ISNAR Agricultural Research Indicator Series

ISNAR agricultural research indicators project
This volume represents the culmination of an effort, begun by ISNAR in late 1984, to compile a global set of commensurable statistics of basic national agricultural research system (NARS) indicators. The unavoidably disparate nature of the data sources and the subject itself, national agricultural research activity, means these statistics are likely to be subject to revision which, in some cases, may be substantial. However, this fully sourced and extensively documented series seeks to establish a basis for informed revision.

While ISNAR does not have a primary vocation for statistical publications or assistance, it is in a unique position, given its on-going contacts at the system level with the NARS of many developing countries, to compile such a series. The significant policy relevance of these statistics at the national, regional, and international level, when combined with the comparative advantage conferred on ISNAR by its mandate to assist developing countries through the provision of research-based services in the areas of agricultural research policy, organization, and management, points to ISNAR's continuing endeavors in this area.

Individuals or agencies with information to hand which may either correct existing inaccuracies or fill existing omissions are requested to forward correspondence on such matters to the following address. Your assistance will be gratefully received.

ISNAR Agricultural Research
Indicators Project
ISNAR
P.O. Box 93375
2509 AJ The Hague
The Netherlands

ISNAR AGRICULTURAL RESEARCH INDICATOR SERIES

A Global Data Base on National Agricultural Research Systems

PHILIP G. PARDEY

JOHANNES ROSEBOOM

Published for the International Service for National Agricultural Research

The right of the
University of Cambridge
to print and sell
all manner of books
was granted by
Henry VIII in 1534.
The University has printed
and published continuously
since 1584.

CAMBRIDGE UNIVERSITY PRESS
Cambridge
New York   Port Chester
Melbourne   Sydney

PUBLISHED BY THE PRESS SYNDICATE OF THE UNIVERSITY OF CAMBRIDGE
The Pitt Building, Trumpington Street, Cambridge, United Kingdom

CAMBRIDGE UNIVERSITY PRESS
The Edinburgh Building, Cambridge CB2 2RU, UK
40 West 20th Street, New York NY 10011–4211, USA
477 Williamstown Road, Port Melbourne, VIC 3207, Australia
Ruiz de Alarcón 13, 28014 Madrid, Spain
Dock House, The Waterfront, Cape Town 8001, South Africa

http://www.cambridge.org

First published 1989
First paperback edition 2003

A catalogue record for this book is available from the British Library

Library of Congress Cataloguing-in-Publication Data
Pardey, Philip G.
ISNAR agricultural research indicator series : a global
data base on national agricultural research systems /
Philip G. Pardey, Johannes Roseboom
        p. cm.
    "Published for the International Service for
National Agricultural Research."
    Includes bibliographical references.
    ISBN 0 521 37368 9 hardback
    1. Agriculture--Research--Statistics. I. Roseboom,
Johannes. II. International Service for National
Agricultural Research. III. Title. IV.Title: National
agricultural research systems.
S541.P37 1989
630´.72--dc20

ISBN 0 521 37368 9 hardback
ISBN 0 521 54333 9 paperback

The International Service for National Agricultural
Research (ISNAR) is one of thirteen centers in the
Consultative Group on International Agricultural
Research (CGIAR) network. ISNAR focuses primarily
on national agricultural research issues and provides
advice to governments, upon request, in the areas of
research policy, organization, and management.
ISNAR is supported by a number of the members
of CGIAR, an informal group of approximately 43
donors, including countries, development banks,
international organizations, and foundations.
The Italian Government generously provided
additional support to the ISNAR Agricultural
Research Indicators Project which facilitated the
research that led to this publication.

CONTENTS

FOREWORD

Technological progress in agriculture is increasingly recognized as a key factor determining the overall economic development in developing countries.

To bring about the kind of technological progress required for promoting overall development, developing countries are investing heavily in the build-up of a national capability for the generation of technology. They are supported in their efforts by donors, development assistance agencies, the international research community, and others.

ISNAR's role and mandate are to assist developing countries in their efforts of strengthening their national research capacities – of making their research systems more productive.

It is essential for successful work in system building and the enhancement of NARS productivity that all those involved in such efforts – the NARS themselves, donors, the international research community, and ISNAR – have access to good information on the policy environment of the NARS, on the organization and structure of NARS, and on the management of the NARS, including the mobilization, availability, and use of resources.

At ISNAR, we have taken up the challenge of providing a data base on NARS that will contribute to this purpose and stimulate analysis of relevant policy and management issues.

We were encouraged by others to take the lead in this important effort. But throughout the project, we collaborated with a broad range of institutions. We acknowledge their contributions.

Naturally, we are proud of the achievement. But we recognize that this was only a first round in what will have to be an ongoing effort – institutionalized at the national or regional level.

We look forward to continuing to play our role in that process, to work with others, to share our lessons, and – above all – to receive your feedback and suggestions.

Alexander von der Osten
Director General
International Service for
National Agricultural Research

ACKNOWLEDGEMENTS

The genesis of this version of the Indicator Series was a survey of national agricultural research systems (NARS) in developing countries, initiated in 1984 by Howard Elliott, with the assistance of Eduardo Trigo and Peter Oram who, at the time, were both working for ISNAR. This 'global' survey, in conjunction with two additional regional surveys, which ISNAR undertook with the assistance of the International Federation of Agricultural Research Systems for Development, the Arab Organization for Agricultural Development, and the Asian Development Bank, respectively, provided a rare opportunity to generate new benchmark data of basic NARS indicators. These surveys were synthesized with data from nearly 900 additional sources to provide the statistical basis for the present series. Howard Elliott's continued substantive and supportive contribution to the project has been invaluable.

Hundreds of individuals, in either a personal or institutional capacity, provided assistance to this project. Many go unnamed here but certainly not unthanked. In most cases their contributions are cited where appropriate. The published work of Bob Evenson and his colleagues provided a solid point of departure for the present effort. We were also most grateful for the large box full of primary data which Bob forwarded to us. ISNAR colleagues tolerated persistent calls for assistance in checking the plausibility of various data sources and clarifying apparent inconsistencies in others. One of our ISNAR colleagues, Paul Bennell, warrants a special mention for supervising the 1987 ISNAR/ADB survey of the South Pacific NARS. University of Minnesota colleagues from the Center for International Food and Agricultural Policy, in particular Ed Schuh, gave freely of their time and expertise to improve the quality of the manuscript. John Dillon also gave valuable comments on an earlier draft of the manuscript.

Alison Young, John Dryden, and their associates from the Scientific, Technological, and Industrial Indicators Division of the Organization for Economic Cooperation and Development provided more assistance than we had a right to expect, to ensure the series for the industrial market economies was as complete as possible. The Comparative Analysis and Data Division of the World Bank also responded generously to our requests for various exchange rate and deflator series as did Robert Summers of the University of Pennsylvannia.

The ISNAR library staff, past and present, not once complained about our repeated requests for obscure documents, and more than once drew our attention to valuable data sources. Sandra Kang, Bonnie Folger, Bob Solinger, Christine Roumagère, and Craig Miller provided accurate and highly competent research and computing assistance throughout the project. Kathleen Sheridan and John Norton undertook a painstaking final check for textual plus numeric accuracy and consistency, while Viviana Galleno and Bob Martin proofed the non-English text. Special thanks go to Arlene Slijk-Holden who typed innumerable drafts of the entire manuscript with dispatch and good cheer.

The Italian government generously provided financial assistance to supplement ISNAR's continuing commitment to this project. We hope the series does justice to their support.

Phil Pardey
Han Roseboom
The Hague

ACRONYMS AND ABBREVIATIONS

Institutional:

AARINENA	Association of Agricultural Research Institutions in the Near East and North Africa
ACIAR	Australian Centre for International Agricultural Research
ADB	Asian Development Bank
AID	Agency for International Development
AOAD	Arab Organization for Agricultural Development
BID	Banco Interamericano de Desarrollo
CAB	Commonwealth Agricultural Bureaux
CARDI	Caribbean Agricultural Research and Development Institute
CARIS	Current Agricultural Research Information System
CGIAR	Consultative Group on International Agricultural Research
CIAT	Centro Internacional de Agricultura Tropical
CIMMYT	Centro Internacional de Mejoramiento de Maíz y Trigo
CIP	Centro Internacional de la Papa
CIRAD	Centre de Coopération Internationale en Recherche Agronomique pour le Développement
DEVRES	Development Research
DSE	Deutsche Stiftung für internationale Entwicklung
ECLAC	Economic Commission for Latin America and Caribbean
ECWA	Economic Commission for Western Asia
EEC	European Economic Community
FAO	Food and Agriculture Organization of the United Nations
GERDAT	Groupement d'Etudes et de Recherches pour le Développement de l'Agronomie Tropicale
IADS	International Agricultural Development Service
IBPGR	International Board for Plant Genetic Resources
IBRD	International Bank for Reconstruction and Development (World Bank)
ICARDA	International Center for Agricultural Research in the Dry Areas
ICRISAT	International Crops Research Institute for the Semi-Arid Tropics
IDRC	International Development Research Centre
IFARD	International Federation of Agricultural Research Systems for Development
IFPRI	International Food Policy Research Institute
IICA	Instituto Interamericano de Cooperación para la Agricultura
IITA	International Institute of Tropical Agriculture
ILCA	International Livestock Center for Africa
ILO	International Labor Organization
ILRAD	International Laboratory for Research on Animal Diseases
IMF	International Monetary Fund
INTERPAKS	International Program for Agricultural Knowledge Systems
IRRI	International Rice Research Institute
ISNAR	International Service for National Agricultural Research
OECD	Organization for Economic Cooperation and Development

ORSTOM	Office de la Recherche Scientifique et Technique Outre-Mer
PCARRD	Philippine Council for Agriculture and Resources Research and Development
SACCAR	Southern African Centre for Cooperation in Agricultural Research
SADCC	Southern African Development Coordination Conference
SAREC	Swedish Agency for Research Cooperation
SOEC	Statistical Office of the European Communities (EUROSTAT)
STIID	Scientific, Technological, and Industrial Indicators Division (OECD)
TAC	Technical Advisory Committee
UNDP	United Nations Development Program
UNESCO	United Nations Educational, Scientific and Cultural Organization
USAID	United States Agency for International Development
USDA	United States Department of Agriculture
WARDA	West Africa Rice Development Association

Other:

AgGDP	Agricultural Gross Domestic Product
CRIS	Current Research Information System
FTE	Full-Time Equivalents
GDP	Gross Domestic Product
ISCED	International Standard Classification of Education
LCU	Local Currency Unit
NARS	National Agricultural Research System
PPP	Purchasing Power Parity
R&D	Research and Development
STI	Science and Technology Indicators
WANA	West Asia and North Africa

PART I: TECHNICAL NOTES

1. Introduction

The ISNAR Agricultural Research Indicator Series is a fully sourced and extensively documented set of research personnel and expenditure indicators for national agricultural research systems (NARS) in 154 developing and developed countries for the 27 years 1960 through to 1986, where possible. Extensive efforts have gone into achieving completeness and commensurability in the series. However, the unavoidably disparate nature of the data sources, plus the subject of the data series itself means that these statistics should be considered indicative rather than definitive. Nevertheless, the series represents a major effort to consolidate and completely restructure previously available data compilations. Further, the scope of the series, in terms of country and time-period coverage plus number of indicators, constitutes a substantial extension of or addition to currently available global compilations.

The Indicator Series attempts to be the most comprehensive time series possible reporting commensurable data within countries over time, across countries, and between the personnel and expenditure series themselves. Several procedures were adopted to ensure commensurability when this series was compiled. These are described in detail in Section 2.

A universally accepted definition of NARS is not feasible, but the NARS concept used here is a practically achievable derivative of the concepts that underpin the basic R&D statistics compiled by other international agencies, such as the Organization for Economic Cooperation and Development, (OECD) the Statistical Office of the European Community (SOEC), and the United Nations Educational, Scientific and Cultural Organization (UNESCO).

A clear understanding of these concepts makes it possible to develop a time series of basic indicators which draws, in a standardized manner, from an underlying data base of documented, multiple, historical observations. The extensive documentation and complete citing of the data included in the series enables users to assess the extent and nature of any deviations from such standards. Certainly, an attempt, where possible, to trace data back to their primary source has enabled this series to work towards an improved cross-country commensurability in institutional coverage over time. Section 3 gives an overview of the various sources used here, including details of the three ISNAR benchmark surveys of NARS in developing countries which formed the immediate point of departure for this series.

The research expenditure data have all been collected in current local currency units. This enabled standardized currency conversion and deflation procedures to be implemented. The conversion options available, and those applied in this instance, are fully described in Section 4. Inappropriate treatment of such matters can have a nontrivial quantitative and qualitative impact on the data; this is also demonstrated in Section 4 in some detail.

The personnel indicators seek to include only research personnel, exclusive of technicians or support staff and, wherever appropriate, attempt to differentiate between local versus expatriate staff. The series also attempts to record (local) researchers according to degree status – PhD, MSc, BSc, or equivalent – and, wherever possible, to measure researchers in full-time equivalent units. This latter issue is particularly important when including personnel from institutions which have, for example, a dual research-extension role, or universities where personnel often hold split research-teaching appointments.

The expenditure series attempts to measure actual research expenditures, not simply appropriations or funds available. Expenditures are total, inclusive of salaries and operating and capital expenses. Where relevant, they also include expenditures funded from development budgets. The series reports expenditures in current and constant local currency units, as well as 1980 Atlas exchange rates and Purchasing Power Parity (PPP) over GDP-converted U.S. dollar volume measures. Again, complete details on the uniform conversion procedures applied to the research expenditures of all countries are given in Section 4.

The format of the Indicator Series country-level files enables them to be used at three levels: the data series themselves can be directly used for aggregate comparative analysis; the data series plus the personnel and expenditure comments are appropriate for more focused analysis at the regional level; while the data, comments, and citations (both sources and additional references) are useful for more targeted analysis at the country-specific or topical level.

The series has not been specifically structured for use as a planning tool by research planners and administrators at the country level. While for numerous NARS these series do indeed represent the most comprehensive set of national-level agricultural research indicators currently available, managers generally need an altogether more detailed set of indicators for use in an administrative capacity.[1] Nevertheless, research managers wishing to track and assess, at a research policy level, the evolution of their national agricultural research systems vis-a-vis other systems should find the Indicator Series particularly useful. The series also provides a basis for improved agricultural research policy analysis and decision making by those operating at the regional and international level.

The need for a comprehensive data base, inclusive of both developing and developed countries, is clear. This is particularly true in view of the important and expanding interdependencies of the agricultural sectors of all economies, both at the factor level (inclusive of agricultural technologies) and at final goods markets. The particular nature of the agricultural research and technology transfer process itself – whereby the output or productivity effect of research expenditures can persist for extended periods of time[2] – points to the need for a relatively long time series, if policy analysis concerning the role of agricultural research in agricultural and economic development is to be meaningful. The process of institutional change is also a time-intensive activity. A basic understanding of recent historical changes in NARS activity is therefore essential if appropriate decisions concerning the capacity for NARS restructuring or adjustments are to be made.

Notes

1. Daniels (1987) begins to sketch the requirements of research administrators with respect to input indicators for use at a management level.
2. Pardey and Craig (1989), for instance, used a long-run research expenditure series for the U.S. to show that the output effect of agricultural research spending may persist for as long as 30 years – around double the prevailing norm assumed to date in the literature.

2. Basic concepts, definitions, and measurement issues

This section describes the conceptual and definitional underpinnings of the ISNAR Agricultural Research Indicator Series. Particular attention is paid to clarifying the concept of a national agricultural research system (NARS) which was used to compile these figures, along with a detailed discussion of measurement issues pertaining to the personnel and expenditure series themselves. It will be clear that the Indicator Series represents a practical, but hopefully transparent, compromise between the conceptual ideal described here and the realities of compiling a global data base on NARS.

One of the major constraints facing this exercise was the nature of the data sources. While benchmark data were obtained from several surveys, which ISNAR conducted in cooperation with other agencies, nearly 900 additional data sources were used to compile the series. The need for a clear understanding of the scope and nature of the institutional coverage which was sought in each country is self-evident if the data are to achieve a tolerable degree of cross-country commensurability over time. Clearly, the heterogeneous nature of the data sources substantially increased the difficulty of achieving complete consistency. It was, therefore, necessary to incorporate explanatory comments and documentation in the country-level data series to alert users as to the likely magnitude of any deviations from the norm.

2.1 Defining a NARS
The ISNAR Agricultural Research Indicator Series is based on the concept of a NARS. While this concept certainly has value as an analytical tool, it is difficult to operationalize as a statistical tool. Moreover, there is not universal agreement as to what constitutes a Nars.

This lack of agreement stems, in part, from the systems characteristic of a NARS. In systems theory, an important distinction is made between open and closed systems. "A closed system has rigid, impenetrable boundaries, whereas an open system has permeable boundaries between itself and a broader supersystem. The boundaries set the domain of the organization's activities. In a physical, mechanical, or biological system the boundaries can be identified. In a social organization, the boundaries are not easily definable and are determined primarily by the functions and activities of the organization" (Kast and Rosenzweig, 1985, 106).

A NARS bears all the hallmarks of an open system and thus has rather vaguely formed boundaries. However, to secure commensurability of NARS indicators on human and financial inputs, it is necessary to delineate the NARS boundary as clearly and as sharply as possible. Thus, a closed rather than open conception of a NARS is required. This section seeks to 'close off' the boundaries of a NARS as much as possible, recognizing that the practicalities of compiling such a data series sometimes forces ad hoc, but hopefully documented, deviations from the notion of a NARS presented here.

A useful beginning is to split the concept of a NARS into its three dimensions, namely 1) national, 2) agriculture, and 3) research, and consider each of these dimensions separately. In many instances, it will be helpful to adopt the approach used in *The Measurement of Scientific and Technical Activities – "Frascati Manual" 1980* (OECD, 1981) and present concrete examples of what lies inside and outside the NARS boundary we seek to establish. While it may not be possible to define exactly when the boundary is crossed, it will hopefully be clear to the user when one is approaching the boundary.

2.1.1 *National*

The notion of what constitutes a 'national' set of statistics on agricultural research is open to several interpretations. One option is to adopt a geographic interpretation and include all agricultural research performed within the boundaries of a country. Another possibility is to pursue a sectoral approach and include domestically targeted research activities funded and/or executed by the public sector of a particular country.[1] This latter approach is adopted here.The Indicator Series attempts to include all agricultural research activities that are financed and/or executed by the public sector, inclusive of private nonprofit agricultural research, while explicitly excluding private, for-profit agricultural research. This sectoral coverage corresponds to the OECD's (1981, 83-91) notion of government, private nonprofit, and higher education sectors, and excludes the business enterprise sector.

Ideally it would be useful to differentiate systematically between the source of resources committed to a national agricultural research program and the execution of the research program itself. At a practical level this distinction is often difficult to achieve, thereby forcing an eclectic approach which depends on the nature of the available data.

The government sector is taken to include those federal or central government agencies, as well as provincial or state and local government agencies, which undertake agricultural R&D. Care is needed to avoid double-counting federal resources which fund agricultural research at the state or provincial level, as well as ensuring that nonresearch activities are excluded. This is a particular problem for research performed by government agencies at the state and local level who, in many instances, also deliver nonresearch services such as rural extension.

The private, nonprofit sector, which in our case also covers nongovernment organizations, generally includes few institutions. Nevertheless, for some countries they can be especially important. Some commodity research centers in developing countries, particularly those concerned with export-oriented estate crops such as tea, coffee, and rubber, are often financed by (industry enforced) export or production levies – with or without matching government contributions – and performed by private or semiprivate nonprofit research institutions. They often operate as pseudopublic-sector research agencies, or at the very least substitute directly for such agencies, so it is appropriate to include them in the Indicator Series.

The higher education sector is fairly readily identified, but does present special problems when compiling agricultural research statistics. As noted earlier, care was taken to isolate research from nonresearch activities (e.g., teaching and extension) and to prorate personnel and expenditure data accordingly.

The 'national' agricultural research statistics reported in the Indicator Series excludes the activities of research institutions with an international or regional mandate such as CIMMYT, IRRI, and WARDA, along with ORSTOM and CIRAD. While their research output may often have substantial impact on the agricultural sectors of their host countries, their mandates direct their research activities towards international and regional, rather than national, applications of their findings. However, all foreign research activities which are either funded and/or executed in collaboration with and/or administered by the national research agencies are included in the series.

2.1.2 *Agriculture*

When measuring science indicators by socioeconomic objective, the OECD (1981, 113) recognizes that two approaches to classification are possible:
(1) according to the *purpose* of an R&D program or project;
(2) according to the general *content* of the R&D program or project.

For example, a research project to improve the fuel efficiency of farm machinery could be placed under 'agriculture' if classified by purpose, but 'energy' if classified by R&D content. The Indicator Series adopts the procedure used by the OECD and classifies research by purpose rather than content, on the grounds that it is generally the purpose for which research is undertaken which has the most policy relevance.

The notion of agricultural research used here includes primary agriculture (crops plus livestock), forestry, and fisheries research. In general terms, this corresponds with the coverage used by both OECD and UNESCO. For policy and analytical purposes, it would be desirable to differentiate agricultural research among commodities, but for many systems this is practically impossible, particularly on a time-series basis back to 1960. Quite a few countries only report national research expenditure data that are not differentiated according to socioeconomic objective, even at this rather aggregate level. For these systems, it was simply not possible, given currently available data, to generate plausible time series at this subsector level.

Prior compilations of NARS indicators have sought to limit their coverage by excluding forestry, fisheries, and even some livestock research (see section 3.2.3 for details). This is extremely difficult to achieve for those systems only reporting aggregate figures, inclusive of all three subsectors. However, attempting to compile statistics that include public-sector agriculture, forestry, and fisheries research is also not without its difficulties. For a significant number of systems there are several individual agencies charged with the responsibility of funding and executing these various areas of research. This complicates the data-collection process. Nevertheless, a substantive argument in favor of adopting this wider definition of agricultural research is that the resulting series is then consistent with the definitions of agricultural GDP, population and so on, as published by the World Bank, United Nations, and FAO.

A further classification difficulty is that a significant amount of agricultural research has an effect at the postharvest stage, while the technology is embodied in inputs that are applied at the farm level. Take, for example, the efforts of plant breeders to improve the storage life of horticultural crops or to alter the baking quality of cereals. These characteristics are embodied in new seed varieties which are adopted by farmers. Furthermore, there is a lack of uniformity in the way research which is applied directly at the

postharvest stage is currently reported. The OECD (1981, 115) classification omits "R&D in favor of the food processing and packaging industries" from their socioeconomic objective of 'agriculture, forestry and fisheries,'[2] while UNESCO (1984, 64) includes "R&D on the processing of food and beverages, their storage and distribution." The Indicator Series sought to implement a variant of these approaches, excluding, where possible, research applied directly at the postharvest stage, while including research that is applied at the preharvest stage but which has an impact at the postharvest stage. Omitting research on food processing and packaging improves the compatibility of these statistics with agricultural GDP and similar figures. Nevertheless, public-sector research targeted directly to food and beverage storage (and in some cases, processing) may in practice be included in this series, although this is more likely to be true of more advanced systems in developed countries.

A final difficulty is to obtain statistics for the higher education sector classified by purpose or 'socioeconomic objective'. The more general case is to find personnel data, and possibly, expenditure data, classified by 'field of science,' where the basis of classification is the nature rather than purpose or objective of the research activity itself.[3] In those cases where it was necessary to rely on 'field of science' data, the series attempts to follow the UNESCO (1984, 77) procedure and consider agronomy, animal husbandry, fisheries, forestry, horticulture, veterinary medicine, and other allied sciences as agricultural sciences, thereby excluding fields such as bacteriology, biochemistry, biology, botany, chemistry, entomology, geology, meteorology, zoology, and other allied sciences. These latter fields are more appropriately classified as natural sciences although in some cases the classification is a little hazy. It was therefore necessary to apply a 'purpose or objective test' to some of these so-called natural science disciplines and to include in this series research undertaken in these areas when the ultimate purpose or objective of that research would have an impact on the agricultural sector.

2.1.3 *Research*

It is possible to identify a whole continuum of research from basic, or upstream research, to applied, or downstream research. Much agricultural research has been characterized as mission-oriented, in the sense that it is problem-solving in orientation, whether or not the solution to those practical problems requires basic or applied research. OECD (1981, 28) states that "the basic criterion for distinguishing R&D from related activities is the presence in R&D of an appreciable element of novelty." For instance, simply monitoring the incidence of plant and animal diseases in and of itself is not considered research and may only be undertaken to enforce quarantine regulations or the like. But, using this information to study the causes or control mechanisms associated with a particular disease is considered research. Of course, some screening of the literature, available plant and animal material, and existing production practices should be included as part of measured research activity, given its importance in the many countries that are undertaking substantial efforts to adapt existing agricultural technology to their local circumstances.

Agricultural research also includes a significant maintenance research component, which seeks to renovate or replace deteriorations in the gains from previous research.[4] Past output gains from much agricultural research are subject to biological degradation as plant and animal pests and pathogens adapt around research-conferred resistance and control mechanisms. The role of maintenance research is not only substantial in many developed countries where current production practices are biological-technology intensive, but also in many developing countries, particularly those situated in the tropics where relatively rapid rates of pest and pathogen adaptation tend to shorten the half-life of previous research-induced gains.

The difficulties of differentiating research from nonresearch activities is especially pertinent in the case of agricultural research, given the dual role of many public-sector agencies charged with agricultural research responsibilities. It is common to find such agencies also involved in additional nonresearch activities such as teaching; extension services; seed certification, multiplication, and distribution; monitoring and eradicating diseases in plants and animals; health maintenance, including veterinary medicine; and fertilizer analyses and certification. In general, it is separating the research component from the joint teaching-research activities (in the case of universities) and the joint extension-research activities of ministerial or department-based agencies which are the most troublesome. If direct measures of expenditure and personnel data were not available at the functional level, then secondary data were often used to estimate the appropriate breakdown of aggregate figures into their research versus nonresearch components.

Even in the case of those institutions whose mandate is ostensibly limited to research, there can be problems in obtaining consistent coverage of research-related activities. For example, general overhead services, including administrative personnel or expenditures required to support research, can be excluded from reported figures for a variety of reasons. In some instances, the institutional relationship between a national research agency and the ministry within which it is located means that overhead services and the like are charged against the ministry and not the research agency. Alternatively, some research agencies report total personnel and expenditure statistics based on an aggregation of project rather than institutional-level data. In such cases, administrative overheads may not be appropriately allocated across projects and thus may be omitted entirely or in part from the agency-level statistics.

A further issue involves identifying the research component of farm operations that are usually undertaken in support of agricultural research. To the extent that such farm operations are necessary to execute a program of research, it seems appropriate that they be included in a measure of national resource commitment to agricultural research. However, some systems undertake farm operations at levels well above those required to support research, with the surplus earnings from farm sales being siphoned off to support research and even various nonresearch activities. In some instances, including all the resources devoted to the farm operations of a NARS substantially overstates the level of support to agricultural research within the system.

There is also the need to make a clear distinction between economic development and experimental development. According to OECD (1981, 25) "experimental development is systematic work, drawing on existing knowledge gained from research and/or practical experience that is directed to producing new materials, products, or devices, to installing new processes, systems, and services, or to improving substantially those already produced or installed." Experimental development is therefore concerned with applying new findings from formal and informal research activity. This contrasts with the notion of economic development, which in general terms, is concerned with improving the well-being or standard of living of members of a society in a particular country or region.

Clearly, while improvements in agricultural productivity that follow from experimental development activities contribute to the process of economic development, they represent only part of the story. Improvements in rural infrastructure, via investments in irrigation, transport, and communication facilities plus improved rural health and education services, also contribute to the economic development of the agricultural sector and, ultimately, to society as a whole (Antle, 1983).

A problem arises when attempting to compile statistics on agricultural research and experimental development activities in developing countries. This occurs when a substantial portion of R&D activity is financed and/or executed as part of an economic development aid package. It is often difficult to identify the experimental versus economic development component of an aid package, particularly given the project orientation of much development aid. For instance, development assistance to establish or upgrade irrigation facilities can often incorporate a research component to evaluate water quality and identify preferred crop varieties as well as agronomic and irrigation practices. However, simply to include all of the project's resources in a measure of NARS capacity could seriously overestimate the level of resource commitment to agricultural research.

Another less obvious difficulty concerns the somewhat transient nature of some of the agricultural research funded through development projects. This research activity tends to be of relatively short duration, between one to five years, and in some cases it is undertaken largely by expatriates at arm's length from the existing national research infrastructure. This type of research presumably contributes to the overall level of national research activity and would be important to capture in a NARS indicator, particularly if one is concerned with measuring sources of growth or technical change within a country. However, to the extent that such research is not integrated into the existing national research infrastructure, it is not a good measure of the 'institutionalized research capacity' of a national system. The strategy pursued here was to include such development-financed research in the indicator series only when the research component could be isolated from the nonresearch component with an acceptable level of precision, and when it appeared to be integrated into the existing agricultural research infrastructure within a country.

2.2 Measurement issues

While some measurement issues that influence the concept of a NARS used here as a frame of reference for the Indicator Series have already been discussed, there are some important issues which remain to be considered.

2.2.1 Construction Methodology

A major objective in compiling the Indicator Series was to improve the consistency, comparability, and coverage of global agricultural research statistics. Considerable effort was invested in attempting to maintain consistent coverage in several dimensions, namely, 1) over time within a country, 2) among countries, and 3) across the personnel and expenditure series.

A key to improving consistency was to identify and track the institutional coverage of the available data, paying particular attention to dates of creation, organizational mergers or divisions, details of name or mandate changes, and the like. This involved gathering quantitative and qualitative data from as many documentable sources as possible, including the benchmark ISNAR data surveys, then proceeding to reconcile and synthesize these multiple observations into a data series that represents as closely as possible the NARS concept identified earlier in this section, while maintaining a consistent institutional coverage.

In all cases, every effort was made to trace data back to primary sources. Simply splicing more recent data onto existing secondary sources was, in many cases, unacceptable. In the first instance, it severely limits the ability to construct a consistent series since many secondary data sources fail to document their institutional coverage in sufficient detail. This results in spurious growth or contraction being incorporated into aggregate statistics simply because of improvements or deterioration in institutional coverage. In other instances, when tracing back to primary sources, it was discovered that some existing 15- to 20-year 'constructed time series' for a given country were, in fact, based on one or two documentable observations. As many observations as possible were traced back to their primary sources in an attempt to break the citation cycle that converts a 'constructed time series' into an 'actual time series'.

The Indicator Series reports data on a calendar-year basis whenever possible. In numerous instances, however, available data are recorded on a fiscal- or academic-year basis. The procedure adopted in such cases was to place the observation in the calendar year that overlaps most with the respective fiscal or academic year. Consequently, a fiscal year running from April 1, 1980, to March 31, 1981, would be placed in calendar year 1980. July 1 to June 30 fiscal years were placed in the year ending June 30.

2.2.2 Personnel Indicators

One possibility for measuring the human resource commitment to NARS is to simply report the total number of personnel employed within a research system, regardless of their qualifications or occupation. However, for a majority of purposes, it is the scientific capacity of NARS personnel that is relevant. Thus, the Indicator Series sought to include only research personnel, i.e., scientists engaged directly in the conception or creation of new knowledge, products, processes, methods, and systems.[5] The series attempts to exclude technicians and support and clerical staff who normally perform scientific and technical tasks under the supervision of a scientist. While separate data on support staff would be useful, simply to include them in a NARS personnel series could seriously distort relativities, both among countries and within countries over time. Such a series may be driven largely by differences in the relative costs of research labor, capital, and other operating expenses – resulting, for example, in quite volatile fluctuations in the ratio of scientists to nonscientists – and would not reflect differences in the underlying scientific capacity the series seeks to measure.

A practical procedure for differentiating scientific from nonscientific staff is to rely principally on educational levels rather than occupational classes. While there are clearly substantial difficulties in standardizing educational levels on a global basis, an international standard classification of education (ISCED) has been developed and is in general use (see UNESCO, 1976). The Indicator Series seeks to include only NARS personnel who hold at least a third-level university degree (ISCED level categories 6 and 7) as researchers.[6] This includes holders of first and postgraduate degrees (or their equivalent) earned at bona fide universities, as well as at specialized institutes of university status.

The series attempts to further classify national research personnel by degree status – PhD, MSc, BSc, or equivalent. This should substantially improve our ability to use the personnel data as an indicator of the human capital or 'quality-adjusted' scientific commitment to national agricultural research. It also attempts to differentiate between local and expatriate scientific staff in order to enhance the information content of the personnel series. As discussed earlier, the series seeks to include only those expatriates who are working directly on domestic issues in an integrated fashion with the national research system.

Personnel who are classed as research managers or administrators present special problems. To the extent they are engaged in the planning and management of the scientific and technical aspects of a researcher's work they should be classified as researchers and included, at least on a prorated basis, in the series. They are usually of a rank equal to or above that of persons directly employed as researchers and will often be former or part-time researchers (OECD, 1981, 67). However, in many cases, it is not at all clear if research managers or administrators maintain a degree of contact with the scientific process itself.

The problems of dealing with research administrators are analogous to those of dealing with other NARS personnel who may hold dual research and nonresearch appointments. This is particularly important when including personnel from institutions of the Ministry or Department of Agriculture, who perform, for example, a dual research-extension function, or universities where personnel often hold split research-teaching appointments. In all cases an attempt has been made to measure researchers in full-time equivalent (FTE) units. If direct measures in FTE units were not available, secondary data, which enabled total researcher figures to be plausibly prorated to FTE units, were used.

2.2.3 Expenditure Indicators

There are several commonly accepted methods of measuring the (annual) financial resource commitment to R&D (OECD, 1981, 72-82):
(1) *performer-based* reporting of the sum of funds received by all relevant R&D agencies for the performance of intramural R&D;
(2) *source-based* reporting of the funds supplied by all relevant agencies for the performance of extramural R&D;
(3) *total* intramural expenditures for R&D performed within a statistical unit or sector of the economy, whatever the source of funds.

The Indicator Series seeks to report actual research expenditures, not simply budgeted

funds, appropriations, or funds available, and so is based on method (3) wherever possible. A substantial number of the major discrepancies in prior compilations were due to large variations – sometimes upward of 30 to 50 percent – between funds budgeted or appropriated and funds actually spent. Some funds allocated to research at the beginning of a fiscal year, for example, may never materialize, especially if governments are forced to trim proposed outlays over the course of the year due to unforeseen budgetary shortfalls. Conversely, some research systems may actually receive more funds than are spent, and thus carry funds over to future budgetary periods. This is particularly true for systems experiencing substantial capital investments where funds are allocated initially in a lump-sum fashion and then are drawn on over a period of time as needed.

The expenditures reported in the Indicator Series are total, inclusive of salaries and operating and capital expenses. While the series reports actual expenses, for some purposes it may be more appropriate to measure resources used rather than funds spent. This would involve explicitly separating capital from noncapital expenditures. Capital expenditures, which represent a measure of capital stock, could then be converted into flow terms by estimating the future service flows derived from the capital expenditures.[7] These capital service flows can then be added back to noncapital expenditures to derive an overall measure of the resources actually used over time for research.

One of the major undertakings in compiling the Indicator Series was to collect all expenditure data in current local currency units. This allowed standardized deflation and currency conversion techniques to be applied to all countries – a critical issue when attempting to use the data for cross-country and/or comparative purposes over time. Section 4 presents complete details of the adjustment procedures that were applied to the raw data. It makes it clear that these procedures can have a significant quantitative and qualitative impact on the data.

Where possible, expenditure data were taken directly from survey and source material in current local currency units. For sources that reported expenditures in other than current local currency units, an attempt was first made to recover the desired figure by applying, in reverse, the specific conversion methodology, deflators, and exchange rates used by these sources. Failing that, annual average exchange rates and implicit GDP deflators were used to recover expenditures in current local currency units.

The base period for all currency units is 1980. If currency units had changed before or after that date, the resulting currency unit was converted to the 1980 base-period unit. This allowed uniform procedures for subsequent currency conversions to be applied to all countries.

Notes

1. Classifying by source of funds is known as a 'funder-based' system of classification as opposed to a 'performer-based system,' which classifies according to the nature of those institutions that actually execute the research. Clearly, these classification systems can give rise to different measures of research capacity, and a preferred approach would be to classify research activity by one or the other method. However, at a practical level, when attempting to construct a global database of agricultural research statistics, it was necessary to adopt an eclectic approach and use an ad hoc combination of both procedures to arrive at a set of statistics.

2. OECD (1981) instead includes it under the socioeconomic objective of 'promotion of industrial development.'

3. Classifying research on the basis of the nature of the R&D activity itself, rather than its principal economic objective, is called a 'functional' approach (OECD, 1981, 53).

4. For additional discussion on maintenance research see Ruttan (1982), Swallow et al. (1985), Miranowski and Carlson (1986), Plucknett and Smith (1986), and Adusei (1987).

5. This corresponds to the OECD (1981, 67) definition of a researcher.

6. An alternative procedure (see OECD, 1981, 67-69) is to first classify researchers, technicians, and other supporting staff on the basis of the ILO (1986) classification scheme and then use ISCED procedures to classify researchers by educational level. Given the rather heavy reliance on secondary data in the Indicator Series, it was not possible to operationalize the ILO classification scheme.

7. See Pardey, Craig, and Hallaway (1989) for details. Unfortunately, there are simply not enough data available at the international level to construct a time series that differentiates research expenditures by factor type.

3. Data sources

The Agricultural Research Indicator Series is certainly not the first attempt to compile global data on national agricultural research expenditures and researchers. Evenson and his colleagues initiated construction of a global series in the late 1960s, which culminated with the 1975 publication of James Boyce and Robert Evenson's *National and International Agricultural Research and Extension Programs*, and after further efforts to update the series, resulted in the 1983 and 1986 publications by Judd, Boyce, and Evenson. The joint 1981 ISNAR/IFPRI publication by Oram and Bindlish also drew heavily on the earlier work by Evenson in an attempt to extend the basic series for 51 developing countries through to 1980. Evenson's pioneering work notwithstanding, much of the existing data sources are fragmented in scope and coverage, difficult to access, uneven in quality, and vary in the degree of documentation. The Indicator Series seeks not only to expand the available data, but also to correct these deficiences, or at least work towards some standardization in both the data collected and the way they are treated.

3.1 Overview

The Indicator Series represents an extensively documented compilation of basic NARS indicators drawing on three benchmark ISNAR surveys, plus information from nearly 900 additional data sources.[1] The character of these additional sources varies enormously. They include official statistical publications and unpublished documents from international, regional, and national agencies; annual reports of various agricultural research agencies; dissertations, mimeos, working papers, and unpublished reports; and a substantial amount of personal communications with individuals working in various international, regional, and national institutions, including universities. Of the 900 or so nonsurvey sources, approximately 28% can be classified as unpublished 'grey' literature.[2]

Tables 3.1 and 3.2 summarize the relative importance, at both a regional and global level, of the sources actually used to construct the personnel and expenditure series, respectively. While only 40% of the nearly 900 references were directly used in the Indicator Series, the quantitative and qualitative information contained in all the references was evaluated before compiling the series. These additional references often proved invaluable when attempting to track changes in organizational structure. This enabled the institutional coverage over time and across countries to be standardized as closely as possible on the NARS concept described above in Section 2.

Published and unpublished material from ISNAR (survey data excluded), United Nations, OECD and EEC, CGIAR, CAB, and DEVRES accounts for 28% of the personnel sources and 25% of the expenditure sources. The three ISNAR benchmark surveys provided a total of 7% and 10% of the personnel and expenditure data, respectively. To the extent that they provided detailed information against which data from alternative sources could be evaluated, these ISNAR surveys had a disproportionately larger impact on the whole series.

Evenson-related data accounted for 11% and 6% of the personnel and expenditure sources, respectively, with over 60% of the personnel and 75% of the expenditure figures obtained from Evenson relating to the pre-1970 period. This apparently small reliance on Evenson-related figures belies the extensive use these data have hitherto received. However, in quite a few instances the Indicator Series drew from the same primary sources that Evenson and his colleagues used or obtained information from additional primary sources that were considered more complete. Given the differences between

Table 3.1: Sources for personnel data by region[a]

	Sub-Saharan Africa	WANA	Caribbean & C. America	South America	Asia	South Pacific	Oceania	Europe	North America	TOTAL
	%	%	%	%	%	%	%	%	%	%
ISNAR										
Surveys[b]	17.2[a]	20.2	7.9	3.5	13.7	5.6	..	..	..	6.7
Other material[c]	12.7	20.2	0.9	..	0.8	2.8	..	0.9	..	5.7
United Nations[d]	13.6	9.6	4.8	1.2	10.5	7.0	2.7	1.4	..	7.1
OECD & SOEC[e]	..	..	..	..	..	..	2.7	16.5	..	2.6
CGIAR[f]	5.7	1.0	1.3	13.3	4.0	..	..	..	..	3.9
CAB	4.1	..	18.9	2.9	12.9	9.9	29.7	4.1	11.3	8.0
DEVRES	3.6	..	..	..	..	..	..	..	..	1.1
NARS reports	1.9	..	6.6	0.6	2.4	..	..	..	..	1.9
Evenson et al.[g]	10.0	18.3	11.4	20.2	10.5	..	5.4	7.8	..	10.8
Oram et al.[h]	1.9	1.0	0.4	7.5	2.4	1.4	..	..	..	1.9
Personal communication	10.8	5.8	7.5	17.3	17.7	9.9	59.5	31.5	..	15.3
Workshops	0.7	4.8	3.5	3.5	8.9	..	..	..	..	2.3
Other										
Published	12.0	14.4	32.9	25.4	16.1	63.4	..	37.4	43.4	24.8
Unpublished[i]	15.8	4.8	3.9	4.6	..	..	..	0.5	45.3	7.9
TOTAL	100.0	100.0	100.0	100.0	100.0	100.0	100.0	100.0	100.0	100.0

See notes to Table 3.2

Evenson's series and the Indicator Series in both coverage and treatment of the data – described in some detail in Sections 2 and 4, and also later in this section – it was often necessary to draw from these alternative data sources in order to improve the cross-country, temporal commensurability of the present series.

For discussion purposes, it is useful to group the references according to their country coverage. This section therefore follows with a brief review of the principal data sources grouped on a global, regional, and country-specific basis.

3.2 Global-level data sources
3.2.1 *ISNAR Surveys*
ISNAR, in cooperation with other agencies, has conducted three benchmark surveys.[3] A 'global' survey, seeking information from 116 developing country NARS, carried out during the period from late 1984 to mid-1985, was undertaken with the cooperation of the International Federation of Agricultural Research Systems for Development (IFARD). The target was the national agricultural research system as conceptualized in Section 2, but the point of entry was usually the publicly funded national agricultural research organization which was requested, in turn, to contact appropriate quasi-public research institutions and universities.

In order to reach member states of the Arab Organization for Agricultural Development (AOAD), ISNAR, in cooperation with AOAD, undertook a similar survey during 1985 at the request of the ministers of agriculture, who constituted the AOAD board. A second regional survey, this time of the South Pacific, was conducted in mid-1987 with the assistance of the Asian Development Bank (ADB). The three surveys yielded usable data for nearly 70 NARS. As expected, the replies were of uneven quality in terms of coverage, internal consistency (between national and institution-specific information), and documentation. Nevertheless, a substantial number of these problems were resolved, or at least nullified, by extensive follow-up activities.

3.2.2 *International Agencies*
While the collection and preparation of basic global indicators of national agricultural research capacity has until now been largely an individual rather than institutionalized effort, several international agencies have collected NARS statistics on a somewhat ad hoc and irregular basis. These include the Food and Agriculture Organization of the United Nations (FAO), Consultative Group on International Agricultural Research (CGIAR), International Food Policy Research Institute (IFPRI), World Bank, and ISNAR.

While some of these international agencies have large statistical offices, they do not compile series of agricultural research indicators on a regular basis. Much of the data-collection activity concerning agricultural research indicators that these organizations undertake occurs in an ad hoc fashion outside their statistical offices. Nevertheless, these activities have proved invaluable for the present series and will be briefly discussed.

The FAO has provided a great deal of published and unpublished material, including agricultural research system reviews, directories of agricultural research institutions, regional overviews of national agricultural research activities, plus workshop and conference reports on agricultural research. An important FAO undertaking is the development of a current agricultural research information system (CARIS). This undertaking seeks to maintain a complete record of agricultural research projects currently underway in developing countries. Each country participating in the project is requested to develop its own agricultural research information system, based on FAO standards, then forward the information on a regular basis to FAO headquarters where it can be integrated into a 'global' agricultural research information system. The unit of record in the CARIS system is a project rather than an institutional or national-agency level. Limitations on the number of countries actively participating in the project and substantial accounting difficulties when attempting to aggregate multi-year, project-level data to system-level data for a particular year, place severe constraints on the usefulness of the CARIS system as a data source for the Indicator Series.[4]

Table 3.2: Sources for expenditure data by region[a]

	Sub-Saharan Africa	WANA	Caribbean & C. America	South America	Asia	South Pacific	Oceania	Europe	North America	TOTAL
	%	%	%	%	%	%	%	%	%	%
ISNAR										
Surveys[b]	12.3[a]	21.1	11.0	2.3	8.2	66.7	..	..	..	19.5
Other material[c]	14.2	2.4	6.2	4.2	0.5	..	3.5	..	..	4.8
United Nations[d]	5.9	2.4	2.7	..	7.7	7.2	..	1.3	..	3.4
OECD & SOEC[e]	..	9.4	..	..	..	..	9.3	29.6	27.7	8.2
CGIAR[f]	8.8	..	3.4	17.8	23.5	..	..	..	..	7.9
CAB	..	..	..	..	..	..	..	..	..	..
DEVRES	1.9	..	..	..	..	..	..	..	..	0.5
NARS reports	..	..	..	12.7	..	..	..	1.3	..	2.0
Evenson et al.[g]	5.9	22.3	4.8	6.1	8.2	..	..	1.0	..	5.5
Oram et al.[h]	7.0	4.7	..	8.0	1.5	1.4	..	..	..	3.3
Personal communication	11.8	15.3	1.4	12.2	7.1	4.3	52.3	43.3	..	18.2
Workshops	2.4	..	8.2	..	15.8	..	..	..	..	3.4
Other										
Published[i]	13.1	14.1	40.4	15.0	16.8	20.3	34.9	22.5	35.4	20.8
Unpublished[j]	16.6	7.1	21.9	21.6	10.8	..	..	1.0	36.9	12.6
TOTAL	100.0	100.0	100.0	100.0	100.0	100.0	100.0	100.0	100.0	100.0

a. The percentages given in Tables 3.1 and 3.2 refer to the proportion of total personnel and expenditure observations within each region derived from a particular source category.

b. ISNAR surveys carried out during 1984-85 with the assistance of the International Federation of Agricultural Research Systems for Development (IFARD) and the Arab Organization for Agricultural Development (AOAD), plus a regional survey of the South Pacific carried out in 1987 with the assistance of the Asian Development Bank (ADB).

c. Includes, inter alia, ISNAR system reviews, monographs, published and unpublished reports.

d. Includes, inter alia, data from UNESCO and FAO sources.

e. Primarily OECD material (published and unpublished) from the Science, Technological and Industrial Indicators Division, and the Development Center.

f. Primarily CGIAR Study Papers and unpublished material from the 1984-85 CGIAR sponsored "Impact Study."

g. Includes data from Evenson and Kislev (1971, 1975, and April 1975), Judd, Boyce, and Evenson (1983 and 1986), and Boyce and Evenson (1975).

h. Includes, inter alia, data from Oram (1986), Oram and Bindlish (1981 and 1984), and Oram and Gieben (1984).

i. Includes, inter alia, dissertations, mimeos, working papers, and unpublished reports.

Sub-Saharan Africa: Angola, Benin, Botswana, Burkina Faso, Burundi, Cameroon, Cape Verde, Central African Republic, Chad, Comoros, Congo, Cote d'Ivoire, Ethiopia, Gabon, Gambia, Ghana, Guinea, Guinea-Bissau, Kenya, Lesotho, Liberia, Madagascar, Malawi, Mali, Mauritania, Mauritius, Mozambique, Niger, Nigeria, Rwanda, Sao Tome & Principe, Senegal, Seychelles, Sierra Leone, Somalia, South Africa, Sudan, Swaziland, Tanzania, Togo, Uganda, Zaire, Zambia, Zimbabwe.
WANA (West Asia & North Africa): Algeria, Cyprus, Egypt, Iran, Iraq, Israel, Jordan, Kuwait, Libya, Morocco, Oman, Qatar, Saudi Arabia, Syria, Tunisia, Turkey, United Arab Emirates, Yemen (Arab Rep.), Yemen (P.D.R.).
Caribbean & Central America: Antigua, Bahamas, Barbados, Bermuda, Belize, Costa Rica, Cuba, Dominica, Dominican Republic, El Salvador, Grenada, Guadeloupe, Guatemala, Haiti, Honduras, Jamaica, Martinique, Mexico, Montserrat, Nicaragua, Panama, Puerto Rico, St.Kitts-Nevis, St.Lucia, St.Vincent, Trinidad & Tobago, Virgin Islands (U.S.).
South America: Argentina, Bolivia, Brazil, Chile, Colombia, Ecuador, Guyana, Paraguay, Peru, Suriname, Uruguay, Venezuela.
Asia: Afghanistan, Bangladesh, Brunei, Burma, China, Hong Kong, India, Indonesia, Korea (Rep. of), Laos, Malaysia, Nepal, Pakistan, Philippines, Singapore, Sri Lanka, Taiwan, Thailand.
South Pacific: Cook Islands, Fiji, French Polynesia, Guam, New Caledonia, Papua New Guinea, Solomon Islands, Tonga, Tuvalu, Vanuatu, Western Samoa.
Oceania: Australia, Japan, New Zealand.
Europe: Austria, Belgium, Denmark, Finland, France, Germany (Fed. Rep. of), Greece, Iceland, Ireland, Italy, Netherlands, Norway, Portugal, Spain, Sweden, Switzerland, United Kingdom.
North America: Canada, United States.

However, in 1978, the CARIS project produced a three-volume publication, of which one volume provides details on agricultural research institutions and the research programs within each institution for 61 developing countries. The entries for this particular CARIS publication were drawn from replies to a special survey that was undertaken at the time, rather than from the project records held within the CARIS system. It includes information on the number of agricultural researchers and research expenditures at an institutional level, a format more suitable for aggregating to the national level.

The World Bank finances many development projects that include components involving agricultural research personnel and infrastructural development. The World Bank's project appraisals and evaluation reports often contain useful descriptive, and occasionally quantitative, information on the state of the existing national agricultural research system. During 1985 and 1986, the World Bank undertook two broad studies of the national agricultural research capacity in Western and South-Eastern Africa, respectively. Unfortunately for the purpose of the Indicator Series, these studies relied heavily on data from existing sources and added little new quantitative information to that already available.

During 1984 and 1985, the CGIAR undertook an extensive assessment of the impact of the international agricultural research centers (IARCs) on national agricultural research systems. The project commissioned a series of study papers which were published for the CGIAR by the World Bank. Many of these provided descriptive and, in several cases, quantitative information on specific national agricultural research systems.[5]

Several other agencies, including the Organization for Economic Cooperation and Development (OECD), the Statistical Office of the European Communities (SOEC), and the United Nations Educational Scientific and Cultural Organization (UNESCO) maintain ongoing efforts to compile science and technology indicators (STIs). Their research expenditure and personnel indicators span the whole spectrum of science and technology, in which the agricultural sciences, in many instances, play a relatively minor role. These STIs seek to achieve commensurability across scientific areas – in addition to the cross-country dimensions over time which are of immediate concern for the present series – hence they can often achieve only a restricted coverage in relation to the NARS concept discussed in Section 2 above. In fact, the published OECD and SOEC data are presented in such a manner that it is difficult to specify the institutional coverage of these series – of particular concern if these data are to be meaningfully integrated with commensurable data from other sources. The Scientific, Technological, and Industrial Indicators Division (STIID) of the OECD gave generous assistance to the present statistical effort, including access to their unpublished source and computer-file material. This substantially improved the commensurability of OECD data used in the Indicator Series.

The OECD and SOEC Science and Technology Indicators (STI) efforts are exclusively oriented towards member countries of the OECD and European Economic Community respectively, although some very useful, but ad hoc, work has also been undertaken on Africa by the OECD Development Center (Kassapu, 1976). The STI series reported by UNESCO for other than OECD countries has been weak or nonexistent, due to the fact that some developing countries have rather weak national statistical offices or in other cases give science and technology indicators relatively low priority.

In addition to STI series, OECD and UNESCO have published numerous studies on science policy. For example, during the 1960s, OECD published an extensive series of *Country Reports on the Organization of Scientific Research* which has continued, since 1970, in the form of *Reviews of National Science Policy*. UNESCO publishes its own series of *Science Policy Studies and Documents* which also contains useful qualitative and quantitative information on NARS.

3.2.3 *Scholars*

The important work by Evenson and his colleagues to compile global statistics on NARS has been mentioned earlier in this chapter. Boyce and Evenson's 1975 publication presented time-series data on national agricultural research and extension expenditures and personnel from the early 1960s (although in some instances 1950 and 1959 figures were also provided) through to 1973. The data were subsequently extended through to 1980 by Judd, Boyce, and Evenson (1983 and 1986). Naturally, this work drew heavily on the 1975 publication, as did the ISNAR/IFPRI 1981 report by Oram and Bindlish, *Resource Allocations to National Agricultural Research: Trends in the 1970s – A Review of Third World Systems*.

The cumulative process, whereby successive research efforts have simply augmented earlier data series, has contributed to some problems. Given the substantial difficulties in obtaining quantitative data on NARS, the primary data base has been rather thin. Consequently, analysts have often been forced to substitute interpolated or shifted observations for missing observations in order to generate constructed time series suitable for analytical purposes.

In some instances, the citation cycle, whereby successive works have included figures from earlier efforts, meant that constructed time series have been converted into actual time series, with the result that a 15- to 20-year constructed time series for some countries has been based on only one or two documentable observations. In developing the Indicator Series, a major effort has gone into tracing data back to their primary source and documenting all observations included in the series.

Previous series have also varied in scope, with Boyce and Evenson's 1975 series seeking to exclude forestry and fisheries research while Oram and Bindlish (1981), like the present series, attempted to include both these categories into their definition of agricultural research. Undocumented changes in institutional coverage over time may also contribute to spurious growth or fluctuations in measured research activity, especially given the gradually improving, but certainly far from ideal, state of the primary data

sources. Additional details concerning differences in measurement and conversion procedures between these previous series and the Indicator Series are developed in Sections 2 and 4.

3.2.4 *Additional Sources*

There are several directories that provide useful country-specific information on a global or semiglobal basis. The first includes a series of directories published initially in 1936, and on an occassional basis since, by the Commonwealth Agricultural Bureaux (CAB).[6] This *List of Research Workers* records the name, degree status, and current position of agricultural researchers working in publicly supported research institutions for most member countries of the Commonwealth. The institutional-level format of the listings makes it possible to distinguish university from nonuniversity staff – a feature of particular assistance when seeking to prorate the part-time research activities of most university personnel into full-time equivalents.

Longman's *Agricultural Research Centres Directory* lists detailed descriptive profiles of about 8,000 research laboratories and locations active in agricultural research and, since 1982, includes information on the research budgets and the number of research personnel of these research units. The eighth edition of Longman's directory, published in 1986, includes information on 132 developed and developing countries. Given that the unit of record is a research laboratory rather than an institute or organization, simply aggregating the often incomplete listings of country-specific research laboratories to yield aggregate NARS indicators can, in some instances, seriously understate national agricultural research capacity.[7]

3.3 Region-specific data sources

3.3.1 *Sub-Saharan Africa*

Regional studies of particular importance for the Indicator Series, especially for the earlier years, include Cooper's (1970) *Agricultural Research in Tropical Africa*; two FAO publications, namely, the *Index of Agricultural Research Institutes* by Brian N. Webster (n.d.) and *Directory of Agricultural Research Institutions and Projects in West Africa – CARIS Pilot Project 1972–73* (1973);

two UNESCO publications, namely, the *Survey on the Scientific and Technical Potential of the Countries of Africa* (1970) and *National Science Policies in Africa* (1973); and Samuel Kassapu's (1976) *Les Dépenses de Recherche Agricole dans les 34 Pays d'Afrique Tropicale*.

Cooper's 1970 study is one of the earliest quantitative assessments of the state of national agricultural research systems in Africa, following independence of many of these countries in the early sixties. The main limitation of this study is that its statistics relate to an exceptionally narrow NARS concept, to the exclusion of forestry, fisheries, and all university-based research. Moreover, agricultural research is taken to include research on crops and livestock but excludes veterinary research.[8] Kassapu's 1976 study provides more detailed institution-specific information which adheres to a broader definition of agricultural research, inclusive of forestry and fisheries research, and also includes university-based research. Kassapu also attempts to distinguish between local and donor funding, paying particular attention to the resources contributed by France to its former colonies. The two FAO-sponsored publications are in a directory-type format, listing agricultural research institutions and stations and briefly describing their current research activities. Both UNESCO publications attempt to report on all research conducted in African countries.

All six of these African reports were extensively used by Boyce and Evenson in their 1975 publication, although they drew from a 1973 draft version of Kassapu's publication, rather than the final 1976 version used here.

For the more recent period, DEVRES, in collaboration with the Southern Africa Development Coordination Conference (SADCC) and the Sahel Institute, undertook a USAID-financed resource assessment of the national agricultural research systems of the countries in the Sahelian region plus member countries of SADCC. These two studies each prepared eight country reports, an overview regional analysis and a strategy report, along with a set of data files on floppy disk. The project yielded a substantial amount of valuable new qualitative and quantitative information. However,

inconsistencies in coverage between the various summary tables presented in the published reports, plus numerous and apparently irreconcilable differences between the data contained in the published reports and on the floppy disks, points to the need for caution when using this data source.

During 1985 and 1986, the World Bank conducted two regional studies on Western and South-Eastern Africa, respectively. Various unpublished reports arising from this study provided useful qualitative but little additional quantitative information.

3.3.2 *Asia and the Pacific*
Overview studies of the Asian and Pacific region are few in number, probably because the relatively large size of many Asian systems makes a country-level approach more appropriate, while the relatively small size and disparate locale of many Pacific NARS ensures that the cost of any regional analysis is often prohibitively high. Nevertheless, four regional studies that proved useful for the Indicator Series are the 1962 publication by Chang, *FAO Directory of Agricultural Research and Experiment Stations in Asia and the Far East*, which contains fairly detailed institutional-level data; Moseman's (1971) *National Agricultural Research Systems in Asia*, which includes a descriptive report with some quantitative data on several research systems in Asia; the 1981 publication by Gamble et al., *South Pacific Agricultural Research Study*, which gives a full description of the agricultural research systems for seven island states plus some brief comments on the agricultural research capacity of seven others (for many island states, this study represents the first description of their NARS available in published form); and a 1986 study published by FAO's Regional Office for Asia and the Pacific entitled *Agricultural Research Systems in the Asia-Pacific Region*, which includes a descriptive, and in many instances a quantitative, overview of the local agricultural research expenditure and personnel commitment for 17 countries in the region.

There have also been several regional workshops and conferences whose proceedings contain useful country-level NARS data. The

Philippine Council for Agriculture and Resources Research and Development (PCARRD) and the International Development Research Center (IDRC) jointly sponsored two workshops in Singapore in 1981 and 1982, the proceedings of which were published in 1983 as *Agriculture and Resources Research Manpower Development in South and Southeast Asia*. This publication contains several country-level reports of national agricultural research capacity, including substantive data on the commitment of human resources to research.

3.3.3 *Latin America and the Caribbean*

Two regional organizations that play an important role in agricultural research in Latin America and the Caribbean are Instituto Interamericano de Cooperacion para la Agricultura (IICA) and the Caribbean Agricultural Research and Development Institute (CARDI). IICA assists NARS by coordinating and facilitating national agricultural research activities in the whole region, while CARDI is directly involved in executing agricultural research in the Caribbean region. CARDI's annual reports proved to be useful sources of information. IICA produced two frequently cited references, namely, the 1979 publication by Segura, *Diagnóstico de la Investigación Agropecuaria en el Istmo Centro Americano*, and the edited volume by Piñeiro and Trigo published in 1983, *Cambio Técnico en el Agro Latinoamericano: Situación y Perspectivas en la Década de 1980*.[9] The Pineiro and Trigo volume includes NARS expenditures and personnel data for 20 countries in the region. Unfortunately, the expenditure data were converted in a nondocumented, nonreplicable manner to constant 1975 U.S. dollars, thereby impeding the use of these data in the Indicator Series.

Additional data sources which each contain information for several NARS in the region are the proceedings of the *Caribbean Workshop on the Organization and Administration of Agricultural Research*, held in Bridgetown, Barbados in 1981; the 1984 publication *Agricultural Research Policy and Management – Volumes I and II*, which includes papers from a workshop sponsored by the Economic Commission for Latin America and the Caribbean (ECLAC), held in Port of Spain, Trinidad, in 1983; and the papers presented at the "Curso-taller sobre la Administración de la Investigación Agricola" held in Panama City in 1986.

3.3.4 *West Asia and North Africa*

The West Asia and North Africa (WANA) region has rather poorly documented or difficult-to-access published material concerning local NARS activities. This problem is particularly acute for the West Asian NARS of Iran, Iraq, Jordan, Kuwait, Lebanon, Oman, Qatar, Saudi Arabia, Syria, Turkey, United Arab Emirates, Arab Republic of Yemen, and People's Democratic Republic of Yemen. Regional studies are virtually nonexistent or incomplete and simply descriptive in nature, for example, the FAO (1979) *Directory of Agricultural Research Institutions in the Near East Region*. However, in some instances specific WANA countries have been included in regional studies of Asia and Africa. For example, the two UNESCO (1970 and 1973) publications on African NARS included information on the NARS of several WANA countries, as did the earlier, 1962, publication by FAO, *Directory of Agricultural Research Institutions and Experiment Stations in Asia and the Far East*.

More recently, useful papers describing agricultural research capacity for several WANA countries have been presented at regional conferences and workshops, such as the "Rainfed Agricultural Information Network Workshop" held in Amman from 17–20 March 1985, "Séminaire sur l'Orientation et l' Organization de la Recherche Agronomique dans les Pays du Bassin Méditerranéen" held in Istanbul from 1–3 December 1986, and the second conference of the Association of Agricultural Research Institutions in the Near East and North Africa (AARINENA) held in Nicosia from 15–17 December 1987.

3.3.5 *Western Europe, North America, and Oceania*

The Scientific, Technological and Industrial Indicators division of the OECD has been the most important single source of information for many of the industrialized countries. Since the late fifties, OECD has published numerous reviews of national science policies, along with various comparative and topical studies of national scientific activity. Agricultural research, however, has generally been analysed in the context of the total research program of a country. Very few OECD studies have concentrated specifically on agricultural research. This contributes to several commensurability problems when attempting to use published OECD data with data from other sources. These and related issues are discussed at some length in Section 3.2.2.

3.3.6 *Eastern Europe*
None of the Eastern European countries are included in this version of the Indicator Series due to a scarcity of usable information. The limited data currently available are difficult to interpret and in a form which makes commensurability with the rest of the series problematic. Moreover, a lack of suitable deflators and exchange rates impedes the possibility of converting research expenditure data from countries in the region into comparable volume measures.[10]

3.4 Country-specific data sources
The discussion to date has focused on sources that provide data on multiple NARS. However, in most cases the Indicator Series drew heavily from country-specific sources, including replies to ISNAR benchmark surveys; ISNAR system reviews, monographs, and published and unpublished reports; (NARS) annual reports; country studies; appraisal and evaluation reports of investments in agricultural research infrastructure and human resource development; monographs and dissertations on national agricultural research or related topics; and personal communication. In fact, personal communication has been the single most important source of information, accounting for 15% and 18% of the personnel and expenditure observations, respectively.

Notes
1. The ISNAR surveys include three surveys that were carried out on a global basis in cooperation with the International Agricultural Research Systems for Development (IFARD) and, on a regional basis, with the assistance of the Arab Organization for Agricultural Development (AOAD) and the Asian Development Bank (ADB). Reference details for all source material cited in this section are included in the alphabetic listing of sources following Part II of this volume.
2. Grey literature was taken to include mimeos, papers presented at seminars, dissertations, and personal communication.
3. See Elliott and Pardey (1987) for a more complete description of these surveys.
4. This is not to be interpreted as a limitation of CARIS per se. It simply reflects the fact that it is often difficult to use data bases structured for one particular use for other purposes. For example, Hallaway (forthcoming) notes similar difficulties when attempting to use the USDA's current research information system (CRIS) to compile a statistical inventory of U.S. national agricultural research capacity.
5. Twenty-four Study Papers have been published to date.
6. Over the last 25 years, the CAB listing has been published in 1966, 1969, 1972, 1975, 1978, and 1981.
7. A careful study of these directories since 1982 also suggests that the quantitative information on research personnel and expenditures may not be subject to systematic updating.
8. See Cooper (1970, 128–129) who observed that if university, forestry, fisheries, and veterinary research were included in his figures "the number of research workers indicated would be immensely larger."
9. These data are also replicated in *Selected Issues in Agricultural Research in Latin America*, a 1984 ISNAR publication edited by Nestle and Trigo.
10. See Marer (1985) for a recent attempt to develop usable exchange rate convertors for these nonmarket economies.

4. Agricultural research expenditure deflators and currency converters

The agricultural research expenditure data contained in the Indicator Series are compiled in current local currency units. The cross-sectional, temporal nature of the data series means these expenditure figures must often be recast into commensurable currency units before they are in a form suitable for comparative analysis.[1] Unfortunately, any international comparisons using expenditure data of this type confront an intractable index number problem whereby international (cross-sectional) and temporal comparisons are generally not independent of choices between alternative weighting schemes.[2] Nevertheless, there are preferred weighting schemes that depend to a large degree on the uses envisaged for the data.

This section describes some of the weighting options that are available, then explores the empirical differences among several of these options. The section concludes with a summary of the approaches adopted for this version of the Indicator Series. The idea was, first, to minimize or at least standardize the currency conversions that were applied to the raw data[3] and, second, to make these conversions as transparent to the user as possible. Like the measurement and definitional issues described in the previous section, the weighting methods used here represent a practical compromise. Nevertheless, there continues to be a substantial effort aimed at achieving more meaningful international comparisons of basic economic indicators which should ultimately prove useful for future versions of the Indicator Series. Compiling the raw data for the Indicator Series in current local currency units will enable future versions of the series, and users of the current version, to take advantage of any developments in this area.

4.1 Alternative methods

In some cases it is possible to make comparative judgments about the data that only involve direct weighting of the expenditure series to account for variations over time – within country rather than between countries – in average price levels. For instance, a cross-country assessment of growth in the 'quantity' or volume of resources committed to national agricultural research activity would only require deflating research expenditures measured in local currency units by a suitable agricultural research expenditure deflator.[4] The quantity or volume growth rate calculation could then be performed on expenditures measured in weighted local currency units, without having to express research expenditures in a numeraire currency.

In other circumstances, it may be possible to avoid using deflation or currency conversion techniques altogether. A comparison of different countries' relative agricultural research expenditures, constructed on the basis of own-price research spending shares, provides an appropriate measure of public support for agricultural research.[5] However, if the overall price of agricultural research goods and services relative to the prices of other goods and services in an economy varies substantially among countries, then research expenditure shares based on own-price figures will give an incorrect measure of the relative volume of resources committed to agricultural research. In this case the preferred alternative is to weight own-price agricultural research expenditure and, for example, agricultural gross domestic product (AgGDP) figures by their respective price indices, to obtain commensurable volume measures before forming the desired volume ratio.

The choice of an appropriate price index entails some conceptual difficulties which will be discussed in detail below. The practical difficulties of obtaining global indices – which enable accurate volume measures to be derived from weighted expenditure figures – means that volume ratios may best be obtained from directly calculating own-price expenditure ratios than by weighting expenditures with inappropriate price indices.

But, in many instances it is not possible to work with weighted local currency units or own-price expenditure ratios. It is often necessary to aggregate country-specific research expenditure data – whether measured in absolute terms or relative to some nonexpenditure figure, such as agricultural population, arable land, or the like – in order to form regional and/or temporal aggregates for use in comparative analyses. There is, consequently, no option but to convert research expenditures measured in current local currency units into some numeraire currency or unit of measurement.

From the various algorithms available for handling price-driven cross-sectional variability over time in the research expenditure data, a practical alternative is to select a two-step procedure which either[6]
(1) first converts local currency expenditure data into a numeraire currency (typically U.S. dollars), then applies an appropriate (U.S.) price deflator to account for price variability over time;
(2) first deflates own-price expenditure data, using appropriate (country-specific) price deflators to account for variability in average price levels over time, then converts into a numeraire currency using a base-year exchange rate.

There are numerous deflators and currency converters that can be incorporated into either algorithm. Six different methods of deriving commensurable volume measures of weighted agricultural research expenditures will be considered here. They compare options used in the past with those which are currently tractable when dealing with a panel data set of 154 countries covering 27 years from 1960 to 1986.

The first three methods use a convert-first algorithm to translate foreign currencies into U.S. dollars using alternative (annual) currency conversion ratios, and then divide every country by a particular U.S. price index to form a volume measure of weighted agricultural research expenditures. The remaining three methods apply a deflate-first algorithm and first deflate research expenditures measured in current local currency units using country-specific implicit GDP deflators, and then translate into a numeraire currency using alternative base-year (1980) currency conversion ratios.

4.1.1 *Specific formulae*
To clarify the discussion of the weighting options that are available it is useful to introduce some notation. The data include

R_t = agricultural research expenditures measured in current local currency units (LCUs);

P_t = local (i.e., country-specific) implicit gross domestic product deflators;

$P_t(\$)$ = implicit gross domestic product deflator for the U.S.;

$WP_t(\$)$ = U.S. wholesale price index;

e_t = annual average market exchange rate, LCUs per U.S. dollar;

e^*_t = World Bank Atlas exchange rate, LCUs per U.S. dollar;

PPP_t = purchasing power parity exchange rate, i.e., the LCU price of a commodity bundle divided by the dollar price of the same bundle.

For this particular application all price deflators are indexed at 1980 = 100. The annual average exchange rate is a yearly official 'market' rate which generally corresponds to the IMF's rf or (inverted) rh rate. The World Bank's 1980 Atlas exchange rate was, for most countries, obtained from unpublished World Bank sources.[7] In several instances, where Atlas rates were not provided, available implicit GDP deflators and annual average exchange rates were used to construct 'Atlas-equivalent' rates in line with the World Bank's current method of construction (see equation 4.1 below). The 1980-based Atlas exchange rate represents an arithmetic average of the actual 1980 average exchange rate and the exchange rates of the previous two years adjusted for relative price changes. It gives the conversion factor for year t = 1980 as follows (where all variables are defined as before).[8]

$$e_t^* = 1/3 \left[e_{t-2} \left[\frac{P_t}{P_{t-2}} \bigg/ \frac{P_t(\$)}{P_{t-2}(\$)} \right] \right.$$

$$\left. + e_{t-1} \left[\frac{P_t}{P_{t-1}} \bigg/ \frac{P_t(\$)}{P_{t-1}(\$)} \right] + e_t \right]$$

(4.1)

The idea is to abstract from short-term fluctuations in exchange rates by constructing an average of two estimated 'real exchange rates' for the single relevant year (1980) and the actual exchange rate for that year. The formula draws implicitly on the Casselian purchasing power parity doctrine that nominal exchange rates adjust fully for relative (i.e., cross-country) price changes over time, such that the 'real exchange rate' – defined as $r_t = e_t \, (P_t(\$)/P_t)$ – is constant over time. In some instances tariffs, subsidies, export or import quotas, government exchange rate policies, and the like cause the (official) 'market' exchange rates used to calculate e^* to deviate egregiously from the rate that applies to the foreign transactions effectively taking place.[9] In such cases, the World Bank draws on secondary data concerning the nature and estimated impact of these distortions to adjust official exchange rates accordingly.[10]

The problem with using exchange rates as a basis for currency conversions, when estimating cross-country volume measures based on own-price expenditure data, is that exchange rates are unreliable measures of the purchasing power of a currency. For example, the per capita GDP for Brazil in 1985 was $ 1,673 when converted at the annual average exchange rate of 6,200 cruzeiros per U.S. dollar. This was only 10.1% of the U.S. per capita GDP figure. However, when converted using the relative 1985 purchasing power of the two currencies given by Summers and Heston (1988), the Brazilian figure was actually $ 3,841, which is 23.2% of the corresponding U.S. figure.

Purchasing power parities (PPP), by definition, measure the domestic cost of buying a bundle of goods and services in a particular country at its own prices relative to the corresponding cost in, say, dollars of the same bundle in the United States.[11] In using PPPs to measure relative price levels, there is clear evidence that, as expected, average price levels are positively associated with per capita income. Moreover, as summarized by Heston and Summers (1988), there is overwhelming evidence that figures converted by annual average exchange rates vary from PPP converted figures in a significant and systematic manner. In particular, the exchange rate deviation index – measuring annual average exchange rates relative to PPPs – is generally greater than unity for low-income countries, often slightly less than unity for high-income countries, and by definition equal to unity for the U.S.

The reason for this pattern of deviations between exchange rates and PPPs lies, to a large degree, in differences across countries in the relative prices of tradable versus nontradable goods and services. In addition to short-run influences on relative price levels across countries, there are several structural factors that cause the prices of nontradables to be lower than tradables, the lower a country's per capita income.[12] Nontradables are generally more labor intensive than tradables, while productivity differences between low- and high-income countries tend to be lower in nontradables than tradables. When combined with the fact that labor is relatively cheap – in opportunity cost not necessarily in terms of labor services – in low-income countries, these labor intensity and productivity differences lead to relatively lower priced nontradables to tradables in low- versus high-income countries.

The overall result is that when measured, for example, by Summers and Heston (1988) as the inverse of the exchange rate deviation index, the average price level for GDP as a whole is relatively low in low- as compared with high-income countries. And because exchange rates are more heavily influenced than PPPs by the price of tradables, this disparity in relative price levels is muted if exchange rates are used to convert own-price expenditure figures into comparable volume measures. Consequently, value measures such as GDP or agricultural research expenditures, when converted by annual exchange rates, tend to understate the relative volume measure in low-income countries compared with PPP-converted figures.

Returning to the notation introduced earlier, we can now write the six formulae used to convert R_t, an own-price agricultural research expenditure figure for year t, to a corresponding volume measure, V^b_t, measured in base-year, b, dollars. Three conversion methods that apply the convert-first algorithm are

$$V^b_t(1) = (R_t/e_t) (WP_b(\$)/WP_t(\$)) \qquad (4.2)$$

and

$$V^b_t(2) = (R_t/e_t) (P_b(\$)/P_t(\$)) \qquad (4.3)$$

where agricultural research expenditures are first converted into U.S. dollars using the annual average (dollar) exchange rate, e_t, and then translated into base-year dollars by applying the U.S. wholesale price index in the case of method (1) and the U.S. implicit GDP deflator in the case of method (2). Method (1) is apparently the approach used by Judd, Boyce and Evenson (1983 and 1986), Boyce and Evenson (1975), and Evenson and Kislev (1971) to convert agricultural research expenditure data into comparable units.[13] The third method uses the Summers and Heston (1988) series on PPP exchange rates to convert to U.S. dollars year by year and then applies the U.S. implicit GDP deflator

$$V^b_t(3) = (R_t/PPP_t) (P_b(\$)/P_t(\$)) \qquad (4.4)$$

The remaining three conversion methods use the deflate-first approach such that

$$V^b_t(4) = (R_t(P_b/P_t))/e_b \qquad (4.5)$$

where agricultural research expenditures are first translated into base-year local currency units using country-specific implicit GDP deflators, and then converted into year b dollars using the base-year annual average (U.S. dollar) exchange rate,

$$V^b_t(5) = (R_t(P_b/P_t))/e^*_b \qquad (4.6)$$

where the base-year Atlas exchange rate, given by equation (4.1), is substituted for the annual average exchange rate, and

$$V^b_t(6) = (R_t(P_b/P_t))/PPP_b \qquad (4.7)$$

where the local currency-to-dollar conversion is now achieved using Summers and Heston's (1988) series on PPP exchange rates.

As Leamer (1988) observed, when discussing various weighting procedures used for international comparisons of capital stock measures, the research expenditure series V(2) can be thought to correct the research expenditure series V(4) for inflation-driven misalignment of the exchange rate. The V(2) series is equal to the V(4) series weighted by an index of the real exchange rate: $V(2) = V(4)/(r_t/r_b)$, where the real exchange rate – also defined earlier with respect to equation 4.1 – is $r_t = e_t (P_t(\$)/P_t)$. These weighted expenditure series, V(2) and V(4), would then be identical if the real exchange rate were constant over time, $r_t = r_b$, that is, if variations in relative prices between countries over time were exactly offset by variation in the nominal exchange rate over time. Furthermore, V(6) can be viewed as a corrected version of the V(4) series, where the weighting factor this time is the reciprocal, $1/P_{Ib}$, of the base-year price index for GDP, $P_{Ib} = PPP_b/e_b$, so that $V(6) = V(4)/P_{Ib}$. Similarly, V(3) can be viewed as a corrected version of V(2) such that $V(3) = V(2)/P_{It}$ where $P_{It} = PPP_t/e_t$.

4.1.2 Sensitivity Analysis

This section presents evidence of the sensitivity of estimating commensurable volume measures from own-price research expenditure data using the six conversion methods described above. Agricultural research expenditures, measured in local currency units, were obtained for 107 developed and developing countries for at least one of the five years 1981 to 1985. Indices for purchasing power parity over GDP were obtained for 121 countries from Summers and Heston (1988), while annual average exchange rates and implicit GDP deflators were generally available from various IMF and World Bank sources. (Appendix A contains more specific details.) The

Table 4.1: Simple correlation coefficients across six agricultural research volume measures

	V(1)	V(2)	V(3)	V(4)	V(5)	V(6)
V(1)	1.000					
V(2)	1.000	1.000				
V(3)	0.954	0.952	1.000			
V(4)	0.995	0.994	0.944	1.000		
V(5)	0.994	0.993	0.947	0.996	1.000	
V(6)	0.964	0.961	0.992	0.964	0.962	1.000

six conversion methods, V(1) through V(6), were applied to the 1981-1985 agricultural research expenditure data to yield comparative estimates of the annual average volume of resources committed to agricultural research for 89 countries.

The simple correlation coefficients across these six volume measures are presented in Table 4.1.

These uniformly high correlation coefficients indicate that country deviations from the mean volume of resources committed to agricultural research are linearly related across the six conversion methodologies under consideration. Consequently, countries that deviate to a large degree from their sample mean using one method are most likely to deviate by a relatively large degree using any of the other methods, and the same is true for countries that deviate by a small degree. Unfortunately, correlation coefficients give no indication of possible differences in the mean volumes implied by each conversion method, nor the relative magnitude across conversion methods of the deviations from such means. A large positive deviation from its sample mean may involve a doubling of the volume of research resources by one method but a tripling of the resources by another. In the same way, a large negative deviation may involve a reduction by half with one method or a reduction by two-thirds by another.

Table 4.2 gives some insights into differences across conversion methods in the mean volume of research resources and relative, country-specific deviations from these means.

Columns 2 to 7 measure the 1981-85 average annual volume of resources committed to agricultural research for 67 developing countries, subdivided into four geographic regions and 21 developed countries, plus the United States, using the six conversion methods under consideration. In this case the regional share is indexed for each conversion method on the 89-country global volume of research resources. Thus the V(1) measure implies that 4.4% of the global volume of resources committed annually to agricultural research over the 1981-85 period was directed to Sub-Saharan Africa, while developed countries as a group, including the United States, accounted for 80.7% of the global total.

For this 1981-85 period at least, the regional shares exhibit nontrivial sensitivity to the conversion method used. Moreover, the regional shares, particularly for the Asia and Pacific region, obtained with the PPP-based V(3) and V(6) methods deviate egregiously from the shares obtained using the alternative conversion methods. Using PPPs rather than exchange rates to convert own-price research expenditures into commensurable volume measures more than doubled the Asia and Pacific region's share of global research resources from around 6.5% to nearly 15%. For the global sample the V(1) method, used widely to date by Evenson and others, gives an annual average research commitment of US$ 5,832 million – nearly a billion dollars (US$ 951 million) less than the corresponding PPP-based estimate, V(3).[14]

In general terms, the difference between methods is even more dramatic for developing

Table 4.2: Six volume measures of annual (1981-85 average) agricultural research resources aggregated by region and indexed on the total sample volume of research resources and measure V(5), respectively

Region	N[a]	Volume of Agricultural Research Resources Indexed on the Total Sample Volume						Volume of Agricultural Research Resources Indexed on Measure V(5)					
		Convert-First			Deflate-First			Convert-First			Deflate-First		
		V(1)	V(2)	V(3)	V(4)	V(5)	V(6)	V(1)	V(2)	V(3)	V(4)	V(5)	V(6)
Column number	1	2	3	4	5	6	7	8	9	10	11	12	13
DEVELOPING COUNTRIES													
Sub-Saharan Africa	30	4.4	4.4	4.8	4.4	4.3	4.8	93.5	88.5	117.2	93.5	100.0	115.4
Asia & Pacific	11	6.3	6.1	14.7	6.7	6.4	14.9	89.9	82.4	243.6	97.2	100.0	244.0
Latin America & Caribbean	17	6.5	6.7	8.8	4.8	5.7	6.5	103.4	100.9	162.8	77.8	100.0	119.1
West Asia & North Africa	9	2.1	2.1	2.8	1.6	1.9	2.3	96.5	92.4	152.0	74.1	100.0	125.0
SUBTOTAL	67	19.3	19.3	31.1	17.4	18.3	28.5	95.6	90.6	179.0	87.8	100.0	162.3
DEVELOPED COUNTRIES													
Developed Countries Exclusive U.S.	21	55.7	55.4	48.3	59.2	60.0	50.6	84.0	79.2	84.8	91.3	100.0	87.6
United States	1	25.0	25.3	20.6	23.4	21.7	20.9	104.3	100.0	100.0	100.0	100.0	100.0
SUBTOTAL	22	80.7	80.7	68.9	82.6	81.7	71.5	89.4	84.7	88.8	93.6	100.0	90.9
TOTAL	89	100.0	100.0	100.0	100.0	100.0	100.0	90.5 (86.7)[b]	85.8 (81.9)	105.3 (106.8)	92.5 (90.5)	100.0 (100.0)	103.9 (105.0)

a. N = number of countries.
b. Figures in parentheses represent total, exclusive of U.S.

versus developed countries. The V(1) method suggests that developing countries account for 19.3% (US$ 1,126 million) of the global volume of research resources, while the V(3) estimate of US$ 2,108 million is 31.1% of the PPP-based global volume estimate – an increase in the developing countries' share of the global total, due simply to differences in conversion methodology, of approximately 61%. The corresponding share for developed countries declined by approximately 15% from 80.7% to 68.9%. Thus, relative price levels in developing countries derived from Summers and Heston's PPPs appear to be much lower than those implied by market exchange rates (especially for the Asia and Pacific region), while the reverse holds, on average, in the case of developed countries.

Indexing regional volumes for each conversion method on a global total reveals the substantial amount of variability that exists across conversion methods in the within-, rather than between-, region dimension. Columns 8 to 13 in Table 4.2 show regional volume measures derived from each conversion method indexed on the deflate-first (Atlas exchange rate-based) V(5) method. The PPP-based methods, V(3) and V(6), increase – in some cases substantially – the estimated volume of research resources relative to other conversion methods for all of the developing country regions, while decreasing the relative volume measure, on average, for the developed countries. In this within-region dimension, the method based on the Atlas exchange rate, V(5), generally comes closest to the

PPP-based measures but still appears to underestimate the volume of resources committed to agricultural research by a nontrivial amount.

An instructive way to summarize the impact of conversion method on the volume of resources committed to agricultural research is given by the volume of research inputs per scientist ratios in Table 4.3. The two PPP-based measures V(3) and V(6) give a ratio of average annual volume per scientist for the 1980-85 period of US$ 69,616. This is 39%, 47%, 37%, and 25% greater than the corresponding V(1), V(2), V(4), and V(5) measures, respectively. Using PPPs rather than annual exchange rates to convert research expenditures increases the minimum observation, decreases the maximum observation, and substantially lowers the coefficient of variation of the ratios for volume per scientist. In fact, the PPP-based conversions increase the minimum ratio of volume per scientist by 65% compared with the corresponding average of the figures converted by exchange rates, while decreasing the maximum ratio of volume per scientist by around 15%. Given that expenditures converted by exchange rates tend to understate the resulting volume measure for low- relative to high-income countries, these results are to be expected.

To date, the sensitivity analysis has focused on data clustered near the 1980 base period. Interactions between various exchange rates (including PPPs) over time and the implicit GDP deflator may introduce spurious variability into

Table 4.3: Agricultural research volume (US$) per scientist ratios[a]

Conversion Method	Mean	Standard Deviation	Coefficient of Variation	Minimum	Maximum
V(1)	50,221	36,627	0.729	6,109	194,840
V(2)	47,259	34,295	0.735	5,787	181,223
V(3)	69,871	34,657	0.496	9,589	149,138
V(4)	50,810	35,053	0.690	5,900	194,817
V(5)	55,819	38,638	0.692	5,382	164,665
V(6)	69,360	37,777	0.545	9,498	162,974

a. Calculated as a simple average of 1981-1985 average ratios for a 74-country sample.

Figure 4.1: Percent deviation of convert-first, V(2), versus deflate-first, V(4), algorithms using annual average exchange rates and implicit GDP deflators[a] (base year = 1980)

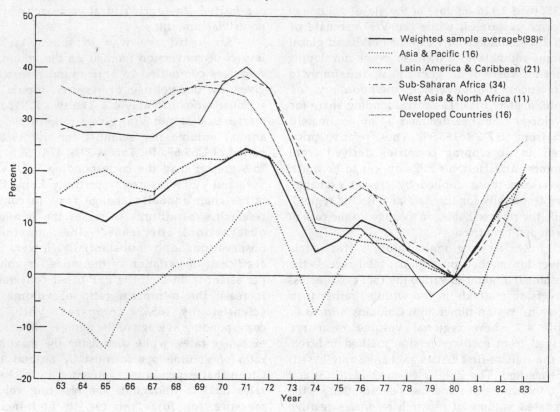

a. Calculated as $([V(4) - V(2)]/V[4]) \times 100 = (1 - [r_b/r_t]) \times 100$ where $r_t = e_t(P_t[\$]/P_t)$.
b. Regional and global averages weighted by proportion of PPP-adjusted 1980 AgGDP for each group accounted for by each country.
c. Figures in parentheses represent number of countries.

the volume measures which is not fully captured by the 1981-85 snapshot. In particular, the volume measure may be sensitive to whether own-price research expenditure data are first deflated or first converted, independent of the specific currency convertors or deflators used.

Figure 4.1 plots the percent deviation of deflate-first, V(4), versus convert-first, V(2), volume measures when annual average exchange rates and implicit GDP deflators are used to derive the respective volume measures.[15] For most regions – Asia and Pacific excepted –

and the sample as a whole, the algorithm, V(4), that first deflates own-price research expenditure figures using country-specific implicit GDP deflators and then converts into U.S. dollars using base-year annual average exchange rates, yields a volume measure that is consistently larger than the volume measure derived from the corresponding conversion algorithm, V(2), which first converts country-specific figures into current U.S. dollars and then deflates all countries using the U.S. implicit GDP deflator. This suggests that, ceteris paribus, either the U.S.

dollar was undervalued with respect to many countries' LCUs in 1980 relative to its historical trend, or in the pre-1980 period local inflation rates were generally higher than the U.S. inflation rate (with the reverse being true for the post-1980 period); or movements in relative inflation rates across countries are imperfectly transmitted to countervailing shifts in relative exchange rates.

Figure 4.2 plots the percent deviation of deflate-first, V(6), versus convert-first, V(3), volume measures when purchasing power parity exchange rates and implicit GDP deflators are used to derive the respective volume measures. Here the pattern of deviations over time are quite muted when compared with the deviations in Figure 4.1. This is not unexpected as, by construction, changes in PPPs over time should

Figure 4.2: Percent deviation of convert-first, V(3), versus deflate-first, V(6), algorithms using PPP over GDP exchange rates and implicit GDP deflators[a] (base year = 1980)

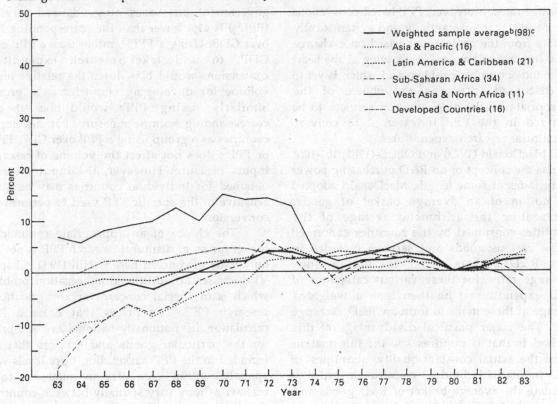

a. Calculated as $([V(6) - V(3)]/V[6]) \times 100 = (1 - [r_b/r_t^*]) \times 100$ where $r_t^* = PPP_t(P_t[\$]/P_t)$.
b. Regional and global averages weighted by proportion of PPP-adjusted 1980 AgGDP for each group accounted for by each country.
c. Figures in parentheses represent number of countries.

counterbalance changes in relative inflation rates among countries. Moreover, for the 1970s and 1980s there appears to be little substantive difference between the deflate-first and convert-first methodologies where PPP exchange rates are used as currency convertors.

4.1.3 *Further Issues*

The choice of conversion algorithm clearly affects the research input volume measure that is derived from an own-price expenditure figure – particularly for data converted by annual exchange rates. Moreover, PPP-based conversions give results that substantially and systematically deviate from the conversions based on exchange rates. But, a fundamental issue, lying at the heart of the index number problem and which is yet to be discussed, concerns the choice of the appropriate basket of goods and services to be included in the PPP index used to convert agricultural research expenditures.

MacDonald (1973) and OECD (1981, 103-108) discuss the concept of an R&D purchasing power parity index at some length. MacDonald adopted the notion of an average basket of goods, representing the arithmetic average of the quantities consumed by the countries concerned of the various goods and services included in their R&D baskets. After calculating 'sub-exchange rates' for these various categories of R&D expenditures, he then took a weighted average of these items to form an 'R&D exchange rate'. The major practical disadvantage of this method is that it requires specific information about the actual constant-quality 'quantities' of R&D inputs in individual countries in order to calculate the average basket of R&D goods and services. As OECD (1981) points out, however, it is possible to modify MacDonald's procedure by substituting price ratios for actual prices and use a weighting scheme based on average expenditures (across countries) by type of cost rather than average quantities.

This is analogous to the method used by Summers and Heston (1988) to derive the PPP indices used in the sensitivity analysis reported above.[16] However, these PPPs were measured with respect to a basket of goods and services that relate to the whole economy. Whether this basket is a reasonable approximation for the mix of goods and services committed on-average to agricultural research is open to question.

Summers and Heston (1988) do present PPP indices for investment and government purchases of goods and services. Table 4.4 presents summary statistics for the 1980 value of both these PPPs measured in relation to the PPP index over gross domestic product (GDP). For developing countries as a group, the PPP over investment goods and services (PPP^I) is 40% larger than the PPP over GDP, while the PPP over government purchases of goods and services (PPP^G) is 41% lower than the corresponding PPP over GDP. Using a PPP^G, rather than a PPP over GDP, to undertake research expenditure conversions would bias down the relative input volume for developing countries as a group. Similarly, using PPP^I would bias up the corresponding volume measure. For developed countries as a group using a PPP over GDP, PPP^I, or PPP^G does not affect the volume of research inputs measure. However, the input measure obtained for individual countries may be quite sensitive to the specific PPP used to perform the conversion.

The choice of an appropriate commodity basket for an agricultural research PPP is not the only issue of concern here.[17] Hill (1982, Chapter V) discusses an additional aggregation problem which is of special concern for an agricultural research PPP. The prices that enter a PPP calculation are notionally national average prices for the particular goods and services that are included in the PPP basket. But, price levels vary spatially within a country in much the same manner as they vary spatially between countries. For many – particularly developing – countries, the prices that enter a PPP calculation are often based on prices obtained from one or two major urban areas. However, in numerous cases national agricultural research agencies support an infrastructure that includes research facilities situated at various urban and rural locations. The average price levels confronting a national research system may therefore vary substantially from the average price levels used to calculate existing PPP indices.

Table 4.4: Purchasing power parity (PPP) indices for investment (I) and government goods and services (G) measured relative to PPP over gross domestic product (year = 1980)

Region[a]	N[b]	PPP^I/PPP			PPP^G/PPP		
		Mean	Min	Max	Mean	Min	Max
DEVELOPING COUNTRIES							
Sub-Saharan Africa	30	164	126	255	57	41	83
West Africa & North Africa	9	148	102	231	66	53	141
Asia & Pacific	11	141	106	210	50	34	82
Latin America & Caribbean	17	117	64	202	78	46	98
SUBTOTAL	67	140	64	255	59	34	141
DEVELOPED COUNTRIES	22	100	85	145	100	60	133
TOTAL	89	112	64	255	88	34	141

a. Regional averages represent a weighted average of country-specific PPP ratios with weights of average 1981-85 agricultural research expenditures expressed in constant 1980 PPPs.
b. N = number of countries.

Analogous problems also confront attempts to calculate appropriate agricultural research expenditure deflators. Pardey, Craig, and Hallaway (1989) calculated an agricultural research deflator for the United States which took account of year-to-year changes in the factor mix of agricultural research expenditures.[18] Even at the national level, the composition of research expenditures shifted quite substantially over time. For the state agricultural experiment stations, capital expenditures (inclusive of land, buildings, and equipment) averaged 8.5% of total expenses during the 1955 to 1974 period. Yet in 1912, some 25 years after the formal establishment of the experiment station system, these expenditures accounted for 28.6% of total expenditures. Failure to account for these shifts in factor mix when constructing an agricultural research expenditure deflator can seriously distort the 'real' research expenditure or volume measure derived from inappropriately weighted expenditure figures.

4.2 Summary
It should be clear by now that the choice of technique used to convert own-price expenditure figures into commensurable volume measures is by no means a trivial matter. The procedure adopted can have substantive quantitative and

qualitative effects on the data. The preferred conversion procedure, in this instance, would employ purchasing power parities and deflators constructed on the basis of a basket of goods and services that relate specifically to agricultural research expenditures. These are not currently available on a global basis, and for practical reasons there is no alternative but to work with proxy PPPs and deflators.

This version of the Indicator Series presents research expenditure data in current and constant local currency units. As discussed earlier, expenditure data in this form can be used for various purposes and in ways that avoid many conversion difficulties altogether. For those cases in which commensurable volume measures are indispensable, two weighted expenditure series are also included.

The first is a V(6) volume measure, where agricultural research expenditures are initially translated into base-year (1980) local currency units using country-specific implicit GDP deflators and then converted into base-year dollars using the Summers and Heston (1988) series on PPP exchange rates. The index of PPPs over GDP was used in lieu of research expenditure-specific PPPs. PPPs at the disaggregated investment and government expenditure levels were available and gave volume measures that were substantially different from the PPP over GDP converted estimates. It is possible that the basket of goods and services in an agricultural research PPP index could be approximated by some appropriately weighted sum of the baskets of goods and services used to calculate investment and government-sector PPPs. Nevertheless, the more broadly defined PPP index was used because the appropriate weighting scheme is unknown and because, as Heston and Summers (1988) concede, their estimated PPPs at the disaggregate level do not reconcile well with the benchmark PPPs currently available.

The second volume measure V(5) also uses the deflate-first algorithm but substitutes the World Bank's 1980 Atlas exchange rate for the 1980 PPP index (Summers and Heston, 1988). Atlas exchange rates represent an arithmetic average of the actual 1980 average exchange rate and the exchange rates of the previous two years adjusted for relative price changes. The Atlas-converted volume measures, at both the regional and global level, were generally more closely aligned with the PPP-converted figures than the estimates derived from alternative conversion procedures.

The Atlas exchange rates have the practical advantage of being derived from readily available annual average exchange rates and country-specific implicit GDP deflators. Their major disadvantage is that, by construction, they imply at least an approximate adherence to the largely discredited Casselian purchasing power parity doctrine. On the other hand, the Summers and Heston (1988) PPP indices, which are conceptually more appealing and substantially improved over previously available PPPs in terms of country coverage and, apparently, precision, are not 'official' estimates. At this point in time, it is left to the informed judgment of the user as to which conversion procedure best suits the circumstances at hand.

Notes

1. By contrast, measurement and definitional issues aside, the agricultural research personnel data are ostensibly in comparable (FTE) units which may be directly aggregated for comparative purposes.

2. See World Bank (1983, i).

3. For instance, in numerous cases Boyce and Evenson (1975) and Judd, Boyce, and Evenson (1983) recorded research expenditure data directly in current or constant U.S. dollars rather than current LCUs. Given the myriad of primary sources underlying their data, it leaves their figures subject to the capricious conversion methodologies of their source authors.

4. Finding a suitable agricultural research deflator is by no means a simple matter and will be discussed later in this section.

5. See Pardey, Kang and Elliott (1988) for an example. They use research spending as a proportion of AgGDP and total government expenditures to address issues relating to the structure of public support for NARS.

6. Khamis (1984) also discusses issues related to aggregating value data measured in current local currency units, where regional or spatio-temporal bilateral and multilateral comparisons are required.

7. For two countries we were forced to use the 1980 annual average rather than the Atlas exchange rate. See Appendix A for details.

8. The World Bank (1986, 244) describes the current Atlas exchange rate method of conversion while the World Bank (1983) gives details of two earlier versions of the Atlas method. The current version is an arithmetic average of three exchange rates. Summers and Heston (1988) argue for a weighted geometric average.

9. Krueger (1983) reviews recent thinking about the determination of exchange rates.

10. See World Bank (1983, 18-20) for more details.

11. See Kravis, et al. (1975), Kravis, Heston and Summers (1978a, 1978b, 1982), Summers and Heston (1984 and 1988) and Kravis (1986) for a comprehensive discussion on purchasing power parity (PPP) indices. An earlier discussion of these conversion issues can be found in Balassa (1961), David (1972) and Balassa and David (1973), while an application of PPPs in a developing-country context is in Salazar-Carrillo (1983). Eurostat (1986 and 1987) provide 1980 PPPs for some 60 countries.

12. According to the specification used by Kravis, Heston and Summers (1982, 191-195), nontradables consist of expenditures on services – defined as goods that cannot be stored – and construction.

13. Unfortunately, it is not clear *exactly* which conversion procedure they employed. Judd, Boyce and Evenson (1983, 3) do state that "(research) expenditures were converted to U.S. dollars using official exchange

rates and were then inflated to 1980 dollars using a general wholesale price index."

14. The deviation of the V(2) from V(3) measures is even larger. It understates the 'global' volume of research resources relative to the PPP-based estimates by US$ 1,256 million.

15. The 1963-1983 period was chosen in order to maximize the sample size, given missing exchange rate and deflator observations for the pre-1963 and post-1983 years.

16. The same procedure was used by the United Nations and World Bank's International Comparisons Project. See footnote 11 for references.

17. See Kravis, Heston and Summers (1982, Appendix Table 6.1) for details of the goods and services coverage of the investment, government goods and services, and GDP categories.

18. See also National Science Foundation (1970), Jaffe (1972), Milton (1972), Sonka and Padberg (1979), Murphy and Kaldor (1981), Mansfield, Romeo and Switzer (1983) and Mansfield (1987) for additional discussions relating to R&D deflators.

References for Part I

Adusei, Edward O. "Evaluation of the Importance and Magnitude of Agricultural Maintenance Research in the United States." PhD diss., Virginia Polytechnic Institute and State University, Blacksburg, October 1987.

Antle, John M. "Infrastructure and Aggregate Agricultural Productivity: International Evidence." *Economic Development and Cultural Change* Vol. 31, No. 3 (April, 1983): 609–619.

Balassa, Bela. "Patterns of Industrial Growth: Comment." *American Economic Review* Vol. LI, No. 3 (June 1961): 394–397.

Balassa, B., and David, P.A. "Just How Misleading Are Official Exchange Rate Conversions? An Interchange." *Economic Journal* Vol. 83, No. 332 (December 1973): 1258–1276.

Boyce, James K., and Robert E. Evenson. *National and International Agricultural Research and Extension Programs.* New York: Agricultural Development Council, Inc., 1975.

Daniels, W.D. "Choosing Input Indicators For Research Managers." *Scientometrics* Vol. 11, No. 1–2 (1987): 17–25.

David, P.A. "Just How Misleading Are Official Exchange Rate Conversions?" *Economic Journal* Vol. 82, No. 327 (September 1972): 979–990.

Elliott, Howard, and Philip Pardey. "Global Data Bases on National Agricultural Research Systems." In *Impact of Research on National Agricultural Development*, proceedings of the second IFARD Global Convention, Brasilia, Brasil, 6–11 October 1986, edited by Brian Webster, Carlos Valverde, and Alan Fletcher, ISNAR, The Hague, July 1987.

Evenson, Robert E., and Yoav Kislev. *Investment in Agricultural Research and Extension: A Survey of International Data.* Economic Growth Center Discussion Paper No. 124. New Haven, Connecticut: Economic Growth Center, Yale University, August 1971.

Evenson, R.E., and Y. Kislev. *Agricultural Research and Productivity.* New Haven: Yale University Press, 1975.

Evenson, Robert E., and Yoav Kislev. "Investment in Agricultural Research and Extension: A Survey of International Data." *Economic Development and Cultural Change* Vol. 23 (April 1975): 507–521.

FAO. *Expert Consultation on Data Needs for Food and Agricultural Analysis and Planning in Developing Countries.* Rome: FAO, 3–6 November 1986.

Hallaway, Michelle L. "The Statistical Evolution of U.S. Agricultural Research: A Study in Institutional Development." MSc thesis, University of Minnesota, St. Paul, Minnesota. (forthcoming)

Heston, Alan, and Robert Summers. "What We Have Learned about Prices and Quantities from International Comparisons: 1987." *American Economic Review* (May 1988): 467–473.

Hill, T. Peter. *Multilateral Measurements of Purchasing Power and Real GDP.* Luxembourg: Eurostat, 1982.

International Labour Office. *International Standard Classification of Occupation.* Revised edition. Geneva: ILO, 1968.

Jaffe, Sidney A. *A Price Index for Deflation of Academic R and D Expenditures.* NSF 72–310. Washington, D.C.: National Science Foundation, May 1972.

Judd, M. Ann, James K. Boyce, and Robert E. Evenson. *Investing in Agricultural Supply.* Economic Growth Center Discussion Paper no. 442. New Haven, Connecticut: Economic Growth Center, Yale University, June 1983.

Judd, M. Ann, James K. Boyce, and Robert E. Evenson. "Investing in Agricultural Supply: The Determinants of Agricultural Research and Extension." *Economic Development and Cultural Change* Vol. 35, No. 1 (October 1986): 77–113.

Kast, Fremont E., and James E. Rosenzweig. *Organization and Management: A Systems and Contingency Approach.* Fourth Edition. New York: McGraw–Hill, 1985.

Khamis, Salem H. "On Aggregation for International Comparisons." *Review of Income and Wealth* Series 30, No. 2 (June 1984): 185–205.

Kravis, Irving B. "The Three Faces of the International Comparison Project." *World Bank Research Observer* Vol. 1, No. 1 (January 1986): 3–26.

Kravis, Irving B., and Robert E. Lipsey. "National Price Levels and the Prices of Tradables and Nontradables." *American Economic Review* (May 1988): 474–478.

Kravis, Irving B., Alan Heston, and Robert Summers. *International Comparisons of Real Product and Purchasing Power.* United Nations International Comparison Project: Phase II. Baltimore: The Johns Hopkins University Press, 1978a.

Kravis, Irving B., Alan Heston, and Robert Summers. "Real GDP Per Capita for More Than One Hundred Countries." *Economic Journal* Vol. 88, No. 350 (June, 1978b): 215–42.

Kravis, Irving B., Alan Heston, and Robert Summers. *World Product and Income: International Comparisons of Real Gross Product.* United Nations International Comparison Project: Phase II. Baltimore: The Johns Hopkins University Press, 1982.

Kravis, Irving B., Zoltan Kenessey, Alan Heston, and Robert Summers. *A System of International Comparisons of Gross Product and Purchasing Power.* United Nations International Comparison Project: Phase One. Baltimore: The John Hopkins University Press, 1975.

Krueger, Anne O. *Exchange-Rate Determination.* Cambridge: Cambridge University Press, 1983.

Leamer, Edward E. "The Sensitivity of International Comparisons of Capital Stock Measures to Different 'Real' Exchange Rates." *American Economic Review* (May 1988): 479–483.

MacDonald, A.S. "Exchange Rates for National Expenditure on Research and Development." *Economic Journal* Vol. 83 (June 1973): 477–494.

Mansfield, Edwin, Anthony Romeo, and Lorne Switzer. "R&D Price Indexes and Real R&D Expenditures in the United States." *Research Policy* Vol. 12 (1983): 105–112.

Mansfield, Edwin. "Price Indexes for R and D Inputs, 1969–1983." *Management Science* Vol. 33, No. 1 (January 1987): 124–129.

Marer, Paul. *Dollar GNPs of the U.S.S.R. and Eastern Europe.* Baltimore: The John Hopkins University Press, 1985.

Milton, Helen S. "Cost-of-Research Index, 1920–1970." *Operations Research* (January-February 1972): 1–18.

Miranowski, J.A., and Carlson, G.A. "Economic Issues in Public and Private Approaches to Preserving Pest Susceptibility." In *Pesticide Resistance: Strategies and Tactics for Management.* Washington, D.C.: National Academy of Sciences, National Academy Press, 1986.

Murphy, Joseph W., and Donald R. Kaldor. "The Changing Cost of Performing Agricultural Research: An Index Approach." In *Evaluation of Agricultural Research,* edited by G.W. Norton, et al., pp. 187–195. Minnesota Agricultural Experiment Station Miscellaneous Publication 8–1981. St. Paul: University of Minnesota, April 1981.

National Science Foundation. *Experimental Input Price Indexes for Research and Development, Fiscal Years 1961–65.* A report by the Bureau of Labor Statistics to the National Science Foundation. NSF 70–7. Washington, D.C.: National Science Foundation, 1970.

O'Connor, John C. Personal communication. Chief, Comparative Analysis & Data Division, Economic Analysis & Projections Department, The World Bank. Washington, D.C., 15 May 1987.

OECD. *The Measurement of Scientific and Technical Activities – "Frascati Manual" 1980.* Paris: OECD, 1981.

OECD. "Purchasing Power Parities and International Comparisons of Price Levels and Real Per Capita GDP in OECD Countries." OECD, Paris, February 1987. Mimeo.

Oram, Peter. "Report on National Agricultural Research in West Africa." IFPRI, Washington, D.C., February 1986. Mimeo.

Oram, Peter A., and Vishva Bindlish. *Resource Allocations to National Agricultural Research: Trends in the 1970's.* The Hague and Washington, D.C.: ISNAR and IFPRI, November 1981.

Oram, Peter A., and Vishva Bindlish. "Investment in Agricultural Research in Developing Countries: Progress, Problems, and the Determination of Priorities." IFPRI, Washington, D.C., January 1984. Mimeo.

Oram, Peter A., and Mark Gieben. "Document Summaries." ISNAR, The Hague, 1984. Mimeo.

Pardey, Philip G., and Barbara Craig. "Causal Relationships between Public Sector Agricultural Research Expenditures and Output." *American Journal of Agricultural Economics* Vol. 71, No. 1 (February 1989):9–19.

Pardey, Philip G., Barbara Craig, and Michelle Hallaway. "US Agricultural Research Deflators: 1890–1985." *Research Policy* Vol. 18 (forthcoming, 1989).

Pardey, Philip G., M. Sandra Kang, and Howard Elliott. "The Structure of Public Support for National Agricultural Research: A Political Economy Perspective." ISNAR Staff Note No. 88–11. The Hague: ISNAR, September 1988.

Pardey, Philip G., Johannes Roseboom, Sandra Kang, and Bonnie Folger. *Agricultural Research Indicator Series: Supplementary Files.* ISNAR Staff Notes No. 88–2. The Hague: ISNAR, June 1988.

Plucknett, D.L., and N.J.H. Smith. "Sustaining Agricultural Yields." *Bioscience* Vol. 36, No. 1 (1986): 40–45.

Ruttan, Vernon W. *Agricultural Research Policy.* Minneapolis: University of Minnesota Press, 1982.

Salazar-Carrillo, Jorge. "Real Product and Price Comparisons for Latin America and Other World Countries." *Economic Development and Cultural Change* Vol. 31, No. 4 (July 1983): 757–774.

Sonka, S.T., and D.I. Padberg. *Estimation of an Academic Research and Development Price Index.* Illinois Agricultural Economics Staff Paper No. 19 E–100. Urbana: Department of Agricultural Economics, University of Illinois, September 1979.

Summers, Robert, and Alan Heston. "Improved International Comparisons of Real Product and its Composition: 1950–1980." *Review of Income and Wealth* Series 30 (June 1984): 207–262.

Summers, Robert, and Alan Heston. "A New Set of International Comparisons of Real Product and Price Levels Estimates for 130 Countries, 1950–1985." *Review of Income and Wealth* Series 34, No. 1 (March 1988): 1–25.

Summers, Robert, Irving B. Kravis, and Alan Heston. "Changes in the World Income Distribution." *Journal of Policy Modeling* Vol. 6, No. 2 (1984): 237–269.

Swallow, Brent M., George W. Norton, Thomas B. Brumback, Jr., and Glenn R. Russ. *Agricultural Research Depreciation and the Importance of Maintenance Research.* Agricultural Economics Report 56. Blacksburg: Virginia Polytechnic Institute and State University, November 1985.

UN, and EUROSTAT. *World Comparisons of Purchasing Power and Real Product for 1980.* Phase IV of the International Comparison Project. Part One: Summary Results for 60 Countries. New York: United Nations and Commission of the European Communities, January 1986.

UN, and EUROSTAT. *World Comparisons of Purchasing Power and Real Product for 1980.* Phase IV of the International Comparison Project. Part Two: Detailed Results for 60 Countries. New York: United Nations and Commission of the European Communities, June 1987.

UNESCO. *International Standard Classification of Education.* Paris: UNESCO – Division of Statistics on Education, Office of Statistics, March 1976.

UNESCO. *Manual for Statistics on Scientific and Technological Activities.* Paris: UNESCO – Division of Statistics on Science and Technology, Office of Statistics, June 1984.

World Bank. *Methodological Problems and Proposals Relating to the Estimation of Internationally Comparable per Capita GNP Figures.* Washington, D.C.: Economic Analysis and Projections Department, The World Bank, November 1983.

World Bank. *World Development Report 1986.* New York: Oxford University Press for The World Bank, 1986.

PART II: INDICATOR SERIES

Country Listing

NOTES ON USING THE STATISTICAL TABLES *

Personnel							EXAMPLE ONLY	
Period/Year	PhD	MSc	BSc	Subtotal	Expat	Total	Source	
1960-64	..	..	..	..	..	9	10/742	— (A)
1965-69	..	..	..	..	..	7	10	— (B)
1970	..	..	..	0	10	10	175/532	— (C)
1971	..	..	..	1	15	16	589	
1972	..	..	..	..	..	..		
1973	..	..	..	1	15	16	742	
1974	..	..	..	..	..	..		
1975	..	..	..	..	..	..		
1976	..	..	..	..	..	23	95	— (D)
1977	..	..	..	..	..	..		
1978	..	..	..	..	..	..		
1979	..	..	..	..	..	..		
1980	.	..	..	..	..	..		
1981	1	- 14 -		15	10	25	235	
1982	..	..	..	26	4	30	23	
1983	2	- 22 -		24	7	31	17	
1984	12	18	11	41	..	41	230	
1985	..	..	..	26	17	43	445	
1986	..	..	..	..	..	..		
1960-64: 1961 & 63				1965-69: 1966 & 67				

(G) — (points to 1981 MSc/BSc "- 14 -")

(E) — (points to 1984 Total 41)

(F) — (points to 1960-64: 1961 & 63)

Notes on Personnel Series Tables:

(A) Total number of researchers for the 1960-64 period estimated as a simple average of two (1961 and 1963) observations. Sources 10 and source 742 may have both been used to compile the 1961 and 1963 observations, or, as is more generally the case, the 1961 observation was taken from source 10 and the 1963 observation from 742. The total number of researchers includes all researchers, both nationals and expatriates.

(B) Total number of researchers for the 1965-69 period estimated as a simple average of two (1966 and 1967) observations, both taken from source 10.

(C) The 1970 observation represents a compilation of figures taken from sources 175 and 532.

(D) If not stated otherwise, the total is inclusive of expatriates. In this particular case, however, the exact number of expatriate researchers is not available, nor could nationals be identified by degree status.

(E) In instances where no expatriate figures are explicitly given, expatriates are included in the degree-specific figures. It is generally the case that expatriates hold a PhD or MSc degree.

(F) The 1960-64 figure is a simple average of 1961 and 1963 observations.

(G) There are 15 local and 10 expatriate reserachers in a total of 25 researchers. One local researcher holds a PhD (or equivalent) and the remaining 14 hold either an MSc or BSc.

* The tables included here are EXAMPLES ONLY, and for expository purposes represent modified versions of the tables which appear in the series to follow.

EXAMPLE ONLY

Expenditure   LCU = *Indian Rupees*

Constant 1980 US$ (millions)

Period/Year	Current LCU (millions)	Constant 1980 LCU (millions)	ATLAS	PPP	Source
1960-64	81.535	308.756	36.982	102.231	876
1965-69	170.447	437.524	52.405	144.866	876/10
1970	267.490	616.336	73.823	204.072	10
1971	303.290	663.654	79.491	219.739	10
1972	372.198	732.673	87.757	242.591	10
1973	382.733	633.664	75.898	209.809	10
1974	..	..	..	..	
1975	582.574	843.088	100.983	279.150	26/781
1976	698.606	949.193	113.692	314.282	26
1977	858.899	1125.687	134.831	372.720	26
1978	..	..	..	..	
1979	..	..	..	..	
1980	1341.518	1341.518	160.683	444.182	778
1981	1443.534	1320.708	158.191	437.292	778
1982	1600.551	1357.550	162.603	449.491	778
1983	1697.568	1321.065	158.233	437.410	778
1984	1898.585	1387.855	166.233	459.525	778
1985	2081.602	1424.804	170.539	471.428	777/778
1986	2356.130	..	..	..	777/778

1960-64: 1960, 61, 62, 63 & 64   1965-69: 1965, 66, 67, 68 & 69

Notes on Expenditure Series Tables:

(A) Deflated to constant 1980 local currency units (LCUs) using country-specific implicit GDP deflators given in Appendix A, Table A.2.

(B) Constant 1980 LCU expenditure converted to constant 1980 dollars using the World Bank's Atlas exchange rate and Summers and Heston's (1988) purchasing power parity (PPP) over gross domestic product annual average exchange rate. The Atlas and PPP exchange rates are given in Appendix A, Table A.1, while complete details of conversion procedures used in this series are given in Chapter 4 of the technical notes.

(C) For guidance on interpreting the sourcing procedure used here, see notes A, B, and C for the personnel series table.

(D) A 1986 current LCU observation is available but cannot be converted to constant LCUs or dollars in the absence of a suitable deflator and/or exchange rate.

NOTES ON USING THE STATISTICAL TABLES (cont.)

Personnel Comments:

The personnel indicator series has been constructed as follows:

	1966	1970	1972	1974	1981	1982	1983	1984	1985
IRAT	10	8	14		-	-	-	-	-
IRCT	7	6	10		-	-	-	-	-
IEMVT	3	2	5		-	-	-	-	-
IDESSAa	-	-	-	104	49	53	51	50	51
IRCA	6	10	10		16	17	17	16	19
CTFT	NA	7	7		7	9	9	8	13
IRHO	19	20	19		22	23	23	24	25
IFAC/IRFA		9	11		16	13	14	15	16
IFCC/IRCC	8	10	25		18	18	19	19	22
ORSTOM	25	69	78	[76]	59	60	70	73	70*
Subtotal	78	141	179	180	187	193	203	205	216
CSRS		3	3		3*				
ITIPAT		5	7						
CRO		19			26**	28*	29*	26*	32*
DPML		3							
CIRES					9*			8	
CIRT					4*			11*	11*
ENSA								11	
Sources	653	532	431	589	86/12* 774**	86 774*	86 774*	86 774*	86 774*

[ ]= estimated or constructed by authors.

a. IRAT, IRCT, and IEMVT merged into IDESSA.

Notes on Personnel Comments:

(A) A dash indicates a figure is not relevant, as the institution is no longer in existence or not formed as yet.

(B) Indicates that a total of 104 researchers are working at the research insitutions indicated.

(C) The subtotals represent the figure that is included in the indicator series. Institutions are listed below the subtotal if only scattered observations were available, or if compatible expenditure data were unavailable.

(D) A blank in the lower half of a table (i.e., below the subtotal) indicates that either note A, G, or H is applicable.

(E) If multiple sources are used to compile a set of institutional figures for a particular year, then a series of superscripts, "*", "**", etc., are used to indicate which figure corresponds to a particular source.

(F) A square bracket indicates that this figure is either estimated or constructed by the authors.

(G) A blank indicates that it is currently not known whether or not a figure is required.

(H) An "NA" indicates that a figure is understood to be required but is not currently available

(I) Institutional acronyms and abbreviations (see definitional listing following comments).

Afghanistan

Personnel

Period/Year	PhD	MSc	BSc	Subtotal	Expat	Total	Source
1960–64	..	..	..	..	..	..	
1965–69	..	..	..	..	..	30	10
1970	..	..	..	..	..	..	
1971	..	..	..	..	..	..	
1972	..	..	..	..	..	44	10
1973	..	..	..	..	..	..	
1974	..	..	..	..	..	..	
1975	..	..	..	..	..	..	
1976	..	..	..	..	..	45	95
1977	..	..	..	..	..	..	
1978	..	..	..	..	..	..	
1979	..	..	..	..	..	..	
1980	3	22	98	123	..	123	590
1981	..	..	..	..	..	..	
1982	..	..	..	..	..	..	
1983	..	..	..	..	..	..	
1984	..	..	..	..	..	..	
1985	..	..	..	..	..	..	
1986	..	..	..	..	..	..	

1960–64:   1965–69: 1967 & 68

Expenditure

LCU = Afghanis

Period/Year	Current LCU (millions)	Constant 1980 LCU (millions)	Constant 1980 US$ (millions)		Source
			Atlas	PPP	
1960–64	..	..	..	..	
1965–69	..	..	..	..	
1970	..	..	..	..	
1971	..	..	..	..	
1972	..	..	..	..	
1973	..	..	..	..	
1974	..	..	..	..	
1975	..	..	..	..	
1976	..	..	..	..	
1977	..	..	..	..	
1978	..	..	..	..	
1979	..	..	..	..	
1980	..	..	..	..	
1981	..	..	..	..	
1982	..	..	..	..	
1983	..	..	..	..	
1984	..	..	..	..	
1985	..	..	..	..	
1986	..	..	..	..	

1960–64:   1965–69:

Personnel Comments:

The 1976 figure has been constructed as follows:

Ministry of Agriculture Research Department	30
Agricultural Research Institute Afghanistan Division of Agronomy	12
Division of Plant Pathology	3
Total	45

According to sources 94 and 173, during the 1970s several other departments at the Ministry of Agriculture conducted some research in addition to the Research Department, such as the Department of Animal Production and Veterinary Services. The number of researchers working in these departments is currently not available.

From the discussion contained in source 590 it is surmised their figures pertain to 1980. They represent the Department of Agricultural Research and Soil Science. It is unclear if these figures are strictly compatible with those of the earlier years.

Sources:

0010 Boyce, James K., and Robert E. Evenson. *National and International Agricultural Research and Extension Programs.* New York: Agricultural Development Council, Inc., 1975.

0095 FAO – CARIS. *Agricultural Research in Developing Countries –Volume 1: Research Institutions.* Rome: FAO – CARIS, 1978.

0590 UNESCO. *Science and Technology in Countries of Asia and the Pacific.* Science Policy Studies and Documents No. 52. Paris: UNESCO, n.d.

Additional References:

0094 FAO – Near East Regional Office. *Directory of Agricultural Research Institutions in the Near East Region.* Cairo: FAO – Near East Regional Office, 1979.

0165 Evenson, R.E., and Y. Kislev. *Agricultural Research and Productivity.* New Haven: Yale University Press, 1975.

0173 FAO. *FAO Near East Regional Studies on Organization and Administration of Agricultural Research.* Rome: FAO, 1974.

0174 Watson, J.M. "Comparative Study of Agricultural Research Organisation and Administration in the Near East Region." Paper presented at the Workshop on Organization and Administration of Agricultural Services in the Arab States, Cairo, 2–15 March 1964.

0852 Evenson, Robert E., and Yoav Kislev. *Investment in Agricultural Research and Extension: A Survey of International Data.* Center Discussion Paper No. 124. New Haven, Connecticut: Economic Growth Center, Yale University, August 1971.

0886 Evenson, Robert E., and Yoav Kislev. "Investment in Agricultural Research and Extension: A Survey of International Data." *Economic Development and Cultural Change* Vol. 23 (April 1975): 507–521.

Algeria

Personnel

Period/Year	PhD	MSc	BSc	Subtotal	Expat	Total	Source
1960– 64	..	..	..	..	..		
1965– 69	..	..	..	..	..		
1970	..	..	..	..	..	62	532
1971	..	..	..	..	..	..	
1972	..	..	..	..	..	..	
1973	..	..	..	..	..	..	
1974	..	..	..	..	..	..	
1975	..	..	..	..	..	..	
1976	..	..	..	..	..	..	
1977	..	..	..	..	..	..	
1978	..	..	..	..	..	..	
1979	..	..	..	..	..	..	
1980	..	..	..	..	..	..	
1981	..	..	..	..	..	..	
1982	..	..	..	..	..	..	
1983	..	..	..	..	..	..	
1984	..	..	..	..	..	..	
1985	44	93	168	305	..	305	408
1986	..	..	..	..	..		

1960– 64: 1965– 69:

Expenditure

LCU = Algerian Dinars

Period/Year	Current LCU (millions)	Constant 1980 LCU (millions)	Constant 1980 US$ (millions)		Source
			Atlas	PPP	
1960– 64	..	..	..	..	
1965– 69	..	..	..	..	
1970	..	..	..	..	
1971	..	..	..	..	
1972	..	..	..	..	
1973	..	..	..	..	
1974	..	..	..	..	
1975	..	..	..	..	
1976	..	..	..	..	
1977	..	..	..	..	
1978	..	..	..	..	
1979	..	..	..	..	
1980	..	..	..	..	
1981	..	..	..	..	
1982	..	..	..	..	
1983	..	..	..	..	
1984	..	..	..	..	
1985	..	..	..	..	
1986	..	..	..	..	

1960– 64: 1965– 69:

Personnel Comments:

The 1970 figure includes the following non-educational institutes: CRZA (18 researchers), CNREF (4 researchers), and INRAA (30 researchers). INRAA comprises CNRA (15 researchers), CNRZ (10 researchers), and CNRESR (5 researchers). In addition to these researchers, source 532 reports, for 1970, 30 graduate staff members at INA, devoting their time to both research and education. It was assumed they spent one-third of their time on research.

The 1985 figure given by source 408 includes those researchers working at institutes under the aegis of the Ministry of Agriculture and Fisheries, and at INA. The Ministry of Agriculture comprises the following institutes: CERP, INRA, INPV, INSA, ITMCI, ITAFV, ITEBO, ITPE, and ITDAS. As for the 1970 observation, it is assumed that staff members at INA spent one-third of their time on research.

Source 408 also reports some agricultural research at two departments of 'Université des Sciences et de la Technologie' (1985: 69 researchers/teachers). To maintain comparability with the 1970 figure, they are excluded from the series.

Acronyms and Abbreviations:

CERP: Centre d'Etudes et de Recherche en Pêches
CNERF: Centre National de Recherche et d'Expérimentation Forestières
CNRA: Centre National de la Recherche Agronomique
CNRESR: Centre National de Recherche en Economie et Sociologie Rurale
CNRZ: Centre National de la Recherche Zootechnique
CRZA: Centre de Recherches sur les Zones Arides
INA: Institut National Agronomique
INPV: Institut National de la Protection des Végétaux
INRA(A): Institut National de la Recherche Agronomique (d'Algérie)
INSA: Institut National de la Santé Animale
ITAFV: Institut Technique de l'Arboriculture Fruitière et de la Viticulture
ITDAS: Institut Technique de Développement de l'Agronomique Saharienne
ITEBO: Institut Technique de l'Elevage Bovin et Ovin
ITMCI: Institut Technique du Maraîchage et des Cultures Industrielles
ITPE: Institut Technique des Petits Elevages

Sources:

0408 Kellou, R. "La Recherche Agricole en Algérie." Paper presented at the Séminaire sur l'Orientation et l'Organisation de la Recherche Agronomique dans les Pays du Bassin Méditerranéen, Centre International de Hautes Etudes Agronomiques Méditerranéennes, Istanbul, 1–3 December 1986.

0532 UNESCO Field Science Office for Africa. *Survey on the Scientific and Technical Potential of the Countries of Africa.* Paris: UNESCO, 1970.

Additional References:

0010 Boyce, James K., and Robert E. Evenson. *National and International Agricultural Research and Extension Programs.* New York: Agricultural Development Council, Inc., 1975.

0027 Harvey, Nigel, ed. *Agricultural Research Centres: A World Directory of Organizations and Programmes.* Seventh Edition, Two Volumes. Harlow, U.K.: Longman, 1983.

0135 FAO, and UNDP. *National Agricultural Research – Report of an Evaluation Study in Selected Countries.* Rome: FAO, and UNDP, 1984.

0175 Cooper, St.G.C. *Agricultural Research in Tropical Africa.* Kampala: East African Literature Bureau, 1970.

0266 UNESCO. *National Science Policies in Africa.* Science Policy Studies and Documents No. 31. Paris: UNESCO, 1974.

0360 Cooper, St.G.C. "Towards Trained Manpower for Agricultural Research in Africa." Paper presented at the Conference on Agricultural Research and Production in Africa, organized by the Association for the Advancement of Agricultural Sciences in Africa (AAASA), Addis Ababa, 29 August– 4 September 1971.

0407 Centre International de Hautes Etudes Agronomiques Méditerranéennes. "Séminaire sur l'Orientation et l'Organisation de la Recherche Agronomique dans les Pays du Bassin Méditerranéen, Istanbul, 1–3 December 1986." Centre International de Hautes Etudes Agronomiques Méditerranéennes, Paris, December 1986. Mimeo.

0422 Casas, Joseph. "Les Ressources Humaines et Financières dans les Pays du Bassin Méditerranéen." Preliminary table. ISNAR, The Hague, 1986. Mimeo.

0653 Webster, B.N. *Index of Agricultural Research Institutions and Stations in Africa.* Rome: FAO, n.d.

0747 ISNAR. *International Workshop on Agricultural Research Management.* Report of a workshop. The Hague: ISNAR, 1987.

0753 Kellou, R. "Agricultural Research in Algeria and Its Human Resources." Paper presented at the International Workshop on Agricultural Research Management, ISNAR, The Hague, 7–11 September 1987.

0852 Evenson, Robert E., and Yoav Kislev. *Investment in Agricultural Research and Extension: A Survey of International Data.* Center Discussion Paper No. 124. New Haven, Connecticut: Economic Growth Center, Yale University, August 1971.

0865 Oram, Peter. "Agricultural Research Objectives and Priorities: Constraints to the Development of Agricultural Research Institutions in Arab Countries." Review paper prepared for the first meeting of the Council for Arab Agricultural Research, organized by the Arab Fund for Economic and Social Development, Kuwait, 23 March 1988.

Angola

Personnel

Period/Year	PhD	MSc	BSc	Subtotal	Expat	Total	Source
1960–64	..	..	..	..	..	..	
1965–69	..	..	..	..	..	39	10
1970	..	..	..	..	..	..	
1971	..	..	..	..	..	..	
1972	..	..	..	..	..	37	10
1973	..	..	..	..	..	..	
1974	..	..	..	..	..	..	
1975	..	..	..	..	..	..	
1976	..	..	..	..	..	..	
1977	..	..	..	..	..	..	
1978	..	..	..	..	..	..	
1979	..	..	..	..	..	..	
1980	..	..	..	..	..	..	
1981	..	..	..	..	..	..	
1982	..	..	..	..	..	..	
1983	0	0	15	15	13	28	17/803
1984	..	..	..	..	..	..	
1985	..	..	..	..	..	..	
1986	..	..	..	..	..	..	

1960–64:   1965–69: 1968

Expenditure   *LCU = Kwanza*

Period/Year	Current LCU (millions)	Constant 1980 LCU (millions)	Constant 1980 US$ (millions)		Source
			Atlas	PPP	
1960–64	..	..	..	..	
1965–69	40.000	279.674	8.299	7.803	10
1970	..	..	..	..	
1971	..	..	..	..	
1972	..	..	..	..	
1973	..	..	..	..	
1974	..	..	..	..	
1975	..	..	..	..	
1976	..	..	..	..	
1977	..	..	..	..	
1978	..	..	..	..	
1979	..	..	..	..	
1980	..	..	..	..	
1981	..	..	..	..	
1982	..	..	..	..	
1983	184.500	155.371	4.611	4.335	17
1984	..	..	..	..	
1985	..	..	..	..	
1986	..	..	..	..	

1960–64:   1965–69: 1968

Personnel Comments:
The 1983 figure represents the Agronomic Research Institute (IIA) and the Veterinary Research Institute (IIV). Both institutes are under the jurisdiction of the Ministry of Agriculture.

Expenditure Comments:
The 1983 expenditure figure refers to IIA and IIV only.

Acronyms and Abbreviations:
IIA: Instituto de Investigação Agronómica (Agronomic Research Institute)
IIV: Instituto de Investigação Veterinária (Veterinary Research Institute)

Sources:

0010 Boyce, James K., and Robert E. Evenson. *National and International Agricultural Research and Extension Programs.* New York: Agricultural Development Council, Inc., 1975.

0017 ISNAR, IFARD & AOAD. Survey of National Agricultural Research Systems: Unpublished Questionnaire Responses. ISNAR, The Hague, 1985.

0803 Branco Marcelino, Fernando. Personal communication. Director, Instituto de Investigação Agronómica. Chianga, Angola, February 1988.

Additional References:

0446 Kyomo, M.L. "Agricultural Research in Eastern and Southern Africa: Issues and Priorities." Southern African Centre for Cooperation in Agricultural Research of SADCC, Gabarone, Botswana, 1986. Mimeo.

Antigua

Personnel

Period/Year	PhD	MSc	BSc	Subtotal	Expat	Total	Source
1960–64	..	..	..	..	..	..	
1965–69	0	2	2	4	..	4	286/285
1970	..	..	..	..	..	..	
1971	1	2	3	6	..	6	279
1972	..	..	..	..	..	..	
1973	..	..	..	..	..	..	
1974	0	2	1	3	..	3	244
1975	..	..	..	..	..	..	
1976	..	..	..	..	..	..	
1977	..	..	..	..	..	..	
1978	1	0	0	1	..	1	341
1979	..	..	..	..	..	..	
1980	1	1	1	3	..	3	343
1981	..	..	..	..	..	7	342
1982	..	..	..	..	..	..	
1983	1	2	2	5	..	5	347/355
1984	1	3	1	5	..	5	354/349
1985	..	..	..	..	..	..	
1986	..	..	..	..	..	..	

1960–64:   1965–69: 1965 & 68

Expenditure

LCU = East Caribbean Dollars

Period/Year	Current LCU (millions)	Constant 1980 LCU(millions)	Constant 1980 US$ (millions)		Source
			Atlas	PPP	
1960–64	..	..	..	..	
1965–69	..	..	..	..	
1970	..	..	..	..	
1971	..	..	..	..	
1972	..	..	..	..	
1973	..	..	..	..	
1974	..	..	..	..	
1975	..	..	..	..	
1976	..	..	..	..	
1977	..	..	..	..	
1978	..	..	..	..	
1979	..	..	..	..	
1980	..	..	..	..	
1981	..	..	..	..	
1982	..	..	..	..	
1983	..	..	..	..	
1984	..	..	..	..	
1985	..	..	..	..	
1986	..	..	..	..	

1960–64:   1965–69:

Personnel Comments:
The pre-1975 figures refer to agricultural researchers identified within the Agricultural Department of the Ministry of Agriculture. This Department includes the Central Cotton Station which in 1982 still reported (source 27) one professional researcher. After 1975 it seems that CARDI (Caribbean Agricultural Research and Development Institute) took over the leading role in agricultural research on Antigua. The figures for the years after 1975 refer to CARDI only.

Expenditure Comments:
The total Ministry of Agriculture recurrent budget was EC$ 5.0 million around 1985 (source 63). The research component of this was not reported.

Sources:
0244 Commonwealth Agricultural Bureaux (CAB). *List of Research Workers in the Agricultural Sciences in the Commonwealth and in the Republic of Ireland 1975.* Slough, England: CAB, 1975.
0279 Commonwealth Agricultural Bureaux (CAB). *List of Research Workers in the Agricultural Sciences in the Commonwealth and in the Republic of Ireland 1972.* Slough, England: CAB, 1972.
0285 Commonwealth Agricultural Bureaux (CAB). *List of Research Workers in the Agricultural Sciences in the Commonwealth and in the Republic of Ireland 1969.* Slough, England: CAB, 1969.
0286 Commonwealth Agricultural Bureaux (CAB). *List of Research Workers in Agriculture, Animal Health and Forestry in the Commonwealth and in the Republic of Ireland 1966.* Slough, England: CAB, 1966.
0341 Caribbean Agricultural Research and Development Institute (CARDI). *Annual Report 1977–1978.* St. Augustine, Trinidad & Tobago: CARDI, November 1978.
0342 Caribbean Agricultural Research and

Development Institute (CARDI). *CARDI 1976–1981: Report of the Chairman.* St. Augustine, Trinidad & Tobago: CARDI, 1982.
0343 Caribbean Agricultural Research and Development Institute (CARDI). *Annual Report 1980.* St. Augustine, Trinidad & Tobago: CARDI, 1981.
0347 Caribbean Agricultural Research and Development Institute (CARDI). *Annual Report 1983 –Antigua Unit.* St. John's, Antigua: CARDI, 1984.
0349 Caribbean Agricultural Research and Development Institute (CARDI). *Annual Report 1983–1984 – Leeward Islands Unit.* CARDI, 1984.
0354 Caribbean Agricultural Research and Development Institute (CARDI). *Annual Report Research and Development 1983–84: Highlights.* St. Augustine, Trinidad: CARDI, 1984.
0355 Caribbean Agricultural Research and Development Institute (CARDI). *Research and Development Summary 1982–83.* St. Augustine, Trinidad: CARDI, 1983.

Additional References:
0027 Harvey, Nigel, ed. *Agricultural Research Centres: A World Directory of Organizations and Programmes.* Seventh Edition, Two Volumes. Harlow, U.K.: Longman, 1983.
0063 ISNAR. *Report to the Board of Directors of CARDI: Analysis, Evaluation and Proposals to Strengthen CARDI's Regional Capacity.* The Hague: ISNAR, 1985.
0287 Commonwealth Agricultural Bureaux (CAB). *List of Research Workers in Agriculture, Animal Health and Forestry in the British Commonwealth, the Republic of Sudan and the Republic of Ireland 1959.* Slough, England: CAB, 1959.
0353 Caribbean Agricultural Research and Development Institute (CARDI). *Paper Presented at a Meeting of the Governing Body of CARDI on January 29, 1982, Basseterre, St. Kitts.* St. Augustine, Trinidad: CARDI, 1982.
0356 Caribbean Agricultural Research and Development Institute (CARDI). *Research and Development Summary 1983–84.* St. Augustine, Trinidad: CARDI, 1984.

Argentina

Personnel

Period/Year	PhD	MSc	BSc	Subtotal	Expat	Total	Source
1960–64	..	..	..	..	..	400	10
1965–69	..	..	..	..	..	634	10
1970	..	..	..	..	..	795	10
1971	..	..	..	..	..	840	10
1972	..	..	..	..	..	820	10
1973	..	..	..	..	..	840	10
1974	..	..	..	..	..	880	14
1975	..	..	..	..	..	953	53
1976	..	..	..	..	..	847	53
1977	..	..	..	..	..	890	53
1978	..	..	..	..	..	880	53
1979	..	..	..	..	..	919	16
1980	..	..	..	..	..	1065	32
1981	..	..	..	..	..	..	
1982	..	..	..	..	..	..	
1983	31	138	836	1005	..	1005	17
1984	..	..	..	..	..	..	
1985	–243–		876	1119	..	1119	655
1986	..	..	..	..	..	..	

1960–64: 1961   1965–69: 1965 & 69

Expenditure   *LCU = Argentine Pesos (1980)*

Period/Year	Current LCU (millions)	Constant 1980 LCU(millions)	Constant 1980 US$ (millions)		Source
			Atlas	PPP	
1960–64	12.069	113867.665	22.550	49.395	711/374
1965–69	38.067	124530.550	24.661	54.021	711/374
1970	60.654	153733.224	30.445	66.688	711/374
1971	70.464	129043.923	25.555	55.978	711/374
1972	126.226	142002.974	28.122	61.600	711/374
1973	253.511	173240.477	34.308	75.151	711/374
1974	365.388	194397.584	38.497	84.328	711/374
1975	831.218	146621.218	29.036	63.603	711/374
1976	4228.607	142862.183	28.292	61.973	711/374
1977	11885.392	154663.872	30.629	67.092	711/374
1978	34205.434	173039.114	34.268	75.063	711/374
1979	88304.747	177504.301	35.152	77.000	711/374
1980	199562.920	199562.920	39.520	86.569	711/374
1981	350615.172	168889.775	33.446	73.263	711/374
1982	743516.250	122530.694	24.265	53.153	711/374
1983	3499464.960	128143.285	25.377	55.588	711/374
1984	23448140.800	199531.975	19.711	43.176	711/374
1985	265607564.800	146802.340	29.072	63.682	711/374
1986	440799609.600	..	..	..	711/374

1960–64: 1960, 61, 62, 63 & 64   1965–69: 1965, 66, 67, 68, & 69

Personnel Comments:

The personnel figures include only researchers within 'Instituto Nacional de Tecnología Agropecuaria' (INTA). Because INTA also has an extension role, its total staff figures are considerably higher. The numbers of researchers given by source 53 were constructed by Piñeiro and Trigo. They considered that 60% of all graduates and postgraduates working within INTA were researchers.

Additional organizations conducting agricultural research, but not included in the indicator series, are: Instituto Forestal Nacional (1985: 153 researchers, source 744), Instituto Nacional de Investigación y Desarollo Pesquero, Instituto Nacional de Vitivinicultura (1985: 29 researchers, source 744), and several universities.

Expenditure Comments:

Argentina has experienced a substantial rate of inflation over the last 15 years or so, and a currency change – from Argentine Pesos to Australs – in 1983. The current local currency expenditure series is based on a 1980 local currency unit, i.e., the Argentine Peso. All expenditure figures expressed in other current units – such as Australes – have been converted to this 1980 currency unit.

The expenditure indicators represent research expenditures by INTA only. Some agricultural research is done in institutes outside INTA, for example, universities and private organizations. Because of a lack of information, they are not included in the indicator series.

In addition to research, INTA is also involved in extension. None of the available sources provide a specific breakdown of total INTA expenditures into research expenditures and extension expenditures. A large general services component in the available total expenditure figures complicates matters considerably. Source 374, however, has made an attempt to split total expenditures for the years 1984–1986 into research and extension.

Under various scenarios the prorated services component of total expenditures for the 1984–86 period averaged 68% to research and 32% to extension. This breakdown was applied to the whole time series, in the absence of similar estimates for the earlier years.

Sources:

0010 Boyce, James K., and Robert E. Evenson. *National and International Agricultural Research and Extension Programs.* New York: Agricultural Development Council, Inc., 1975.

0014 Judd, M. Ann, James K. Boyce, and Robert E. Evenson. "Investing in Agricultural Supply." Economic Growth Center, Yale University, New Haven, Connecticut, 1983. Mimeo.

0016 Oram, Peter A., and Vishva Bindlish. *Resource Allocations to National Agricultural Research: Trends in the 1970s.* The Hague and Washington, D.C.: ISNAR and IFPRI, November 1981.

0017 ISNAR, IFARD & AOAD. Survey of National Agricultural Research Systems: Unpublished Questionnaire Responses. ISNAR, The Hague, 1985.

0032 Castronovo, Alfonso J.P. "Diagnóstico Descriptivo de la Situación Actual del Sistema Nacional de Investigación y Extensión en Argentina, Chile, Paraguay y Uruguay." Buenos Aires, December 1980. Mimeo.

0053 Piñeiro, Martín, and Eduardo Trigo, eds. *Cambio Técnico en el Agro Latino Americano – Situación y Perspectivas en la Década de 1980.* San José: IICA, 1983.

0374 Instituto Nacional de Tecnología Agropecuaria (INTA). "Elementos para Priorización Institucional." Documento Preliminar para Discusión. INTA, Buenos Aires, August 1987. Mimeo.

0655 Marzocca, Angel. "Proceso de Formación y Evolución del INTA, en Argentina." In *Organización y Administración de la Generación y Transferencia de Tecnología Agropecuaria,* Serie: Ponencias, Resultados y Recomendaciones de Eventos Técnicos No. A4/UY-86-001, edited by Horacio H. Stagno and Mario Allegri. Montevideo, Uruguay: IICA and Centro de Investigaciones Agrícolas 'Alberto Boerger', 1986.

0711 Instituto Nacional de Tecnología Agropecuaria (INTA). "Cuadro de Erogaciones de Ejecución del Presupuesto y Cuadro de Recursos Presupuestarios, 1956–1986 (En Valores Históricos y en Valores Actualizados – Año Base 1986)." INTA, Buenos Aires, n.d. Mimeo.

Additional References:

0001 Trigo, Eduardo J., and Martín E. Piñeiro. "Funding Agricultural Research." In *Report of a Conference: Selected Issues in Agricultural Research in Latin America,* edited by Barry

60

Nestel and Eduardo J. Trigo. The Hague: ISNAR, March 1984.

0022 UNESCO. *Statistical Yearbook 1983.* Paris: UNESCO, 1983.

0026 Oram, Peter A., and Mark Gieben. "Document Summaries." ISNAR, The Hague, 1984. Mimeo.

0027 Harvey, Nigel, ed. *Agricultural Research Centres: A World Directory of Organizations and Programmes.* Seventh Edition, Two Volumes. Harlow, U.K.: Longman, 1983.

0029 Morras, Héctor José María. Personal communication. Director (Int.) de Desarrollo de Recursos Humanos, Instituto Nacional de Tecnología Agropecuaria. Buenos Aires, Argentina, March 1988.

0073 Oram, Peter A., and Vishva Bindlish. "Investment in Agricultural Research in Developing Countries: Progress, Problems, and the Determination of Priorities." IFPRI, Washington, D.C., January 1984. Mimeo.

0080 Trigo, Eduardo, Martín Piñeiro, and Jorge Ardila. *Organización de la Investigación Agropecuaria en América Latina.* San José: IICA, 1982.

0081 Instituto Nacional de Tecnología Agropecuaria (INTA). *Memoria y Balance Anual 1973.* Buenos Aires: INTA, 1974.

0095 FAO – CARIS. *Agricultural Research in Developing Countries – Volume 1: Research Institutions.* Rome: FAO – CARIS, 1978.

0096 Ardila, Jorge, Eduardo Trigo, and Martín Piñeiro. "Human Resources in Agricultural Research: Three Cases in Latin America." IICA – PROTAAL, San José, March 1981. Mimeo.

0097 Ardila, Jorge, N. Reichart, and A. Rincón. *Sistemas Nacionales de Investigación Agropecuaria en América Latina: Análisis Comparativo de los Recursos Humanos en Países Seleccionados – El Caso del INTA en Argentina.* Documento PROTAAL 48. Costa Rica: IICA, 1980.

0165 Evenson, R.E., and Y. Kislev. *Agricultural Research and Productivity.* New Haven: Yale University Press, 1975.

0172 Trigo, Eduardo. Background material. ISNAR, The Hague, n.d.

0288 Instituto Nacional Tecnología Agropecuaria (INTA). *Memoria y Balance Anual 1978.* Buenos Aires: INTA, 1979.

0289 Instituto Nacional Tecnología Agropecuaria (INTA). *Memoria y Balance Anual 1979.* Buenos Aires: INTA, 1980.

0290 Instituto Nacional Tecnología Agropecuaria (INTA). *Memoria y Balance Anual 1980.* Buenos Aires: INTA, 1981.

0291 Instituto Nacional Tecnología Agropecuaria (INTA). *Memoria y Balance Anual 1983.* Buenos Aires: INTA, 1984.

0305 Ardila, Jorge, Eduardo Trigo, and Martín Piñeiro. "Human Resources in Agricultural Research: Three Cases in Latin America." In *Resource Allocation to Agricultural Research.* Proceedings of a workshop held in Singapore 8–10 June 1981, edited by Douglas Daniels and Barry Nestel, pp. 151–167. Ottawa: IDRC, 1981.

0852 Evenson, Robert E., and Yoav Kislev. *Investment in Agricultural Research and Extension: A Survey of International Data.* Center Discussion Paper No. 124. New Haven, Connecticut: Economic Growth Center, Yale University, August 1971.

0886 Evenson, Robert E., and Yoav Kislev. "Investment in Agricultural Research and Extension: A Survey of International Data." *Economic Development and Cultural Change* Vol. 23 (April 1975): 507–521.

Australia

Personnel

Period/Year	PhD	MSc	BSc	Subtotal	Expat	Total	Source
1960-64	..	..	..	..	..	..	
1965-69	612	382	1396	2390	..	2390	286/285
1970	..	..	..	..	..	..	
1971	914	497	1864	3275	..	3275	279
1972	..	..	..	..	..	..	
1973	..	..	..	..	..	..	
1974	1129	550	2083	3762	..	3762	244
1975	..	..	..	..	..	..	
1976	..	..	..	..	..	..	
1977	1435	632	2086	4153	..	4153	243
1978	..	..	..	..	..	..	
1979	..	..	..	..	..	..	
1980	1603	725	2027	4355	..	4355	242
1981	..	..	..	..	..	..	
1982	..	..	..	..	..	..	
1983	..	..	..	..	..	..	
1984	..	..	..	..	..	..	
1985	..	..	..	..	..	..	
1986	..	..	..	..	..	..	

1960-64:   1965-69: 1965 & 68

Expenditure

LCU = Australian Dollars

Period/Year	Current LCU (millions)	Constant 1980 LCU (millions)	Constant 1980 US$ (millions)		Source
			Atlas	PPP	
1960-64	28.660	105.368	118.308	98.789	852
1965-69	55.463	171.272	192.307	160.578	852/10/685
1970	..	..	..	..	
1971	..	..	..	..	
1972	96.115	239.092	268.456	224.164	565
1973	109.732	243.308	273.190	228.117	565
1974	132.925	251.276	282.137	235.587	565
1975	..	..	..	..	
1976	..	..	..	..	
1977	154.635	202.402	227.260	189.765	687
1978	..	..	..	..	
1979	..	..	..	..	
1980	..	..	..	..	
1981	..	..	..	..	
1982	306.054	252.312	283.300	236.558	610
1983	..	..	..	..	
1984	..	..	..	..	
1985	..	..	..	..	
1986	..	..	..	..	

1960-64: 1961   1965-69: 1965, 68 & 69

Personnel Comments:

These figures include CSIRO personnel and agricultural scientists working for other federal (national) government research organizations, or those sponsored by the federal government, along with researchers working in the state departments of agriculture (and affiliated research institutes) plus those working in universities and colleges. The higher education researchers were prorated to full-time equivalents using a 0.5 weight. The estimates were derived from a detailed compilation of researchers at the institute level. Some sources give far higher estimates for the number of researchers (expressed in FTEs) working in the university sector. This is caused by the fact that they include research students and professional administrators (source 686), while this count explicitly excludes them.

The table below gives a breakdown of the prorated number of researchers by the non-university and university sectors.

	1965	1968	1971	1974	1977	1980
Non-university	1861	2378	2938	3376	3679	3864
University	257	284	337	386	474	491
Total	2118	2662	3275	3762	4153	4355
Source	286	285	279	244	243	242

Expenditure Comments:

The table below gives an overview of construction details for the post-1969 expenditure estimates.

	1969	1972	1973	1974	1977	1978	1981	1982
Commonwealth Government	25.186	48.405	53.640	67.334	54.583	NA	NA	122.589
State Governments	27.165	39.723	47.042	54.997	87.624	NA	NA	153.765
Private nonprofit		1.706	1.988	2.412	0.095	NA	NA	
Universities	5.778	6.281	7.062	8.182	12.333	20.600	29.700	[29.700]
Total	58.129	96.115	109.732	132.925	154.635	NA	NA	306.054
Source	685	565	565	565	687	611	611	610

[ ] = estimated or constructed by authors.

Acronyms and Abbreviations:
CSIRO: Commonwealth Scientific and Industrial Research Organization

Sources:
0010 Boyce, James K., and Robert E. Evenson. *National and International Agricultural Research and Extension Programs.* New York: Agricultural Development Council, Inc., 1975.

0242 Commonwealth Agricultural Bureaux (CAB). *List of Research Workers in Agricultural Sciences in the Commonwealth 1981.* Slough, England: CAB, 1981.

0243 Commonwealth Agricultural Bureaux (CAB). *List of Research Workers in the Agricultural Sciences in the Commonwealth and in the Republic of Ireland 1978.* Slough, England: CAB, 1978.

0244 Commonwealth Agricultural Bureaux (CAB). *List of Research Workers in the Agricultural Sciences in the Commonwealth and in the Republic of Ireland 1975.* Slough, England: CAB, 1975.

63

0279 Commonwealth Agricultural Bureaux (CAB). *List of Research Workers in the Agricultural Sciences in the Commonwealth and in the Republic of Ireland 1972.* Slough, England: CAB, 1972.

0285 Commonwealth Agricultural Bureaux (CAB). *List of Research Workers in the Agricultural Sciences in the Commonwealth and in the Republic of Ireland 1969.* Slough, England: CAB, 1969.

0286 Commonwealth Agricultural Bureaux (CAB). *List of Research Workers in Agriculture, Animal Health and Forestry in the Commonwealth and in the Republic of Ireland 1966.* Slough, England: CAB, 1966.

0565 Industries Assistance Commission. *Financing Rural Research.* Canberra: Australian Government Publishing Service, June 1976.

0610 Cameron, R.J. *Research and Experimental Development – General Government Organisations, Australia, 1981-82.* Catalogue No. 8109.0. Canberra: Australian Bureau of Statistics, January 1984.

0685 Department of Science. *Project SCORE Report 5: Summary of All Expenditures by Australia on Research and Development during 1968-69.* Canberra: Australian Government Publishing Service, 1973.

0687 Department of Science and the Environment. *Project SCORE Research and Development in Australia 1976-77 – All-Sector Results.* Canberra: Australian Government Publishing Service, 1980.

0852 Evenson, Robert E., and Yoav Kislev. *Investment in Agricultural Research and Extension: A Survey of International Data.* Center Discussion Paper No. 124. New Haven, Connecticut: Economic Growth Center, Yale University, August 1971.

Additional References:

0014 Judd, M. Ann, James K. Boyce, and Robert E. Evenson. "Investing in Agricultural Supply." Economic Growth Center, Yale University, New Haven, Connecticut, 1983. Mimeo.

0027 Harvey, Nigel, ed. *Agricultural Research Centres: A World Directory of Organizations and Programmes.* Seventh Edition, Two Volumes. Harlow, U.K.: Longman, 1983.

0165 Evenson, R.E., and Y. Kislev. *Agricultural Research and Productivity.* New Haven: Yale University Press, 1975.

0287 Commonwealth Agricultural Bureaux (CAB). *List of Research Workers in Agriculture, Animal Health and Forestry in the British Commonwealth, the Republic of Sudan and the Republic of Ireland 1959.* Slough, England: CAB, 1959.

0473 OECD. *Science and Technology Indicators: Basic Statistical Series – Volume A: The Objectives of Government R&D Funding 1974-1985.* Paris: OECD, May 1983.

0474 OECD. *Science and Technology Indicators: Basic Statistical Series - Volume A: The Objectives of Government R&D, 1969-1981.* Paris: OECD, July 1981.

0512 OECD. *Survey of the Resources Devoted to R&D by OECD Member Countries, International Statistical Year 1973 – Volume 5: Total Tables, Statistical Tables and Notes.* Paris: OECD, September 1976.

0530 OECD. *OECD Science and Technology Indicators No. 2 – R&D, Invention and Competitiveness.* Paris: OECD, 1986.

0566 Tisdell, C.A. "Research and Development in Australia – A Summary of Expenditure and the Allocation of Funds." In *Science and Industry Forum Report No. 5: The Influence of Research and Development on Economic Growth,* edited by the Australian Academy of Science, pp. 28-34. Canberra: Australian Academy of Science, June 1972.

0567 Underwood, E.J. "The Organisation of Agricultural Research in Australia." *Agricultural Administration* Vol. 1, No. 1 (1974): 73-81.

0568 Jarrett, F.G., and R.K. Lindner. "Rural Research in Australia." In *Agriculture in the Australian Economy* (second edition), edited by D.B. Williams, pp. 94-98. Sydney: Sydney University Press, 1980.

0569 Lindner, R.K. "Accountability in Research through Industry Funding." In *Evaluation of Agricultural Research.* Miscellaneous publication 8-1981, edited by G.W. Norton, et al., pp. 177-181. St. Paul, Minnesota: Minnesota Agricultural Experiment Station, University of Minnesota, April 1981.

0570 Underwood, E.J. "Agricultural Research in Australia: A Critical Appraisal." *Search* Vol. 4, No. 5 (May 1973): 155-160.

0571 Hastings, Trevor. "The Impact of Scientific Research on Australian Rural Productivity." *Australian Journal of Agricultural Economics* Vol. 25, No. 1 (April 1981): 48-59.

0572 OECD. *Reviews of National Science Policy: Australia.* Paris: OECD, 1977.

0573 OECD. *Reviews of National Science and Technology Policy: Australia.* Paris: OECD, 1986.

0601 OECD – STIID Data Bank. "Public Funding of R&D by Socio-Economic Objective." Tables. OECD, Paris, April 1987. Mimeo.

0602 OECD – STIID Data Bank. "R&D Expenditure in the Higher Education and Private Non-Profit

64

Sectors." Tables. OECD, Paris, April 1987. Mimeo.

0611 Cameron, R.J. *Research and Experimental Development, Higher Education Organisations, Australia, 1981.* Canberra: Australian Bureau of Statistics, October 1983.

0629 Commonwealth Scientific & Industrial Research Organization (CSIRO). *Twelfth Annual Report 1959-60.* Australia: CSIRO, n.d.

0630 Commonwealth Scientific and Industrial Research Organization (CSIRO). *Seventeenth Annual Report 1964-65.* Australia: CSIRO, n.d.

0686 Department of Science. *Project SCORE Research and Development in Australia 1973-74 – Volume 2: All-Sector Results, Business Enterprise Sector, and Higher Education Sector.* Canberra: Australian Government Publishing Service, 1977.

0708 OECD – STIID Data Bank. "Gross Domestic Expenditure on R&D: All Fields of Sciences – Agriculture, Forestry and Fishing." Tables. OECD, Paris, June 1987. Mimeo.

0732 FAO – Regional Office for Asia and the Pacific. *Agricultural Research Systems in the Asia-Pacific Region.* Bangkok: FAO – Regional Office for Asia and the Pacific, 1986.

0744 *Agricultural Research Centres: A World Directory of Organizations and Programmes.* Eighth Edition, Two Volumes. Harlow, U.K.: Longman, 1986.

0886 Evenson, Robert E., and Yoav Kislev. "Investment in Agricultural Research and Extension: A Survey of International Data." *Economic Development and Cultural Change* Vol. 23 (April 1975): 507-521.

Austria

Personnel

Period/Year	PhD	MSc	BSc	Subtotal	Expat	Total	Source
1960–64	..	..	..	..	..	145	514
1965–69	..	..	..	..	..	219	515
1970	..	..	..	..	..	225	10
1971	..	..	..	..	..	..	
1972	..	..	..	..	..	..	
1973	..	..	..	..	..	..	
1974	..	..	..	..	..	..	
1975	..	..	..	..	..	..	
1976	..	..	..	..	..	..	
1977	..	..	..	..	..	..	
1978	..	..	..	..	..	..	
1979	..	..	..	..	..	..	
1980	..	..	..	..	..	..	
1981	..	..	..	..	..	..	
1982	..	..	..	..	..	..	
1983	..	..	..	..	..	..	
1984	..	..	..	..	..	..	
1985	..	..	..	..	..	..	
1986	..	..	..	..	..	..	

1960–64: 1964   1965–69: 1967

Expenditure   LCU = A. Schillings

Period/Year	Current LCU (millions)	Constant 1980 LCU(millions)	Constant 1980 US$ (millions)		Source
			Atlas	PPP	
1960–64	43.856	102.948	7.893	6.420	514
1965–69	80.650	169.812	13.019	10.590	527/510
1970	139.600	257.565	19.747	16.062	10
1971	..	..	..	..	
1972	..	..	..	..	
1973	..	..	..	..	
1974	..	..	..	..	
1975	160.935	206.327	15.819	12.867	704
1976	215.630	261.687	20.064	16.320	704
1977	221.148	254.779	19.534	15.889	704
1978	237.089	259.681	19.910	16.194	704
1979	256.511	269.728	20.680	16.821	704
1980	273.733	273.733	20.987	17.071	704
1981	293.402	276.013	21.162	17.213	704
1982	331.300	293.446	22.498	18.300	704
1983	349.562	298.007	22.848	18.585	704
1984	361.783	294.133	22.551	18.343	677
1985	390.600	307.559	23.581	19.180	677
1986	..	..	..	..	

1960–64: 1964   1965–69: 1966 & 67

Personnel Comments:

The 1964 figure from source 514 includes 98 full-time equivalents (FTE) in the general government sector and 47 FTE in the higher education sector. Source 10 incorrectly placed this figure in 1963. The 1967 figure from source 515 measures agricultural researchers by field of science rather than by objective, which is more appropriate.

Expenditure Comments:

The expenditure indicators for the early years were constructed as follows:

	1964	1966	1967
Central Government	29.952	44.800	74.299
Higher Education	13.904	[19.200]	23.000
Total	43.856	64.000	97.299
Source	514	527	510

[ ] = estimated or constructed by authors.

The post-1974 expenditure indicators are based on the following OECD statistical series: "Public Funding of R & D by Socio-Economic Objective: Agriculture, Forestry and Fisheries – Version 2." This version differs from the standard version in that the Advancement of Knowledge category (including General University Funds) is further broken down into socio-economic objectives – although still leaving a considerable amount of expenditures under Advancement of Knowledge. A complete breakdown of this item into socioeconomic objectives would certainly result in some increase in the agricultural research expenditure estimates.

Sources:

0010 Boyce, James K., and Robert E. Evenson. *National and International Agricultural Research and Extension Programs.* New York: Agricultural Development Council, Inc., 1975.

0510 OECD. *International Survey of the Resources Devoted to R&D in 1967 by OECD Member*

Countries, Statistical Tables and Notes – Volume 5: Total Tables. Paris: OECD, August 1970.

0514 OECD. *A Study of Resources Devoted to R&D in OECD Member Countries in 1963/64, International Statistical Year for Research and Development – Volume 2: Statistical Tables and Notes.* Paris: OECD, 1968.

0515 UNESCO. *Statistical Yearbook 1972.* Paris: UNESCO, 1973.

0527 OECD. *Reviews of National Science Policy: Austria.* Paris: OECD, 1971.

0677 Bundesministerium für Wissenschaft und Forschung. *Bericht 1986 der Bundesregierung an den Nationalrat.* Wien: Bundesministerium für Wissenschaft und Forschung, 1986.

0704 OECD – Directorate for Science, Technology and Industry. "Public Funding of R&D by Socio-Economic Objectives in Austria." Table 0.1, version 2. OECD, Paris, January 1985. Mimeo.

Additional References:

0021 OECD. *OECD Science and Technology Indicators.* Paris: OECD, 1984.

0027 Harvey, Nigel, ed. *Agricultural Research Centres: A World Directory of Organizations and Programmes.* Seventh Edition, Two Volumes. Harlow, U.K.: Longman, 1983.

0165 Evenson, R.E., and Y. Kislev. *Agricultural Research and Productivity.* New Haven: Yale University Press, 1975.

0509 OECD. *Survey of the Resources Devoted to R&D by OECD Member Countries, International Statistical Year 1971 – Volume 5: Total Tables, Statistical Tables and Notes.* Paris: OECD, August 1974.

0511 OECD. *International Survey of the Resources Devoted to R&D in 1969 by OECD Member Countries, Statistical Tables and Notes – Volume 5: Total Tables.* Paris: OECD, June 1973.

0512 OECD. *Survey of the Resources Devoted to R&D by OECD Member Countries, International Statistical Year 1973 – Volume 5: Total Tables, Statistical Tables and Notes.* Paris: OECD, September 1976.

0526 OECD. *Country Reports on the Organisation of Scientific Research: Austria.* Paris: OECD, January 1964.

0530 OECD. *OECD Science and Technology Indicators No. 2 – R&D, Invention and Competitiveness.* Paris: OECD, 1986.

0601 OECD – STIID Data Bank. "Public Funding of R&D by Socio-Economic Objective." Tables. OECD, Paris, April 1987. Mimeo.

0602 OECD – STIID Data Bank. "R&D Expenditure in the Higher Education and Private Non-Profit

Sectors." Tables. OECD, Paris, April 1987. Mimeo.

0674 Bundesministerium für Wissenschaft und Forschung. *Bericht 1977 der Bundesregierung an den Nationalrat.* Wien: Bundesministerium für Wissenschaft und Forschung, 1977.

0675 Bundesministerium für Wissenschaft und Forschung. *Bericht 1980 der Bundesregierung an den Nationalrat.* Wien: Bundesministerium für Wissenschaft und Forschung, 1980.

0676 Bundesministerium für Wissenschaft und Forschung. *Bericht 1985 der Bundesregierung an den Nationalrat.* Wien: Bundesministerium für Wissenschaft und Forschung, 1985.

0678 Bundesministerium für Land- und Forstwirtschaft. *Das Forschungs- und Versuchswesen im Bereich des Bundesministeriums für Land- und Forstwirtschaft.* Wien: Bundesministerium für Land- und Forstwirtschaft, March 1980.

0679 Messmann, Karl. "Forschung und Experimentelle Entwicklung in Österreich 1966/67 bis 1985." *Statistische Nachrichten* Vol. 40, No. 5 (1985): 291–300.

0680 Bundesministerium für Wissenschaft und Forschung. *Bericht 1972 der Bundesregierung an den Nationalrat.* Wien: Bundesministerium für Wissenschaft und Forschung, April 1972.

0705 OECD – Directorate for Science, Technology and Industry. "Public R&D Funding by Socio-Economic Objective in Austria." Table 0.1, version 2. OECD, Paris, January 1986. Mimeo.

0707 OECD – STIID Data Bank. "R&D Expenditure in the Higher Education and Private Non-Profit Sectors: Higher Education, Agricultural Sciences, General Universities Funds." Table. OECD, Paris, June 1987. Mimeo.

0708 OECD – STIID Data Bank. "Gross Domestic Expenditure on R&D: All Fields of Sciences – Agriculture, Forestry and Fishing." Tables. OECD, Paris, June 1987. Mimeo.

0852 Evenson, Robert E., and Yoav Kislev. *Investment in Agricultural Research and Extension: A Survey of International Data.* Center Discussion Paper No. 124. New Haven, Connecticut: Economic Growth Center, Yale University, August 1971.

0886 Evenson, Robert E., and Yoav Kislev. "Investment in Agricultural Research and Extension: A Survey of International Data." *Economic Development and Cultural Change* Vol. 23 (April 1975): 507–521.

Bahamas

Personnel

Period/Year	PhD	MSc	BSc	Subtotal	Expat	Total	Source
1960–64	..	..	..	..	..	..	
1965–69	0	3	3	6	..	6	286/285
1970	..	..	..	..	..		
1971	..	..	..	..	..	0	279
1972	..	..	..	..	..		
1973	..	..	..	..	..	..	
1974	..	..	..	..	..	0	244
1975	..	..	..	..	..		
1976	..	..	..	..	..		
1977	..	..	..	..	..	0	243
1978	..	..	..	..	..		
1979	..	..	..	..	..	..	
1980	1	11	15	27	..	27	242
1981	..	..	..	..	..		
1982	..	..	..	..	..	..	
1983	0	12	10	22	..	22	115
1984	..	..	..	..	..	..	
1985	..	..	..	..	..	..	
1986	..	..	..	..	..	..	

1960-64:   1965-69: 1965 & 68

Expenditure

LCU = Bahama Dollars

Period/Year	Current LCU (millions)	Constant 1980 LCU(millions)	Constant 1980 US$ (millions) Atlas	Constant 1980 US$ (millions) PPP	Source
1960–64	..	..	..	..	
1965–69	..	..	..	..	
1970	..	..	..	..	
1971	..	..	..	..	
1972	..	..	..	..	
1973	..	..	..	..	
1974	..	..	..	..	
1975	..	..	..	..	
1976	..	..	..	..	
1977	..	..	..	..	
1978	..	..	..	..	
1979	..	..	..	..	
1980	..	..	..	..	
1981	..	..	..	..	
1982	..	..	..	..	
1983	..	..	..	..	
1984	..	..	..	..	
1985	..	..	..	..	
1986	..	..	..	..	

1960-64:   1965-69:

Personnel Comments:

For the earlier years (1965 and 1968) some Ministry of Agriculture personnel are included in the list of research workers of the Commonwealth Agricultural Bureaux (CAB) (sources 285 and 286). Although their function titles are not explicitly research oriented, they were included in the indicator series. Sources 243, 244, and 279 state that the Ministry of Agriculture is only involved in agricultural administration and extension, and report no researchers for the years 1971, 1974, and 1977. For the more recent period fisheries development and agriculture have become important areas of action (source 115), and a considerable increase in agricultural research effort has taken place in the last seven years.

At present the main agricultural research stations within the Ministry of Agriculture are Gladstone Road Agricultural Research Complex (GRARC), the Bahamas Agricultural Research Centre (BARC), and Coppice Island Research Station (source 115).

Sources:

0115 Economic Commission for Latin America and the Caribbean (ECLAC). *Agricultural Research Policy and Management – Volumes I and II.* Papers presented at the Workshop on Agricultural Research Policy and Management, 26–30 September 1983, Port of Spain, Trinidad. Port of Spain, Trinidad: ECLAC, November 1984.

0242 Commonwealth Agricultural Bureaux (CAB). *List of Research Workers in Agricultural Sciences in the Commonwealth 1981.* Slough, England: CAB, 1981.

0243 Commonwealth Agricultural Bureaux (CAB). *List of Research Workers in the Agricultural Sciences in the Commonwealth and in the Republic of Ireland 1978.* Slough, England: CAB, 1978.

0244 Commonwealth Agricultural Bureaux (CAB). *List of Research Workers in the Agricultural Sciences in the Commonwealth and in the Republic of Ireland 1975.* Slough, England: CAB, 1975.

0279 Commonwealth Agricultural Bureaux (CAB). *List of Research Workers in the Agricultural Sciences in the Commonwealth and in the Republic of Ireland 1972.* Slough, England: CAB, 1972.

0285 Commonwealth Agricultural Bureaux (CAB). *List of Research Workers in the Agricultural Sciences in the Commonwealth and in the Republic of Ireland 1969.* Slough, England: CAB, 1969.

0286 Commonwealth Agricultural Bureaux (CAB). *List of Research Workers in Agriculture, Animal Health and Forestry in the Commonwealth and in the Republic of Ireland 1966.* Slough, England: CAB, 1966.

Bangladesh

Personnel

Period/Year	PhD	MSc	BSc	Subtotal	Expat	Total	Source
1960–64	..	..	..	..	..	..	
1965–69	57	187	165	409	..	409	454
1970	..	..	..	..	..		
1971	..	..	..	..			
1972	..	..	..	..			
1973	..	..	..	..			
1974	..	..	..	..			
1975	..	..	..	..	..	635	16
1976	..	..	..	921	22	943	476
1977	..	..	..	952	32	984	476
1978	..	..	..	1190	44	1234	476
1979	66	933	241	1240		1240	100
1980	..	..	..	1256	46	1302	476
1981	..	..	..	..			
1982	..	..	..	..			
1983	81	534	312	927		927	99
1984	..	..	..	..			
1985	..	..	..	..			
1986	131	732	289	1152	..	1152	854

1960–64: 1965–69: 1968

Expenditure

LCU = Taka

Period/Year	Current LCU (millions)	Constant 1980 LCU (millions)	Constant 1980 US$ (millions)		Source
			Atlas	PPP	
1960–64	..	..	..	..	
1965–69	..	..	..	..	
1970	..	..	..	..	
1971	..	..	..	..	
1972	..	..	..	..	
1973	..	..	..	..	
1974	..	..	..	..	
1975	..	..	..	..	
1976	103.500	169.394	10.690	34.687	476
1977	99.800	139.776	8.821	28.622	476
1978	164.500	204.094	12.880	41.793	476
1979	192.600	212.583	13.416	43.531	476
1980	379.900	379.900	23.975	77.793	476
1981	..	..	..	..	
1982	..	..	..	..	
1983	..	..	..	..	
1984	508.360	317.924	20.064	65.102	17
1985	..	..	..	..	
1986	..	..	..	..	

1960–64:. 1965–69:

Personnel Comments:

All post-1974 figures claim to cover the whole agricultural research system of Bangladesh, except for the university system. For the years 1979, 1983, and 1986, the following institutes are included: BARI, BINA, BJRI, BRRI, BTRI, FRI, SRTI, and livestock and fisheries research. The 1986 figure also includes 49 researchers at BARC. In addition, source 99 reports, for 1983, 362 (possibly part-time) researchers at BAU and 92 researchers at other universities.

Several sources (95, 590, 111, 795, and 114) give institution-specific data. For the data provided by these sources it is difficult to distinguish between established post and those which are actually filled. In some cases filled positions account for less than 50% of established posts.

Expenditure Comments:

The expenditure figures given by source 476 are consistent with the personnel figures. All figures include donor contributions, in addition to funds from domestic sources.

Acronyms and Abbreviations:

BARC: Bangladesh Agricultural Research Council
BARI: Bangladesh Agricultural Research Institute
BAU: Bangladesh Agricultural University
BINA: Bangladesh Institute of Nuclear Agriculture
BJRI: Bangladesh Jute Research Institute
BRRI: Bangladesh Rice Research Institute
BTRI: Bangladesh Tea Research Institute
BWDB: Bangladesh Water Development Board
FRI: Forestry Research Institute
SRTI: Sugarcane Research and Training Institute

Sources:

0016 Oram, Peter A., and Vishva Bindlish. *Resource Allocations to National Agricultural Research: Trends in the 1970s.* The Hague and Washington, D.C.: ISNAR and IFPRI, November 1981.

0017 ISNAR, IFARD & AOAD. Survey of National Agricultural Research Systems: Unpublished Questionnaire Responses. ISNAR, The Hague, 1985.

0099 Bangladesh Agricultural Research Council (BARC). *National Agricultural Research Plan 1984–1989.* Bangladesh: BARC, May 1984.

0100 Wennergren, Boyd E., Charles H. Antholt, and Morris D. Whitaker. *Agricultural Development*

in Bangladesh. Boulder, Colorado: Westview Press, 1984.

0454 Wahid, Abdul. "State of Current Agricultural Research and Education in Pakistan." *Agriculture Pakistan* Vol. 19, No. 3 (1968): 261–285.

0476 Gill, Gerard J. *Operational Funding Constraints on Agricultural Research in Bangladesh.* BARC Agricultural Economics and Rural Social Science Papers No. 9. Dacca: Bangladesh Agricultural Research Council (BARC), April 1981.

0854 Bangladesh Agricultural Research Council (BARC). *Manpower Planning and Development in Agricultural Research.* Dhaka: BARC, September 1987.

Additional References:

0010 Boyce, James K., and Robert E. Evenson. *National and International Agricultural Research and Extension Programs.* New York: Agricultural Development Council, Inc., 1975.

0014 Judd, M. Ann, James K. Boyce, and Robert E. Evenson. "Investing in Agricultural Supply." Economic Growth Center, Yale University, New Haven, Connecticut, 1983. Mimeo.

0018 Daniels, Douglas, and Barry Nestel, eds. *Resource Allocation to Agricultural Research.* Proceedings of a workshop held in Singapore 8–10 June 1981. Ottowa: IDRC, 1981.

0019 Bengtsson, Bo, and Tedla Getachew, eds. *Strengthening National Agricultural Research.* Report from a SAREC workshop, 10–17 September 1979. Part 1: Background Documents; Part II: Summary and Conclusions. Sweden: SAREC, 1980.

0026 Oram, Peter A., and Mark Gieben. "Document Summaries." ISNAR, The Hague, 1984. Mimeo.

0027 Harvey, Nigel, ed. *Agricultural Research Centres: A World Directory of Organizations and Programmes.* Seventh Edition, Two Volumes. Harlow, U.K.: Longman, 1983.

0050 Pray, Carl E., and Jock R. Anderson. *Bangladesh and the CGIAR Centers: A Study of Their Collaboration in Agricultural Research.* CGIAR Study Paper Number 8. Washington, D.C.: World Bank, 1985.

0073 Oram, Peter A., and Vishva Bindlish. "Investment in Agricultural Research in Developing Countries: Progress, Problems, and the Determination of Priorities." IFPRI, Washington, D.C., January 1984. Mimeo.

0093 Bangladesh Agricultural Research Council (BARC), ISNAR, and Winrock International. *Management of Human Resources in Agricultural Research.* Report of the International Workshop on Management of Human Resources in

Agricultural Research held March 3–5, 1986 in Dhaka, Bangladesh. Dhaka: BARC, ISNAR, and Winrock International, 1986.

0095 FAO – CARIS. *Agricultural Research in Developing Countries – Volume 1: Research Institutions.* Rome: FAO – CARIS, 1978.

0101 Ahsan, Ekramul. *Bangladesh: Resources Allocation in Agricultural Research.* Bangladesh: Bangladesh Agricultural Research Council, n.d.

0111 Moseman, A.H., et al. *Bangladesh Agricultural Research System.* Bangladesh: Ministry of Agriculture and Fisheries, July 1980.

0112 Elias, S.M. "Manpower Requirements and Developments for Agricultural Research in Bangladesh." Bangladesh Agricultural Research Council, Bangladesh, 1980. Mimeo.

0113 Moseman, A.H. "Agricultural Research in Bangladesh." FAO/DANIDA, Rome/Dacca, n.d. Mimeo.

0114 Bangladesh Agricultural Research Council (BARC). *Agricultural Research in Bangladesh.* Dhaka: BARC, 1983.

0163 CGIAR. "National Agricultural Research." CGIAR, Washington, D.C., 1985. Mimeo.

0179 The Philippine Council for Agriculture and Resources Research and Development (PCARRD), and IDRC. *Agriculture and Resources Research Manpower Development in South and Southeast Asia.* Proceedings of the Workshop on Agriculture and Resources Research Manpower Development in South and Southeast Asia, 21–23 October 1981 and 4–6 February 1982, Singapore. Los Baños, Philippines: PCARRD, 1983.

0193 Joint Pakistan–American Agricultural Research Review Team. *Report of the Joint Pakistan–American Agricultural Research Review Team.* Dacca: Bangladesh Agricultural Research Council, April 1968.

0242 Commonwealth Agricultural Bureaux (CAB). *List of Research Workers in Agricultural Sciences in the Commonwealth 1981.* Slough, England: CAB, 1981.

0243 Commonwealth Agricultural Bureaux (CAB). *List of Research Workers in the Agricultural Sciences in the Commonwealth and in the Republic of Ireland 1978.* Slough, England: CAB, 1978.

0244 Commonwealth Agricultural Bureaux (CAB). *List of Research Workers in the Agricultural Sciences in the Commonwealth and in the Republic of Ireland 1975.* Slough, England: CAB, 1975.

0285 Commonwealth Agricultural Bureaux (CAB). *List of Research Workers in the Agricultural Sciences in the Commonwealth and in the Republic of Ireland 1969.* Slough, England: CAB, 1969.

0286 Commonwealth Agricultural Bureaux (CAB). *List of Research Workers in Agriculture, Animal Health and Forestry in the Commonwealth and in the Republic of Ireland 1966.* Slough, England: CAB, 1966.

0287 Commonwealth Agricultural Bureaux (CAB). *List of Research Workers in Agriculture, Animal Health and Forestry in the British Commonwealth, the Republic of Sudan and the Republic of Ireland 1959.* Slough, England: CAB, 1959.

0302 Ahsan, Ekramul. "Resource Allocation to Agricultural Research in Bangladesh." In *Resource Allocation to Agricultural Research.* Proceedings of a workshop held in Singapore 8–10 June 1981, edited by Douglas Daniels and Barry Nestel, pp. 129–136. Ottawa: IDRC, 1981.

0306 Elias, S.M. "Manpower Developments for Agricultural Research in Bangladesh." In *Resource Allocation to Agricultural Research.* Proceedings of a workshop held in Singapore 8–10 June 1981, edited by Douglas Daniels and Barry Nestel, pp. 168–170. Ottawa: IDRC, 1981.

0315 Rahman, M.M. "Agricultural Research in Bangladesh." In *Strengthening National Agricultural Research.* Report from a SAREC workshop, 10–17 September 1979, edited by Bo Bengtsson and Getachew Tedla, pp. 77–88. Sweden: SAREC, 1980.

0326 Ahsan, Ekramul. "Keynote Address: Management of Human Resources in Agricultural Research." In *Management of Human Resources in Agricultural Research.* Report of the International Workshop on Management of Human Resources in Agricultural Research held 3–5 March 1986, edited by Theodore Hutchcroft, pp.27–31. Dhaka: Bangladesh Agricultural Research Council, ISNAR, and Winrock International, 1986.

0434 Anh, Do. "Agricultural Research Management Systems in the Philippines, India, and Bangladesh." In *Proceedings of the Research Extension Linkage Workshop Held in Hanoi, Viet Nam, 9–13 June 1986.* Rome: FAO, 1986.

0590 UNESCO. *Science and Technology in Countries of Asia and the Pacific.* Science Policy Studies and Documents No. 52. Paris: UNESCO, n.d.

0732 FAO – Regional Office for Asia and the Pacific. *Agricultural Research Systems in the Asia–Pacific Region.* Bangkok: FAO – Regional Office for Asia and the Pacific, 1986.

0770 Ruttan, Vernon W. *Agricultural Research Policy and Development.* FAO Research and Technology Paper 2. Rome: FAO, 1987.

0795 Ahsan, Ekramul. "Upgrading Manpower Resources for Agricultural Research in Bangladesh." In *Agriculture and Resources Research Manpower Development in South and Southeast Asia*, pp. 201–206. Los Baños, Philippines: Philippine Council for Agriculture and Resources Research and Development, 1983.

0804 Bangladesh Agricultural Research Council (BARC). "Progress Report Manpower Planning in Agricultural Research." BARC, Dhaka, July 1986. Mimeo.

0807 Cushing, R.L. *Master Plan for Bangladesh Agricultural Research Council Phase II*. Dhaka: Bangladesh Agricultural Research Council, November 1982.

0830 Islam, Mohammad Zahirul. *Directory of Agricultural Scientists and Technologists in Bangladesh*. Dhaka: Bangladesh Agricultural Research Council, 1985.

0869 Boyce, James K. "Agricultural Research in Indonesia, The Philippines, Bangladesh, South Korea and India: A Documentary History." University of Minnesota Asian Agricultural Research Review Project, July 1980. Mimeo.

0878 Pray, C.E. "The Economics of Agricultural Research in British Punjab and Pakistani Punjab, 1905–1975." PhD diss., University of Pennsylvania, Pennsylvania, 1978.

0879 Pray, C.E. "The Economics of Agricultural Research in Bangladesh." *Bangladesh Journal of Agricultural Economics* Vol. 2 (December 1979): 1–36.

Barbados

Personnel

Period/Year	PhD	MSc	BSc	Subtotal	Expat	Total	Source
1960–64	..	..	..	..	..	..	
1965–69	1	5	3	9	..	9	286/285
1970	2	5	19	26	..	26	17/800
1971	2	6	19	27	..	27	17/800
1972	2	6	19	27	..	27	17/800
1973	2	6	20	28	..	28	17/800
1974	2	6	22	30	..	30	17/800
1975	3	7	24	34	..	34	17/800
1976	3	8	25	36	..	36	17/800
1977	3	8	25	36	..	36	17/800
1978	3	8	26	37	..	37	17/800
1979	3	8	27	38	..	38	17/800
1980	3	9	28	40	..	40	17/800
1981	3	9	28	40	..	40	17/800
1982	3	10	28	41	12	53	17/800
1983	3	11	29	43	10	53	17/800
1984	3	11	29	43	10	53	800
1985	3	11	29	43	6	49	800
1986	..	..	..	..	..	..	

1960–64:   1965–69: 1965 & 68

Expenditure

LCU = Barbados Dollars

Period/Year	Current LCU (millions)	Constant 1980 LCU(millions)	Constant 1980 US$ (millions)		Source
			Atlas	PPP	
1960–64	..	..	..	..	
1965–69	..	..	..	..	
1970	0.700	2.157	1.000	1.383	17
1971	0.690	2.024	0.938	1.297	17
1972	0.810	2.315	1.073	1.484	17
1973	0.750	1.830	0.848	1.173	17
1974	0.770	1.398	0.648	0.896	17
1975	1.890	3.001	1.391	1.923	17
1976	2.000	3.226	1.495	2.067	17
1977	2.440	3.782	1.753	2.424	17
1978	2.550	3.578	1.659	2.293	17
1979	2.720	3.421	1.586	2.193	17
1980	2.830	2.830	1.312	1.814	17
1981	2.920	2.577	1.185	1.639	17
1982	4.250	3.411	1.581	2.186	17
1983	4.190	3.172	1.471	2.033	17
1984	2.538	1.823	0.845	1.169	800
1985	4.818	3.234	1.499	2.072	800
1986	..	..	..	..	

1960–64:   1965–69:

Personnel Comments:

The figures from source 800 include the Research Division of the Ministry of Agriculture, Food and Consumer Affairs only. According to source 17 the following organizations also conduct agricultural research in Barbados: the Barbados Sugar Industries Ltd. (1985: 8 researchers, source 744), the West Indies Central Sugar Cane Breeding Station, and the Barbados unit of the Caribbean Agricultural Research and Development Institute (CARDI).

CARDI was created in 1975. The following professional staff development of the Barbados unit of CARDI was reconstructed on the basis of several annual reports and source 800.

	1975	1978	1980	1981	1983	1984	1985
PhD	NA	3	4	2	0	0	0
MSc	NA	5	5	4	4	3	2
BSc	NA	1	2	1	2	1	2
Total	4	9	11	7	6	4	4
Source	342	341	343	800	355	800	348

Expenditure Comments:

The expenditure indicators represent Ministry of Agriculture, Food and Consumer Affairs research expenditures only. The fluctuations in the early eighties are caused by relatively large capital investments in certain years.

Sources:

0017   ISNAR, IFARD & AOAD. Survey of National Agricultural Research Systems: Unpublished Questionnaire Responses. ISNAR, The Hague, 1985.

0285   Commonwealth Agricultural Bureaux (CAB). *List of Research Workers in the Agricultural Sciences in the Commonwealth and in the Republic of Ireland 1969.* Slough, England: CAB, 1969.

0286   Commonwealth Agricultural Bureaux (CAB). *List of Research Workers in Agriculture, Animal Health and Forestry in the Commonwealth and in the Republic of Ireland 1966.* Slough, England: CAB, 1966.

0800   Jeffers, J.P.W. Personal communication. Deputy Chief Agricultural Officer (Research), Ministry of Agriculture, Food and Consumer Affairs. St. George, Barbados, February 1988.

Additional References:

0001   Trigo, Eduardo J., and Martín E. Piñeiro. "Funding Agricultural Research." In *Report of a Conference: Selected Issues in Agricultural Research in Latin America*, edited by Barry Nestel and Eduardo J. Trigo. The Hague: ISNAR, March 1984.

0010   Boyce, James K., and Robert E. Evenson. *National and International Agricultural Research and Extension Programs.* New York: Agricultural Development Council, Inc., 1975.

0014   Judd, M. Ann, James K. Boyce, and Robert E. Evenson. "Investing in Agricultural Supply." Economic Growth Center, Yale University, New Haven, Connecticut, 1983. Mimeo.

0016   Oram, Peter A., and Vishva Bindlish. *Resource Allocations to National Agricultural Research: Trends in the 1970s.* The Hague and Washington, D.C.: ISNAR and IFPRI, November 1981.

0053   Piñeiro, Martín, and Eduardo Trigo, eds. *Cambio Técnico en el Agro Latino Americano – Situación y Perspectivas en la Década de 1980.* San José: IICA, 1983.

0063   ISNAR. *Report to the Board of Directors of CARDI: Analysis, Evaluation and Proposals to Strengthen CARDI's Regional Capacity.* The Hague: ISNAR, 1985.

0073   Oram, Peter A., and Vishva Bindlish. "Investment in Agricultural Research in Developing Countries: Progress, Problems, and the Determination of Priorities." IFPRI, Washington, D.C, January 1984. Mimeo.

0102   Pinchinat, Antonio. *Barbados Agricultural Research System – A Condensed Report.* Santo Domingo: IICA, December 1980.

0103   Jeffers, J.P.W. "Agricultural Research Profile, Barbados." Ministry of Agriculture, Food and Consumer Affairs, Barbados, 1977. Mimeo.

0115 Economic Commission for Latin America and the Caribbean (ECLAC). *Agricultural Research Policy and Management – Volumes I and II.* Papers presented at the Workshop on Agricultural Research Policy and Management, 26–30 September 1983, Port of Spain, Trinidad. Trinidad: ECLAC, November 1984.

0129 Caribbean Agricultural Research and Development Institute (CARDI) – Barbados Unit. *Annual Report 1985.* St. Michael, Barbados: CARDI, 1985.

0163 CGIAR. "National Agricultural Research." CGIAR, Washington, D.C., 1985. Mimeo.

0172 Trigo, Eduardo. Background material. ISNAR. The Hague, n.d.

0219 Department of Agriculture, Natural Resources and Rural Development (DARNDR), and IICA. "Réunion Technique Régionale sur les Systèmes de Recherche Agricole dans les Antilles." DARNDR and IICA, Port-au-Prince, Haiti, November/December 1977. Mimeo.

0244 Commonwealth Agricultural Bureaux (CAB). *List of Research Workers in the Agricultural Sciences in the Commonwealth and in the Republic of Ireland 1975.* Slough, England: CAB, 1975.

0279 Commonwealth Agricultural Bureaux (CAB). *List of Research Workers in the Agricultural Sciences in the Commonwealth and in the Republic of Ireland 1972.* Slough, England: CAB, 1972.

0287 Commonwealth Agricultural Bureaux (CAB). *List of Research Workers in Agriculture, Animal Health and Forestry in the British Commonwealth, the Republic of Sudan and the Republic of Ireland 1959.* Slough, England: CAB, 1959.

0341 Caribbean Agricultural Research and Development Institute (CARDI). *Annual Report 1977–1978.* St. Augustine, Trinidad & Tobago: CARDI, November 1978.

0342 Caribbean Agricultural Research and Development Institute (CARDI). *CARDI 1976–*

1981: Report of the Chairman. St. Augustine, Trinidad & Tobago: CARDI, 1982.

0343 Caribbean Agricultural Research and Development Institute (CARDI). *Annual Report 1980.* St. Augustine, Trinidad & Tobago: CARDI, 1981.

0344 Caribbean Agricultural Research and Development Institute (CARDI). *Annual Report 1980 – Barbados Unit.* Barbados: CARDI, 1982.

0345 Caribbean Agricultural Research and Development Institute (CARDI). *Annual Report 1981 – Barbados Unit.* Barbados: CARDI, 1982.

0348 Caribbean Agricultural Research and Development Institute (CARDI). *Annual Report 1985 – Barbados Unit.* Barbados: CARDI, 1985.

0350 Caribbean Agricultural Research and Development Institute (CARDI). *Annual Report 1984 – Barbados Unit.* Barbados: CARDI, 1984.

0353 Caribbean Agricultural Research and Development Institute (CARDI). *Paper Presented at a Meeting of the Governing Body of CARDI on January 29, 1982, Basseterre, St. Kitts.* St. Augustine, Trinidad: CARDI, 1982.

0354 Caribbean Agricultural Research and Development Institute (CARDI). *Annual Report Research and Development 1983–84: Highlights.* St. Augustine, Trinidad: CARDI, 1984.

0355 Caribbean Agricultural Research and Development Institute (CARDI). *Research and Development Summary 1982–83.* St. Augustine, Trinidad: CARDI, 1983.

0356 Caribbean Agricultural Research and Development Institute (CARDI). *Research and Development Summary 1983–84.* St. Augustine, Trinidad: CARDI, 1984.

0365 Trigo, Eduardo, and Guillermo E. Gálvez. "Trip Report of the CIAT and IICA Mission to the Caribbean English-Speaking Countries for UNDP." 1979. Mimeo.

0744 *Agricultural Research Centres: A World Directory of Organizations and Programmes.* Eighth Edition, Two Volumes. Harlow, U.K.: Longman, 1986.

Belgium

Personnel

Period/Year	PhD	MSc	BSc	Subtotal	Expat	Total	Source
1960–64	..	..	..	..	..	478	282/852
1965–69	..	..	..	..	..	648	852/10
1970	..	..	..	..	..		
1971	..	..	..	..	..		
1972	..	..	..	..	..		
1973	..	..	..	..	..		
1974	..	..	..	..	..		
1975	..	..	..	..	..		
1976	..	..	..	..	..	546	283
1977	..	..	..	..	..		
1978	..	..	..	..	..		
1979	..	..	..	..	..		
1980	..	..	..	..	..		
1981	..	..	..	..	..	496	376
1982	..	..	..	..	..		
1983	..	..	..	..	..		
1984	..	..	..	..	..		
1985	..	..	..	..	..		
1986	..	..	..	..	..		

1960–64: 1960 & 63   1965–69: 1965, 67 & 69

Expenditure   *LCU = Belgium Francs*

Period/Year	Current LCU (millions)	Constant 1980 LCU(millions)	Constant 1980 US$ (millions) Atlas	Constant 1980 US$ (millions) PPP	Source
1960–64	274.017	724.370	25.099	19.213	282/398
1965–69	..	..	..	..	
1970	..	..	..	..	
1971	538.500	1016.038	32.205	26.949	509
1972	..	..	..	..	
1973	..	..	..	..	
1974	..	..	..	..	
1975	964.800	1264.482	43.813	33.539	513
1976	..	..	..	..	
1977	..	..	..	..	
1978	..	..	..	..	
1979	..	..	..	..	
1980	..	..	..	..	
1981	1652.000	1571.836	54.463	41.691	376
1982	..	..	..	..	
1983	..	..	..	..	
1984	..	..	..	..	
1985	..	..	..	..	
1986	..	..	..	..	

1960–64: 1960 & 63   1965–69:

78

Personnel Comments

The main institutions undertaking agricultural research in Belgium are the Ministry of Agriculture, IWONL/IRSIA (Institute for the Advancement of Scientific Research in Industry and Agriculture), and the universities. IWONL/IRSIA executes research within its own organizations, as well as funding agricultural research in external agencies. Statistical information on agricultural research in Belgium is rather weak. The indicators should therefore be used with some caution. However, as far as it was possible to check, the personnel indicators include the three main institutions noted above.

Expenditure Comments:

The expenditure series has been constructed in an attempt to ensure commensurability with the personnel figures.

Acronyms and Abbreviations:

IWONL/IRSIA: (Institute for the Advancement of Scientific Research in Industry and Agriculture)

Sources:

0010 Boyce, James K., and Robert E. Evenson. *National and International Agricultural Research and Extension Programs.* New York: Agricultural Development Council, Inc., 1975.

0282 EEG. *De Organisatie van het Landbouwkundig Onderzoek in de Landen van de EEG.* EEG Studies, Serie Landbouw 9. Brussel: EEG, 1963.

0283 European Economic Community (EEC). "Monograph on the Organization of Agricultural Research." EEC, Brussels, 1982. Mimeo.

0376 FAO. *Fourteenth FAO Regional Conference for Europe, Reykjavik, Iceland, 17–21 September 1984: Research in Support of Agricultural Policies in Europe.* Rome: FAO, 1984.

0398 UNESCO. *La Politique Scientifique et l'Organisation de la Recherche Scientifique en Belgique.* Etudes et Documents de Politique Scientifique No. 1. Paris: UNESCO, 1965.

0509 OECD. *Survey of the Resources Devoted to R&D by OECD Member Countries, International Statistical Year 1971 – Volume 5: Total Tables, Statistical Tables and Notes.* Paris: OECD, August 1974.

0513 OECD. *International Survey of the Resources Devoted to R&D by OECD Member Countries, International Statistical Year 1975 – International Volume: Statistical Tables and Notes.* Paris: OECD, March 1979.

0852 Evenson, Robert E., and Yoav Kislev. *Investment in Agricultural Research and Extension: A Survey of International Data.* Center Discussion Paper No. 124. New Haven, Connecticut: Economic Growth Center, Yale University, August 1971.

Additional References:

0021 OECD. *OECD Science and Technology Indicators.* Paris: OECD, 1984.

0027 Harvey, Nigel, ed. *Agricultural Research Centres: A World Directory of Organizations and Programmes.* Seventh Edition, Two Volumes. Harlow, U.K.: Longman, 1983.

0165 Evenson, R.E., and Y. Kislev. *Agricultural Research and Productivity.* New Haven: Yale University Press, 1975.

0450 Ministerie van Landbouw en Visserij – Directie Akkerbouw en Tuinbouw. "Indrukken van een Orientatiereis naar Belgie 25 t/m 29 Juli 1977." Ministerie van Landbouw en Visserij, The Hague, March 1978. Mimeo.

0463 Instituut tot Aanmoediging van het Wetenschappelijk Onderzoek in Nijverheid en Landbouw (IWONL). *De Centra voor Landbouwkundig Onderzoek gesubsidieerd door het IWONL.* Belgium: IWONL, n.d.

0473 OECD. *Science and Technology Indicators: Basic Statistical Series – Volume A: The Objectives of Government R&D Funding 1974–1985.* Paris: OECD, May 1983.

0474 OECD. *Science and Technology Indicators: Basic Statistical Series – Volume A: The Objectives of Government R&D, 1969–1981.* Paris: OECD, July 1981.

0510 OECD. *International Survey of the Resources Devoted to R&D in 1967 by OECD Member Countries, Statistical Tables and Notes – Volume 5: Total Tables.* Paris: OECD, August 1970.

0511 OECD. *International Survey of the Resources Devoted to R&D in 1969 by OECD Member Countries, Statistical Tables and Notes – Volume 5: Total Tables.* Paris: OECD, June 1973.

0512 OECD. *Survey of the Resources Devoted to R&D by OECD Member Countries, International Statistical Year 1973 – Volume 5: Total Tables, Statistical Tables and Notes.* Paris: OECD, September 1976.

0514 OECD. *A Study of Resources Devoted to R&D in OECD Member Countries in 1963/64, International Statistical Year for Research and Development – Volume 2: Statistical Tables and Notes.* Paris: OECD, 1968.

0530 OECD. *OECD Science and Technology Indicators No. 2 – R&D, Invention and Competitiveness.* Paris: OECD, 1986.

0558 OECD. *Country Reports on the Organisation of Scientific Research: Belgium.* Paris: OECD, October 1963.

0559 OECD. *Reviews of National Science Policy: Belgium.* Paris: OECD, 1966.

0574 OECD. *Intellectual Investment in Agriculture for Economic and Social Development.* Documentation in Food and Agriculture No. 60. Paris: OECD, 1962.

0601 OECD – STIID Data Bank. "Public Funding of R&D by Socio-Economic Objective." Tables. OECD, Paris, April 1987. Mimeo.

0602 OECD – STIID Data Bank. "R&D Expenditure in the Higher Education and Private Non-Profit Sectors." Tables. OECD, Paris, April 1987. Mimeo.

0606 EUROSTAT. *Government Financing of Research and Development 1975–1984.* Luxembourg: Office des Publications Officielles des Communautés Européennes, February 1985.

0706 EUROSTAT. *Government Financing of Research and Development 1975–1985.* Luxembourg: Office for Official Publications of the European Communities, 1987.

0707 OECD – STIID Data Bank. "R&D Expenditure in the Higher Education and Private Non-Profit Sectors: Higher Education, Agricultural Sciences, General Universities Funds." Table. OECD, Paris, June 1987. Mimeo.

0886 Evenson, Robert E., and Yoav Kislev. "Investment in Agricultural Research and Extension: A Survey of International Data." *Economic Development and Cultural Change* Vol. 23 (April 1975): 507–521.

Belize

Personnel

Period/Year	PhD	MSc	BSc	Subtotal	Expat	Total	Source
1960–64	..	..	..	..	..	..	
1965–69	..	..	..	..	..	..	
1970	..	..	..	..	..	..	
1971	..	..	..	..	..	..	
1972	..	..	..	..	..	..	
1973	..	..	..	..	..	..	
1974	..	..	..	..	..	..	
1975	..	..	..	..	..	..	
1976	..	..	..	..	..	..	
1977	..	..	..	..	..	..	
1978	..	..	..	..	..	..	
1979	..	..	..	..	..	..	
1980	..	..	..	..	..	13	242
1981	..	..	..	..	..		
1982	..	..	..	..	..	16	27/115
1983	..	..	..	..	..		
1984	..	..	..	..	..		
1985	..	..	..	..	..		
1986	..	..	..	..	..		

1960–64:   1965–69:

Expenditure   *LCU = Belize Dollars*

Period/Year	Current LCU (millions)	Constant 1980 LCU(millions)	Constant 1980 US$ (millions)		Source
			Atlas	PPP	
1960–64	..	..	..	..	
1965–69	..	..	..	..	
1970	..	..	..	..	
1971	..	..	..	..	
1972	..	..	..	..	
1973	..	..	..	..	
1974	..	..	..	..	
1975	..	..	..	..	
1976	..	..	..	..	
1977	..	..	..	..	
1978	..	..	..	..	
1979	..	..	..	..	
1980	..	..	..	..	
1981	..	..	..	..	
1982	..	..	..	..	
1983	..	..	..	..	
1984	..	..	..	..	
1985	..	..	..	..	
1986	..	..	..	..	

1960–64:   1965–69:

Personnel Comments:
The personnel figures have been constructed as follows:

	1977	1978	1980	1981	1982	1983	1984
MNR							
DOA/RD	4	NA	5	NA	10	NA	NA
TRDP – Rice Research			4	NA	3*	NA	NA
CARDI		5	[4]	3	[3]	3	3
Total	NA	NA	13	NA	16	NA	NA
Source	243	341	242	342	27 115*	355	354/ 352

[ ] = estimated or constructed by authors.

In addition to the research institutes included in the table above, two private agricultural research institutes operate in Belize, namely, the Sugar Cane Research Station of the Belize Sugar Industries Ltd. (1 researcher in 1980, source 242, and 3 researchers in 1983, source 17) and Caricom Farms Ltd. (1 researcher in 1983, source 17).

Expenditure Comments:
Although source 17 provided expenditure data, it was not possible to derive an expenditure figure consistent with the personnel figures.

Acronyms and Abbreviations:
CARDI: Caribbean Agricultural Research and
  Development Institute
DOA/RD: Department of Agriculture – Research
  Division
MNR: Ministry of Natural Resources
TRDP: Toledo Rural Development Project

Sources:
0027 Harvey, Nigel, ed. *Agricultural Research Centres: A World Directory of Organizations and Programmes. Seventh Edition, Two Volumes.* Harlow, U.K.: Longman, 1983.
0115 Economic Commission for Latin America and the Caribbean (ECLAC). Agricultural Research Policy and Management – Volumes I and II. Papers presented at the Workshop on Agricultural Research Policy and Management, 26–30 September 1983, Port of Spain, Trinidad. Port of Spain, Trinidad: ECLAC, November 1984.
0242 Commonwealth Agricultural Bureaux (CAB). *List of Research Workers in Agricultural Sciences*

in the Commonwealth 1981. Slough, England: CAB, 1981.

Additional References:
0017 ISNAR, IFARD & AOAD. Survey of National Agricultural Research Systems: Unpublished Questionnaire Responses. ISNAR, The Hague, 1985.
0063 ISNAR. *Report to the Board of Directors of CARDI: Analysis, Evaluation and Proposals to Strengthen CARDI's Regional Capacity.* The Hague: ISNAR, 1985.
0243 Commonwealth Agricultural Bureaux (CAB). *List of Research Workers in the Agricultural Sciences in the Commonwealth and in the Republic of Ireland 1978.* Slough, England: CAB, 1978.
0287 Commonwealth Agricultural Bureaux (CAB). *List of Research Workers in Agriculture, Animal Health and Forestry in the British Commonwealth, the Republic of Sudan and the Republic of Ireland 1959.* Slough, England: CAB, 1959.
0341 Caribbean Agricultural Research and Development Institute (CARDI). *Annual Report 1977–1978.* St. Augustine, Trinidad & Tobago: CARDI, November 1978.
0342 Caribbean Agricultural Research and Development Institute (CARDI). *CARDI 1976–1981: Report of the Chairman.* St. Augustine, Trinidad & Tobago: CARDI, 1982.
0343 Caribbean Agricultural Research and Development Institute (CARDI). *Annual Report 1980.* St. Augustine, Trinidad & Tobago: CARDI, 1981.
0352 Caribbean Agricultural Research and Development Institute (CARDI). *Annual Report*

82

1983–1984 – Belize Unit. St. Augustine, Trinidad: CARDI, 1984.

0353 Caribbean Agricultural Research and Development Institute (CARDI). *Paper Presented at a Meeting of the Governing Body of CARDI on January 29, 1982, Basseterre, St. Kitts.* St. Augustine, Trinidad: CARDI, 1982.

0354 Caribbean Agricultural Research and Development Institute (CARDI). *Annual Report*

Research and Development 1983–84: Highlights. St. Augustine, Trinidad: CARDI, 1984.

0355 Caribbean Agricultural Research and Development Institute (CARDI). *Research and Development Summary 1982–83.* St. Augustine, Trinidad: CARDI, 1983.

0356 Caribbean Agricultural Research and Development Institute (CARDI). *Research and Development Summary 1983–84.* St. Augustine, Trinidad: CARDI, 1984.

Benin

Personnel

Period/Year	PhD	MSc	BSc	Subtotal	Expat	Total	Source
1960–64	..	..	..	..	..		
1965–69	..	..	..	..	..	13	653
1970	..	..	..	..	..	15	17
1971	..	..	..	..	..	15	17
1972	..	..	..	..	..	19	17
1973	..	..	..	..	..	21	17
1974	..	..	..	..	..	26	17
1975	..	..	..	..	..	29	17
1976	..	..	..	..	..	30	17
1977	..	..	..	..	..	32	17
1978	..	..	..	..	..	35	17
1979	..	..	..	..	..	36	17
1980	..	..	..	..	..	40	17
1981	..	..	..	..	..	42	17
1982	..	..	..	..	..	42	17
1983	6	31	15	52	4	56	17
1984	..	..	..	..	..		
1985	..	..	..	..	..		
1986	..	..	..	..	..		

1960–64:   1965–69: 1966

Expenditure

LCU = CFA Francs

Period/Year	Current LCU (millions)	Constant 1980 LCU (millions)	Constant 1980 US$ (millions)		Source
			Atlas	PPP	
1960–64	..	..	..	..	
1965–69	..	..	..	..	
1970	273.000	649.308	2.938	5.092	17
1971	283.700	648.625	2.935	5.087	17
1972	316.800	689.948	3.122	5.411	17
1973	290.400	594.098	2.689	4.659	17
1974	285.800	511.285	2.314	4.010	17
1975	379.200	626.818	2.837	4.916	17
1976	395.100	571.892	2.588	4.485	17
1977	425.500	568.595	2.573	4.459	17
1978	406.200	513.340	2.323	4.026	17
1979	474.700	545.674	2.469	4.279	17
1980	362.600	362.600	1.641	2.844	17
1981	388.300	354.619	1.605	2.781	17
1982	341.108	275.722	1.248	2.162	466
1983	319.924	222.667	1.008	1.746	466
1984	..	..	..	..	
1985	..	..	..	..	
1986	..	..	..	..	

1960–64:   1965–69:

Personnel Comments:

The personnel figures given by source 17 represent all agricultural researchers under the responsibility of the 'Département de la Recherche Agronomique' (DRA), which at the time of the questionnaire was in transition from the 'Ministère des Enseignements Moyen et Superieur' to the 'Ministère du Développement Rural et de l'Action Cooperative'. It is understood that DRA covers nearly all agricultural research in Benin, so DRA's personnel figures seem a reasonable measure of the total number of researchers within the NARS.

As in many former French colonies, the agricultural research system was not nationalized, and subsequently reorganized, until the early seventies.

According to source 653 the following institutes were involved in agricultural research in 1966: IRAT/IFAC (5 researchers), IRHO (7 researchers), and IRCT (1 researcher). Source 431 indicates that the following organizations were conducting agricultural research in Benin in 1972: IRAT, IRCT, IRHO, IFAC and 'Laboratoire des Pêches'. The total number of researchers (17) working at these institutes, according to source 431, approximates the number given by source 17 for the same year, namely 19. The same can be said for the expenditure figures given by both sources.

Expenditure Comments:

The 1970–81 expenditure figures are based on the ISNAR & IFARD questionnaire (source 17). The expenditure figures represent all agricultural research under the responsibility of DRA and include external funding.

Acronyms and Abbreviations:

DRA: Département de la Recherche Agronomique
IFAC: Institut Francais de Recherches Fruitières Outre-Mer
IRAT: Institut de Recherches Agronomiques Tropicales et des Cultures Vivrières
IRCT: Institut de Recherche du Coton et des Textiles Exotiques

IRHO: Institut de Recherches pour les Huiles et Oléagineux

Sources:

0017 ISNAR, IFARD & AOAD. Survey of National Agricultural Research Systems: Unpublished Questionnaire Responses. ISNAR, The Hague, 1985.

0466 World Bank – West Africa Agricultural Research Review Team. "National Agricultural Research Systems (NARS): A Regional Appraisal and Selected Issues." World Bank, Washington, D.C., August 1986. Mimeo.

0653 Webster, B.N. *Index of Agricultural Research Institutions and Stations in Africa*. Rome: FAO, n.d.

Additional References:

0010 Boyce, James K., and Robert E. Evenson. *National and International Agricultural Research and Extension Programs*. New York: Agricultural Development Council, Inc., 1975.

0016 Oram, Peter A., and Vishva Bindlish. *Resource Allocations to National Agricultural Research: Trends in the 1970s*. The Hague and Washington, D.C.: ISNAR and IFPRI, November 1981.

0023 Bennell, Paul. *Agricultural Researchers in Sub-Saharan Africa: An Overview*. Working Paper No. 4. The Hague: ISNAR, October 1985.

0027 Harvey, Nigel, ed. *Agricultural Research Centres: A World Directory of Organizations and Programmes*. Seventh Edition, Two Volumes. Harlow, U.K.: Longman, 1983.

0068 Conte, Stephan. "West Africa Agricultural Research Data Base Building Task: Termination Report." World Bank, Washington, D.C., 1985. Mimeo.

0073 Oram, Peter A., and Vishva Bindlish. "Investment in Agricultural Research in Developing Countries: Progress, Problems, and the Determination of Priorities." IFPRI, Washington, D.C., January 1984. Mimeo.

0086 World Bank. "West Africa Agricultural Research Review – Country Studies." World Bank, Washington, D.C., 1985. Mimeo.

0088 Swanson, Burton E., and Wade H. Reeves. "Agricultural Research West Africa: Manpower and Training." World Bank, Washington, D.C., November 1985. Mimeo.

0089 Oram, Peter. "Report on National Agricultural Research in West Africa." IFPRI, Washington, D.C., February 1986. Mimeo.

0163 CGIAR. "National Agricultural Research." CGIAR, Washington, D.C., 1985. Mimeo.

0165 Evenson, R.E., and Y. Kislev. *Agricultural Research and Productivity*. New Haven: Yale University Press, 1975.

0175 Cooper, St.G.C. *Agricultural Research in Tropical Africa*. Kampala: East African Literature Bureau, 1970.

0266 UNESCO. *National Science Policies in Africa*. Science Policy Studies and Documents No. 31. Paris: UNESCO, 1974.

0360 Cooper, St.G.C. "Towards Trained Manpower for Agricultural Research in Africa." Paper presented at the Conference on Agricultural Research and Production in Africa, organized by the Association for the Advancement of Agricultural Sciences in Africa (AAASA), Addis Ababa, 29 August–4 September 1971.

0373 Groupement d'Etudes et de Recherches pour le Développement de l'Agronomie Tropicale (GERDAT). *Rapport Général d'Activité pour 1982*. Paris: GERDAT, 1983.

0400 UNESCO. *The Promotion of Scientific Activity in Tropical Africa*. Science Policy Studies and Documents No. 11. Paris: UNESCO, 1969.

0431 Current Agricultural Research Information System (CARIS). *Directory of Agricultural Research Institutions and Projects in West Africa – Pilot Project 1972–73*. Rome: FAO, 1973.

0532 UNESCO Field Science Office for Africa. *Survey on the Scientific and Technical Potential of the Countries of Africa*. Paris: UNESCO, 1970.

0589 Kassapu, Samuel. *Les Dépenses de Recherche Agricole dans 34 Pays d'Afrique Tropicale*. Paris: Centre de Développement de l'OCDE, 1976.

0654 Office de la Recherche Scientifique et Technique Outre-Mer (ORSTOM). *Rapport d'Activité 1967*. Paris: ORSTOM, 1969.

0852 Evenson, Robert E., and Yoav Kislev. *Investment in Agricultural Research and Extension: A Survey of International Data*. Center Discussion Paper No. 124. New Haven, Connecticut: Economic Growth Center, Yale University, August 1971.

0886 Evenson, Robert E., and Yoav Kislev. "Investment in Agricultural Research and Extension: A Survey of International Data." *Economic Development and Cultural Change* Vol. 23 (April 1975): 507–521.

Bermuda

Personnel

Period/Year	PhD	MSc	BSc	Subtotal	Expat	Total	Source
1960–64	..	..	..	..	..	..	
1965–69	1	2	1	4	..	4	286/285
1970	..	..	..	..	..	..	
1971	1	3	1	5	..	5	279
1972	..	..	..	..	..	..	
1973	..	..	..	..	..	..	
1974	2	2	0	4	..	4	244
1975	..	..	..	..	..	..	
1976	..	..	..	..	..	..	
1977	2	2	1	5	..	5	243
1978	..	..	..	..	..	..	
1979	..	..	..	..	..	..	
1980	3	3	1	7	..	7	242
1981	..	..	..	..	..	..	
1982	..	..	..	..	..	..	
1983	..	..	..	..	..	..	
1984	..	..	..	..	..	..	
1985	..	..	..	..	..	4	744
1986	..	..	..	..	..	..	

1960–64:   1965–69: 1965 & 68

Expenditure

LCU = Bermuda Dollars

Period/Year	Current LCU (millions)	Constant 1980 LCU(millions)	Constant 1980 US$ (millions)		Source
			Atlas	PPP	
1960–64	..	..	..	..	
1965–69	..	..	..	..	
1970	..	..	..	..	
1971	..	..	..	..	
1972	..	..	..	..	
1973	..	..	..	..	
1974	..	..	..	..	
1975	..	..	..	..	
1976	..	..	..	..	
1977	..	..	..	..	
1978	..	..	..	..	
1979	..	..	..	..	
1980	..	..	..	..	
1981	..	..	..	..	
1982	..	..	..	..	
1983	..	..	..	..	
1984	..	..	..	..	
1985	..	..	..	..	
1986	..	..	..	..	

1960–64:   1965–69:

Personnel Comments:

These figures refer to the Department of Agriculture and Fisheries.

In addition to the Department of Agriculture and Fisheries, there is the Bermuda Biological Station for Research, Incorporated, an independent research center involved in both research and education. Total graduate staff: 12 in 1985 (source 744).

Sources:

0242 Commonwealth Agricultural Bureaux (CAB). *List of Research Workers in Agricultural Sciences in the Commonwealth 1981.* Slough, England: CAB, 1981.

0243 Commonwealth Agricultural Bureaux (CAB). *List of Research Workers in the Agricultural Sciences in the Commonwealth and in the Republic of Ireland 1978.* Slough, England: CAB, 1978.

0244 Commonwealth Agricultural Bureaux (CAB). *List of Research Workers in the Agricultural Sciences in the Commonwealth and in the Republic of Ireland 1975.* Slough, England: CAB, 1975.

0279 Commonwealth Agricultural Bureaux (CAB). *List of Research Workers in the Agricultural Sciences in the Commonwealth and in the Republic of Ireland 1972.* Slough, England: CAB, 1972.

0285 Commonwealth Agricultural Bureaux (CAB). *List of Research Workers in the Agricultural Sciences in the Commonwealth and in the Republic of Ireland 1969.* Slough, England: CAB, 1969.

0286 Commonwealth Agricultural Bureaux (CAB). *List of Research Workers in Agriculture, Animal Health and Forestry in the Commonwealth and in the Republic of Ireland 1966.* Slough, England: CAB, 1966.

0744 *Agricultural Research Centres: A World Directory of Organizations and Programmes.* Eighth Edition, Two Volumes. Harlow, U.K.: Longman, 1986.

Additional References:

0027 Harvey, Nigel, ed. *Agricultural Research Centres: A World Directory of Organizations and Programmes.* Seventh Edition, Two Volumes. Harlow, U.K.: Longman, 1983.

0287 Commonwealth Agricultural Bureaux (CAB). *List of Research Workers in Agriculture, Animal Health and Forestry in the British Commonwealth, the Republic of Sudan and the Republic of Ireland 1959.* Slough, England: CAB, 1959.

Bolivia

Personnel

Period/Year	PhD	MSc	BSc	Subtotal	Expat	Total	Source
1960–64	..	..	..	..	..	129	10
1965–69	..	..	..	..	..	..	
1970	..	..	..	..	..	61	10
1971	..	..	..	..	..	60	10
1972	..	..	..	..	..	48	10
1973	..	..	..	..	..	49	10
1974	..	..	..	..	..	..	
1975	..	..	..	..	..	48	16
1976	..	..	..	..	..	50	95
1977	..	..	..	50	7	57	721
1978	..	..	..	67	7	74	721
1979	..	..	..	75	7	82	721
1980	..	..	..	101	13	114	721
1981	..	..	..			..	
1982	..	..	..		..	103	27
1983	2	31	54	87	17	104	17
1984	..	..	..	..	..	..	
1985	..	..	..	..	..	..	
1986	..	..	..	..	..	..	

1960–64: 1962   1965–69:

Expenditure   *LCU = Bolivian Pesos (1980)*

Period/Year	Current LCU (millions)	Constant 1980 LCU(millions)	Constant 1980 US$ (millions)		Source
			Atlas	PPP	
1960–64	2.881	25.955	0.605	1.717	10
1965–69	..	..	..	..	
1970	10.440	70.067	1.634	4.636	1
1971	10.913	69.955	1.631	4.629	1
1972	10.511	55.910	1.304	3.699	1
1973	15.208	57.173	1.333	3.783	1
1974	24.535	58.417	1.362	3.865	1
1975	24.820	55.402	1.292	3.666	1
1976	24.880	51.405	1.199	3.401	721
1977	46.480	86.555	2.018	5.727	721
1978	55.780	91.593	2.136	6.061	721
1979	56.780	78.861	1.839	5.218	721
1980	68.020	68.020	1.586	4.501	721
1981	..	..	..	..	
1982	..	..	..	..	
1983	459.556	33.608	0.784	2.224	17
1984	..	..	..	..	
1985	..	..	..	..	
1986	..	..	..	..	

1960–64: 1962   1965–69:

Personnel Comments:

It is presumed that the numbers of researchers presented here represent those working at the 'Instituto Boliviano de Tecnología Agropecuaria' (IBTA) only. Additional institutions undertaking agricultural research but not included in the indicator series were reported by source 17 as Centro de Investigaciones de Agricultura Tropical (CIAT) (1982: 32 researchers, source 27), Centro de Investigaciones Fitoecogenéticas, Estación Experimental Abapo Izozog de las Fuerzas Armadas, and several universities (1985: 18 researchers, source 744).

Expenditure Comments:

The expenditure figures represent IBTA only.

Acronyms and Abbreviations:

CIAT: Centro de Investigaciones de Agricultura Tropical

IBTA: Instituto Boliviano de Tecnología Agropecuaria

Sources:

0001 Trigo, Eduardo J., and Martín E. Piñeiro. "Funding Agricultural Research." In *Report of a Conference: Selected Issues in Agricultural Research in Latin America*, edited by Barry Nestel and Eduardo J. Trigo. The Hague: ISNAR, March 1984.

0010 Boyce, James K., and Robert E. Evenson. *National and International Agricultural Research and Extension Programs*. New York: Agricultural Development Council, Inc., 1975.

0016 Oram, Peter A., and Vishva Bindlish. *Resource Allocations to National Agricultural Research: Trends in the 1970s*. The Hague and Washington, D.C.: ISNAR and IFPRI, November 1981.

0017 ISNAR, IFARD & AOAD. Survey of National Agricultural Research Systems: Unpublished Questionnaire Responses. ISNAR, The Hague, 1985.

0027 Harvey, Nigel, ed. *Agricultural Research*

Centres: A World Directory of Organizations and Programmes. Seventh Edition, Two Volumes. Harlow, U.K.: Longman, 1983.

0095 FAO – CARIS. *Agricultural Research in Developing Countries – Volume 1: Research Institutions*. Rome: FAO – CARIS, 1978.

0721 Ministerio de Asuntos Campesinos y Agropecuarios. "Gastos para la Investigación en Agricultura, Forestal y Pesca en Bolivia." Table. Bolivia, n.d. Mimeo.

Additional References:

0014 Judd, M. Ann, James K. Boyce, and Robert E. Evenson. "Investing in Agricultural Supply." Economic Growth Center, Yale University, New Haven, Connecticut, 1983. Mimeo.

0053 Piñeiro, Martín, and Eduardo Trigo, eds. *Cambio Técnico en el Agro Latino Americano – Situación y Perspectivas en la Década de 1980*. San José: IICA, 1983.

0073 Oram, Peter A., and Vishva Bindlish. "Investment in Agricultural Research in Developing Countries: Progress, Problems, and the Determination of Priorities." IFPRI, Washington, D.C, January 1984. Mimeo.

0165 Evenson, R.E., and Y. Kislev. *Agricultural Research and Productivity*. New Haven: Yale University Press, 1975.

0172 Trigo, Eduardo. Background material. ISNAR. The Hague, n.d.

0744 *Agricultural Research Centres: A World Directory of Organizations and Programmes*. Eighth Edition, Two Volumes. Harlow, U.K.: Longman, 1986.

0852 Evenson, Robert E., and Yoav Kislev. *Investment in Agricultural Research and Extension: A Survey of International Data*. Center Discussion Paper No. 124. New Haven, Connecticut: Economic Growth Center, Yale University, August 1971.

0886 Evenson, Robert E., and Yoav Kislev. "Investment in Agricultural Research and Extension: A Survey of International Data." *Economic Development and Cultural Change* Vol. 23 (April 1975): 507–521.

Botswana

Personnel

Period/Year	PhD	MSc	BSc	Subtotal	Expat	Total	Source
1960–64	..	..	..	..	..	1	10
1965–69	..	..	..	..	..	9	286/10
1970	..	..	..	..	..	17	532
1971	5	5	12	22	..	22	279
1972	..	..	..	1	23	24	589
1973	..	..	..	..	..	..	
1974	3	8	14	25	..	25	244
1975	..	..	..	..	..	..	
1976	..	..	..	..	..	..	
1977	6	7	13	26	..	26	243
1978	..	..	..	..	..	..	
1979	..	..	..	..	..	..	
1980	1	2	9	12	28	40	104
1981	..	..	..	..	..	..	
1982	..	..	..	..	..	48	17
1983	0	7	14	21	33	54	235
1984	0	11	18	29	32	61	3
1985	0	11	20	31	31	62	445
1986	1	8	6	15	19	34	720

1960–64: 1961   1965–69: 1965 & 67

Expenditure   *LCU = Pula*

Period/Year	Current LCU (millions)	Constant 1980 LCU(millions)	Constant 1980 US$ (millions)		Source
			Atlas	PPP	
1960–64	0.016	0.071	0.077	0.123	10
1965–69	0.137	0.503	0.548	0.870	589/10
1970	0.288	0.949	1.034	1.642	17
1971	0.319	0.968	1.055	1.676	17
1972	..	..	..	..	
1973	0.511	1.007	1.098	1.743	17
1974	0.863	1.469	1.600	2.541	17
1975	0.534	0.819	0.892	1.416	17
1976	0.730	0.990	1.079	1.713	17
1977	0.795	1.114	1.214	1.927	17
1978	1.071	1.176	1.281	2.034	17
1979	1.561	1.452	1.583	2.513	17
1980	3.027	3.027	3.298	5.238	17
1981	3.439	3.362	3.664	5.818	17
1982	4.473	4.262	4.645	7.375	17
1983	3.667	3.285	3.579	5.683	17
1984	3.021	2.608	2.842	4.512	3
1985	4.469	3.739	4.074	6.470	720
1986	5.528	..	..	..	720

1960–64: 1961   1965–69: 1966 & 69

Personnel Comments:

Agricultural research in Botswana falls under the responsibility of the Department of Agricultural Research (DAR) of the Ministry of Agriculture (MOA). DAR consists of four administrative units: the Arable Crop Research Division (ACRD), the Animal Production and Range Research Unit (APRU), the Estate Management Unit, and the Laboratory Services Unit. A 1980 figure of 26 researchers (7 PhD, 8 MSc, and 11 BSc) is reported by source 242. This figure was excluded, however, because it appears to cover only established posts and omits project posts. For the same year, source 104 gives a total of 40 researchers, of which 15 are on project postings. According to source 105, all project postings are filled by expatriates. They were added to the 13 established post expatriates (sources 104 and 105) to give a total of 28 expatriates. Until the early 1970s, all agricultural researchers were expatriates. It is only during the last 10 years that nationals have been appointed to research positions.

The 1986 personnel figure indicates a dramatic decline in the number of researchers – both nationals and expatriates – because, according to source 720: 1) quite a few nationals are away on training, 2) there has been a reduction of the number of expatriate posts, 3) some expatriate posts are vacant, and 4) there have been delays in project implementation.

Expenditure Comments:

The expenditure figures given by source 17 are, according to this source, only estimates and may differ from actual expenditures. All expenditure figures refer to the Department of Agricultural Research of the Ministry of Agriculture.

Sources:

0003 SADCC, and DEVRES, Inc. *Agricultural Research Resource Assessment in the SADCC Countries, Volume II: Country Report Botswana.* Washington, D.C.: DEVRES, Inc., January 1985.

0010 Boyce, James K., and Robert E. Evenson. *National and International Agricultural Research and Extension Programs.* New York: Agricultural Development Council, Inc., 1975.

0017 ISNAR, IFARD & AOAD. Survey of National Agricultural Research Systems: Unpublished Questionnaire Responses. ISNAR, The Hague, 1985.

0104 Oland, Kristian. "Agricultural Research in Botswana." Gaborone, August 1980. Mimeo.

0235 FAO. "Trained Agricultural Manpower Assessment (TAMA): Original Answer Sheets." FAO, Rome, 1982. Mimeo.

0243 Commonwealth Agricultural Bureaux (CAB). *List of Research Workers in the Agricultural Sciences in the Commonwealth and in the Republic of Ireland 1978.* Slough, England: CAB, 1978.

0244 Commonwealth Agricultural Bureaux (CAB). *List of Research Workers in the Agricultural Sciences in the Commonwealth and in the Republic of Ireland 1975.* Slough, England: CAB, 1975.

0279 Commonwealth Agricultural Bureaux (CAB). *List of Research Workers in the Agricultural Sciences in the Commonwealth and in the Republic of Ireland 1972.* Slough, England: CAB, 1972.

0286 Commonwealth Agricultural Bureaux (CAB). *List of Research Workers in Agriculture, Animal Health and Forestry in the Commonwealth and in the Republic of Ireland 1966.* Slough, England: CAB, 1966.

0445 Swanson, Burton E., and Wade H. Reeves. "Agricultural Research Eastern and Southern Africa: Manpower and Training." World Bank, Washington, D.C., August 1986. Mimeo.

0532 UNESCO Field Science Office for Africa. *Survey on the Scientific and Technical Potential of the Countries of Africa.* Paris: UNESCO, 1970.

0589 Kassapu, Samuel. *Les Dépenses de Recherche Agricole dans 34 Pays d'Afrique Tropicale.* Paris: Centre de Développement de l'OCDE, 1976.

0720 Gollifer, D.E. Personal communication. Director of Research, Department of Agricultural Research. Gaborone, Botswana, October 1986.

Additional References:

0002 SADCC, and DEVRES, Inc. *Agricultural Research Resource Assessment in the SADCC Countries, Volume I: Regional Analysis and Strategy.* Washington, D.C.: DEVRES, Inc., January 1985.

0014 Judd, M. Ann, James K. Boyce, and Robert E. Evenson. "Investing in Agricultural Supply." Economic Growth Center, Yale University, New Haven, Connecticut, 1983. Mimeo.

0016 Oram, Peter A., and Vishva Bindlish. *Resource Allocations to National Agricultural Research: Trends in the 1970s.* The Hague and Washington, D.C.: ISNAR and IFPRI, November 1981.

0019 Bengtsson, Bo, and Tedla Getachew, eds. *Strengthening National Agricultural Research.* Report from a SAREC workshop, 10–17 September 1979. Part 1: Background Documents; Part II: Summary and Conclusions. Sweden: SAREC, 1980.

0023 Bennell, Paul. *Agricultural Researchers in Sub-Saharan Africa: An Overview.* Working Paper No. 4. The Hague: ISNAR, October 1985.

0027 Harvey, Nigel, ed. *Agricultural Research Centres: A World Directory of Organizations and Programmes.* Seventh Edition, Two Volumes. Harlow, U.K.: Longman, 1983.

0105 ICRISAT. "Botswana." ICRISAT, 1980. Mimeo.

0164 Association for the Advancement of Agricultural Sciences in Africa (AAASA). *Proceedings of the Workshop on Agricultural Research Administration, Nairobi, Kenya, 27–30 June 1977.* Proceedings Series PE-4. Addis Ababa, Ethiopia: AAASA and IDRC, August 1979.

0175 Cooper, St.G.C. *Agricultural Research in Tropical Africa.* Kampala: East African Literature Bureau, 1970.

0234 SADCC. *SADCC Agricultural Research Conference Gaborone Botswana.* Sebele, Botswana: SADCC, April 1984.

0242 Commonwealth Agricultural Bureaux (CAB). *List of Research Workers in Agricultural Sciences in the Commonwealth 1981.* Slough, England: CAB, 1981.

0307 Oland, K. "Agricultural Research in Botswana." In *Strengthening National Agricultural Research.* Report from a SAREC workshop, 10–17 September 1979, edited by Bo Bengtsson and Getachew Tedla, pp. 7–19. Sweden: SAREC, 1980.

0336 Ministry of Agriculture – Animal Production Research Unit. *Ten Years of Animal Production and Range Research in Botswana.* Gaborone, Botswana: Ministry of Agriculture, 1980.

0360 Cooper, St.G.C. "Towards Trained Manpower for Agricultural Research in Africa." Paper presented at the Conference on Agricultural Research and Production in Africa, organized by the Association for the Advancement of Agricultural Sciences in Africa (AAASA), Addis Ababa, 29 August–4 September 1971.

0385 SADCC, and DEVRES, Inc. *SADCC Region Agricultural Research Resource Assessment – Data Base Management Information System: Diskettes and User's Guide.* Printouts. Washington, D.C.: SADCC, and DEVRES, Inc, n.d.

0446 Kyomo, M.L. "Agricultural Research in Eastern and Southern Africa: Issues and Priorities." Southern African Centre for Cooperation in Agricultural Research of SADCC, Gaborone, Botswana, 1986. Mimeo.

0467 DEVRES, Inc. *The Agricultural Research Resource Assessment Pilot Report for Botswana, Malawi and Swaziland.* Washington, D.C.: DEVRES, Inc., November 1983.

0653 Webster, B.N. *Index of Agricultural Research Institutions and Stations in Africa.* Rome: FAO, n.d.

0847 Kimura, J.H. "Financial and Administrative Management of Research Institutions in Eastern and Southern Africa: Report on Responses to a Questionnaire." In *Promotion of Technology Policy and Science Management in Africa,* edited by Karl Wolfgang Menck and Wolfgang Gmelin. Bonn: Deutsche Stiftung für internationale Entwicklung (DSE), 1986.

0852 Evenson, Robert E., and Yoav Kislev. *Investment in Agricultural Research and Extension: A Survey of International Data.* Center Discussion Paper No. 124. New Haven, Connecticut: Economic Growth Center, Yale University, August 1971.

Brazil

Personnel

Period/Year	PhD	MSc	BSc	Subtotal	Expat	Total	Source
1960–64	..	..	..	..	..	..	
1965–69	..	..	..	..	..	..	
1970	..	..	..	..	..	..	
1971	..	..	..	..	..	1920	108
1972	..	..	..	..	..	..	
1973	..	..	..	..	..	..	
1974	..	..	..	..	..	..	
1975	..	..	..	..	..	2442	14
1976	..	..	..	..	..	2529	16
1977	..	..	..	..	..	2496	16
1978	..	..	..	..	..	2554	16
1979	210	1145	1372	2727	..	2727	26
1980	..	..	..	..	..	2935	303
1981	..	..	..	..	..	3426	26
1982	..	..	..	..	..	3714	264
1983	481	1763	1900	4144	31	4175	17
1984	..	..	..	..	..	..	
1985	651	1896	1314	3861	..	3861	725
1986	..	..	..	..	..	..	

1960–64: 1965–69:

Expenditure *LCU = Cruzeiros*

Period/Year	Current LCU (millions)	Constant 1980 LCU(millions)	Constant 1980 US$ (millions) Atlas	PPP	Source
1960–64	..	..	..	..	
1965–69	..	..	..	..	
1970	..	..	..	..	
1971	..	..	..	..	
1972	..	..	..	..	
1973	..	..	..	..	
1974	382.200	4851.051	98.834	150.001	106
1975	745.400	6994.309	142.500	216.274	106
1976	1371.600	8774.202	178.763	271.311	106
1977	2193.700	9696.252	197.549	299.822	106
1978	3809.700	11682.131	238.008	361.228	106
1979	6478.277	12604.209	256.795	389.740	172
1980	..	..	..	..	
1981	18973.000	9719.775	198.028	300.549	26
1982	..	..	..	..	
1983	..	..	..	..	
1984	..	..	..	..	
1985	..	..	..	..	
1986	..	..	..	..	

1960–64: 1965–69:

Personnel Comments:

EMBRAPA is the major agricultural research organization in Brazil. It comprises nearly all agricultural research activities at the federal level, except some important commodity research institutes for coffee and sugar. In addition to agricultural research at the federal level, a significant amount of agricultural research is also performed at the state level. In São Paulo, Paraná, and Rio Grande do Sul, state-level agricultural research systems existed long before EMBRAPA was created in 1974.

Some of the figures can be broken down as follows:

	1979	1980	1981	1982	1983	1984	1985	1986
EMBRAPA	1448	1553	1576	1597	1610	1619	1650	NA
State Organizations	674	765	887	954	1216	NA		
Integrated Programs	605	617	963	1163	1349	NA	[2211]	2211
Total	2727	2935	3426	3714	4175	NA	3861	NA
Source	26	303	26	264	17	725	725	725

[ ] = estimated or constructed by authors.

A detailed breakdown of EMBRAPA research personnel is provided below.

	'74	'75	'76	'77	'78	'79	'80	'81	'82	'83	'84	'85
PhD	15	27	36	37	89	123	162	196	226	268	298	363
MSc	133	178	194	188	699	777	882	941	968	986	1001	1012
BSc	724	832	1098	1086	548	548	509	439	403	355	320	275
Total	872	1037	1328	1311	1336	1448	1553	1576	1597	1609	1619	1650
Source	70	70	70	70	70	70	70	70	70	726	726	726

Expenditure Comments:

The 1981 expenditure figure (in Cr. 1,000) is constructed as follows:

	Federal Government	State Government	Total
EMBRAPA	14,794,979		14,794,979
State Organizations	701,165	1,943,349	2,644,514
Integrated Programs	364,102	1,113,829	1,477,931
Special Projects	55,472		55,472
Total	15,915,718	3,057,178	18,972,896

Agricultural research outside EMBRAPA is poorly documented. This 1981 estimate should be treated with some care as the EMBRAPA figure it includes conflicts with the estimate given by source 71.

	1974	1975	1976	1977	1978	1979	1980	1981	1982	1983	1984	1985
EMBRAPA	151	411	807	1277	2028	3988	8004	16386	41716	74559	214152	769707
Source	71	71	71	71	71	71	71	71	71	726	726	726

EMBRAPA's 1974 to 1985 research expenditures (in millions of current LCUs) is provided above.

Acronyms and Abbreviations:
EMBRAPA: Empresa Brasileira de Pesquisa Agropecuária (Brazilian Agricultural Research Corporation)

Sources:
0014 Judd, M. Ann, James K. Boyce, and Robert E. Evenson. "Investing in Agricultural Supply." Economic Growth Center, Yale University, New Haven, Connecticut, 1983. Mimeo.

0016 Oram, Peter A., and Vishva Bindlish. *Resource Allocations to National Agricultural Research: Trends in the 1970s.* The Hague and Washington, D.C.: ISNAR and IFPRI, November 1981.

0017 ISNAR, IFARD & AOAD. Survey of National Agricultural Research Systems: Unpublished Questionnaire Responses. ISNAR, The Hague, 1985.

0026 Oram, Peter A., and Mark Gieben. "Document Summaries." ISNAR, The Hague, 1984. Mimeo.

0106 Sanches da Fonseca, Maria A., and Roberto Mendonça de Barros. "A Preliminary Attempt to Evaluate the Agricultural Research System in Brazil." Brazil, n.d. Mimeo.

0108 Alves, Eliseu Roberto de Andrade, et al. *Pesquisa Agropecuária: Perspectiva Histórica e Desenvolvimento Institucional.* Brasília: Empresa Brasileira de Pesquisa Agropecuária, 1985.

0172 Trigo, Eduardo. *Background material.* ISNAR. The Hague, n.d.

0264 Empresa Brasileira de Pesquisa Agropecuária (EMBRAPA). *EMBRAPA Ano 11 – Destaque dos Resultados da Pesquisa de 1983.* Brasília: EMBRAPA, 1984.

0303 Sanches da Fonseca, Maria Aparecida, and José Roberto Mendonça de Barros. "A Preliminary Attempt to Evaluate the Agricultural Research System in Brazil." In *Resource Allocation to Agricultural Research.* Proceedings of a workshop held in Singapore 8–10 June 1981, edited by Douglas Daniels and Barry Nestel, pp. 137–144. Ottawa: IDRC, 1981.

0725 Alves, Eliseu, and Elisio Contini. "A Modernização da Agricultura Brasileira." EMBRAPA, Brazil, 13 March 1987. Mimeo.

Additional References:
0001 Trigo, Eduardo J., and Martín E. Piñeiro. "Funding Agricultural Research." In *Report of a Conference: Selected Issues in Agricultural Research in Latin America,* edited by Barry Nestel and Eduardo J. Trigo. The Hague: ISNAR, March 1984.

0010 Boyce, James K., and Robert E. Evenson. *National and International Agricultural Research and Extension Programs.* New York: Agricultural Development Council, Inc., 1975.

0018 Daniels, Douglas, and Barry Nestel, eds. *Resource Allocation to Agricultural Research.* Proceedings of a workshop held in Singapore 8–10 June 1981. Ottawa: IDRC, 1981.

0027 Harvey, Nigel, ed. *Agricultural Research Centres: A World Directory of Organizations and Programmes.* Seventh Edition, Two Volumes. Harlow, U.K.: Longman, 1983.

0053 Piñeiro, Martín, and Eduardo Trigo, eds. *Cambio Técnico en el Agro Latino Americano – Situación y Perspectivas en la Década de 1980.* San José: IICA, 1983.

0070 Brazilian Agricultural Research Corporation (EMBRAPA). *Formation of Human Capital and Returns on Investment in Manpower Training by EMBRAPA.* Brasília: EMBRAPA, 1984.

0071 Brazilian Agricultural Research Corporation (EMBRAPA). *The Socio-Economic Impact of Investments in Research by EMBRAPA: Results Obtained, Profitability and Future Prospects.* Brasília: EMBRAPA, 1985.

0073 Oram, Peter A., and Vishva Bindlish. "Investment in Agricultural Research in Developing Countries: Progress, Problems, and the Determination of Priorities." IFPRI, Washington, D.C., January 1984. Mimeo.

0082 Homem de Melo, Fernando. *Brazil and the CGIAR Centers: A Study of Their Collaboration in Agricultural Research.* CGIAR Study Paper Number 9. Washington, D.C.: World Bank, 1986.

0095 FAO – CARIS. *Agricultural Research in Developing Countries – Volume 1: Research Institutions.* Rome: FAO – CARIS, 1978.

96

0107 Yeganiantz, Levon, ed. *Brazilian Agriculture and Agricultural Research*. Brasília: Brazilian Agricultural Research Corporation – EMBRAPA, 1984.

0109 World Bank. *Brazil: Agricultural Research II Project*. Washington, D.C.: World Bank, 1981.

0135 FAO, and UNDP. *National Agricultural Research – Report of an Evaluation Study in Selected Countries*. Rome: FAO, and UNDP, 1984.

0163 CGIAR. "National Agricultural Research." CGIAR, Washington, D.C., 1985. Mimeo.

0165 Evenson, R.E., and Y. Kislev. *Agricultural Research and Productivity*. New Haven: Yale University Press, 1975.

0321 Lopes, J.R.B. "Agricultural Research in Brazil." In *Strengthening National Agricultural Research*. Report from a SAREC workshop, 10–17 September 1979, edited by Bo Bengtsson and Getachew Tedla, pp. 144–154. Sweden: SAREC, 1980.

0726 Webster, Brian, Carlos Valverde, and Alan Fletcher, eds. *The Impact of Research on National Agricultural Development*. Report on the First International Meeting of National Agricultural Research Systems and the Second IFARD Global Convention. The Hague: ISNAR, July 1987.

0766 Freitas Rivaldo, Ormuz. "Strategies for Strengthening the Brazilian Agricultural Research System." In *The Impact of Research on National Agricultural Development*, edited by Brian Webster, Carlos Valverde, and Alan Fletcher, pp. 161–181. The Hague: ISNAR, July 1987.

0770 Ruttan, Vernon W. *Agricultural Research Policy and Development*. FAO Research and Technology Paper 2. Rome: FAO, 1987.

0852 Evenson, Robert E., and Yoav Kislev. *Investment in Agricultural Research and Extension: A Survey of International Data*. Center Discussion Paper No. 124. New Haven, Connecticut: Economic Growth Center, Yale University, August 1971.

0886 Evenson, Robert E., and Yoav Kislev. "Investment in Agricultural Research and Extension: A Survey of International Data." *Economic Development and Cultural Change* Vol. 23 (April 1975): 507–521.

Brunei

Personnel

Period/Year	PhD	MSc	BSc	Subtotal	Expat	Total	Source
1960–64	..	..	..	..	..	..	
1965–69	0	1	2	3	..	3	286/285
1970	..	..	..	..	..	..	
1971	0	3	6	9	..	9	279
1972	..	..	..	..	..	..	
1973	..	..	..	..	..	..	
1974	0	3	7	10	..	10	244
1975	..	..	..	..	..	..	
1976	..	..	..	..	..	..	
1977	..	..	..	..	..	..	
1978	..	..	..	..	..	..	
1979	..	..	..	..	..	..	
1980	..	..	..	..	..	..	
1981	..	..	..	..	..	..	
1982	..	..	..	..	..	20	27
1983	..	..	..	..	..	..	
1984	..	..	..	..	..	..	
1985	..	..	..	..	..	..	
1986	..	..	..	..	..	..	

1960–64:   1965–69: 1965 & 68

Expenditure   *LCU = Brunei Dollars*

Period/Year	Current LCU (millions)	Constant 1980 LCU(millions)	Constant 1980 US$ (millions)		Source
			Atlas	PPP	
1960–64	..	..	..	..	
1965–69	..	..	..	..	
1970	..	..	..	..	
1971	..	..	..	..	
1972	..	..	..	..	
1973	..	..	..	..	
1974	..	..	..	..	
1975	..	..	..	..	
1976	..	..	..	..	
1977	..	..	..	..	
1978	..	..	..	..	
1979	..	..	..	..	
1980	..	..	..	..	
1981	..	..	..	..	
1982	..	..	..	..	
1983	..	..	..	..	
1984	..	..	..	..	
1985	..	..	..	..	
1986	..	..	..	..	

1960–64:   1965–69:

Personnel Comments:

The 1982 figure has been constructed as follows:

Agricultural Department	
Research Division	
Agronomy Department	7
Entomology Department	2
Livestock Husbandry Department	[3]
Plant Pathology Department	4
Soil Department	4
Total	20

[ ] = estimated or constructed by authors.

Sources:

0027 Harvey, Nigel, ed. *Agricultural Research Centres: A World Directory of Organizations and Programmes.* Seventh Edition, Two Volumes. Harlow, U.K.: Longman, 1983.

0244 Commonwealth Agricultural Bureaux (CAB). *List of Research Workers in the Agricultural Sciences in the Commonwealth and in the Republic of Ireland 1975.* Slough, England: CAB, 1975.

0279 Commonwealth Agricultural Bureaux (CAB). *List of Research Workers in the Agricultural Sciences in the Commonwealth and in the Republic of Ireland 1972.* Slough, England: CAB, 1972.

0285 Commonwealth Agricultural Bureaux (CAB). *List of Research Workers in the Agricultural Sciences in the Commonwealth and in the Republic of Ireland 1969.* Slough, England: CAB, 1969.

0286 Commonwealth Agricultural Bureaux (CAB). *List of Research Workers in Agriculture, Animal Health and Forestry in the Commonwealth and in the Republic of Ireland 1966.* Slough, England: CAB, 1966.

Burkina Faso

Personnel

Period/Year	PhD	MSc	BSc	Subtotal	Expat	Total	Source
1960–64	..	..	..	..		6	10
1965–69	..	..	..	..		9	10
1970						9	532
1971						10	10
1972						11	431
1973							
1974	..	..	..	1	14	15	589
1975						33	16
1976	..	..	..				
1977	..	..	..				
1978							
1979							
1980	..	..	..	33	64	97	746
1981							
1982	..	..	..	70	66	136	24
1983	..	..	..	78	35	113	83
1984	..	..	..	55	55	110	88
1985				..		..	
1986				..		..	

1960–64: 1961   1965–69: 1966 & 67

Expenditure

LCU = CFA Francs

Period/Year	Current LCU (millions)	Constant 1980 LCU(millions)	Constant 1980 US$ (millions)		Source
			Atlas	PPP	
1960–64	45.800	152.667	0.707	1.249	10
1965–69	75.900	227.246	1.053	1.859	589
1970	..	..	..	..	
1971	82.300	203.713	0.944	1.667	589
1972	..	..	..	..	
1973	187.950	406.818	1.885	3.328	589
1974	..	..	..	..	
1975	..	..	..	..	
1976	..	..	..	..	
1977	..	..	..	..	
1978	..	..	..	..	
1979	..	..	..	..	
1980	1584.000	1584.000	7.340	12.958	746
1981	..	..	..	..	
1982	3040.000	2461.538	11.406	20.137	24
1983	2185.700	1663.394	7.708	13.608	83
1984	..	..	..	..	
1985	..	..	..	..	
1986	..	..	..	..	

1960–64: 1961   1965–69: 1966

Personnel Comments:

Agricultural research in Burkina Faso is dispersed over many national, regional, and international research institutes and over an even larger number of projects funded by foreign donors. Although an attempt has been made by the government of Burkina Faso to improve the coordination and management of agricultural research by the creation, in 1981, of INERA (previously IVRAZ and then IBRAZ), no clear improvements of the situation have been reported. This means that centralized information on the agricultural research system of Burkina Faso is still non-existent. Since the early eighties, three major studies on the organization of agricultural research in Burkina Faso have been undertaken by ISNAR (jointly with World Bank and FAO), DEVRES/INSAH, and the World Bank. None of them completely quantified the complex structure of agricultural research in Burkina Faso. Therefore, the total number of researchers given here have to be considered incomplete. They may indicate no more than the fact that, in the early eighties, the number of agricultural researchers in Burkina Faso fluctuated between 100 and 150, of which nearly half were expatriate researchers.

In the sixties and the early seventies agricultural research in Burkina Faso was completely dominated by the French. Most sources do not specify for these years which organizations or institutes are taken into account when constructing estimates of the total number of researchers.

The 1970 and the 1972 figures, however, can be broken down as follows:

	1970	1972
IRAT	4	
IRCT	2	3
IRHO	2	4
CTFT	1	2
LDRV	0	1
IEMVT		1
Total	9	11

Expenditure Comments:

The 1966 and 1971 expenditure figures include IRHO, IRAT, and LDRV. The 1973 expenditure figure includes CTFT, IEMVT, IRAT, IRCT, and IRHO. As mentioned above, no centralized information on NARS is available. Therefore the expenditure figures presented here for 1982 and 1983 are no more than estimates. The 1982 figure seems to be the most complete.

Acronyms and Abbreviations:

CTFT: Centre Technique Forestier Tropical
IBRAZ: Institut Burkinabé de Recherche Agronomique et Zootechnique
IEMVT: Institut d'Elevage et de Médecine Vétérinaire des Pays Tropicaux
INERA: Institut National d'Etudes et de Recherches Agricoles
IRAT: Institut de Recherches Agronomiques Tropicales et des Cultures Vivrières
IRCT: Institut de Recherches du Coton et des Textiles Exotiques
IRHO: Institut de Recherches pour les Huiles et Oléagineux
IVRAZ: Institut Voltaique de Recherches Agronomiques et Zootechniques
LDRV: Laboratoire de Diagnostic et des Recherches Vétérinaires

Sources:

0010 Boyce, James K., and Robert E. Evenson. *National and International Agricultural Research and Extension Programs.* New York: Agricultural Development Council, Inc., 1975.

0016 Oram, Peter A., and Vishva Bindlish. *Resource Allocations to National Agricultural Research: Trends in the 1970s.* The Hague and Washington, D.C.: ISNAR and IFPRI, November 1981.

0024 World Bank, FAO, and ISNAR. *Agricultural and Livestock Research: Upper Volta.* Report prepared for the Government of the Republic of Upper Volta on the joint mission of World Bank-FAO-ISNAR. The Hague: ISNAR, March 1983.

0083 Institut du Sahel, and DEVRES, Inc. *Bilan des Ressources de la Recherche Agricole dans les Pays du Sahel, Volume II: Résumés des Rapports Nationaux.* Washington, D.C.: DEVRES, Inc., August 1984.

0088 Swanson, Burton E., and Wade H. Reeves. "Agricultural Research West Africa: Manpower and Training." World Bank, Washington, D.C., November 1985. Mimeo.

0431 Current Agricultural Research Information System (CARIS). *Directory of Agricultural Research Institutions and Projects in West Africa – Pilot Project 1972–73*. Rome: FAO, 1973.

0532 UNESCO Field Science Office for Africa. *Survey on the Scientific and Technical Potential of the Countries of Africa*. Paris: UNESCO, 1970.

0589 Kassapu, Samuel. *Les Dépenses de la Recherche Agricole dans 34 Pays d'Afrique Tropicale*. Paris: Centre de Développement de l'OCDE, 1976.

0746 Ouali, Ibrahim Firmin. *Burkina Faso and the CGIAR Centers: A Study of Their Collaboration in Agricultural Research*. CGIAR Study Paper Number 23. Washington D.C.: World Bank, 1987.

Additional References:

0023 Bennell, Paul. *Agricultural Researchers in Sub-Saharan Africa: An Overview*. Working Paper No. 4. The Hague: ISNAR, October 1985.

0027 Harvey, Nigel, ed. *Agricultural Research Centres: A World Directory of Organizations and Programmes*. Seventh Edition, Two Volumes. Harlow, U.K.: Longman, 1983.

0045 Institut du Sahel, and DEVRES, Inc. *Bilan des Ressources de la Recherche Agricole dans les Pays du Sahel, Volume I: Analyse et Stratégie Régionale*. Washington, D.C.: DEVRES, Inc., 1984.

0068 Conte, Stephan. "West Africa Agricultural Research Data Base Building Task: Termination Report." World Bank, Washington, D.C., 1985. Mimeo.

0073 Oram, Peter A., and Vishva Bindlish. "Investment in Agricultural Research in Developing Countries: Progress, Problems, and the Determination of Priorities." IFPRI, Washington, D.C, January 1984. Mimeo.

0086 World Bank. "West Africa Agricultural Research Review – Country Studies." World Bank, Washington, D.C., 1985. Mimeo.

0089 Oram, Peter. "Report on National Agricultural Research in West Africa." IFPRI, Washington, D.C., February 1986. Mimeo.

0095 FAO – CARIS. *Agricultural Research in Developing Countries – Volume 1: Research Institutions*. Rome: FAO – CARIS, 1978.

0110 Development Alternatives, Inc. *Agricultural Sector Assistance Strategy for Upper Volta*. Washington, D.C.: Development Alternatives, Inc., March 1982.

0163 CGIAR. "National Agricultural Research." CGIAR, Washington, D.C., 1985. Mimeo.

0165 Evenson, R.E., and Y. Kislev. *Agricultural Research and Productivity*. New Haven: Yale University Press, 1975.

0175 Cooper, St.G.C. *Agricultural Research in Tropical Africa*. Kampala: East African Literature Bureau, 1970.

0228 CIRAD. *Coopération Francaise en Recherche Agronomique: Activités du CIRAD dans les Pays du Sahel*. Paris: CIRAD, March 1985.

0258 "Effort Francais en Faveur de la Recherche Agronomique dans les Pays du Sahel 1984–1985." 1985. Mimeo.

0266 UNESCO. *National Science Policies in Africa*. Science Policy Studies and Documents No. 31. Paris: UNESCO, 1974.

0360 Cooper, St.G.C. "Towards Trained Manpower for Agricultural Research in Africa." Paper presented at the Conference on Agricultural Research and Production in Africa, organized by the Association for the Advancement of Agricultural Sciences in Africa (AAASA), Addis Ababa, 29 August–4 September 1971.

0372 Gabas, Jean-Jacques. *La Recherche Agricole dans les Pays Membres du CILSS*. Club du Sahel, September 1986.

0373 Groupement d'Etudes et de Recherches pour le Développement de l'Agronomie Tropicale (GERDAT). *Rapport Général d'Activité pour 1982*. Paris: GERDAT, 1983.

0384 Institut du Sahel, and DEVRES, Inc. *Sahel Region Agricultural Research Resource Assessment – Data Base Management Information System: Diskettes and User's Guide*. Printouts. Washington, D.C.: Institut du Sahel, and DEVRES, Inc., n.d.

0389 Ministère de l'Enseignement Supérieur et de la Recherche Scientifique (MESRS). "Organigramme du Ministère de l'Enseignement Supérieur et de la Recherche Scientifique." MESRS, n.d. Mimeo.

0390 Trouchaud, Jean-Pierre. "Rapport de Mission d'Appui a l'Organisation de la Recherche Scientifique et Technologique en Haute-Volta." 1979. Mimeo.

0391 Institut d'Etudes et de Recherches Agricoles. "Lignes Directrices d'Organisation et d'Administration des Recherches Agricoles." MESRS, CNRST, and INERA, Ougadougou, n.d. Mimeo.

0392 ISNAR. "INERA – Masterplan." ISNAR, The Hague, 1986. Mimeo.

0400 UNESCO. *The Promotion of Scientific Activity in Tropical Africa*. Science Policy Studies and Documents No. 11. Paris: UNESCO, 1969.

0466 World Bank – West Africa Agricultural Research Review Team. "National Agricultural Research Systems (NARS): A Regional Appraisal and Selected Issues." World Bank, Washington, D.C., August 1986. Mimeo.

0654 Office de la Recherche Scientifique et Technique Outre-Mer (ORSTOM). *Rapport d'Activité 1967.* Paris: ORSTOM, 1969.

0852 Evenson, Robert E., and Yoav Kislev. *Investment in Agricultural Research and Extension: A Survey of International Data.* Center Discussion Paper No. 124. New Haven, Connecticut: Economic Growth Center, Yale University, August 1971.

0886 Evenson, Robert E., and Yoav Kislev. "Investment in Agricultural Research and Extension: A Survey of International Data." *Economic Development and Cultural Change* Vol. 23 (April 1975): 507–521.

Burma

Personnel

Period/Year	PhD	MSc	BSc	Subtotal	Expat	Total	Source
1960–64	..	..	..	..	..	14	715/716
1965–69	..	..	..	..	..	25	10
1970	..	..	..	..	..	..	
1971	..	..	..	..	..	..	
1972	..	..	..	..	..	..	
1973	..	..	..	..	..	..	
1974	..	..	..	..	..	..	
1975	..	..	..	..	..	..	
1976	..	..	..	..	..	..	
1977	..	..	..	..	..	..	
1978	..	..	..	..	..	..	
1979	..	..	..	..	..	..	
1980	..	..	..	..	..	..	
1981	..	..	..	..	..	..	
1982	..	..	..	..	..	..	
1983	..	..	..	..	..	..	
1984	8	24	235	267	..	267	57
1985	..	..	..	..	..	..	
1986	..	..	..	..	..	..	

1960–64: 1960   1965–69: 1966

Expenditure   *LCU = Kyats*

Period/Year	Current LCU (millions)	Constant 1980 LCU(millions)	Constant 1980 US$ (millions)		Source
			Atlas	PPP	
1960–64	0.555	1.126	0.178	0.471	715/716
1965–69	0.952	2.065	0.327	0.864	10
1970	..	..	..	..	
1971	..	..	..	..	
1972	..	..	..	..	
1973	..	..	..	..	
1974	..	..	..	..	
1975	..	..	..	..	
1976	..	..	..	..	
1977	..	..	..	..	
1978	..	..	..	..	
1979	..	..	..	..	
1980	..	..	..	..	
1981	..	..	..	..	
1982	..	..	..	..	
1983	..	..	..	..	
1984	..	..	..	..	
1985	..	..	..	..	
1986	..	..	..	..	

1960–64: 1960   1965–69: 1966

Personnel Comments:

The 1984 figure is constructed as follows:

	PhD	MSc	BSc	Total
Agricultural Research Institute (ARI)	2	10	82	94
Applied Research Division (ARD)	6	14	153	173
Total	8	24	235	267

Sources:

0010 Boyce, James K., and Robert E. Evenson. *National and International Agricultural Research and Extension Programs.* New York: Agricultural Development Council, Inc., 1975.

0057 Zin, Kyaw. *Burma and the CGIAR Centers: A Study of Their Collaboration in Agricultural Research.* CGIAR Study Paper Number 19. Washington, D.C.: World Bank, November 1986.

0715 Chang, C.W. *FAO Directory of Agricultural Research Institutes and Experiment Stations in Asia and the Far East.* Bangkok, Thailand: FAO – Regional Office for Asia and the Far East, September 1962.

0716 Chang, C.W. *FAO Directory of Agricultural Research Institutes and Experiment Stations in Asia and the Far East, Supplement.* Bangkok, Thailand: FAO – Regional Office for Asia and the Far East, May 1963.

Additional References:

0135 FAO, and UNDP. *National Agricultural Research – Report of an Evaluation Study in Selected Countries.* Rome: FAO and UNDP, 1984.

0163 CGIAR. "National Agricultural Research." CGIAR, Washington, D.C., 1985. Mimeo.

0165 Evenson, R.E., and Y. Kislev. *Agricultural Research and Productivity.* New Haven: Yale University Press, 1975.

0732 FAO – Regional Office for Asia and the Pacific. *Agricultural Research Systems in the Asia-Pacific Region.* Bangkok: FAO – Regional Office for Asia and the Pacific, 1986.

0852 Evenson, Robert E., and Yoav Kislev. *Investment in Agricultural Research and Extension: A Survey of International Data.* Center Discussion Paper No. 124. New Haven, Connecticut: Economic Growth Center, Yale University, August 1971.

0886 Evenson, Robert E., and Yoav Kislev. "Investment in Agricultural Research and Extension: A Survey of International Data." *Economic Development and Cultural Change* Vol. 23 (April 1975): 507–521.

Burundi

Personnel

Period/Year	PhD	MSc	BSc	Subtotal	Expat	Total	Source
1960–64	..	..	..	..	..	11	10
1965–69	..	..	..	..	..	14	10
1970				3	18	21	175
1971				3	17	20	589
1972				..	..	..	
1973				..	..	..	
1974				..	..	..	
1975				..	..	..	
1976				..	..	..	
1977				..	..	24	17
1978				..	..	26	17
1979				..	..	37	17
1980				..	..	43	17
1981				..	..	54	17
1982				..	..	52	17
1983	1	23	9	33	26	59	17
1984	..	..	..	32	25	57	87
1985				..	..	..	
1986				..	..	..	

1960–64: 1963   1965–69: 1966

Expenditure

LCU = Burundi Francs

Period/Year	Current LCU (millions)	Constant 1980 LCU(millions)	Constant 1980 US$ (millions)		Source
			Atlas	PPP	
1960–64	11.000	40.786	0.406	0.698	10
1965–69	26.000	89.926	0.896	1.539	589/10
1970	..	..	..	..	
1971	29.250	90.058	0.897	1.541	589
1972	..	..	..	..	
1973	..	..	..	..	
1974	..	..	..	..	
1975	..	..	..	..	
1976	..	..	..	..	
1977	..	..	..	..	
1978	..	..	..	..	
1979	..	..	..	..	
1980	..	..	..	..	
1981	..	..	..	..	
1982	..	..	..	..	
1983	281.600	256.000	2.550	4.381	17
1984	..	..	..	..	
1985	..	..	..	..	
1986	..	..	..	..	

1960–64: 1963   1965–69: 1966 & 67

Personnel Comments:
These figures refer to 'Institut des Sciences Agronomiques du Burundi' (ISABU) only. Other research institutes not included are 'Laboratoire Vétérinaire' (LV), 'Institut de Recherche Agronomique et Zootechnique' (IRAZ) – an intergovernmental institute of Burundi, Rwanda, and Zaire – and the Faculty of Agriculture.

Expenditure Comments:
These expenditure figures refer to ISABU only.

Sources:
0010 Boyce, James K., and Robert E. Evenson. *National and International Agricultural Research and Extension Programs.* New York: Agricultural Development Council, Inc., 1975.

0017 ISNAR, IFARD & AOAD. Survey of National Agricultural Research Systems: Unpublished Questionnaire Responses. ISNAR, The Hague, 1985.

0087 FAO. *Assistance a l'Institut des Sciences Agronomiques Burundi.* Rome: FAO, 1984.

0175 Cooper, St.G.C. *Agricultural Research in Tropical Africa.* Kampala: East African Literature Bureau, 1970.

0589 Kassapu, Samuel. *Les Dépenses de Recherche Agricole dans 34 Pays d'Afrique Tropicale.* Paris: Centre de Développement de l'OCDE, 1976.

Additional References:
0014 Judd, M. Ann, James K. Boyce, and Robert E. Evenson. "Investing in Agricultural Supply." Economic Growth Center, Yale University, New Haven, Connecticut, 1983. Mimeo.

0016 Oram, Peter A., and Vishva Bindlish. *Resource Allocations to National Agricultural Research: Trends in the 1970s.* The Hague and Washington, D.C.: ISNAR and IFPRI, November 1981.

0023 Bennell, Paul. *Agricultural Researchers in Sub-Saharan Africa: An Overview.* Working Paper No. 4. The Hague: ISNAR, October 1985.

0073 Oram, Peter A., and Vishva Bindlish. "Investment in Agricultural Research in Developing Countries: Progress, Problems, and the Determination of Priorities." IFPRI, Washington, D.C., January 1984. Mimeo.

0090 Institut de Recherche Agronomique et Zootechnique – Communauté Economique des Pays des Grands Lacs. *Répertoire des Recherches Agronomiques en Cours; Burundi-Rwanda-Zaire: 1984.* Gitega, Burundi: Institut de Recherche Agronomique et Zootechnique, 1984.

0163 CGIAR. "National Agricultural Research." CGIAR, Washington, D.C., 1985. Mimeo.

0225 Ministère de l'Agriculture et de l'Elevage. *Institut des Sciences Agronomiques du Burundi (ISABU) – Rapport des Activités de Recherches 1983.* Bujumbura, Burundi: Ministère de l'Agriculture et de l'Elevage, 1984.

0266 UNESCO. *National Science Policies in Africa.* Science Policy Studies and Documents No. 31. Paris: UNESCO, 1974.

0360 Cooper, St.G.C. "Towards Trained Manpower for Agricultural Research in Africa." Paper presented at the Conference on Agricultural Research and Production in Africa, organized by the Association for the Advancement of Agricultural Sciences in Africa (AAASA), Addis Ababa, 29 August–4 September 1971.

0400 UNESCO. *The Promotion of Scientific Activity in Tropical Africa.* Science Policy Studies and Documents No. 11. Paris: UNESCO, 1969.

0445 Swanson, Burton E., and Wade H. Reeves. "Agricultural Research Eastern and Southern Africa: Manpower and Training." World Bank, Washington, D.C., August 1986. Mimeo.

0532 UNESCO Field Science Office for Africa. *Survey on the Scientific and Technical Potential of the Countries of Africa.* Paris: UNESCO, 1970.

0650 Focan, A., and E. Sebatutisi. "La Recherche Agronomique au Burundi et ses Applications." In *Papers presented to the Conference on Agricultural Research and Production in Africa, September 1971 – Part 3,* supplement to volume 2 of the Journal of the Association for the Advancement of Agricultural Sciences in Africa. Addis Ababa, Ethiopia: Association for the Advancement of Agricultural Sciences in Africa, June 1975.

0653 Webster, B.N. *Index of Agricultural Research Institutions and Stations in Africa.* Rome: FAO, n.d.

0852 Evenson, Robert E., and Yoav Kislev. *Investment in Agricultural Research and Extension: A Survey of International Data.* Center Discussion Paper No. 124. New Haven, Connecticut: Economic Growth Center, Yale University, August 1971.

Cameroon

Personnel

Period/Year	PhD	MSc	BSc	Subtotal	Expat	Total	Source
1960–64	..	..	..	..	..	..	
1965–69	..	..	..	8	73	81	84
1970	..	..	..	11	77	88	84
1971	..	..	..	13	84	97	84
1972	..	..	..	..	..	..	
1973	..	..	..	..	..	..	
1974	..	..	..	..	..	..	
1975	..	..	..	44	..	..	69
1976	..	..	..	44	..	..	69
1977	..	..	..	51	..	..	69
1978	..	..	..	51	..	..	69
1979	..	..	..	54	..	..	69
1980	..	..	..	54	..	..	69
1981	..	..	..	72	48	120	69
1982	..	..	..	91	62	153	757
1983	16	93	0	109	67	176	17
1984	..	..	..	132	67	199	757
1985	..	..	..	169	62	231	757
1986	..	..	..	185	60	245	757

1960–64:   1965–69: 1966, 67, 68 & 69

Expenditure   *LCU = CFA Francs*

Period/Year	Current LCU (millions)	Constant 1980 LCU(millions)	Constant 1980 US$ (millions) Atlas	PPP	Source
1960–64	..	..	..	..	
1965–69	171.543	527.466	2.426	2.361	84
1970	253.200	656.508	3.020	2.939	84
1971	276.000	694.334	3.194	3.108	84
1972	309.350	721.027	3.317	3.228	84
1973	..	..	..	..	84
1974	332.080	654.697	3.012	2.931	84
1975	506.898	841.329	3.870	3.766	84
1976	746.725	1142.227	5.254	5.113	84
1977	1059.151	1466.813	6.747	6.567	84
1978	884.904	1113.998	5.124	4.987	84
1979	1023.840	1146.641	5.274	5.133	84
1980	1578.533	1578.533	7.261	7.067	84
1981	2058.184	1919.948	8.832	8.595	84
1982	3398.949	2834.820	13.040	12.691	837
1983	5055.425	3717.224	17.099	16.641	837
1984	7202.227	4833.709	22.235	21.639	837
1985	7963.082	5312.263	24.436	23.782	837
1986	9227.820	..	..	..	837

1960–64:   1965–69: 1965, 66, 67, 68 & 69

Personnel Comments:

Before the nationalization and reorganization of agricultural research in Cameroon in 1973, it was dominated by French institutes, funding, and researchers. The personnel figures for the period 1965–72 represent the number of researchers working at the French agricultural research institutes and ORSTOM in Cameroon at that time. Source 532 reports, for the year 1970, a total of 79 researchers at the following institutes: CTFT (2), IEMVT (5), IRAT (5), IRCT (3), IRHO (2), IRCA (2), IFAC (4), IFCC (13), and ORSTOM (38). This is slightly less than the 88 researchers reported by source 84 and included in the indicator series. Besides these French research institutes, source 532 also reports on the following organizations executing agricultural research: CDC (7), PPCL (3), TDC (1), 'Secrétariat d'Etat au Développement Rural – Direction des Eaux et Forêts et des Chasses': 'Station Botanique de Yaounde' (1), 'Station Centrale de Pisciculture' (0), and EFSA (14 teacher/ researchers). In 1973 two new agricultural research institutes were created, namely, 'Institut de la Recherche Agronomique' (IRA) and 'Institut des Recherches Zootechniques' (IRZ), substituting for the old French agricultural research institutes, except ORSTOM. During the years directly after the reorganization only Cameroonian agricultural researchers were counted in the statistics, which makes these statistics inconsistent with those of other years. In the more recent years expatriates are again included, providing a more complete measure of the research potential. The figures presented for the years after 1973 relate to IRA and IRZ only.

Expenditure Comments:

The 1965–72 expenditure figures refer to the governmental budget contributions to the following institutes: CTFT, IEMVT, IFAC, IFCC, IRAT, IRCT, and IRHO. In 1973 the system was nationalized and reorganized into IRA and IRZ. From 1974 on, the expenditure figures represent the approved budget contributions of the Cameroonian government to these two institutions.

Source 757 demonstrates that over the 1982–1985 period there is a large discrepancy between the approved budget figures and actual expenditures on agricultural research by the Cameroonian government. These figures (in current LCUs) are presented in the table below. Source 757 also provides estimates of contributions to agricultural research by non-Cameroonian sources.

	1982	1983	1984	1985
Budgeted Expenses				
Cameroonian Sources	3638	5545	7412	7963
Actual Expenses				
Cameroonian Sources (a)	2481	3511	4659	5518
Non-Cameroonian				
Sources (b)	756	970	999	1656
Total (a+b)	3237	4481	5658	7174

Although the indicator series attempts to measure actual expenditures, inclusive of donor contributions, we opted here to include the approved budget estimates for 1982–1985 in order to maintain consistency with the earlier figures. Comparing the total expenditure estimates given by source 757 with the approved budget figures from Cameroonian sources shows that it only slightly distorts the quantitative picture in this particular case.

Acronyms and Abbreviations:

CDC: Cameroon Development Corporation
CTFT: Centre Technique Forestier Tropical
EFSA: Ecole Fédérale Supérieure d'Agriculture
IEMVT: Institut d'Elevage et de Médecine Vétérinaire des Pays Tropicaux
IFAC: Institut Français de Recherches Fruitières Outre-Mer
IFCC: Institut Français du Café, Cacao et autres plantes stimulantes
IRA: Institut de la Recherche Agronomique
IRAT: Institut de Recherche Agronomique Tropical et des Cultures Vivrières
IRCA: Institut de Recherches sur le Caoutchouc en Afrique
IRCT: Institut de Recherche du Coton et Textiles Exotiques
IRHO: Institut de Recherches pour les Huiles et Oléagineux
IRZ: Institut des Recherches Zootechniques
ORSTOM: Office de la Recherche Scientifique et Technique Outre-Mer

PPCL: Plantations Pamol du Cameroun Limited
TDC: Timber Development Centre

Sources:

0017 ISNAR, IFARD & AOAD. Survey of National
Agricultural Research Systems: Unpublished
Questionnaire Responses. ISNAR, The Hague,
1985.

0069 Dikoumé, Cosme, Oscar Cordeiro, and Mathieu
Gracia. *Affectation des Ressources à la
Recherche Agricole au Cameroun.* Douala,
Cameroon: Institut Panafricain pour le
Développement, May 1984.

0084 Ngatchou, Nya J. *Evolution de la Recherche
Scientifique et Technique au Cameroun.* Yaounde:
Délégation Générale à la Recherche Scientifique
et Technique (DGRST), March 1982.

0757 ISNAR. *An Analysis of Structure and
Management of the Institute of Agricultural
Research (IRA) and the Institute of Animal
Research (IRZ) of Cameroon.* Report to the
Ministry of Higher Education and Scientific
Research of the Republic of Cameroon. The
Hague: ISNAR, December 1987.

0837 Ayuk-Takem, J.A. Personal communication.
Interim Director of the Institute of Agronomic
Research. Yaounde, Cameroon, March 1988.

Additional References:

0010 Boyce, James K., and Robert E. Evenson. *National
and International Agricultural Research and
Extension Programs.* New York: Agricultural
Development Council, Inc., 1975.

0016 Oram, Peter A., and Vishva Bindlish. *Resource
Allocations to National Agricultural Research:
Trends in the 1970s.* The Hague and Washington,
D.C.: ISNAR and IFPRI, November 1981.

0023 Bennell, Paul. *Agricultural Researchers in Sub-
Saharan Africa: An Overview.* Working Paper
No. 4. The Hague: ISNAR, October 1985.

0026 Oram, Peter A., and Mark Gieben. "Document
Summaries." ISNAR, The Hague, 1984. Mimeo.

0027 Harvey, Nigel, ed. *Agricultural Research
Centres: A World Directory of Organizations and
Programmes.* Seventh Edition, Two Volumes.
Harlow, U.K.: Longman, 1983.

0060 Lyonga, S.N., and E.T. Pamo. "The Impact of the
Collaboration between the International
Agricultural Research System and the National
Agricultural Research System in Cameroon."
CGIAR, Washington, D.C., March 1985. Mimeo.

0068 Conte, Stephan. "West Africa Agricultural
Research Data Base Building Task: Termination
Report." World Bank, Washington, D.C., 1985.
Mimeo.

0073 Oram, Peter A., and Vishva Bindlish.
"Investment in Agricultural Research in
Developing Countries: Progress, Problems, and
the Determination of Priorities." IFPRI,
Washington, D.C., January 1984. Mimeo.

0074 Institut de la Recherche Agronomique (IRA).
"Note Succincte sur l'Institut de la Recherche
Agronomique." IRA, Cameroon, April 1985.
Mimeo.

0085 Institut de la Recherche Agronomique (IRA).
Rapport d'Activités Techniques 1981. Yaounde:
IRA/Délégation Générale a la Recherche
Scientifique et Technique (DGRST), 1981.

0086 World Bank. "West Africa Agricultural
Research Review – Country Studies." World
Bank, Washington, D.C., 1985. Mimeo.

0088 Swanson, Burton E., and Wade H. Reeves.
"Agricultural Research West Africa: Manpower
and Training." World Bank, Washington, D.C.,
November 1985. Mimeo.

0089 Oram, Peter. "Report on National Agricultural
Research in West Africa." IFPRI, Washington,
D.C., February 1986. Mimeo.

0117 ISNAR, and Institut Panafricain pour le
Développement – Afrique Centrale (IPD/AC).
*Rapport au Ministère de l'Enseignement
Supérieur et de la Recherche Scientifique du
Cameroun: L'Amélioration de la Gestion de la
Recherche Agricole au Cameroun.* The Hague:
ISNAR, June 1984.

0118 Ministère de l'Agriculture. "Bilan Diagnostic du
Secteur Agricole de 1960 à 1980." Ministère de
l'Agriculture, Yaounde, 1980. Mimeo.

0119 World Bank – Western Africa Projects
Department. "Staff Appraisal Report Cameroon
– National Agricultural Research Project."
World Bank, Washington, D.C., 1984. Mimeo.

0163 CGIAR. "National Agricultural Research."
CGIAR, Washington, D.C., 1985. Mimeo.

0165 Evenson, R.E., and Y. Kislev. *Agricultural
Research and Productivity.* New Haven: Yale
University Press, 1975.

0170 Institut des Recherches Zootechniques (IRZ).
Annual Report, 1981–1982. Yaounde, Cameroon:
IRZ, 1982.

0175 Cooper, St.G.C. *Agricultural Research in
Tropical Africa.* Kampala: East African
Literature Bureau, 1970.

0266 UNESCO. *National Science Policies in Africa.*
Science Policy Studies and Documents No. 31.
Paris: UNESCO, 1974.

0360 Cooper, St.G.C. "Towards Trained Manpower for
Agricultural Research in Africa." Paper
presented at the Conference on Agricultural
Research and Production in Africa, organized by
the Association for the Advancement of

110

Agricultural Sciences in Africa (AAASA), Addis Ababa, 29 August–4 September 1971.

0373 Groupement d'Etudes et de Recherches pour le Développement de l'Agronomie Tropicale (GERDAT). *Rapport Général d'Activité pour 1982*. Paris: GERDAT, 1983.

0400 UNESCO. *The Promotion of Scientific Activity in Tropical Africa*. Science Policy Studies and Documents No. 11. Paris: UNESCO, 1969.

0406 Institut des Recherches Zootechniques (IRZ). *Annual Report, 1982–1983*. Yaounde, Cameroon: IRZ, 1983.

0466 World Bank – West Africa Agricultural Research Review Team. "National Agricultural Research Systems (NARS): A Regional Appraisal and Selected Issues." World Bank, Washington, D.C., August 1986. Mimeo.

0532 UNESCO Field Science Office for Africa. *Survey on the Scientific and Technical Potential of the Countries of Africa*. Paris: UNESCO, 1970.

0589 Kassapu, Samuel. *Les Dépenses de Recherche Agricole dans 34 Pays d'Afrique Tropicale*. Paris: Centre de Développement de l'OCDE, 1976.

0653 Webster, B.N. *Index of Agricultural Research Institutions and Stations in Africa*. Rome: FAO, n.d.

0654 Office de la Recherche Scientifique et Technique Outre-Mer (ORSTOM). *Rapport d'Activité 1967*. Paris: ORSTOM, 1969.

0852 Evenson, Robert E., and Yoav Kislev. *Investment in Agricultural Research and Extension: A Survey of International Data*. Center Discussion Paper No. 124. New Haven, Connecticut: Economic Growth Center, Yale University, August 1971.

0886 Evenson, Robert E., and Yoav Kislev. "Investment in Agricultural Research and Extension: A Survey of International Data." *Economic Development and Cultural Change* Vol. 23 (April 1975): 507–521.

Canada

Personnel

Period/Year	PhD	MSc	BSc	Subtotal	Expat	Total	Source
1960–64	..	..	..	..	..	..	
1965–69	1052	632	355	2039	..	2039	286/285
1970	..	..	..	..	..	..	
1971	1229	659	334	2222	..	2222	279
1972	..	..	..	..	..	..	
1973	..	..	..	..	..	..	
1974	1400	585	297	2282	..	2282	244
1975	..	..	..	..	..	..	
1976	..	..	..	..	..	..	
1977	1530	604	271	2405	..	2405	243
1978	..	..	..	..	..	..	
1979	..	..	..	..	..	..	
1980	1490	503	236	2229	..	2229	242
1981	..	..	..	..	..	..	
1982	..	..	..	..	..	..	
1983	..	..	..	..	..	..	
1984	..	..	..	..	..	..	
1985	..	..	..	..	..	..	
1986	..	..	..	..	..	..	

1960–64:   1965–69: 1965 & 68

Expenditure

LCU = Canadian Dollars

Period/Year	Current LCU (millions)	Constant 1980 LCU(millions)	Constant 1980 US$ (millions)		Source
			Atlas	PPP	
1960–64	56.773	167.967	141.377	151.040	514
1965–69	128.645	322.227	271.217	289.755	510/511
1970	..	..	..	..	
1971	145.600	329.412	277.264	296.215	509
1972	137.500	295.699	248.888	265.900	601
1973	127.500	251.479	211.669	226.136	601
1974	177.500	303.419	255.386	272.842	601
1975	210.000	322.086	271.098	289.627	601
1976	217.000	306.065	257.613	275.221	601
1977	231.500	303.010	255.042	272.474	601
1978	305.000	374.233	314.990	336.519	601
1979	356.250	395.833	333.171	355.943	601
1980	395.000	395.000	332.470	355.194	601
1981	452.500	409.502	344.676	368.234	601
1982	508.750	418.036	351.859	375.908	601
1983	592.500	462.891	389.613	416.242	601
1984	677.500	500.739	421.470	450.277	601
1985	708.750	506.974	426.718	455.884	601
1986	..			..	

1960–64: 1963   1965–69: 1967 & 69

Personnel Comments:

The main institutions undertaking agricultural research in Canada are the Canadian Department of Agriculture, the Canadian Department of Fisheries, the Canadian Department of Environment (Forestry), the Provinces of Canada, and the universities. Most sources only give information about the Canadian Department of Agriculture or only include crop and livestock research. Fisheries and forestry research, however, are relatively important research areas in Canada, and not counting them substantially distorts the picture.

The personnel indicators were constructed by an aggregation of detailed research personnel figures provided by the Commonwealth Agricultural Bureaux (sources 242, 243, 244, 279, 285, & 286) and compiled at the institute level. In addition to the 5 major entities mentioned above, agricultural researchers within the following organizations were included in the indicator series: the Canadian Department of Industry, Trade and Commerce; the National Research Council of Canada; the Ministry of Natural Resources; and Statistics Canada. Researchers at universities were prorated at a 50% rate.

Expenditure Comments:

Most Canadian agricultural research expenditure data exclude forestry and fisheries as well as university research. OECD statistics do include forestry and fisheries research but only a portion of relevant university research expenditures. For the years 1972–1985 an estimate was made of the relevant General University Fund component, as this was not included in the original OECD time series – Public Funding of R & D by Socio-Economic Objective: Agriculture, Forestry and Fisheries – given by source 601. Using source 509, the relevant General University Fund component was estimated at roughly 25% of the time series given by source 601, and increased the time series accordingly.

Sources:

0242 Commonwealth Agricultural Bureaux (CAB). *List of Research Workers in Agricultural Sciences in the Commonwealth 1981.* Slough, England: CAB, 1981.

0243 Commonwealth Agricultural Bureaux (CAB). *List of Research Workers in the Agricultural Sciences in the Commonwealth and in the Republic of Ireland 1978.* Slough, England: CAB, 1978.

0244 Commonwealth Agricultural Bureaux (CAB). *List of Research Workers in the Agricultural Sciences in the Commonwealth and in the Republic of Ireland 1975.* Slough, England: CAB, 1975.

0279 Commonwealth Agricultural Bureaux (CAB). *List of Research Workers in the Agricultural Sciences in the Commonwealth and in the Republic of Ireland 1972.* Slough, England: CAB, 1972.

0285 Commonwealth Agricultural Bureaux (CAB). *List of Research Workers in the Agricultural Sciences in the Commonwealth and in the Republic of Ireland 1969.* Slough, England: CAB, 1969.

0286 Commonwealth Agricultural Bureaux (CAB). *List of Research Workers in Agriculture, Animal Health and Forestry in the Commonwealth and in the Republic of Ireland 1966.* Slough, England: CAB, 1966.

0509 OECD. *Survey of the Resources Devoted to R&D by OECD Member Countries, International Statistical Year 1971 – Volume 5: Total Tables, Statistical Tables and Notes.* Paris: OECD, August 1974.

0510 OECD. *International Survey of the Resources Devoted to R&D in 1967 by OECD Member Countries, Statistical Tables and Notes – Volume 5: Total Tables.* Paris: OECD, August 1970.

0511 OECD. *International Survey of the Resources Devoted to R&D in 1969 by OECD Member Countries, Statistical Tables and Notes – Volume 5: Total Tables.* Paris: OECD, June 1973.

0514 OECD. *A Study of Resources Devoted to R&D in OECD Member Countries in 1963/64, International Statistical Year for Research and Development – Volume 2: Statistical Tables and Notes.* Paris: OECD, 1968.

0601 OECD – STIID Data Bank. "Public Funding of R&D by Socio-Economic Objective." Tables. OECD, Paris, April 1987. Mimeo.

Additional References:

0010 Boyce, James K., and Robert E. Evenson. *National and International Agricultural Research and Extension Programs.* New York: Agricultural Development Council, Inc., 1975.

0014 Judd, M. Ann, James K. Boyce, and Robert E. Evenson. "Investing in Agricultural Supply." Economic Growth Center, Yale University, New Haven, Connecticut, 1983. Mimeo.

0027 Harvey, Nigel, ed. *Agricultural Research Centres: A World Directory of Organizations and Programmes.* Seventh Edition, Two Volumes. Harlow, U.K.: Longman, 1983.

0165 Evenson, R.E., and Y. Kislev. *Agricultural Research and Productivity.* New Haven: Yale University Press, 1975.

0287 Commonwealth Agricultural Bureaux (CAB). *List of Research Workers in Agriculture, Animal Health and Forestry in the British Commonwealth, the Republic of Sudan and the Republic of Ireland 1959.* Slough, England: CAB, 1959.

0473 OECD. *Science and Technology Indicators: Basic Statistical Series – Volume A: The Objectives of Government R&D Funding 1974–1985.* Paris: OECD, May 1983.

0474 OECD. *Science and Technology Indicators: Basic Statistical Series – Volume A: The Objectives of Government R&D, 1969–1981.* Paris: OECD, July 1981.

0512 OECD. *Survey of the Resources Devoted to R&D by OECD Member Countries, International Statistical Year 1973 – Volume 5: Total Tables, Statistical Tables and Notes.* Paris: OECD, September 1976.

0513 OECD. *International Survey of the Resources Devoted to R&D by OECD Member Countries, International Statistical Year 1975 – International Volume: Statistical Tables and Notes.* Paris: OECD, March 1979.

0530 OECD. *OECD Science and Technology Indicators No. 2 – R&D, Invention and Competitiveness.* Paris: OECD, 1986.

0597 Hamilton, D.G. *Evaluation of Research and Development in Agriculture and Food in Canada.* Ontario: Canadian Agricultural Research Council, January 1980.

0598 McGlaughlin, Glen R. "A Study Regarding Agricultural Research in Canada." Research Division, Saskatchewan Wheat Pool, January 1977. Mimeo.

0599 Canada Grains Council. *Agricultural Research in Canada.* Manitoba, Canada: Canada Grains Council, 1986.

0600 Christie, Leonard A. *Agricultural Research in Canada: Priorities, Funding and Manpower.* Ottawa: Library of Parliament – Science and Technology Division, Research Branch, April 1980.

0602 OECD – STIID Data Bank. "R&D Expenditure in the Higher Education and Private Non-Profit Sectors." Tables. OECD, Paris, April 1987. Mimeo.

0607 Brinkman, George L. *An Analysis of Sources of Multifactor Productivity Growth in Canadian Agriculture, 1961 to 1980, with Projections to 2000.* Report prepared for the Development Policy Directorate of the Regional Development Branch of Agriculture Canada. Ontario: School of Agricultural Economics and Extension Education, University of Guelph, December 1984.

0608 Zentner, Robert Paul. "An Economic Evaluation of Public Wheat Research Expenditures in Canada." PhD diss., University of Minnesota, Minneapolis, January 1982.

0649 OECD. *Reviews of National Science Policy: Canada.* Paris: OECD, 1969.

0852 Evenson, Robert E., and Yoav Kislev. *Investment in Agricultural Research and Extension: A Survey of International Data.* Center Discussion Paper No. 124. New Haven, Connecticut: Economic Growth Center, Yale University, August 1971.

0874 Klein, K.K., and W.H. Furtan, eds. *Economics of Agricultural Research in Canada.* Papers presented at a conference held September 1983. Calgary, Alberta, Canada: The University of Calgary Press, 1985.

0886 Evenson, Robert E., and Yoav Kislev. "Investment in Agricultural Research and Extension: A Survey of International Data." *Economic Development and Cultural Change* Vol. 23 (April 1975): 507–521.

Cape Verde

Personnel

Period/Year	PhD	MSc	BSc	Subtotal	Expat	Total	Source
1960–64	..	..	..	..	..	..	
1965–69	..	..	..	..	..	..	
1970	..	..	..	..	..	..	
1971	..	..	..	..	..	..	
1972	..	..	..	..	..	..	
1973	..	..	..	..	..	..	
1974	..	..	..	..	..	..	
1975	..	..	..	..	..	..	
1976	..	..	..	..	..	..	
1977	..	..	..	..	..	..	
1978	..	..	..	..	..	..	
1979	..	..	..	..	..	..	
1980	..	..	..	..	..	..	
1981	..	..	..	..	..	..	
1982	..	..	..	..	..	..	
1983	1	4	6	11	3	14	46
1984	..	..	..	..	..	..	
1985						18	744
1986	..	..	..	..	..	..	

1960–64:   1965–69:

Expenditure   *LCU = Cape Verdian Escudos*

Period/Year	Current LCU (millions)	Constant 1980 LCU(millions)	Constant 1980 US$ (millions)		Source
			Atlas	PPP	
1960–64	..	..	..	..	
1965–69	..	..	..	..	
1970	..	..	..	..	
1971	..	..	..	..	
1972	..	..	..	..	
1973	..	..	..	..	
1974	..	..	..	..	
1975	..	..	..	..	
1976	..	..	..	..	
1977	..	..	..	..	
1978	..	..	..	..	
1979	..	..	..	..	
1980	..	..	..	..	
1981	..	..	..	..	
1982			..	..	
1983	9.000	6.845	0.189	0.320	46
1984	..	..	..	..	
1985	..	..	..	..	
1986	..	..	..	..	

1960–64:   1965–69:

Personnel Comments:

According to source 46 there were no research institutions in the country before independence (1975). In 1977 CEA (Center for Agrarian Studies) was created within the Ministry of Rural Development and entrusted with building up an agricultural research system. In 1984 it was proposed that CEA be given a more independent status and change its name to 'Instituto Nacional de Investigação Agraría Amilcar Cabral' (INIAAC). Also in that year a livestock research institute was created. Therefore some expansion of the total number of researchers is expected.

The 1985 figure given by source 744 only includes INIAAC.

Expenditure Comments:

See comments on personnel figures. These expenditure figures represent the operating budget only. Because the Cape Verdian agricultural research system is in its first phase of development, capital investments are likely to be considerable. An estimate of capital investment, however, is not available, because nearly all such investments are funded by foreign donors and kept outside the official budget figures.

Sources:

0046 Institut du Sahel, and DEVRES, Inc. *Assessment of Agricultural Research Resources in the Sahel, Volume III: National Report Cape Verde.* Washington, D.C.: DEVRES, Inc., August 1984.

0744 *Agricultural Research Centres: A World Directory of Organizations and Programmes.* Eighth Edition, Two Volumes. Harlow, U.K.: Longman, 1986.

Additional References:

0021 OECD. *OECD Science and Technology Indicators.* Paris: OECD, 1984.

0045 Institut du Sahel, and DEVRES, Inc. *Bilan des Ressources de la Recherche Agricole dans les Pays du Sahel, Volume I: Analyse et Stratégie Régionale.* Washington, D.C.: DEVRES, Inc., 1984.

0068 Conte, Stephan. "West Africa Agricultural Research Data Base Building Task: Termination Report." World Bank, Washington, D.C., 1985. Mimeo.

0083 Institut du Sahel, and DEVRES, Inc. *Bilan des Ressources de la Recherche Agricole dans les Pays du Sahel, Volume II: Résumés des Rapports Nationaux.* Washington, D.C.: DEVRES, Inc., August 1984.

0088 Swanson, Burton E., and Wade H. Reeves. "Agricultural Research West Africa: Manpower and Training." World Bank, Washington, D.C., November 1985. Mimeo.

0089 Oram, Peter. "Report on National Agricultural Research in West Africa." IFPRI, Washington, D.C., February 1986. Mimeo.

0163 CGIAR. "National Agricultural Research." CGIAR, Washington, D.C., 1985. Mimeo.

0258 "Effort Francais en Faveur de la Recherche Agronomique dans les Pays du Sahel 1984–1985." 1985. Mimeo.

0372 Gabas, Jean-Jacques. *La Recherche Agricole dans les Pays Membres du CILSS.* Club du Sahel, September 1986.

0384 Institut du Sahel, and DEVRES, Inc. *Sahel Region Agricultural Research Resource Assessment – Data Base Management Information System: Diskettes and User's Guide.* Printouts. Washington, D.C.: Institut du Sahel, and DEVRES, Inc., n.d.

0466 World Bank – West Africa Agricultural Research Review Team. "National Agricultural Research Systems (NARS): A Regional Appraisal and Selected Issues." World Bank, Washington, D.C., August 1986. Mimeo.

Central African Republic

Personnel

Period/Year	PhD	MSc	BSc	Subtotal	Expat	Total	Source
1960–64	..	..	..	..	..	..	
1965–69	..	..	..	..	..	21	653
1970	..	..	..	..	..	27	532
1971	..	..	..	..	..	..	
1972	..	..	..	..	..	..	
1973	..	..	..	..	..	..	
1974	..	..	..	..	..	..	
1975	..	..	..	..	..	..	
1976	..	..	..	..	..	..	
1977	..	..	..	..	..	..	
1978	..	..	..	..	..	..	
1979	..	..	..	..	..	..	
1980	..	..	..	..	..	..	
1981	..	..	..	..	..	..	
1982	..	..	..	..	..	..	
1983	..	..	..	..	..	..	
1984	..	..	..	..	..	..	
1985	..	..	..	..	..	..	
1986	..	..	..	..	..	..	

1960–64:   1965–69: 1966

Expenditure   *LCU = CFA Francs*

Period/Year	Current LCU (millions)	Constant 1980 LCU(millions)	Constant 1980 US$ (millions) Atlas	Constant 1980 US$ (millions) PPP	Source
1960–64	..	..		..	
1965–69	183.900	451.843	1.970	2.985	266
1970	..	..	..	..	
1971	..	..	..	..	
1972	..	..	..	..	
1973	..	..	..	..	
1974	..	..	..	..	
1975	..	..	..	..	
1976	..	..	..	..	
1977	..	..	..	..	
1978	..	..	..	..	
1979	..	..	..	..	
1980	..	..	..	..	
1981	..	..	..	..	
1982	..	..	..	..	
1983	..	..	..	..	
1984	..	..	..	..	
1985	..	..	..	..	
1986	..	..	..	..	

1960–64:   1965–69: 1969

Personnel Comments:

The 1966 figure includes the following institutes: IRAT (3 researchers), IFCC (6 researchers), IRCT (9 researchers), IRHO (1 researcher), and ORSTOM (2 researchers). Source 653 does not give the number of researchers at IEMVT for the year 1966.

The 1970 figure includes the following institutes: IRAT (2 researchers), IRCT (5 researchers), IFCC (5 researchers), IEMVT (5 researchers), and ORSTOM (10 researchers). Source 589 reports that by the end of 1970 the relations between almost all French research institutes and the CAR government were disrupted, resulting in a massive withdrawal of French agricultural research resources. Only ORSTOM continued its activities. It took several years before contacts were re-established, and French agricultural research institutes returned to CAR.

The total number of agricultural researchers in the early eighties differs considerably from source to source. The WAARR study (sources 86 & 88) gives a number of 28 researchers, however the preparation study of WAARR (source 68) mentions a number of 93, and Oram gives a figure of 54 agricultural researchers in his West Africa study (source 89). On the basis of the available information, it was not possible to decide which is the most plausible estimate.

Expenditure Comments:

The 1965–69 average is based on a 1969 expenditure figure. This expenditure figure includes the expenditures of the following institutes: IRAT, IFCC, IEMVT, IRCT, and ORSTOM.

The WAARR study estimates the budget for the agricultural research system of CAR at US$ 42 million. It is not stated which institute(s) and projects are included in this expenditure figure and to which year(s) it applies. Because no specific sources are mentioned, it is impossible to check this estimated expenditure figure. When comparing this expenditure figure to the estimated number of researchers (28) in the same report, it seems clear that one or both of these figures are unrealistic.

Acronyms and Abbreviations:

IEMVT: Institut d'Elevage et de Médecine Vétérinaire des Pays Tropicaux
IFCC: Institut Français du Café, Cacao et autres plantes stimulantes
IRAT: Institut de Recherche Agronomique Tropicale et des Cultures Vivrières
IRCT: Institut de Recherche du Coton et Textiles Exotiques
IRHO: Institut de Recherches pour les Huiles et Oléagineux
ORSTOM: Office de la Recherche Scientifique et Technique Outre-Mer
WAARR: West African Agricultural Research Review

Sources:

0266 UNESCO. *National Science Policies in Africa.* Science Policy Studies and Documents No. 31. Paris: UNESCO, 1974.
0532 UNESCO Field Science Office for Africa. *Survey on the Scientific and Technical Potential of the Countries of Africa.* Paris: UNESCO, 1970.
0653 Webster, B.N. *Index of Agricultural Research Institutions and Stations in Africa.* Rome: FAO, n.d.

Additional References:

0010 Boyce, James K., and Robert E. Evenson. *National and International Agricultural Research and Extension Programs.* New York: Agricultural Development Council, Inc., 1975.
0017 ISNAR, IFARD & AOAD. Survey of National Agricultural Research Systems: Unpublished Questionnaire Responses. ISNAR, The Hague, 1985.
0023 Bennell, Paul. *Agricultural Researchers in Sub-Saharan Africa: An Overview.* Working Paper No. 4. The Hague: ISNAR, October 1985.
0027 Harvey, Nigel, ed. *Agricultural Research Centres: A World Directory of Organizations and Programmes.* Seventh Edition, Two Volumes. Harlow, U.K.: Longman, 1983.
0068 Conte, Stephan. "West Africa Agricultural Research Data Base Building Task: Termination Report." World Bank, Washington, D.C., 1985. Mimeo.
0086 World Bank. "West Africa Agricultural Research Review – Country Studies." World Bank, Washington, D.C., 1985. Mimeo.
0088 Swanson, Burton E., and Wade H. Reeves. "Agricultural Research West Africa: Manpower and Training." World Bank, Washington, D.C., November 1985. Mimeo.
0089 Oram, Peter. "Report on National Agricultural

118

Research in West Africa." IFPRI, Washington, D.C., February 1986. Mimeo.

0175 Cooper, St.G.C. *Agricultural Research in Tropical Africa.* Kampala: East African Literature Bureau, 1970.

0360 Cooper, St.G.C. "Towards Trained Manpower for Agricultural Research in Africa." Paper presented at the Conference on Agricultural Research and Production in Africa, organized by the Association for the Advancement of Agricultural Sciences in Africa (AAASA), Addis Ababa, 29 August–4 September 1971.

0373 Groupement d'Etudes et de Recherches pour le Développement de l'Agronomie Tropicale (GERDAT). *Rapport Général d'Activité pour 1982.* Paris: GERDAT, 1983.

0466 World Bank – West Africa Agricultural Research Review Team. "National Agricultural Research Systems (NARS): A Regional Appraisal and Selected Issues." World Bank, Washington, D.C., August 1986. Mimeo.

0589 Kassapu, Samuel. *Les Dépenses de Recherche Agricole dans 34 Pays d'Afrique Tropicale.* Paris: Centre de Développement de l'OCDE, 1976.

0654 Office de la Recherche Scientifique et Technique Outre-Mer (ORSTOM). *Rapport d'Activité 1967.* Paris: ORSTOM, 1969.

0852 Evenson, Robert E., and Yoav Kislev. *Investment in Agricultural Research and Extension: A Survey of International Data.* Center Discussion Paper No. 124. New Haven, Connecticut: Economic Growth Center, Yale University, August 1971.

Chad

Personnel

Period/Year	PhD	MSc	BSc	Subtotal	Expat	Total	Source
1960–64	..	..	..	..	..	..	
1965–69	..	..	..	..	..	20	10/175
1970	..	..	..	..	..	20	532
1971	..	..	..	..	..	..	
1972	..	..	..	..	..	..	
1973	..	..	..	..	..	..	
1974	..	..	..	..	..	..	
1975	..	..	..	..	..	..	
1976	..	..	..	..	..	..	
1977	..	..	..	..	..	..	
1978	..	..	..	..	..	..	
1979	..	..	..	..	..	..	
1980	..	..	..	..	..	..	
1981	..	..	..	..	..	..	
1982	..	..	..	..	..	..	
1983	..	..	..	..	..	..	
1984	..	..	..	20	8	28	92/17
1985	..	..	..	..	..	..	
1986	..	..	..	..	..	..	

1960–64:   1965–69: 1966 & 67

Expenditure

LCU = CFA Francs

	Current LCU (millions)	Constant 1980 LCU(millions)	Constant 1980 US$ (millions)		Source
Period/Year			Atlas	PPP	
1960–64	..	..	..	..	
1965–69	..	..	..	..	
1970	..	..	..	..	
1971	..	..	..	..	
1972	213.000	432.927	2.008	3.222	266
1973	294.800	578.039	2.681	4.302	589
1974	..	..	..	..	
1975	..	..	..	..	
1976	..	..	..	..	
1977	..	..	..	..	
1978	..	..	..	..	
1979	..	..	..	..	
1980	..	..	..	..	
1981	..	..	..	..	
1982	..	..	..	..	
1983	286.690	198.952	0.923	1.481	17
1984	..	..	..	..	
1985	..	..	..	..	
1986	..	..	..	..	

1960–64:   1965–69:

Personnel Comments:

The 1970 figure includes, according to source 532, IEMVT/LRVZ (10 researchers, partly involved in teaching at IEZVAC) and IRCT (10 researchers). The same source reports 24 researchers working for ORSTOM. An unknown number of them are involved in non-agricultural research, so this ORSTOM figure was not included in the indicator series. Agricultural research in Chad is currently conducted by three institutions, namely, 'Division de la Recherche Agronomique', IRCT, and IEMVT-LRVZ. The last two institutions are still part of the old French colonial research system and, as such, are managed by France. The number of researchers reported here for 1984 is based on the ISNAR, IFARD & AOAD questionnaire response (source 17) for DRA and IRCT, and the DEVRES report (source 92) for IEMVT/LRVZ.

Expenditure Comments:

Nearly all agricultural research in Chad is financed by France or other foreign donors. This makes it very difficult to get an overview of total agricultural research expenditures in Chad, especially because there is – as far as can be discerned – no coordinating body which collects this kind of data. The 1972 and 1973 expenditure figures include IRCT and IEMVT/LRVZ. According to the same source, another 200 million FCFA is spent by ORSTOM. Because it is not clear how much of ORSTOM's activities are agriculturally related, ORSTOM expenditures were excluded from the indicator series. The ISNAR, IFARD & AOAD questionnaire response (source 17) gives a 1983 figure of FCFA 286.69 million (of which 275 goes to IRCT and 11.69 to DRA). However, the DRA figures exclude salaries. There are also no data for the veterinary institute (IEMVT/LRVZ), as it temporarily stopped operations due to the civil war.

Acronyms and Abbreviations:

DRA: Département de la Recherche Agronomique
IRCT: Institut de Recherche du Coton et Textiles Exotiques
IEMVT/LRVZ: Institut d'Elevage et de Médecine Vétérinaire des Pays Tropicaux – Laboratoire de Recherches Vétérinaires et Zootechniques de Farcha

IEZVAC: Institut d'Enseignement Zootechnique et Vétérinaire d'Afrique Centrale
ORSTOM: Office de la Recherche Scientifique et Technique Outre-Mer

Sources:

0010 Boyce, James K., and Robert E. Evenson. *National and International Agricultural Research and Extension Programs.* New York: Agricultural Development Council, Inc., 1975.

0017 ISNAR, IFARD & AOAD. Survey of National Agricultural Research Systems: Unpublished Questionnaire Responses. ISNAR, The Hague, 1985.

0092 Institut du Sahel, and DEVRES, Inc. *Assessment of Agricultural Research Resources in the Sahel, Volume III: National Report Chad.* Washington, D.C.: DEVRES, Inc., August 1984.

0175 Cooper, St.G.C. *Agricultural Research in Tropical Africa.* Kampala: East African Literature Bureau, 1970.

0266 UNESCO. *National Science Policies in Africa.* Science Policy Studies and Documents No. 31. Paris: UNESCO, 1974.

0532 UNESCO Field Science Office for Africa. *Survey on the Scientific and Technical Potential of the Countries of Africa.* Paris: UNESCO, 1970.

0589 Kassapu, Samuel. *Les Dépenses de Recherche Agricole dans 34 Pays d'Afrique Tropicale.* Paris: Centre de Développement de l'OCDE, 1976.

Additional References:

0016 Oram, Peter A., and Vishva Bindlish. *Resource Allocations to National Agricultural Research: Trends in the 1970s.* The Hague and Washington, D.C.: ISNAR and IFPRI, November 1981.

0023 Bennell, Paul. *Agricultural Researchers in Sub-Saharan Africa: An Overview.* Working Paper No. 4. The Hague: ISNAR, October 1985.

0045 Institut du Sahel, and DEVRES, Inc. *Bilan des Ressources de la Recherche Agricole dans les Pays du Sahel, Volume I: Analyse et Stratégie Régionale.* Washington, D.C.: DEVRES, Inc., 1984.

0068 Conte, Stephan. "West Africa Agricultural Research Data Base Building Task: Termination Report." World Bank, Washington, D.C., 1985. Mimeo.

0073 Oram, Peter A., and Vishva Bindlish. "Investment in Agricultural Research in Developing Countries: Progress, Problems, and the Determination of Priorities." IFPRI, Washington, D.C., January 1984. Mimeo.

0083 Institut du Sahel, and DEVRES, Inc. *Bilan des Ressources de la Recherche Agricole dans les Pays du Sahel, Volume II: Résumés des Rapports*

Nationaux. Washington, D.C.: DEVRES, Inc., August 1984.

0086 World Bank. "West Africa Agricultural Research Review – Country Studies." World Bank, Washington, D.C., 1985. Mimeo.

0088 Swanson, Burton E., and Wade H. Reeves. "Agricultural Research West Africa: Manpower and Training." World Bank, Washington, D.C., November 1985. Mimeo.

0089 Oram, Peter. "Report on National Agricultural Research in West Africa." IFPRI, Washington, D.C., February 1986. Mimeo.

0095 FAO – CARIS. *Agricultural Research in Developing Countries – Volume 1: Research Institutions.* Rome: FAO – CARIS, 1978.

0131 FAO. *Rapport d'une Mission de Consultation: La Recherche Agronomique au Tchad.* Rome: FAO, December 1984.

0163 CGIAR. "National Agricultural Research." CGIAR, Washington D.C., 1985. Mimeo.

0228 CIRAD. *Coopération Francaise en Recherche Agronomique: Activités du CIRAD dans les Pays du Sahel.* Paris: CIRAD, March 1985.

0258 "Effort Francais en Faveur de la Recherche Agronomique dans les Pays du Sahel 1984–1985." 1985. Mimeo.

0360 Cooper, St.G.C. "Towards Trained Manpower for Agricultural Research in Africa." Paper presented at the Conference on Agricultural Research and Production in Africa, organized by the Association for the Advancement of Agricultural Sciences in Africa (AAASA), Addis Ababa, 29 August–4 September 1971.

0372 Gabas, Jean-Jacques. *La Recherche Agricole dans les Pays Membres du CILSS.* Club du Sahel, September 1986.

0373 Groupement d'Etudes et de Recherches pour le Développement de l'Agronomie Tropicale (GERDAT). *Rapport Général d'Activité pour 1982.* Paris: GERDAT, 1983.

0384 Institut du Sahel, and DEVRES, Inc. *Sahel Region Agricultural Research Resource Assessment – Data Base Management Information System: Diskettes and User's Guide.* Printouts. Washington, D.C.: Institut du Sahel, and DEVRES, Inc., n.d.

0466 World Bank – West Africa Agricultural Research Review Team. "National Agricultural Research Systems (NARS): A Regional Appraisal and Selected Issues." World Bank, Washington, D.C., August 1986. Mimeo.

0653 Webster, B.N. *Index of Agricultural Research Institutions and Stations in Africa.* Rome: FAO, n.d.

0654 Office de la Recherche Scientifique et Technique Outre-Mer (ORSTOM). *Rapport d'Activité 1967.* Paris: ORSTOM, 1969.

0852 Evenson, Robert E., and Yoav Kislev. *Investment in Agricultural Research and Extension: A Survey of International Data.* Center Discussion Paper No. 124. New Haven, Connecticut: Economic Growth Center, Yale University, August 1971.

Chile

Personnel

Period/Year	PhD	MSc	BSc	Subtotal	Expat	Total	Source
1960–64	..	..	..	..	..	121	10/25
1965–69	12	29	134	175	..	175	25
1970	19	41	147	207	..	207	25
1971	24	44	149	217	..	217	25
1972	..	..	..	..	..	..	
1973	..	..	..	..	..	..	
1974	26	43	161	230	..	230	25
1975	28	42	166	236	..	236	25
1976	29	45	173	247	..	247	25
1977	30	46	177	253	..	253	25
1978	..	..	..	..	..	..	
1979	35	61	192	288	..	288	25
1980	36	56	172	264	..	264	25
1981	35	61	172	268	..	268	25
1982	37	66	161	264	..	264	25
1983	41	64	169	274	..	274	25
1984	42	66	168	276	..	276	25
1985				..	..	..	
1986	..	..	..	..	..	..	

1960–64: 1964   1965–69: 1965, 66, 67, 68 & 69

Expenditure

LCU = Chilean Pesos

Period/Year	Current LCU (millions)	Constant 1980 LCU(millions)	Constant 1980 US$ (millions)		Source
			Atlas	PPP	
1960–64	0.003	261.160	5.859	11.518	25
1965–69	0.020	383.412	8.601	16.910	25
1970	0.052	443.680	9.953	19.568	25
1971	0.103	742.976	16.668	32.769	25
1972	0.182	699.702	15.697	30.860	25
1973	0.689	511.040	11.464	22.539	25
1974	5.793	541.302	12.143	23.874	25
1975	27.583	582.557	13.069	25.693	25
1976	107.653	648.365	14.545	28.596	25
1977	219.588	649.659	14.574	28.653	25
1978	298.882	564.859	12.672	24.913	25
1979	413.507	534.230	11.985	23.562	25
1980	570.950	570.950	12.808	25.182	25
1981	752.513	670.689	15.046	29.581	25
1982	728.581	573.234	12.860	25.282	25
1983	903.356	561.356	12.593	24.758	25
1984	1165.501	633.769	14.218	27.952	25
1985	..	..	..	..	
1986	..	..	..	..	

1960–64: 1960, 61, 62, 63 & 64   1965–69: 1965, 66, 67, 68 & 69

Personnel Comments:

These figures include the 'Instituto de Investigaciones Agropecuarias' (INIA), and the Faculties of Agriculture of the University of Chile, the Catholic University, the University of Concepcion, and the Austral University. University staff were assumed to spend one-third of their time on research. The figures explicitly exclude forestry, fisheries, and veterinary research; basic research conducted in universities that may pertain to agriculture outside the faculties of agriculture; and nutrition and food technology research outside agricultural institutions. The following organizations are definitely excluded: Fundación Chile, Instituto de Investigaciones del Desarrollo Rural, Instituto de Investigaciones Oceanológicas, Instituto de la Patagonia, Instituto Forestal, and Instituto Nacional de Investigación de Recursos Naturales (source 744).

Expenditure Comments:

These figures correspond to the personnel figures. The pre-1965 expenditure figures for INIA refer to the Ministry of Agriculture and the Rockefeller Foundation program.

Expenditure data in source 25 was reported in constant 1984 pesos after deflating by the CPI index given in Table A-15, p. 149. The procedure used to construct the indicator series was first to recover a current LCU estimate of the expenditure data using the CPI index cited by source 25, and then to sum across each research institution.

Sources:

0010 Boyce, James K., and Robert E. Evenson. *National and International Agricultural Research and Extension Programs.* New York: Agricultural Development Council, Inc., 1975.

0025 Venezian, Eduardo. *Chile and the CGIAR Centers: A Study of Their Collaboration in Agricultural Research.* CGIAR Study Paper Number 20. Washington, D.C.: World Bank, April 1987.

Additional References:

0001 Trigo, Eduardo J., and Martín E. Piñeiro. "Funding Agricultural Research." In *Report of a Conference: Selected Issues in Agricultural*

Research in Latin America, edited by Barry Nestel and Eduardo J. Trigo. The Hague: ISNAR, March 1984.

0014 Judd, M. Ann, James K. Boyce, and Robert E. Evenson. "Investing in Agricultural Supply." Economic Growth Center, Yale University, New Haven, Connecticut, 1983. Mimeo.

0016 Oram, Peter A., and Vishva Bindlish. *Resource Allocations to National Agricultural Research: Trends in the 1970s.* The Hague and Washington, D.C.: ISNAR and IFPRI, November 1981.

0017 ISNAR, IFARD & AOAD. Survey of National Agricultural Research Systems: Unpublished Questionnaire Responses. ISNAR, The Hague, 1985.

0022 UNESCO. *Statistical Yearbook 1983.* Paris: UNESCO, 1983.

0026 Oram, Peter A., and Mark Gieben. "Document Summaries." ISNAR, The Hague, 1984. Mimeo.

0027 Harvey, Nigel, ed. *Agricultural Research Centres: A World Directory of Organizations and Programmes.* Seventh Edition, Two Volumes. Harlow, U.K.: Longman, 1983.

0032 Castronovo, Alfonso J.P. "Diagnóstico Descriptivo de la Situación Actual del Sistema Nacional de Investigación y Extensión en Argentina, Chile, Paraguay y Uruguay." Buenos Aires, December 1980. Mimeo.

0053 Piñeiro, Martín, and Eduardo Trigo, eds. *Cambio Técnico en el Agro Latino Americano – Situación y Perspectivas en la Década de 1980.* San José: IICA, 1983.

0064 Instituto de Investigaciones Agropecuarias (INIA). *Memoria del INIA, 1980, 1981, 1982, 1983, 1984, 1985.* Santiago: INIA, various years.

0073 Oram, Peter A., and Vishva Bindlish. "Investment in Agricultural Research in Developing Countries: Progress, Problems, and the Determination of Priorities." IFPRI, Washington, D.C., January 1984. Mimeo.

0095 FAO – CARIS. *Agricultural Research in Developing Countries – Volume 1: Research Institutions.* Rome: FAO – CARIS, 1978.

0163 CGIAR. "National Agricultural Research." CGIAR, Washington, D.C., 1985. Mimeo.

0165 Evenson, R.E., and Y. Kislev. *Agricultural Research and Productivity.* New Haven: Yale University Press, 1975.

0172 Trigo, Eduardo. Background material. ISNAR. The Hague, n.d.

0575 Instituto de Investigaciones Agropecuarias (INIA). *Decimocuarta Memoria del INIA 1977/1979.* Santiago, Chile: INIA, 1981.

0576 Instituto de Investigaciones Agropecuarias (INIA). *Memoria del INIA.* Years: 1964/1965 – 1976/1977. Santiago, Chile: INIA, various years.

124

0585 Departamento de Economía Agraria, Universidad Católica de Chile. "La Investigación Agropecuaria en Chile." In *Panorama Económico de la Agricultura*, edited by Paul Aldunate V, et al., pp. 2–8. Santiago, Chile: Departamento de Economía Agraria – Universidad Católica de Chile, March 1985.

0646 Instituto de Investigaciones Agropecuarias (INIA). *Investigación Agropecuaria*. Santiago, Chile: INIA, 1970.

0652 Bonilla, Sergio. "Proceso de Formación y Evolución del INIA en Chile." In *Organización y Administración de la Generación y Transferencia de Tecnología Agropecuaria*, Serie: Ponencias, Resultados y Recomendaciones de Eventos Técnicos No. A4/UY-86-001, edited by Horacio H. Stagno and Mario Allegri. Montevideo,

Uruguay: IICA and Centro de Investigaciones Agrícolas 'Alberto Boerger', 1986.

0744 *Agricultural Research Centres: A World Directory of Organizations and Programmes*. Eighth Edition, Two Volumes. Harlow, U.K.: Longman, 1986.

0852 Evenson, Robert E., and Yoav Kislev. *Investment in Agricultural Research and Extension: A Survey of International Data*. Center Discussion Paper No. 124. New Haven, Connecticut: Economic Growth Center, Yale University, August 1971.

0886 Evenson, Robert E., and Yoav Kislev. "Investment in Agricultural Research and Extension: A Survey of International Data." *Economic Development and Cultural Change* Vol. 23 (April 1975): 507–521.

China, P. R.

Personnel

Period/Year	PhD	MSc	BSc	Subtotal	Expat	Total	Source
1960–64	..	..	..	..	..	..	
1965–69	..	..	..	..	..	..	
1970	..	..	..	..	..	..	
1971	..	..	..	..	..	..	
1972	..	..	..	..	..	..	
1973	..	..	..	..	..	..	
1974	..	..	..	..	..	..	
1975	..	..	..	..	..	..	
1976	..	..	..	..	..	..	
1977	..	..	..	..	..	..	
1978	..	..	..	..	..	19539	883
1979	..	..	..	..	..	20676	883
1980	..	..	..	..	..	20790	883
1981	..	..	..	..	..	23729	883
1982	..	..	..	..	..	31226	883
1983	..	..	..	..	..		
1984	..	..	..	..	..	31956	883
1985	..	..	..	..	..	33454	371/588
1986	..	..	..	..	..	..	

1960–64:   1965–69:

Expenditure

LCU = Yuan

Period/Year	Current LCU (millions)	Constant 1980 LCU (millions)	Constant 1980 US$ (millions)		Source
			Atlas	PPP	
1960–64	..	..	..	..	
1965–69	..	..	..	..	
1970	..	..	..	..	
1971	..	..	..	..	
1972	..	..	..	..	
1973	810.000	880.435	586.697	1103.676	554
1974	..	..	..	..	
1975	..	..	..	..	
1976	..	..	..	..	
1977	..	..	..	..	
1978	..	..	..	..	
1979	..	..	..	..	
1980	..	..	..	..	
1981	..	..	..	..	
1982	..	..	..	..	
1983	..	..	..	..	
1984	..	..	..	..	
1985	..	..	..	..	
1986	..	..	..	..	

1960–64:   1965–69:

Personnel Comments:

Statistical indicators of agricultural research in the PRC are extremely difficult to estimate. In addition to the paucity of data, there are major problems in deriving a researcher count where the classification and qualification levels of a researcher/scientist are standardized over time and roughly compatible with other countries. This situation appears to have improved of late.

The source 371 figure reported here appears the most acceptable to date. It was compiled on the basis of an Autumn 1985 survey which used UNESCO standard definitions and concepts. It includes only scientists possessing a university graduate, or 'higher cultural level,' and others having the senior or intermediate academic title working in R & D institutions subordinate to government departments above the county (i.e., at the provincial and central) level (4690 institutions, source 371). In the areas of agriculture, forestry, animal husbandry, and water management, 37,799 scientists and engineers are engaged in science and technology activities at and above the county level (source 371, Table I–2). An estimated 4,345 scientists and engineers who work at the county level (67,782 county-level personnel x 6.4%, the total number of county- level personnel possessing a university education or higher cultural level) were excluded from the total. According to sources 588 and 754, they more appropriately can be classified as extension workers rather than researchers.

The 1978–84 figures from source 883 were obtained directly from the Ministry of Agriculture, Animal Husbandry and Fisheries of China. The 1984 figure consists of 499 MSc-equivalent researchers and 17,965 BSc-equivalent researchers. The degree-level breakdown provided by source 883 indicates there were no PhD-equivalent researchers working in agricultural research in 1984. This is corroborated by source 882 who states that the Chinese have trained very few, probably less than 25, PhDs in agriculture, and, in spite of the significant number currently seeking training abroad (US) to the PhD level, probably no more than 10 foreign-trained PhD-level agricultural researchers had returned to China in the last few years. Source

883 also indicated that the remaining 13,492 researchers in the 1984 figure do not possess a PhD, MSc, or BSc degree and have generally had less than 4 (probably 2 or 3) years of university training. The series may best be considered an upper-bound estimate of the total number of agricultural researchers in China, when evaluated with respect to the concept of a researcher that was generally used to compile these statistics. Finally, source 883 confirmed that the 1981–82 jump in the series was largely due to an influx of newly trained research personnel into the system.

Efforts to estimate a consistent long-run time series from available sources have been unsuccessful. Source 10 gives a 1965 estimate of 8,000 researchers. It was calculated on the assumption that 15% of the total figure for research and development personnel given by source 367 – 53,000 scientists and engineers – are oriented to agriculture. Unfortunately source 367 (p. 397) states that its total figure explicitly excludes research personnel in the life sciences, medicine, and agriculture. The 1,200 research *and* technical personnel reported by source 10 for 1958 was taken from source 369 (p. 253) and includes Academy of Agricultural Science personnel only. The Academy of Agricultural Science personnel represent an (unknown) fraction of the total agricultural science personnel in the PRC. For example, source 367 (p. 398) states that in 1957 the Chinese Academy of Science received only 31% of the national science budget.

Source 483 reports a 1966 estimate of 150,000 agricultural scientists and a mid-1975 estimate of 180,000. Source 483 (p. 99) points out that although the later figure almost certainly includes graduates of state-run (national and provincial) institutions "the entrants into universities in 1970 and 1971 were probably the least qualified and most poorly trained of all." Thus, there is a tremendous range of competence among these research professionals, which suggests the latter two estimates significantly overstate the number of agricultural research personnel when considered in relation to the UNESCO standard which was used to compile the 1985 figure.

Finally source 484 gives an alternative indicator, the number of agricultural (research plus nonresearch) institutes in the PRC as follows:

Year	Extension Stations	Animal Breeding Stations	Vet. Stations	Seed Stations	Seed Demonstration Sites
1950	10	148	251		
1952	232	389	1,005		
1957	13,669	821	2,930	1,390	1,899
1979	17,622	1,174	8,495	2,369	2,418
1980	15,114	533	5,530	2,436	2,404
1981	15,415	566	6,778	2,370	2,392
1982	17,300	547	6,358	2,787	2,421
1983	14,694	669	7,689	2,548	2,271

Source 370 (p. 26) gives a breakdown of the total number of research (agricultural plus non-agricultural) institutes in 1983 as follows:

Level	Independent	Attached	Total
Academy of Science	1,119	11	130
Ministry and Commission	738	1,249	1,987
Province Autonomous Region Municipality directly under central government	1,439	1,288	2,727
City	2,151	2,346	4,497
Total	4,447	4,894	9,341

Here independent research institutes receive their financial allocations directly from the state budget (Ministry of Finance), while the remainder are financed through other organizations – often with funds from several sources. The bulk of the research is carried out by the independent research institutes belonging to the Academy and to the ministries.

Expenditure Comments:
Source 10 estimates the total agricultural research expenditures for 1965 at 135 million Yuan (US$ 55.0 million). This was calculated as 10% of a total research expenditure estimate of 1350 million Yuan given by source 367. Unfortunately source 367's estimate explicitly excluded agricultural (plus medical and life science) research expenditures. Source 485 (p. 632) gives a 'budget for agricultural research' of 'roughly' US$ 330 million for around 1965, but gives no indication

of how it was derived. The 1973 (source 554) estimate includes R&D expenditures on "agriculture and natural resources, excluding energy" and accounts for 18% of estimated total R&D spending for that year of Y4,590 million.

Acronyms and Abbreviations:
PRC: People's Republic of China

Sources:
0371 State Science and Technology Commission of the People's Republic of China. *Statistical Data on Science and Technology of 1985*. Beijing, China: State Science and Technology Commission of the People's Republic of China, 1986.

0554 Billgren, Boel, and Jon Sigurdson. *An Estimate of Research and Development Expenditures in the People's Republic of China in 1973*. Industry and Technology Occasional Paper No. 16. Paris: OECD Development Centre, July 1977.

0588 Kong, Guasong. Personal communication. Second Secretary of Science and Technology of the

128

Embassy of the People's Republic of China in the Netherlands. The Hague, April 1987.

0883 Kong, Guasong. Personal communication. Second Secretary of Science and Technology of the Embassy of the People's Republic of China in the Netherlands. The Hague, March 1988.

Additional References:

0010 Boyce, James K., and Robert E. Evenson. *National and International Agricultural Research and Extension Programs.* New York: Agricultural Development Council, Inc., 1975.

0014 Judd, M. Ann, James K. Boyce, and Robert E. Evenson. "Investing in Agricultural Supply." Economic Growth Center, Yale University, New Haven, Connecticut, 1983. Mimeo.

0027 Harvey, Nigel, ed. *Agricultural Research Centres: A World Directory of Organizations and Programmes.* Seventh Edition, Two Volumes. Harlow, U.K.: Longman, 1983.

0178 IRRI. *Rice Research and Production in China: An IRRI Team's View.* Los Baños, Philippines: IRRI, 1978.

0277 Berner, Boel. *The Organization and Economy of Pest Control in China.* Discussion Paper Series 128. Sweden: Research Policy Institute, University of Lund, July 1979.

0278 Berner, Boel. *The Organization and Planning of Scientific Research in China Today.* Discussion Paper Series 134. Sweden: Research Policy Institute, University of Lund, November 1979.

0367 Orleans, Leo A. "Research and Development in Communist China." *Science* Vol. 157 (28 July 1967): 392–400.

0369 Wu, Yuan-li, and Robert B. Sheeks. *The Organization and Support of Scientific Research and Development in Mainland China.* New York: Praeger Publishers Inc., 1970.

0370 Sigurdson, Jon, and Zhang Wei. "Technological Transformation in the People's Republic of China – Achievements and Problems Faced." Paper prepared for the World Institute for Development Economics Research (WIDER), Helsinki, Finland. University of Lund and Tsinghua University, Sweden and China, August 1986. Mimeo.

0483 Orleans, Leo A. *Science and Technology in the People's Republic of China.* Paris: OECD, 1977.

0484 *Statistical Yearbook of China 1984.* Beijing, China, 1984.

0485 Stavis, Benedict. "Agricultural Research and Extension Services in China." *World Development* Vol. 6 (1978): 631–645.

0486 Saich, Tony. "Linking Research to the Productive Sector: Reforms of the Civilian Science and Technology System in Post-Mao China." *Development and Change* Vol. 17 (1986): 3–33.

0487 OECD. *Science and Technology in the People's Republic of China.* Paris: OECD, 1977.

0590 UNESCO. *Science and Technology in Countries of Asia and the Pacific.* Science Policy Studies and Documents No. 52. Paris: UNESCO, n.d.

0732 FAO – Regional Office for Asia and the Pacific. *Agricultural Research Systems in the Asia-Pacific Region.* Bangkok: FAO – Regional Office for Asia and the Pacific, 1986.

0747 ISNAR. *International Workshop on Agricultural Research Management.* Report of a workshop. The Hague: ISNAR, 1987.

0754 Zhou, Fang. "Organization and Structure of the National Agricultural Research System in China." Paper presented at the International Workshop on Agricultural Research Management, ISNAR, The Hague, 7–11 September 1987.

0882 Orleans, Leo A. Personal communication. Library of Congress, Science and Technology Division. Washington D.C., March 1987.

Colombia

Personnel

Period/Year	PhD	MSc	BSc	Subtotal	Expat	Total	Source
1960–64	..	..	..	..	..	227	10/729
1965–69	..	..	..	..	..	326	10/729
1970	..	..	..	..	..	..	
1971	..	..	..	..	..	..	
1972	..	..	..	..	..	..	
1973	..	..	..	..	..	..	
1974	35	91	433	559	..	559	18/729
1975	..	..	..	..	..	..	
1976	33	170	231	434	..	434	18/729
1977	..	..	..	..	..	..	
1978		–182–		194	376	..	376	30/729
1979	40	160	166	366	..	366	18/729
1980	36	157	231	424	..	424	91/729
1981	42	186	203	431	..	431	55/729
1982	..	..	..	..	..	..	
1983	41	188	195	424	..	424	17/729
1984	42	213	201	456	..	456	55/729
1985	47	196	262	505	..	505	735/729
1986	..	..	..	..	..	..	

1960–64: 1960   1965–69: 1965

Expenditure

LCU = Colombian Pesos

Period/Year	Current LCU (millions)	Constant 1980 LCU(millions)	Constant 1980 US$ (millions)		Source
			Atlas	PPP	
1960–64	..	..	..	..	
1965–69	..	..	..	..	
1970	160.763	1118.121	22.485	46.793	55/729
1971	173.852	1091.231	21.944	45.668	55/729
1972	183.419	1019.070	20.493	42.648	55/729
1973	211.325	977.133	19.649	40.893	55/729
1974	228.154	841.703	16.926	35.225	55/729
1975	299.091	898.170	18.062	37.588	55/729
1976	335.379	802.504	16.138	33.585	55/729
1977	421.174	780.732	15.700	32.673	55/729
1978	574.302	909.135	18.282	38.047	55/729
1979	707.970	903.158	18.162	37.797	55/729
1980	877.978	877.978	17.656	36.743	55/729
1981	1212.083	987.038	19.849	41.307	55/729
1982	1372.507	895.892	18.016	37.493	55/729
1983	2353.662	1276.389	25.667	53.417	17
1984	2408.336	1068.946	21.496	44.735	736
1985	4166.100	1501.297	30.190	62.829	736/730
1986	5181.000	1490.078	29.964	62.359	736/730

1960–64:   1965–69:

Personnel Comments:
The indicator series includes ICA and CENICAFE only. The table below gives a breakdown of the number of researchers at the institutional level.

	1960	1965	1974	1976	1978	1979	1980	1981	1983	1984	1985
ICA	210	300	517	392	333	321	378	381	373	405	454
CENICAFE	17*	26*	42*	42*	43*	45*	46*	50*	[51]*	[51]*	[51]*
Total	227	326	559	434	376	366	424	431	424	456	505
Source	10	10	18	18	30	18	91	55	17	55	735
	729*	729*	729*	729*	729*	729*	729*	729*	729*	729*	729*

[ ] = estimated or constructed by authors.

In addition to research, ICA is also involved in extension and development activities. In all instances, except possibly for 1974, these nonresearch activities have been excluded from the indicator series.

According to source 55, the following institutes also conduct agricultural research in Colombia, namely, CONIF, CARC, CENICANA, ASOCOFLORES, INDERENA, and universities. In addition to the 456 agricultural researchers working at ICA and CENICAFE in 1984, source 55 reports another 215 agricultural researchers working at the institutes mentioned above.

Expenditure Comments:
The table below gives an overview of how the expenditure figures have been constructed, plus some scattered observations for institutes which have not been included in the indicator series.

Acronyms and Abbreviations:
ASOCOFLORES: Asociación Colombiana de Productores de Flores
CENICAFE: Nacional de Investigaciones del Café
CENICANA: Centro Nacional de Investigaciones de la Caña
CONIF: Corporación Nacional de Investigación y Fomento Forestal
CARC: Corporación Autónoma Regional del Cauca
ICA: Instituto Colombiano Agropecuario
INDERENA: Instituto Nacional de los Recursos Naturales Renovables y del Ambiente

	1970	1971	1972	1973	1974	1975
ICA	142.900	152.600	155.800	176.700	182.200	242.300
CENICAFE	17.863*	21.252*	27.619*	34.625*	45.954*	56.791*
Subtotal	160.763	173.852	183.419	211.325	228.154	299.091
INDERENA				9.047	9.481	9.503
CONIF						3.053
CARC						1.928
Universities			4.576	6.776	7.430	10.812
Source	55	55	55	55	55	55
	729*	729*	729*	729*	729*	729*

	1976	1977	1978	1979	1980	1981
ICA	265.600	330.000	463.300	558.000	673.800	924.100
CENICAFE	69.779*	91.174*	111.002*	149.970*	204.178*	287.983*
Subtotal	335.379	421.174	574.302	707.970	877.978	1212.083
INDERENA	9.023				211.600	
CONIF	2.813				10.800	
CARC	3.136				6.000	
Universities	13.401				14.000	
CENICANA					55.100	
ASOCOFLORES					20.000	
Source	55	55	55	55	55	55
	729*	729*	729*	729*	729*	729*

	1982	1983	1984	1985	1986
ICA	1059.000	1847.991	1710.500	3276.100	3836.100
CENICAFE	313.507*	[505.671]	[697.836]	890.000*	1345.000*
Subtotal	1372.507	2353.662	2408.336	4166.100	5181.000
Source	55	17	736	736	736
	729*	730*	730*		

[ ] = estimated or constructed by authors.

Sources:

0010 Boyce, James K., and Robert E. Evenson. *National and International Agricultural Research and Extension Programs.* New York: Agricultural Development Council, Inc., 1975.

0017 ISNAR, IFARD & AOAD. Survey of National Agricultural Research Systems: Unpublished Questionnaire Responses. ISNAR, The Hague, 1985.

0018 Daniels, Douglas, and Barry Nestel, eds. *Resource Allocation to Agricultural Research.* Proceedings of a workshop held in Singapore 8–10 June 1981. Ottawa: IDRC, 1981.

0030 Ardila V, Jorge. "Colombia: Análisis de la Cooperación entre los Centros Nacionales y los Centros Internacionales de Investigación Agropecuaria." BID, OWA & IICA, 1980. Mimeo.

0055 Weersma-Haworth, Teresa. "Colombian Case Study." CGIAR Impact Study. CGIAR, Washington, D.C., October 1984. Mimeo.

0091 Instituto Colombiano Agropecuario (ICA). *Plan Nacional de Investigación Agropecuaria del ICA.* Tomo 1. Colombia: ICA, January 1981.

0729 Federación Nacional de Cafeteros de Colombia. *Informe del Gerente General al XLI Congreso Nacional de Cafeteros.* Anexo 2. Bogotá: Federación Nacional de Cafeteros de Colombia, November 1982.

0730 Federación Nacional de Cafeteros de Colombia. *Informe del Gerente General al XLIV Congreso Nacional de Cafeteros.* Anexo. Bogotá: Federación Nacional de Cafeteros de Colombia, November 1986.

0735 Gómez Moncayo, Fernando. *Instituto Colombiano Agropecuario – Informe de Gerencia 1985.* Bogotá: Instituto Colombiano Agropecuario, n.d.

0736 Montes Llamas, Gabriel. *Instituto Colombiano Agropecuario – Informe de Gerencia 1986.* Bogotá: Instituto Colombiano Agropecuario, n.d.

Additional References:

0001 Trigo, Eduardo J., and Martín E. Piñeiro. "Funding Agricultural Research." In *Report of a Conference: Selected Issues in Agricultural Research in Latin America,* edited by Barry Nestel and Eduardo J. Trigo. The Hague: ISNAR, March 1984.

132

0014 Judd, M. Ann, James K. Boyce, and Robert E. Evenson. "Investing in Agricultural Supply." Economic Growth Center, Yale University, New Haven, Connecticut, 1983. Mimeo.

0016 Oram, Peter A., and Vishva Bindlish. *Resource Allocations to National Agricultural Research: Trends in the 1970s*. The Hague and Washington, D.C.: ISNAR and IFPRI, November 1981.

0026 Oram, Peter A., and Mark Gieben. "Document Summaries." ISNAR, The Hague, 1984. Mimeo.

0027 Harvey, Nigel, ed. *Agricultural Research Centres: A World Directory of Organizations and Programmes*. Seventh Edition, Two Volumes. Harlow, U.K.: Longman, 1983.

0053 Piñeiro, Martín, and Eduardo Trigo, eds. *Cambio Técnico en el Agro Latino Americano – Situación y Perspectivas en la Década de 1980*. San José: IICA, 1983.

0073 Oram, Peter A., and Vishva Bindlish. "Investment in Agricultural Research in Developing Countries: Progress, Problems, and the Determination of Priorities." IFPRI, Washington, D.C., January 1984. Mimeo.

0080 Trigo, Eduardo, Martín Piñeiro, and Jorge Ardila. *Organización de la Investigación Agropecuaria en América Latina*. San José: IICA, 1982.

0095 FAO – CARIS. *Agricultural Research in Developing Countries – Volume 1: Research Institutions*. Rome: FAO – CARIS, 1978.

0120 Chaparro, Fernando, et al. "Research Priorities and Resource Allocation in Agriculture: The Case of Colombia." Paper presented at the International Workshop on Resources Allocation in National Agricultural Research Systems, IDRC-IFARD, Singapore, 8–10 June 1981.

0121 Ardila V, Jorge, "Organización Institucional para la Investigación Agropecuaria en Colombia, en Relación al Plan Nacional de Investigaciones Agropecuarias." ICA, Cali, April 1982. Mimeo.

0163 CGIAR. "National Agricultural Research." CGIAR, Washington, D.C., 1985. Mimeo.

0165 Evenson, R.E., and Y. Kislev. *Agricultural Research and Productivity*. New Haven: Yale University Press, 1975.

0172 Trigo, Eduardo. Background material. ISNAR. The Hague, n.d.

0194 Ardila, Jorge, and Eduardo Trigo, et al. *Sistemas Nacionales de Investigación Agropecuaria en América Latina: Análisis Comparativo de los Recursos Humanos en Países Seleccionados*. Colombia: IICA – Oficina en Colombia, February 1980.

0195 Instituto Colombiano Agropecuario (ICA). "Informe de Gerencia 1982." ICA, Colombia, n.d. Mimeo.

0298 Chaparro, Fernando, et al. "Research Priorities and Resource Allocation in Agriculture: The Case of Colombia." In *Resource Allocation to Agricultural Research*. Proceedings of a workshop held in Singapore 8–10 June 1981, edited by Douglas Daniels and Barry Nestel, pp. 68–96. Ottawa: IDRC, 1981.

0305 Ardila, Jorge, Eduardo Trigo, and Martín Piñeiro. "Human Resources in Agricultural Research: Three Cases in Latin America." In *Resource Allocation to Agricultural Research*. Proceedings of a Workshop held in Singapore 8–10 June 1981, edited by Douglas Daniels and Barry Nestel, pp. 151–167. Ottawa: IDRC, 1981.

0733 Bernal C., Fernando. "Análisis de la Estructura de Generación y Tranferencia de Tecnología: La Subgerencia de Investigación y Transferencia del ICA." Bogotá, October 1987. Mimeo.

0734 Alarcón Millan, Enrique. *El Modelo Institucional para la Investigación Agropecuaria: Problemática y Planteamiento para su Cambio*. Bogotá: Instituto Colombiano Agropecuario – Subgerencia de Investigación y Transferencia, May 1986.

0744 *Agricultural Research Centres: A World Directory of Organizations and Programmes*. Eighth Edition, Two Volumes. Harlow, U.K.: Longman, 1986.

0863 Lindarte, Eduardo. "Proyecto Asignación de Recursos para Investigación Agrícola en América Latina (ARIAL) – Colombia: Estudio de Caso." Unidad de Apoyo a la Programación – PSU, March 1979. Mimeo.

0886 Evenson, Robert E., and Yoav Kislev. "Investment in Agricultural Research and Extension: A Survey of International Data." *Economic Development and Cultural Change* Vol. 23 (April 1975): 507–521.

Comoros

Personnel

Period/Year	PhD	MSc	BSc	Subtotal	Expat	Total	Source
1960–64	..	..	..	..	..	..	
1965–69	..	..	..	..	..	..	
1970	..	..	..	..	..	..	
1971	..	..	..	..	..	..	
1972	..	..	..	..	..	..	
1973	..	..	..	..	..	..	
1974	..	..	..	..	..	..	
1975	..	..	..	..	..	..	
1976	..	..	..	..	..	..	
1977	..	..	..	..	..	..	
1978	..	..	..	..	..	..	
1979	..	..	..	..	..	..	
1980	..	..	..	..	..	..	
1981	..	..	..	..	..	..	
1982	..	..	..	..	..	..	
1983	..	..	..	..	..	..	
1984	..	..	..	..	..	..	
1985	0	0	7	7	7	14	445
1986	..	..	..	..	..		

1960–64:   1965–69:

Expenditure

LCU = Comorian Francs

Period/Year	Current LCU (millions)	Constant 1980 LCU(millions)	Constant 1980 US$ (millions)		Source
			Atlas	PPP	
1960–64	..	..	..	..	
1965–69	..	..	..	..	
1970	..	..	..	..	
1971	..	..	..	..	
1972	..	..	..	..	
1973	..	..	..	..	
1974	..	..	..	..	
1975	..	..	..	..	
1976	..	..	..	..	
1977	..	..	..	..	
1978	..	..	..	..	
1979	..	..	..	..	
1980	..	..	..	..	
1981	..	..	..	..	
1982	..	..	..	..	
1983	..	..	..	..	
1984	..	..	..	..	
1985	..	..	..	..	
1986	..	..	..	..	

1960–64:   1965–69:

134

Personnel Comments:
Agricultural Research in Comoros is conducted under the authority of the Research and Development Unit of the Federal Center of Rural Development (CEFADER). The personnel figure for 1985 includes, in addition to the Research and Development Unit of CEFADER, 1 researcher who works for IRHO.

Acronyms and Abbreviations:
CEFADER: Centre Fédéral d'Appui au Développement Rural
IRHO: Institut de Recherches pour les Huiles et Oléagineux

Sources:
0445 Swanson, Burton E., and Wade H. Reeves. "Agricultural Research Eastern and Southern Africa: Manpower and Training." World Bank, Washington, D.C., August 1986. Mimeo.

Additional References:
0027 Harvey, Nigel, ed. *Agricultural Research Centres: A World Directory of Organizations and Programmes.* Seventh Edition, Two Volumes. Harlow, U.K.: Longman, 1983.
0122 World Bank – Eastern Africa Regional Office. *The Comoros: Problems and Prospects of a Small, Island Economy.* Washington, D.C.: World Bank, July 1979.

Congo

Personnel

Period/Year	PhD	MSc	BSc	Subtotal	Expat	Total	Source
1960–64	..	..	..	..		..	
1965–69	..	..	..	..	..		
1970	..	..	..	..	..	27	532
1971	..	..	..	0	24	24	589
1972	..	..	..	..			
1973	..	..	..	..	..	..	
1974	..	..	..	..		..	
1975	..	..	..	..			
1976	..	..	..	..		39	95/856
1977	..	..	..	..		..	
1978	..	..	..	..		..	
1979	..	..	..	..		..	
1980	..	..	..	..		..	
1981	..	..	..	..		..	
1982	..	..	..	..	..		
1983	..	..	..	..	..	72	68/856
1984	..	..	..	40	34	74	68/856
1985	..	..	..	..			
1986	..	..	..	..		..	

1960–64:   1965–69:

Expenditure

LCU = CFA Francs

Period/Year	Current LCU (millions)	Constant 1980 LCU(millions)	Constant 1980 US$ (millions)		Source
			Atlas	PPP	
1960–64	..	..	..	..	
1965–69	100.641	371.369	1.563	1.622	589
1970	..	..	..		
1971	116.923	335.023	1.410	1.463	589
1972	..	..	..		
1973	..	..	..		
1974	..	..	..		
1975	..	..	..		
1976	..	..	..		
1977	..	..	..		
1978	..	..	..		
1979	..	..	..		
1980	..	..	..		
1981	..	..	..		
1982	..	..	..		
1983	..	..	..		
1984	956.000	571.087	2.404	2.494	86/17/856
1985	..	..	..		
1986	..	..	..		

1960–64:   1965–69: 1966

Personnel Comments:

According to source 86, following independence in 1963 all agricultural research was abandoned except for some effort in support of development activities, ORSTOM's basic research in soils, phytopathology, and entomology, and a small CTFT unit. It was not until the mid-seventies that the Congolese government allocated some funds to agricultural research. With the financial support of France it was possible to reactivate the old research station at Loudina in 1976. Since then, however, financial support by the Congelese government has been very limited.

At the same time CRVZ (formerly Laboratoire Vétérinaire Sanitaire) was established with the help of the Soviet Union.

The table below gives details of how the personnel figures have been constructed.

	1967	1970	1971	1974	1976	1983	1984	1985
ORSTOM[a]	25	25	22	NA	16	27	30	NA
CTFT	NA	2	[2]	2	5	12	10	NA
CRAL	–	–	–	–	5	[9]	[10]	10
CRVZ	–	–	–	–	13*	24*	[24]*	NA
Total	NA	27	24	NA	39	72	74	NA
Source	654	532	589	589	95 856*	68 856*	86 856*	68

[ ] = estimated or constructed by authors.
a. ORSTOM-Brazzaville only. ORSTOM-Pointe Noire not included because it is involved in oceanography research only.

Expenditure Comments:

The table below gives an overview of how the expenditure indicators have been constructed.

	1966	1971	1973	1977	1979	1980	1981	1982	1983	1984	1985
ORSTOM[a]	78.803	79.330	NA	NA	NA	NA	NA	NA	NA	800.000	710.000
CTFT	21.838	37.593	25.000	80.000	63.000	90.000	125.000	142.000	NA	120.000*	NA
CRAL	–	–	–	NA	NA	NA	NA	NA	10.000	[8.000]	4.600*
CRVZ	–	–	–	NA	8.273*	11.266*	11.406*	25.800*	27.350*	[28.000]**	NA
Total	100.641	116.923	NA	NA	NA	NA	NA	NA	NA	956.000	NA
Source	589	589	589	68	68 856*	68 856*	68 856*	68 856*	86 856*	86 17* 856**	656 86*

[ ] = estimated or constructed by authors.
a. ORSTOM-Brazzaville only. ORSTOM-Pointe Noire not included because it is involved in oceanography only.

Acronyms and Abbreviations:

CRAL: Centre de Recherche Agronomique de Loudima
CRVZ: Centre de Recherche Vétérinaire et Zootechnique
CTFT: Centre Technique Forestier Tropical
ORSTOM: Office de la Recherche Scientifique et Technique Outre-Mer

Sources:

0017 ISNAR, IFARD & AOAD. Survey of National Agricultural Research Systems: Unpublished Questionnaire Responses. ISNAR, The Hague, 1985.
0068 Conte, Stephan. "West Africa Agricultural Research Data Base Building Task: Termination

Report." World Bank, Washington, D.C., 1985. Mimeo.

0086 World Bank. "West Africa Agricultural Research Review – Country Studies." World Bank, Washington, D.C., 1985. Mimeo.

0095 FAO – CARIS. *Agricultural Research in Developing Countries – Volume 1: Research Institutions.* Rome: FAO – CARIS, 1978.

0532 UNESCO Field Science Office for Africa. *Survey on the Scientific and Technical Potential of the Countries of Africa.* Paris: UNESCO, 1970.

0589 Kassapu, Samuel. *Les Dépenses de Recherche Agricole dans 34 Pays d'Afrique Tropicale.* Paris: Centre de Développement de l'OCDE, 1976.

0856 Doulou, V. Personal communication. Directeur des Activités Scientifiques et Technologiques, Direction Générale de la Recherche Scientifique et Technique, Ministère de la Recherche Scientifique et de l'Environnement. Brazzaville, Congo, April 1988.

Additional References:

0010 Boyce, James K., and Robert E. Evenson. National and International Agricultural Research and Extension Programs. New York: Agricultural Development Council, Inc., 1975.

0023 Bennell, Paul. *Agricultural Researchers in Sub-Saharan Africa: An Overview.* Working Paper No. 4. The Hague: ISNAR, October 1985.

0027 Harvey, Nigel, ed. *Agricultural Research Centres: A World Directory of Organizations and Programmes.* Seventh Edition, Two Volumes. Harlow, U.K.: Longman, 1983.

0088 Swanson, Burton E., and Wade H. Reeves. "Agricultural Research West Africa: Manpower and Training." World Bank, Washington, D.C., November 1985. Mimeo.

0089 Oram, Peter. "Report on National Agricultural Research in West Africa." IFPRI, Washington, D.C., February 1986. Mimeo.

0165 Evenson, R.E., and Y. Kislev. *Agricultural Research and Productivity.* New Haven: Yale University Press, 1975.

0175 Cooper, St.G.C. *Agricultural Research in Tropical Africa.* Kampala: East African Literature Bureau, 1970.

0266 UNESCO. *National Science Policies in Africa.* Science Policy Studies and Documents No. 31. Paris: UNESCO, 1974.

0360 Cooper, St.G.C. "Towards Trained Manpower for Agricultural Research in Africa." Paper presented at the Conference on Agricultural Research and Production in Africa, organized by the Association for the Advancement of Agricultural Sciences in Africa (AAASA), Addis Ababa, 29 August–4 September 1971.

0373 Groupement d'Etudes et de Recherches pour le Développement de l'Agronomie Tropicale (GERDAT). *Rapport Général d'Activité pour 1982.* Paris: GERDAT, 1983.

0400 UNESCO. *The Promotion of Scientific Activity in Tropical Africa.* Science Policy Studies and Documents No. 11. Paris: UNESCO, 1969.

0466 World Bank – West Africa Agricultural Research Review Team. "National Agricultural Research Systems (NARS): A Regional Appraisal and Selected Issues." World Bank, Washington, D.C., August 1986. Mimeo.

0653 Webster, B.N. *Index of Agricultural Research Institutions and Stations in Africa.* Rome: FAO, n.d.

0654 Office de la Recherche Scientifique et Technique Outre-Mer (ORSTOM). *Rapport d'Activité 1967.* Paris: ORSTOM, 1969.

0656 Office de la Recherche Scientifique et Technique Outre-Mer (ORSTOM) *Rapport d'Activité 1985.* Paris: ORSTOM, n.d.

0852 Evenson, Robert E., and Yoav Kislev. *Investment in Agricultural Research and Extension: A Survey of International Data.* Center Discussion Paper No. 124. New Haven, Connecticut: Economic Growth Center, Yale University, August 1971.

0886 Evenson, Robert E., and Yoav Kislev. "Investment in Agricultural Research and Extension: A Survey of International Data." *Economic Development and Cultural Change* Vol. 23 (April 1975): 507–521.

Cook Islands

Personnel

Period/Year	PhD	MSc	BSc	Subtotal	Expat	Total	Source
1960–64	..	..	..	..	..	..	
1965–69	..	..	..	..	..	..	
1970	..	..	..	..	..	..	
1971	..	..	..	..	..	..	
1972	..	..	..	..	..	..	
1973	..	..	..	..	..	..	
1974	..	..	..	..	..	..	
1975	..	..	..	..	..	..	
1976	..	..	..	..	..	..	
1977	..	..	..	..	..	..	
1978	..	..	..	..	..	..	
1979	..	..	..	..	..	..	
1980	..	..	..	1	3	4	38
1981	..	..	..	..	..	..	
1982	..	..	..	..	..	..	
1983	..	..	..	..	..	4	723
1984	..	..	..	..	..	3	723
1985	..	..	..	..	..	5	723
1986	0	1	1	2	2	4	723

1960–64:   1965–69:

Expenditure

LCU = New Zealand Dollars

Period/Year	Current LCU (millions)	Constant 1980 LCU (millions)	Constant 1980 US$ (millions)		Source
			Atlas	PPP	
1960–64	..	..	..	..	
1965–69	..	..	..	..	
1970	..	..	..	..	
1971	..	..	..	..	
1972	..	..	..	..	
1973	..	..	..	..	
1974	..	..	..	..	
1975	..	..	..	..	
1976	..	..	..	..	
1977	..	..	..	..	
1978	..	..	..	..	
1979	..	..	..	..	
1980	0.214	0.214	0.208	0.295	38
1981	..	..	..	..	
1982	..	..	..	..	
1983	0.212	0.149	0.144	0.205	723
1984	0.171	0.113	0.110	0.156	723
1985	0.209	0.121	0.118	0.167	723
1986	0.301	..	..	..	723

1960–64:   1965–69:

Personnel Comments:

Agricultural research on the Cook Islands is performed mainly by the New Zealand Department of Scientific and Industrial Research. The number of researchers presented here for 1980, in standard scientist years, was derived from a descriptive overview by source 38 of agricultural research on the Cook Islands. Most of the research is conducted at the Totokoitu Research Station on Rarotonga Island or on farmers' land on that island.

Expenditure Comments:

New Zealand plays a dominant role in agricultural research on the Cook Islands. Source 38 indicates that more than 75% of the estimated expenditures are funded by New Zealand.

Sources:

0038 Gamble, W.K., R.M. Bourke, and C.W. Brookson. *South Pacific Agricultural Research Study.* Consultants' Report to the Asian Development Bank (two volumes). Manila: Asian Development Bank, June 1981.

0723 Fernando, L.H. "South Pacific NARS Resource Commitment Survey." ADB and ISNAR, The Hague, 1987. Mimeo.

Additional References:

0771 Fernando, L.H. "ISNAR/IRETA Workshop on Planning and Management of Agricultural Research in the South Pacific: Survey of Agricultural Research Resources." ISNAR, The Hague, n.d. Mimeo.

Costa Rica

Personnel

Period/Year	PhD	MSc	BSc	Subtotal	Expat	Total	Source
1960–64	..	..	..	..	..	44	10
1965–69	..	..	..	..	..	54	10
1970	..	..	..	..	..	43	10
1971	..	..	..	..	..	55	10
1972	..	..	..	..	..	59	10
1973	..	..	..	..	..	60	10
1974	..	..	..	..	..	65	10
1975	..	..	..	..	..	..	
1976	..	..	..	..	..	..	
1977	..	..	..	..	..	..	
1978	..	..	..	..	..	90	124
1979	..	..	..	..	..	..	
1980	..	..	..	..	..	..	
1981	..	..	..	..	..	114	31
1982	..	..	..	..	..	..	
1983	..	..	..	..	..	..	
1984	..	..	..	..	..	..	
1985	..	..	..	..	..	..	
1986	..	..	..	..	..	..	

1960–64: 1960 & 62   1965–69: 1965

Expenditure

LCU = Costa Rica Colones

Period/Year	Current LCU (millions)	Constant 1980 LCU(millions)	Constant 1980 US$ (millions)		Source
			Atlas	PPP	
1960–64	..	..	..	..	
1965–69	..	..	..	..	
1970	..	..	..	..	
1971	..	..	..	..	
1972	..	..	..	..	
1973	..	..	..	..	
1974	..	..	..	..	
1975	..	..	..	..	
1976	16.347	26.755	2.940	4.396	208
1977	..	..	..	..	
1978	..	..	..	..	
1979	21.196	25.173	2.766	4.137	31
1980	26.268	26.268	2.886	4.316	31
1981	30.679	21.743	2.389	3.573	31
1982	..	..	..	..	
1983	..	..	..	..	
1984	47.200	12.072	1.327	1.984	713
1985	..	..	..	..	
1986	..	..	..	..	

1960–64:   1965–69:

Personnel Comments:
The figures refer to the Ministry of Agriculture and Livestock (MAG). Within the ministry there is a special department for agricultural research ('Dirección General de Investigaciones Agrícolas' – DIA). This department conducts mainly crop research. Livestock research and forestry research is done by research divisions of the Livestock Department and the Forestry Department of the Ministry of Agriculture and Livestock. In many sources only DIA is identified as doing research at the ministry. However, the indicator series also includes researchers working at the research divisions of the various departments within MAG. Source 31 indicates that a considerable amount of agricultural research is also done by organizations other than the Ministry of Agriculture and Livestock. The exact number of researchers working for these organizations is not available.

	1974	1975	1976	1977	1978	1980	1981	1983
DIA	NA	56	64	59	60	75	91	82
GANADERIA	NA	NA	NA	NA	NA	8	8	NA
SANIDAD ANIMAL	NA	NA	NA	NA	NA	NA	2	NA
FORESTAL	NA	NA	NA	NA	NA	NA	3	NA
PESCA Y FAUNA	NA	NA	NA	NA	NA	NA	3	NA
AUTROS	NA	NA	NA	NA	NA	NA	7	NA
TOTAL MAG[a]	65	NA	NA	NA	90	NA	114	NA
Source	10	127	127	127	124	53	31	17

a. Source 10 and 124 do not give a detailed breakdown of the total MAG figure.

Expenditure Comments:
In many publications only DIA is identified as doing research within the Ministry of Agriculture and Livestock (see, for example, source 1). Hence, research expenditures have generally been underestimated. The figures presented here cover all agricultural research conducted by the Ministry of Agriculture and Livestock. There are, however, at least 10 other organizations in Costa Rica that to some extent are also involved in agricultural research (see, for example, source 31). Expenditure figures on agricultural research by these organizations are scarce and are therefore not included in the figures presented here. To give some idea of the importance of these organizations, in 1976 total agricultural research expenditures were estimated (source 208) at C 29,775,502, of which 55% was spent by the Ministry of Agriculture and Livestock, 19% by 'Asociacion Nacional Bananera', 17% by the University, and 9% by a group of small organizations.

Acronyms and Abbreviations:
DIA: Dirección General de Investigaciones Agrícolas
MAG: Ministerio de Agricultura y Ganadería

Sources:
0010 Boyce, James K., and Robert E. Evenson. *National and International Agricultural Research and Extension Programs.* New York: Agricultural Development Council, Inc., 1975.
0031 ISNAR. *Informe al Gobierno de Costa Rica: El Sistema de Investigación Agropecuaria y Transferencia de Tecnología en Costa Rica.* The Hague: ISNAR, June 1981.
0124 Consejo Nacional de Investigaciones Científicas y Tecnológicas (CNICT). *Un Analisis del Desarrollo Científico Tecnológico del Sector Agropecuario de Costa Rica.* Volumes 1, 2 and 3. San José: CNICT, 1980.
0208 Segura, M. "Diagnóstico de la Investigación Agropecuaria en el Istmo Centroamericano." IICA, Guatemala, 1979. Mimeo.
0713 Carranza, Germán. "Sector Agropecuario de Costa Rica." Ministerio de Agricultura y Ganadería, San José, Costa Rica, March 1987. Mimeo.

142

Additional References:

0001 Trigo, Eduardo J., and Martín E. Piñeiro. "Funding Agricultural Research." In *Report of a Conference: Selected Issues in Agricultural Research in Latin America*, edited by Barry Nestel and Eduardo J. Trigo. The Hague: ISNAR, March 1984.

0014 Judd, M. Ann, James K. Boyce, and Robert E. Evenson. "Investing in Agricultural Supply." Economic Growth Center, Yale University, New Haven, Connecticut, 1983. Mimeo.

0016 Oram, Peter A., and Vishva Bindlish. *Resource Allocations to National Agricultural Research: Trends in the 1970s.* The Hague and Washington, D.C.: ISNAR and IFPRI, November 1981.

0017 ISNAR, IFARD & AOAD. Survey of National Agricultural Research Systems: Unpublished Questionnaire Responses. ISNAR, The Hague, 1985.

0026 Oram, Peter A., and Mark Gieben. "Document Summaries." ISNAR, The Hague, 1984. Mimeo.

0027 Harvey, Nigel, ed. *Agricultural Research Centres: A World Directory of Organizations and Programmes.* Seventh Edition, Two Volumes. Harlow, U.K.: Longman, 1983.

0053 Piñeiro, Martín, and Eduardo Trigo, eds. *Cambio Técnico en el Agro Latino Americano – Situación y Perspectivas en la Década de 1980.* San José: IICA, 1983.

0073 Oram, Peter A., and Vishva Bindlish. "Investment in Agricultural Research in Developing Countries: Progress, Problems, and the Determination of Priorities." IFPRI, Washington, D.C., January 1984. Mimeo.

0095 FAO – CARIS. *Agricultural Research in Developing Countries – Volume 1: Research Institutions.* Rome: FAO – CARIS, 1978.

0098 Stewart, Rigoberto. *Costa Rica and the CGIAR Centers: A Study of Their Collaboration in Agricultural Research.* CGIAR Study Paper Number 4. Washington, D.C.: World Bank, 1985.

0123 World Bank. "Agricultural Research and Farmer Advisory Services in Central America and Panama." World Bank, Washington, D.C., 1979. Mimeo.

0125 Ministerio de Agricultura y Ganadería – Dirección General de Investigaciones Agrícolas. *Memoria Anual de la Investigación Agrícola 1983.* San José: Ministerio de Agricultura y Ganadería, 1984.

0126 IICA. "Reunión de Programación de Investigación Agrícola del Istmo Centro-Americano." IICA, Guatemala, 1979. Mimeo.

0127 Secretaría Ejecutiva de Planificación Sectorial Agropecuaria (SEPSA). "Proyecto de Investigación y Transferencia de Tecnología: Información Preparada para la Misión FAO-Banco Mundial." SEPSA, San José, 1981. Mimeo.

0163 CGIAR. "National Agricultural Research." CGIAR, Washington, D.C., 1985. Mimeo.

0165 Evenson, R.E., and Y. Kislev. *Agricultural Research and Productivity.* New Haven: Yale University Press, 1975.

0172 Trigo, Eduardo. Background material. ISNAR. The Hague, n.d.

0207 Rosales, Franklin E. "Situación del Sistema Nacional de Investigación Agronómica en Costa Rica." IICA – OEA, San José, December 1980. Mimeo.

0322 Barquero, Vargas A. "Costa Rica – Some Facts and Figures about Agriculture and Agricultural Research." In *Strengthening National Agricultural Research.* Report from a SAREC workshop, 10–17 September 1979, edited by Bo Bengtsson and Getachew Tedla, pp. 155–166. Sweden: SAREC, 1980.

0380 Alfaro, R. "La Investigación Agrícola en el Ministerio de Agricultura y Ganadería de Costa Rica." Paper presented at Curso-Taller sobre la Administración de la Investigación Agrícola, Panama City, 14–25 July 1986.

0744 *Agricultural Research Centres: A World Directory of Organizations and Programmes.* Eighth Edition, Two Volumes. Harlow, U.K.: Longman, 1986.

0775 Hernández López, Jesus. Personal Communication. Director General Investigación y Extensión Agrícola, Ministerio de Agricultura y Ganadería. San José, Costa Rica, January 1988.

0852 Evenson, Robert E., and Yoav Kislev. *Investment in Agricultural Research and Extension: A Survey of International Data.* Center Discussion Paper No. 124. New Haven, Connecticut: Economic Growth Center, Yale University, August 1971.

0886 Evenson, Robert E., and Yoav Kislev. "Investment in Agricultural Research and Extension: A Survey of International Data." *Economic Development and Cultural Change* Vol. 23 (April 1975): 507–521.

Cote D'Ivoire

Personnel

Period/Year	PhD	MSc	BSc	Subtotal	Expat	Total	Source
1960–64	..	..	..	..	..	..	
1965–69	..	..	..	..	..	85	653
1970	..	..	..	..	..	141	532
1971	..	..	..	..	..	..	
1972	..	..	..	..	..	179	431
1973	..	..	..	..	..	..	
1974	..	..	..	..	..	180	589
1975	..	..	..	..	..	..	
1976	..	..	..	..	..	..	
1977	..	..	..	..	..	..	
1978	..	..	..	..	..	..	
1979	..	..	..	..	..	..	
1980	..	..	..	..	..	..	
1981	..	..	..	36	151	187	86
1982	..	..	..	43	150	193	86
1983	..	..	..	53	150	203	86
1984	..	..	..	65	140	205	86
1985	..	..	..	71	145	216	86/774
1986	..	..	..	..		..	

1960–64: 1965–69: 1966

Expenditure *LCU = CFA Francs*

Period/Year	Current LCU (millions)	Constant 1980 LCU (millions)	Constant 1980 US$ (millions)		Source
			Atlas	PPP	
1960–64	..	..	..	..	
1965–69	..	..	..	..	
1970	..	..	..	..	
1971	..	..	..	..	
1972	1942.377	5640.364	28.239	23.432	431
1973	..	..	..	..	
1974	..	..	..	..	
1975	..	..	..	..	
1976	..	..	..	..	
1977	..	..	..	..	
1978	..	..	..	..	
1979	..	..	..	..	
1980	..	..	..	..	
1981	6906.600	6844.995	34.271	28.436	86
1982	7512.530	7073.945	35.417	29.387	86
1983	8304.390	7362.048	36.859	30.584	86
1984	8322.130	6426.355	32.175	26.697	86
1985	8633.930	6390.770	31.997	26.549	86
1986	..	..	..	..	

1960–64: 1965–69:

Personnel Comments:

The personnel indicator series has been constructed as follows:

	1966	1970	1972	1974	1981	1982	1983	1984	1985
IRAT	10	8	14		–	–	–	–	–
IRCT	7	6	10		–	–	–	–	–
IEMVT	3	2	5		–	–	–	–	–
IDESSA[a]	–	–	–	104[b]	49	53	51	50	51
IRCA	6	10	10		16	17	17	16	19
CTFT	NA	7	7		7	9	9	8	13
IRHO	19	20	19		22	23	23	24	25
IRFA(IFAC)	7	9	11		16	13	14	15	16
IRCC(IFCC)	8	10	25		18	18	19	19	22
ORSTOM	25	69	78	[76]	59	60	70	73	70*
Subtotal	85	141	179	180	187	193	203	205	216
CSRS		3	3		3*				
CIRT (ITIPAT)		5	7		4*			11*	11*
CRO		19			26**	28*	29*	26*	32*
DPML		3							
CIRES					9*			8	
ENSA								11	
Sources	653	532	431	589	86/12* 774**	86 774*	86 774*	86 774*	86 774*

[ ] = estimated or constructed by authors.
a. IRAT, IRCT, and IEMVT merged into IDESSA.
b. Comprises all GERDAT institutes.

Expenditure Comments:

France is still heavily involved in agricultural research in Côte d'Ivoire, by means of both human resources and funding. In addition to performing research, some research institutes are also involved in agricultural production, the sale of which is used to finance part of their research. An exceptional case in this matter is IRHO, for which sales of produce represent more than 85% of total financial resources and expenditures

	1972	1981	1982	1983	1984	1985
CTFT	72.956	275.080	340.150	379.000	376.210	372.290
IDESSA	349.694[a]	1332.180	1476.750	1670.240	1686.480	1713.100
IRCA	170.950	380.250	471.040	509.510	582.760	537.670
IRCC(IFCC)	256.688	1094.140	1099.020	1050.600	1198.390	1045.200
IRFA(IFAC)	179.892	488.440	516.990	561.170	490.510	508.900
IRHO	186.697	756.810	769.480	978.570	787.780	1156.770
ORSTOM	[725.500]	2579.700	2839.100	3155.300	[3200.000]	[3300.000]
Total	1942.377	6906.600	7512.530	8304.390	8322.130	8633.930
Source	431	86	86	86	86	86

[ ] = estimated or constructed by authors.
a. In 1972 the IDESSA figure comprises IRAT, IRCT, and IEMVT.

(source 86). Including this large figure would seriously bias upwards a total research expenditure estimate. For example, if all revenue from sales were included in research funds available, then research expenditures per scientist for IRHO would be as high as US$ 675,000. To correct for this distortion, it was assumed, for all organizations, that 10% of sales of produce was used to fund agricultural research.

In the table above an institutional breakdown of the expenditure series is given.

Acronyms and Abbreviations:
CIRES: Centre Ivoirien de Recherche Economique et
    Sociale
CIRT: Centre Ivoirien de Recherches Technologiques
CRO: Centre de Recherches Océanographiques
CSRS: Centre Suisse de Recherches Scientifiques en
    Côte d'Ivoire
CTFT: Centre Technique Forestier Tropical
DPML: Direction des Pêches Maritimes et Lagunaires
ENSA: Ecole Nationale Supérieure Agronomique
GERDAT: Groupement d'Etudes et de Recherche pour le
    Développement de l'Agronomie Tropicale
IDESSA: Institut des Savannes
IEMVT: Institut d'Elevage et de Médecine Vétérinaire
    des Pays Tropicaux
IFAC: Institut Francais de Recherches Fruitières Outre-
    Mer
IFCC: Institut Francais du Café, du Cacao et Autres
    Plantes Stimulantes
IRAT: Institut de Recherches Agronomiques Tropicales
    et des Cultures Vivrières
IRCA: Institut de Recherche sur le Caoutchouc en
    Afrique
IRCC: Institut de Recherche du Café et du Cacao et
    autres plantes stimulantes
IRCT: Institut de Recherche du Coton et des Textiles
    Exotiques
IRFA: Institut de Recherche sur les Fruits et Agrumes
IRHO: Institut de Recherche sur les Huiles et les
    Oléagineux
ITIPAT: Institut pour la Technologie et
    l'Industrialisation des Produits Agricoles
    Tropicaux
ORSTOM: Office de Recherche Scientifique et
    Technique d'Outre-Mer

Sources:
0086 World Bank. "West Africa Agricultural Research Review – Country Studies." World Bank, Washington, D.C., 1985. Mimeo.

0431 Current Agricultural Research Information System (CARIS). *Directory of Agricultural Research Institutions and Projects in West Africa – Pilot Project 1972–73.* Rome: FAO, 1973.

0532 UNESCO Field Science Office for Africa. *Survey on the Scientific and Technical Potential of the Countries of Africa.* Paris: UNESCO, 1970.

0589 Kassapu, Samuel. *Les Dépenses de Recherche Agricole dans 34 Pays d'Afrique Tropicale.* Paris: Centre de Développement de l'OCDE, 1976.

0653 Webster, B.N. *Index of Agricultural Research Institutions and Stations in Africa.* Rome: FAO, n.d.

0774 Alassane Salif, N'Diaye. Personal communication. Le Ministre de la Recherche Scientifique, Ministère de la Recherche Scientifique. Abidjan, Côte d'Ivoire, January 1988.

Additional References:
0010 Boyce, James K., and Robert E. Evenson. *National and International Agricultural Research and Extension Programs.* New York: Agricultural Development Council, Inc., 1975.

0012 ISNAR. *Report to the Government of Ivory Coast: Agricultural Research in Ivory Coast.* The Hague: ISNAR, December 1984.

0023 Bennell, Paul. *Agricultural Researchers in Sub-Saharan Africa: An Overview.* Working Paper No. 4. The Hague: ISNAR, October 1985.

0026 Oram, Peter A., and Mark Gieben. "Document Summaries." ISNAR, The Hague, 1984. Mimeo.

0027 Harvey, Nigel, ed. *Agricultural Research Centres: A World Directory of Organizations and Programmes.* Seventh Edition, Two Volumes. Harlow, U.K.: Longman, 1983.

0068 Conte, Stephan. "West Africa Agricultural Research Data Base Building Task: Termination Report." World Bank, Washington, D.C., 1985. Mimeo.

0073 Oram, Peter A., and Vishva Bindlish. "Investment in Agricultural Research in Developing Countries: Progress, Problems, and the Determination of Priorities." IFPRI, Washington, D.C., January 1984. Mimeo.

0088 Swanson, Burton E., and Wade H. Reeves. "Agricultural Research West Africa: Manpower and Training." World Bank, Washington, D.C., November 1985. Mimeo.

0089 Oram, Peter. "Report on National Agricultural Research in West Africa." IFPRI, Washington, D.C., February 1986. Mimeo.

0146 Hocombe, S.D. "FAO/World Bank Cooperative Programme – Project Brief Cote d'Ivoire: National Agricultural Research Project." Report No.: 4/86 CP–IVC 12 PB. FAO and the World Bank, Rome/Washington, D.C., January 1986. Mimeo.

146

0147 Ellinger, K.R. "Ivory Coast Agricultural Research." World Bank – Regional Mission in Western Africa, Abidjan, 1982. Mimeo.

0148 Ministère de la Coopération et du Développement de France. "Procès-Verbal de la IXeme Commission Mixte Franco-Ivoirienne de Recherche." Ministère de la Coopération et du Développement de France, Paris, January 1982. Mimeo.

0149 Ministère du Plan. *Plan Quinquennial de Développement Economique, Social et Culturel (1981–1985): Politique Sectionnelle.* Abidjan: Ministère du Plan, 1981.

0150 ORSTOM. *Activités de l'ORSTOM en Cote d'Ivoire.* Abidjan: ORSTOM, 1980.

0161 Plessix, C.J. du, et al. "La Programmation de la Recherche Agronomique en Côte d'Ivoire." Extrait du Compte rendu du 10e Colloque de l'Institut International de la Potasse, Abidjan, Côte d'Ivoire, December 1973.

0169 ISNAR. *Agrotechnical Research in Ivory Coast.* The Hague: ISNAR, 1982.

0175 Cooper, St.G.C. *Agricultural Research in Tropical Africa.* Kampala: East African Literature Bureau, 1970.

0266 UNESCO. *National Science Policies in Africa.* Science Policy Studies and Documents No. 31. Paris: UNESCO, 1974.

0360 Cooper, St.G.C. "Towards Trained Manpower for Agricultural Research in Africa." Paper presented at the Conference on Agricultural Research and Production in Africa, organized by the Association for the Advancement of Agricultural Sciences in Africa (AAASA), Addis Ababa, 29 August–4 September 1971.

0373 Groupement d'Etudes et de Recherches pour le Développement de l'Agronomie Tropicale (GERDAT). *Rapport Général d'Activité pour 1982.* Paris: GERDAT, 1983.

0400 UNESCO. *The Promotion of Scientific Activity in Tropical Africa.* Science Policy Studies and Documents No. 11. Paris: UNESCO, 1969.

0466 World Bank – West Africa Agricultural Research Review Team. "National Agricultural Research Systems (NARS): A Regional Appraisal and Selected Issues." World Bank, Washington, D.C., August 1986. Mimeo.

0654 Office de la Recherche Scientifique et Technique Outre-Mer (ORSTOM). *Rapport d'Activité 1967.* Paris: ORSTOM, 1969.

0852 Evenson, Robert E., and Yoav Kislev. *Investment in Agricultural Research and Extension: A Survey of International Data.* Center Discussion Paper No. 124. New Haven, Connecticut: Economic Growth Center, Yale University, August 1971.

Cuba

Personnel

Period/Year	PhD	MSc	BSc	Subtotal	Expat	Total	Source
1960–64	..	..	..	..	..	..	
1965–69	..	..	..	..	..	274	10
1970	..	..	..	..	..		
1971	..	..	..	..	..		
1972	..	..	..	..	..		
1973	..	..	..	..	..		
1974	..	..	..	..	..		
1975	..	..	..	..	..		
1976	..	..	..	..	..		
1977	..	..	..	..	..		
1978	..	..	..	..	..	1257	22
1979	..	..	..	..	..		
1980	..	..	..	..	..		
1981	..	..	..	..	..		
1982	..	..	..	..	..		
1983	..	..	..	..	..		
1984	..	..	..	..	..	2191	334
1985	..	..	..	..	..	..	
1986	..	..	..	..	..		

1960–64:   1965–69: 1969

Expenditure   *LCU = Cuban Pesos*

Period/Year	Current LCU (millions)	Constant 1980 LCU(millions)	Constant 1980 US$ (millions)		Source
			Atlas	PPP	
1960–64	..	..	..	..	
1965–69	..	..	..	..	
1970	..	..	..	..	
1971	..	..	..	..	
1972	..	..	..	..	
1973	..	..	..	..	
1974	..	..	..	..	
1975	..	..	..	..	
1976	..	..	..	..	
1977	..	..	..	..	
1978	23.868	..	..	..	22
1979	..	..	..	..	
1980	..	..	..	..	
1981	..	..	..	..	
1982	..	..	..	..	
1983	45.000	..	..	..	65
1984	62.000	..	..	..	334
1985	..	..	..	..	
1986	..	..	..	..	

1960–64:   1965–69:

Personnel Comments:

The 1969 and the 1978 personnel figures are both derived from UNESCO figures. It is understood that the UNESCO figures more or less include the whole agricultural research system of Cuba, as does the 1984 figure. In the table below a breakdown of this figure has been given:

	1984
Ministry of Agriculture[a]	1053
Ministry of Sugar Industry[b]	415
Ministry of Higher Education[c]	393
Academy of Sciences[d]	330
Total	2191

a. Includes 17 institutes.
b. Includes ICINAZ, ICIDCA, & CEDIC.
c. Includes CENSA, ICA, INCA, & Estación Indio Hatuey.
d. Source 334 assumed that about a third of all research of the Academy is related to agriculture.

Expenditure Comments:

The expenditure indicators are more or less commensurable with the personnel figures.

The 1978 expenditure figure includes the total agricultural research expenditure in the productive sector (table 5.10 a, source 22) and higher education sector (table 5.10 b, source 22).

Acronyms and Abbreviations:

CEDIC: Centro para el Desarrollo Integrado de la Caña de Azúcar
CENSA: Centro Nacional de Sanidad Agropecuaria
ICA: Instituto de Ciencia Animal
ICIDCA: Instituto Cubano de Investigación en Derivados de la Caña
ICINAZ: Instituto Cubano de Investigaciones Azucareras
INCA: Instituto de Ciencia Agrícola

Sources:

0010 Boyce, James K., and Robert E. Evenson. *National and International Agricultural Research and Extension Programs.* New York: Agricultural Development Council, Inc., 1975.
0022 UNESCO. *Statistical Yearbook 1983.* Paris: UNESCO, 1983.
0065 Casas, Joseph. "Cuba: A Small Country, a Large Agricultural Research Potential." Paper presented at the Workshop on Agricultural Research Policy, Minneapolis, Minnesota, April 1985.
0334 Sánchez, Pedro A., and Grant M. Scobie. *Cuba and the CGIAR Centers: A Study of Their Collaboration in Agricultural Research.* CGIAR Study Paper No. 14. Washington, D.C.: World Bank, 1986.

Additional References:

0027 Harvey, Nigel, ed. *Agricultural Research Centres: A World Directory of Organizations and Programmes.* Seventh Edition, Two Volumes. Harlow, U.K.: Longman, 1983.
0116 Sussman, Jean. "Agricultural Research in Cuba." University of Minnesota, St. Paul, July 1982. Mimeo.
0163 CGIAR. "National Agricultural Research." CGIAR, Washington, D.C., 1985. Mimeo.
0378 Rodríguez, P. "Importancia de la Investigación Agrícola en la República de Cuba." Paper presented at Curso-Taller sobre la Administración de la Investigación Agrícola, Panama City, 14–25 July 1986.

Cyprus

Personnel

Period/Year	PhD	MSc	BSc	Subtotal	Expat	Total	Source
1960–64	..	..	..		..		
1965–69	..	..	..		..	24	10
1970	..	..	..		..	39	17
1971	..	..	..		..	42	17
1972	..	..	..		..	42	17
1973	..	..	..		..	45	17
1974	..	..	..		..	47	17
1975	..	..	..		..	47	17
1976	..	..	..		..	47	17
1977	..	..	..		..	47	17
1978	..	..	..		..	50	17
1979	..	..	..		..	54	17
1980	..	..	..		..	56	17
1981	..	..	..		..	57	17
1982	..	..	..		..	59	17
1983	15	26	16	57	..	57	17
1984	..	..	..		..	55	744
1985	..	..	..		..	..	
1986	..	..	..		..	..	

1960–64:   1965–69: 1967 & 69

Expenditure

LCU = Cyprus Pounds

Period/Year	Current LCU (millions)	Constant 1980 LCU (millions)	Constant 1980 US$ (millions)		Source
			Atlas	PPP	
1960–64	..	..	..	..	
1965–69	0.160	0.404	1.070	1.463*	10
1970	0.241	0.568	1.505	2.058	17
1971	0.273	0.629	1.666	2.277	17
1972	0.290	0.681	1.803	2.465	17
1973	0.350	0.707	1.873	2.560	17
1974	0.340	0.618	1.637	2.238	17
1975	0.350	0.611	1.618	2.211	17
1976	0.359	0.573	1.519	2.076	17
1977	0.505	0.730	1.933	2.642	17
1978	0.593	0.777	2.058	2.814	17
1979	0.685	0.783	2.073	2.834	17
1980	0.852	0.852	2.256	3.084	17
1981	0.990	0.888	2.352	3.215	17
1982	1.165	0.948	2.511	3.432	17
1983	1.305	1.007	2.667	3.646	17
1984	1.517	1.075	2.848	3.893	744
1985	..	..	..	..	
1986	..	..	..	..	

1960–64:   1965–69: 1967 & 69

150

Personnel Comments:
The 1983 figure includes the Agricultural Research Institute (ARI), and the Research Divisions of the Fisheries and Forestry Departments of the Ministry of Agriculture. All other figures given by source 17 are likely to include the same organizations. The organizational coverage of the 1984 figure is compatible with the earlier figures, assuming the Forestry Department continues to employ 1 researcher. Source 27 notes that the Veterinary Services Department also undertakes research. In 1982 it employed 7 researchers. These researchers are not included in the figures presented here, for reasons of consistency with the expenditure figures.

Expenditure Comments:
The 1983 expenditure figure includes ARI and the Research Divisions of the Forestry and Fisheries Departments of the Ministry of Agriculture. The expenditure figures for the other years are likely to include the same organizations.

Sources:
0010 Boyce, James K., and Robert E. Evenson. *National and International Agricultural Research and Extension Programs.* New York: Agricultural Development Council, Inc., 1975.

0017 ISNAR, IFARD & AOAD. Survey of National Agricultural Research Systems: Unpublished Questionnaire Responses. ISNAR, The Hague, 1985.

0744 *Agricultural Research Centres: A World Directory of Organizations and Programmes.* Eighth Edition, Two Volumes. Harlow, U.K.: Longman, 1986.

Additional References:
0014 Judd, M. Ann, James K. Boyce, and Robert E. Evenson. "Investing in Agricultural Supply." Economic Growth Center, Yale University, New Haven, Connecticut, 1983. Mimeo.

0016 Oram, Peter A., and Vishva Bindlish. *Resource Allocations to National Agricultural Research: Trends in the 1970s.* The Hague and Washington, D.C.: ISNAR and IFPRI, November 1981.

0026 Oram, Peter A., and Mark Gieben. "Document Summaries." ISNAR, The Hague, 1984. Mimeo.

0027 Harvey, Nigel, ed. *Agricultural Research Centres: A World Directory of Organizations and Programmes.* Seventh Edition, Two Volumes. Harlow, U.K.: Longman, 1983.

0073 Oram, Peter A., and Vishva Bindlish. "Investment in Agricultural Research in Developing Countries: Progress, Problems, and the Determination of Priorities." IFPRI, Washington, D.C., January 1984. Mimeo.

0094 FAO – Near East Regional Office. *Directory of Agricultural Research Institutions in the Near East Region.* Cairo: FAO – Near East Regional Office, 1979.

0095 FAO – CARIS. *Agricultural Research in Developing Countries – Volume 1: Research Institutions.* Rome: FAO – CARIS, 1978.

0165 Evenson, R.E., and Y. Kislev. *Agricultural Research and Productivity.* New Haven: Yale University Press, 1975.

0173 FAO. *FAO Near East Regional Studies on Organization and Administration of Agricultural Research.* Rome: FAO, 1974.

0187 Cyprus Agricultural Research Institute (CARI). *Cyprus Agricultural Research Institute, Annual Report for 1983.* Nicosia, Cyprus: Ministry of Agriculture and Natural Resources, March 1984.

0242 Commonwealth Agricultural Bureaux (CAB). *List of Research Workers in Agricultural Sciences in the Commonwealth 1981.* Slough, England: CAB, 1981.

0243 Commonwealth Agricultural Bureaux (CAB). *List of Research Workers in the Agricultural Sciences in the Commonwealth and in the Republic of Ireland 1978.* Slough, England: CAB, 1978.

0244 Commonwealth Agricultural Bureaux (CAB). *List of Research Workers in the Agricultural Sciences in the Commonwealth and in the Republic of Ireland 1975.* Slough, England: CAB, 1975.

0279 Commonwealth Agricultural Bureaux (CAB). *List of Research Workers in the Agricultural Sciences in the Commonwealth and in the Republic of Ireland 1972.* Slough, England: CAB, 1972.

0285 Commonwealth Agricultural Bureaux (CAB). *List of Research Workers in the Agricultural Sciences in the Commonwealth and in the Republic of Ireland 1969.* Slough, England: CAB, 1969.

0286 Commonwealth Agricultural Bureaux (CAB). *List of Research Workers in Agriculture, Animal Health and Forestry in the Commonwealth and in the Republic of Ireland 1966.* Slough, England: CAB, 1966.

0287 Commonwealth Agricultural Bureaux (CAB). *List of Research Workers in Agriculture, Animal*

Health and Forestry in the British Commonwealth, the Republic of Sudan and the Republic of Ireland 1959. Slough, England: CAB, 1959.

0442 UNDP, and FAO. *Agricultural Research Institute, Nicosia, Cyprus – Final Report.* Rome: FAO, 1968.

0831 FAO. *Report of an Expert Consultation on Agricultural Extension and Research Linkages in the Near East, Amman, Jordan, 22–26 July 1985.* Rome: FAO, 1985.

0852 Evenson, Robert E., and Yoav Kislev. *Investment in Agricultural Research and Extension: A Survey of International Data.* Center Discussion Paper No. 124. New Haven, Connecticut: Economic Growth Center, Yale University, August 1971.

0886 Evenson, Robert E., and Yoav Kislev. "Investment in Agricultural Research and Extension: A Survey of International Data." *Economic Development and Cultural Change* Vol. 23 (April 1975): 507–521.

Denmark

Personnel

Period/Year	PhD	MSc	BSc	Subtotal	Expat	Total	Source
1960–64	..	..	..	..	..	..	
1965–69	..	..	..	..	..	458	621
1970	..	..	..	..	..	429	505
1971	..	..	..	..	..	422	505
1972	..	..	..	..	..	415	505
1973	..	..	..	..	..	408	505
1974	..	..	..	..	..	406	505
1975	..	..	..	..	..	403	505
1976	..	..	..	..	..	400	505
1977	..	..	..	..	..	399	505
1978	..	..	..	..	..	401	505
1979	..	..	..	..	..	401	505
1980	..	..	..	..	..	424	505
1981	..	..	..	..	..	446	505
1982	..	..	..	..	..	469	505
1983	..	..	..	..	..	..	
1984	..	..	..	..	..	..	
1985	..	..	..	..	..	..	
1986	..	..	..	..	..	..	

1960–64: 1965–69: 1967

Expenditure LCU = *Danish Kroner*

Period/Year	Current LCU (millions)	Constant 1980 LCU(millions)	Constant 1980 US$ (millions) Atlas	Constant 1980 US$ (millions) PPP	Source
1960–64	..	..	..	..	
1965–69	75.623	233.404	42.941	30.704	510
1970	90.200	224.938	41.383	29.590	505
1971	96.000	222.222	40.884	29.233	505
1972	101.800	216.136	39.764	28.432	505
1973	107.600	208.932	38.438	27.485	505
1974	116.400	200.000	36.795	26.310	505
1975	125.200	191.145	35.166	25.145	505
1976	134.200	187.955	34.579	24.725	505
1977	144.500	184.783	33.996	24.308	505
1978	154.900	180.326	33.176	23.722	505
1979	165.400	179.004	32.933	23.548	505
1980	218.000	218.000	40.107	28.678	505
1981	270.700	245.867	45.234	32.344	505
1982	323.500	265.818	48.904	34.968	505
1983	..	..	..	..	
1984	..	..	..	..	
1985	..	..	..	..	
1986	..	..	..	..	

1960–64: 1965–69: 1967

Personnel Comments:

These figures, measured in full-time equivalents (FTEs), include all public-sector researchers in the agricultural sciences, inclusive of university (or more specifically the higher education sector) personnel. During the 1980–82 period, FTE researchers at universities accounted for an average 28.6% of the total, compared with 27.5% for the 1970–72 period. Government-financed research is performed by institutions which fall within the Ministry of Agriculture (for example, National Institute for Soil and Crop Research, National Institute of Animal Science, National Dairy Research Institute, National Engineering Institute, Institute of Agricultural Economics, National Forestry Research Station, National Veterinary Serum Laboratory, National Veterinary Virus Research Institute, and the Poultry Disease Institute) and the Ministry of Education (for example, Royal Veterinary and Agricultural University, University of Advanced Technology, and the National Agricultural and Veterinary Research Council) (see sources 283 and 27).

Expenditure Comments:

These figures are compatible with the personnel figures and include total (university plus other public sector) R&D expenditures in the agricultural sciences. Based on a slightly higher 1978 estimate than included here of DKR 180 million, source 283 reports that 63% of this budget comes from the Ministry of Agriculture, 7% from the Ministry of Education, and 25% from the Royal (Veterinary and) Agricultural University. Source 505 indicates that over the 1980–82 period the higher education sector accounted for 18.6% of the total expenditure reported here, compared with a 28.4% average for the 1970–72 period.

Sources:

0505 Forskningssekretariatet. *Forskningsstatistik 1982. Ressource forbruget ved forskning og udviklingsarbejde i den offentlige sektor.* København: Forskningssekretariatet, 1986.

0510 OECD. *International Survey of the Resources Devoted to R&D in 1967 by OECD Member Countries, Statistical Tables and Notes – Volume 5: Total Tables.* Paris: OECD, August 1970.

0621 NORDFORSK. *Forsknings Virksomhet i Norden i 1967 – Utgifter og Personale.* Nordisk Statistisk Skriftserie 18. Oslo: NORDFORSK, 1970.

Additional References:

0010 Boyce, James K., and Robert E. Evenson. *National and International Agricultural Research and Extension Programs.* New York: Agricultural Development Council, Inc., 1975.

0021 OECD. *OECD Science and Technology Indicators.* Paris: OECD, 1984.

0027 Harvey, Nigel, ed. *Agricultural Research Centres: A World Directory of Organizations and Programmes.* Seventh Edition, Two Volumes. Harlow, U.K.: Longman, 1983.

0165 Evenson, R.E., and Y. Kislev. *Agricultural Research and Productivity.* New Haven: Yale University Press, 1975.

0283 European Economic Community (EEC). "Monograph on the Organization of Agricultural Research." EEC, Brussels, 1982. Mimeo.

0376 FAO. *Fourteenth FAO Regional Conference for Europe, Reykjavik, Iceland, 17–21 September 1984: Research in Support of Agricultural Policies in Europe.* Rome: FAO, 1984.

0452 Knudsen, P.H., ed. *Agriculture in Denmark.* Copenhagen: Agricultural Council of Denmark, 1977.

0473 OECD. *Science and Technology Indicators: Basic Statistical Series – Volume A: The Objectives of Government R&D Funding 1974–1985.* Paris: OECD, May 1983.

0474 OECD. *Science and Technology Indicators: Basic Statistical Series – Volume A: The Objectives of Government R&D, 1969–1981.* Paris: OECD, July 1981.

0506 OECD. *Country Reports on the Organisation of Scientific Research: Denmark.* Paris: OECD, July 1964.

0509 OECD. *Survey of the Resources Devoted to R&D by OECD Member Countries, International Statistical Year 1971 – Volume 5: Total Tables, Statistical Tables and Notes.* Paris: OECD, August 1974.

0511 OECD. *International Survey of the Resources Devoted to R&D in 1969 by OECD Member Countries, Statistical Tables and Notes – Volume 5: Total Tables.* Paris: OECD, June 1973.

0512 OECD. *Survey of the Resources Devoted to R&D by OECD Member Countries, International Statistical Year 1973 – Volume 5: Total Tables, Statistical Tables and Notes.* Paris: OECD, September 1976.

0513 OECD. *International Survey of the Resources Devoted to R&D by OECD Member Countries, International Statistical Year 1975 –*

154

International Volume: Statistical Tables and Notes. Paris: OECD, March 1979.

0530 OECD. *OECD Science and Technology Indicators No. 2 – R&D, Invention and Competitiveness.* Paris: OECD, 1986.

0536 Rasmussen, Ole Kjeldsen. Personal communication. Landbrugets Samråd for Forskning og Forsøg. Copenhagen, January – March 1987.

0601 OECD – STIID Data Bank. "Public Funding of R&D by Socio-Economic Objective." Tables. OECD, Paris, April 1987. Mimeo.

0602 OECD – STIID Data Bank. "R&D Expenditure in the Higher Education and Private Non-Profit Sectors." Tables. OECD, Paris, April 1987. Mimeo.

0606 EUROSTAT. *Government Financing of Research and Development 1975–1984.* Luxembourg: Office des Publications Officielles des Communautés Européennes, February 1985.

0706 EUROSTAT. *Government Financing of Research and Development 1975–1985.* Luxembourg: Office for Official Publications of the European Communities, 1987.

0707 OECD – STIID Data Bank. "R&D Expenditure in the Higher Education and Private Non-Profit Sectors: Higher Education, Agricultural Sciences, General Universities Funds." Table. OECD, Paris, June 1987. Mimeo.

0708 OECD – STIID Data Bank. "Gross Domestic Expenditure on R&D: All Fields of Sciences – Agriculture, Forestry and Fishing." Tables. OECD, Paris, June 1987. Mimeo.

0852 Evenson, Robert E., and Yoav Kislev. *Investment in Agricultural Research and Extension: A Survey of International Data.* Center Discussion Paper No. 124. New Haven, Connecticut: Economic Growth Center, Yale University, August 1971.

0886 Evenson, Robert E., and Yoav Kislev. "Investment in Agricultural Research and Extension: A Survey of International Data." *Economic Development and Cultural Change* Vol. 23 (April 1975): 507–521.

Dominica

Personnel

Period/Year	PhD	MSc	BSc	Subtotal	Expat	Total	Source
1960–64	..	..	..	..	..	..	
1965–69	1	1	1	3	..	3	285
1970					..	..	
1971	2	2	4	8	..	8	279
1972				..	..	..	
1973				..	..	..	
1974	1	2	2	5	..	5	244
1975	..	..	..		..	..	
1976	..	..	..		..	..	
1977	..	..	..		..	..	
1978	..	..	..		..	..	
1979	..	..	..		..	..	
1980	..	..	..		..	..	
1981	..	..	..		..	..	
1982	..	..	..		..	..	
1983	0	2	4	6	..	6	115
1984	..	..			..	..	
1985	..	..	..		..	..	
1986	..	..	..		..	..	

1960–64: 1965–69: 1968

Expenditure *LCU = East Caribbean Dollars*

Period/Year	Current LCU (millions)	Constant 1980 LCU(millions)	Constant 1980 US$ (millions)		Source
			Atlas	PPP	
1960–64	..	..	..	..	
1965–69	..	..	..	..	
1970	..	..	..	..	
1971	..	..	..	..	
1972	..	..	..	..	
1973	..	..	..	..	
1974	..	..	..	..	
1975	..	..	..	..	
1976	..	..	..	..	
1977	..	..	..	..	
1978	..	..	..	..	
1979	..	..	..	..	
1980	..	..	..	..	
1981	..	..	..	..	
1982	0.429	0.403	0.138	0.211	115
1983	0.440	0.391	0.134	0.205	115
1984	0.230	0.197	0.067	0.103	115
1985	..	..	..	..	
1986	..	..	..	..	

1960–64: 1965–69:

Personnel Comments:

The number of researchers given by the Commonwealth Agricultural Bureaux (CAB) 'Lists of Research Workers' (sources 285, 279, & 244) for the years 1968, 1971, and 1974 refer to Ministry of Agriculture personnel who appear to perform agricultural research in Dominica. Care is required when using CAB sources in this instance as functional names and divisional or unit affiliations vary from one CAB listing to another and do not always clearly imply that the listed individuals are involved in agricultural research.

The 1983 figure includes the Ministry of Agriculture, CARDI (Caribbean Agricultural Research and Development Institute), and the French Technical Cooperation Group. These three organizations work together in a national agricultural research program. One researcher comes from the Ministry, three from CARDI, and two from the French group. In addition, 3 MSc's and 9 BSc's were reported by source 355 as "other technical staff in the Ministry of Agriculture."

CARDI was created in 1975. The following professional staff at CARDI's Dominica unit have been reconstructed on the basis of CARDI annual reports:

	1975	1980	1981	1983	1984
PhD	0	0	NA	0	0
MSc	0	0	NA	1	1
BSc	0	2	NA	1	2
Total	0	2	3	2	3
Source	342	43	342	355	354

Expenditure Comments:

These expenditure figures refer to the national agricultural research program in which the Ministry of Agriculture, CARDI, and the French Technical Cooperation Group work together.

Sources:

0115 Economic Commission for Latin America and the Caribbean (ECLAC). *Agricultural Research Policy and Management – Volumes I and II.* Papers presented at the Workshop on Agricultural Research Policy and Management, 26–30 September 1983, Port of Spain, Trinidad. Port of Spain, Trinidad: ECLAC, November 1984.

0244 Commonwealth Agricultural Bureaux (CAB). *List of Research Workers in the Agricultural Sciences in the Commonwealth and in the Republic of Ireland 1975.* Slough, England: CAB, 1975.

0279 Commonwealth Agricultural Bureaux (CAB). *List of Research Workers in the Agricultural Sciences in the Commonwealth and in the Republic of Ireland 1972.* Slough, England: CAB, 1972.

0285 Commonwealth Agricultural Bureaux (CAB). *List of Research Workers in the Agricultural Sciences in the Commonwealth and in the Republic of Ireland 1969.* Slough, England: CAB, 1969.

Additional References:

0063 ISNAR. *Report to the Board of Directors of CARDI: Analysis, Evaluation and Proposals to Strengthen CARDI's Regional Capacity.* The Hague: ISNAR, 1985.

0287 Commonwealth Agricultural Bureaux (CAB). *List of Research Workers in Agriculture, Animal Health and Forestry in the British Commonwealth, the Republic of Sudan and the Republic of Ireland 1959.* Slough, England: CAB, 1959.

0342 Caribbean Agricultural Research and Development Institute (CARDI). *CARDI 1976–1981: Report of the Chairman.* St. Augustine, Trinidad & Tobago: CARDI, 1982.

0343 Caribbean Agricultural Research and Development Institute (CARDI). *Annual Report 1980.* St. Augustine, Trinidad & Tobago: CARDI, 1981.

0353 Caribbean Agricultural Research and Development Institute (CARDI). *Paper Presented at a Meeting of the Governing Body of CARDI on January 29, 1982, Basseterre, St. Kitts.* St. Augustine, Trinidad: CARDI, 1982.

0354 Caribbean Agricultural Research and Development Institute (CARDI). *Annual Report Research and Development 1983–84: Highlights.* St. Augustine, Trinidad: CARDI, 1984.

0355 Caribbean Agricultural Research and Development Institute (CARDI). *Research and Development Summary 1982–83.* St. Augustine, Trinidad: CARDI, 1983.

0356 Caribbean Agricultural Research and Development Institute (CARDI). *Research and Development Summary 1983–84.* St. Augustine, Trinidad: CARDI, 1984.

Dominican Republic

Personnel

Period/Year	PhD	MSc	BSc	Subtotal	Expat	Total	Source
1960–64	..	..	..	..	..	..	
1965–69	..	..	..	..	..	..	
1970	..	..	..	..	..	..	
1971	..	..	..	..	..	12	10
1972	..	..	..	..	..	..	
1973	..	..	..	..	..	..	
1974	..	..	..	..	..	..	
1975	..	..	..	..	..	..	
1976	..	..	..	..	..	62	219
1977	0	9	51	60	..	60	220
1978	..	..	..	..	..	..	
1979	0	8	59	67	..	67	220
1980	3	12	91	106	..	106	220
1981	..	..	..	..	..	..	
1982	..	..	..	..	..	..	
1983	2	29	105	136	..	136	28
1984	..	..	..	..	..	..	
1985	..	..	..	..	..	..	
1986	..	..	..	..	..	..	

1960–64:   1965–69:

Expenditure

LCU = Dominican Pesos

Period/Year	Current LCU (millions)	Constant 1980 LCU(millions)	Constant 1980 US$ (millions) Atlas	PPP	Source
1960–64	..	..	..	..	
1965–69	..	..	..	..	
1970	..	..	..	..	
1971	..	..	..	..	
1972	..	..	..	..	
1973	..	..	..	..	
1974	..	..	..	..	
1975	..	..	..	..	
1976	1.263	1.784	1.713	2.797	219
1977	2.780	3.550	3.409	5.566	221
1978	2.780	3.510	3.370	5.503	221
1979	2.214	2.516	2.416	3.944	28
1980	2.216	2.216	2.128	3.474	28
1981	2.119	2.012	1.932	3.155	28
1982	..	..	..	..	
1983	3.575	3.040	2.919	4.766	17
1984	..	..	..	..	
1985	..	..	..	..	
1986	..	..	..	..	

1960–64:   1965–69:

Personnel Comments:

These figures represent agricultural research under the jurisdiction of the Ministry of Agriculture only. They can be broken down as follows:

	1977	1979	1980	1982	1983
CESDA	30	37	65	NA	51
CENDA	24	24	35	NA	28
CEDIA	6	6	6	NA	10
CENIP				NA	32
CIAZA				NA	8
DIA				NA	7
Subtotal SEA	60	67	106	NA	136
CEA/Duquesa			22		
CEA/CEDIPCA			8		
UASD[a]					100*
Source	220	220	220	27	28
					17*

a. Number of teachers/researchers.

Expenditure Comments:

These expenditure figures refer to 'Departamento de Investigación Agropecuaria' which finances all the research institutes under the jurisdiction of the Ministry of Agriculture (SEA).

Acronyms and Abbreviations:

CEA/CEDIPCA: Consejo Estatal de Azúcar – Centro Dominicano de Investigación Pecuaria con Caña de Azúcar

CEA/Duquesa: Consejo Estatal de Azúcar – Centro de Investigaciones en Caña de Azúcar

CEDIA: Centro de Investigaciones Arroceras

CENDA: Centro Norte de Desarrollo Agropecuario

CENIP: Centro de Investigaciones Pecuarias

CESDA: Centro Sur de Desarrollo Agropecuario

CIAZA: Centro de Investigaciones Aplicadas a Zonas Aridas

DIA: Departamento de Investigación Agropecuaria de Secretaría de Estado de Agricultura (SEA)

SEA: Secretaria de Estado de Agricultura (Ministry of Agriculture)

UASD: Universidad Autónoma Santo Domingo

Sources:

0010 Boyce, James K., and Robert E. Evenson. *National and International Agricultural Research and Extension Programs.* New York: Agricultural Development Council, Inc., 1975.

0017 ISNAR, IFARD & AOAD. Survey of National Agricultural Research Systems: Unpublished Questionnaire Responses. ISNAR, The Hague, 1985.

0028 ISNAR. *Informe al Gobierno de la República Dominicana; El Sistema de Investigación Agropecuaria en la República Dominicana.* The Hague: ISNAR, August 1983.

0219 Department of Agriculture, Natural Resources and Rural Development (DARNDR), and IICA. "Réunion Technique Régionale sur les Systèmes de Recherche Agricole dans les Antilles." DARNDR and IICA, Port-au-Prince, Haiti, November/December 1977. Mimeo.

0220 Secretariado Técnico de la Presidencia – Departamento de Ciencia y Tecnología. *Estudio de Base del Sector Agropecuario y Forestal.* Santo Domingo: Secretariado Técnico de la Presidencia, August 1982.

0221 Stagno, H., and R. Pineda. *Situación Actual del Sistema Nacional de Investigación y Extensión, y Descripción de los Programas Principales en Cultivos Alimenticios.* Informe para el Proyecto BID-IICA-Iowa State University. Santo Domingo: IICA, 1980.

Additional References:

0014 Judd, M. Ann, James K. Boyce, and Robert E. Evenson. "Investing in Agricultural Supply." Economic Growth Center, Yale University, New Haven, Connecticut, 1983. Mimeo.

0016 Oram, Peter A., and Vishva Bindlish. *Resource Allocations to National Agricultural Research: Trends in the 1970s.* The Hague and Washington, D.C.: ISNAR and IFPRI, November 1981.

0027 Harvey, Nigel, ed. *Agricultural Research Centres: A World Directory of Organizations and Programmes.* Seventh Edition, Two Volumes. Harlow, U.K.: Longman, 1983.

0053 Piñeiro, Martín, and Eduardo Trigo, eds. *Cambio Técnico en el Agro Latino Americano – Situación y Perspectivas en la Década de 1980.* San José: IICA, 1983.

0073 Oram, Peter A., and Vishva Bindlish. "Investment in Agricultural Research in Developing Countries: Progress, Problems, and the Determination of Priorities." IFPRI, Washington, D.C., January 1984. Mimeo.

159

0172 Trigo, Eduardo. Background material. ISNAR. The Hague, n.d.
0379 Tejada, S. "La Planificación de las Investigaciones Agropecuarias en la República Dominicana." Paper presented at Curso-Taller sobre la Administración de la Investigación Agrícola, Panama City, 14–25 July 1986.

Ecuador

Personnel

Period/Year	PhD	MSc	BSc	Subtotal	Expat	Total	Source
1960–64	..	..	..	..	..	..	
1965–69	..	..	..	..	..	34	10
1970	..	..	..	..	..	54	10
1971	..	..	..	..	..	94	10
1972	..	..	..	..	..	123	10
1973	..	..	..	..	..	158	10
1974	5	38	114	157	..	157	758
1975	5	38	115	158	..	158	758
1976	5	38	117	160	..	160	758
1977	5	39	134	178	..	178	758
1978	5	36	147	188	..	188	758
1979	6	51	119	176	..	176	758
1980	4	49	147	200	..	200	758
1981	5	54	117	176	..	176	758
1982	5	55	117	177	..	177	758
1983	5	58	169	232	..	232	758
1984	5	67	163	235	..	235	758
1985	4	67	162	233	..	233	758
1986	5	67	153	225	..	225	758

1960–64:   1965–69: 1965

Expenditure

LCU = Ecuadoran Sucres

Period/Year	Current LCU (millions)	Constant 1980 LCU(millions)	Constant 1980 US$ (millions)		Source
			Atlas	PPP	
1960–64	7.470	33.558	1.233	2.264	758
1965–69	20.146	87.183	3.204	5.883	758
1970	36.361	129.861	4.772	8.763	758
1971	51.515	171.146	6.290	11.548	758
1972	62.803	203.906	7.494	13.759	758
1973	82.289	251.648	9.248	16.980	758
1974	118.739	259.823	9.549	17.532	758
1975	164.330	326.700	12.006	22.045	758
1976	155.740	274.190	10.077	18.501	758
1977	166.103	248.657	9.138	16.778	758
1978	183.075	253.918	9.332	17.133	758
1979	191.223	228.462	8.396	15.416	758
1980	267.392	267.392	9.827	18.043	758
1981	263.169	230.043	8.454	15.522	758
1982	291.755	216.596	7.960	14.615	758
1983	332.873	178.102	6.545	12.018	758
1984	468.285	180.040	6.617	12.148	758
1985	607.722	178.427	6.557	12.040	758
1986	661.727	162.626	5.977	10.973	758

1960–64: 1961, 62, 63 & 64   1965–69: 1965, 66, 67, 68 & 69

Personnel Comments:
These figures refer to 'Instituto Nacional de Investigaciones Agropecuarias' (INIAP) only. Source 27 also mentions a small forestry research institute (5 researchers), a publicly funded veterinary laboratory (42 researchers), and the Universidad Técnica de Machala with a faculty of Agronomy and Veterinary Science of 11 graduate research staff in 1982.

Expenditure Comments:
These expenditure figures refer to INIAP only. Other sources (1, 17, and 368) give time series which differ from year to year with this time series. In general, however, they give the same trend. The figures presented are actual expenditures, not simply funds available.

Sources:
0010 Boyce, James K., and Robert E. Evenson. *National and International Agricultural Research and Extension Programs.* New York: Agricultural Development Council, Inc., 1975.
0758 Larrea Herrera, Pablo. Personal communication. Director General, Instituto Nacional de Investigaciones Agropecuarias. Quito, Ecuador, January 1988.

Additional References:
0001 Trigo, Eduardo J., and Martín E. Piñeiro. "Funding Agricultural Research." In *Report of a Conference: Selected Issues in Agricultural Research in Latin America*, edited by Barry Nestel and Eduardo J. Trigo. The Hague: ISNAR, March 1984.
0014 Judd, M. Ann, James K. Boyce, and Robert E. Evenson. "Investing in Agricultural Supply." Economic Growth Center, Yale University, New Haven, Connecticut, 1983. Mimeo.
0016 Oram, Peter A., and Vishva Bindlish. *Resource Allocations to National Agricultural Research: Trends in the 1970s.* The Hague and Washington, D.C.: ISNAR and IFPRI, November 1981.
0017 ISNAR, IFARD & AOAD. Survey of National Agricultural Research Systems: Unpublished Questionnaire Responses. ISNAR, The Hague, 1985.
0027 Harvey, Nigel, ed. *Agricultural Research Centres: A World Directory of Organizations and Programmes.* Seventh Edition, Two Volumes. Harlow, U.K.: Longman, 1983.
0053 Piñeiro, Martín, and Eduardo Trigo, eds. *Cambio Técnico en el Agro Latino Americano – Situación y Perspectivas en la Década de 1980.* San José: IICA, 1983.
0073 Oram, Peter A., and Vishva Bindlish. "Investment in Agricultural Research in Developing Countries: Progress, Problems, and the Determination of Priorities." IFPRI, Washington, D.C., January 1984. Mimeo.
0095 FAO – CARIS. *Agricultural Research in Developing Countries – Volume 1: Research Institutions.* Rome: FAO – CARIS, 1978.
0163 CGIAR. "National Agricultural Research." CGIAR, Washington, D.C., 1985. Mimeo.
0165 Evenson, R.E., and Y. Kislev. *Agricultural Research and Productivity.* New Haven: Yale University Press, 1975.
0172 Trigo, Eduardo. Background material. ISNAR, The Hague, n.d.
0196 Solis, Rómulo V. "Documento Sobre Investigación Agropecuaria, Extensión Agrícola y Rural y Principales Programas de Investigación que Reciben." IICA, Quito, 1980. Mimeo.
0368 Posada Torres, Rafael. *Ecuador and the CGIAR Centers – A Study of Their Collaboration in Agricultural Research.* CGIAR Study Paper Number 11. Washington, D.C.: World Bank, 1986.
0744 *Agricultural Research Centres: A World Directory of Organizations and Programmes.* Eighth Edition, Two Volumes. Harlow, U.K.: Longman, 1986.
0852 Evenson, Robert E., and Yoav Kislev. *Investment in Agricultural Research and Extension: A Survey of International Data.* Center Discussion Paper No. 124. New Haven, Connecticut: Economic Growth Center, Yale University, August 1971.
0886 Evenson, Robert E., and Yoav Kislev. "Investment in Agricultural Research and Extension: A Survey of International Data." *Economic Development and Cultural Change* Vol. 23 (April 1975): 507–521.

Egypt

Personnel

Period/Year	PhD	MSc	BSc	Subtotal	Expat	Total	Source
1960–64	..	..	..	..	..	569	10
1965–69	..	..	..	..	..	1257	175/10
1970	..	..	..	1780	..	1780	175
1971	..	..	..	..	..	..	
1972	..	..	..	..	..	..	
1973	..	..	..	..	..	..	
1974	580	–1490–		2070	..	2070	173
1975	..	..	..	..	..	..	
1976	..	..	..	..	..	..	
1977	..	..	..	..	..	..	
1978	..	..	..	..	..	..	
1979	..	..	..	..	..	..	
1980	..	..	..	..	..	..	
1981	..	..	..	..	..	..	
1982	..	..	..	..	..	3556	163
1983	944	1548	1689	4181	..	4181	249
1984	..	..	..	..	..	..	
1985	800	1200	3000	5000	..	5000	248
1986	..	..	..	..	..	..	

1960–64: 1963   1965–69: 1966 & 68

Expenditure   *LCU = Egyptian Pounds*

Period/Year	Current LCU (millions)	Constant 1980 LCU(millions)	Constant 1980 US$ (millions)		Source
			Atlas	PPP	
1960–64	1.620	5.906	7.625	13.768	10
1965–69	4.000	12.341	15.934	28.769	175
1970	..	..	..	..	
1971	..	..	..	..	
1972	..	..	..	..	
1973	4.200	10.446	13.487	24.352	10
1974	..	..	..	..	
1975	..	..	..	..	
1976	..	..	..	..	
1977	..	..	..	..	
1978	..	..	..	..	
1979	..	..	..	..	
1980	13.650	13.650	17.624	31.821	249
1981	15.930	14.249	18.397	33.217	249
1982	21.480	17.143	22.134	39.964	249
1983	24.110	16.848	21.754	39.277	249
1984	28.100	16.442	21.229	38.331	249
1985	..	..	..	..	
1986	..	..	..	..	

1960–64: 1960   1965–69: 1966

Personnel Comments:

The 1966 and 1970 figures, both given by source 175, explicitly exclude university research and forestry, fisheries, or veterinary research. No indication is given as to which institutes are included in these figures, but it is likely that they include, more or less, the same institutes as the Agricultural Research Center, which was established in 1971. The 1974, 1983, and 1985 figures represent the Agricultural Research Center (ARC) only. ARC, which is under the aegis of the Ministry of Agriculture, is the principal agricultural research organization in Egypt. A considerable amount of agricultural research, however, is conducted by other organizations. The most important of these are the Food and Agriculture Council of the Academy of Scientific Research and Technology (this organization only funds research – it does not execute agricultural research), the Agricultural Research Division of the National Research Center (1983: 600 researchers including 200 with a PhD degree, source 249), the Water Research Center of the Ministry of Irrigation, the Desert Research Institute under the responsibility of the Ministry of Land Reclamation (1970: 150 researchers, source 532; 1985: 100 researchers, source 248), and the Institute of Oceanography and Fisheries (1970: 88 researchers, source 532; 1985: 136 researchers, source 248).

Expenditure Comments:

The figures from source 249 refer to the research division of ARC only. The 1980–83 figures are actual expenditures – which differ little from funds available – while the 1984 figure represents funds allocated. For the earlier observations, no indication is given as to which institutes are included.

Sources:

0010 Boyce, James K., and Robert E. Evenson. *National and International Agricultural Research and Extension Programs*. New York: Agricultural Development Council, Inc., 1975.

0163 CGIAR. "National Agricultural Research." CGIAR, Washington, D.C., 1985. Mimeo.

0173 FAO. *FAO Near East Regional Studies on Organization and Administration of Agricultural Research*. Rome: FAO, 1974.

0175 Cooper, St.G.C. *Agricultural Research in Tropical Africa*. Kampala: East African Literature Bureau, 1970.

0248 FAO. *Review of the Organization of Agricultural and Fisheries Research, Technology and Development in Egypt*. Rome: FAO, 1985.

0249 United States Team of Consultants to the Ministry of Agriculture of the Arab Republic of Egypt. "Increasing Egyptian Agricultural Production through Strengthened Research and Extension Programs." Ministry of Agriculture, Cairo, September 1983. Mimeo.

0852 Evenson, Robert E., and Yoav Kislev. *Investment in Agricultural Research and Extension: A Survey of International Data*. Center Discussion Paper No. 124. New Haven, Connecticut: Economic Growth Center, Yale University, August 1971.

Additional References:

0014 Judd, M. Ann, James K. Boyce, and Robert E. Evenson. "Investing in Agricultural Supply." Economic Growth Center, Yale University, New Haven, Connecticut, 1983. Mimeo.

0016 Oram, Peter A., and Vishva Bindlish. *Resource Allocations to National Agricultural Research: Trends in the 1970s*. The Hague and Washington, D.C.: ISNAR and IFPRI, November 1981.

0019 Bengtsson, Bo, and Tedla Getachew, eds. *Strengthening National Agricultural Research*. Report from a SAREC workshop, 10–17 September 1979. Part 1: Background Documents; Part II: Summary and Conclusions. Sweden: SAREC, 1980.

0026 Oram, Peter A., and Mark Gieben. "Document Summaries." ISNAR, The Hague, 1984. Mimeo.

0027 Harvey, Nigel, ed. *Agricultural Research Centres: A World Directory of Organizations and Programmes*. Seventh Edition, Two Volumes. Harlow, U.K.: Longman, 1983.

0073 Oram, Peter A., and Vishva Bindlish. "Investment in Agricultural Research in Developing Countries: Progress, Problems, and the Determination of Priorities." IFPRI, Washington, D.C., January 1984. Mimeo.

0094 FAO – Near East Regional Office. *Directory of Agricultural Research Institutions in the Near East Region*. Cairo: FAO – Near East Regional Office, 1979.

0095 FAO – CARIS. *Agricultural Research in Developing Countries – Volume 1: Research Institutions*. Rome: FAO – CARIS, 1978.

0135 FAO, and UNDP. *National Agricultural Research – Report of an Evaluation Study in Selected Countries*. Rome: FAO and UNDP, 1984.

164

0165 Evenson, R.E., and Y. Kislev. *Agricultural Research and Productivity*. New Haven: Yale University Press, 1975.

0257 Ministry of Agriculture Egypt. *Allocation of Resources in Agricultural Research*. Cairo: Ministry of Agriculture, November 1983.

0266 UNESCO. *National Science Policies in Africa*. Science Policy Studies and Documents No. 31. Paris: UNESCO, 1974.

0308 Oteifa, B.A. "Status of Agricultural Research in Egypt." In *Strengthening National Agricultural Research*. Report from a SAREC workshop, 10–17 September 1979, edited by Bo Bengtsson and Getachew Tedla, pp. 20–24. Sweden: SAREC, 1980.

0360 Cooper, St.G.C. "Towards Trained Manpower for Agricultural Research in Africa." Paper presented at the Conference on Agricultural Research and Production in Africa, organized by the Association for the Advancement of Agricultural Sciences in Africa (AAASA), Addis Ababa, 29 August–4 September 1971.

0395 Oteifa, Bakir A. "Role of Universities and Private Sector Organizations in the National Agricultural Research System of Egypt." Paper presented at the 1st International Meeting of National Agricultural Research Systems and 2nd IFARD Global Convention, IFARD/EMBRAPA, Brasília, Brasil, 6–11 October 1986.

0407 Centre International de Hautes Etudes Agronomiques Méditerranéennes. "Séminaire sur l'Orientation et l'Organisation de la Recherche Agronomique dans les Pays du Bassin Méditerranéen, Istanbul, 1–3 December 1986." Centre International de Hautes Etudes Agronomiques Méditerranéennes, Paris, December 1986. Mimeo.

0409 El–Sharkawy, Ahmed. "Les Systèmes de Recherche Agricole en Egypte." Paper presented at the Séminaire sur l'Orientation et l'Organisation de la Recherche Agronomique dans les Pays du Bassin Méditerranéen, Centre International de Hautes Etudes Agronomiques Méditerranéennes, Istanbul, 1–3 December 1986.

0422 Casas, Joseph. "Les Ressources Humaines et Financières dans les Pays du Bassin Méditerranéen." Preliminary table. ISNAR, The Hague, 1986. Mimeo.

0532 UNESCO Field Science Office for Africa. *Survey on the Scientific and Technical Potential of the Countries of Africa*. Paris: UNESCO, 1970.

0744 *Agricultural Research Centres: A World Directory of Organizations and Programmes*. Eighth Edition, Two Volumes. Harlow, U.K.: Longman, 1986.

0747 ISNAR. *International Workshop on Agricultural Research Management*. Report of a workshop. The Hague: ISNAR, 1987.

0751 Shehata, A.H. "Agricultural Research System in Egypt." Paper presented at the International Workshop on Agricultural Research Management, ISNAR, The Hague, August 1987.

0831 FAO. *Report of an Expert Consultation on Agricultural Extension and Research Linkages in the Near East, Amman, Jordan, 22–26 July 1985*. Rome: FAO, 1985.

0865 Oram, Peter. "Agricultural Research Objectives and Priorities: Constraints to the Development of Agricultural Research Institutions in Arab Countries." Review paper prepared for the first meeting of the Council for Arab Agricultural Research, organized by the Arab Fund for Economic and Social Development, Kuwait, 23 March 1988.

0886 Evenson, Robert E., and Yoav Kislev. "Investment in Agricultural Research and Extension: A Survey of International Data." *Economic Development and Cultural Change* Vol. 23 (April 1975): 507–521.

El Salvador

Personnel

Period/Year	PhD	MSc	BSc	Subtotal	Expat	Total	Source
1960–64	..	..	..	..	..	..	
1965–69	..	..	..	..	..	49	10
1970	..	..	..	..	..	67	10
1971	..	..	..	..	..	77	10
1972	..	..	..	..	..	75	10
1973	..	..	..	..	..	80	10
1974	..	..	..	..	..	..	
1975	..	..	..	..	..	..	
1976	..	..	..	..	..	..	
1977	..	..	..	..	..	107	14
1978	..	..	..	..	..	..	
1979	..	..	..	..	..	..	
1980	..	..	..	..	..	106	209
1981	..	..	..	..	..	..	
1982	..	..	..	..	..	..	
1983	..	..	..	..	..	..	
1984	..	..	..	..	..	..	
1985	..	..	..	..	..	..	
1986	..	..	..	..	..	..	

1960–64:   1965–69: 1965 & 1966

Expenditure

LCU = Salvadoran Colones

Period/Year	Current LCU (millions)	Constant 1980 LCU(millions)	Constant 1980 US$ (millions)		Source
			Atlas	PPP	
1960–64	..		..		
1965–69	1.459	3.819	1.465	2.897	10
1970	..	..	..		
1971	..	..	..		
1972	1.733	4.311	1.654	3.270	10
1973	2.156	4.878	1.871	3.700	10
1974	..	..	..		
1975	..	..	..		
1976	3.380	5.216	2.001	3.956	209
1977	4.133	5.382	2.065	4.082	209
1978	5.363	6.929	2.658	5.255	209
1979	5.433	6.160	2.363	4.672	209
1980	5.873	5.873	2.253	4.454	209
1981	..	..	..		
1982	..	..	..		
1983	..	..	..		
1984	..	..	..		
1985	..	..	..		
1986	..	..	..		

1960–64:   1965–69: 1965

166

Personnel Comments:

'Centro Nacional de Tecnología Agropecuaria' (CENTA), or more precisely its Research Division, is the key agricultural research unit within the Ministry of Agriculture and Livestock (MAG). According to source 835, CENTA was established in 1972. The research activities which were performed by MAG's former Directorate of Agricultural Research were integrated into CENTA, along with other MAG activities such as seed production, extension, and agricultural education. In 1977, education was separated from CENTA. In 1982 there was an attempt to integrate research under the other directorates of MAG with CENTA into a new organization called 'Instituto Salvadoreño de Investigaciones Agrarias y Pesqueras' (ISIAP). However, according to source 835, this attempt failed; only the name of CENTA was (slightly) changed into 'Centro de Tecnología Agrícola,' still using the same abbreviation. Then, in 1986 the agricultural extension division was separated from CENTA, leaving only research and seed production in its mandate.

Although CENTA is the key research institute within MAG, several other directorates of MAG conduct research, or have a research institute. According to source 381, the following research institutes are linked with MAG: ISIC, CENDEPESCA, CENREN, and CDG. In addition, several non-MAG organizations are mentioned by source 208 to be involved in agricultural research. These include the Faculty of Agriculture of the El Salvador University, Cooperativa Algodenera, and INAZUCAR.

The indicator series only includes CENTA and ISIC, for which reasonable time-series data are available. The table below gives an overview of how these figures have been constructed.

	1965	1966	1970	1971	1972	1973	1977	1980	1982	1983	1986
CENTA[a]	33	32	46	55	52	56	78	74	65	85	80
ISIC	16	[17]	[21]	[22]	[23]	[24]	[29]	32	NA	NA	NA
Total	49	49	67	77	75	80	107	106	NA	NA	NA
Source	10	10	10	10	10	10	14	209	27	17	381

[ ] = estimated or constructed by authors.
a. Before 1972 the figures represent the Research Directorate of MAG.

Expenditure Comments:

The institutional coverage of the expenditure series corresponds with the personnel figures.

The table below gives an overview of how the expenditure figures have been constructed.

	1965	1972	1973	1976	1977	1978	1979	1980	1983
CENTA[a]	0.978	1.230	1.650	2.865	3.108	3.548	3.647	3.960	5.070
ISIC	0.481	[0.503]	[0.506]	0.515	1.025	1.815	1.786	1.913	NA
Total	1.459	1.733	2.156	3.380	4.133	5.363	5.433	5.873	NA
Source	10	10	10	209	209	209	209	209	17

[ ] = estimated or constructed by authors.
a. Before 1972 the figures represent the Research Directorate of MAG.

Acronyms and Abbreviations:
CDG: Centro de Desarrollo Ganadero
CENDEPESCA: Centro de Desarrollo Pesquero
CENREN: Centro de Recursos Naturales
CENTA: Centro Nacional de Tecnología Agropecuaria
    (Centro de Tecnología Agrícola)
INAZUCAR: unknown
ISIC: Instituto Salvadoreño de Investigaciones del Café
MAG: Ministerio de Agricultura y Ganadería

Sources:

0010 Boyce, James K., and Robert E. Evenson. *National and International Agricultural Research and Extension Programs.* New York: Agricultural Development Council, Inc., 1975.

0014 Judd, M. Ann, James K. Boyce, and Robert E. Evenson. "Investing in Agricultural Supply." Economic Growth Center, Yale University, New Haven, Connecticut, 1983. Mimeo.

0209 Centro Nacional de Tecnología Agropecuaria – Ministerio de Agricultura y Ganadería El Salvador. "Informes Sobre Actividades de Investigación y Extensión Agropecuarias realizadas por Instituciones del Sector Público Agropecuario." Documento para el Proyecto BID-IICA-Iowa State University. IICA, San Andrés, Colombia, 1980. Mimeo.

Additional References:

0001 Trigo, Eduardo J., and Martín E. Piñeiro. "Funding Agricultural Research." In *Report of a Conference: Selected Issues in Agricultural Research in Latin America,* edited by Barry Nestel and Eduardo J. Trigo. The Hague: ISNAR, March 1984.

0016 Oram, Peter A., and Vishva Bindlish. *Resource Allocations to National Agricultural Research: Trends in the 1970s.* The Hague and Washington, D.C.: ISNAR and IFPRI, November 1981.

0017 ISNAR, IFARD & AOAD. Survey of National Agricultural Research Systems: Unpublished Questionnaire Responses. ISNAR, The Hague, 1985.

0027 Harvey, Nigel, ed. *Agricultural Research Centres: A World Directory of Organizations and Programmes.* Seventh Edition, Two Volumes. Harlow, U.K.: Longman, 1983.

0053 Piñeiro, Martín, and Eduardo Trigo, eds. *Cambio Técnico en el Agro Latino Americano – Situación y Perspectivas en la Década de 1980.* San José: IICA, 1983.

0073 Oram, Peter A., and Vishva Bindlish. "Investment in Agricultural Research in Developing Countries: Progress, Problems, and the Determination of Priorities." IFPRI, Washington, D.C., January 1984. Mimeo.

0095 FAO – CARIS. *Agricultural Research in Developing Countries – Volume 1: Research Institutions.* Rome: FAO – CARIS, 1978.

0165 Evenson, R.E., and Y. Kislev. *Agricultural Research and Productivity.* New Haven: Yale University Press, 1975.

0172 Trigo, Eduardo. Background material. ISNAR, The Hague, n.d.

0208 Segura, M. "Diagnóstico de la Investigación Agropecuaria en el Istmo Centroamericano." IICA, Guatemala, 1979. Mimeo.

0381 Marroquín, Víctor René. "La Investigación Agrícola Salvadoreña." Paper presented at Curso-Taller sobre la Administración de la Investigación Agrícola, Panama City, 14–25 July 1986.

0835 García Berríos, Carlos Mario. Personal communication. Centro de Tecnología Agrícola. San Andrés, El Salvador, March 1988.

0836 García Berríos, Carlos Mario, and Víctor Manuel Rodríguez A. "Estructura, Logros y Objetivos." Centro de Tecnología Agrícola – División de Investigación Agrícola, San Andrés, El Salvador, December 1987. Mimeo.

0852 Evenson, Robert E., and Yoav Kislev. *Investment in Agricultural Research and Extension: A Survey of International Data.* Center Discussion Paper No. 124. New Haven, Connecticut: Economic Growth Center, Yale University, August 1971.

0886 Evenson, Robert E., and Yoav Kislev. "Investment in Agricultural Research and Extension: A Survey of International Data." *Economic Development and Cultural Change* Vol. 23 (April 1975): 507–521.

Ethiopia

Personnel

Period/Year	PhD	MSc	BSc	Subtotal	Expat	Total	Source
1960–64	..	..	..	..	..	10	10
1965–69	..	..	..	..	..	33	10/175/618
1970	1	6	10	17	23	40	772
1971	1	5	15	21	26	47	772
1972	4	4	20	28	28	56	772
1973	4	6	19	29	28	57	772
1974	5	9	16	30	29	59	772
1975	4	13	9	26	22	48	772
1976	3	7	18	28	19	47	772
1977	3	4	25	32	9	41	772
1978	5	19	33	57	22	79	772
1979	5	20	37	62	15	77	772
1980	5	22	31	58	17	75	772
1981	7	26	51	84	15	99	772
1982	9	35	65	109	12	121	772
1983	12	44	74	130	2	132	772
1984	15	52	87	154	2	156	772
1985	15	55	102	172	1	173	772
1986	17	60	153	230	10	240	383

1960–64: 1961   1965–69: 1966, 68 & 69

Expenditure   *LCU = Birr*

Period/Year	Current LCU (millions)	Constant 1980 LCU (millions)	Constant 1980 US$ (millions)		Source
			Atlas	PPP	
1960–64	..	..	..	..	
1965–69	1.914	3.049	1.564	4.222	589/48
1970	1.725	2.544	1.305	3.523	772
1971	1.764	2.575	1.321	3.566	772
1972	2.964	4.505	2.310	6.237	772
1973	2.347	3.472	1.780	4.807	772
1974	3.413	4.612	2.365	6.386	772
1975	3.154	4.291	2.201	5.942	772
1976	5.067	6.488	3.327	8.983	772
1977	6.211	7.034	3.607	9.739	772
1978	6.695	7.047	3.614	9.758	772
1979	7.180	7.238	3.712	10.022	772
1980	6.926	6.926	3.552	9.590	772
1981	7.948	7.650	3.923	10.593	772
1982	8.172	7.759	3.979	10.744	772
1983	8.169	7.464	3.827	10.334	772
1984	10.414	9.364	4.802	12.966	772
1985	11.866	9.903	5.078	13.712	772
1986	..	..	..	..	

1960–64:   1965–69: 1966 & 69

Personnel Comments:

The figures given by source 772 represent the 'Institute of Agricultural Research' (IAR) only. For the observations in earlier years (except 1969 which includes IAR only) it is not clear which institutes are included, but they seem to be in line with the IAR-only figures given by source 772 for the later years. The 1968 and 1970 figures explicitly exclude university research and fisheries, forestry, and veterinary research. IAR is the principal agricultural research institute in the country. There are, however, a considerable number of other institutes which undertake some agricultural research. The following were identified: Forestry Resources Institute (12 researchers in 1982, source 27), Tendaho Agricultural Development Enterprise Research Section (4 researchers in 1982, source 27), IEMVT – Imperial Veterinary Institute (5 researchers in 1969/1970, source 532, and 8 researchers in 1985, source 228), Arsi Rural Development Unit (8 researchers in 1976, source 95), and Agronomic Research and Advisory Unit of the State Farms Development Authority (1 researcher in 1976, source 95).

Expenditure Comments:

The figures given by source 589, 48, and 772 refer to IAR only.

Acronyms and Abbreviations:

IAR: Institute of Agricultural Research
IEMVT: Institut d'Elevage et de Médecine Vétérinaire des Pays Tropicaux

Sources:

0010 Boyce, James K., and Robert E. Evenson. *National and International Agricultural Research and Extension Programs*. New York: Agricultural Development Council, Inc., 1975.

0048 Shawel, Hailu, and Akalu Negewo. "The Impact of the Collaboration between the International Agricultural Research System and the National Agricultural Research System in Ethiopia." CGIAR, Washington, D.C., December 1984. Mimeo.

0175 Cooper, St.G.C. *Agricultural Research in Tropical Africa*. Kampala: East African Literature Bureau, 1970.

0383 ISNAR. "Tables Prepared by the Ethiopia Review Team." ISNAR, The Hague, 1986. Mimeo.

0589 Kassapu, Samuel. *Les Dépenses de Recherche Agricole dans 34 Pays d'Afrique Tropicale*. Paris: Centre de Développement de l'OCDE, 1976.

0618 Institute of Agricultural Research (IAR). *Report for the Period April 1969 to March 1970*. Addis Ababa: IAR, 1970.

0772 Debela, Seme. Personal communication. General Manager, Institute of Agricultural Research. Addis Ababa, January 1988.

Additional References:

0014 Judd, M. Ann, James K. Boyce, and Robert E. Evenson. "Investing in Agricultural Supply." Economic Growth Center, Yale University, New Haven, Connecticut, 1983. Mimeo.

0016 Oram, Peter A., and Vishva Bindlish. *Resource Allocations to National Agricultural Research: Trends in the 1970s*. The Hague and Washington, D.C.: ISNAR and IFPRI, November 1981.

0017 ISNAR, IFARD & AOAD. Survey of National Agricultural Research Systems: Unpublished Questionnaire Responses. ISNAR, The Hague, 1985.

0019 Bengtsson, Bo, and Tedla Getachew, eds. *Strengthening National Agricultural Research*. Report from a SAREC workshop, 10–17 September 1979. Part 1: Background Documents; Part II: Summary and Conclusions. Sweden: SAREC, 1980.

0023 Bennell, Paul. *Agricultural Researchers in Sub-Saharan Africa: An Overview*. Working Paper No. 4. The Hague: ISNAR, October 1985.

0026 Oram, Peter A., and Mark Gieben. "Document Summaries." ISNAR, The Hague, 1984. Mimeo.

0027 Harvey, Nigel, ed. *Agricultural Research Centres: A World Directory of Organizations and Programmes*. Seventh Edition, Two Volumes. Harlow, U.K.: Longman, 1983.

0073 Oram, Peter A., and Vishva Bindlish. "Investment in Agricultural Research in Developing Countries: Progress, Problems, and the Determination of Priorities." IFPRI, Washington, D.C., January 1984. Mimeo.

0095 FAO – CARIS. *Agricultural Research in Developing Countries – Volume 1: Research Institutions*. Rome: FAO – CARIS, 1978.

0163 CGIAR. "National Agricultural Research." CGIAR, Washington, D.C., 1985. Mimeo.

0164 Association for the Advancement of Agricultural Sciences in Africa (AAASA). *Proceedings of the Workshop on Agricultural Research*

170

Administration, Nairobi, Kenya, 27–30 June 1977.
Proceedings Series PE-4. Addis Ababa, Ethiopia:
AAASA and IDRC, August 1979.

0165 Evenson, R.E., and Y. Kislev. *Agricultural
Research and Productivity.* New Haven: Yale
University Press, 1975.

0174 Watson, J.M. "Comparative Study of
Agricultural Research Organisation and
Administration in the Near East Region." Paper
presented at the Workshop on Organization and
Administration of Agricultural Services in the
Arab States, Cairo, 2–15 March 1964.

0226 World Bank. "Ethiopia Agricultural Research
Project." World Bank, Washington, D.C., 1984.
Mimeo.

0227 Australian Centre for International Agricultural
Research (ACIAR). *Proceedings of the Eastern
Africa-ACIAR Consultation on Agricultural
Research, 18–22 July 1983, Nairobi, Kenya.*
Australia: ACIAR, 1984.

0228 CIRAD. *Coopération Francaise en Recherche
Agronomique: Activités du CIRAD dans les Pays
du Sahel.* Paris: CIRAD, March 1985.

0266 UNESCO. *National Science Policies in Africa.*
Science Policy Studies and Documents No. 31.
Paris: UNESCO, 1974.

0309 Gebre, H. "Agricultural Research in Ethiopia."
In *Strengthening National Agricultural
Research.* Report from a SAREC workshop, 10–17
September 1979, edited by Bo Bengtsson and
Getachew Tedla, pp. 25–35. Sweden: SAREC,
1980.

0360 Cooper, St.G.C. "Towards Trained Manpower for
Agricultural Research in Africa." Paper
presented at the Conference on Agricultural
Research and Production in Africa, organized by
the Association for the Advancement of
Agricultural Sciences in Africa (AAASA), Addis
Ababa, 29 August–4 September 1971.

0400 UNESCO. *The Promotion of Scientific Activity
in Tropical Africa.* Science Policy Studies and
Documents No. 11. Paris: UNESCO, 1969.

0445 Swanson, Burton E., and Wade H. Reeves.
"Agricultural Research Eastern and Southern
Africa: Manpower and Training." World Bank,
Washington, D.C., August 1986. Mimeo.

0465 ISNAR. "Ethiopia: Agricultural Sector Study –

Agricultural Research." ISNAR, The Hague,
1986. Mimeo.

0532 UNESCO Field Science Office for Africa. *Survey
on the Scientific and Technical Potential of the
Countries of Africa.* Paris: UNESCO, 1970.

0653 Webster, B.N. *Index of Agricultural Research
Institutions and Stations in Africa.* Rome: FAO,
n.d.

0712 ISNAR. *Review of Research Program
Management and Manpower Planning at the
Institute of Agricultural Research in Ethiopia.*
The Hague: ISNAR, March 1987.

0744 *Agricultural Research Centres: A World
Directory of Organizations and Programmes.*
Eighth Edition, Two Volumes. Harlow, U.K.:
Longman, 1986.

0745 FAO. *Institute of Agricultural Research Ethiopia
– Project Findings and Recommendations.* Rome:
FAO, 1982.

0760 Abebe, M., M. Mekuria, and T. Gebremeskel.
"Status of Agricultural Research in Ethiopia." In
*Proceedings of the Eastern Africa-ACIAR
Consultation on Agricultural Research,* pp. 9–15.
Australia: ACIAR, 1984.

0847 Kimura, J.H. "Financial and Administrative
Management of Research Institutions in Eastern
and Southern Africa: Report on Responses to a
Questionnaire." In *Promotion of Technology
Policy and Science Management in Africa,* edited
by Karl Wolfgang Menck and Wolfgang Gmelin.
Bonn: Deutsche Stiftung für internationale
Entwicklung (DSE), 1986.

0852 Evenson, Robert E., and Yoav Kislev. *Investment
in Agricultural Research and Extension: A Survey
of International Data.* Center Discussion Paper
No. 124. New Haven, Connecticut: Economic
Growth Center, Yale University, August 1971.

0873 Abdalla, Abdalla Ahmed. "Agricultural
Research in the IGADD Sub-Region and Related
Manpower Training." Inter-Governmental
Authority on Drought and Development
(IGADD) and FAO, 1987. Mimeo.

0886 Evenson, Robert E., and Yoav Kislev. "Investment
in Agricultural Research and Extension: A Survey
of International Data." *Economic Development
and Cultural Change* Vol. 23 (April 1975): 507–
521.

Fiji

Personnel

Period/Year	PhD	MSc	BSc	Subtotal	Expat	Total	Source
1960–64	..			..		..	
1965–69	2	3	16	21	..	21	286/285
1970	..			..		..	
1971	1	10	20	31	..	31	279
1972	..	..		..		..	
1973	..	..		..	..	..	
1974	5	8	18	31	..	31	244
1975	..	..		..	..	..	
1976	..			..		26	95
1977	4	7	16	27		27	243
1978	..	..		..		..	
1979	..	..		..		..	
1980	2	5	22	29	7	36	776
1981	1	7	22	30	7	37	776
1982	1	8	18	27	9	36	776
1983	1	8	19	28	7	35	776
1984		8	22	32	6	38	776
1985	2	9	25	36	6	42	776
1986	2	11	27	40	10	50	776

1960–64:   1965–69: 1965 & 68

Expenditure

LCU = Fiji Dollars

Period/Year	Current LCU (millions)	Constant 1980 LCU(millions)	Constant 1980 US$ (millions)		Source
			Atlas	PPP	
1960–64	..	..	..	..	
1965–69	..	..	..	..	
1970	..	..	..	..	
1971	..	..	..	..	
1972	..	..	..	..	
1973	..	..	..	..	
1974	..	..	..	..	
1975	..	..	..	..	
1976	0.991	1.313	1.503	2.600	95
1977	..	..	..	..	
1978	..	..	..	..	
1979	..	..	..	..	
1980	2.200	2.200	2.518	4.357	723
1981	2.360	2.338	2.675	4.629	723
1982	2.900	2.655	3.039	5.258	723
1983	2.820	2.467	2.823	4.886	723
1984	3.200	2.713	3.104	5.372	723
1985	3.730	2.923	3.345	5.789	723
1986	5.010	..	..	..	723

1960–64:   1965–69:

Personnel Comments:
The time series includes research executed by the Research Division, the Animal Health and Production Division, and the Fisheries Division of the Ministry of Agriculture and Fisheries, plus research executed by the Ministry of Forestry. Not included are the Sugar Cane Research Institute (6, 8, and 9 researchers in 1980, 1983, and 1985, sources 242, 17, and 744, respectively), the Fiji College of Agriculture, and the University of the South Pacific. Source 242 reports, for 1980, a total of 25 teacher-researchers at these two educational establishments.

Expenditure Comments:
The institutional coverage of the expenditure indicators corresponds with the personnel series. The Sugar Cane Research Institute, which for consistency was excluded, spent F$ 900,000 in 1983, source 17; and F$ 1,000,000 in 1985, source 723.

Sources:
0095 FAO – CARIS. *Agricultural Research in Developing Countries – Volume 1: Research Institutions.* Rome: FAO – CARIS, 1978.
0243 Commonwealth Agricultural Bureaux (CAB). *List of Research Workers in the Agricultural Sciences in the Commonwealth and in the Republic of Ireland 1978.* Slough, England: CAB, 1978.
0244 Commonwealth Agricultural Bureaux (CAB). *List of Research Workers in the Agricultural Sciences in the Commonwealth and in the Republic of Ireland 1975.* Slough, England: CAB, 1975.
0279 Commonwealth Agricultural Bureaux (CAB). *List of Research Workers in the Agricultural Sciences in the Commonwealth and in the Republic of Ireland 1972.* Slough, England: CAB, 1972.
0285 Commonwealth Agricultural Bureaux (CAB). *List of Research Workers in the Agricultural Sciences in the Commonwealth and in the Republic of Ireland 1969.* Slough, England: CAB, 1969.
0286 Commonwealth Agricultural Bureaux (CAB). *List of Research Workers in Agriculture, Animal Health and Forestry in the Commonwealth and in the Republic of Ireland 1966.* Slough, England: CAB, 1966.
0723 Fernando, L.H. "South Pacific NARS Resource Commitment Survey." ADB and ISNAR, The Hague, 1987. Mimeo.

0776 Sivan, P. Personal communication. Director of Research, Ministry of Primary Industries. Nausori, Fiji, January 1988.

Additional References:
0016 Oram, Peter A., and Vishva Bindlish. *Resource Allocations to National Agricultural Research: Trends in the 1970s.* The Hague and Washington, D.C.: ISNAR and IFPRI, November 1981.
0017 ISNAR, IFARD & AOAD. Survey of National Agricultural Research Systems: Unpublished Questionnaire Responses. ISNAR, The Hague, 1985.
0026 Oram, Peter A., and Mark Gieben. "Document Summaries." ISNAR, The Hague, 1984. Mimeo.
0027 Harvey, Nigel, ed. *Agricultural Research Centres: A World Directory of Organizations and Programmes.* Seventh Edition, Two Volumes. Harlow, U.K.: Longman, 1983.
0038 Gamble, W.K., R.M. Bourke, and C.W. Brookson. *South Pacific Agricultural Research Study.* Consultants Report to the Asian Development Bank (two volumes). Manila: Asian Development Bank, June 1981.
0042 ISNAR. *Report to the Ministry of Agriculture and Fisheries Fiji: A Review of the Agricultural Research Division.* The Hague: ISNAR, September 1982.
0073 Oram, Peter A., and Vishva Bindlish. "Investment in Agricultural Research in Developing Countries: Progress, Problems, and the Determination of Priorities." IFPRI, Washington, D.C., January 1984. Mimeo.
0163 CGIAR. "National Agricultural Research." CGIAR, Washington, D.C., 1985. Mimeo.
0186 ISNAR. *Agricultural Research Plan Fiji.* The Hague: ISNAR, June 1985.
0242 Commonwealth Agricultural Bureaux (CAB). *List of Research Workers in Agricultural Sciences in the Commonwealth 1981.* Slough, England: CAB, 1981.
0287 Commonwealth Agricultural Bureaux (CAB). *List of Research Workers in Agriculture, Animal Health and Forestry in the British Commonwealth, the Republic of Sudan and the Republic of Ireland 1959.* Slough, England: CAB, 1959.
0732 FAO – Regional Office for Asia and the Pacific. *Agricultural Research Systems in the Asia-Pacific Region.* Bangkok: FAO – Regional Office for Asia and the Pacific, 1986.
0771 Fernando, L.H. "ISNAR/IRETA Workshop on Planning and Management of Agricultural Research in the South Pacific: Survey of Agricultural Research Resources." ISNAR, The Hague, n.d. Mimeo.

Finland

Personnel

Period/Year	PhD	MSc	BSc	Subtotal	Expat	Total	Source
1960–64	..	..	..	..	..	..	
1965–69	..	..	..	..	..	284	10
1970	..	..	..	..	..	..	
1971	..	..	..	..	..	320	609
1972	..	..	..	..	..	..	
1973	..	..	..	..	..	355	609
1974	..	..	..	..	..	..	
1975	..	..	..	..	..	349	609
1976	..	..	..	..	..	..	
1977	..	..	..	..	..	365	609
1978	..	..	..	..	..	..	
1979	..	..	..	..	..	410	609
1980	..	..	..	..	..	..	
1981	..	..	..	..	..	398	609
1982	..	..	..	..	..	..	
1983	..	..	..	..	..	411	609
1984	..	..	..	..	..	..	
1985	..	..	..	..	..	..	
1986	..	..	..	..	..	..	

1960–64:   1965–69: 1969

Expenditure   *LCU = Markka*

Period/Year	Current LCU (millions)	Constant 1980 LCU(millions)	Constant 1980 US$ (millions)		Source
			Atlas	PPP	
1960–64	..	..	..	..	
1965–69	31.920	97.615	24.989	20.968	10
1970	..	..	..	..	
1971	31.120	85.260	21.827	18.314	601/609
1972	37.085	93.649	23.974	20.116	601/609
1973	43.650	96.571	24.722	20.744	601/609
1974	50.360	91.067	23.313	19.562	601/609
1975	65.670	103.744	26.558	22.285	601/609
1976	81.420	114.194	29.233	24.529	601/609
1977	108.870	138.688	35.504	29.791	601/609
1978	103.780	122.671	31.404	26.350	601/609
1979	128.190	139.945	35.826	30.061	601/609
1980	141.700	141.700	36.275	30.438	601/609
1981	163.710	146.957	37.621	31.567	601/609
1982	189.070	155.613	39.837	33.426	601/609
1983	195.140	147.721	37.817	31.731	601/609
1984	213.610	148.443	38.001	31.886	601
1985	232.860	152.595	39.064	32.778	601
1986	267.880	167.530	42.887	35.986	601

1960–64:   1965–69: 1969

Personnel Comments:
These figures represent scientist years and include both university and non-university personnel. The figures are – according to source 507 – exclusive of veterinary research but do include forestry research, which constitutes a significant proportion of the total agricultural research activity in Finland.

The table below shows how the time series has been constructed. The number of researchers has been broken down into graduates and postgraduates. This latter category includes doctors, which are equivalent to PhD, and licentiates, which have (according to source 507) a degree equivalent to something between MSc and PhD. The graduate category includes only candidates which are equivalent to a MSc (source 507). No BSc level is provided for in the statistics. Because the reported categories could not readily be made commensurable with the standard indicator series categories, a degree-level breakdown was not attempted.

	1971			1973			1975			1977		
	P-G	G	Tot.	P-G	G	Tot.	P-G	G	Tot.	P-G	G	Tot.
Government & PNP[a]	82	180	262	91	206	297	94	197	291	105	201	306
Universities[a]	[29]	[29]	58	[29]	[29]	[58]	[29]	[29]	[58]	[29]	[30]	[59]
Total	111	209	320	120	235	355	123	226	349	134	23	365

	1979			1981			1983		
	P-G	G	Tot.	P-G	G	Tot.	P-G	G	Tot.
Government & PNP[a]	116	235	351	106	233	339	102	250	352
Universities[a]	[29]	[30]	[59]	[29]	[30]	[59]	[29]	[30]	[59]
Total	145	265	410	135	263	398	131	280	411

[ ] = estimated or constructed by authors.
PNP = Private Nonprofit.
P-G = Postgraduates (Doctors and Licentiates).
G = Graduates (Candidates).
a. Source 609.

Source 609 provides, for the years 1973, 1975, and 1977, the total number of researchers at universities involved in agricultural research. FTEs for these years were constructed using the proportion of total researchers to FTEs as given for the year 1971 (i.e., 38%). The university figures for the years 1979, 1981, and 1983 are extrapolations.

Expenditure Comments:
These expenditure figures are based on two sources. One is "Public Funding of R & D by Socio-Economic Objective: Agriculture, Forestry and Fishing" (source 601) which includes all public funding of agricultural research except for universities. General university funds are, by convention, placed under the objective 'Advancement of Science' and no breakdown into objectives like 'Industries, Agriculture, Forestry, and Fishing' is provided. To calculate the university component, another source was used, namely, "R & D Expenditures in the Higher Education and Private Nonprofit Sectors: Higher Education – Agricultural Sciences" (sources 602 and 609). Although the definitions used by sources 601 and 602/609 differ to some degree and

therefore incomplete coverage or double-counting of certain items may occur, it is probably the best estimate that is possible.

The table below gives an overview of how the expenditure indicators have been constructed.

	1971	1972	1973	1974	1975	1976	1977	1978
Public Funding	28.60	33.70	39.40	44.90	59.00	73.30	99.30	92.70
Universities	2.52	[3.39]	4.25	[5.46]	6.67	[8.12]	9.57	[11.08]
Total	31.12	37.09	43.65	50.36	65.67	81.42	108.87	103.78
Source	601 609	601 609	601 609	601 609	601 609	601 609	601 609	601 609

	1979	1980	1981	1982	1983	1984	1985	1986
Public Funding	115.60	126.80	146.50	169.55	173.30	189.70	206.80	237.90
Universities	12.59	[14.90]	[17.21]	[19.52]	21.84	[23.91]	[26.06]	[29.98]
Total	128.19	141.7	163.71	189.07	195.14	213.61	232.86	267.88
Source	601 609	601 609	601 609	601 609	601 602	601	601	601

[ ] = estimated or constructed by authors.

Sources:

0010 Boyce, James K., and Robert E. Evenson. *National and International Agricultural Research and Extension Programs.* New York: Agricultural Development Council, Inc., 1975.

0601 OECD – STIID Data Bank. "Public Funding of R&D by Socio-Economic Objective." Tables. OECD, Paris, April 1987. Mimeo.

0609 Central Statistical Office of Finland. *Research Activity.* Official Statistics of Finland XXXVIII:1–7, bi-annually 1971–1983. Helsinki: Central Statistical Office of Finland, various years.

Additional References:

0021 OECD. *OECD Science and Technology Indicators.* Paris: OECD, 1984.

0022 UNESCO. *Statistical Yearbook 1983.* Paris: UNESCO, 1983.

0027 Harvey, Nigel, ed. *Agricultural Research Centres: A World Directory of Organizations and Programmes.* Seventh Edition, Two Volumes. Harlow, U.K.: Longman, 1983.

0165 Evenson, R.E., and Y. Kislev. *Agricultural Research and Productivity.* New Haven: Yale University Press, 1975.

0376 FAO. *Fourteenth FAO Regional Conference for Europe, Reykjavik, Iceland, 17–21 September 1984: Research in Support of Agricultural Policies in Europe.* Rome: FAO, 1984.

0429 Sumelius, John. "The Returns to Investments in Agricultural Research in Finland 1950–1984." *Journal of Agricultural Science in Finland* Vol. 59, No. 4 (1987): 251–354.

0473 OECD. *Science and Technology Indicators: Basic Statistical Series – Volume A: The Objectives of Government R&D Funding 1974–1985.* Paris: OECD, May 1983.

0474 OECD. *Science and Technology Indicators: Basic Statistical Series – Volume A: The Objectives of Government R&D, 1969–1981.* Paris: OECD, July 1981.

0507 Sumelius, John. Personal communication. Agricultural Economics Research Institute. Helsinki, Finland, December 1986.

0508 Poutiainen, Esko. Personal communication. Director General of the Agricultural Research Centre. Jokioinen, Finland, January 1987.

0509 OECD. *Survey of the Resources Devoted to R&D by OECD Member Countries, International Statistical Year 1971 – Volume 5: Total Tables, Statistical Tables and Notes.* Paris: OECD, August 1974.

0510 OECD. *International Survey of the Resources Devoted to R&D in 1967 by OECD Member Countries, Statistical Tables and Notes – Volume 5: Total Tables.* Paris: OECD, August 1970.

0511 OECD. *International Survey of the Resources Devoted to R&D in 1969 by OECD Member Countries, Statistical Tables and Notes – Volume 5: Total Tables.* Paris: OECD, June 1973.

0512 OECD. *Survey of the Resources Devoted to R&D by OECD Member Countries, International Statistical Year 1973 – Volume 5: Total Tables, Statistical Tables and Notes.* Paris: OECD, September 1976.

0513 OECD. *International Survey of the Resources Devoted to R&D by OECD Member Countries, International Statistical Year 1975 – International Volume: Statistical Tables and Notes.* Paris: OECD, March 1979.

0530 OECD. *OECD Science and Technology Indicators No. 2 – R&D, Invention and Competitiveness.* Paris: OECD, 1986.

0602 OECD – STIID Data Bank. "R&D Expenditure in the Higher Education and Private Non-Profit Sectors." Tables. OECD, Paris, April 1987. Mimeo.

0605 Academy of Finland. *Kertomus Suomen Akatemian Toiminnasta 1984.* Helsinki: Academy of Finland, 1985. 0621 NORDFORSK. *Forsknings Virksomhet i Norden i 1967 – Utgifter og Personale.* Nordisk Statistisk Skriftserie 18. Oslo: NORDFORSK, 1970.

0707 OECD – STIID Data Bank. "R&D Expenditure in the Higher Education and Private Non-Profit Sectors: Higher Education, Agricultural Sciences, General Universities Funds." Table. OECD, Paris, June 1987. Mimeo.

0852 Evenson, Robert E., and Yoav Kislev. *Investment in Agricultural Research and Extension: A Survey of International Data.* Center Discussion Paper No. 124. New Haven, Connecticut: Economic Growth Center, Yale University, August 1971.

0886 Evenson, Robert E., and Yoav Kislev. "Investment in Agricultural Research and Extension: A Survey of International Data." *Economic Development and Cultural Change* Vol. 23 (April 1975): 507–521.

France

Personnel

Period/Year	PhD	MSc	BSc	Subtotal	Expat	Total	Source
1960–64	..	..	..	..	..	507	282
1965–69	..	..	..	..	..	1282	405/444
1970	..	..	..	..	..	1398	404
1971	..	..	..	..	..	1461	404
1972	..	..	..	..	..	1521	404
1973	..	..	..	..	..	1570	404
1974	..	..	..	..	..	1602	404
1975	..	..	..	..	..	1635	404
1976	..	..	..	..	..	1730	404
1977	..	..	..	..	..	1810	404
1978	..	..	..	..	..	1871	404
1979	..	..	..	..	..	1928	404
1980	..	..	..	..	..	2017	404
1981	..	..	..	..	..	2068	404
1982	..	..	..	..	..	2166	404
1983	..	..	..	..	..	2386	404
1984	..	..	..	..	..	2592	404
1985	..	..	..	..	..	2593	404
1986	..	..	..	..	..	2750	853

1960–64: 1960   1965–69: 1966 & 68

Expenditure

LCU = French Francs

Period/Year	Current LCU (millions)	Constant 1980 LCU(millions)	Constant 1980 US$ (millions)		Source
			Atlas	PPP	
1960–64	76.270	259.633	58.473	48.856	282/552/533
1965–69	169.897	499.487	112.492	93.990	533/444
1970	264.600	654.950	147.504	123.244	404
1971	368.700	863.466	194.465	162.481	404
1972	367.900	812.141	182.906	152.823	404
1973	456.000	932.515	210.016	175.475	404
1974	528.300	972.928	219.118	183.079	404
1975	621.900	1009.578	227.372	189.976	404
1976	721.800	1067.751	240.473	200.922	404
1977	798.300	1078.784	242.958	202.998	404
1978	874.700	1073.252	241.712	201.957	404
1979	1031.600	1148.775	258.721	216.169	404
1980	1183.500	1183.500	266.542	222.703	404
1981	1367.500	1227.558	276.464	230.994	404
1982	1561.500	1255.225	282.695	236.200	404
1983	1780.400	1303.367	293.538	245.259	404
1984	1905.278	1298.758	292.499	244.392	404
1985	2064.461	1330.194	299.579	250.307	405
1986	2230.000	1373.153	309.264	258.391	853

1960–64: 1960, 61, 62, 63 & 64   1965–69: 1965, 66 & 68

Personnel Comments:
These figures refer to 'Institut National de la Recherche Agronomique' (INRA) only. They include both "chercheurs" and "ingénieurs". In several sources only "chercheurs" are classified and counted as researchers, while other sources indicate that "ingénieurs" should also be classified as researchers and counted as such. INRA is the main agricultural research institute in France and in most sources the only institute which is mentioned and for which statistics are provided. There are, however, several other organizations which conduct agricultural research.

Following INRA, the most important institute is CEMAGREF. CEMAGREF was created in 1982, or there about, through the amalgamation of CNEEMA and CERAFER (formerly CREGR). In 1986 CEMAGREF employed, according to source 537, 460 "chercheurs et ingénieurs". Earlier observations (source 283 and 444) for CNEEMA and CERAFER give only the number of "chercheurs" working in the institute. In contrast to INRA, which has a somewhat more independent status, CEMAGREF (and previously CNEEMA and CERAFER) is under the direct control of the Ministry of Agriculture. INRA is formally under the control of both the Ministry of Agriculture and the Ministry of Science and Education.

In addition to researchers within these two public, noneducational, agricultural research organizations, one can identify, according to source 422, about 500 full-time equivalent (FTE) agricultural researchers within the higher education system in 1986. There are relatively strong links between the higher education system and INRA. Around 1980, according to source 283, 59 of the 121 agricultural and veterinary faculties were associated with INRA and employed INRA research workers. On the basis of the figures provided by source 283, there was an estimated 417 FTE agricultural researchers working in the higher education system around 1980.

INRA, CEMAGREF, and the higher education system are involved in research activities in the strict sense of the word. There are, however, a large number of (private nonprofit) organizations which are involved in so-called "Recherche et Développement". Two key associations, namely ACTA and ACTIA, consist of 17 and 15 organizations, respectively. Very limited quantitative information concerning the research executed by these organizations is available. Source 423 gives an estimated 900 "researchers" within ACTA organizations and 110 researchers within ACTIA organizations (circa 1985).

Two major organizations in France are involved in international or tropical agricultural research, namely, CIRAD (formerly GERDAT) and ORSTOM. Because of their nondomestic mandate, they are not included in the Indicator Series for France. During the early eighties, both organizations employed about 700 researchers each (sources 656 and 373), of which more than half were stationed overseas.

Expenditures Comments:
The expenditure figures refer to INRA only. CEMAGREF was reported to have spent FF13.270 million in 1969 (CNEEMA and CERAFER – source 444) and FF228.200 million in 1985 (source 411). No expenditure figures are mentioned for the higher education sector in the sources consulted for this series. Source 423 estimates a total of FF600 million spent by ACTA and ACTIA organizations (circa 1985).

Acronyms and Abbreviations:
ACTA: Association de Coordination Technique Agricole
ACTIA: Association de Coordination Technique des Industries Agro-alimentaires
CEMAGREF: Centre National du Machinisme Agricole, du Génie Rural, des Eaux et Forêts
CERAFER: Centre National d'Etudes Techniques et de Recherches Techniques pour l'Agriculture, les Forêts et l'Equipement Rural
CIRAD: Centre de Cooperation Internationale en Recherche Agronomique pour le Développement
CNEEMA: Centre National d'Etudes et d'Expérimentation du Machinisme Agricole
CREGR: Centre de Recherche et d'Expérimentation du Génie Rural
GERDAT: Groupement d'Etudes et de Recherches pour le Développement de l'Agronomie Tropicale
INRA: Institut National de la Recherche Agronomique
ORSTOM: Office de la Recherche Scentifique et Technique Outre-Mer

Sources:

0282 EEG. *De Organisatie van het Landbouwkundig Onderzoek in de Landen van de EEG.* EEG Studies, Serie Landbouw 9. Brussel: EEG, 1963.

0404 Institut National de la Recherche Agronomique (INRA). *Analyse Retrospective des Moyens de l'INRA de 1970 a 1985.* Paris: INRA, 1986.

0405 Institut National de la Recherche Agronomique (INRA). *INRA 1986.* Paris: INRA, 1986.

0444 UNESCO. *La Politique Scientifique et l'Organisation de la Recherche en France.* Etudes et Documents de Politique Scientifique No. 24. Paris: UNESCO, 1971.

0533 Institut National de la Recherche Agronomique (INRA). *L'Institut National de la Recherche Agronomique, Edition du 20e Anniversaire 1946–1966.* Paris: INRA, 1966.

0552 OECD. *Country Reports on the Organisation of Scientific Research: France.* Paris: OECD, May 1964.

0853 Casas, Joseph, and Alfred Conesa. "Bref Aperçu du Système Francais de Recherche Agricole et Agro-Alimentaire." Centre International des Hautes Etudes Agronomiques Méditerranéennes (CIHEAM), Paris, September 1987. Mimeo.

Additional References:

0010 Boyce, James K., and Robert E. Evenson. *National and International Agricultural Research and Extension Programs.* New York: Agricultural Development Council, Inc., 1975.

0014 Judd, M. Ann, James K. Boyce, and Robert E. Evenson. "Investing in Agricultural Supply." Economic Growth Center, Yale University, New Haven, Connecticut, 1983. Mimeo.

0022 UNESCO. *Statistical Yearbook 1983.* Paris: UNESCO, 1983.

0027 Harvey, Nigel, ed. *Agricultural Research Centres: A World Directory of Organizations and Programmes.* Seventh Edition, Two Volumes. Harlow, U.K.: Longman, 1983.

0095 FAO – CARIS. *Agricultural Research in Developing Countries – Volume 1: Research Institutions.* Rome: FAO – CARIS, 1978.

0165 Evenson, R.E., and Y. Kislev. *Agricultural Research and Productivity.* New Haven: Yale University Press, 1975.

0283 European Economic Community (EEC). "Monograph on the Organization of Agricultural Research." EEC, Brussels, 1982. Mimeo.

0373 Groupement d'Etudes et de Recherches pour le Développement de l'Agronomie Tropicale (GERDAT). *Rapport Général d'Activité pour 1982.* Paris: GERDAT, 1983.

0376 FAO. *Fourteenth FAO Regional Conference for Europe, Reykjavik, Iceland, 17–21 September 1984: Research in Support of Agricultural Policies in Europe.* Rome: FAO, 1984.

0403 Institut National de la Recherche Agronomique (INRA). *Rapport d'Activité 1985.* Paris: INRA, 1985.

0407 Centre International de Hautes Etudes Agronomiques Méditerranéennes. "Séminaire sur l'Orientation et l'Organisation de la Recherche Agronomique dans les Pays du Bassin Méditerranéen, Istanbul, 1–3 December 1986." Centre International de Hautes Etudes Agronomiques Méditerranéennes, Paris, December 1986. Mimeo.

0411 Conesa, A.P. "Quelques Aspects du Dispositif de Recherche Agricole et Agro-Alimentaire en France." Paper presented at the Séminaire sur l'Orientation et l'Organisation de la Recherche Agronomique dans les Pays du Bassin Méditerranéen, Centre International de Hautes Etudes Agronomiques Méditerranéennes, Istanbul, 1–3 December 1986.

0422 Casas, Joseph. "Les Ressources Humaines et Financières dans les Pays du Bassin Méditerranéen." Preliminary table. ISNAR, The Hague, 1986. Mimeo.

0423 Casas, Joseph. "Dispositif de Recherche-Développement du Secteur Agricole – Les Budgets des Organismes." ISNAR, The Hague, 1986. Mimeo.

0473 OECD. *Science and Technology Indicators: Basic Statistical Series – Volume A: The Objectives of Government R&D Funding 1974–1985.* Paris: OECD, May 1983.

0474 OECD. *Science and Technology Indicators: Basic Statistical Series – Volume A: The Objectives of Government R&D, 1969–1981.* Paris: OECD, July 1981.

0509 OECD. *Survey of the Resources Devoted to R&D by OECD Member Countries, International Statistical Year 1971 – Volume 5: Total Tables, Statistical Tables and Notes.* Paris: OECD, August 1974.

0510 OECD. *International Survey of the Resources Devoted to R&D in 1967 by OECD Member Countries, Statistical Tables and Notes – Volume 5: Total Tables.* Paris: OECD, August 1970.

0511 OECD. *International Survey of the Resources Devoted to R&D in 1969 by OECD Member Countries, Statistical Tables and Notes – Volume 5: Total Tables.* Paris: OECD, June 1973.

0512 OECD. *Survey of the Resources Devoted to R&D by OECD Member Countries, International Statistical Year 1973 – Volume 5: Total Tables, Statistical Tables and Notes.* Paris: OECD, September 1976.

0513 OECD. *International Survey of the Resources Devoted to R&D by OECD Member Countries,*

180

International Statistical Year 1975 – International Volume: Statistical Tables and Notes. Paris: OECD, March 1979.

0514 OECD. *A Study of Resources Devoted to R&D in OECD Member Countries in 1963/64, International Statistical Year for Research and Development – Volume 2: Statistical Tables and Notes.* Paris: OECD, 1968.

0530 OECD. *OECD Science and Technology Indicators No. 2 – R&D, Invention and Competitiveness.* Paris: OECD, 1986.

0535 Commissariat Général du Plan d'Equipement et de la Productivité. *5e Plan de Développement Economique et Social – La Recherche Scientifique et Technique, Tome 1.* Paris: Commissariat Général du Plan d'Equipement et de la Productivité, 1970.

0537 Centre National du Machinisme Agricole, du Génie Rural, des Eaux et des Forêts (CEMAGREF). *Programmation des Actions CEMAGREF 1986–1988.* Antony, France: CEMAGREF, October 1986.

0553 OECD. *Reviews of National Science Policy: France.* Paris: OECD, 1966.

0574 OECD. *Intellectual Investment in Agriculture for Economic and Social Development.* Documentation in Food and Agriculture No. 60. Paris: OECD, 1962.

0601 OECD – STIID Data Bank. "Public Funding of R&D by Socio-Economic Objective." Tables. OECD, Paris, April 1987. Mimeo.

0602 OECD – STIID Data Bank. "R&D Expenditure in the Higher Education and Private Non-Profit Sectors." Tables. OECD, Paris, April 1987. Mimeo.

0606 EUROSTAT. *Government Financing of Research and Development 1975–1984.* Luxembourg: Office des Publications Officielles des Communautés Européennes, February 1985.

0703 Martin, Ben R., and John Irvine. *An International Comparison of Government Funding of Academic and Academically Related Research.* Advisory Board for the Research Council (ABRC) Science Policy Study No. 2. Brighton: ABRC – Science Policy Research Unit, University of Sussex, October 1986.

0706 EUROSTAT. *Government Financing of Research and Development 1975–1985.* Luxembourg: Office for Official Publications of the European Communities, 1987.

0852 Evenson, Robert E., and Yoav Kislev. *Investment in Agricultural Research and Extension: A Survey of International Data.* Center Discussion Paper No. 124. New Haven, Connecticut: Economic Growth Center, Yale University, August 1971.

0875 Bouchet, Frederic C. "An Analysis of Sources of Growth in French Agriculture 1960–1984." PhD diss., Virginia Polytechnic Institute and State University, Blacksburg, Virginia, August 1987.

0886 Evenson, Robert E., and Yoav Kislev. "Investment in Agricultural Research and Extension: A Survey of International Data." *Economic Development and Cultural Change* Vol. 23 (April 1975): 507–521.

French Polynesia

Personnel

Period/Year	PhD	MSc	BSc	Subtotal	Expat	Total	Source
1960–64	..	..	..	..	..	..	
1965–69	..	..	..	..	..	..	
1970	..	..	..	..	..	..	
1971	..	..	..	..	..	..	
1972	..	..	..	..	..	..	
1973	..	..	..	..	..	..	
1974	..	..	..	..	..	..	
1975	..	..	..	..	..	..	
1976	..	..	..	..	..	..	
1977	..	..	..	..	..	..	
1978	..	..	..	..	..	..	
1979	..	..	..	..	..	..	
1980	..	..	..	..	..	..	
1981	..	..	..	..	..	3	38
1982	..	..	..	..	..	..	
1983	..	..	..	..	..	..	
1984	..	..	..	..	..	..	
1985	..	..	..	..	..	..	
1986	..	..	..	..	..	..	

1960–64:   1965–69:

Expenditure   LCU = CFP Francs

Period/Year	Current LCU (millions)	Constant 1980 LCU(millions)	Constant 1980 US$ (millions)		Source
			Atlas	PPP	
1960–64	..	..	..	..	
1965–69	..	..	..	..	
1970	..	..	..	..	
1971	..	..	..	..	
1972	..	..	..	..	
1973	..	..	..	..	
1974	..	..	..	..	
1975	..	..	..	..	
1976	..	..	..	..	
1977	..	..	..	..	
1978	..	..	..	..	
1979	..	..	..	..	
1980	..	..	..	..	
1981	..	..	..	..	
1982	..	..	..	..	
1983	..	..	..	..	
1984	..	..	..	..	
1985	..	..	..	..	
1986	..	..	..	..	

1960–64:   1965–69:

Personnel Comments:

Agricultural Research is conducted in French Polynesia by 'Institut de Recherches Agronomiques Tropicales et des Cultures Vivrières' (IRAT). Its staff consisted in 1981 of an agronomist, a soil scientist, and a plant pathologist (source 38).

Sources:

0038 Gamble, W.K., R.M. Bourke, and C.W. Brookson. *South Pacific Agricultural Research Study.* Consultants' Report to the Asian Development Bank (two volumes). Manila: Asian Development Bank, June 1981.

Additional References:

0654 Office de la Recherche Scientifique et Technique Outre-Mer (ORSTOM). *Rapport d'Activité 1967.* Paris: ORSTOM, 1969.

Gabon

Personnel

Period/Year	PhD	MSc	BSc	Subtotal	Expat	Total	Source
1960–64	..	..	..	..	..	..	
1965–69	..	..	..	..	..	..	
1970	..	..	..	0	7	7	532/10
1971	..	..	..	0	7	7	589/10
1972	..	..	..		..	..	
1973	..	..	..	0	7	7	589/10
1974	..	..	..	0	8	8	589/10
1975	..	..	..		..		
1976	..	..	..	..	..		
1977	..	..	..		..	..	
1978	..	..	..	..	..	..	
1979	..	..	..	..	..	..	
1980	..	..	..		..	..	
1981	..	..	..	..	..		
1982	..	..	..	..	..		
1983	..	..	..	..			
1984	..	..	..	..			
1985	–3–		7	10	14	24	86
1986	..	..	..	..	..	..	

1960–64:　　　　　　　　　　1965–69:

Expenditure　　　　　　　　　　　　　　　　　　　　*LCU = CFA Francs*

Period/Year	Current LCU (millions)	Constant 1980 LCU(millions)	Constant 1980 US$ (millions)		Source
			Atlas	PPP	
1960–64	..	..	..	..	
1965–69	58.343	334.191	1.273	1.204	589/10
1970	69.292	316.202	1.205	1.139	589/10
1971	73.346	322.864	1.230	1.163	589/10
1972	..	..	..	..	
1973	84.879	309.041	1.177	1.113	589/10
1974	..	..	..	..	
1975	..	..	..	..	
1976	..	..	..	..	
1977	..	..	..	..	
1978	..	..	..	..	
1979	..	..	..	..	
1980	..	..	..	..	
1981	..	..	..	..	
1982	..	..	..	..	
1983	..	..	..	..	
1984	..	..	..	..	
1985	1180.000	692.488	2.638	2.494	68
1986	..	..	..	..	

1960–64:　　　　　　　　1965–69: 1966

Personnel Comments:

As in most former French colonies, agricultural research was dominated by France long after independence. Information contained in sources 448 and 589, indicates that the French research institutes, CTFT, IRAT, and ORSTOM were permanently based in Gabon during the period 1960–1975. IRAT, however, abandoned its research station in Gabon in 1970 or 1971. The other French research agencies, like IRCA and IRHO, only occasionaly send a research mission to Gabon. Only the pedology component of ORSTOM's research activities has been included in the indicator series. The table below gives an overview of how the indicators have been constructed.

	1960	1965	1966	1970	1971	1973	1974
ORSTOM	2	3	NA	[3]	[3]	2	[2]
IRAT	NA	NA	1	1	–	–	–
CTFT	NA	NA	NA	3	4	[5]	6
Total	NA	NA	NA	7	7	7	8
Source	10	10	653	532	589	10	589

[ ] = estimated or constructed by authors.

For the period after 1975 there is a considerable lack of documented information concerning agricultural research activities in Gabon. For 1985, source 68 gives the following institutions: CMCE (1 researcher), CIAM (8 researchers), IRCA (2 researchers), OGAPROV (2 researchers), 'Ferme Porcine' (3 researchers), and IRAF (8 researchers). Except for IRAF, which is under the responsibility of the 'Ministère de l'Enseignement Supérieur et de la Recherche Scientifique,' all other institutions are under the responsibility of the 'Ministère de l'Agriculture, de l'Elevage et de l'Economie Rurale'.

Expenditure Comments:

The pre-1975 expenditure indicators correspond with the personnel indicators and have been constructed as follows:

	1960	1965	1966	1970	1971	1973
ORSTOM	3.100	4.005	4.783	7.895	[9.990]	14.179
IRAT	NA	NA	NA	NA	–	–
CTFT	NA	NA	53.560	[61.397]	63.356	70.700*
Total	NA	NA	58.343	69.292	73.346	84.879
Source	10	10	589	10	589	10/589*

[ ] = estimated or constructed by authors.

The 1985 expenditure figure includes the same agricultural research organizations as the personnel figure, except for IRAF. About one-third of the identified researchers are located in IRAF, so it is possible that 1985 agricultural research expenditures may be up to 50% (approximately 590 million FCFA) higher than the figure included in the indicator series. Source 68, however, states that IRAF staff are all Gabonese and that there is a high staff turnover. This may indicate that 1) the foreign donor component is low, and 2) salaries are low. This being the case, IRAF's 1985 expenditures may be much lower than the estimated 590 million FCFA figure. In the absence of a reliable estimate for IRAF's expenditures, they were excluded from the 1985 indicator series estimate.

Acronyms and Abbreviations:

CIAM: Centre d'Introduction, d'Adaptation et de Multiplication de Matériel Végétal Vivrier et Fruitier

CMCE: Centre de Multiplication des Cacaoyers d'Elite

CTFT: Centre Technique Forestier Tropical

IRAF: Institut des Recherches Agronomiques et Forestières

IRAT: Institut de la Recherche Agronomique Tropicale

IRCA: Institut de Recherches sur le Caoutchouc

IRHO: Institut de Recherche sur les Huiles et les Oléagineux

OGAPROV: Office Gabonais d'Amélioration et de Production de Viande

ORSTOM: Organisation de Recherche Scientifique et Technique Outre Mer

Sources:

0010 Boyce, James K., and Robert E. Evenson. *National and International Agricultural Research and Extension Programs.* New York: Agricultural Development Council, Inc., 1975.

0068 Conte, Stephan. "West Africa Agricultural Research Data Base Building Task: Termination Report." World Bank, Washington, D.C., 1985. Mimeo.

0086 World Bank. "West Africa Agricultural Research Review – Country Studies." World Bank, Washington, D.C., 1985. Mimeo.

0532 UNESCO Field Science Office for Africa. *Survey on the Scientific and Technical Potential of the Countries of Africa.* Paris: UNESCO, 1970.

0589 Kassapu, Samuel. *Les Dépenses de Recherche Agricole dans 34 Pays d'Afrique Tropicale.* Paris: Centre de Développement de l'OCDE, 1976.

Additional References:

0088 Swanson, Burton E., and Wade H. Reeves. "Agricultural Research West Africa: Manpower and Training." World Bank, Washington, D.C., November 1985. Mimeo.

0089 Oram, Peter. "Report on National Agricultural Research in West Africa." IFPRI, Washington, D.C., February 1986. Mimeo.

0175 Cooper, St.G.C. *Agricultural Research in Tropical Africa.* Kampala: East African Literature Bureau, 1970.

0266 UNESCO. *National Science Policies in Africa.* Science Policy Studies and Documents No. 31. Paris: UNESCO, 1974.

0360 Cooper, St.G.C. "Towards Trained Manpower for Agricultural Research in Africa." Paper presented at the Conference on Agricultural Research and Production in Africa, organized by the Association for the Advancement of Agricultural Sciences in Africa (AAASA), Addis Ababa, 29 August–4 September 1971.

0373 Groupement d'Etudes et de Recherches pour le Développement de l'Agronomie Tropicale (GERDAT). *Rapport Général d'Activité pour 1982.* Paris: GERDAT, 1983.

0400 UNESCO. *The Promotion of Scientific Activity in Tropical Africa.* Science Policy Studies and Documents No. 11. Paris: UNESCO, 1969.

0466 World Bank – West Africa Agricultural Research Review Team. "National Agricultural Research Systems (NARS): A Regional Appraisal and Selected Issues." World Bank, Washington, D.C., August 1986. Mimeo.

0653 Webster, B.N. *Index of Agricultural Research Institutions and Stations in Africa.* Rome: FAO, n.d.

0654 Office de la Recherche Scientifique et Technique Outre-Mer (ORSTOM). *Rapport d'Activité 1967.* Paris: ORSTOM, 1969.

0852 Evenson, Robert E., and Yoav Kislev. *Investment in Agricultural Research and Extension: A Survey of International Data.* Center Discussion Paper No. 124. New Haven, Connecticut: Economic Growth Center, Yale University, August 1971.

Gambia

Personnel

Period/Year	PhD	MSc	BSc	Subtotal	Expat	Total	Source
1960–64	..	..	..	..	..	..	
1965–69	..	..	..	..	..	..	
1970	..	..	..	..	..	..	
1971	..	..	..	..	..	..	
1972	..	..	..	..	..	..	
1973	..	..	..	..	..	..	
1974	..	..	..	..	..	..	
1975	..	..	..	..	..	..	
1976	..	..	..	..	..	..	
1977	..	..	..	..	..	..	
1978	..	..	..	..	..	..	
1979	..	..	..	..	..	..	
1980	..	..	..	..	..	..	
1981	..	..	..	..	..	..	
1982	..	..	..	..	..	..	
1983	..	..	..	..	..	..	
1984	..	..	..	45	17	62	140
1985	..	..	..	..	..	..	
1986	..	..	..	..	..	..	

1960–64:   1965–69:

Expenditure

LCU = Dalasis

Period/Year	Current LCU (millions)	Constant 1980 LCU(millions)	Constant 1980 US$ (millions)		Source
			Atlas	PPP	
1960–64	..	..	..	..	
1965–69	..	..	..	..	
1970	..	..	..	..	
1971	..	..	..	..	
1972	..	..	..	..	
1973	..	..	..	..	
1974	..	..	..	..	
1975	..	..	..	..	
1976	..	..	..	..	
1977	..	..	..	..	
1978	..	..	..	..	
1979	..	..	..	..	
1980	..	..	..	..	
1981	..	..	..	..	
1982	..	..	..	..	
1983	..	..	..	..	
1984	..	..	..	..	
1985	..	..	..	..	
1986	..	..	..	..	

1960–64:   1965–69:

Personnel Comments:
Sources 68, 86, and 88 give a figure for 1983/84 (year not exactly identifiable) of 23, 23, and 27 researchers, respectively. There is a confusing cycle of internal citation among these sources but it is surmised that all these observations originated from two Devres reports – source 139, which gives a figure of 23 researchers, and source 45, which gives a figure of 28 researchers.

However, these only include researchers working in the Department of Agriculture – Agricultural Research Services and the Department of Animal Health. The indicator series includes a more extensive number of research entities and projects. The table below gives an overview of the data which have been compiled to date. However, no consistent time series could be constructed.

	'65	'66	'68	'70	'71	'72	'74	'77	'80	'81	'82	'83	'84
MOA													
DOA	6	7	5	7	5	4[c]	6	10	13				
ARS												11	17
CPS													10
AES													3
VD	1			1		1							
DOAH&P[a]							3				15	22[b]	21[b]
MOWR&E													
WRD													5
FD										5		6	5
DOF													1
Source	286	175	285	175	279	431	244	243	242	68	27	384	140

a. Created in 1973, the Veterinary Department was incorporated into this department.
b. These figures explicitly include the Mixed Farming and Resources Management Program.
c. Yundum Experimental Station only.

Expenditure Comments:
Source 384 gives an average expenditure figure of D. 2,866,000 (US$ 1,433,000) for the early eighties (to be more specific: the average is based on the most recent three years recorded, probably 1980–1982). It includes the Agricultural Research Services, the Department of Animal Health and Production, and the Fisheries Department. However, the 1985 expenditure figure given by source 140 totals D.1,190,000 but includes even more institutes than the figure given by source 384.

Acronyms and Abbreviations:
AES: Agricultural Engineering Service
ARS: Agricultural Research Service
CPS: Crop Protection Service
DOA: Department of Agriculture
DOAH&P: Department of Animal Health & Production
DOF: Department of Forestry
FD: Fisheries Department

MOA: Ministry of Agriculture
MOWR&E : Ministry of Water Resources & Environment
VD: Veterinary Department
WRD: Water Resources Department

Sources:
0140 University of Wisconsin. "Gambia Agricultural Research and Diversification Project." Discussion paper. University of Wisconsin, Madison, 1985. Mimeo.

Additional References:
0010 Boyce, James K., and Robert E. Evenson. *National and International Agricultural Research and Extension Programs.* New York: Agricultural Development Council, Inc., 1975.
0014 Judd, M. Ann, James K. Boyce, and Robert E. Evenson. "Investing in Agricultural Supply." Economic Growth Center, Yale University, New Haven, Connecticut, 1983. Mimeo.
0016 Oram, Peter A., and Vishva Bindlish. *Resource Allocations to National Agricultural Research:*

Trends in the 1970s. The Hague and Washington, D.C.: ISNAR and IFPRI, November 1981.

0023 Bennell, Paul. *Agricultural Researchers in Sub-Saharan Africa: An Overview.* Working Paper No. 4. The Hague: ISNAR, October 1985.

0027 Harvey, Nigel, ed. *Agricultural Research Centres: A World Directory of Organizations and Programmes.* Seventh Edition, Two Volumes. Harlow, U.K.: Longman, 1983.

0045 Institut du Sahel, and DEVRES, Inc. *Bilan des Ressources de la Recherche Agricole dans les Pays du Sahel, Volume I: Analyse et Stratégie Régionale.* Washington, D.C.: DEVRES, Inc., 1984.

0068 Conte, Stephan. "West Africa Agricultural Research Data Base Building Task: Termination Report." World Bank, Washington, D.C., 1985. Mimeo.

0083 Institut du Sahel, and DEVRES, Inc. *Bilan des Ressources de la Recherche Agricole dans les Pays du Sahel, Volume II: Résumés des Rapports Nationaux.* Washington, D.C.: DEVRES, Inc., August 1984.

0086 World Bank. "West Africa Agricultural Research Review – Country Studies." World Bank, Washington, D.C., 1985. Mimeo.

0088 Swanson, Burton E., and Wade H. Reeves. "Agricultural Research West Africa: Manpower and Training." World Bank, Washington, D.C., November 1985. Mimeo.

0089 Oram, Peter. "Report on National Agricultural Research in West Africa." IFPRI, Washington, D.C., February 1986. Mimeo.

0136 Snyder, Monteze M. *The Organizational Context and Manpower Implications for the Ministry of Agriculture The Gambia: Part II.* Banjul, The Gambia: AID, November 1984.

0137 IBRD – Western Africa Projects Department Agriculture Division 4. "The Gambia Agricultural Development Project II." IBRD, Washington, D.C., April 1984. Mimeo.

0138 McLean, Diana. "Agricultural Research Analysis." AID, Banjul, The Gambia, September 1984. Mimeo.

0139 Institut du Sahel, and DEVRES, Inc. *Assessment of Agricultural Research Resources in the Sahel, Volume III: National Report The Gambia.* Washington, D.C.: DEVRES, Inc., August 1984.

0163 CGIAR. "National Agricultural Research." CGIAR, Washington, D.C., 1985. Mimeo.

0175 Cooper, St.G.C. *Agricultural Research in Tropical Africa.* Kampala: East African Literature Bureau, 1970.

0242 Commonwealth Agricultural Bureaux (CAB). *List of Research Workers in Agricultural Sciences in the Commonwealth 1981.* Slough, England: CAB, 1981.

0243 Commonwealth Agricultural Bureaux (CAB). *List of Research Workers in the Agricultural Sciences in the Commonwealth and in the Republic of Ireland 1978.* Slough, England: CAB, 1978.

0244 Commonwealth Agricultural Bureaux (CAB). *List of Research Workers in the Agricultural Sciences in the Commonwealth and in the Republic of Ireland 1975.* Slough, England: CAB, 1975.

0273 Sompo-Ceesay, M.S. "Organization and Structure of Agricultural Research in The Gambia: Past, Present and Future." Paper presented at the Workshop on Improving Agricultural Research Organization and Management: Progress and Issues for the Future, ISNAR, The Hague, 8–12 September 1986.

0279 Commonwealth Agricultural Bureaux (CAB). *List of Research Workers in the Agricultural Sciences in the Commonwealth and in the Republic of Ireland 1972.* Slough, England: CAB, 1972.

0285 Commonwealth Agricultural Bureaux (CAB). *List of Research Workers in the Agricultural Sciences in the Commonwealth and in the Republic of Ireland 1969.* Slough, England: CAB, 1969.

0286 Commonwealth Agricultural Bureaux (CAB). *List of Research Workers in Agriculture, Animal Health and Forestry in the Commonwealth and in the Republic of Ireland 1966.* Slough, England: CAB, 1966.

0360 Cooper, St.G.C. "Towards Trained Manpower for Agricultural Research in Africa." Paper presented at the Conference on Agricultural Research and Production in Africa, organized by the Association for the Advancement of Agricultural Sciences in Africa (AAASA), Addis Ababa, 29 August–4 September 1971.

0372 Gabas, Jean-Jacques. *La Recherche Agricole dans les Pays Membres du CILSS.* Club du Sahel, September 1986.

0384 Institut du Sahel, and DEVRES, Inc. *Sahel Region Agricultural Research Resource Assessment – Data Base Management Information System: Diskettes and User's Guide.* Printouts. Washington, D.C.: Institut du Sahel, and DEVRES, Inc, n.d.

0386 "Manpower Analysis Gambia." n.d. Mimeo.

0431 Current Agricultural Research Information System (CARIS). *Directory of Agricultural Research Institutions and Projects in West Africa – Pilot Project 1972–73.* Rome: FAO, 1973.

0466 World Bank – West Africa Agricultural Research Review Team. "National Agricultural Research Systems (NARS): A Regional Appraisal and

Selected Issues." World Bank, Washington, D.C., August 1986. Mimeo.

0532 UNESCO Field Science Office for Africa. *Survey on the Scientific and Technical Potential of the Countries of Africa.* Paris: UNESCO, 1970.

0589 Kassapu, Samuel. *Les Dépenses de Recherche Agricole dans 34 Pays d'Afrique Tropicale.* Paris: Centre de Développement de l'OCDE, 1976.

0653 Webster, B.N. *Index of Agricultural Research Institutions and Stations in Africa.* Rome: FAO, n.d.

0852 Evenson, Robert E., and Yoav Kislev. *Investment in Agricultural Research and Extension: A Survey of International Data.* Center Discussion Paper No. 124. New Haven, Connecticut: Economic Growth Center, Yale University, August 1971.

Germany, Fed. Rep. of

Personnel

Period/Year	PhD	MSc	BSc	Subtotal	Expat	Total	Source
1960–64	..	..	..	..	..	..	
1965–69	..	..	..	..	..	2592	672
1970	..	..	..	..	..	..	
1971	..	..	..	..	..	2681	672
1972	..	..	..	..	..	..	
1973	..	..	..	..	..	2780	672
1974	..	..	..	..	..	..	
1975	..	..	..	..	..	2468	672
1976	..	..	..	..	..	..	
1977	..	..	..	..	..	2283	673
1978	..	..	..	..	..	..	
1979	..	..	..	..	..	2019	673
1980	..	..	..	..	..	..	
1981	..	..	..	..	..	2070	673
1982	..	..	..	..	..	..	
1983	..	..	..	..	..	2179	673
1984	..	..	..	..	..	..	
1985	..	..	..	..	..	..	
1986	..	..	..	..	..	..	

1960–64:   1965–69: 1969

Expenditure   LCU = D–Marken

Period/Year	Current LCU (millions)	Constant 1980 LCU(millions)	Constant 1980 US$ (millions) Atlas	Constant 1980 US$ (millions) PPP	Source
1960–64	174.000	363.257	201.756	147.881	799
1965–69	258.500	479.766	266.466	195.311	799
1970	..	..	..	..	
1971	406.000	630.435	350.149	256.648	799
1972	..	..	..	..	
1973	497.000	689.320	382.854	280.620	799
1974	..	..	..	..	
1975	635.000	776.284	431.154	316.022	799
1976	..	..	..	..	
1977	555.000	630.682	350.286	256.748	799
1978	..	..	..	..	
1979	599.000	627.883	348.731	255.609	799
1980	..	..	..	..	799
1981	633.000	608.654	338.051	247.781	799
1982	..	..	..	..	799
1983	700.000	624.442	346.820	254.208	799
1984	..	..	..	..	799
1985	741.000	634.418	352.361	258.269	799
1986	..	..	..	..	

1960–64: 1964   1965–69: 1967 & 69

Personnel Comments:

Agricultural research in the Federal Republic of Germany (FRG) has a complex organizational structure involving universities, autonomous institutes, and several ministries at both federal as well as state level.

The table below gives a breakdown of the total number of researchers into the university and non-university sector. The numbers represent researchers in agricultural sciences and not researchers classified by the socioeconomic objective category: 'agriculture, forestry, and fisheries'. There is considerable overlap between the two categories, though they are not identical (source 714).

	1969	1971	1973	1975	1977	1979	1981	1983
Non-university sector	1564	1391	1335	1303	1123	1070	1079	1179
University sector	1028	1290	1445	1165	1160	949	991	1000
Total	2592	2681	2780	2468	2283	2019	2070	2179
Source	672	672	672	672	673	673	673	673

Expenditure Comments:

The table below shows how the expenditure indicators have been constructed. The figures represent actual expenditures in millions of current local currency units.

	1964	1967	1969	1971	1973	1975	1977	1979	1981	1983	1985
Non-university	54	93	119	186	224	290	287	270	307	346	365
University	120	130	175	220	273	345	268	329	326	354	376
Total	174	223	294	406	497	635	555	599	633	700	741
Source	799	799	799	799	799	799	799	799	799	799	799

Sources:

0672 Der Bundesminister für Forschung und Technologie. *Faktenbericht 1981 zum Bundesbericht Forschung.* Reihe: Berichte und Dokumentationen. Bonn: Der Bundesminister für Forschung und Technologie, 1982.

0673 Der Bundesminister für Forschung und Technologie. *Faktenbericht 1986 zum Bundesbericht Forschung.* Bonn: Der Bundesminister für Forschung und Technologie, May 1986.

0799 Rost, E. Personal communication. Bundesministerium für Forschung und Technologie. Bonn, February 1988.

Additional References:

0010 Boyce, James K., and Robert E. Evenson. *National and International Agricultural Research and Extension Programs.* New York: Agricultural Development Council, Inc., 1975.

0021 OECD. *OECD Science and Technology Indicators.* Paris: OECD, 1984.

0022 UNESCO. *Statistical Yearbook 1983.* Paris: UNESCO, 1983.

0027 Harvey, Nigel, ed. *Agricultural Research Centres: A World Directory of Organizations and Programmes.* Seventh Edition, Two Volumes. Harlow, U.K.: Longman, 1983.

0165 Evenson, R.E., and Y. Kislev. *Agricultural Research and Productivity*. New Haven: Yale University Press, 1975.

0282 EEG. *De Organisatie van het Landbouwkundig Onderzoek in de Landen van de EEG*. EEG Studies, Serie Landbouw 9. Brussel: EEG, 1963.

0283 European Economic Community (EEC). "Monograph on the Organization of Agricultural Research." EEC, Brussels, 1982. Mimeo.

0376 FAO. *Fourteenth FAO Regional Conference for Europe, Reykjavik, Iceland, 17–21 September 1984: Research in Support of Agricultural Policies in Europe*. Rome: FAO, 1984.

0449 Ministerie van Landbouw en Visserij – Directie Akkerbouw en Tuinbouw. "Verslag van een Studiereis naar de Bondsrepubliek Duitsland in de Periode 11 t/m 15 Augustus 1975." Ministerie van Landbouw en Visserij, The Hague, 1975. Mimeo.

0473 OECD. *Science and Technology Indicators: Basic Statistical Series – Volume A: The Objectives of Government R&D Funding 1974–1985*. Paris: OECD, May 1983.

0474 OECD. *Science and Technology Indicators: Basic Statistical Series – Volume A: The Objectives of Government R&D, 1969–1981*. Paris: OECD, July 1981.

0509 OECD. *Survey of the Resources Devoted to R&D by OECD Member Countries, International Statistical Year 1971 – Volume 5: Total Tables, Statistical Tables and Notes*. Paris: OECD, August 1974.

0510 OECD. *International Survey of the Resources Devoted to R&D in 1967 by OECD Member Countries, Statistical Tables and Notes – Volume 5: Total Tables*. Paris: OECD, August 1970.

0511 OECD. *International Survey of the Resources Devoted to R&D in 1969 by OECD Member Countries, Statistical Tables and Notes – Volume 5: Total Tables*. Paris: OECD, June 1973.

0512 OECD. *Survey of the Resources Devoted to R&D by OECD Member Countries, International Statistical Year 1973 – Volume 5: Total Tables, Statistical Tables and Notes*. Paris: OECD, September 1976.

0513 OECD. *International Survey of the Resources Devoted to R&D by OECD Member Countries, International Statistical Year 1975 – International Volume: Statistical Tables and Notes*. Paris: OECD, March 1979.

0514 OECD. *A Study of Resources Devoted to R&D in OECD Member Countries in 1963/64, International Statistical Year for Research and Development – Volume 2: Statistical Tables and Notes*. Paris: OECD, 1968.

0530 OECD. *OECD Science and Technology Indicators No. 2 – R&D, Invention and Competitiveness*. Paris: OECD, 1986.

0562 OECD. *Country Reports on the Organisation of Scientific Research: Germany*. Paris: OECD, November 1963.

0601 OECD – STIID Data Bank. "Public Funding of R&D by Socio-Economic Objective." Tables. OECD, Paris, April 1987. Mimeo.

0602 OECD – STIID Data Bank. "R&D Expenditure in the Higher Education and Private Non-Profit Sectors." Tables. OECD, Paris, April 1987. Mimeo.

0606 EUROSTAT. *Government Financing of Research and Development 1975–1984*. Luxembourg: Office des Publications Officielles des Communautés Européennes, February 1985.

0670 Der Bundesminister für Forschung und Technologie. *Fünfter Forschungsbericht der Bundesregierung*. Bonn: Der Bundesminister für Forschung und Technologie, 1975.

0671 Der Bundesminister für Forschung und Technologie. *Bundesbericht Forschung VI*. Reihe: Berichte und Dokumentationen, Band 4. Bonn: Der Bundesminister für Forschung und Technologie, 1979.

0703 Martin, Ben R., and John Irvine. *An International Comparison of Government Funding of Academic and Academically Related Research*. Advisory Board for the Research Council (ABRC) Science Policy Study No. 2. Brighton: ABRC – Science Policy Research Unit, University of Sussex, October 1986.

0706 EUROSTAT. *Government Financing of Research and Development 1975–1985*. Luxembourg: Office for Official Publications of the European Communities, 1987.

0714 OECD. Personal communication. Background information from the OECD Science and Technology Indicators Division. Paris, Summer 1987.

0852 Evenson, Robert E., and Yoav Kislev. *Investment in Agricultural Research and Extension: A Survey of International Data*. Center Discussion Paper No. 124. New Haven, Connecticut: Economic Growth Center, Yale University, August 1971.

0886 Evenson, Robert E., and Yoav Kislev. "Investment in Agricultural Research and Extension: A Survey of International Data." *Economic Development and Cultural Change* Vol. 23 (April 1975): 507–521.

Ghana

Personnel

Period/Year	PhD	MSc	BSc	Subtotal	Expat	Total	Source
1960–64	..	..	..	..	..	..	
1965–69	21	30	25	76	..	76	286/285
1970	..	..	..	..	..	115	532
1971	..	..	..	..	..		
1972	..	..	..	..	..	112	431
1973	..	..	..	..	..		
1974	32	46	27	105	..	105	244
1975	..	..	..	..	..	116	164
1976	..	..	..	..	..	129	164
1977	35	64	31	130	..	130	243
1978	..	..	..	..	..	..	
1979	..	..	..	..	..	..	
1980	35	63	22	120	..	120	242
1981	..	..	..	..	..	..	
1982	..	..	..	..	..	..	
1983	28	66	43	137	6	143	17
1984	..	..	..	..	..	..	
1985	31	66	42	139	12	151	86
1986	..	..	..	..	..	..	

1960–64: 1965–69: 1965 & 68

Expenditure

LCU = Cedis

Period/Year	Current LCU (millions)	Constant 1980 LCU(millions)	Constant 1980 US$ (millions)		Source
			Atlas	PPP	
1960–64	..	..	..	..	
1965–69	2.069	48.476	4.926	6.335	175
1970	3.079	56.071	5.698	7.327	26
1971	..	..	..	..	
1972	..	..	..	..	
1973	..	..	..	..	
1974	3.423	34.075	3.463	4.453	16
1975	4.557	35.021	3.559	4.577	16
1976	9.447	56.675	5.759	7.406	16
1977	10.881	39.039	3.967	5.102	16
1978	13.500	27.949	2.840	3.652	16
1979	23.916	35.897	3.648	4.691	16
1980	27.850	27.850	2.830	3.639	16
1981	..	..	..	..	
1982	..	..	..	..	
1983	109.125	23.322	2.370	3.048	17
1984	..	..	..	..	
1985	..	..	..	..	
1986	..	..	..	..	

1960–64: 1965–69: 1967

Personnel Comments:

These figures can be broken down as follows:

	'65	'68	'70	'71	'72	'74	'75	'76	'77	'80	'81	'83	'85
ARI	20	13	16	7	10	11	12	20	18	17	NA	14	18
CRI[a]	17	22	35	28	31	28	30	29	32	29	NA	54	61
FRI	9	16	25	10	25	13	19	19	19	21	NA	21	22
IAB			10	[10]	[10]	5	13	13	15	11	NA	16	16
SRI	9	11	8	9	16	17	18	20	18	19	NA	8	6
CCRI	15	19	21	20	20	31	24	28	28	23	NA	30	28
Subtotal	70	81	115	84	112	105	116	129	130	120	NA	143	151
FPRI	16	13	18	20		19	21	22	26	24			6*
FRU		4	7	6		6			6		5		
CDB						4			3	3	2		
WRI													14
VRA	1		5							9			
MOA[b]	10	16		28		27			28	52			
MOF	3	3		1		1			1	1			
Universities	45	75	113	78	21	136			98	65			
GAEC										7			
DVS/MOA			2		3								
Source	286	285	532	279	431	244	164	164	243	242	68	17	86 744*

[ ] = estimated or constructed by authors.
a. Inclusive Oil Research Institute which was split out of the Crop Research Institute in 1981 or 1982.
b. Exclusive of the Fishery Research Institute.

The lists of Research Workers published by the Commonwealth Agricultural Bureaux over the period 1965–80 (sources 286, 285, 279, 244, and 242) also identify agricultural researchers within the ministries and universities. They are not included in our indicator time series because 1) personnel identified within the ministries held positions such as head of a division, etc.; it seems reasonable not to classify them as researchers. 2) For those agricultural researchers working at the universities, it is not clear what proportion of their time is spent on research. The far lower number of researchers given by source 431 suggests that the time spent on research by university personnel is low.

Expenditure Comments:

For the 1983 figure, the following research institutes are included: CRI, ARI, FRI, IAB, SRI, OPRI, and CCRI. For all other expenditure figures, it is not clear which institutes are included. Sources for the later years (sources 68, 86, and 310) report considerable differences between budgeted and actual expenditures. Only 70–80% of the budget was released in these years. For these sources in particular, budgeted amounts versus expenditures were not clearly identified, making it difficult to select the actual expenditure figures.

Acronyms and Abbreviations:
ARI: Animal Research Institute
CCRI: Cocoa Research Institute
CDB: Cotton Development Board
CRI: Crops Research Institute
DVS: Division of Veterinary Services
FPRI: Forest Productions Research Institute
FRI: Fishery Research Institute
FRU: Fisheries Research Unit
GAEC: Ghana Atomic Energy Committee
IAB: Institute of Aquatic Biology
MOA: Ministry of Agriculture
MOF: Ministry of Forestry
OPRI: Oil Palm Research Institute

SRI: Soil Research Institute
VRA : Volta River Authority
WRI: Water Resources Institute

Sources:

0016 Oram, Peter A., and Vishva Bindlish. *Resource Allocations to National Agricultural Research: Trends in the 1970s.* The Hague and Washington, D.C.: ISNAR and IFPRI, November 1981.

0017 ISNAR, IFARD & AOAD. Survey of National Agricultural Research Systems: Unpublished Questionnaire Responses. ISNAR, The Hague, 1985.

0026 Oram, Peter A., and Mark Gieben. "Document Summaries." ISNAR, The Hague, 1984. Mimeo.

0086 World Bank. "West Africa Agricultural Research Review – Country Studies." World Bank, Washington, D.C., 1985. Mimeo.

0164 Association for the Advancement of Agricultural Sciences in Africa (AAASA). *Proceedings of the Workshop on Agricultural Research Administration, Nairobi, Kenya, 27–30 June 1977.* Proceedings Series PE-4. Addis Ababa, Ethiopia: AAASA and IDRC, August 1979.

0175 Cooper, St.G.C. *Agricultural Research in Tropical Africa.* Kampala: East African Literature Bureau, 1970.

0242 Commonwealth Agricultural Bureaux (CAB). *List of Research Workers in Agricultural Sciences in the Commonwealth 1981.* Slough, England: CAB, 1981.

0243 Commonwealth Agricultural Bureaux (CAB). *List of Research Workers in the Agricultural Sciences in the Commonwealth and in the Republic of Ireland 1978.* Slough, England: CAB, 1978.

0244 Commonwealth Agricultural Bureaux (CAB). *List of Research Workers in the Agricultural Sciences in the Commonwealth and in the Republic of Ireland 1975.* Slough, England: CAB, 1975.

0285 Commonwealth Agricultural Bureaux (CAB). *List of Research Workers in the Agricultural Sciences in the Commonwealth and in the Republic of Ireland 1969.* Slough, England: CAB, 1969.

0286 Commonwealth Agricultural Bureaux (CAB). *List of Research Workers in Agriculture, Animal Health and Forestry in the Commonwealth and in the Republic of Ireland 1966.* Slough, England: CAB, 1966.

0431 Current Agricultural Research Information System (CARIS). *Directory of Agricultural Research Institutions and Projects in West Africa – Pilot Project 1972–73.* Rome: FAO, 1973.

0532 UNESCO Field Science Office for Africa. *Survey on the Scientific and Technical Potential of the Countries of Africa.* Paris: UNESCO, 1970.

Additional References:

0010 Boyce, James K., and Robert E. Evenson. *National and International Agricultural Research and Extension Programs.* New York: Agricultural Development Council, Inc., 1975.

0014 Judd, M. Ann, James K. Boyce, and Robert E. Evenson. "Investing in Agricultural Supply." Economic Growth Center, Yale University, New Haven, Connecticut, 1983. Mimeo.

0019 Bengtsson, Bo, and Tedla Getachew, eds. *Strengthening National Agricultural Research.* Report from a SAREC workshop, 10–17 September 1979. Part 1: Background Documents; Part II: Summary and Conclusions. Sweden: SAREC, 1980.

0023 Bennell, Paul. *Agricultural Researchers in Sub-Saharan Africa: An Overview.* Working Paper No. 4. The Hague: ISNAR, October 1985.

0027 Harvey, Nigel, ed. *Agricultural Research Centres: A World Directory of Organizations and Programmes.* Seventh Edition, Two Volumes. Harlow, U.K.: Longman, 1983.

0068 Conte, Stephan. "West Africa Agricultural Research Data Base Building Task: Termination Report." World Bank, Washington, D.C., 1985. Mimeo.

0073 Oram, Peter A., and Vishva Bindlish. "Investment in Agricultural Research in Developing Countries: Progress, Problems, and the Determination of Priorities." IFPRI, Washington, D.C., January 1984. Mimeo.

0088 Swanson, Burton E., and Wade H. Reeves. "Agricultural Research West Africa: Manpower and Training." World Bank, Washington, D.C., November 1985. Mimeo.

0089 Oram, Peter. "Report on National Agricultural Research in West Africa." IFPRI, Washington, D.C., February 1986. Mimeo.

0095 FAO – CARIS. *Agricultural Research in Developing Countries – Volume 1: Research Institutions.* Rome: FAO – CARIS, 1978.

0141 IAR&T Consultancy Group – University of Ife. *A Consultancy Report on Agricultural Research Delivery Systems in West Africa.* Rome: FAO, December 1979.

0165 Evenson, R.E., and Y. Kislev. *Agricultural Research and Productivity.* New Haven: Yale University Press, 1975.

0266 UNESCO. *National Science Policies in Africa.* Science Policy Studies and Documents No. 31. Paris: UNESCO, 1974.

0279 Commonwealth Agricultural Bureaux (CAB). *List of Research Workers in the Agricultural Sciences in the Commonwealth and in the Republic of Ireland 1972.* Slough, England: CAB, 1972.

0287 Commonwealth Agricultural Bureaux (CAB). *List of Research Workers in Agriculture, Animal Health and Forestry in the British Commonwealth, the Republic of Sudan and the Republic of Ireland 1959.* Slough, England: CAB, 1959.

0310 Agble, W.K. "Agricultural Research in Ghana." In *Strengthening National Agricultural Research.* Report from a SAREC workshop, 10–17 September 1979, edited by Bo Bengtsson and Getachew Tedla, pp. 36–43. Sweden: SAREC, 1980.

0360 Cooper, St.G.C. "Towards Trained Manpower for Agricultural Research in Africa." Paper presented at the Conference on Agricultural Research and Production in Africa, organized by the Association for the Advancement of Agricultural Sciences in Africa (AAASA), Addis Ababa, 29 August–4 September 1971.

0400 UNESCO. *The Promotion of Scientific Activity in Tropical Africa.* Science Policy Studies and Documents No. 11. Paris: UNESCO, 1969.

0462 Soil Research Institute. *Annual Report of the Soil Research Institute January 1965–March 1966.* Accra, Ghana: Ghana Academy of Sciences, 1966.

0466 World Bank – West Africa Agricultural Research Review Team. "National Agricultural Research Systems (NARS): A Regional Appraisal and Selected Issues." World Bank, Washington, D.C., August 1986. Mimeo.

0589 Kassapu, Samuel. *Les Dépenses de Recherche Agricole dans 34 Pays d'Afrique Tropicale.* Paris: Centre de Développement de l'OCDE, 1976.

0653 Webster, B.N. *Index of Agricultural Research Institutions and Stations in Africa.* Rome: FAO, n.d.

0744 *Agricultural Research Centres: A World Directory of Organizations and Programmes.* Eighth Edition, Two Volumes. Harlow, U.K.: Longman, 1986.

0852 Evenson, Robert E., and Yoav Kislev. *Investment in Agricultural Research and Extension: A Survey of International Data.* Center Discussion Paper No. 124. New Haven, Connecticut: Economic Growth Center, Yale University, August 1971.

0886 Evenson, Robert E., and Yoav Kislev. "Investment in Agricultural Research and Extension: A Survey of International Data." *Economic Development and Cultural Change* Vol. 23 (April 1975): 507–521.

Greece

Personnel

Period/Year	PhD	MSc	BSc	Subtotal	Expat	Total	Source
1960–64	..	..	..	..	..	235	574/10/514
1965–69	..	..	..	..	..	281	10/55
1970	..	..	..	..	..	..	
1971	..	..	..	..	..	..	
1972	..	..	..	..	..	..	
1973	..	..	..	..	..	371	10
1974	..	..	..	..	..	..	
1975	..	..	..	..	..	..	
1976	..	..	..	..	..	..	
1977	..	..	..	..	..	..	
1978	..	..	..	..	..	..	
1979	..	..	..	..	..	..	
1980	..	..	..	..	..	425	283
1981	..	..	..	..	..	..	
1982	..	..	..	..	..	..	
1983	..	..	..	..	..	..	
1984	..	..	..	..	..	..	
1985	..	..	..	..	..	460	422
1986	..	..	..	..	..	..	

1960–64: 1960, 62 & 64   1965–69: 1966, 67 & 69

Expenditure

LCU = Drachmas

Period/Year	Current LCU (millions)	Constant 1980 LCU(millions)	Constant 1980 US$ (millions) Atlas	PPP	Source
1960–64	..	..		..	
1965–69	91.708	356.481	8.489	8.807	510/10
1970	..	..	..	..	
1971	..	..	..	..	
1972	..	..	..	..	
1973	..	..	..	..	
1974	..	..	..	..	
1975	..	..	..	..	
1976	..	..	..	..	
1977	584.400	921.767	21.950	22.773	474
1978	..	..	..	..	
1979	809.100	951.882	22.667	23.517	601
1980	723.000	723.000	17.217	17.862	601
1981	1275.600	1065.664	25.377	26.328	601
1982	1188.000	793.057	18.885	19.593	601
1983	1441.770	808.620	19.256	19.978	601
1984	2399.880	1118.304	26.630	27.629	601
1985	3245.610	1284.373	30.585	31.732	601
1986	..	..	..	..	

1960–64:   1965–69: 1966 & 69

Personnel Comments:

The 1964 full-time equivalents (FTEs) figure from source 514 – which source 10 incorrectly placed in 1963 – includes 273 FTEs working in the general government sector and 20 working in the higher education sector. For all other observations it is not clear whether the university sector is included, except for the 1985 figure which explicitly excludes the university sector.

Expenditure Comments:

The 1966 figure, from source 510, represents expenditures of D66.906 million by the general government sector and D14.110 million by the higher education sector. From 1977 on, no higher education research expenditures are included. A 1961 estimate of D63.0 million from source 529, which source 10 cites, plus another estimate of D137.2 million for 1973, also cited by source 10, are omitted from the series as they both exclude salary expenses.

Sources:

0010 Boyce, James K., and Robert E. Evenson. *National and International Agricultural Research and Extension Programs.* New York: Agricultural Development Council, Inc., 1975.

0283 European Economic Community (EEC). "Monograph on the Organization of Agricultural Research." EEC, Brussels, 1982. Mimeo.

0422 Casas, Joseph. "Les Ressources Humaines et Financières dans les Pays du Bassin Méditerranéen." Preliminary table. ISNAR, The Hague, 1986. Mimeo.

0474 OECD. *Science and Technology Indicators: Basic Statistical Series – Volume A: The Objectives of Government R&D, 1969–1981.* Paris: OECD, July 1981.

0510 OECD. *International Survey of the Resources Devoted to R&D in 1967 by OECD Member Countries, Statistical Tables and Notes – Volume 5: Total Tables.* Paris: OECD, August 1970.

0514 OECD. *A Study of Resources Devoted to R&D in OECD Member Countries in 1963/64, International Statistical Year for Research and Development – Volume 2: Statistical Tables and Notes.* Paris: OECD, 1968.

0574 OECD. *Intellectual Investment in Agriculture for Economic and Social Development.* Documentation in Food and Agriculture No. 60. Paris: OECD, 1962.

0601 OECD – STIID Data Bank. "Public Funding of R&D by Socio-Economic Objective." Tables. OECD, Paris, April 1987. Mimeo.

Additional References:

0021 OECD. *OECD Science and Technology Indicators.* Paris: OECD, 1984.

0022 UNESCO. *Statistical Yearbook 1983.* Paris: UNESCO, 1983.

0027 Harvey, Nigel, ed. *Agricultural Research Centres: A World Directory of Organizations and Programmes.* Seventh Edition, Two Volumes. Harlow, U.K.: Longman, 1983.

0165 Evenson, R.E., and Y. Kislev. *Agricultural Research and Productivity.* New Haven: Yale University Press, 1975.

0376 FAO. *Fourteenth FAO Regional Conference for Europe, Reykjavik, Iceland, 17–21 September 1984: Research in Support of Agricultural Policies in Europe.* Rome: FAO, 1984.

0407 Centre International de Hautes Etudes Agronomiques Méditerranéennes. "Séminaire sur l'Orientation et l'Organisation de la Recherche Agronomique dans les Pays du Bassin Méditerranéen, Istanbul, 1–3 December 1986." Centre International de Hautes Etudes Agronomiques Méditerranéennes, Paris, December 1986. Mimeo.

0412 Liacos, Leonidas. "La Recherche Vétérinaire en Grêce." Paper presented at the Séminaire sur l'Orientation et l'Organisation de la Recherche Agronomique dans les Pays du Bassin Méditerranéen, Centre International de Hautes Etudes Agronomiques Méditerranéennes, Istanbul, 1–3 December 1986.

0413 Papanastasis, Vasilios. "La Recherche Forestière en Grêce." Paper presented at the Séminaire sur l'Orientation et l'Organisation de la Recherche Agronomique dans les Pays du Bassin Méditerranéen, Centre International de Hautes Etudes Agronomiques Méditerranéennes, Istanbul, 1–3 December 1986.

0414 Papanastasis, Vasilios. "Quelques Détails sur la Recherche Forestière." Paper presented at the Séminaire sur l'Orientation et l'Organisation de la Recherche Agronomique dans les Pays du Bassin Méditerranéen, Centre International de Hautes Etudes Agronomiques Méditerranéennes, Istanbul, 1–3 December 1986.

0415 Sficas, A.G. "L'Organisation et l'Orientation de la Recherche." Paper presented at the Séminaire sur l'Orientation et l'Organisation de la Recherche Agronomique dans les Pays du Bassin Méditerranéen, Centre International de Hautes

Etudes Agronomiques Méditerranéennes, Istanbul, 1–3 December 1986.

0473 OECD. *Science and Technology Indicators: Basic Statistical Series – Volume A: The Objectives of Government R&D Funding 1974–1985.* Paris: OECD, May 1983.

0509 OECD. *Survey of the Resources Devoted to R&D by OECD Member Countries, International Statistical Year 1971 – Volume 5: Total Tables, Statistical Tables and Notes.* Paris: OECD, August 1974.

0511 OECD. *International Survey of the Resources Devoted to R&D in 1969 by OECD Member Countries, Statistical Tables and Notes – Volume 5: Total Tables.* Paris: OECD, June 1973.

0528 OECD. *Country Reports on the Organisation of Scientific Research: Greece.* Paris: OECD, November 1963.

0529 OECD. *Reviews of National Science Policy: Greece.* Paris: OECD, September 1965.

0530 OECD. *OECD Science and Technology Indicators No. 2 – R&D, Invention and Competitiveness.* Paris: OECD, 1986.

0602 OECD – STIID Data Bank. "R&D Expenditure in the Higher Education and Private Non-Profit Sectors." Tables. OECD, Paris, April 1987. Mimeo.

0606 EUROSTAT. *Government Financing of Research and Development 1975–1984.* Luxembourg: Office des Publications Officielles des Communautés Européennes, February 1985.

0706 EUROSTAT. *Government Financing of Research and Development 1975–1985.* Luxembourg: Office for Official Publications of the European Communities, 1987.

0852 Evenson, Robert E., and Yoav Kislev. *Investment in Agricultural Research and Extension: A Survey of International Data.* Center Discussion Paper No. 124. New Haven, Connecticut: Economic Growth Center, Yale University, August 1971.

0886 Evenson, Robert E., and Yoav Kislev. "Investment in Agricultural Research and Extension: A Survey of International Data." *Economic Development and Cultural Change* Vol. 23 (April 1975): 507–521.

Grenada

Personnel

Period/Year	PhD	MSc	BSc	Subtotal	Expat	Total	Source
1960–64	..	..	..		..	..	
1965–69	0	1	1	2	..	2	286/285
1970	..	..	..	..	..	..	
1971	0	1	1	2	..	2	279
1972	..	..	..	..		..	
1973	..	..	..	..		..	
1974	0	0	0	0	..	0	244
1975	0	0	0	0	..	0	342
1976	..	..	..	..		..	
1977	..	..	..	..		..	
1978	..	..	..	..		..	
1979	..	..	..	..		..	
1980	0	1	1	2	..	2	343
1981	..	..	..	..	..	5	342
1982	..	..	..	..		..	
1983	0	1	1	2	..	2	355
1984	0	1	0	1	..	1	354
1985	..	..	..	..		..	
1986	..	..	..	..		..	

1960–64:   1965–69: 1965 & 68

Expenditure   *LCU = East Caribbean Dollars*

Period/Year	Current LCU (millions)	Constant 1980 LCU(millions)	Constant 1980 US$ (millions)		Source
			Atlas	PPP	
1960–64	..	..	..	..	
1965–69	..	..	..	..	
1970	..	..	..	..	
1971	..	..	..	..	
1972	..	..	..	..	
1973	..	..	..	..	
1974	..	..	..	..	
1975	..	..	..	..	
1976	..	..	..	..	
1977	..	..	..	..	
1978	..	..	..	..	
1979	..	..	..	..	
1980	..	..	..	..	
1981	..	..	..	..	
1982	..	..	..	..	
1983	..	..	..	..	
1984	..	..	..	..	
1985	..	..	..	..	
1986	..	..	..	..	

1960–64:   1965–69:

Personnel Comments:
The figures prior to 1975 include the Ministry of Agriculture only. In 1975 CARDI (Caribbean Agricultural Research and Development Institute) was created. According to source 63, CARDI presently plays a major role within the Agricultural Research System of Grenada. The post-1975 figures refer to the Grenada unit of CARDI only.

Sources:
0244 Commonwealth Agricultural Bureaux (CAB). *List of Research Workers in the Agricultural Sciences in the Commonwealth and in the Republic of Ireland 1975.* Slough, England: CAB, 1975.
0279 Commonwealth Agricultural Bureaux (CAB). *List of Research Workers in the Agricultural Sciences in the Commonwealth and in the Republic of Ireland 1972.* Slough, England: CAB, 1972.
0285 Commonwealth Agricultural Bureaux (CAB). *List of Research Workers in the Agricultural Sciences in the Commonwealth and in the Republic of Ireland 1969.* Slough, England: CAB, 1969.
0286 Commonwealth Agricultural Bureaux (CAB). *List of Research Workers in Agriculture, Animal Health and Forestry in the Commonwealth and in the Republic of Ireland 1966.* Slough, England: CAB, 1966.
0342 Caribbean Agricultural Research and Development Institute (CARDI). *CARDI 1976–1981: Report of the Chairman.* St. Augustine, Trinidad & Tobago: CARDI, 1982.
0343 Caribbean Agricultural Research and Development Institute (CARDI). *Annual Report 1980.* St. Augustine, Trinidad & Tobago: CARDI, 1981.
0354 Caribbean Agricultural Research and Development Institute (CARDI). *Annual Report Research and Development 1983–84: Highlights.* St. Augustine, Trinidad: CARDI, 1984.
0355 Caribbean Agricultural Research and Development Institute (CARDI). *Research and Development Summary 1982–83.* St. Augustine, Trinidad: CARDI, 1983.

Additional References:
0053 Piñeiro, Martín, and Eduardo Trigo, eds. *Cambio Tecnico en el Agro Latino Americano – Situacion y Perspectivas en la Decada de 1980.* San José: IICA, 1983.
0063 ISNAR. *Report to the Board of Directors of CARDI: Analysis, Evaluation and Proposals to Strengthen CARDI's Regional Capacity.* The Hague: ISNAR, 1985.
0287 Commonwealth Agricultural Bureaux (CAB). *List of Research Workers in Agriculture, Animal Health and Forestry in the British Commonwealth, the Republic of Sudan and the Republic of Ireland 1959.* Slough, England: CAB, 1959.
0353 Caribbean Agricultural Research and Development Institute (CARDI). *Paper Presented at a Meeting of the Governing Body of CARDI on January 29, 1982, Basseterre, St. Kitts.* St. Augustine, Trinidad: CARDI, 1982.
0356 Caribbean Agricultural Research and Development Institute (CARDI). *Research and Development Summary 1983–84.* St. Augustine, Trinidad: CARDI, 1984.

Guadeloupe

Personnel

Period/Year	PhD	MSc	BSc	Subtotal	Expat	Total	Source
1960–64	..	..	..	..	..	..	
1965–69	..	..	..	..	..	..	
1970	..	..	..	..	..	..	
1971	..	..	..	..	..	..	
1972	..	..	..	..	..	..	
1973	..	..	..	..	..	..	
1974	..	..	..	..	..	..	
1975	..	..	..	..	..	..	
1976	..	..	..	..	..	..	
1977	..	..	..	..	..	..	
1978	..	..	..	..	..	..	
1979	..	..	..	..	..	..	
1980	..	..	..	..	..	..	
1981	..	..	..	..	..	..	
1982	..	..	..	..	..	14	27
1983	..	..	..	..	..	..	
1984	..	..	..	..	..	..	
1985	..	..	..	..	..	..	
1986	..	..	..	..	..	..	

1960–64:   1965–69:

Expenditure

LCU = French Francs

Period/Year	Current LCU (millions)	Constant 1980 LCU(millions)	Constant 1980 US$ (millions)		Source
			Atlas	PPP	
1960–64	..	..	..	..	
1965–69	..	..	..	..	
1970	..	..	..	..	
1971	..	..	..	..	
1972	..	..	..	..	
1973	..	..	..	..	
1974	..	..	..	..	
1975	..	..	..	..	
1976	..	..	..	..	
1977	..	..	..	..	
1978	..	..	..	..	
1979	..	..	..	..	
1980	..	..	..	..	
1981	..	..	..	..	
1982	..	..	..	..	
1983	..	..	..	..	
1984	..	..	..	..	
1985	..	..	..	..	
1986	..	..	..	..	

1960–64:   1965–69:

Personnel Comments:
The 1982 figure includes only 'Office de la Recherche Scientifique et Technique Outre-Mer' (ORSTOM). This particular ORSTOM institute serves all the French Antilles.

Sources:
0027 Harvey, Nigel, ed. *Agricultural Research Centres: A World Directory of Organizations and Programmes.* Seventh Edition, Two Volumes. Harlow, U.K.: Longman, 1983.

Guam

Personnel

Period/Year	PhD	MSc	BSc	Subtotal	Expat	Total	Source
1960–64	..	..	..	..	..	..	
1965–69	..	..	..	..	..	..	
1970	..	..	..	..	..	..	
1971	..	..	..	..	..	..	
1972	..	..	..	..	..	..	
1973	..	..	..	..	..	1	492
1974	..	..	..	..	..	2	493
1975	..	..	..	..	..	2	488
1976	..	..	..	..	..	2	488
1977	..	..	..	..	..	2	488
1978	..	..	..	..	..	8	488
1979	..	..	..	..	..	12	488
1980	..	..	..	..	..	9	488
1981	..	..	..	..	..	9	488
1982	..	..	..	..	..	12	488
1983	..	..	..	..	..	11	488
1984	..	..	..	..	..	11	488
1985	..	..	..	..	..	9	488
1986	..	..	..	..	..	..	

1960–64:   1965–69:

Expenditure   LCU = U.S. Dollars

Period/Year	Current LCU (millions)	Constant 1980 LCU(millions)	Constant 1980 US$ (millions)		Source
			Atlas	PPP	
1960–64	..	..	..	..	
1965–69	..	..	..	..	
1970	..	..	..	..	
1971	..	..	..	..	
1972	..	..	..	..	
1973	0.070	0.141	0.133	0.158	492
1974	0.051	0.089	0.084	0.100	493
1975	0.192	0.307	0.288	0.344	488
1976	0.439	0.688	0.646	0.771	488
1977	0.480	0.746	0.700	0.835	488
1978	0.903	0.826	0.774	0.924	488
1979	0.698	0.834	0.782	0.933	488
1980	0.755	0.755	0.708	0.845	488
1981	0.877	0.728	0.683	0.815	488
1982	1.105	..	..	..	488
1983	1.159	..	..	..	488
1984	1.225	..	..	..	488
1985	1.455	..	..	..	488
1986	..	..	..	..	

1960–64:   1965–69:

Personnel Comments:
The 1973 and 1974 researcher figures from sources 492 and 493, respectively, are measured in full-time equivalent units. The 1975–85 figures are measured in scientific years and represent an aggregation of project-level data obtained directly from the USDA's Current Research Information System (source 488).

Expenditure Comments:
The 1973–74 figures are research expenditures, not simply funds available, while the 1975–85 figures from source 488 represent research obligations.

Sources:
0488 US Department of Agriculture – Cooperative State Research Service. *Inventory of Agricultural Research – Fiscal Years: 1970–1985.* Washington, D.C.: US Government Printing Office, various years.

0492 US Department of Agriculture – Cooperative State Research Service. *Funds for Research at State Agricultural Experiment Stations and Other State Institutions.* Annual issues 1964 to 1974. Washington, D.C.: US Government Printing Office, various years.

0493 US Department of Agriculture – Cooperative State Research Service. *Funds for Research at State Agricultural Experiment Stations and Other Cooperating Institutions.* Annual issues 1974 and 1975. Washington, D.C.: US Government Printing Office, various years.

Additional References:
0038 Gamble, W.K., R.M. Bourke, and C.W. Brookson. *South Pacific Agricultural Research Study.* Consultants' Report to the Asian Development Bank (two volumes). Manila: Asian Development Bank, June 1981.

Guatemala

Personnel

Period/Year	PhD	MSc	BSc	Subtotal	Expat	Total	Source
1960–64	..	..	..	..	..	..	
1965–69	..	..	..	..	..	22	53/211
1970	..	..	..	..	..	..	
1971	..	..	..	..	..	..	
1972	..	..	..	..	..	50	53
1973	..	..	..	..	..	..	
1974	..	..	..	..	..	..	
1975	2	12	55	69	7	76	66
1976	4	4	86	94	..	94	211
1977	..	..	..	..	..	..	
1978	..	..	..	..	..	..	
1979	3	15	103	121	..	121	211
1980	2	15	103	120	..	120	211
1981	..	..	..	..	..	..	
1982	..	..	..	..	..	..	
1983	2	21	150	173	3	176	17
1984	2	21	147	170	2	172	66
1985	2	25	101	128	4	132	17
1986	..	..	..	..	..	..	

1960–64: 1965–69: 1966 & 69

Expenditure

LCU = Quetzales

Period/Year	Current LCU (millions)	Constant 1980 LCU(millions)	Constant 1980 US$ (millions)		Source
			Atlas	PPP	
1960–64	..	..	..	..	
1965–69	..	..	..	..	
1970	..	..	..	..	
1971	..	..	..	..	
1972	..	..	..	..	
1973	1.189	2.546	2.532	4.581	208
1974	2.009	3.720	3.700	6.694	208
1975	2.331	3.815	3.794	6.864	208
1976	2.494	3.662	3.643	6.589	208
1977	3.267	4.120	4.098	7.412	208
1978	3.757	4.489	4.464	8.076	208
1979	5.140	5.655	5.624	10.174	208
1980	4.254	4.254	4.231	7.654	66
1981	4.967	4.578	4.553	8.237	66
1982	4.368	3.835	3.814	6.900	66
1983	5.009	4.126	4.104	7.424	17
1984	4.778	3.780	3.760	6.801	66
1985	..	..	..	..	
1986	..	..	..	..	

1960–64: 1965–69:

Personnel Comments:
The figures presented here refer to 'Instituto de Ciencia y Tecnología Agrícola' (ICTA) only. The BSc figure in 1983 and 1984 most likely includes some 'Ingenieros Agrónomos,' 25 of whom were identified by source 17 as holding a degree lower than a BSc and were consequently excluded from the indicator series. Besides ICTA, the following organizations – exclusive of universities – could be identified from the ISNAR, IFARD, & AOAD questionnaire (1983 figures) as performing some agricultural research: 'Comisión MOSCAMED' (20 scientists), ANACAFE (15 scientists), INF (4 scientists), 'Comisión México-Guatemala para el Control de la Roya del Cafeto' (20 scientists), and INCAP (58 researchers in 1976, source 95).

Source 744 reports IIA as an educational establishment with R&D capability, and 44 graduate research staff in 1985.

Expenditure Comments:
These expenditure figures refer to 'Instituto de Ciencia y Tecnología Agrícola' (ICTA) only. According to the ISNAR & IFARD questionnaire, research expenditures by other research organizations ('Comisión MOSCAMED', ANACAFE, INF, 'Comisión México-Guatemala para el Control de la Roya del Cafeto') in 1983 nearly equaled the research expenditures by ICTA. As this information was unavailable for the other years, these institutes were not included in the expenditure series.

Acronyms and Abbreviations:
ANACAFE: Asociación Nacional del Café
ICTA: Instituto de Ciencia y Tecnología Agrícola
IIA: Instituto de Investigaciones Agronómicas
INCAP: Instituto de Nutrición de Centro América y Panamá
INF: Instituto Nacional Forestal
MOSCAMED: unknown

Sources:
0017 ISNAR, IFARD & AOAD. Survey of National Agricultural Research Systems: Unpublished Questionnaire Responses. ISNAR, The Hague, 1985.

0053 Piñeiro, Martín, and Eduardo Trigo, eds. *Cambio Técnico en el Agro Latino Americano – Situación y Perspectivas en la Década de 1980.* San José: IICA, 1983.

0066 Stewart, Rigoberto. *Guatemala and the CGIAR Centers: A Study of Their Collaboration in Agricultural Research.* CGIAR Study Paper Number 5. Washington, D.C.: World Bank, 1985.

0208 Segura, M. "Diagnóstico de la Investigación Agropecuaria en el Istmo Centroamericano." IICA, Guatemala, 1979. Mimeo.

0211 McDermott, J.K., and David Bathrick. *Guatemala: Development of the Institute of Agricultural Science and Technology (ICTA) and Its Impact on Agricultural and Farm Productivity.* AID Project Impact Evaluation No. 30. Washington, D.C.: AID, February 1982.

Additional References:
0001 Trigo, Eduardo J., and Martín E. Piñeiro. "Funding Agricultural Research." In *Report of a Conference: Selected Issues in Agricultural Research in Latin America*, edited by Barry Nestel and Eduardo J. Trigo. The Hague: ISNAR, March 1984.

0010 Boyce, James K., and Robert E. Evenson. *National and International Agricultural Research and Extension Programs.* New York: Agricultural Development Council, Inc., 1975.

0014 Judd, M. Ann, James K. Boyce, and Robert E. Evenson. "Investing in Agricultural Supply." Economic Growth Center, Yale University, New Haven, Connecticut, 1983. Mimeo.

0016 Oram, Peter A., and Vishva Bindlish. *Resource Allocations to National Agricultural Research: Trends in the 1970s.* The Hague and Washington, D.C.: ISNAR and IFPRI, November 1981.

0026 Oram, Peter A., and Mark Gieben. "Document Summaries." ISNAR, The Hague, 1984. Mimeo.

0027 Harvey, Nigel, ed. *Agricultural Research Centres: A World Directory of Organizations and Programmes.* Seventh Edition, Two Volumes. Harlow, U.K.: Longman, 1983.

0078 Instituto de Ciencia y Tecnología Agrícolas (ICTA). *Objectivos, Programas de Acción, Realizaciones, Metas.* Guatemala: ICTA, August 1983.

0095 FAO – CARIS. *Agricultural Research in Developing Countries – Volume 1: Research Institutions.* Rome: FAO – CARIS, 1978.

0163 CGIAR. "National Agricultural Research." CGIAR, Washington, D.C., 1985. Mimeo.

0165 Evenson, R.E., and Y. Kislev. *Agricultural Research and Productivity.* New Haven: Yale University Press, 1975.

0172 Trigo, Eduardo. Background material. ISNAR, The Hague, n.d.

0210 Soikes, Raúl. "Situación del Sistema Nacional de Investigación y Extensión de Guatemala." IICA, Costa Rica, 1980. Mimeo.

208

0744 *Agricultural Research Centres: A World Directory of Organizations and Programmes.* Eighth Edition, Two Volumes. Harlow, U.K.: Longman, 1986.

0852 Evenson, Robert E., and Yoav Kislev. *Investment in Agricultural Research and Extension: A Survey of International Data.* Center Discussion Paper No. 124. New Haven, Connecticut: Economic Growth Center, Yale University, August 1971.

0886 Evenson, Robert E., and Yoav Kislev. "Investment in Agricultural Research and Extension: A Survey of International Data." *Economic Development and Cultural Change* Vol. 23 (April 1975): 507–521.

Guinea

Personnel

Period/Year	PhD	MSc	BSc	Subtotal	Expat	Total	Source
1960–64	..	..	..	..	..	..	
1965–69	..	..	..	..	..	..	
1970	..	..	..	9	8	17	175
1971	..	..	..	..	..	..	
1972	..	..	..	..	..	..	
1973	..	..	..	..	..	..	
1974	..	..	..	..	..	..	
1975	..	..	..	..	..	..	
1976	..	..	..	..	..	..	
1977	..	..	..	..	..	..	
1978	..	..	..	..	..	..	
1979	..	..	..	..	..	..	
1980	..	..	..	..	..	..	
1981	..	..	..	..	..	..	
1982	..	..	..	..	..	..	
1983	..	..	..	..	..	..	
1984	..	..	..	..	..	..	
1985	..	..	..	..	..	177	132
1986	..	..	..	..	..	..	

1960–64:   1965–69:

Expenditure

LCU = Guinean Sylis

Period/Year	Current LCU (millions)	Constant 1980 LCU(millions)	Constant 1980 US$ (millions)		Source
			Atlas	PPP	
1960–64	..	..	..	..	
1965–69	..	..	..	..	
1970	..	..	..	..	
1971	..	..	..	..	
1972	..	..	..	..	
1973	..	..	..	..	
1974	..	..	..	..	
1975	..	..	..	..	
1976	..	..	..	..	
1977	..	..	..	..	
1978	..	..	..	..	
1979	..	..	..	..	
1980	..	..	..	..	
1981	..	..	..	..	
1982	..	..	..	..	
1983	..	..	..	..	
1984	..	..	..	..	
1985	116.900	77.111	4.146	5.605	132
1986	..	..	..	..	

1960–64:   1965–69:

210

Personnel Comments:
The 1970 figure given by source 175 explicitly excludes university research and forestry, fisheries, and veterinary research. Source 132 (p.48) gives figures of 168 'enseignants-chercheurs' (teacher- researchers), 101 'ingénieurs de recherche et de coordination,' and 110 'ingénieurs de recherche' for 1985. The first 2 categories were prorated as 25% researchers and added to the 'ingénieurs de recherche' figure. This source gives no indication of qualification levels so it may well be that this figure is inclusive of technical personnel and therefore overestimates the total number of researchers. Source 27, for instance, gives a figure of only 22 graduate research staff in 1982, but this number refers to 'Institut National de Recherches Agronomiques de Foulaya' (INRAF) personnel only.

Expenditure Comments:
Source 132 warns that their total expenditure figure is incomplete and only indicative. We include this figure with serious reservations, as sources 86 and 132 indicate a high proportion of research is 'self-funded' so it is not entirely clear what this figure represents.

Sources:
0132 FAO. *Assistance à la Réorganisation et au Renforcement de la Recherche Agricole Guinée.* Rome: FAO, 1986.
0175 Cooper, St.G.C. *Agricultural Research in Tropical Africa.* Kampala: East African Literature Bureau, 1970.

Additional References:
0010 Boyce, James K., and Robert E. Evenson. *National and International Agricultural Research and Extension Programs.* New York: Agricultural Development Council, Inc., 1975.
0023 Bennell, Paul. *Agricultural Researchers in Sub-Saharan Africa: An Overview.* Working Paper No. 4. The Hague: ISNAR, October 1985.
0027 Harvey, Nigel, ed. *Agricultural Research Centres: A World Directory of Organizations and Programmes.* Seventh Edition, Two Volumes. Harlow, U.K.: Longman, 1983.
0068 Conte, Stephan. "West Africa Agricultural Research Data Base Building Task: Termination Report." World Bank, Washington, D.C., 1985. Mimeo.
0086 World Bank. "West Africa Agricultural Research Review – Country Studies." World Bank, Washington, D.C., 1985. Mimeo.
0088 Swanson, Burton E., and Wade H. Reeves. "Agricultural Research West Africa: Manpower and Training." World Bank, Washington, D.C., November 1985. Mimeo.
0089 Oram, Peter. "Report on National Agricultural Research in West Africa." IFPRI, Washington, D.C., February 1986. Mimeo.
0266 UNESCO. *National Science Policies in Africa.* Science Policy Studies and Documents No. 31. Paris: UNESCO, 1974.
0360 Cooper, St.G.C. "Towards Trained Manpower for Agricultural Research in Africa." Paper presented at the Conference on Agricultural Research and Production in Africa, organized by the Association for the Advancement of Agricultural Sciences in Africa (AAASA), Addis Ababa, 29 August–4 September 1971.
0466 World Bank – West Africa Agricultural Research Review Team. "National Agricultural Research Systems (NARS): A Regional Appraisal and Selected Issues." World Bank, Washington, D.C., August 1986. Mimeo.
0532 UNESCO Field Science Office for Africa. *Survey on the Scientific and Technical Potential of the Countries of Africa.* Paris: UNESCO, 1970.

Guinea-Bissau

Personnel

Period/Year	PhD	MSc	BSc	Subtotal	Expat	Total	Source
1960–64	..	..	..				
1965–69	..	..	..				
1970	..	..	..				
1971	..	..	..				
1972	..	..	..				
1973	..	..	..				
1974	..	..	..	..	..	..	
1975	..	..	..	..	..	..	
1976	..	..	..	..	..	..	
1977	..	..	..	..	..	..	
1978	..	..	..	..	..	..	
1979	..	..	..	..	..	..	
1980	..	..	..	..	..	..	
1981	..	..	..	..	..	..	
1982	..	..	..	..	..	..	
1983	0	5	2	7	1	8	17
1984	..	..	..	..	..	..	
1985	..	..	..	..	..	..	
1986	..	..	..	..	..	..	

1960–64:   1965–69:

Expenditure

LCU = Guinea-Bissau Pesos

Period/Year	Current LCU (millions)	Constant 1980 LCU(millions)	Constant 1980 US$ (millions) Atlas	PPP	Source
1960–64	..	..	..	..	
1965–69	..	..	..	..	
1970	..	..	..	..	
1971	..	..	..	..	
1972	..	..	..	..	
1973	..	..	..	..	
1974	..	..	..	..	
1975	..	..	..	..	
1976	..	..	..	..	
1977	..	..	..	..	
1978	..	..	..	..	
1979	..	..	..	..	
1980	..	..	..	..	
1981	..	..	..	..	
1982	..	..	..	..	
1983	21.049	16.322	0.349	0.907	17
1984	..	..	..	..	
1985	..	..	..	..	
1986	..	..	..	..	

1960–64:   1965–69:

Personnel Comments:

The figure presented here refers to the number of researchers in agronomic research only.

Expenditure Comments:

The 1983 expenditure figure is based on the ISNAR, IFARD, & AOAD questionnaire response. At the time of independence in 1974 no agricultural research system existed (source 311). In the last few years some agricultural research institutes have been created. The 1983 expenditure figure includes only agronomic research. There is some zootechnic and veterinary research (source 68) but it is considered negligible at present.

Sources:

0017 ISNAR, IFARD & AOAD. Survey of National Agricultural Research Systems: Unpublished Questionnaire Responses. ISNAR, The Hague, 1985.

Additional References:

0019 Bengtsson, Bo, and Tedla Getachew, eds. *Strengthening National Agricultural Research.* Report from a SAREC workshop, 10–17 September 1979. Part 1: Background Documents; Part II: Summary and Conclusions. Sweden: SAREC, 1980.

0068 Conte, Stephan. "West Africa Agricultural Research Data Base Building Task: Termination Report." World Bank, Washington, D.C., 1985. Mimeo.

0088 Swanson, Burton E., and Wade H. Reeves. "Agricultural Research West Africa: Manpower and Training." World Bank, Washington, D.C., November 1985. Mimeo.

0311 Britos Dos Santos, F.L. "Some Information Regarding Agricultural Investigations in Guinea Bissau." In *Strengthening National Agricultural Research.* Report from a SAREC workshop, 10–17 September 1979, edited by Bo Bengtsson and Getachew Tedla, pp. 44–47. Sweden: SAREC, 1980.

0466 World Bank – West Africa Agricultural Research Review Team. "National Agricultural Research Systems (NARS): A Regional Appraisal and Selected Issues." World Bank, Washington, D.C., August 1986. Mimeo.

Guyana

Personnel

Period/Year	PhD	MSc	BSc	Subtotal	Expat	Total	Source
1960–64	..	..	..	..	..	..	
1965–69	2	6	5	13	..	13	286/285
1970	..	..	..		..		
1971	5	9	9	23	..	23	279
1972	..	..	..	..	..	..	
1973	6	12	24	42	..	42	323
1974	3	14	26	43	..	43	244
1975	5	13	25	43	..	43	197
1976	4	13	24	41	..	41	197
1977	4	15	30	49	..	49	243/197/341
1978	..	..	..	..	..	..	
1979	..	..	..	..	..	..	
1980	2	21	33	56	..	56	242/358/342
1981	..	..	..	..	..	..	
1982	..	..	..	..	..	..	
1983	..	..	..	..	..	50	115/355/27
1984	..	..	..	..	..	..	
1985	..	..	..	..	..	..	
1986	..	..	..	..	..	..	

1960–64:　　　　　　　　　　1965–69: 1965 & 68

Expenditure

LCU = Guyana Dollars

Period/Year	Current LCU (millions)	Constant 1980 LCU(millions)	Constant 1980 US$ (millions)		Source
			Atlas	PPP	
1960–64	..	..	..	..	
1965–69	..	..	..	..	
1970	..	..	..	..	
1971	..	..	..	..	
1972	..	..	..	..	
1973	2.630	5.602	2.178	4.776	197
1974	2.680	4.185	1.627	3.568	197
1975	3.135	4.155	1.615	3.542	197
1976	2.670	3.579	1.392	3.051	197
1977	2.965	4.015	1.561	3.423	197
1978	..	..	..	..	
1979	..	..	..	..	
1980	..	..	..	..	
1981	..	..	..	..	
1982	..	..	..	..	
1983	..	..	..	..	
1984	..	..	..	..	
1985	..	..	..	..	
1986	..	..	..	..	

1960–64:　　　　　　　　　　1965–69:

Personnel Comments:

The indicator figures are constructed as follows:

	1965	1968	1971	1973	1974	1975	1976	1977	1980	1983
MOA[a]	7	18	23	25	30	28	27	31	34	20
GSC[b]				14	8	12	11	10*	12	18*
GRB				3	5	3	3	5*	7	6
CARDI[c]	–	–	–	–	–	–	–	3**	3*	6**
Total	7	18	23	42	43	43	41	49	56	50
Sources	286	285	279	323	244	323	323	243	242	115
								197*	342*	27*
								341**		355**

a. MOA includes the Crop Science Division, the Soil Science Division, and the Veterinary and Livestock Division. Some other divisions (e.g., extension, planning) and other departments (e.g., forestry and fisheries) are mentioned in several sources as performing agricultural research. Quantitative information about them, however, is very limited so they are not included in the indicator series.

b. The 1983 figure of GSC possibly includes the Field Equipment Department within which is a so-called 'Field Equipment Experimental Unit.' It seems that the research component of the Field Equipment Department is rather limited, and although not possible in this particular instance, it would be desirable on grounds of consistency to exclude it from the 1983 figure. (For 1977 and 1980 sources 243 and 242 report, respectively, 15 and 12 'graduate staff positions' within the Field Equipment Department.)

c. CARDI was created in 1975. The CARDI unit in Guyana was created in 1977.

In 1984, research by the Ministry of Agriculture and by the Rice Board were merged into the National Agricultural Research Institute (NARI).

Expenditure Comments:

The expenditure figures have been constructed as follows:

	1973	1974	1975	1976	1977
MOA	1.130	1.180	1.635	1.170	0.675
GSC	1.500	1.500	1.500	1.500	1.500
GRB	NA	NA	NA	NA	0.790
CARDI	–	–	–	–	NA
Total	2.630	2.680	3.135	2.670	2.965

Acronyms and Abbreviations:

CARDI: Caribbean Agricultural Research and Development Institute
GRB: Guyana Rice Board
GSC: Guyana Sugar Corporation
MOA: Ministry of Agriculture
NARI: National Agricultural Research Institute

Sources:

0027 Harvey, Nigel, ed. *Agricultural Research Centres: A World Directory of Organizations and Programmes.* Seventh Edition, Two Volumes. Harlow, U.K.: Longman, 1983.

0115 Economic Commission for Latin America and the Caribbean (ECLAC). *Agricultural Research Policy and Management – Volumes I and II.* Papers presented at the Workshop on Agricultural Research Policy and Management, 26–30 September 1983, Port of Spain, Trinidad. Port of Spain, Trinidad: ECLAC, November 1984.

0197 Fletcher, R., et al. "Agricultural Research in Guyana." March 1978. Mimeo.

0242 Commonwealth Agricultural Bureaux (CAB). *List of Research Workers in Agricultural Sciences in the Commonwealth 1981.* Slough, England: CAB, 1981.

0243 Commonwealth Agricultural Bureaux (CAB). *List of Research Workers in the Agricultural Sciences in the Commonwealth and in the Republic of Ireland 1978.* Slough, England: CAB, 1978.

0244 Commonwealth Agricultural Bureaux (CAB). *List of Research Workers in the Agricultural Sciences in the Commonwealth and in the Republic of Ireland 1975.* Slough, England: CAB, 1975.

0279 Commonwealth Agricultural Bureaux (CAB). *List of Research Workers in the Agricultural Sciences in the Commonwealth and in the Republic of Ireland 1972.* Slough, England: CAB, 1972.

0285 Commonwealth Agricultural Bureaux (CAB). *List of Research Workers in the Agricultural Sciences in the Commonwealth and in the Republic of Ireland 1969.* Slough, England: CAB, 1969.

0286 Commonwealth Agricultural Bureaux (CAB). *List of Research Workers in Agriculture, Animal Health and Forestry in the Commonwealth and in the Republic of Ireland 1966.* Slough, England: CAB, 1966.

0323 Chesney, H.A.D. "Agricultural Research in Guyana." In *Strengthening National Agricultural Research.* Report from a SAREC Workshop 10–17 September 1979, edited by Bo Bengtsson and Getachew Tedla, pp. 167–174. Sweden: SAREC, 1980.

0341 Caribbean Agricultural Research and Development Institute (CARDI). *Annual Report 1977–1978.* St. Augustine, Trinidad & Tobago: CARDI, November 1978.

0342 Caribbean Agricultural Research and Development Institute (CARDI). *CARDI 1976–1981: Report of the Chairman.* St. Augustine, Trinidad & Tobago: CARDI, 1982.

0355 Caribbean Agricultural Research and Development Institute (CARDI). *Research and Development Summary 1982–83.* St. Augustine, Trinidad: CARDI, 1983.

0358 Persaud, H.B., and M. Granger. "Resources for Agricultural Research in Guyana – Central Agricultural Station." Ministry of Agriculture, Georgetown, 1981. Mimeo.

Additional References:

0001 Trigo, Eduardo J., and Martín E. Piñeiro. "Funding Agricultural Research." In *Report of a Conference: Selected Issues in Agricultural Research in Latin America,* edited by Barry Nestel and Eduardo J. Trigo. The Hague: ISNAR, March 1984.

0010 Boyce, James K., and Robert E. Evenson. *National and International Agricultural Research and Extension Programs.* New York: Agricultural Development Council, Inc., 1975.

0014 Judd, M. Ann, James K. Boyce, and Robert E. Evenson. "Investing in Agricultural Supply." Economic Growth Center, Yale University, New Haven, Connecticut, 1983. Mimeo.

0016 Oram, Peter A., and Vishva Bindlish. *Resource Allocations to National Agricultural Research: Trends in the 1970s.* The Hague and Washington, D.C.: ISNAR and IFPRI, November 1981.

0053 Piñeiro, Martín, and Eduardo Trigo, eds. *Cambio Técnico en el Agro Latino Americano – Situación y Perspectivas en la Década de 1980.* San José: IICA, 1983.

0063 ISNAR. *Report to the Board of Directors of CARDI: Analysis, Evaluation and Proposals to Strengthen CARDI's Regional Capacity.* The Hague: ISNAR, 1985.

0073 Oram, Peter A., and Vishva Bindlish. "Investment in Agricultural Research in Developing Countries: Progress, Problems, and the Determination of Priorities." IFPRI, Washington, D.C., January 1984. Mimeo.

0163 CGIAR. "National Agricultural Research." CGIAR, Washington, D.C., 1985. Mimeo.

0172 Trigo, Eduardo. Background material. ISNAR. The Hague, n.d.

0198 IICA. "Strengthening the Agricultural Research Capability of the Ministry of Agriculture of Guyana." IICA, n.d. Mimeo.

0199 South-East Consortium for International Development, and Tuskegee Institute. *Baseline Study of Agricultural Research, Education, and Extension in Guyana.* Chapel Hill, North Carolina: South-East Consortium for International Development, June 1981.

0200 Downer. "Proposal for a National Agricultural Research Service in Guyana." 1979. Mimeo.

0201 ISNAR. *Guyana: The Agricultural Research System.* The Hague: ISNAR, March 1982.

0287 Commonwealth Agricultural Bureaux (CAB). *List of Research Workers in Agriculture, Animal Health and Forestry in the British Commonwealth, the Republic of Sudan and the Republic of Ireland 1959.* Slough, England: CAB, 1959.

0343 Caribbean Agricultural Research and Development Institute (CARDI). *Annual Report 1980.* St. Augustine, Trinidad & Tobago: CARDI, 1981.

0353 Caribbean Agricultural Research and Development Institute (CARDI). *Paper Presented at a Meeting of the Governing Body of CARDI on January 29, 1982, Basseterre, St. Kitts.* St. Augustine, Trinidad: CARDI, 1982.

0354 Caribbean Agricultural Research and Development Institute (CARDI). *Annual Report Research and Development 1983–84: Highlights.* St. Augustine, Trinidad: CARDI, 1984.

0356 Caribbean Agricultural Research and Development Institute (CARDI). *Research and Development Summary 1983–84.* St. Augustine, Trinidad: CARDI, 1984.

0365 Trigo, Eduardo, and Guillermo E. Galvez. "Trip Report of the CIAT and IICA Mission to the Caribbean English-Speaking Countries for UNDP." 1979. Mimeo.

0654 Office de la Recherche Scientifique et Technique Outre-Mer (ORSTOM). *Rapport d'Activité 1967.* Paris: ORSTOM, 1969.

Haiti

Personnel

Period/Year	PhD	MSc	BSc	Subtotal	Expat	Total	Source
1960–64	..	..	..	..	..	..	
1965–69	..	..	..	..	..	..	
1970	..	..	..	..	..	..	
1971	..	..	..	..	..	..	
1972	..	..	..	..	..	..	
1973	..	..	..	..	..	..	
1974	..	..	..	..	..	..	
1975	..	..	..	..	..	..	
1976	..	..	..	..	..	..	
1977	..	..	..	..	..	..	
1978	..	..	..	..	..	..	
1979	..	..	..	..	..	..	
1980	..	..	..	..	..	..	
1981	..	..	..	..	..	..	
1982	..	..	..	..	..	..	
1983	7	23	2	32	..	32	115
1984	..	..	..	..	..	..	
1985	..	..	..	..	..	..	
1986	..	..	..	..	..	..	

1960–64:   1965–69:

Expenditure   *LCU = Gourdes*

Period/Year	Current LCU (millions)	Constant 1980 LCU(millions)	Constant 1980 US$ (millions)		Source
			Atlas	PPP	
1960–64	0.720	2.182	0.412	1.097	10
1965–69	..	..	..	..	
1970	..	..	..	..	
1971	..	..	..	..	
1972	..	..	..	..	
1973	..	..	..	..	
1974	..	..	..	..	
1975	..	..	..	..	
1976	1.070	1.436	0.271	0.722	222
1977	1.325	1.604	0.303	0.806	222
1978	1.235	1.517	0.286	0.763	222
1979	..	..	..	..	
1980	..	..	..	..	
1981	..	..	..	..	
1982	..	..	..	..	
1983	3.875	3.229	0.609	1.623	115
1984	..	..	..	..	
1985	..	..	..	..	
1986	..	..	..	..	

1960–64: 1960   1965–69:

Personnel Comments:

For 1980, source 222 identified 38 persons with a university degree (1 PhD, 12 MSc, and 25 BSc) who were involved in agricultural research and/or extension.

Expenditure Comments:

When compared with other observations, the expenditure figures in source 222 appear to have been incorrectly reported in units of 1,000 US dollars rather than 10,000 US dollars, and accordingly scaled upwards by a factor of 10. The 1983 figure refers to the Agricultural Research and Documentation Center (CDRA). Source 222 indicates that a substantial proportion (unspecified) of CDRA's budget is externally funded.

Sources:

0010 Boyce, James K., and Robert E. Evenson. *National and International Agricultural Research and Extension Programs.* New York: Agricultural Development Council, Inc., 1975.

0115 Economic Commission for Latin America and the Caribbean (ECLAC). *Agricultural Research Policy and Management – Volumes I and II.* Papers presented at the Workshop on Agricultural Research Policy and Management, 26–30 September 1983, Port of Spain, Trinidad. Port of Spain, Trinidad: ECLAC, November 1984.

0222 Department of Agriculture, Natural Resources and Rural Development (DARNDR). "Diagnosis of the Situation of the Agricultural Research and Extension System in Haiti (A Summary)." Paper presented at the Caribbean Workshop on the Organization and Administration of Agricultural Research, Bridgetown, Barbados, February 1981.

Additional References:

0014 Judd, M. Ann, James K. Boyce, and Robert E. Evenson. "Investing in Agricultural Supply." Economic Growth Center, Yale University, New Haven, Connecticut, 1983. Mimeo.

0016 Oram, Peter A., and Vishva Bindlish. *Resource Allocations to National Agricultural Research: Trends in the 1970s.* The Hague and Washington, D.C.: ISNAR and IFPRI, November 1981.

0053 Piñeiro, Martín, and Eduardo Trigo, eds. *Cambio Técnico en el Agro Latino Americano – Situación y Perspectivas en la Década de 1980.* San José: IICA, 1983.

0095 FAO – CARIS. *Agricultural Research in Developing Countries – Volume 1: Research Institutions.* Rome: FAO – CARIS, 1978.

0163 CGIAR. "National Agricultural Research." CGIAR, Washington, D.C., 1985. Mimeo.

0165 Evenson, R.E., and Y. Kislev. *Agricultural Research and Productivity.* New Haven: Yale University Press, 1975.

0172 Trigo, Eduardo. Background material. ISNAR, The Hague, n.d.

0219 Department of Agriculture, Natural Resources and Rural Development (DARNDR), and IICA. "Réunion Technique Régionale sur les Systèmes de Recherche Agricole dans les Antilles." DARNDR and IICA, Port-au-Prince, Haiti, November/December 1977. Mimeo.

0886 Evenson, Robert E., and Yoav Kislev. "Investment in Agricultural Research and Extension: A Survey of International Data." *Economic Development and Cultural Change* Vol. 23 (April 1975): 507–521.

Honduras

Personnel

Period/Year	PhD	MSc	BSc	Subtotal	Expat	Total	Source
1960–64	..	..	..	..	..	..	
1965–69	..	..	..	..	..	..	
1970	..	..	..	..	..	..	
1971	..	..	..	..	..	..	
1972	..	..	..	..	..	56	208
1973	..	..	..	..	..	..	
1974	..	..	..	..	..	..	
1975	..	..	..	..	..	..	
1976	..	..	..	..	..	38	95
1977	2	2	33	37	3	40	208
1978	..	..	..	..	..	..	
1979	..	..	..	..	..	..	
1980	1	6	53	60	11	71	212
1981	..	..	..	..	..	..	
1982	..	..	..	..	..	65	27
1983	..	..	..	..	..	..	
1984	..	..	..	..	..	..	
1985	..	..	..	..	..	..	
1986	..	..	..	..	..	..	

1960–64:   1965–69:

Expenditure   *LCU = Lempiras*

Period/Year	Current LCU (millions)	Constant 1980 LCU(millions)	Constant 1980 US$ (millions)		Source
			Atlas	PPP	
1960–64	..	..	..	..	
1965–69	..	..	..	..	
1970	..	..	..	..	
1971	..	..	..	..	
1972	..	..	..	..	
1973	..	..	..	..	
1974	..	..	..	..	
1975	..	..	..	..	
1976	0.827	1.211	0.603	0.965	208
1977	1.238	1.481	0.737	1.180	208
1978	1.594	1.983	0.987	1.580	208
1979	1.617	1.835	0.914	1.463	208
1980	1.950	1.950	0.971	1.554	208
1981	..	..	..	..	
1982	..	..	..	..	
1983	..	..	..	..	
1984	..	..	..	..	
1985	..	..	..	..	
1986	..	..	..	..	

1960–64:   1965–69:

Personnel Comments:

These figures refer to 'Programa Nacional de Investigación Agropecuaria' (PNIA) only. Although PNIA is the principal public agricultural research institution in Honduras, source 172 suggests that a roughly equivalent level of resources may be committed to agricultural research by other (semi-) public and private research agencies. However, quantifiable information on these is scarce, so it was not possible to estimate the total number of researchers working in these institutes. In addition to PNIA the following organizations were identified as conducting agricultural research: IHCAFE, COHBANA, INA, THSA, SIATSA, SFC, IA, and COHDEFOR.

Expenditure Comments:

These expenditure figures refer to PNIA only.

Acronyms and Abbreviations:

COHBANA: Corporación Hondureña del Banano
COHDEFOR: Corporación Hondureña para el
    Desarrollo Forestal
IA: Ingenios Azucareros
IHCAFE: Instituto Hondureño del Café
INA: Instituto Nacional Agrario
PNIA: Programa Nacional de Investigación
    Agropecuaria
SFC: Standard Fruit Company
SIATSA: Servicios Internacionales de Agricultura
    Tropical S.A.
THSA: Tabacalera Hondureña S.A.

Sources:

0027 Harvey, Nigel, ed. *Agricultural Research Centres: A World Directory of Organizations and Programmes*. Two Volumes. Harlow: Longman, 1983.

0095 FAO – CARIS. *Agricultural Research in Developing Countries – Volume 1: Research Institutions*. Rome: FAO – CARIS, 1978.

0208 Segura, M. "Diagnóstico de la Investigación Agropecuaria en el Istmo Centro-americano." IICA, Guatemala, 1979. Mimeo.

0212 Rosales, Franklin E. "Situación del Sistema Nacional de Investigación Agronómica en Honduras." IICA – OEA, San José, December 1980. Mimeo.

Additional References:

0010 Boyce, James K., and Robert E. Evenson. *National and International Agricultural Research and Extension Programs*. New York: Agricultural Development Council, Inc., 1975.

0014 Judd, M. Ann, James K. Boyce, and Robert E. Evenson. "Investing in Agricultural Supply." Economic Growth Center, Yale University, New Haven, Connecticut, 1983. Mimeo.

0016 Oram, Peter A., and Vishva Bindlish. *Resource Allocations to National Agricultural Research: Trends in the 1970s*. The Hague and Washington, D.C.: ISNAR and IFPRI, November 1981.

0017 ISNAR, IFARD & AOAD. Survey of National Agricultural Research Systems: Unpublished Questionnaire Responses. ISNAR, The Hague, 1985.

0026 Oram, Peter A., and Mark Gieben. "Document Summaries." ISNAR, The Hague, 1984. Mimeo.

0053 Piñeiro, Martín, and Eduardo Trigo, eds. *Cambio Técnico en el Agro Latino Americano – Situación y Perspectivas en la Década de 1980*. San José: IICA, 1983.

0073 Oram, Peter A., and Vishva Bindlish. "Investment in Agricultural Research in Developing Countries: Progress, Problems, and the Determination of Priorities." IFPRI, Washington, D.C., January 1984. Mimeo.

0172 Trigo, Eduardo. Background material. ISNAR, The Hague, n.d.

0852 Evenson, Robert E., and Yoav Kislev. *Investment in Agricultural Research and Extension: A Survey of International Data*. Center Discussion Paper No. 124. New Haven, Connecticut: Economic Growth Center, Yale University, August 1971.

Hong Kong

Personnel

Period/Year	PhD	MSc	BSc	Subtotal	Expat	Total	Source
1960–64	..	..	..	..	..	19	715/716
1965–69	..	..	..	..	..	18	286/285
1970	..	..	..	..	..	..	
1971	..	..	..	..	..	21	279
1972	..	..	..	..	..	..	
1973	..	..	..	..	..	..	
1974	..	..	..	..	..	26	244
1975	..	..	..	..	..	..	
1976	..	..	..	..	..	22	644
1977	..	..	..	..	..	22	644
1978	..	..	..	..	..	23	644
1979	..	..	..	..	..	23	644
1980	..	..	..	..	..	24	644
1981	..	..	..	..	..	23	644
1982	..	..	..	..	..	21	644
1983	4	5	7	16	6	22	644/779
1984	4	4	10	18	4	22	644/779
1985	4	5	11	20	2	22	644/779
1986	4	5	12	21	1	22	779

1960–64: 1960   1965–69: 1965 & 68

Expenditure   *LCU = Hong Kong Dollars*

Period/Year	Current LCU (millions)	Constant 1980 LCU (millions)	Constant 1980 US$ (millions)		Source
			Atlas	PPP	
1960–64	..	..	..	..	
1965–69	..	..	..	..	
1970	..	..	..	..	
1971	..	..	..	..	
1972	..	..	..	..	
1973	..	..	..	..	
1974	..	..	..	..	
1975	..	..	..	..	
1976	..	..	..	..	
1977	..	..	..	..	
1978	..	..	..	..	
1979	..	..	..	..	
1980	..	..	..	..	
1981	..	..	..	..	
1982	..	..	..	..	
1983	..	..	..	..	
1984	..	..	..	..	
1985	..	..	..	..	
1986	..	..	..	..	

1960–64:   1965–69:

Personnel Comments:

The personnel indicators have been constructed as follows:

	1960	1965	1968	1971	1974	1976	1977	1978
Agriculture and Fisheries Department								
Administration Branch		4	2	2	2	2	2	2
Agricultural Branch	17	5	3	9	11	10	10	10
Fisheries Branch		10	8	8	[9]	9	9	10
Conservation and Country Parks Branch	2*	2	2	2	4	1	1	1
Total	19	21	15	21	26	22	22	23
Source	715 716*	286	285	279	244	644	644	644

	1979	1980	1981	1982	1983	1984	1985	1986
Agriculture and Fisheries Department								
Administration Branch	2	2	2	0	0	1	1	1
Agricultural Branch	10	11	10	10	11	11	11	11
Fisheries Branch	10	10	10	10	10	9	9	9
Conservation and Country Parks Branch	1	1	1	1	1	1	1	1
Total	23	24	23	21	22	22	22	22
Source	644	644	644	644	644	644	644	779

[ ] = estimated or constructed by authors.

Sources:

0244 Commonwealth Agricultural Bureaux (CAB). *List of Research Workers in the Agricultural Sciences in the Commonwealth and in the Republic of Ireland 1975.* Slough, England: CAB, 1975.

0279 Commonwealth Agricultural Bureaux (CAB). *List of Research Workers in the Agricultural Sciences in the Commonwealth and in the Republic of Ireland 1972.* Slough, England: CAB, 1972.

0285 Commonwealth Agricultural Bureaux (CAB). *List of Research Workers in the Agricultural Sciences in the Commonwealth and in the Republic of Ireland 1969.* Slough, England: CAB, 1969.

0286 Commonwealth Agricultural Bureaux (CAB). *List of Research Workers in Agriculture, Animal Health and Forestry in the Commonwealth and in the Republic of Ireland 1966.* Slough, England: CAB, 1966.

0644 Shin, Joseph. Personal communication. Agriculture and Fisheries Department. Hong Kong, May 1987.

0715 Chang, C.W. *FAO Directory of Agricultural Research Institutes and Experiment Stations in Asia and the Far East.* Bangkok, Thailand: FAO – Regional Office for Asia and the Far East, September 1962.

0716 Chang, C.W. *FAO Directory of Agricultural Research Institutes and Experiment Stations in Asia and the Far East, Supplement.* Bangkok, Thailand: FAO – Regional Office for Asia and the Far East, May 1963.

0779 Chan, T.K. Personal communication. Agriculture and Fisheries Department. Hong Kong, January 1988.

Additional References:

0010 Boyce, James K., and Robert E. Evenson. *National and International Agricultural Research and Extension Programs.* New York: Agricultural Development Council, Inc., 1975.

0014 Judd, M. Ann, James K. Boyce, and Robert E. Evenson. "Investing in Agricultural Supply." Economic Growth Center, Yale University, New Haven, Connecticut, 1983. Mimeo.

0027 Harvey, Nigel, ed. *Agricultural Research Centres: A World Directory of Organizations and Programmes.* Seventh Edition, Two Volumes. Harlow, U.K.: Longman, 1983.

0242 Commonwealth Agricultural Bureaux (CAB). *List of Research Workers in Agricultural Sciences in the Commonwealth 1981.* Slough, England: CAB, 1981.

0243 Commonwealth Agricultural Bureaux (CAB). *List of Research Workers in the Agricultural Sciences in the Commonwealth and in the Republic of Ireland 1978. Slough, England: CAB, 1978.*

0287 Commonwealth Agricultural Bureaux (CAB). *List of Research Workers in Agriculture, Animal Health and Forestry in the British Commonwealth, the Republic of Sudan and the Republic of Ireland 1959.* Slough, England: CAB, 1959.

0852 Evenson, Robert E., and Yoav Kislev. *Investment in Agricultural Research and Extension: A Survey of International Data.* Center Discussion Paper No. 124. New Haven, Connecticut: Economic Growth Center, Yale University, August 1971.

Iceland

Personnel

Period/Year	PhD	MSc	BSc	Subtotal	Expat	Total	Source
1960–64	..	..	..	..	..	37	503
1965–69	..	..	..	..	..	40	503
1970	..	..	..	..	..	48	503
1971	..	..	..	..	..	..	
1972	..	..	..	..	..	..	
1973	..	..	..	..	..	..	
1974	..	..	..	..	..	..	
1975	..	..	..	..	..	42	14
1976	..	..	..	..	..	..	
1977	..	..	..	..	..	48	14
1978	..	..	..	..	..	49	14
1979	..	..	..	..	..	49	14
1980	..	..	..	..	..	51	14
1981	..	..	..	..	..	..	
1982	..	..	..	..	..	..	
1983	..	..	..	..	..	..	
1984	..	..	..	..	..	..	
1985	..	..	..	..	..	77	744
1986	..	..	..	..	..	..	

1960–64: 1960   1965–69: 1965, 66 & 69

Expenditure   *LCU = Kronur*

Period/Year	Current LCU (millions)	Constant 1980 LCU(millions)	Constant 1980 US$ (millions)		Source
			Atlas	PPP	
1960–64	0.200	9.400	1.787	1.438	503
1965–69	0.528	14.635	2.782	2.238	503
1970	0.970	18.664	3.547	2.855	503
1971	1.090	18.607	3.536	2.846	503
1972	..	..	..	..	
1973	..	..	..	..	
1974	..	..	..	..	
1975	3.820	21.776	4.139	3.331	601/602
1976	..	..	..	..	
1977	9.280	29.357	5.579	4.490	601/602
1978	..	..	..	..	
1979	22.760	34.798	6.614	5.322	601/602
1980	..	..	..	..	
1981	..	..	..	..	
1982	..	..	..	..	
1983	96.400	23.090	4.388	3.532	601/602
1984	..	..	..	..	
1985	..	..	..	..	
1986	..	..	..	..	

1960–64: 1961, 62, 63 & 64   1965–69: 1965, 66, 67, 68 & 69

Personnel Comments:
These figures include fisheries research. Fisheries research accounts for half or more of all agricultural research in Iceland. Boyce and Evenson (source 10) systematically excluded fisheries research.

The 1985 personnel figure can be broken down as follows:

Agricultural Research Institute	35
Marine Research Institute	30
Icelandic Soil Conservation Service	2
Iceland Forestry Research Station	2
Biological Institute (FTEs)	8
Total	**77**

Expenditure Comments:

	1975	1977	1979	1981	1983	1984	1985
Non-university	3.15	8.41	20.49	NA	81.70	134.60	190.10
University	0.67*	0.87*	2.27*	4.60	14.70*	NA	NA
Total	3.82	9.28	22.76	NA	96.40	NA	NA
Source	601	601	601	602	601	601	601
	602*	602*	602*	602*			

The post-1975 expenditure figures have been constructed as above.

It appears there was a currency change in the early seventies. To get the pre-1971 figures in line with the later figures, and with the deflator series, they were divided by 100.

Sources:

0014 Judd, M. Ann, James K. Boyce, and Robert E. Evenson. "Investing in Agricultural Supply." Economic Growth Center, Yale University, New Haven, Connecticut, 1983. Mimeo.

0503 OECD. *Reviews of National Science Policy: Iceland*. Paris: OECD, 1972.

0601 OECD – STIID Data Bank. "Public Funding of R&D by Socio-Economic Objective." Tables. OECD, Paris, April 1987. Mimeo.

0602 OECD – STIID Data Bank. "R&D Expenditure in the Higher Education and Private Non-Profit Sectors." Tables. OECD, Paris, April 1987. Mimeo.

0744 *Agricultural Research Centres: A World Directory of Organizations and Programmes.* Eighth Edition, Two Volumes. Harlow, U.K.: Longman, 1986.

Additional References:

0010 Boyce, James K., and Robert E. Evenson. *National and International Agricultural Research and Extension Programs.* New York: Agricultural Development Council, Inc., 1975.

0021 OECD. *OECD Science and Technology Indicators.* Paris: OECD, 1984.

0022 UNESCO. *Statistical Yearbook 1983.* Paris: UNESCO, 1983.

0027 Harvey, Nigel, ed. *Agricultural Research Centres: A World Directory of Organizations and Programmes.* Seventh Edition, Two Volumes. Harlow, U.K.: Longman, 1983.

0376 FAO. *Fourteenth FAO Regional Conference for Europe, Reykjavik, Iceland, 17–21 September 1984: Research in Support of Agricultural Policies in Europe.* Rome: FAO, 1984.

0473 OECD. *Science and Technology Indicators: Basic Statistical Series – Volume A: The Objectives of Government R&D Funding 1974–1985.* Paris: OECD, May 1983.

0504 OECD. *Country Reports on the Organisation of Scientific Research: Iceland.* Paris: OECD, July 1963.

0509 OECD. *Survey of the Resources Devoted to R&D by OECD Member Countries, International Statistical Year 1971 – Volume 5: Total Tables, Statistical Tables and Notes.* Paris: OECD, August 1974.

0512 OECD. *Survey of the Resources Devoted to R&D by OECD Member Countries, International Statistical Year 1973 – Volume 5: Total Tables, Statistical Tables and Notes.* Paris: OECD, September 1976.

0513 OECD. *International Survey of the Resources Devoted to R&D by OECD Member Countries, International Statistical Year 1975 –*

226

International Volume: Statistical Tables and Notes. Paris: OECD, March 1979.

0530 OECD. *OECD Science and Technology Indicators No. 2 – R&D, Invention and Competitiveness.* Paris: OECD, 1986.

0707 OECD – STIID Data Bank. "R&D Expenditure in the Higher Education and Private Non-Profit Sectors: Higher Education, Agricultural Sciences, General Universities Funds." Table. OECD, Paris, June 1987. Mimeo.

0708 OECD – STIID Data Bank. "Gross Domestic Expenditure on R&D: All Fields of Sciences – Agriculture, Forestry and Fishing." Tables. OECD, Paris, June 1987. Mimeo.

0852 Evenson, Robert E., and Yoav Kislev. *Investment in Agricultural Research and Extension: A Survey of International Data.* Center Discussion Paper No. 124. New Haven, Connecticut: Economic Growth Center, Yale University, August 1971.

India

Personnel

Period/Year	PhD	MSc	BSc	Subtotal	Expat	Total	Source
1960–64	..	..	..	..	..		
1965–69	..	..	..	..	..		
1970	..	..	..	..	..		
1971	..	..	..	..	..		
1972	..	..	..	..	..		
1973	..	..	..	..	..		
1974	..	..	..	..	..		
1975	..	..	..	..	..	5666	781
1976	..	..	..	..	..		
1977	..	..	..	..	..		
1978	..	..	..	..	..		
1979	..	..	..	..	..	6910	26
1980	..	..	..	..	..		
1981	..	..	..	..	..		
1982	..	..	..	..	..		
1983	..	..	..	..	..		
1984	..	..	..	..	..		
1985	..	..	..	..	..	8389	778
1986	..	..	..	..	..		

1960–64: 1965–69:

Expenditure

LCU = Indian Rupees

Period/Year	Current LCU (millions)	Constant 1980 LCU(millions)	Constant 1980 US$ (millions)		Source
			Atlas	PPP	
1960–64	81.535	308.758	36.982	102.231	876
1965–69	170.447	437.524	52.405	144.866	876/10
1970	267.490	616.336	73.823	204.072	10
1971	303.290	663.654	79.491	219.739	10
1972	372.198	732.673	87.757	242.591	10
1973	382.733	633.664	75.898	209.809	10
1974	..	..	..	..	
1975	582.574	843.088	100.983	279.150	26/781
1976	698.606	949.193	113.692	314.282	26
1977	858.899	1125.687	134.831	372.720	26
1978	..	..	..	..	
1979	..	..	..	..	
1980	1341.518	1341.518	160.683	444.182	778
1981	1443.534	1320.708	158.191	437.292	778
1982	1600.551	1357.550	162.603	449.491	778
1983	1697.568	1321.065	158.233	437.410	778
1984	1898.585	1387.855	166.233	459.525	778
1985	2081.602	1423.804	170.539	471.428	777/778
1986	2356.130	..	..	..	777/778

1960–64: 1960, 61, 62, 63 & 64   1965–69: 1965, 66, 67, 68 & 69

228

Personnel Comments:

Agricultural research in India is conducted at both the national and state level.

At the national level is the Indian Council for Agricultural Research (ICAR) which, according to source 778, in 1986 or so comprised more than 40 national agricultural research institutes, 14 national research centers, more than 70 All-India Coordinated Research Projects, and 12 general (national-level) universities.

At the state level, agricultural research is conducted within State Agricultural Universities (SAUs), which have been adapted from the model of the US land-grant system. The State Agricultural Universities and ICAR collaborate closely through the more than 70 All-India Coordinated Research Projects and other ad hoc research schemes. There are also close budgetary links between SAUs and ICAR. According to source 778, p. 111, ICAR assistance constitutes 15 to 30% of the annual expenditure of several SAUs.

Because of the large and relatively complex organizational structure of the research system, it has been difficult to construct a time series of aggregate personnel indicators. Two aspects need to be taken into account, namely:

(1) Total numbers of scientists at SAUs do not reflect the total number of person-years spent on research. SAUs perform three tasks, namely, education, extension, and research. It is assumed that at the SAUs on average, one-third of total person-years are spent on research.

(2) Source 777 indicates that there is a large discrepancy between the total number of posts sanctioned and those actually filled. This is corroborated by source 778, p.190, who states, "... every third position in ICAR institutions and every fourth position in the state research system continues to be vacant."

Where possible, the series included only the number of filled rather than sanctioned research positions. Prorating only filled positions gives considerably lower estimates of agricultural researchers (roughly expressed in full-time equivalents (FTEs)) than some other sources (e.g., 93 & 433).

The table below gives more detailed information on how the estimates have been constructed.

	1975	1979	1985	1987[a]
ICAR	[3440]	3707	4109	4052
SAUs	2126*	[3203]	4280	5800
Total	5666	6910	8389	9852
Sources	26/781*	26	778	777

[ ] = estimated or constructed by authors.
a. Not included in the indicator series.

Most of the pre-1975 figures given by Boyce and Evenson (source 10) originate from the CAB Lists of Research Workers (sources 242, 243, 244, 279, 285, 286, & 287). A close examination of these sources gives cause for concern about the coverage of these lists in this particular instance. It appears that only senior research positions have been included because few BSc-level researchers are reported. Consequently totals are considerably lower than those reported by other sources. Therefore, to be consistent within the time series and between countries, these CAB-based figures have not been included here.

Expenditure Comments:

The pre-1974 figures have been derived from sources 876 and 10. The figures given by source 876 (1960-68) have been estimated by Mohan, Jha, and Evenson on the basis of several sources such as the Directory of Scientific Research Institutions in India (1969) and Annual Budget Reports of each state, and on a considerable number of assumptions based on inside knowledge of the Indian NARS. According to source 10, the 1969, '70, '72, and '73 figures are estimates made on the basis of data presented in source 876, plus publications listed in Indian Science Abstracts of the relevant years. An expenditure per publication ratio was calculated for the base period 1965–68, and later expenditure figures were derived by applying this ratio to a suitable publication series. The post-1974 expenditure

	1975	1976	1977	1980	1981	1982	1983	1984	1985	1986
ICAR	415.141	488.156	605.432	959.000	1018.000	1132.000	1186.000	1344.000	1484.000	1578.000
SAUs	167.433*	[210.450]	[253.467]	[382.518]	[425.534]	[468.551]	[511.568]	[544.585]	597.602*	778.130*
Total	582.574	698.606	858.899	1341.518	1443.534	1600.551	1697.568	1898.585	2081.602	2356.130
Source	26 781*	26	26	778	778	778	778	778	778 777*	778 777*

[ ] = estimated or constructed by authors.

indicators have been constructed as shown in the table above.

ICAR expenditures include the ICAR contributions to SAUs. The SAUs research expenditures represent the state government contributions only. It is assumed that a third of all state government contributions to the SAUs are spent on research.

Acronyms and Abbreviations:
ICAR: Indian Council for Agricultural Research
SAUs: State Agricultural Universities

Sources:
0026 Oram, Peter A., and Mark Gieben. "Document Summaries." ISNAR, The Hague, 1984. Mimeo.
0777 Acharya, R.M. Personal communication. Deputy Director General, Indian Council of Agricultural Research. New Delhi, January 1988.
0778 Randhawa, N.S. *Agricultural Research in India: An Overview of Its Organization, Management and Operations.* FAO Research and Technology Paper 3. Rome: FAO, 1987.
0781 Indian Council of Agricultural Research (ICAR). *Report of the Review Committee on Agricultural Universities.* New Delhi: ICAR, October 1978.
0876 Mohan, Rakesh, D. Jha, and Robert Evenson. "The Indian Agricultural Research System." *Economic and Political Weekly* Vol. VIII, No. 13 (March 1973): 223–228.

Additional References:
0010 Boyce, James K., and Robert E. Evenson. *National and International Agricultural Research and Extension Programs.* New York: Agricultural Development Council, Inc., 1975.
0014 Judd, M. Ann, James K. Boyce, and Robert E. Evenson. "Investing in Agricultural Supply." Economic Growth Center, Yale University, New Haven, Connecticut, 1983. Mimeo.
0016 Oram, Peter A., and Vishva Bindlish. *Resource Allocations to National Agricultural Research: Trends in the 1970s.* The Hague and Washington, D.C.: ISNAR and IFPRI, November 1981.
0019 Bengtsson, Bo, and Tedla Getachew, eds. *Strengthening National Agricultural Research.* Report from a SAREC workshop, 10–17 September 1979. Part 1: Background Documents; Part II: Summary and Conclusions. Sweden: SAREC, 1980.
0027 Harvey, Nigel, ed. *Agricultural Research Centres: A World Directory of Organizations and Programmes.* Seventh Edition, Two Volumes. Harlow, U.K.: Longman, 1983.
0073 Oram, Peter A., and Vishva Bindlish. "Investment in Agricultural Research in Developing Countries: Progress, Problems, and the Determination of Priorities." IFPRI, Washington, D.C., January 1984. Mimeo.
0093 Bangladesh Agricultural Research Council (BARC), ISNAR, and Winrock International. *Management of Human Resources in Agricultural Research.* Report of the International Workshop on Management of Human Resources in Agricultural Research held March 3–5, 1986 in Dhaka, Bangladesh. Dhaka: BARC, ISNAR, and Winrock International, 1986.
0163 CGIAR. "National Agricultural Research." CGIAR, Washington, D.C., 1985. Mimeo.
0165 Evenson, R.E., and Y. Kislev. *Agricultural Research and Productivity.* New Haven: Yale University Press, 1975.
0176 Moseman, Albert, ed. *National Agricultural Research Systems in Asia.* Report of the Regional Seminar held at the India International Centre, New Delhi, India, March 8–13, 1971. New York: ADC, Inc., 1971.
0242 Commonwealth Agricultural Bureaux (CAB). *List of Research Workers in Agricultural Sciences in the Commonwealth 1981.* Slough, England: CAB, 1981.
0243 Commonwealth Agricultural Bureaux (CAB). *List of Research Workers in the Agricultural*

230

Sciences in the Commonwealth and in the Republic of Ireland 1978. Slough, England: CAB, 1978.

0244 Commonwealth Agricultural Bureaux (CAB). *List of Research Workers in the Agricultural Sciences in the Commonwealth and in the Republic of Ireland 1975.* Slough, England: CAB, 1975.

0279 Commonwealth Agricultural Bureaux (CAB). *List of Research Workers in the Agricultural Sciences in the Commonwealth and in the Republic of Ireland 1972.* Slough, England: CAB, 1972.

0285 Commonwealth Agricultural Bureaux (CAB). *List of Research Workers in the Agricultural Sciences in the Commonwealth and in the Republic of Ireland 1969.* Slough, England: CAB, 1969.

0286 Commonwealth Agricultural Bureaux (CAB). *List of Research Workers in Agriculture, Animal Health and Forestry in the Commonwealth and in the Republic of Ireland 1966.* Slough, England: CAB, 1966.

0287 Commonwealth Agricultural Bureaux (CAB). *List of Research Workers in Agriculture, Animal Health and Forestry in the British Commonwealth, the Republic of Sudan and the Republic of Ireland 1959.* Slough, England: CAB, 1959.

0316 Jain, H.K. "Reorganization and Intensification of Agricultural Research in India." In *Strengthening National Agricultural Research.* Report from a SAREC workshop, 10–17 September 1979, edited by Bo Bengtsson and Getachew Tedla, pp. 89–97. Sweden: SAREC, 1980.

0328 Raman, K.V. "Management of Human Resources in Agricultural Research: Indian Efforts and Experiences." In *Management of Human Resources in Agricultural Research.* Report of the International Workshop on Management of Human Resources in Agricultural Research held 3–5 March 1986, edited by Theodore Hutchcroft, pp.87–96. Dhaka: Bangladesh Agricultural Research Council, ISNAR, and Winrock International, 1986.

0337 Randhawa, N.S., K.V. Raman, and M. Rajagopalan. *National Agricultural Research System in India and Its Impact on Agricultural Production and Productivity.* New Delhi: Indian Council of Agricultural Research, October 1986.

0377 World Bank. *India Staff Appraisal Report of the National Agricultural Research Project II.* Report No. 5469-IN. Washington, D.C.: World Bank, September 1985.

0433 Raman, K.V., and C. Prasad. "Trends in Agricultural Research and Extension in India." Paper presented at the regional seminar organised by the Commonwealth Association of Scientific Agricultural Societies and the Agricultural Institute of Malaysia, Kuala Lumpur, November 1986.

0434 Anh, Do. "Agricultural Research Management Systems in the Philippines, India, and Bangladesh." In *Proceedings of the Research Extension Linkage Workshop Held in Hanoi, Viet Nam, 9–13 June 1986.* Rome: FAO, 1986.

0684 Morehouse, Ward, and Brijen Gupta. "India: Success and Failure." Chapter 8 in *Learning by Doing – Science and Technology in the Developing World,* edited by Aaron Segal, pp. 189–212. Boulder and London: Westview Press, 1987.

0732 FAO – Regional Office for Asia and the Pacific. *Agricultural Research Systems in the Asia-Pacific Region.* Bangkok: FAO – Regional Office for Asia and the Pacific, 1986.

0770 Ruttan, Vernon W. *Agricultural Research Policy and Development.* FAO Research and Technology Paper 2. Rome: FAO, 1987.

0852 Evenson, Robert E., and Yoav Kislev. *Investment in Agricultural Research and Extension: A Survey of International Data.* Center Discussion Paper No. 124. New Haven, Connecticut: Economic Growth Center, Yale University, August 1971.

0869 Boyce, James K. "Agricultural Research in Indonesia, The Philippines, Bangladesh, South Korea and India: A Documentary History." University of Minnesota Asian Agricultural Research Review Project, July 1980. Mimeo.

0886 Evenson, Robert E., and Yoav Kislev. "Investment in Agricultural Research and Extension: A Survey of International Data." *Economic Development and Cultural Change* Vol. 23 (April 1975): 507–521.

Indonesia

Personnel

Period/Year	PhD	MSc	BSc	Subtotal	Expat	Total	Source
1960–64	..	..	..	..	..	..	
1965–69	..	..	..	..	..	..	
1970	..	..	..	..	..	..	
1971	..	..	..	..	..	..	
1972	..	..	..	..	..	..	
1973	..	..	..	..	..	..	
1974	19	167	250	436		436	459
1975	16	204	243	463		463	181
1976	..	..	..	..	..	..	
1977	..	..	..	..	..	..	
1978	..	..	..	..	..	..	
1979	27	–978–		1005		1005	869
1980	..	..	..	..	..	..	
1981	..	..	..	..	..	..	
1982	..	..	..	..	..	..	
1983	..	..	..	..	..	..	
1984	90	279	965	1334		1334	181
1985	102	296	965	1363		1363	47
1986	129	235	1008	1372		1372	180

1960–64:　　　　　　　　　　1965–69:

Expenditure

LCU = Indonesian New Rupiahs

Period/Year	Current LCU (millions)	Constant 1980 LCU(millions)	Constant 1980 US$ (millions)		Source
			Atlas	PPP	
1960–64	..	..	..	..	
1965–69	..	..	..	..	
1970	..	..	..	..	
1971	..	..	..	..	
1972	..	..	..	..	
1973	..	..	..	..	
1974	6934.000	19154.410	28.459	65.611	47
1975	9129.000	22417.562	33.307	76.789	47
1976	13776.000	29564.633	43.926	101.270	47
1977	17907.000	34000.739	50.517	116.466	47
1978	20074.000	34348.657	51.034	117.657	47
1979	22846.000	29501.864	43.833	101.055	47
1980	35890.000	35890.000	53.324	122.937	47
1981	42830.000	36143.460	53.701	123.805	47
1982	40240.000	32295.345	47.983	110.624	47
1983	63902.000	45320.567	67.336	155.240	47
1984	67164.000	42562.738	63.238	145.794	47
1985	68745.000	40462.036	60.117	138.598	811
1986	89910.000	..	..	..	811

1960–64:　　　　　　　　　　1965–69:

Personnel Comments:

The personal indicators represent the Agency for Agricultural Research and Development (AARD) only. AARD was established in 1974 with the statutory responsibility to undertake research and development in agriculture according to the policy stated by the Ministry of Agriculture and to manage all technical executive units in agricultural research and development within the Ministry of Agriculture. Prior to 1974, agricultural research was conducted separately within each of the Directorate Generalates of the Ministry of Agriculture, such as the Directorate Generalate of Food Crops, Estate Crops, Forestry, Fisheries, and Animal Husbandry, all of which had few trained researchers and limited research budgets.

In 1979 an internal reorganization of AARD took place, followed by another reorganization in 1983, which resulted in the removal of forestry research to the Agency for Forestry Research and Development (AFRD) of the newly created Ministry of Forestry. According to source 774, AFRD accounted for 127 researchers in 1985. Although most agricultural research in Indonesia is performed by AARD, some work on germ plasm conservation is done by the National Biological Institute (LBN) and some on oceanography by the National Oceanographic Institute (LON), both of which are part of the Ministry of Science and Technology. Sources 27 and 744 note 117 and 184 researchers at LBN for the years 1982 and 1985, respectively. Universities also carry out some agricultural research, although budgetary constraints severely limit the scale of such activities (source 47).

Expenditure Comments:

The expenditure indicators represent the total budget, inclusive of donor contributions, of AARD only. Until 1983, the year that AFRD was created, forestry research expenditures were included in the AARD expenditures. According to source 732, the proposed budget of AFRD for 1983 was 62.53 million Rupees.

Since the Indonesian fiscal year runs from April 1 to March 31, according to the standard practice in this series, an expenditure figure for, say, the 1975–76 period would be placed in 1975.

Acronyms and Abbreviations:

AARD: Agency for Agricultural Research and Development
AFRD: Agency for Forestry Research and Development
LBN: (National Biological Institute)
LON: (National Oceanographic Institute)

Sources:

0047 Nestel, Barry. *Indonesia and the CGIAR Centers: A Study of Their Collaboration in Agricultural Research.* CGIAR Study Paper Number 10. Washington, D.C.: World Bank, 1985.

0180 Retzlaff, Ralph H. "Human Resources: Lessons from AARD in Indonesia." In *Management of Human Resources in Agricultural Research.* Report of the International Workshop on Management of Human Resources in Agricultural Research held 3–5 March 1986, edited by Theodore Hutchcroft, pp. 58–65. Dhaka: Bangladesh Agricultural Research Council, ISNAR, and Winrock International, 1986.

0181 IRRI. *Education for Agriculture.* Proceedings of the Symposium on Education for Agriculture, 12–16 November 1984. Manila: IRRI, 1985.

0459 World Bank – East Asia and Pacific Projects Department. *Appraisal of Agricultural Research and Extension Project Indonesia.* Washington, D.C.: World Bank, April 1975.

0811 Agency for Agricultural Research and Development (AARD). *AARD Reviews 1984–1987: An Analysis of their Recommendations and their Implementation.* Jakarta, Indonesia: AARD, August 1988.

0869 Boyce, James K. "Agricultural Research in Indonesia, The Philippines, Bangladesh, South Korea and India: A Documentary History." University of Minnesota Asian Agricultural Research Review Project, July 1980. Mimeo.

Additional References:

0010 Boyce, James K., and Robert E. Evenson. *National and International Agricultural Research and Extension Programs.* New York: Agricultural Development Council, Inc., 1975.

0014 Judd, M. Ann, James K. Boyce, and Robert E. Evenson. "Investing in Agricultural Supply." Economic Growth Center, Yale University, New Haven, Connecticut, 1983. Mimeo.

0016 Oram, Peter A., and Vishva Bindlish. *Resource Allocations to National Agricultural Research: Trends in the 1970s.* The Hague and Washington, D.C.: ISNAR and IFPRI, November 1981.

0018 Daniels, Douglas, and Barry Nestel, eds. *Resource Allocation to Agricultural Research.* Proceedings of a workshop held in Singapore 8–10 June 1981. Ottawa: IDRC, 1981.

0019 Bengtsson, Bo, and Tedla Getachew, eds. *Strengthening National Agricultural Research.* Report from a SAREC workshop, 10–17 September 1979. Part 1: Background Documents; Part II: Summary and Conclusions. Sweden: SAREC, 1980.

0026 Oram, Peter A., and Mark Gieben. "Document Summaries." ISNAR, The Hague, 1984. Mimeo.

0027 Harvey, Nigel, ed. *Agricultural Research Centres: A World Directory of Organizations and Programmes.* Seventh Edition, Two Volumes. Harlow, U.K.: Longman, 1983.

0073 Oram, Peter A., and Vishva Bindlish. "Investment in Agricultural Research in Developing Countries: Progress, Problems, and the Determination of Priorities." IFPRI, Washington, D.C., January 1984. Mimeo.

0093 Bangladesh Agricultural Research Council (BARC), ISNAR, and Winrock International. *Management of Human Resources in Agricultural Research.* Report of the International Workshop on Management of Human Resources in Agricultural Research held March 3–5, 1986 in Dhaka, Bangladesh. Dhaka: BARC, ISNAR, and Winrock International, 1986.

0095 FAO – CARIS. *Agricultural Research in Developing Countries – Volume 1: Research Institutions.* Rome: FAO – CARIS, 1978.

0135 FAO, and UNDP. *National Agricultural Research – Report of an Evaluation Study in Selected Countries.* Rome: FAO and UNDP, 1984.

0159 Agency for Agricultural Research and Development (AARD). *An Evaluation of the Industrial Crop Research Program of AARD.* Two volumes. Indonesia: AARD, August 1985.

0163 CGIAR. "National Agricultural Research." CGIAR, Washington, D.C., 1985. Mimeo.

0165 Evenson, R.E., and Y. Kislev. *Agricultural Research and Productivity.* New Haven: Yale University Press, 1975.

0176 Moseman, Albert, ed. *National Agricultural Research Systems in Asia.* Report of the regional seminar held at the India International Centre, New Delhi, India, March 8–13, 1971. New York: ADC, Inc., 1971.

0179 The Philippine Council for Agriculture and Resources Research and Development (PCARRD), and IDRC. *Agriculture and Resources Research Manpower Development in South and Southeast Asia.* Proceedings of the Workshop on Agriculture and Resources Research Manpower Development in South and Southeast Asia, 21–23 October 1981 and 4–6 February 1982, Singapore. Los Baños, Philippines: PCARRD, 1983.

0182 Agency for Agricultural Research and Development (AARD). *This is AARD.* Indonesia: AARD, 1985.

0183 World Bank – East Asia and the Pacific Regional Office. *Indonesia National Agricultural Research Project: Staff Appraisal Report.* Washington, D.C.: World Bank, April 1980.

0270 Karama, A. Syarifuddin. "Planning, Programming and Evaluation of Research and Development Indonesia Experiences." Paper presented at the Workshop on Improving Agricultural Research Organizations and Management: Progress and Issues for the Future, ISNAR, The Hague, 8–12 September 1986.

0317 Birowo, A.T. "Policy Issues for Agricultural Research in Indonesia." In *Strengthening National Agricultural Research.* Report from a SAREC workshop, 10–17 September 1979, edited by Bo Bengtsson and Getachew Tedla, pp. 98–111. Sweden: SAREC, 1980.

0435 Badruddoza, K.M. "Agricultural Research System in Indonesia." In *Proceedings of the Research Extension Linkage Workshop Held in Hanoi, Viet Nam, 9–13 June 1986.* Rome: FAO, 1986.

0441 Joint Agriculture Research Survey Team. *Report of Survey and Recommendations on Organizations, Systems, and Requirements for Research in Agriculture and Related Industries in Indonesia.* Djakarta, Indonesia: Ministry of Agriculture, July 1969.

0447 Agency for Agricultural Research and Development (AARD). *5 Years (1976–1980) of Agricultural Research and Development for Indonesia.* Indonesia: AARD, n.d.

0460 World Bank – Projects Department, East Asia and Pacific Regional Office. *Indonesia: Technical and Professional Manpower in Agriculture.* Two vols. Washington, D.C.: World Bank, March 1981.

0731 Salmon, David Christopher. "An Evaluation of Investment in Agricultural Research in Indonesia: 1965–1977." PhD diss., University of Minnesota, Minneapolis, April 1984.

0732 FAO – Regional Office for Asia and the Pacific. *Agricultural Research Systems in the Asia-Pacific Region.* Bangkok: FAO – Regional Office for Asia and the Pacific, 1986.

0744 *Agricultural Research Centres: A World Directory of Organizations and Programmes.* Eighth Edition, Two Volumes. Harlow, U.K.: Longman, 1986.

0852 Evenson, Robert E., and Yoav Kislev. *Investment in Agricultural Research and Extension: A Survey of International Data.* Center Discussion Paper No. 124. New Haven, Connecticut: Economic Growth Center, Yale University, August 1971.

0886 Evenson, Robert E., and Yoav Kislev. "Investment in Agricultural Research and Extension: A Survey of International Data." *Economic Development and Cultural Change* Vol. 23 (April 1975): 507–521.

Iran

Personnel

Period/Year	PhD	MSc	BSc	Subtotal	Expat	Total	Source
1960–64	..	..	..	..	..	..	
1965–69	..	..	..	..	..	300	10
1970	..	..	..	..	..	509	10
1971	..	..	..	..	..	589	10
1972	..	..	..	..	..	..	
1973	..	..	..	..	..	..	
1974	..	..	..	..	..	537	10
1975	..	..	..	..	..	..	
1976	..	..	..	..	..	518	95
1977	..	..	..	..	..	..	
1978	..	..	..	..	..	..	
1979	..	..	..	..	..	..	
1980	..	..	..	..	..	..	
1981	..	..	..	..	..	..	
1982	..	..	..	..	..	493	27
1983	..	..	..	..	..	..	
1984	..	..	..	..	..	..	
1985	..	..	..	..	..	..	
1986	..	..	..	..	..	..	

1960–64:   1965–69: 1967

Expenditure

LCU = Iranian Rials

Period/Year	Current LCU (millions)	Constant 1980 LCU(millions)	Constant 1980 US$ (millions) Atlas	PPP	Source
1960–64	200.000	1290.323	14.668	23.397	10
1965–69	..	..	..	..	
1970	704.000	4789.116	54.440	86.840	10
1971	757.500	4618.902	52.505	83.754	10
1972	1072.600	6094.318	69.277	110.507	10
1973	1147.600	4842.194	55.043	87.802	10
1974	..	..	..	..	
1975	..	..	..	..	
1976	1864.000	3915.966	44.515	71.007	95
1977	..	..	..	..	
1978	..	..	..	..	
1979	..	..	..	..	
1980	..	..	..	..	
1981	..	..	..	..	
1982	..	..	..	..	
1983	..	..	..	..	
1984	..	..	..	..	
1985	..	..	..	..	
1986	..	..	..	..	

1960–64: 1962   1965–69:

Personnel Comments:

The 1976 and 1982 figures have been constructed as follows:

Institute	1976	1982
Forest and Rangeland Research Institute	59	50
Animal Husbandry Research Institute	23	
Plant Pests and Diseases Research Institute	130	120
Razi State Serum and Vaccine Institute	61	43
Safiabad Agricultural Research Center	29	15
Soil Institute of Iran	156	191
Sugar Beet Seed Research Institute	20	50
Seed and Plant Improvement Institute	40	
Institute of Horticulture (University of Isfahan)		24
Total	518	493

The institutes included here operate at the national level. Some of them have regional substations. In addition, source 94 identifies 8 regional research centers which conduct agricultural and animal research. Each of these regional research centers have 4 to 9 substations per region. The total number of researchers working at these centers may be substantial but there is currently no quantifiable information available.

Expenditure Comments:

The 1976 expenditure figure includes the following institutes: Forest and Rangeland Research Institute, Animal Husbandry Research Institute, Plant Pests and Diseases Research Institute, Razi State Serum and Vaccine Institute

and Soil Institute of Iran. Total research expenditures are probably substantially larger than the figure presented in the table.

Sources:

0010 Boyce, James K., and Robert E. Evenson. *National and International Agricultural Research and Extension Programs.* New York: Agricultural Development Council, Inc., 1975.

0027 Harvey, Nigel, ed. *Agricultural Research Centres: A World Directory of Organizations and Programmes.* Seventh Edition, Two Volumes. Harlow, U.K.: Longman, 1983.

0095 FAO – CARIS. *Agricultural Research in Developing Countries – Volume 1: Research Institutions.* Rome: FAO – CARIS, 1978.

Additional References:

0094 FAO – Near East Regional Office. *Directory of Agricultural Research Institutions in the Near East Region.* Cairo: FAO – Near East Regional Office, 1979.

0173 FAO. *FAO Near East Regional Studies on Organization and Administration of Agricultural Research.* Rome: FAO, 1974.

0174 Watson, J.M. "Comparative Study of Agricultural Research Organisation and Administration in the Near East Region." Paper presented at the Workshop on Organization and Administration of Agricultural Services in the Arab States, Cairo, 2–15 March 1964.

0517 Aresvik, Oddvar. *The Agricultural Development of Iran.* Praeger Special Studies in International Economics and Developement. New York: Praeger Publishers, 1976.

0852 Evenson, Robert E., and Yoav Kislev. *Investment in Agricultural Research and Extension: A Survey of International Data.* Center Discussion Paper No. 124. New Haven, Connecticut: Economic Growth Center, Yale University, August 1971.

Iraq

Personnel

Period/Year	PhD	MSc	BSc	Subtotal	Expat	Total	Source
1960–64	..	..	..	..	..	..	
1965–69	..	..	..	..	..	..	
1970	..	..	..	..	..	..	
1971	..	..	..	..	..	..	
1972	..	..	..	..	..	..	
1973	..	..	..	..	..	..	
1974	..	..	..	..	..	..	
1975	..	..	..	..	..	..	
1976	..	..	..	..	..	..	
1977	..	..	..	..	..	..	
1978	..	..	..	..	..	..	
1979	..	..	..	..	..	..	
1980	..	..	..	..	..	..	
1981	..	..	..	..	..	..	
1982	..	..	..	..	..	..	
1983	93	75	374	542	..	542	188
1984	..	..	..	..	..	..	
1985	..	..	..	..	..	..	
1986	..	..	..	..	..	..	

1960–64:   1965–69:

Expenditure   *LCU = Iraqi Dinars*

Period/Year	Current LCU (millions)	Constant 1980 LCU(millions)	Constant 1980 US$ (millions)		Source
			Atlas	PPP	
1960–64	..	..	..	..	
1965–69	..	..	..	..	
1970	..	..	..	..	
1971	..	..	..	..	
1972	..	..	..	..	
1973	..	..	..	..	
1974	..	..	..	..	
1975	..	..	..	..	
1976	..	..	..	..	
1977	..	..	..	..	
1978	..	..	..	..	
1979	..	..	..	..	
1980	..	..	..	..	
1981	..	..	..	..	
1982	..	..	..	..	
1983	..	..	..	..	
1984	..	..	..	..	
1985	..	..	..	..	
1986	..	..	..	..	

1960–64:   1965–69:

Personnel Comments:

The 1983 figure includes two institutes, namely, the State Board for Applied Agricultural Research (SBAAR) and the Agriculture and Water Resources Research Centre (AWRRC). According to source 188, SBAAR is under the administrative control of the Ministry of Agriculture and Agrarian Reform, while AWRRC is under the Scientific Research Council, which is under the direct control of the Prime Minister's Secretariat. SBAAR was organized in its present form in 1979 and has been built around the nucleus of the former Field Crops Directorate, by attaching to it research units of other Directorates of the Ministry of Agriculture, along with other Ministries such as Irrigation and autonomous bodies such as the State Organization in Soils and Land Reclamation. Because of the large dispersion of agricultural research in the past, the limited additional data available appears to significally understate the national research capacity.

Sources:

0188 FAO. *Report of an FAO Mission: Organization and Administration of Applied Agricultural Research in Iraq.* Rome: FAO, February 1984.

Additional References:

0010 Boyce, James K., and Robert E. Evenson. *National and International Agricultural Research and Extension Programs.* New York: Agricultural Development Council, Inc., 1975.

0016 Oram, Peter A., and Vishva Bindlish. *Resource Allocations to National Agricultural Research: Trends in the 1970s.* The Hague and Washington, D.C.: ISNAR and IFPRI, November 1981.

0027 Harvey, Nigel, ed. *Agricultural Research Centres: A World Directory of Organizations and Programmes.* Seventh Edition, Two Volumes. Harlow, U.K.: Longman, 1983.

0094 FAO – Near East Regional Office. *Directory of Agricultural Research Institutions in the Near East Region.* Cairo: FAO – Near East Regional Office, 1979.

0095 FAO – CARIS. *Agricultural Research in Developing Countries – Volume 1: Research Institutions.* Rome: FAO – CARIS, 1978.

0173 FAO. *FAO Near East Regional Studies on Organization and Administration of Agricultural Research.* Rome: FAO, 1974.

0174 Watson, J.M. "Comparative Study of Agricultural Research Organisation and Administration in the Near East Region." Paper presented at the Workshop on Organization and Administration of Agricultural Services in the Arab States, Cairo, 2–15 March 1964.

0189 FAO. *Report to the Government of Iraq: Organization and Administration of Agricultural Research.* Rome: FAO, 1964.

0831 FAO. *Report of an Expert Consultation on Agricultural Extension and Research Linkages in the Near East, Amman, Jordan, 22–26 July 1985.* Rome: FAO, 1985.

0865 Oram, Peter. "Agricultural Research Objectives and Priorities: Constraints to the Development of Agricultural Research Institutions in Arab Countries." Review paper prepared for the first meeting of the Council for Arab Agricultural Research, organized by the Arab Fund for Economic and Social Development, Kuwait, 23 March 1988.

Ireland

Personnel

Period/Year	PhD	MSc	BSc	Subtotal	Expat	Total	Source
1960–64	..	..	..		..		
1965–69	58	113	128	299	..	299	286/285
1970	..	..	..		..		
1971	86	148	104	338	..	338	279
1972	..	..	..		..		
1973	..	..	..		..		
1974	111	139	101	351	..	351	244
1975	..	..	..		..		
1976	..	..	..		..		
1977	..	..	..		..	393	283
1978	..	..	..		..		
1979	..	..	..		..		
1980	..	..	..		..		
1981	..	..	..		..		
1982	..	..	..		..		
1983	..	..	..		..		
1984	..	..	..		..		
1985	..	..	..		..		
1986	..	..	..		..		

1960–64:   1965–69: 1965 & 68

Expenditure

LCU = Irish Pounds

Period/Year	Current LCU (millions)	Constant 1980 LCU(millions)	Constant 1980 US$ (millions) Atlas	PPP	Source
1960–64	1.686	9.918	18.621	18.753	514
1965–69	2.533	11.265	21.151	21.301	510/474/511
1970	3.290	12.230	22.963	23.126	474
1971	3.725	12.500	23.469	23.636	601/509
1972	4.531	13.405	25.169	25.347	601
1973	5.347	13.781	25.874	26.058	601
1974	5.143	12.453	23.381	23.546	601
1975	7.059	13.951	26.193	26.378	601/513
1976	9.014	14.705	27.609	27.804	601
1977	11.209	16.151	30.325	30.540	601
1978	11.364	14.816	27.818	28.015	601
1979	13.030	14.943	28.055	28.254	601
1980	11.875	11.875	22.296	22.454	601
1981	15.050	12.819	24.069	24.239	601/602
1982	18.875	13.899	26.096	26.281	601
1983	19.280	12.751	23.941	24.111	601/602
1984	20.010	12.460	23.393	23.559	601/602
1985	22.720	13.468	25.286	25.465	601
1986	..	..	..	..	

1960–64: 1963   1965–69: 1967 & 69

Personnel Comments:

The personnel series has been constructed as below.

The non–university component includes researchers at the Department of Agriculture and Fisheries, the Department of Lands (Forest and Wildlife Service), and the Agricultural Institute 'An Foras Taluntais'. The latter is the main agricultural research institute in the country. The university component includes agricultural researchers at the University College, Dublin; at the University College, Cork; and at the University of Dublin, Trinity College. On the basis of source 632, which gives the total number of researchers for universities expressed in full-time equivalents for 1967, it was estimated that researchers at these universities spend no more than 30% of their time on research. This percentage has been applied to the total number of agricultural researchers at universities for all other years.

	1965				1968			
	PhD	MSc	BSc	Total	PhD	MSc	BSc	Total
Non-university	39	72	134	245	58	114	111	283
University	7	18	6	31	11	22	5	38
Total	46	90	140	276	69	136	116	321
Source				286				285

	1971				1974				1977
	PhD	MSc	BSc	Total	PhD	MSc	BSc	Total	Total
Non-university	70	129	99	298	88	122	99	309	346
University	16	19	5	40	23	17	2	42	[47]
Total	86	148	104	338	111	139	101	351	393
Source				279				244	283

[ ] = estimated or constructed by authors.

Expenditure Comments:

The expenditure series has been constructed as below:

The non-university and university institutions included here correspond, more or less, to the institutions included in the personnel series.

	1963	1967	1969	1970	1971	1972	1973	1974	1975	1976
Non-university	1.651	2.225	2.582	3.084	3.480	4.220	4.970	4.700	6.550	8.350
University	0.035	0.091	0.167*	0.206	0.245*	[0.311]	[0.377]	[0.443]	0.509*	0.664
Total	1.686	2.316	2.749	3.290	3.725	4.531	5.347	5.143	7.059	9.014
Source	514	510	474 511*	474	601 509*	601	601	601	601 513*	601

	1977	1978	1979	1980	1981	1982	1983	1984	1985
Non-university	10.390	10.390	11.900	10.590	13.610	17.520	18.010	18.790	21.500
University	[0.819]	[0.974]	[1.130]	[1.285]	1.440*	[1.355]	1.270*	1.220*	[1.220]
Total	11.209	11.364	13.030	11.875	15.050	18.875	19.280	20.010	22.720
Source	601	601	601	601	601 602*	601	601 602*	601 602*	601

[ ] = estimated or constructed by authors.

Sources:

0244 Commonwealth Agricultural Bureaux (CAB). *List of Research Workers in the Agricultural Sciences in the Commonwealth and in the Republic of Ireland 1975.* Slough, England: CAB, 1975.

0279 Commonwealth Agricultural Bureaux (CAB). *List of Research Workers in the Agricultural Sciences in the Commonwealth and in the Republic of Ireland 1972.* Slough, England: CAB, 1972.

0283 European Economic Community (EEC). "Monograph on the Organization of Agricultural Research." EEC, Brussels, 1982. Mimeo.

0285 Commonwealth Agricultural Bureaux (CAB). *List of Research Workers in the Agricultural Sciences in the Commonwealth and in the Republic of Ireland 1969.* Slough, England: CAB, 1969.

0286 Commonwealth Agricultural Bureaux (CAB). *List of Research Workers in Agriculture, Animal Health and Forestry in the Commonwealth and in the Republic of Ireland 1966.* Slough, England: CAB, 1966.

0474 OECD. *Science and Technology Indicators: Basic Statistical Series – Volume A: The Objectives of Government R&D, 1969–1981.* Paris: OECD, July 1981.

0509 OECD. *Survey of the Resources Devoted to R&D by OECD Member Countries, International Statistical Year 1971 – Volume 5: Total Tables, Statistical Tables and Notes.* Paris: OECD, August 1974.

0510 OECD. *International Survey of the Resources Devoted to R&D in 1967 by OECD Member Countries, Statistical Tables and Notes – Volume 5: Total Tables.* Paris: OECD, August 1970.

0511 OECD. *International Survey of the Resources Devoted to R&D in 1969 by OECD Member Countries, Statistical Tables and Notes – Volume 5: Total Tables.* Paris: OECD, June 1973.

0513 OECD. *International Survey of the Resources Devoted to R&D by OECD Member Countries,*
International Statistical Year 1975 – International Volume: Statistical Tables and Notes. Paris: OECD, March 1979.

0514 OECD. *A Study of Resources Devoted to R&D in OECD Member Countries in 1963/64, International Statistical Year for Research and Development – Volume 2: Statistical Tables and Notes.* Paris: OECD, 1968.

0601 OECD – STIID Data Bank. "Public Funding of R&D by Socio-Economic Objective." Tables. OECD, Paris, April 1987. Mimeo.

0602 OECD – STIID Data Bank. "R&D Expenditure in the Higher Education and Private Non-Profit Sectors." Tables. OECD, Paris, April 1987. Mimeo.

Additional References:

0010 Boyce, James K., and Robert E. Evenson. *National and International Agricultural Research and Extension Programs.* New York: Agricultural Development Council, Inc., 1975.

0014 Judd, M. Ann, James K. Boyce, and Robert E. Evenson. "Investing in Agricultural Supply." Economic Growth Center, Yale University, New Haven, Connecticut, 1983. Mimeo.

0021 OECD. *OECD Science and Technology Indicators.* Paris: OECD, 1984.

0022 UNESCO. *Statistical Yearbook 1983.* Paris: UNESCO, 1983.

0027 Harvey, Nigel, ed. *Agricultural Research Centres: A World Directory of Organizations and Programmes.* Seventh Edition, Two Volumes. Harlow, U.K.: Longman, 1983.

0165 Evenson, R.E., and Y. Kislev. *Agricultural Research and Productivity.* New Haven: Yale University Press, 1975.

0287 Commonwealth Agricultural Bureaux (CAB). *List of Research Workers in Agriculture, Animal Health and Forestry in the British Commonwealth, the Republic of Sudan and the Republic of Ireland 1959.* Slough, England: CAB, 1959.

0376 FAO. *Fourteenth FAO Regional Conference for Europe, Reykjavik, Iceland, 17–21 September 1984: Research in Support of Agricultural Policies in Europe.* Rome: FAO, 1984.

0473 OECD. *Science and Technology Indicators: Basic Statistical Series – Volume A: The Objectives of Government R&D Funding 1974–1985.* Paris: OECD, May 1983.

0512 OECD. *Survey of the Resources Devoted to R&D by OECD Member Countries, International Statistical Year 1973 – Volume 5: Total Tables, Statistical Tables and Notes.* Paris: OECD, September 1976.

0530 OECD. *OECD Science and Technology Indicators No. 2 – R&D, Invention and Competitiveness.* Paris: OECD, 1986.

0574 OECD. *Intellectual Investment in Agriculture for Economic and Social Development.* Documentation in Food and Agriculture No. 60. Paris: OECD, 1962.

0606 EUROSTAT. *Government Financing of Research and Development 1975–1984.* Luxembourg: Office des Publications Officielles des Communautés Européennes, February 1985.

0631 An Foras Taluntais. *Annual Report.* Years: 1964/65 – 1969/70. Dublin: An Foras Taluntais, various years.

0632 Murphy, Diarmuid. *Research and Development in Ireland 1967.* Dublin: National Science Council, October 1969.

0633 Department of Agriculture and Fisheries. *Annual Report of the Veterinary Research Laboratory 1968–69.* Dublin: Stationary Office, n.d.

0706 EUROSTAT. *Government Financing of Research and Development 1975–1985.* Luxembourg: Office for Official Publications of the European Communities, 1987.

0707 OECD – STIID Data Bank. "R&D Expenditure in the Higher Education and Private Non-Profit Sectors: Higher Education, Agricultural Sciences, General Universities Funds." Table. OECD, Paris, June 1987. Mimeo.

0708 OECD – STIID Data Bank. "Gross Domestic Expenditure on R&D: All Fields of Sciences – Agriculture, Forestry and Fishing." Tables. OECD, Paris, June 1987. Mimeo.

0744 *Agricultural Research Centres: A World Directory of Organizations and Programmes.* Eighth Edition, Two Volumes. Harlow, U.K.: Longman, 1986.

0852 Evenson, Robert E., and Yoav Kislev. *Investment in Agricultural Research and Extension: A Survey of International Data.* Center Discussion Paper No. 124. New Haven, Connecticut: Economic Growth Center, Yale University, August 1971.

0886 Evenson, Robert E., and Yoav Kislev. "Investment in Agricultural Research and Extension: A Survey of International Data." *Economic Development and Cultural Change* Vol. 23 (April 1975): 507–521.

Israel

Personnel

Period/Year	PhD	MSc	BSc	Subtotal	Expat	Total	Source
1960–64	..	..	..	..	..	..	
1965–69	..	..	..	..	..	357	10
1970	..	..	..	..	..	430	852
1971	..	..	..	..	..	..	
1972	..	..	..	..	..	..	
1973	..	..	..	..	..	..	
1974	..	..	..	..	..	..	
1975	..	..	..	..	..	..	
1976	..	..	..	..	..	416	724
1977	..	..	..	..	..	..	
1978	..	..	..	..	..	..	
1979	..	..	..	..	..	..	
1980	..	..	..	..	..	500	724
1981	..	..	..	..	..	..	
1982	..	..	..	..	..	..	
1983	..	..	..	..	..	..	
1984	..	..	..	..	..	..	
1985	..	..	..	..	..	..	
1986	..	..	..	..	..	..	

1960–64:   1965: 1967, 68 & 69

Expenditure

LCU = Shekels (1980)

Period/Year	Current LCU (millions)	Constant 1980 LCU(millions)	Constant 1980 US$ (millions) Atlas	PPP	Source
1960–64	0.779	48.688	8.343	10.509	10
1965–69	1.840	71.552	12.261	15.444	10
1970	2.928	100.966	17.302	21.792	724
1971	..	..	..	..	
1972	4.150	106.410	18.235	22.967	724
1973	..	..	..	..	
1974	..	..	..	..	
1975	8.758	99.523	17.054	21.481	724
1976	..	..	..	..	
1977	24.116	148.864	25.509	32.130	724
1978	..	..	..	..	
1979	94.427	211.246	36.199	45.595	724
1980	211.245	211.245	36.199	45.595	724
1981	464.567	199.214	34.137	42.998	724
1982	1045.973	212.812	36.468	45.933	724
1983	2807.167	238.097	40.801	51.390	724
1984	..	..	..	..	
1985	..	..	..	..	
1986	..	..	..	..	

1960–64: 1960   1965–69: 1965, 67, 68 & 69

Personnel Comments:

The 1976 and 1980 figures are constructed on the basis of source 724 and include the following institutes: the Agricultural Research Organization (ARO), the Institute for Plant Protection, the Kidron Veterinary Institute, the Soil Erosion Research Institute, the Hydrological Service (all of the Ministry of Agriculture), the Israel Center for Water Works Appliances, Mekorot, Tahal Water Planning for Israel Ltd., the Biological Control Institute, the Oceanographic and Limnological Research Company Ltd., and the Faculty of Agriculture of the Hebrew University. Using secondary information from source 724, university full-time time equivalent (FTE) academic staff were prorated to FTE researchers, assuming that 30% of their academic time was spent on research. Source 10 does not indicate which institutes are included in the pre-1971 figures.

Expenditure Comments:

The pre-1970 figures include the Agricultural Research Organization (ARO) only. Source 10 estimates that at that time ARO represented about 85% of all agricultural research in the country. The post-1969 figures are constructed on the basis of source 724 and include both the university and non-university sector. Construction details are given in the table below.

| | Public Expenditure | Universities | | Total |
		Operating Budget	Specially Budgeted R&D	
1970	2.600	0.190	0.138	2.928
1972	3.600	0.340	0.210	4.150
1975	8.000	0.560	0.198	8.758
1977	22.400	1.260	0.456	4.116
1979	90.000	3.660	0.767	94.427
1980	196.900	[12.325]	2.020	211.245
1981	437.400	20.990	[6.177]	464.567
1982	994.700	40.940	10.333	1045.973
1983	2634.200	143.300	29.667	2807.167

[ ] = estimated or constructed by authors.

The public expenditure figure represents public expenditures on R&D for the socioeconomic objective of 'agriculture, forestry and fisheries'. It explicitly excludes research expenditures paid from general university funds. They are included here under the operating budget of the university. The specially budgeted R&D of the universities includes only that portion of the special budget which is not covered by public expenditures on R&D. From 1970 to 1974, university expenditures were reported on an April-to-March fiscal-year basis, and hence are recorded here in the first calendar year. The later years were reported on an October-to-September academic-year basis, and hence are recorded here in the second calender year.

These expenditure figures should be consistent with the personnel figures given by source 724.

Sources:

0010 Boyce, James K., and Robert E. Evenson. *National and International Agricultural Research and Extension Programs.* New York: Agricultural Development Council, Inc., 1975.

0724 Herskovic, Shlomo. Personal communication. Head, Department of R&D Economics, National Council for Research and Development. Jerusalem, 10 July 1987.

0852 Evenson, Robert E., and Yoav Kislev. *Investment in Agricultural Research and Extension: A Survey of International Data.* Center Discussion Paper No. 124. New Haven, Connecticut: Economic Growth Center, Yale University, August 1971.

Additional References:

0027 Harvey, Nigel, ed. *Agricultural Research Centres: A World Directory of Organizations and Programmes.* Seventh Edition, Two Volumes. Harlow, U.K.: Longman, 1983.

0165 Evenson, R.E., and Y. Kislev. *Agricultural Research and Productivity.* New Haven: Yale University Press, 1975.

0744 *Agricultural Research Centres: A World Directory of Organizations and Programmes.* Eighth Edition, Two Volumes. Harlow, U.K.: Longman, 1986.

0886 Evenson, Robert E., and Yoav Kislev. "Investment in Agricultural Research and Extension: A Survey of International Data." *Economic Development and Cultural Change* Vol. 23 (April 1975): 507–521.

Italy

funded by three major organizations, namely the Ministry of Agriculture and Forestry, the Ministry of Education (represented the universities) and the National Council of Research). In addition to these three main organizations, several ...

Personnel

Period/Year	PhD	MSc	BSc	Subtotal	Expat	Total	Source
1960–64	..	..	..	..	..	995	520/881
1965–69	..	..	..	..	..	1016	714/881
1970	..	..	..	..	..	..	
1971	..	..	..	..	..	934	714/881
1972	..	..	..	..	..	..	
1973	..	..	..	..	..	1219	714/881
1974	..	..	..	..	..	..	
1975	..	..	..	..	..	1447	714/881
1976	..	..	..	..	..	1470	714/881
1977	..	..	..	..	..	1683	714/881
1978	..	..	..	..	..	1755	881
1979	..	..	..	..	..	1964	681/881
1980	..	..	..	..	..	2092	681/881
1981	..	..	..	..	..	2135	681/881
1982	..	..	..	..	..	2225	681/881
1983	..	..	..	..	..	2349	681/881
1984	..	..	..	..	..	2473	682/081
1985	..	..	..	..	..	2455	860/881
1986	..	..	..	..	..	2512	881

1960–64: 1963   1965–69: 1969

Expenditure

LCU = Lira

Period/Year	Current LCU (millions)	Constant 1980 LCU(millions)	Constant 1980 US$ (millions)		Source
			Atlas	PPP	
1960–64	..	..	..		
1965–69	9599.750	48578.211	52.506	57.961	881
1970	14887.000	67977.169	73.474	81.107	881
1971	12548.000	53395.745	57.713	63.709	881
1972	16472.000	65888.000	71.216	78.614	881
1973	23893.000	85637.993	92.562	102.179	881
1974	19262.000	58369.697	63.089	69.644	881
1975	24624.000	63627.907	68.773	75.917	881
1976	30026.000	65702.407	71.015	78.393	881
1977	38646.000	70910.092	76.644	84.606	881
1978	55966.000	90267.742	97.567	107.703	881
1979	58403.000	81228.095	87.796	96.917	881
1980	85294.000	85294.000	92.191	101.768	881
1981	152460.000	128658.228	139.061	155.508	881
1982	188411.000	136827.160	147.891	163.255	881
1983	243972.000	153634.761	166.057	183.309	881
1984	262756.000	150146.286	162.287	179.147	881
1985	359310.000	188713.235	203.972	225.163	881
1986	454761.000	220972.303	238.890	263.652	881

1960–64:   1965–69: 1966, 67, 68 & 69

Personnel Comments:

Agricultural research in Italy is executed and funded by three major organizations, namely, the Ministry of Agriculture and Forestry, the Ministry of Education – representing the universities – and the National Council of Research (CNR). In addition to these three major organizations, several other independent organizations are also involved in agricultural research: the Agricultural Department of the National Agency for Alternative Energy (ENEA) and the Italian National Agency for the Development of the South of Italy (CASMEZ).

The table below gives a breakdown of the total personnel figures into their respective university and non-university components.

	1963	1969	1971	1973	1975	1976	1977	1978	1979
Non-university	445	534	379	488	598	624	806	[844]	885
University	550	482	555	731	849	846	877	911	1079
Total	955	1016	934	1219	1447	1470	1683	1755	1964
Source	714	714	714	714	714	714	714	881	681
	881	881	881	881	881	881	881		881

	1980	1981	1982	1983	1984	1985	1986	1987[a]
Non-university	981	991	1047	1136	1151	1083	1087	1091
University	[1111]	[1144]	[1178]	1213	1322	[1372]	[1425]	1479
Total	2092	2135	2225	2349	2473	2455	2512	2570
Source	681	681	681	681	682	860	881	881
	881	881	881	881	881	881		

[ ] = estimated by source 881.
a. Not included in the indicator series.

Expenditure Comments:

The Italian expenditure series has been based on the table on the next page – provided by source 881 – which reports public funds for agricultural and forestry research in Italy measured in millions of current Lira. According to source 881, these are the most complete Italian figures presently available representing public funds allocated for research on the following activities: agriculture and forestry, postharvest technology, processing of agricultural and forestry products, nutritional research related to agricultural production, and social and economic research for agricultural systems. The table originates from INEA (National Institute for Agricultural Economics) data. This institute undertakes an annual survey of the Italian agricultural research system.

Acronyms and Abbreviations:

CASMEZ: (Italian National Agency for the Development of the South of Italy)
CNR: Consiglio Nazionale delle Ricerche (National Research Council)
ENEA: (National Agency for Alternative Energy)
INEA: (National Institute for Agricultural Economics)
MAF: (Ministry of Agriculture and Forestry)
MPI: (Ministry of Education)

	MPI	CNR	MAF	ENEA	OTHERS	REGIONS	CASMEZ	INTERN[a]	TOTAL
1966	1600	1993	1324	63	560		0	463	6003
1967	2260	2382	2236	534	270		0	106	7788
1968	2447	2710	4853	511	472		0	106	11099
1969	2749	2543	5025	589	868		1605	130	13509
1970	3026	3735	6026	597	1373		0	130	14887
1971	3371	3890	2220	815	1286		827	139	12548
1972	4577	4305	4556	910	1985		0	139	16472
1973	5542	5192	8046	1307	3663		0	143	23893
1974	5983	3944	6411	998	1731		0	195	19262
1975	7376	5035	7424	1845	1893	1037	0	14	24624
1976	8435	9311	8359	2420	689	812	0	0	30026
1977	8321	14529	8893	2585	4134	184	0	0	38646
1978	10834	17460	15819	2441	5535	2210	0	1667	55966
1979	11819	20243	15004	4127	5540	0	0	1670	58403
1980	17665	23184	19633	2656	5382	0	15103	1671	85294
1981	60209	26082	25071	0	5593	10797	22710	1998	152460
1982	75966	27952	28059	12644	8328	20074	13190	2198	188411
1983	83252	45677	30026	11690	9796	34290	24332	5809	243972
1984	88185	64582	41984	16750	12840	30850	1537	6108	262756
1985	108276	70678	56029	14750	12420	31083	58166	7908	359310
1986	164287	82148	51940	14112	26554	26940	78252	10528	454761
1987	226846	77437	67580	14258	21648	10599	78611	8072	505051

a. INTERN = Italian contribution to international research organizations.

Sources:

0520 OECD. *Reviews of National Science Policy: Italy.* Paris: OECD, 1969.

0681 Istituto Centrale di Statistica (ICS). *Indagine Statistica sulla Ricerca Scientifica 1975–1983.* Supplemento al Bollettino Mensile di Statistica 1978:1&22, 1980:1, 1981:13, 1982:19, 1983:17&27, 1984:11, 1985:19. Roma: ICS, various years.

0682 Istituto Centrale di Statistica (ICS). *Indagine Statistica sulla Ricerca Scientifica – Consuntivo 1984 Previsione 1985 e 1986.* Collana d'Informazione Anno 1987, No. 1. Roma: ICS, 1987.

0714 OECD. Personal communication. Background information from the OECD Science and Technology Indicators Division. Paris, Summer 1987.

0860 Instituto Centrale di Statistica (ICS). Personal communication. Director General, ICS. Rome, April 1988.

0881 Galante, Ennio, and Cesare Sala. Personal communication. Istituto Biosintesi Vegetali, Consiglio Nazionale delle Ricerche. Milano, Italy, July 1988.

Additional References:

0010 Boyce, James K., and Robert E. Evenson. *National and International Agricultural Research and Extension Programs.* New York: Agricultural Development Council, Inc., 1975.

0014 Judd, M. Ann, James K. Boyce, and Robert E. Evenson. "Investing in Agricultural Supply." Economic Growth Center, Yale University, New Haven, Connecticut, 1983. Mimeo.

0021 OECD. *OECD Science and Technology Indicators.* Paris: OECD, 1984.

0022 UNESCO. *Statistical Yearbook 1983.* Paris: UNESCO, 1983.

0027 Harvey, Nigel, ed. *Agricultural Research Centres: A World Directory of Organizations and Programmes.* Seventh Edition, Two Volumes. Harlow, U.K.: Longman, 1983.

0165 Evenson, R.E., and Y. Kislev. *Agricultural Research and Productivity.* New Haven: Yale University Press, 1975.

0282 EEG. *De Organisatie van het Landbouwkundig Onderzoek in de Landen van de EEG.* EEG Studies, Serie Landbouw 9. Brussel: EEG, 1963.

0283 European Economic Community (EEC). "Monograph on the Organization of Agricultural Research." EEC, Brussels, 1982. Mimeo.

0376 FAO. *Fourteenth FAO Regional Conference for Europe, Reykjavik, Iceland, 17–21 September 1984: Research in Support of Agricultural Policies in Europe.* Rome: FAO, 1984.

0407 Centre International de Hautes Etudes Agronomiques Méditerranéennes. "Séminaire sur l'Orientation et l'Organisation de la Recherche Agronomique dans les Pays du Bassin Méditerranéen, Istanbul, 1–3 December 1986." Centre International de Hautes Etudes Agronomiques Méditerranéennes, Paris, December 1986. Mimeo.

0416 Fideghelli, C. "La Recherche Agricole en Italie." Paper presented at the Séminaire sur l'Orientation et l'Organisation de la Recherche Agronomique dans les Pays du Bassin Méditerranéen, Centre International de Hautes Etudes Agronomiques Méditerranéennes, Istanbul, 1–3 December 1986.

0422 Casas, Joseph. "Les Ressources Humaines et Financières dans les Pays du Bassin Méditerranéen." Preliminary table. ISNAR, The Hague, 1986. Mimeo.

0473 OECD. *Science and Technology Indicators: Basic Statistical Series – Volume A: The Objectives of Government R&D Funding 1974–1985*. Paris: OECD, May 1983.

0474 OECD. *Science and Technology Indicators: Basic Statistical Series – Volume A: The Objectives of Government R&D, 1969–1981*. Paris: OECD, July 1981.

0509 OECD. *Survey of the Resources Devoted to R&D by OECD Member Countries, International Statistical Year 1971 – Volume 5: Total Tables, Statistical Tables and Notes*. Paris: OECD, August 1974.

0510 OECD. *International Survey of the Resources Devoted to R&D in 1967 by OECD Member Countries, Statistical Tables and Notes – Volume 5: Total Tables*. Paris: OECD, August 1970.

0511 OECD. *International Survey of the Resources Devoted to R&D in 1969 by OECD Member Countries, Statistical Tables and Notes – Volume 5: Total Tables*. Paris: OECD, June 1973.

0513 OECD. *International Survey of the Resources Devoted to R&D by OECD Member Countries, International Statistical Year 1975 – International Volume: Statistical Tables and Notes*. Paris: OECD, March 1979.

0514 OECD. *A Study of Resources Devoted to R&D in OECD Member Countries in 1963/64, International Statistical Year for Research and Development – Volume 2: Statistical Tables and Notes*. Paris: OECD, 1968.

0519 OECD. *Country Reports on the Organisation of Scientific Research: Italy*. Paris: OECD, November 1964.

0530 OECD. *OECD Science and Technology Indicators No. 2 – R&D, Invention and Competitiveness*. Paris: OECD, 1986.

0601 OECD – STIID Data Bank. "Public Funding of R&D by Socio-Economic Objective." Tables. OECD, Paris, April 1987. Mimeo.

0602 OECD – STIID Data Bank. "R&D Expenditure in the Higher Education and Private Non-Profit Sectors." Tables. OECD, Paris, April 1987. Mimeo.

0606 EUROSTAT. *Government Financing of Research and Development 1975–1984*. Luxembourg: Office des Publications Officielles des Communautés Européennes, February 1985.

0683 Istituto Centrale di Statistica (ICS). *Statistiche sulla Ricerca Scientifica Anni 1967–1974*. Estratto da: Bollettino Mensile di Statistica 1973:10, 1975:1, 1976:5, 1977:1. Roma: ICS, various years.

0706 EUROSTAT. *Government Financing of Research and Development 1975–1985*. Luxembourg: Office for Official Publications of the European Communities, 1987.

0708 OECD – STIID Data Bank. "Gross Domestic Expenditure on R&D: All Fields of Sciences – Agriculture, Forestry and Fishing." Tables. OECD, Paris, June 1987. Mimeo.

0818 EUROSTAT. *Government Financing of Research and Development 1970–1978*. Luxembourg: Office for Official Publications of the European Communities, 1978.

0819 EUROSTAT. *Government Financing of Research and Development 1970–1977*. Luxembourg: Office for Official Publications of the European Communities, 1977.

0820 EUROSTAT. *Government Financing of Research and Development 1970–1979*. Luxembourg: Office for Official Publications of the European Communities, 1980.

0821 EUROSTAT. *Government Financing of Research and Development 1970–1980*. Luxembourg: Office for Official Publications of the European Communities, 1981.

0822 EUROSTAT. *Government Financing of Research and Development 1975–1981*. Luxembourg: Office for Official Publications of the European Communities, 1982.

0823 EUROSTAT. *Government Financing of Research and Development 1975–1982*. Luxembourg: Office for Official Publications of the European Communities, 1983.

0824 EUROSTAT. *Government Financing of Research and Development 1975–1983*. Luxembourg: Office for Official Publications of the European Communities, 1984.

0825 Scarda, Anna Maria. Personal communication. Consiglio Nazionale delle Ricerche – Istituto di Studi sulla Ricerca e Documentazione Scientifica. Roma, February 1988.

0826 Comitato per la Ricerca Scientifica e Tecnica (CREST). *Il Finanziamento Pubblico della Ricerca e dello Sviluppo nei Paesi della Comunità: Analisi per Categorie di Obiettivi 1969–1973*. Luxembourg: Comunità Europee, n.d.

0827 Committee on Scientific and Technical Research (CREST). *Public Expenditure on Research and Development in the Community Countries: Summary Report 1974–1976.* Luxembourg: European Communities, 1976.

0828 Consiglio Nazionale delle Ricerche (CNR). *Il Finanziamento Pubblico della Ricerca e dello Sviluppo nei Paesi della Comunità: Analisi per Obiettivi 1967–1970.* Roma: CNR, 1970.

0829 Comitato per la Ricerca Scientifica e Tecnica (CREST). *Il Finanziamento Pubblico della Ricerca e dello Sviluppo nei Paesi della Comunita: Analisi Sommaria per Obiettivi Principali di Ricerca e Sviluppo 1974.* Luxembourg: Comunita Europee, April 1975.

0852 Evenson, Robert E., and Yoav Kislev. *Investment in Agricultural Research and Extension: A Survey of International Data.* Center Discussion Paper No. 124. New Haven, Connecticut: Economic Growth Center, Yale University, August 1971.

0886 Evenson, Robert E., and Yoav Kislev. "Investment in Agricultural Research and Extension: A Survey of International Data." *Economic Development and Cultural Change* Vol. 23 (April 1975): 507–521.

Jamaica

Personnel

Period/Year	PhD	MSc	BSc	Subtotal	Expat	Total	Source
1960–64	..	..	..	..	..	40	880
1965–69	10	22	23	55	..	55	286/285
1970	..	..	..	..	..	..	
1971	12	29	40	81	..	81	279
1972	..	..	..	..	..	..	
1973	..	..	..	..	..	..	
1974	10	25	41	76	..	76	244
1975	..	..	..	..	..	..	
1976	..	..	..	..	..	..	
1977	6	24	30	60	..	60	324/243/341
1978	..	..	..	..	..	..	
1979	..	..	..	..	..	..	
1980	4	23	22	49	..	49	242
1981	..	..	..	..	..	..	
1982	..	..	..	..	..	..	
1983	..	..	..	..	..	..	
1984	..	..	..	..	..	..	
1985	..	..	..	..	..	..	
1986	..	..	..	..	..	..	

1960–64: 1960   1965–69: 1965 & 68

Expenditure   *LCU = Jamaican Dollars*

Period/Year	Current LCU (millions)	Constant 1980 LCU(millions)	Constant 1980 US$ (millions)		Source
			Atlas	PPP	
1960–64	..	..	..	..	
1965–69	..	..	..	..	
1970	..	..	..	..	
1971	1.345	5.558	3.057	4.733	63
1972	..	..	..	..	
1973	..	..	..	..	
1974	..	..	..	..	
1975	..	..	..	..	
1976	..	..	..	..	
1977	..	..	..	..	
1978	..	..	..	..	
1979	..	..	..	..	
1980	..	..	..	..	
1981	3.048	2.817	1.550	2.399	63
1982	..	..	..	..	
1983	..	..	..	..	
1984	..	..	..	..	
1985	..	..	..	..	
1986	..	..	..	..	

1960–64:   1965–69:

Personnel Comments:
The table below gives a breakdown at the research unit level.

	1965	1968	1971	1974	1977	1980	1981	1982	1983
Banana Industries Board	287	211	212	211	225	210	NA	NA	8
Coconut Industries Board	3	3	3	3	3	3	NA	NA	3
MOA – Crop & Soils Research	14	26	39	36	29*	12	40	NA	NA
MOA – Livestock Research	11	15	17	18	12	8	NA	NA	NA
Sugar Industries Research	9	10	10	8	6	10	NA	15	NA
CARDI	–	–	–	–	5**	6	8*	NA	8*
Subtotal	44	65	81	76	60	49	NA	NA	NA
MOA – Fisheries Research					2	1			
MOA – Forestry Research		3	2	2	4	4		5	
MOA – Socioeconomic Research	4	14	17	20	2	1			
Scientific Research Council						8			
Sources	286	285	279	244	243	242	223	27	115
					324*		342*		355*
					341**		343*		346*

According to source 223, agricultural research within the Ministry of Agriculture virtually collapsed at the end of the seventies because of political instability, and it seems to have been in a state of reorganization since that time.

Expenditure Comments:
These expenditure figures refer to total crop and livestock research expenditures by the Ministry of Agriculture. Many other references give only the crop-related research expenditures of the ministry, to the exclusion of livestock research.

Acronyms and Abbreviations:
CARDI: Caribbean Agricultural Research and
Development Institute
MOA: Ministry of Agriculture

Sources:
0063 ISNAR. *Report to the Board of Directors of CARDI: Analysis, Evaluation and Proposals to Strengthen CARDI's Regional Capacity.* The Hague: ISNAR, 1985.

0242 Commonwealth Agricultural Bureaux (CAB). *List of Research Workers in Agricultural Sciences in the Commonwealth 1981.* Slough, England: CAB, 1981.

0243 Commonwealth Agricultural Bureaux (CAB). *List of Research Workers in the Agricultural Sciences in the Commonwealth and in the Republic of Ireland 1978.* Slough, England: CAB, 1978.

0244 Commonwealth Agricultural Bureaux (CAB). *List of Research Workers in the Agricultural Sciences in the Commonwealth and in the Republic of Ireland 1975.* Slough, England: CAB, 1975.

0279 Commonwealth Agricultural Bureaux (CAB). *List of Research Workers in the Agricultural Sciences in the Commonwealth and in the Republic of Ireland 1972.* Slough, England: CAB, 1972.

0285 Commonwealth Agricultural Bureaux (CAB). *List of Research Workers in the Agricultural Sciences in the Commonwealth and in the Republic of Ireland 1969.* Slough, England: CAB, 1969.

0286 Commonwealth Agricultural Bureaux (CAB). *List of Research Workers in Agriculture, Animal Health and Forestry in the Commonwealth and in the Republic of Ireland 1966.* Slough, England: CAB, 1966.

0287 Commonwealth Agricultural Bureaux (CAB). *List of Research Workers in Agriculture, Animal Health and Forestry in the British Commonwealth, the Republic of Sudan and the Republic of Ireland 1959.* Slough, England: CAB, 1959.

0324 Barker, G., L.A. Bell, and A. Wahab. "Agricultural Research in Jamaica." In *Strengthening National Agricultural Research.* Report from a SAREC workshop, 10–17 September 1979, edited by Bo Bengtsson and

Getachew Tedla, pp. 175–181. Sweden: SAREC, 1980.

0341 Caribbean Agricultural Research and Development Institute (CARDI). *Annual Report 1977–1978*. St. Augustine, Trinidad & Tobago: CARDI, November 1978.

0880 Edwards, D.T. "An Economic View of Jamaican Agricultural Research." *Social and Economic Studies* Vol. 10, No. 3 (September 1961): 306–339.

Additional References:

0001 Trigo, Eduardo J., and Martín E. Piñeiro. "Funding Agricultural Research." In *Report of a Conference: Selected Issues in Agricultural Research in Latin America*, edited by Barry Nestel and Eduardo J. Trigo. The Hague: ISNAR, March 1984.

0010 Boyce, James K., and Robert E. Evenson. *National and International Agricultural Research and Extension Programs*. New York: Agricultural Development Council, Inc., 1975.

0014 Judd, M. Ann, James K. Boyce, and Robert E. Evenson. "Investing in Agricultural Supply." Economic Growth Center, Yale University, New Haven, Connecticut, 1983. Mimeo.

0016 Oram, Peter A., and Vishva Bindlish. *Resource Allocations to National Agricultural Research: Trends in the 1970s*. The Hague and Washington, D.C.: ISNAR and IFPRI, November 1981.

0026 Oram, Peter A., and Mark Gieben. "Document Summaries." ISNAR, The Hague, 1984. Mimeo.

0027 Harvey, Nigel, ed. *Agricultural Research Centres: A World Directory of Organizations and Programmes*. Seventh Edition, Two Volumes. Harlow, U.K.: Longman, 1983.

0053 Piñeiro, Martín, and Eduardo Trigo, eds. *Cambio Técnico en el Agro Latino Americano – Situación y Perspectivas en la Década de 1980*. San José: IICA, 1983.

0073 Oram, Peter A., and Vishva Bindlish. "Investment in Agricultural Research in Developing Countries: Progress, Problems, and the Determination of Priorities." IFPRI, Washington, D.C, January 1984. Mimeo.

0095 FAO – CARIS. *Agricultural Research in Developing Countries – Volume 1: Research Institutions*. Rome: FAO – CARIS, 1978.

0115 Economic Commission for Latin America and the Caribbean (ECLAC). *Agricultural Research Policy and Management – Volumes I and II*. Papers presented at the Workshop on Agricultural Research Policy and Management, 26–30 September 1983, Port of Spain, Trinidad. Port of Spain, Trinidad: ECLAC, November 1984.

0163 CGIAR. "National Agricultural Research." CGIAR, Washington, D.C., 1985. Mimeo.

0172 Trigo, Eduardo. Background material. ISNAR, The Hague, n.d.

0219 Department of Agriculture, Natural Resources and Rural Development (DARNDR), and IICA. "Réunion Technique Régionale sur les Systèmes de Recherche Agricole dans les Antilles." DARNDR and IICA, Port-au-Prince, Haiti, November/December 1977. Mimeo.

0223 Wahab, Abdul H., and Noel Singh. *Agricultural Research in Jamaica*. Miscellaneous Publication No. 274. Jamaica: IICA, 1981.

0342 Caribbean Agricultural Research and Development Institute (CARDI). *CARDI 1976–1981: Report of the Chairman*. St. Augustine, Trinidad & Tobago: CARDI, 1982.

0343 Caribbean Agricultural Research and Development Institute (CARDI). *Annual Report 1980*. St. Augustine, Trinidad & Tobago: CARDI, 1981.

0346 Caribbean Agricultural Research and Development Institute (CARDI). *Annual Report 1983 – Jamaica Unit*. Kingston, Jamaica: CARDI, 1983.

0351 Caribbean Agricultural Research and Development Institute (CARDI). *Annual Report 1984 – Jamaica Unit*. St. Augustine, Trinidad: CARDI, 1985.

0354 Caribbean Agricultural Research and Development Institute (CARDI). *Annual Report Research and Development 1983–84: Highlights*. St. Augustine, Trinidad: CARDI, 1984.

0355 Caribbean Agricultural Research and Development Institute (CARDI). *Research and Development Summary 1982–83*. St. Augustine, Trinidad: CARDI, 1983.

0356 Caribbean Agricultural Research and Development Institute (CARDI). *Research and Development Summary 1983–84*. St. Augustine, Trinidad: CARDI, 1984.

0359 Cook, Theodore W., Gaylord L. Walker, and Carl. J. Metzger. *Final Report Agricultural Research Project Jamaica*. Arlington, USA: Multinational Agribusiness Systems Incorporated, September 1983.

0365 Trigo, Eduardo, and Guillermo E. Gálvez. "Trip Report of the CIAT and IICA Mission to the Caribbean English-Speaking Countries for UNDP." 1979. Mimeo.

0740 Sands, Carolyn Marie. "The Theoretical and Empirical Basis for Analyzing Agricultural Technology Systems." PhD thesis, University of Illinois, Urbana-Champaign, Illinois, 1988.

0852 Evenson, Robert E., and Yoav Kislev. *Investment in Agricultural Research and Extension: A Survey of International Data*. Center Discussion Paper No. 124. New Haven, Connecticut: Economic Growth Center, Yale University, August 1971.

Japan

Personnel

Period/Year	PhD	MSc	BSc	Subtotal	Expat	Total	Source
1960–64	..	..	..	..	..	12535	612/10
1965–69	..	..	..	..	..	13085	612
1970	..	..	..	..	..	13275	612
1971	..	..	..	..	..	13461	612
1972	..	..	..	..	..	13850	612/10
1973	..	..	..	..	..	13955	612
1974	..	..	..	..	..	13872	612
1975	..	..	..	..	..	13850	612
1976	..	..	..	..	..	13706	612
1977	..	..	..	..	..	13753	612
1978	..	..	..	..	..	13718	612
1979	..	..	..	..	..	13737	612
1980	..	..	..	..	..	13822	612
1981	..	..	..	..	..	14800	612/22
1982	..	..	..	..	..	14825	612
1983	..	..	..	..	..	14657	612/602
1984	..	..	..	..	..	14820	612
1985	..	..	..	..	..	14793	612
1986	..	..	..	..	..	14759	612

1960–64: 1964   1965–69: 1966, 67, 68 & 69

Expenditure   *LCU = Yen*

Period/Year	Current LCU (millions)	Constant 1980 LCU (millions)	Constant 1980 US$ (millions)		Source
			Atlas	PPP	
1960–64	32740.000	89103.658	425.925	358.030	612/514
1965–69	52969.800	130056.779	621.685	522.586	612/510
1970	82197.000	171243.750	818.563	688.080	612
1971	92080.000	181617.357	868.150	729.763	612
1972	103240.000	192611.940	920.706	773.941	612/10
1973	120974.000	199957.025	955.816	803.454	612
1974	143960.000	196935.705	941.374	791.314	612
1975	157355.000	199942.821	955.748	803.397	612/513
1976	169798.000	201421.115	962.815	809.337	612
1977	189530.000	212477.578	1015.666	853.763	612/602
1978	205498.000	219783.957	1050.591	883.121	612
1979	227450.000	236188.993	1129.009	949.039	612/602
1980	239291.000	239291.000	1143.837	961.503	612/602
1981	255642.000	247715.116	1184.105	995.353	612/602
1982	260700.000	248048.525	1185.699	996.696	612
1983	268789.000	253813.975	1213.258	1019.859	612/602
1984	277626.000	258979.478	1237.950	1040.614	612
1985	285559.000	262221.304	1253.446	1053.640	612
1986	..	..	..	..	

1960–64: 1963 & 64   1965–69: 1965, 66, 67, 68 & 69

Personnel Comments:

In Japan three entities play an important role in public agricultural research, namely, the national government, the prefectural governments, and the universities. The table below gives construction details for the indicator series.

	1963	1964	1966	1967	1968	1969	1970
National Govt.	NA	3127	3112	3145	3162	3090	3151
Prefectures	NA	6798	6854	7130	6949	6748	6782
Universities[a]	2488	[2610]	[2854]	[2976]	[3098]	[3220]	[3342]
Total	NA	12535	12820	13251	13209	13058	13275
Source	10	612	612	612	612	612	612

	1971	1972	1973	1974	1975	1976	1977	1978
National Govt.	3092	3084	3108	3099	3123	3098	3099	3091
Prefectures	6905	7179	7241	7149	7084	6947	6974	6929
Universities[b]	[3464]	3587*	[3606]	[3624]	[3643]	[3661]	[3680]	[3698]
Total	13461	13850	13955	13872	13850	13706	13753	13718
Source	612	612/10*	612	612	612	612	612	612

	1979	1980	1981	1982	1983	1984	1985	1986
National Govt.	3091	3132	3164	3174	3165	3139	3103	3080
Prefectures	6929	6955	7882*	7971	7887	8076	8085	8074
Universities[c]	[3717]	[3735]	3754*	[3680]	3605*	[3605]	[3605]	[3605]
Total	13737	13822	14800	14825	14657	14820	14793	14759
Source	612	612	612/22*	612	612/602*	612	612	612

[ ] = estimated or constructed by authors.

a. According to source 10, the 1963 figure is measured in scientific person-years.

b. Source 10 gives a total university researcher estimate of 7174 for 1972, which was prorated here to full-time equivalent (FTE) units at a 50% rate.

c. Source 22 (p. V-53) reports a 1981 figure of 7507 full- plus part-time scientists and engineers. It was prorated to FTEs at a 50% rate, as was the 7210 figure given by source 602, representing the total number of researchers in the higher education sector in 1983.

Expenditure Comments:
The indicator series measured in current Yen has been constructed as follows:

	1960	1961	1962	1963	1964	1965	1966	1967	1968
National Govt.	4969	6071	5830	7129	7883	8540	9578	10641	11776
Prefectures	NA	NA	12192	15393	17782	19047	22095	25781	28976
Universities	NA	NA	NA	7718*	[9575]	[11440]	[13297]	15144*	[18955]
Total	NA	NA	NA	30240	35240	39027	44970	51566	59707
Source	612	612	612	612/514*	612	612	612	612/510*	612

	1969	1970	1971	1972	1973	1974	1975	1976
National Govt.	13168	14747	16957	18951	22808	27994	32122	34432
Prefectures	33645	40782	44734	50089	60165	74163	79629	84764
Universities	[22766]	[26578]	[30389]	34200*	[38001]	[41803]	45604*	[50602]
Total	69579	82107	92080	103240	120974	143960	157355	169798
Source	612	612	612	612/10*	612	612	612/513*	612

	1977	1978	1979	1980	1981	1982	1983	1984	1985
National Govt.	37900	40090	45266	47469	49380	48760	50790	53509	54078
Prefectures	96030	102879	112725	120970	134017	135481	137321	140790	145706
Universities	55600*	[62530]	69459*	[70852]	72245*	[76459]	80673*	[83327]	[85775]
Total	189530	205499	227450	23929	1255642	260700	268784	277626	285559
Source	612/602*	612	612/602*	612	612/602*	612	612/602*	612	612

[ ] = estimated or constructed by authors.

Sources:
0010 Boyce, James K., and Robert E. Evenson. *National and International Agricultural Research and Extension Programs.* New York: Agricultural Development Council, Inc., 1975.

0022 UNESCO. *Statistical Yearbook 1983.* Paris: UNESCO, 1983.

0510 OECD. *International Survey of the Resources Devoted to R&D in 1967 by OECD Member Countries, Statistical Tables and Notes – Volume 5: Total Tables.* Paris: OECD, August 1970.

0513 OECD. *International Survey of the Resources Devoted to R&D by OECD Member Countries, International Statistical Year 1975 – International Volume: Statistical Tables and Notes.* Paris: OECD, March 1979.

0514 OECD. *A Study of Resources Devoted to R&D in OECD Member Countries in 1963/64, International Statistical Year for Research and Development – Volume 2: Statistical Tables and Notes.* Paris: OECD, 1968.

0602 OECD – STIID Data Bank. "R&D Expenditure in the Higher Education and Private Non-Profit Sectors." Tables. OECD, Paris, April 1987. Mimeo.

0612 Ohashi, Tetsuro. Personal communication. Director, International Research Division –

Agriculture, Forestry and Fisheries Research Council. Tokyo, April 1987.

Additional References:

0021 OECD. *OECD Science and Technology Indicators.* Paris: OECD, 1984.

0027 Harvey, Nigel, ed. *Agricultural Research Centres: A World Directory of Organizations and Programmes.* Seventh Edition, Two Volumes. Harlow, U.K.: Longman, 1983.

0163 CGIAR. "National Agricultural Research." CGIAR, Washington, D.C., 1985. Mimeo.

0165 Evenson, R.E., and Y. Kislev. *Agricultural Research and Productivity.* New Haven: Yale University Press, 1975.

0176 Moseman, Albert, ed. *National Agricultural Research Systems in Asia.* Report of the Regional Seminar held at the India International Centre, New Delhi, India, March 8–13, 1971. New York: ADC, Inc., 1971.

0333 Hayami, Yujiro, and Masakatsu Akino. "Organization and Productivity of Agricultural Research Systems in Japan." In *Resource Allocation and Productivity in National and International Agricultural Research*, edited by Thomas M. Arndt, Dana G. Dalrymple, and Vernon W. Ruttan, pp. 29–43. Minneapolis, Minnesota: University of Minnesota Press, 1977.

0375 Ministry of Agriculture, Forestry and Fisheries. *Present Situation of Research Relating to Agriculture, Forestry and Fisheries 1984.* Tokyo, Japan: Ministry of Agriculture, Forestry and Fisheries, 1984.

0473 OECD. *Science and Technology Indicators: Basic Statistical Series – Volume A: The Objectives of Government R&D Funding 1974–1985.* Paris: OECD, May 1983.

0474 OECD. *Science and Technology Indicators: Basic Statistical Series – Volume A: The Objectives of Government R&D, 1969–1981.* Paris: OECD, July 1981.

0509 OECD. *Survey of the Resources Devoted to R&D by OECD Member Countries, International Statistical Year 1971 – Volume 5: Total Tables, Statistical Tables and Notes.* Paris: OECD, August 1974.

0511 OECD. *International Survey of the Resources Devoted to R&D in 1969 by OECD Member Countries, Statistical Tables and Notes – Volume 5: Total Tables.* Paris: OECD, June 1973.

0512 OECD. *Survey of the Resources Devoted to R&D by OECD Member Countries, International Statistical Year 1973 – Volume 5: Total Tables, Statistical Tables and Notes.* Paris: OECD, September 1976.

0530 OECD. *OECD Science and Technology Indicators No. 2 – R&D, Invention and Competitiveness.* Paris: OECD, 1986.

0590 UNESCO. *Science and Technology in Countries of Asia and the Pacific.* Science Policy Studies and Documents No. 52. Paris: UNESCO, n.d.

0595 OECD. *Reviews of National Science Policy: Japan.* Paris: OECD, 1967.

0601 OECD – STIID Data Bank. "Public Funding of R&D by Socio-Economic Objective." Tables. OECD, Paris, April 1987. Mimeo.

0703 Martin, Ben R., and John Irvine. *An International Comparison of Government Funding of Academic and Academically Related Research.* Advisory Board for the Research Council (ABRC) Science Policy Study No. 2. Brighton: ABRC – Science Policy Research Unit, University of Sussex, October 1986.

0708 OECD – STIID Data Bank. "Gross Domestic Expenditure on R&D: All Fields of Sciences – Agriculture, Forestry and Fishing." Tables. OECD, Paris, June 1987. Mimeo.

0732 FAO – Regional Office for Asia and the Pacific. *Agricultural Research Systems in the Asia-Pacific Region.* Bangkok: FAO – Regional Office for Asia and the Pacific, 1986.

0852 Evenson, Robert E., and Yoav Kislev. *Investment in Agricultural Research and Extension: A Survey of International Data.* Center Discussion Paper No. 124. New Haven, Connecticut: Economic Growth Center, Yale University, August 1971.

0886 Evenson, Robert E., and Yoav Kislev. "Investment in Agricultural Research and Extension: A Survey of International Data." *Economic Development and Cultural Change* Vol. 23 (April 1975): 507–521.

Jordan

Personnel

Period/Year	PhD	MSc	BSc	Subtotal	Expat	Total	Source
1960–64	..	..	..	..	..	..	
1965–69	..	..	..	..	..	..	
1970	..	..	..	..	..	..	
1971	..	..	..	..	..	..	
1972	..	..	..	..	..	..	
1973	..	..	..	..	..	..	
1974	3	4	20	27	..	27	75
1975	..	..	..	..	..	..	
1976	8	8	26	42	..	42	75
1977	..	..	..	..	..	..	
1978	18	13	28	59	..	59	75
1979	..	..	..	..	..	..	
1980	24	20	28	72	..	72	75
1981	..	..	..	..	..	..	
1982	16	14	24	54	..	54	75
1983	..	..	..	..	..	..	
1984	18	11	30	59	..	59	739
1985	..	..	..	..	..	..	
1986	..	..	..	..	..	..	

1960–64:   1965–69:

Expenditure

LCU = Jordan Dinars

Period/Year	Current LCU (millions)	Constant 1980 LCU(millions)	Constant 1980 US$ (millions) Atlas	PPP	Source
1960–64	..	..	..	..	
1965–69	..	..	..	..	
1970	..	..	..	..	
1971	..	..	..	..	
1972	..	..	..	..	
1973	0.116	0.269	0.847	1.485	75
1974	0.133	0.258	0.813	1.425	75
1975	0.198	0.343	1.080	1.893	75
1976	0.240	0.373	1.175	2.059	75
1977	0.286	0.388	1.221	2.141	75
1978	0.317	0.402	1.264	2.217	75
1979	0.290	0.322	1.014	1.778	75
1980	0.337	0.337	1.060	1.859	75
1981	0.298	0.277	0.871	1.527	75
1982	0.316	0.273	0.859	1.507	75
1983	..	..	..	..	
1984	0.329	0.261	0.820	1.438	739
1985	..	..	..	..	
1986	..	..	..	..	

1960–64:   1965–69:

Personnel Comments:

These figures include the Department of Agricultural Research and Extension (DARE) within the Ministry of Agriculture, plus the Faculty of Agriculture at the University of Jordan. They are given in scientist-year equivalents and are therefore a more precise measure of research personnel than other sources (e.g., source 95) whose higher figures represent total (full- and part-time) research personnel.

Expenditure Comments:

These expenditure figures include DARE and the Faculty of Agriculture, University of Jordan, and represent only recurrent expenditures.

Sources:

0075 Qasem, Subhi. "Technology Transfer and Feedback: Agricultural Research and Extension." Part IV, Chapter 11 in *The Agricultural Sector of Jordan: Policy and Systems Studies*, edited by A.B. Zahlan and Subhi Qasem, pp. 343–375. London: Ithaca Press, 1985.

0739 Qasem, Subhi. "Study on Small Country Research: Research in Jordan." IDRC, Ottawa, n.d. Mimeo.

Additional References:

0010 Boyce, James K., and Robert E. Evenson. *National and International Agricultural Research and Extension Programs.* New York: Agricultural Development Council, Inc., 1975.

0014 Judd, M. Ann, James K. Boyce, and Robert E. Evenson. "Investing in Agricultural Supply." Economic Growth Center, Yale University, New Haven, Connecticut, 1983. Mimeo.

0016 Oram, Peter A., and Vishva Bindlish. *Resource Allocations to National Agricultural Research: Trends in the 1970s.* The Hague and Washington, D.C.: ISNAR and IFPRI, November 1981.

0022 UNESCO. *Statistical Yearbook 1983.* Paris: UNESCO, 1983.

0026 Oram, Peter A., and Mark Gieben. "Document Summaries." ISNAR, The Hague, 1984. Mimeo.

0027 Harvey, Nigel, ed. *Agricultural Research Centres: A World Directory of Organizations and Programmes.* Seventh Edition, Two Volumes. Harlow, U.K.: Longman, 1983.

0073 Oram, Peter A., and Vishva Bindlish. "Investment in Agricultural Research in Developing Countries: Progress, Problems, and the Determination of Priorities." IFPRI, Washington, D.C., January 1984. Mimeo.

0094 FAO – Near East Regional Office. *Directory of Agricultural Research Institutions in the Near East Region.* Cairo: FAO – Near East Regional Office, 1979.

0095 FAO – CARIS. *Agricultural Research in Developing Countries – Volume 1: Research Institutions.* Rome: FAO – CARIS, 1978.

0173 FAO. *FAO Near East Regional Studies on Organization and Administration of Agricultural Research.* Rome: FAO, 1974.

0174 Watson, J.M. "Comparative Study of Agricultural Research Organisation and Administration in the Near East Region." Paper presented at the Workshop on Organization and Administration of Agricultural Services in the Arab States, Cairo, 2–15 March 1964.

0722 USAID – Highlands Agricultural Development Project. Title not known. Jordan, n.d. Mimeo.

0747 ISNAR. *International Workshop on Agricultural Research Management.* Report of a workshop. The Hague: ISNAR, 1987.

0752 Bilbeisi, Usama. "The Jordanian Experiment in Agricultural Research and Extension Linkage: An Analytic Overview." Paper presented at the International Workshop on Agricultural Research Management, ISNAR, The Hague, 7–11 September 1987.

0831 FAO. *Report of an Expert Consultation on Agricultural Extension and Research Linkages in the Near East, Amman, Jordan, 22–26 July 1985.* Rome: FAO, 1985.

0865 Oram, Peter. "Agricultural Research Objectives and Priorities: Constraints to the Development of Agricultural Research Institutions in Arab Countries." Review paper prepared for the first meeting of the Council for Arab Agricultural Research, organized by the Arab Fund for Economic and Social Development, Kuwait, 23 March 1988.

Kenya

Personnel

Period/Year	PhD	MSc	BSc	Subtotal	Expat	Total	Source
1960–64	4	3	15	22	103	125	229
1965–69	5	18	30	53	134	187	229
1970	8	29	54	91	160	251	229
1971	12	37	74	123	172	295	229
1972	11	56	92	159	189	348	229
1973	14	65	82	161	177	338	229
1974	19	69	92	180	154	334	229
1975	17	75	99	191	155	346	229
1976	16	76	101	193	120	313	229
1977	16	77	116	209	106	315	229
1978	15	100	127	242	90	332	229
1979	..	..	..	..	..	..	
1980	..	..	..	..	..	..	
1981	..	..	..	390	72	462	51
1982	..	..	..	..	..	..	
1983	..	..	..	..	..	..	
1984	..	..	..	..	..	..	
1985	..	..	..	..	..	..	
1986	17	212	275	504		504	780

1960–64: 1963 & 64   1965–69: 1965, 66, 67, 68 & 69

Expenditure

LCU = Kenyan Shillings

Period/Year	Current LCU (millions)	Constant 1980 LCU(millions)	Constant 1980 US$ (millions)		Source
			Atlas	PPP	
1960–64	21.962	58.306	7.732	12.199	229/518
1965–69	33.094	86.255	11.438	18.047	229/589
1970	45.060	99.958	13.255	20.914	229
1971	48.018	117.616	15.596	24.609	229/589
1972	54.869	132.426	17.560	27.707	229
1973	63.925	140.294	18.604	29.354	229
1974	64.547	121.840	16.157	25.492	229
1975	74.985	126.720	16.804	26.513	229
1976	82.926	117.872	15.630	24.662	229
1977	108.345	131.710	17.465	27.558	229
1978	151.537	178.675	23.693	37.384	229
1979	177.825	196.707	26.084	41.157	229
1980	152.057	152.057	20.163	31.815	292
1981	165.739	149.990	19.889	31.382	798
1982	179.416	146.942	19.485	30.744	798
1983	171.288	129.274	17.142	27.048	798
1984	146.917	100.353	13.307	20.997	798
1985	..	..	..	..	
1986	..	..	..	..	

1960–64: 1964   1965–69: 1965, 66, 67, 68 & 69

260

Personnel Comments:

The indicator series represents all research institutes under the jurisdiction of the Scientific Research Division of the Ministry of Agriculture, the Veterinary Services Department of the Ministry of Agriculture, the Coffee Research Foundation, the Tea Research Institute, the National Irrigation Board, the Kenya Agricultural Research Institute, and the Kenya Veterinary Research Institute. During 1987 all public (non-university) agricultural research institutes were finally integrated into the Kenyan Agricultural Research Institute (KARI). Explicitly not included in the indicator series are universities and international agricultural research institutes.

Source 33, Table 2, indicates that over 100 graduates were recruited into the system in 1981. Source 33 gives the following figures for 1982: 15 PhD's, 242 MSc's, 284 BSc's, and 72 expatriates, which totals to 613 researchers. According to the same source, this figure included a) 25 researchers for whom the academic degree was not known and b) an unknown number of researchers who were on study leave. Available evidence suggests that this unknown number is likely to be high. For these reasons, the 1982 figure given by source 33 was not included in the indicator series.

Expenditure Comments:

The expenditure indicators up to 1980 are based on source 229 and include the Scientific Research Division and the Division of Veterinary Services of the Ministry of Agriculture, the Coffee Research Foundation, the Tea Research Institute, the National Irrigation Board, the Kenyan (formerly East African) Agriculture and Forestry Research Organization, and the Kenyan (formerly East African) Veterinary Research Organization. From 1981 on, the figures from source 798 include a compatible set of organizations.

Sources:

0051 ISNAR. *A Report to the Government of Kenya: Kenya's National Agricultural Research System.* The Hague: ISNAR, September 1981.

0229 Jamieson, Barbara M. "Resource Allocation to Agricultural Research in Kenya from 1963 to 1978." PhD diss., University of Toronto, Canada, 1981.

0292 Wang'ati, F.J. "Allocation of Resources to Agricultural Research: An Inventory of the Current Situation in Kenya." In *Resource Allocation to Agricultural Research.* Proceedings of a workshop held in Singapore 8–10 June 1981, edited by Douglas Daniels and Barry Nestel, pp. 27–31. Ottawa: IDRC, 1981.

0518 East African Academy. *Research Services in East Africa.* Nairobi: East African Publishing House, 1966.

0589 Kassapu, Samuel. *Les Dépenses de la Recherche Agricole dans 34 Pays d'Afrique Tropicale.* Paris: Centre de Développement de l'OCDE, 1976.

0780 Muturi, S.N. Personal communication. Director of Research Development, Ministry of Research, Science and Technology. Nairobi, January 1988.

0798 Muturi, Stachys N. *Training in the CGIAR System: The Case of Kenya.* Rome: CGIAR – TAC, October 1984.

Additional References:

0010 Boyce, James K., and Robert E. Evenson. *National and International Agricultural Research and Extension Programs.* New York: Agricultural Development Council, Inc., 1975.

0014 Judd, M. Ann, James K. Boyce, and Robert E. Evenson. "Investing in Agricultural Supply." Economic Growth Center, Yale University, New Haven, Connecticut, 1983. Mimeo.

0016 Oram, Peter A., and Vishva Bindlish. *Resource Allocations to National Agricultural Research: Trends in the 1970s.* The Hague and Washington, D.C.: ISNAR and IFPRI, November 1981.

0018 Daniels, Douglas, and Barry Nestel, eds. *Resource Allocation to Agricultural Research.* Proceedings of a workshop held in Singapore 8–10 June 1981. Ottawa: IDRC, 1981.

0019 Bengtsson, Bo, and Tedla Getachew, eds. *Strengthening National Agricultural Research.* Report from a SAREC workshop, 10–17 September 1979. Part 1: Background Documents; Part II: Summary and Conclusions. Sweden: SAREC, 1980.

0023 Bennell, Paul. *Agricultural Researchers in Sub-Saharan Africa: An Overview.* Working Paper No. 4. The Hague: ISNAR, October 1985.

0026 Oram, Peter A., and Mark Gieben. "Document Summaries." ISNAR, The Hague, 1984. Mimeo.

0027 Harvey, Nigel, ed. *Agricultural Research Centres: A World Directory of Organizations and Programmes.* Seventh Edition, Two Volumes. Harlow, U.K.: Longman, 1983.

0033 National Council for Science and Technology (NCST), and ISNAR. *A Manpower and Training Plan for the Agricultural Research System in Kenya 1983–1987.* The Hague: ISNAR, November 1982.

0073 Oram, Peter A., and Vishva Bindlish.
"Investment in Agricultural Research in
Developing Countries: Progress, Problems, and
the Determination of Priorities." IFPRI,
Washington, D.C., January 1984. Mimeo.

0135 FAO, and UNDP. *National Agricultural
Research – Report of an Evaluation Study in
Selected Countries.* Rome: FAO and UNDP, 1984.

0163 CGIAR. "National Agricultural Research."
CGIAR, Washington, D.C., 1985. Mimeo.

0164 Association for the Advancement of Agricultural
Sciences in Africa (AAASA). *Proceedings of the
Workshop on Agricultural Research
Administration, Nairobi, Kenya, 27–30 June 1977.*
Proceedings Series PE–4. Addis Ababa, Ethiopia:
AAASA and IDRC, August 1979.

0165 Evenson, R.E., and Y. Kislev. *Agricultural
Research and Productivity.* New Haven: Yale
University Press, 1975.

0175 Cooper, St.G.C. *Agricultural Research in
Tropical Africa.* Kampala: East African
Literature Bureau, 1970.

0227 Australian Centre for International Agricultural
Research (ACIAR). *Proceedings of the Eastern
Africa-ACIAR Consultation on Agricultural
Research, 18–22 July 1983, Nairobi, Kenya.*
Australia: ACIAR, 1984.

0242 Commonwealth Agricultural Bureaux (CAB).
*List of Research Workers in Agricultural Sciences
in the Commonwealth 1981.* Slough, England:
CAB, 1981.

0243 Commonwealth Agricultural Bureaux (CAB).
*List of Research Workers in the Agricultural
Sciences in the Commonwealth and in the
Republic of Ireland 1978.* Slough, England: CAB,
1978.

0244 Commonwealth Agricultural Bureaux (CAB).
*List of Research Workers in the Agricultural
Sciences in the Commonwealth and in the
Republic of Ireland 1975.* Slough, England: CAB,
1975.

0266 UNESCO. *National Science Policies in Africa.*
Science Policy Studies and Documents No. 31.
Paris: UNESCO, 1974.

0269 Wapakala, William W. "Improving
Agricultural Research Organisation and
Management – A Historical Perspective of the
Kenyan Situation." Paper presented at the
Workshop on Improving Agricultural Research
Organisation and Management: Progress and
Issues for the Future, ISNAR, The Hague, 8–12
September 1986.

0274 Hill, Jeffery Max. "Agricultural Research
Useability Bias toward Small or Large Scale
Farmers: An Examination of Crop Production
Research in Kenya." MSc thesis, University of
California, Davis, California, 1981.

0279 Commonwealth Agricultural Bureaux (CAB).
*List of Research Workers in the Agricultural
Sciences in the Commonwealth and in the
Republic of Ireland 1972.* Slough, England: CAB,
1972.

0280 ISNAR. *Kenya Agricultural Research Strategy
and Plan, Volumes 1 & 2.* The Hague: ISNAR,
December 1985.

0285 Commonwealth Agricultural Bureaux (CAB).
*List of Research Workers in the Agricultural
Sciences in the Commonwealth and in the
Republic of Ireland 1969.* Slough, England: CAB,
1969.

0286 Commonwealth Agricultural Bureaux (CAB).
*List of Research Workers in Agriculture, Animal
Health and Forestry in the Commonwealth and
in the Republic of Ireland 1966.* Slough, England:
CAB, 1966.

0287 Commonwealth Agricultural Bureaux (CAB).
*List of Research Workers in Agriculture, Animal
Health and Forestry in the British
Commonwealth, the Republic of Sudan and the
Republic of Ireland 1959.* Slough, England: CAB,
1959.

0301 Muturi, S.N. "The System of Resource Allocation
to Agricultural Research in Kenya." In *Resource
Allocation to Agricultural Research.* Proceedings
of a workshop held in Singapore 8–10 June 1981,
edited by Douglas Daniels and Barry Nestel, pp.
123–128. Ottawa: IDRC, 1981.

0312 Thairu, D.M. "Agricultural Research System in
Kenya." In *Strengthening National Agricultural
Research.* Report from a SAREC workshop, 10–17
September 1979, edited by Bo Bengtsson and
Getachew Tedla, pp. 48–54. Sweden: SAREC,
1980.

0360 Cooper, St.G.C. "Towards Trained Manpower for
Agricultural Research in Africa." Paper
presented at the Conference on Agricultural
Research and Production in Africa, organized by
the Association for the Advancement of
Agricultural Sciences in Africa (AAASA), Addis
Ababa, 29 August–4 September 1971.

0445 Swanson, Burton E., and Wade H. Reeves.
"Agricultural Research Eastern and Southern
Africa: Manpower and Training." World Bank,
Washington, D.C., August 1986. Mimeo.

0532 UNESCO Field Science Office for Africa. *Survey
on the Scientific and Technical Potential of the
Countries of Africa.* Paris: UNESCO, 1970.

0640 East African Agriculture and Forestry Research
Organization. *Record of Research – Annual
Report 1969.* Nairobi: East African Community,
1970.

0641 East African Veterinary Research Organization.
Record of Research – Annual Report 1968. Kenya:
East African Community, 1969.

262

0642 Keen, B.A. *The East African Agriculture and Forestry Research Organization – Its Origin and Objects*. Nairobi: East African Community, n.d.

0643 Binns, H.R. *The East African Veterinary Research Organisation – Its Development, Objects and Scientific Activities*. Nairobi: East African Community, n.d.

0653 Webster, B.N. *Index of Agricultural Research Institutions and Stations in Africa*. Rome: FAO, n.d.

0763 Wang'ati, F.J. "Agricultural Research in Kenya." In *Proceedings of the Eastern Africa-ACIAR Consultation on Agricultural Research*, pp. 31–43. Australia: ACIAR, 1984.

0770 Ruttan, Vernon W. *Agricultural Research Policy and Development*. FAO Research and Technology Paper 2. Rome: FAO, 1987.

0847 Kimura, J.H. "Financial and Administrative Management of Research Institutions in Eastern and Southern Africa: Report on Responses to a Questionnaire." In *Promotion of Technology Policy and Science Management in Africa*, edited by Karl Wolfgang Menck and Wolfgang Gmelin. Bonn: Deutsche Stiftung für internationale Entwicklung (DSE), 1986.

0852 Evenson, Robert E., and Yoav Kislev. *Investment in Agricultural Research and Extension: A Survey of International Data*. Center Discussion Paper No. 124. New Haven, Connecticut: Economic Growth Center, Yale University, August 1971.

0873 Abdalla, Abdalla Ahmed. "Agricultural Research in the IGADD Sub-Region and Related Manpower Training." Inter-Governmental Authority on Drought and Development (IGADD) and FAO, 1987. Mimeo.

0886 Evenson, Robert E., and Yoav Kislev. "Investment in Agricultural Research and Extension: A Survey of International Data." *Economic Development and Cultural Change* Vol. 23 (April 1975): 507–521.

Korea, Republic of

Personnel

Period/Year	PhD	MSc	BSc	Subtotal	Expat	Total	Source
1960–64	..	..	..	..	..	521	176/10
1965–69	..	..	..	..	..	747	10
1970	..	..	..	..	..	864	17
1971	..	..	..	..	..	757	17
1972	..	..	..	..	..	831	17
1973	..	..	..	..	..	969	17
1974	..	..	..	..	..	..	
1975	..	..	..	..	..	992	17
1976	..	..	..	..	..	998	17
1977	..	..	..	..	..	911	17
1978	..	..	..	..	..	1029	17
1979	..	..	..	..	..	1027	17
1980	..	..	..	..	..	1296	17
1981	..	..	..	..	..	1334	17
1982	..	..	..	..	..	1361	17
1983	155	398	583	1136	237	1373	17
1984	..	..	..	..	..	..	
1985	..	..	..	..	..	..	
1986	..	..	..	..	..	..	

1960–64: 1962 & 63   1965–69: 1966, 68 & 69

Expenditure   *LCU = Won*

Period/Year	Current LCU (millions)	Constant 1980 LCU(millions)	Constant 1980 US$ (millions)		Source
			Atlas	PPP	
1960–64	187.550	3388.517	5.694	8.650	10
1965–69	1083.000	8330.769	13.999	21.265	10
1970	1652.300	10457.595	17.573	26.694	17
1971	2193.200	12184.444	20.475	31.102	17
1972	2427.800	11560.952	19.427	29.510	17
1973	2125.300	8929.832	15.006	22.794	17
1974	2765.600	8979.221	15.089	22.920	17
1975	3697.700	9530.155	16.015	24.327	17
1976	4836.100	10377.897	17.439	26.491	17
1977	5830.100	9081.153	15.260	23.181	17
1978	7022.300	10623.752	17.852	27.118	17
1979	9142.900	11414.357	19.181	29.136	17
1980	12012.600	12012.600	20.186	30.663	17
1981	14057.900	12150.303	20.418	31.015	17
1982	19098.800	15414.689	25.903	39.347	17
1983	27145.400	21290.510	35.777	54.346	17
1984	28103.473	21114.555	35.481	53.897	842
1985	38501.750	27899.819	46.883	71.217	842
1986	46435.270	32932.816	55.341	84.064	842

1960–64: 1960 & 64   1965–69: 1968

Personnel Comments:

The 1983 figure (source 17) has been constructed as follows:

	PhD	MSc	BSc	Sub	Exp	Total
MAF/RDA	–	–	–	–	221	221
Agricultural Science Institute	25	88	69	182	–	182
Farm Management Bureau	10	19	28	57	–	57
Veterinary Research Institute	12	24	15	51	–	51
Agricultural Mechanization Institute	1	10	17	28	–	28
Wheat and Barley Research Institute	7	19	2	28	–	28
Agr. Chemicals Research Institute	8	13	12	33	–	33
Crops Experiment Station	0	26	9	55	–	55
Honam Crop Experiment Station	6	23	42	71	–	71
Yeongam Crop Experiment Station	10	1	25	56	–	56
Horticultural Experiment Station	13	22	40	75	–	75
Sericultural Experiment Station	8	10	10	28	–	28
Livestock Experiment Station	18	26	21	0	–	65
Alpine Experiment Station	0	9	0	22	–	2
Cheju Experiment Station	0	9	7	16	–	16
Rural Nutrition Institute	2	8	6	16	–	16
Subtotal MAF/RDA	141	339	303	783	221	1004
Fisheries Research Development Center (MAF/NFA)	6	23	50	79	0	79
Agricultural Engineering Research Center (MAF/ADC)	0	8	18	26	0	26
Forest Research Institute (MHA/FA)	3	24	155	182	16	198
Institute of Forest Genetics (MHA/FA)	5	4	57	66	0	66
Total	155	398	583	1136	237	1373

Some effort was made to resolve apparent internal inconsistencies in the responses to various parts of the ISNAR, IFARD & AOAD questionnaire. The most likely reconstruction has been presented here. Part A of the questionnaire included only those institutes under the Rural Development Administration (RDA) of the Ministry of Agriculture and Fisheries. In part C, the number of expatriate researchers working at RDA institutes were not indicated, while part A notes 221 expatriate researchers. Adding these 221 expatriate researchers to the total sum of part C gives a total of 1373 researchers which corresponds, as it should, to the 1374 researchers listed in part B.

The pre-1983 figures given by source 17 seem to include RDA plus four additional institutes as indicated in the table above.

Source 842 provided the following breakdown of the total number of RDA-only researchers:

	PhD	MSc	BSc	Subtotal	Expat	Total
1970	7	61	559	627	178	805
1975	52	123	620	795	64	859
1980	81	190	520	791	69	860
1984	141	419	425	985	19	1004
1987	199	539	232	970	61	1031

Source 842 also provided a 1987 figure of 1594 researchers, covering RDA plus the four other institutes. This figure was broken down as follows: 224 PhD's, 711 MSc's, 519 BSc's, and 140 expatriates.

Expenditure Comments:
The 1983 expenditure figure is constructed as follows:

	(1000 Won)
MAF/RDA	
Agricultural Science Institute	2,237,585
Farm Management Bureau	253,750
Veterinary Research Institute	1,273,642
Agricultural Mechanization Institute	490,000
Wheat and Barley Research Institute	653,000
Agr. Chemicals Research Institute	777,253
Crops Experiment Station	1,623,000
Honam Crop Experiment Station	1,436,984
Yeongam Crop Experiment Station	1,168,574
Horticultural Experiment Station	1,535,617
Sericultural Experiment Station	615,697
Livestock Experiment Station	1,629,000
Alpine Experiment Station	865,000
Cheju Experiment Station	487,593
Rural Nutrition Institute	269,080
Subtotal MAF/RDA	15,315,775
Fisheries Research and Development Agency (MAF/NFA)	5,676,000
Agricultural Engineering Research Center (MAF/ADC)	820,000
Subtotal MAF	21,811,775
Institute of Forest Genetics (MHA/FA)	1,620,648
Forest Research Institute (MHA/FA)	3,713,000
Total	27,145,423

Parts A, B, and C of the ISNAR, IFARD & AOAD questionnaire give as total expenditures by NARS 20195, 22221, and 27145 million Won, respectively. Possible explanations for the differences are 1) in part A, only the expenditures of those institutes under the Ministry of Agriculture and Fisheries are included; 2) the expenditure figure given by part A represents only the government contribution to NARS, while in part C other sources like sale of products are included; 3) part B refers to recurrent expenditures only. Because of the detailed information given in this particular case by the institutional-level responses contained in part C of the questionnaire, the sum total of these institutional-level responses is the most likely aggregate estimate.

Acronyms and Abbreviations:
MAF/RDA: Ministry of Agriculture and Fisheries – Rural Development Administration
MAF/NFA: Ministry of Agriculture and Fisheries – National Fisheries Administration
MAF/ADC: Ministry of Agriculture and Fisheries – Agricultural Development Corporation
MHA/FA: Ministry of Home Affairs – Forestry Administration

Sources:
0010 Boyce, James K., and Robert E. Evenson. *National and International Agricultural Research and Extension Programs.* New York: Agricultural Development Council, Inc., 1975.
0017 ISNAR, IFARD & AOAD. Survey of National Agricultural Research Systems: Unpublished Questionnaire Responses. ISNAR, The Hague, 1985.
0176 Moseman, Albert, ed. *National Agricultural Research Systems in Asia.* Report of the Regional Seminar held at the India International Centre, New Delhi, India, March 8–13, 1971. New York: ADC, Inc., 1971.
0842 Yun Cho, Chae. Personal communication. Director General, International Technical Cooperation Center – Rural Development Administration. Suweon, Republic of Korea, March 1988.

Additional References:
0014 Judd, M. Ann, James K. Boyce, and Robert E. Evenson. "Investing in Agricultural Supply." Economic Growth Center, Yale University, New Haven, Connecticut, 1983. Mimeo.
0016 Oram, Peter A., and Vishva Bindlish. *Resource Allocations to National Agricultural Research: Trends in the 1970s.* The Hague and Washington, D.C.: ISNAR and IFPRI, November 1981.
0022 UNESCO. *Statistical Yearbook 1983.* Paris: UNESCO, 1983.
0026 Oram, Peter A., and Mark Gieben. "Document Summaries." ISNAR, The Hague, 1984. Mimeo.
0027 Harvey, Nigel, ed. *Agricultural Research Centres: A World Directory of Organizations and*

Programmes. Seventh Edition, Two Volumes. Harlow, U.K.: Longman, 1983.

0073 Oram, Peter A., and Vishva Bindlish. "Investment in Agricultural Research in Developing Countries: Progress, Problems, and the Determination of Priorities." IFPRI, Washington, D.C, January 1984. Mimeo.

0095 FAO – CARIS. *Agricultural Research in Developing Countries – Volume 1: Research Institutions.* Rome: FAO – CARIS, 1978.

0165 Evenson, R.E., and Y. Kislev. *Agricultural Research and Productivity.* New Haven: Yale University Press, 1975.

0177 Office of Rural Development. *Annual Research Report for 1982.* South Korea: Office of Rural Development, 1983.

0393 Kim, Young Sang. "The Progress and Contribution of Research Projects for Agricultural Development in Korea." Paper presented at the 1st International Meeting of National Agricultural Research Systems and 2nd IFARD Global Convention, IFARD/EMBRAPA, Brasília, Brasil, 6–11 October 1986.

0437 Sharma, S.K. "Research Management System in Republic of Korea." In *Proceedings of the Research Extension Linkage Workshop Held in Hanoi, Viet Nam, 9–13 June 1986.* Rome: FAO, 1986.

0715 Chang, C.W. *FAO Directory of Agricultural Research Institutes and Experiment Stations in Asia and the Far East.* Bangkok, Thailand: FAO – Regional Office for Asia and the Far East, September 1962.

0716 Chang, C.W. *FAO Directory of Agricultural Research Institutes and Experiment Stations in Asia and the Far East, Supplement.* Bangkok, Thailand: FAO – Regional Office for Asia and the Far East, May 1963.

0726 Webster, Brian, Carlos Valverde, and Alan Fletcher, eds. *The Impact of Research on National Agricultural Development.* Report on the First International Meeting of National Agricultural Research Systems and the Second IFARD Global Convention. The Hague: ISNAR, July 1987.

0732 FAO – Regional Office for Asia and the Pacific. *Agricultural Research Systems in the Asia–Pacific Region.* Bangkok: FAO – Regional Office for Asia and the Pacific, 1986.

0768 Kim, Young Sang. "The Progress and Contribution of Research Projects for Agricultural Development in Korea." In *The Impact of Research on National Agricultural Development,* edited by Brian Webster, Carlos Valverde and Alan Fletcher, pp. 211–224. The Hague: ISNAR, July 1987.

0869 Boyce, James K. "Agricultural Research in Indonesia, The Philippines, Bangladesh, South Korea and India: A Documentary History." University of Minnesota Asian Agricultural Research Review Project, July 1980. Mimeo.

0886 Evenson, Robert E., and Yoav Kislev. "Investment in Agricultural Research and Extension: A Survey of International Data." *Economic Development and Cultural Change* Vol. 23 (April 1975): 507–521.

Kuwait

Personnel

Period/Year	PhD	MSc	BSc	Subtotal	Expat	Total	Source
1960–64	..	..	..	..	..	..	
1965–69	..	..	..	..	..	..	
1970	..	..	..	..	..	..	
1971	..	..	..	..	..	..	
1972	..	..	..	..	..	..	
1973	..	..	..	..	..	..	
1974	..	..	..	..	..	..	
1975	..	..	..	..	..	..	
1976	..	..	..	..	..	..	
1977	..	..	..	..	..	..	
1978	..	..	..	..	..	..	
1979	..	..	..	..	..	..	
1980	..	..	..	..	..	..	
1981	..	..	..	..	..	..	
1982	..	..	..	..	..	..	
1983	0	1	0	1	7	8	17
1984	..	..	..	..	..	..	
1985	..	..	..	..	..	..	
1986	2	1	9	12	..	12	843

1960–64:   1965–69:

Expenditure

LCU = *Kuwaiti Dinars*

Period/Year	Current LCU (millions)	Constant 1980 LCU (millions)	Constant 1980 US$ (millions)		Source
			Atlas	PPP	
1960–64	..	..	..	..	
1965–69	..	..	..	..	
1970	..	..	..	..	
1971	..	..	..	..	
1972	..	..	..	..	
1973	..	..	..	..	
1974	..	..	..	..	
1975	..	..	..	..	
1976	..	..	..	..	
1977	..	..	..	..	
1978	..	..	..	..	
1979	..	..	..	..	
1980	..	..	..	..	
1981	..	..	..	..	
1982	..	..	..	..	
1983	..	..	..	..	
1984	..	..	..	..	
1985	..	..	..	..	
1986	..	..	..	..	

1960–64:   1965–69:

Personnel Comments:

The 1983 and 1986 figures represent the Agriculture Affairs and Fish Resources Authority (AAFRA) within the Ministry of Public Works. The Kuwait Institute for Scientific Research is mentioned as also doing some agricultural research but no data are available. For 1982, source 27 gives a total of 239 graduate research staff in the "Agricultural Experiment Station" in the Ministry of Public Works. In the light of other information which is available, this figure appears dubious.

Sources:

0017 ISNAR, IFARD & AOAD. Survey of National Agricultural Research Systems: Unpublished Questionnaire Responses. ISNAR, The Hague, 1985.

0843 Agriculture Affairs and Fish Resources Authority (AAFRA). Handout submitted during the Rainfed Agricultural Information Network Workshop, Amman, Jordan, 17–20 March 1985.

Additional References:

0027 Harvey, Nigel, ed. *Agricultural Research Centres: A World Directory of Organizations and Programmes.* Seventh Edition, Two Volumes. Harlow, U.K.: Longman, 1983.

0094 FAO – Near East Regional Office. *Directory of Agricultural Research Institutions in the Near East Region.* Cairo: FAO – Near East Regional Office, 1979.

0173 FAO. *FAO Near East Regional Studies on Organization and Administration of Agricultural Research.* Rome: FAO, 1974.

0865 Oram, Peter. "Agricultural Research Objectives and Priorities: Constraints to the Development of Agricultural Research Institutions in Arab Countries." Review paper prepared for the first meeting of the Council for Arab Agricultural Research, organized by the Arab Fund for Economic and Social Development, Kuwait, 23 March 1988.

Laos, P.D.R.

Personnel

Period/Year	PhD	MSc	BSc	Subtotal	Expat	Total	Source
1960–64	..	..	..	..	..		
1965–69	..	..	..	..	..		
1970	..	..	..	..	..	29	10
1971	..	..	..	..	..		
1972	..	..	..	..	..		
1973	..	..	..	..	..		
1974	..	..	..	..	..		
1975	..	..	..	..	..		
1976	..	..	..	..	..		
1977	..	..	..	..	..		
1978	..	..	..	..	..		
1979	..	..	..	..	..		
1980	..	..	..	..	..		
1981	..	..	..	..	..		
1982	..	..	..	..	..		
1983	..	..	..	..	..		
1984	..	..	..	..	..		
1985	..	..	..	..	..		
1986	..	..	..	..	..		

1960–64:   1965–69:

Expenditure   *LCU = Kip*

Period/Year	Current LCU (millions)	Constant 1980 LCU(millions)	Constant 1980 US$ (millions)		Source
			Atlas	PPP	
1960–64	..	..	..	..	
1965–69	..	..	..	..	
1970	..	..	..	..	
1971	..	..	..	..	
1972	..	..	..	..	
1973	..	..	..	..	
1974	..	..	..	..	
1975	..	..	..	..	
1976	..	..	..	..	
1977	..	..	..	..	
1978	..	..	..	..	
1979	..	..	..	..	
1980	..	..	..	..	
1981	..	..	..	..	
1982	..	..	..	..	
1983	..	..	..	..	
1984	..	..	..	..	
1985	..	..	..	..	
1986	..	..	..	..	

1960–64:   1965–69:

Personnel Comments:

Source 590 reports, for the year 1980, 82 scientists and engineers with an educational training in agricultural sciences. The number working in agricultural research is not given, but is thought to be low.

Source 817 reports 62 researchers at Centre National de Recherche Agronomique (CNRA) in 1987, of which 3 held a PhD degree, 18 a MSc, and 31 a BSc, and 10 were expatriates.

Expenditure Comments:

The only expenditure figure available is from source 10 and is given directly in US dollars (1972: US$ 467,000). A lack of exchange rates for Laos means this expenditure figure cannot be converted into local currency units.

Total budget of CNRA in 1987 was estimated at US$ 750,000 in 1987 (source 817).

Sources:
0010 Boyce, James K., and Robert E. Evenson. *National and International Agricultural Research and Extension Programs.* New York: Agricultural Development Council, Inc., 1975.

Additional References:
0590 UNESCO. *Science and Technology in Countries of Asia and the Pacific.* Science Policy Studies and Documents No. 52. Paris: UNESCO, n.d.

0817 Giacich, Walter. "Document de Projet Révisé UNDP/FAO/Lao/82/011." Vientiane, Lao, June 1987. Mimeo.

Lebanon

Personnel

Period/Year	PhD	MSc	BSc	Subtotal	Expat	Total	Source
1960–64	..	..	..	..	..	..	
1965–69	..	..	..	..	..	71	10
1970	..	..	..	..	..	..	
1971	..	..	..	..	..	..	
1972	..	..	..	..	..	..	
1973	..	..	..	..	..	..	
1974	..	..	..	..	..	..	
1975	..	..	..	..	..	..	
1976	..	..	..	..	..	..	
1977	..	..	..	..	..	..	
1978	..	..	..	..	..	..	
1979	..	..	..	..	..	..	
1980	..	..	..	..	..	..	
1981	..	..	..	..	..	..	
1982	..	..	..	..	..	..	
1983	..	..	..	..	..	..	
1984	..	..	..	..	..	..	
1985	..	..	..	..	..	..	
1986	..	..	..	..	..	..	

1960–64:   1965–69: 1967

Expenditure

LCU = Lebanese Pounds

Period/Year	Current LCU (millions)	Constant 1980 LCU(millions)	Constant 1980 US$ (millions)		Source
			Atlas	PPP	
1960–64	..	..	..	..	
1965–69	..	..	..	..	
1970	..	..	..	..	
1971	..	..	..	..	
1972	..	..	..	..	
1973	..	..	..	..	
1974	..	..	..	..	
1975	..	..	..	..	
1976	..	..	..	..	
1977	..	..	..	..	
1978	..	..	..	..	
1979	..	..	..	..	
1980	..	..	..	..	
1981	..	..	..	..	
1982	..	..	..	..	
1983	..	..	..	..	
1984	..	..	..	..	
1985	..	..	..	..	
1986	..	..	..	..	

1960–64:   1965–69:

Personnel Comments:

Source 27 lists 28 graduate research staff in the Faculty of Agricultural and Food Science, American University of Beirut, in 1982, and source 744 lists 75 graduate research staff for 1985.

Sources:

0010 Boyce, James K., and Robert E. Evenson. *National and International Agricultural Research and Extension Programs.* New York: Agricultural Development Council, Inc., 1975.

Additional References:

0027 Harvey, Nigel, ed. *Agricultural Research Centres: A World Directory of Organizations and Programmes.* Seventh Edition, Two Volumes. Harlow, U.K.: Longman, 1983.

0094 FAO – Near East Regional Office. *Directory of Agricultural Research Institutions in the Near East Region.* Cairo: FAO – Near East Regional Office, 1979.

0173 FAO. *FAO Near East Regional Studies on Organization and Administration of Agricultural Research.* Rome: FAO, 1974.

0174 Watson, J.M. "Comparative Study of Agricultural Research Organisation and Administration in the Near East Region." Paper presented at the Workshop on Organization and Administration of Agricultural Services in the Arab States, Cairo, 2–15 March 1964.

0407 Centre International de Hautes Etudes Agronomiques Méditerranéennes.

"Séminaire sur l'Orientation et l'Organisation de la Recherche Agronomique dans les Pays du Bassin Méditerranéen, Istanbul, 1–3 December 1986." Centre International de Hautes Etudes Agronomiques Méditerranéennes, Paris, December 1986. Mimeo.

0417 Haidar, Wael. "La Recherche Agronomique au Liban." Paper presented at the Séminaire sur l'Orientation et l'Organisation de la Recherche Agronomique dans les Pays du Bassin Méditerranéen, Centre International de Hautes Etudes Agronomiques Méditerranéennes, Istanbul, 1–3 December 1986.

0422 Casas, Joseph. "Les Ressources Humaines et Financières dans les Pays du Bassin Meditérranéen." Preliminary table. ISNAR, The Hague, 1986. Mimeo.

0744 *Agricultural Research Centres: A World Directory of Organizations and Programmes.* Eighth Edition, Two Volumes. Harlow, U.K.: Longman, 1986.

0831 FAO. *Report of an Expert Consultation on Agricultural Extension and Research Linkages in the Near East, Amman, Jordan, 22–26 July 1985.* Rome: FAO, 1985.

0865 Oram, Peter. "Agricultural Research Objectives and Priorities: Constraints to the Development of Agricultural Research Institutions in Arab Countries." Review paper prepared for the first meeting of the Council for Arab Agricultural Research, organized by the Arab Fund for Economic and Social Development, Kuwait, 23 March 1988.

Lesotho

Personnel

Period/Year	PhD	MSc	BSc	Subtotal	Expat	Total	Source
1960–64	..	..	..	..		1	10
1965–69	..	..	..	..	..	3	286/10
1970	..	..	..	..		3	266
1971	..	..	..	..		7	589
1972	..	..	..	..			
1973	..	..	..	..			
1974	..	..	..	..		10	16
1975	..	..	..	..			
1976	..	..	..	..			
1977	..	..	..	..			
1978	..	..	..	..			
1979	..	..	..	..		..	
1980	..	..	..	..		..	
1981	..	..	..	..		..	
1982	..	..	..	..		..	
1983	..	..	..	..		..	
1984	0	3	6	9	9	18	4
1985	..	..	..	..		..	
1986	..	..	..	..		..	

1960–64: 1961   1965–69: 1965 & 66

Expenditure

LCU = Maloti/South African Rands

Period/Year	Current LCU (millions)	Constant 1980 LCU(millions)	Constant 1980 US$ (millions)		Source
			Atlas	PPP	
1960–64	..	..	..		
1965–69	0.029	0.093	0.108	0.269	589
1970	..	..	..		
1971	0.097	0.260	0.300	0.750	589
1972	..	..	..		
1973	..	..	..		
1974	..	..	..		
1975	..	..	..		
1976	..	..	..		
1977	..	..	..		
1978	..	..	..		
1979	..	..	..		
1980	..	..	..		
1981	..	..	..		
1982	..	..	..		
1983	..	..	..		
1984	3.185	2.093	2.415	6.043	4
1985	..	..	..		
1986	..	..	..		

1960–64:   1965–69: 1966

Personnel Comments:
The 1984 figure refers to the Research Division of the Ministry of Agriculture. The Research Division in its present form was established in 1979, having previously been part of the Crops Division since 1952.

Expenditure Comments:
The 1984 expenditure figure refers to the Research Division of the Ministry of Agriculture. The figures given by source 589 refer to the Agricultural Research Station only. According to information contained in source 4, it appears that the Agricultural Research Station accounts for most of the research activity within the Research Division of the Ministry of Agriculture.

Sources:

0004　SADCC, and DEVRES, Inc. *Agricultural Research Resource Assessment in the SADCC Countries, Volume II: Country Report Lesotho.* Washington, D.C.: DEVRES, Inc., January 1985.

0010　Boyce, James K., and Robert E. Evenson. *National and International Agricultural Research and Extension Programs.* New York: Agricultural Development Council, Inc., 1975.

0016　Oram, Peter A., and Vishva Bindlish. *Resource Allocations to National Agricultural Research: Trends in the 1970s.* The Hague and Washington, D.C.: ISNAR and IFPRI, November 1981.

0266　UNESCO. *National Science Policies in Africa.* Science Policy Studies and Documents No. 31. Paris: UNESCO, 1974.

0286　Commonwealth Agricultural Bureaux (CAB). *List of Research Workers in Agriculture, Animal Health and Forestry in the Commonwealth and in the Republic of Ireland 1966.* Slough, England: CAB, 1966.

0589　Kassapu, Samuel. *Les Dépenses de la Recherche Agricole dans 34 Pays d'Afrique Tropicale.* Paris: Centre de Développement de l'OCDE, 1976.

Additional References:

0002　SADCC, and DEVRES, Inc. *Agricultural Research Resource Assessment in the SADCC Countries, Volume I: Regional Analysis and Strategy.* Washington, D.C.: DEVRES, Inc., January 1985.

0023　Bennell, Paul. *Agricultural Researchers in Sub-Saharan Africa: An Overview.* Working Paper No. 4. The Hague: ISNAR, October 1985.

0027　Harvey, Nigel, ed. *Agricultural Research Centres: A World Directory of Organizations and Programmes.* Seventh Edition, Two Volumes. Harlow, U.K.: Longman, 1983.

0073　Oram, Peter A., and Vishva Bindlish. "Investment in Agricultural Research in Developing Countries: Progress, Problems, and the Determination of Priorities." IFPRI, Washington, D.C, January 1984. Mimeo.

0163　CGIAR. "National Agricultural Research." CGIAR, Washington, D.C., 1985. Mimeo.

0175　Cooper, St.G.C. *Agricultural Research in Tropical Africa.* Kampala: East African Literature Bureau, 1970.

0236　Setai, Bethuel. "Assessment of Future Trained Manpower Requirements in the Agricultural Sector of Lesotho." FAO, Rome, 1983. Mimeo.

0360　Cooper, St.G.C. "Towards Trained Manpower for Agricultural Research in Africa." Paper presented at the Conference on Agricultural Research and Production in Africa, organized by the Association for the Advancement of Agricultural Sciences in Africa (AAASA), Addis Ababa, 29 August–4 September 1971.

0385　SADCC, and DEVRES, Inc. *SADCC Region Agricultural Research Resource Assessment – Data Base Management Information System: Diskettes and User's Guide.* Printouts. Washington, D.C.: SADCC, and DEVRES, Inc., n.d.

0445　Swanson, Burton E., and Wade H. Reeves. "Agricultural Research Eastern and Southern Africa: Manpower and Training." World Bank, Washington, D.C., August 1986. Mimeo.

0446　Kyomo, M.L. "Agricultural Research in Eastern and Southern Africa: Issues and Priorities." Southern African Centre for Cooperation in Agricultural Research of SADCC, Gabarone, Botswana, 1986. Mimeo.

0532　UNESCO Field Science Office for Africa. *Survey on the Scientific and Technical Potential of the Countries of Africa.* Paris: UNESCO, 1970.

0653　Webster, B.N. *Index of Agricultural Research Institutions and Stations in Africa.* Rome: FAO, n.d.

Liberia

Personnel

Period/Year	PhD	MSc	BSc	Subtotal	Expat	Total	Source
1960–64	..	..	..				
1965–69	..	..	..			11	653
1970	..	..	..			5	532
1971	..	..	..				
1972	..	..	..			15	431
1973	..	..	..				
1974	..	..	..				
1975	..	..	..			19	16
1976	..	..	..				
1977	..	..	..				
1978	..	..	..				
1979	..	..	..				
1980	..	..	..				
1981	..	..	..				
1982	..	..	..				
1983	..	..	..			31	86
1984	5	8	10	23	9	32	133
1985	..	..	..			36	466
1986	..	..	..				

1960–64:   1965–69: 1966

Expenditure

LCU = Liberian Dollars

Period/Year	Current LCU (millions)	Constant 1980 LCU(millions)	Constant 1980 US$ (millions) Atlas	PPP	Source
1960–64	..	..	..		
1965–69	0.160	0.371	0.371	0.515	175
1970	0.139	0.317	0.316	0.440	443
1971	0.144	0.327	0.327	0.455	443
1972	0.134	0.292	0.292	0.406	443
1973	0.155	0.313	0.312	0.435	443
1974	0.354	0.597	0.596	0.829	443
1975	0.441	0.611	0.610	0.849	16
1976	..	..	..		
1977	..	..	..		
1978	..	..	..		
1979	..	..	..		
1980	..	..	..		
1981	..	..	..		
1982	..	..	..		
1983	3.867	3.844	3.838	5.340	133
1984	3.856	3.711	3.706	5.155	133
1985	..	..	..		
1986	..	..	..		

1960–64:   1965–69: 1965

Personnel Comments:

The indicator series includes only the Central Agricultural Experiment Station (CAES), also cited as the Central Agricultural Research Institute (CARI). According to source 443, around 1970 CAES "was more or less in a state of suspended animation." At that time FAO assisted in a reorganization and revival of CAES. This may explain the dip in the time series for 1970. In addition to CAES or CARI, the following Liberian agricultural research units can be identified: 1) the Botanical Research Department of the Firestone Plantations Company. This privately funded research unit (about US$ 250,000 a year) not only undertakes rubber-related research but has also research programs on other commercial crops. The number of researchers at the Botanical Research Department has declined over the years from 5 researchers in 1970 to 3 in 1985. 2) The College of Agriculture and Forestry of the University of Liberia, where the number of teacher-researchers has increased from 19 in 1970 to 31 in 1985 (sources 532 and 86). No indication is given of the amount of time spent on research by these teacher-researchers. 3) WARDA (West African Rice Development Association), which, according to source 86, had a scientific staff of 51 in 1982. A large proportion of its research is decentralized and conducted in different West African countries. For consistency reasons, the first two research units were excluded from the series, while WARDA was not included because of its international or regional character.

Expenditure Comments:

The expenditure series refer to CAES or CARI only.

Sources:

0016 Oram, Peter A., and Vishva Bindlish. Resource Allocations to National Agricultural Research: Trends in the 1970s. The Hague and Washington, D.C.: ISNAR and IFPRI, November 1981.

0086 World Bank. "West Africa Agricultural Research Review – Country Studies." World Bank, Washington, D.C., 1985. Mimeo.

0133 FAO. Report of a Review Mission: Agricultural Research in Liberia. Rome: FAO, June 1984.

0175 Cooper, St.G.C. Agricultural Research in Tropical Africa. Kampala: East African Literature Bureau, 1970.

0431 Current Agricultural Research Information System (CARIS). Directory of Agricultural Research Institutions and Projects in West Africa – Pilot Project 1972–73. Rome: FAO, 1973.

0443 UNDP, and FAO. Report to the Government of Liberia: Agricultural Research Organization. Project Report No. TA 3299. Rome: FAO, 1974.

0466 World Bank – West Africa Agricultural Research Review Team. "National Agricultural Research Systems (NARS): A Regional Appraisal and Selected Issues." World Bank, Washington, D.C., August 1986. Mimeo.

0532 UNESCO Field Science Office for Africa. Survey on the Scientific and Technical Potential of the Countries of Africa. Paris: UNESCO, 1970.

0653 Webster, B.N. Index of Agricultural Research Institutions and Stations in Africa. Rome: FAO, n.d.

Additional References:

0010 Boyce, James K., and Robert E. Evenson. National and International Agricultural Research and Extension Programs. New York: Agricultural Development Council, Inc., 1975.

0023 Bennell, Paul. Agricultural Researchers in Sub-Saharan Africa: An Overview. Working Paper No. 4. The Hague: ISNAR, October 1985.

0027 Harvey, Nigel, ed. Agricultural Research Centres: A World Directory of Organizations and Programmes. Seventh Edition, Two Volumes. Harlow, U.K.: Longman, 1983.

0068 Conte, Stephan. "West Africa Agricultural Research Data Base Building Task: Termination Report." World Bank, Washington, D.C., 1985. Mimeo.

0073 Oram, Peter A., and Vishva Bindlish. "Investment in Agricultural Research in Developing Countries: Progress, Problems, and the Determination of Priorities." IFPRI, Washington, D.C., January 1984. Mimeo.

0088 Swanson, Burton E., and Wade H. Reeves. "Agricultural Research West Africa: Manpower and Training." World Bank, Washington, D.C., November 1985. Mimeo.

0089 Oram, Peter. "Report on National Agricultural Research in West Africa." IFPRI, Washington, D.C., February 1986. Mimeo.

0095 FAO – CARIS. Agricultural Research in Developing Countries – Volume 1: Research Institutions. Rome: FAO – CARIS, 1978.

0163 CGIAR. "National Agricultural Research." CGIAR, Washington, D.C., 1985. Mimeo.

0165 Evenson, R.E., and Y. Kislev. *Agricultural Research and Productivity.* New Haven: Yale University Press, 1975.

0266 UNESCO. *National Science Policies in Africa.* Science Policy Studies and Documents No. 31. Paris: UNESCO, 1974.

0360 Cooper, St.G.C. "Towards Trained Manpower for Agricultural Research in Africa." Paper presented at the Conference on Agricultural Research and Production in Africa, organized by the Association for the Advancement of Agricultural Sciences in Africa (AAASA), Addis Ababa, 29 August–4 September 1971.

0400 UNESCO. *The Promotion of Scientific Activity in Tropical Africa.* Science Policy Studies and Documents No. 11. Paris: UNESCO, 1969.

0589 Kassapu, Samuel. *Les Dépenses de la Recherche Agricole dans 34 Pays d'Afrique Tropicale.* Paris: Centre de Développement de l'OCDE, 1976.

0852 Evenson, Robert E., and Yoav Kislev. *Investment in Agricultural Research and Extension: A Survey of International Data.* Center Discussion Paper No. 124. New Haven, Connecticut: Economic Growth Center, Yale University, August 1971.

0886 Evenson, Robert E., and Yoav Kislev. "Investment in Agricultural Research and Extension: A Survey of International Data." *Economic Development and Cultural Change* Vol. 23 (April 1975): 507–521.

Libyan Arab Republic

Personnel

Period/Year	PhD	MSc	BSc	Subtotal	Expat	Total	Source
1960–64	..	..	..	..	..	..	
1965–69	..	..	..	..	..	..	
1970	..	..	..	..	..	..	
1971	..	..	..	..	..	..	
1972	..	..	..	..	..	..	
1973	..	..	..	..	..	..	
1974	..	..	..	..	..	..	
1975	..	..	..	..	..	..	
1976	..	..	..	..	..	112	95
1977	..	..	..	..	..	..	
1978	..	..	..	..	..	..	
1979	..	..	..	..	..	..	
1980	..	..	..	..	..	..	
1981	..	..	..	..	..	..	
1982	..	..	..	..	..	..	
1983	..	..	..	..	..	..	
1984	..	..	..	..	..	..	
1985	..	..	..	..	..	..	
1986	..	..	..	..	..	..	

1960–64:   1965–69:

Expenditure   *LCU = Libyan Dinars*

Period/Year	Current LCU (millions)	Constant 1980 LCU(millions)	Constant 1980 US$ (millions)		Source
			Atlas	PPP	
1960–64	..	..	..	..	
1965–69	..	..	..	..	
1970	..	..	..	..	
1971	..	..	..	..	
1972	..	..	..	..	
1973	..	..	..	..	
1974	..	..	..	..	
1975	..	..	..	..	
1976	..	..	..	..	
1977	..	..	..	..	
1978	..	..	..	..	
1979	..	..	..	..	
1980	..	..	..	..	
1981	..	..	..	..	
1982	..	..	..	..	
1983	..	..	..	..	
1984	..	..	..	..	
1985	..	..	..	..	
1986	..	..	..	..	

1960–64:   1965–69:

Personnel Comments:

The 1976 figure includes the Agricultural Research Center, the Arab Development Institute, and the Council for Nutrition Affairs and Marine Wealth.

Sources:

0095 FAO – CARIS. *Agricultural Research in Developing Countries – Volume 1: Research Institutions.* Rome: FAO – CARIS, 1978.

Additional References:

0010 Boyce, James K., and Robert E. Evenson. *National and International Agricultural Research and Extension Programs.* New York: Agricultural Development Council, Inc., 1975.

0016 Oram, Peter A., and Vishva Bindlish. *Resource Allocations to National Agricultural Research: Trends in the 1970s.* The Hague and Washington, D.C.: ISNAR and IFPRI, November 1981.

0027 Harvey, Nigel, ed. *Agricultural Research Centres: A World Directory of Organizations and Programmes.* Seventh Edition, Two Volumes. Harlow, U.K.: Longman, 1983.

0094 FAO – Near East Regional Office. *Directory of Agricultural Research Institutions in the Near East Region.* Cairo: FAO – Near East Regional Office, 1979.

0165 Evenson, R.E., and Y. Kislev. *Agricultural Research and Productivity.* New Haven: Yale University Press, 1975.

0173 FAO. *FAO Near East Regional Studies on Organization and Administration of Agricultural Research.* Rome: FAO, 1974.

0174 Watson, J.M. "Comparative Study of Agricultural Research Organisation and Administration in the Near East Region."

Paper presented at the Workshop on Organization and Administration of Agricultural Services in the Arab States, Cairo, 2–15 March 1964.

0175 Cooper, St.G.C. *Agricultural Research in Tropical Africa.* Kampala: East African Literature Bureau, 1970.

0266 UNESCO. *National Science Policies in Africa.* Science Policy Studies and Documents No. 31. Paris: UNESCO, 1974.

0360 Cooper, St.G.C. "Towards Trained Manpower for Agricultural Research in Africa." Paper presented at the Conference on Agricultural Research and Production in Africa, organized by the Association for the Advancement of Agricultural Sciences in Africa (AAASA), Addis Ababa, 29 August–4 September 1971.

0532 UNESCO Field Science Office for Africa. *Survey on the Scientific and Technical Potential of the Countries of Africa.* Paris: UNESCO, 1970.

0852 Evenson, Robert E., and Yoav Kislev. *Investment in Agricultural Research and Extension: A Survey of International Data.* Center Discussion Paper No. 124. New Haven, Connecticut: Economic Growth Center, Yale University, August 1971.

0865 Oram, Peter. "Agricultural Research Objectives and Priorities: Constraints to the Development of Agricultural Research Institutions in Arab Countries." Review paper prepared for the first meeting of the Council for Arab Agricultural Research, organized by the Arab Fund for Economic and Social Development, Kuwait, 23 March 1988.

0886 Evenson, Robert E., and Yoav Kislev. "Investment in Agricultural Research and Extension: A Survey of International Data." *Economic Development and Cultural Change* Vol. 23 (April 1975): 507–521.

Madagascar

Personnel

Period/Year	PhD	MSc	BSc	Subtotal	Expat	Total	Source
1960–64	..	..	..	..	..	..	
1965–69	..	..	..	..	..	60	175
1970	..	..	..	..	..	67	532
1971	..	..	..	..	..	70	266
1972	..	..	..	..	..	..	
1973	..	..	..	12	60	72	37
1974	..	..	..	..	..	..	
1975	..	..	..	..	..	..	
1976	..	..	..	..	..	46	95
1977	..	..	..	..	..	..	
1978	..	..	..	..	..	..	
1979	..	..	..	..	..	..	
1980	..	..	..	..	..	51	237
1981	..	..	..	..	..	..	
1982	–13–		56	69	6	75	37
1983	9	13	39	61	7	68	17
1984	..	..	..	..	..	..	
1985	–43–		48	91	12	103	271
1986	–51–		47	98	19	117	271

1960–64:   1965–69: 1967

Expenditure   LCU = Malagasy Francs

Period/Year	Current LCU (millions)	Constant 1980 LCU (millions)	Constant 1980 US$ (millions)		Source
			Atlas	PPP	
1960–64	330.100	1144.164	5.059	8.514	266
1965–69	665.500	1824.123	8.066	13.573	175/37
1970	695.263	1742.329	7.705	12.965	266
1971	..	..	..	..	
1972	..	..	..	..	
1973	976.725	2049.977	9.065	15.254	589
1974	618.000	1055.697	4.668	7.855	17
1975	856.000	1397.483	6.180	10.399	17
1976	1029.000	1527.650	6.755	11.367	17
1977	1071.000	1464.226	6.475	10.895	17
1978	1344.000	1721.034	7.610	12.806	17
1979	1249.000	1436.436	6.352	10.689	17
1980	1351.000	1351.000	5.974	10.053	17
1981	1007.000	816.045	3.609	6.072	17
1982	1128.000	706.767	3.125	5.259	17
1983	1840.000	962.343	4.255	7.161	17
1984	2717.000	1296.279	5.732	9.646	17
1985	1776.145	768.561	3.399	5.719	17
1986	1584.751	..	..	..	816

1960–64: 1960   1965–69: 1967 & 69

Personnel Comments:

The table below presents construction details for the pre-1974 indicators. It is clear that French overseas agricultural research organizations played a predominant role in agricultural research in Madagascar during this period.

	1966	1967	1970	1971	1973
CTFT	NA		8	9	
IEMVT	NA		14	13	
IFAC	3		5	4	
IFCC	2		7	8	
IRAM	15		22	30	
IRCT	6		9	5	
IRHO	1		2	1	
Total	NA	60	67	70	72
Source	653	175	532	448	37

According to source 532, ORSTOM's research effort in Madagascar was focused in two stations and involved a total of 67 researchers in Madagascar in 1970. However, it is not clear which part of ORSTOM's research activities were agriculturally related. Because of this uncertainty, and with access to only limited information, ORSTOM's activities were excluded from the time series.

The post-1974 personnel figures refer to FOFIFA (the National Center for Applied Research on Rural Development). FOFIFA was established in 1974 to centralize all agricultural research in a single national organization. It took over all the physical facilities of the research centers previously operated by the French research institutions and by the Ministry of Rural Development. FOFIFA falls under the Ministry of Rural Development and Agrarian Reform and under the Ministry of Higher Education and Scientific Research. Within FOFIFA the following research departments can be identified: 'Département de Recherches Agronomiques', 'Département de Recherches Zootechniques et Vétérinaires' and 'Département de Recherches Chimie, Technologie, Machinisme et Hydraulique'.

Expenditure Comments:

The post-1974 figures represent FOFIFA only. The table below gives construction details for the pre-1974 figures.

	1960	1967	1969	1970	1973
CTFT	22.000			91.275	107.750
IEMVT	60.000			134.500	227.500
IFAC	2.000			40.600	56.350
IFCC	31.000			79.100	132.200
IRAM	180.000			285.300	366.000
IRCT	35.100			52.865	69.125
IRHO				11.623	17.800
Total	330.100	626.000	705.000	695.263	976.725
Source	448	175	37	448	589

Acronyms and Abbreviations:

CTFT: Centre Technique Forestier Tropical
FOFIFA: (National Center for Applied Research on Rural Development)
IEMVT: Institut d'Elevage et de Médicine Vétérinaire des Pays Tropicaux
IFAC: Institut Français de Recherches Fruitières Outre-Mer
IFCC: Institut Français du Café, du Cacao et autres plantes stimulantes
IRAM: Institut de Recherches Agronomiques à Madagascar
IRCT: Institut de Recherches du Coton et des Textiles Exotiques
IRHO: Institut de Recherches pour les Huiles et Oléagineux
ORSTOM: Office de la Recherche Scientifique et Technique Outre-Mer

Sources:

0017 ISNAR, IFARD & AOAD. Survey of National Agricultural Research Systems: Unpublished Questionnaire Responses. ISNAR, The Hague, 1985.

0037 ISNAR. *Rapport au Gouvernement de la Repoblika Demokratika Malagasy: La Recherche Agricole a Madagascar – Bilan et Perspectives du FOFIFA.* The Hague: ISNAR, August 1983.

0095 FAO – CARIS. *Agricultural Research in Developing Countries – Volume 1: Research Institutions.* Rome: FAO – CARIS, 1978.

282

0175 Cooper, St.G.C. *Agricultural Research in Tropical Africa.* Kampala: East African Literature Bureau, 1970.

0237 World Bank. "Madagascar Agricultural Research Subsector Review." Report No. 3188-MAG. World Bank, Washington, D.C., October 1980. Mimeo.

0266 UNESCO. *National Science Policies in Africa.* Science Policy Studies and Documents No. 31. Paris: UNESCO, 1974.

0271 Raharinosy, V., and C.P. Ravohitrarivo. "Planification et Programmation de la Recherche a Madagascar (La Recherche Agricole)." Paper presented at the Workshop on Improving Agricultural Research Organizations and Management: Progress and Issues for the Future, ISNAR, The Hague, 8–12 September 1986.

0532 UNESCO Field Science Office for Africa. *Survey on the Scientific and Technical Potential of the Countries of Africa.* Paris: UNESCO, 1970.

0589 Kassapu, Samuel. *Les Dépenses de Recherche Agricole dans 34 Pays d'Afrique Tropicale.* Paris: Centre de Développement de l'OCDE, 1976.

0816 Saint-Clair, P.M. "Réflexions sur l'Organisation et le Fonctionnement Actuels de la Recherche au FOFIFA." ISNAR, Madagascar, April 1987. Mimeo.

Additional References:

0010 Boyce, James K., and Robert E. Evenson. *National and International Agricultural Research and Extension Programs.* New York: Agricultural Development Council, Inc., 1975.

0014 Judd, M. Ann, James K. Boyce, and Robert E. Evenson. "Investing in Agricultural Supply." Economic Growth Center, Yale University, New Haven, Connecticut, 1983. Mimeo.

0016 Oram, Peter A., and Vishva Bindlish. *Resource Allocations to National Agricultural Research: Trends in the 1970s.* The Hague and Washington, D.C.: ISNAR and IFPRI, November 1981.

0023 Bennell, Paul. *Agricultural Researchers in Sub-Saharan Africa: An Overview.* Working Paper No. 4. The Hague: ISNAR, October 1985.

0026 Oram, Peter A., and Mark Gieben. "Document Summaries." ISNAR, The Hague, 1984. Mimeo.

0027 Harvey, Nigel, ed. *Agricultural Research Centres: A World Directory of Organizations and Programmes.* Seventh Edition, Two Volumes. Harlow, U.K.: Longman, 1983.

0073 Oram, Peter A., and Vishva Bindlish. "Investment in Agricultural Research in Developing Countries: Progress, Problems, and the Determination of Priorities." IFPRI, Washington, D.C, January 1984. Mimeo.

0165 Evenson, R.E., and Y. Kislev. *Agricultural Research and Productivity.* New Haven: Yale University Press, 1975.

0360 Cooper, St.G.C. "Towards Trained Manpower for Agricultural Research in Africa." Paper presented at the Conference on Agricultural Research and Production in Africa, organized by the Association for the Advancement of Agricultural Sciences in Africa (AAASA), Addis Ababa, 29 August–4 September 1971.

0373 Groupement d'Etudes et de Recherches pour le Développement de l'Agronomie Tropicale (GERDAT). *Rapport Général d'Activité pour 1982.* Paris: GERDAT, 1983.

0400 UNESCO. *The Promotion of Scientific Activity in Tropical Africa.* Science Policy Studies and Documents No. 11. Paris: UNESCO, 1969.

0445 Swanson, Burton E., and Wade H. Reeves. "Agricultural Research Eastern and Southern Africa: Manpower and Training." World Bank, Washington, D.C., August 1986. Mimeo.

0653 Webster, B.N. *Index of Agricultural Research Institutions and Stations in Africa.* Rome: FAO, n.d.

0654 Office de la Recherche Scientifique et Technique Outre-Mer (ORSTOM). *Rapport d'Activité 1967.* Paris: ORSTOM, 1969.

0852 Evenson, Robert E., and Yoav Kislev. *Investment in Agricultural Research and Extension: A Survey of International Data.* Center Discussion Paper No. 124. New Haven, Connecticut: Economic Growth Center, Yale University, August 1971.

0886 Evenson, Robert E., and Yoav Kislev. "Investment in Agricultural Research and Extension: A Survey of International Data." *Economic Development and Cultural Change* Vol. 23 (April 1975): 507–521.

Malawi

Personnel

Period/Year	PhD	MSc	BSc	Subtotal	Expat	Total	Source
1960–64	3	3	15	21	..	21	871
1965–69	0	0	6	6	20	26	11
1970	–2–		9	11	21	32	11
1971	–2–		9	11	24	35	11
1972	–3–		11	14	24	38	11
1973	–4–		14	18	20	38	11
1974	–5–		18	23	7	30	11
1975	–8–		27	35	13	48	11
1976	–8–		34	42	10	52	11
1977	–13–		36	49	9	58	11
1978	–17–		41	58	7	65	11
1979	–22–		48	70	6	76	11
1980	–22–		50	72	1	73	11
1981	–22–		50	72	3	75	11
1982	4	10	53	67	9	76	11
1983	3	17	64	84	1	85	241
1984	..	..	..	82	10	92	5
1985	..	..	..	..		..	
1986	..	..	..	..		..	

1960–64: 1960, 61 & 64   1965–69: 1965, 66, 67, 68 & 69

Expenditure

LCU = Kwacha

Period/Year	Current LCU (millions)	Constant 1980 LCU (millions)	Constant 1980 US$ (millions)		Source
			Atlas	PPP	
1960–64	..	..	..	..	
1965–69	0.446	0.860	1.005	2.174	239
1970	0.877	1.534	1.792	3.876	239
1971	0.746	1.168	1.305	2.952	239
1972	0.711	1.165	1.361	2.944	239
1973	0.457	0.728	0.850	1.840	239
1974	1.057	1.477	1.726	3.733	239
1975	1.079	1.363	1.592	3.444	239
1976	0.843	0.983	1.148	2.483	439
1977	1.013	1.111	1.298	2.808	439
1978	1.081	1.183	1.382	2.990	439
1979	1.366	1.422	1.661	3.593	439
1980	1.657	1.657	1.936	4.187	439
1981	1.876	1.572	1.837	3.973	439
1982	2.206	1.847	2.158	4.668	439
1983	3.220	2.427	2.835	6.132	439
1984	3.225	2.003	2.340	5.061	439
1985	3.487	2.135	2.494	5.394	439
1986	..	..	..	..	

1960–64:   1965–69: 1966, 67, 68 & 69

Personnel Comments:

These figures refer to the Department of Agricultural Research within the Ministry of Agriculture only. Other organizations which conduct agricultural research are the Department of Veterinary Services within the Ministry of Agriculture, the Tobacco Research Authority, the Tea Research Foundation, the Fisheries Research Station, the Forestry Research Institute, and the Sugar Corporation of Malawi. These organizations together employed approximately 55 full-time equivalent researchers in 1984 (source 5).

Most of the figures in source 11 originated in source 239.

Expenditure Comments:

These expenditure figures refer to the Department of Agricultural Research of the Ministry of Agriculture only. Expenditures by other research organizations (i.e., Department of Veterinary Services, Tobacco Research Authority, Tea Research Foundation, and Sugar Corporation of Malawi) amounted to K$ 1.893 million in 1984 (source 5).

Sources:

0005 SADCC, and DEVRES, Inc. *Agricultural Research Resource Assessment in the SADCC Countries, Volume II: Country Report Malawi.* Washington, D.C.: DEVRES, Inc., January 1985.

0011 ISNAR. *Report to the Government of Malawi: A Review of the Agricultural Research System of Malawi.* The Hague: ISNAR, August 1982.

0239 "Development and Administration of Agricultural Research and Its Contribution to Agricultural Development in Malawi." n.d. Mimeo.

0241 Department of Agricultural Research – Ministry of Agriculture. *Malawi Agricultural Research Strategy Plan.* Malawi: Ministry of Agriculture, September 1983.

0439 Swanson, Burton E., et al. "An INTERPAKS Case Study of the Agricultural Technology System in Malawi." INTERPAKS – University of Illinois, Urbana, Illinois, March 1986. Mimeo.

0871 Department of Agriculture Nyasaland/Malawi. *Annual Report of the Department of Agriculture for the Years 1959/60, 1960/61, 1963/64, and 1967/68.* Part II. Zomba, Nyasaland/Malawi: The Government Printer, various years.

Additional References:

0002 SADCC, and DEVRES, Inc. *Agricultural Research Resource Assessment in the SADCC Countries, Volume I: Regional Analysis and Strategy.* Washington, D.C.: DEVRES, Inc., January 1985.

0010 Boyce, James K., and Robert E. Evenson. *National and International Agricultural Research and Extension Programs.* New York: Agricultural Development Council, Inc., 1975.

0014 Judd, M. Ann, James K. Boyce, and Robert E. Evenson. "Investing in Agricultural Supply." Economic Growth Center, Yale University, New Haven, Connecticut, 1983. Mimeo.

0016 Oram, Peter A., and Vishva Bindlish. *Resource Allocations to National Agricultural Research: Trends in the 1970s.* The Hague and Washington, D.C.: ISNAR and IFPRI, November 1981.

0022 UNESCO. *Statistical Yearbook 1983.* Paris: UNESCO, 1983.

0023 Bennell, Paul. *Agricultural Researchers in Sub-Saharan Africa: An Overview.* Working Paper No. 4. The Hague: ISNAR, October 1985.

0026 Oram, Peter A., and Mark Gieben. "Document Summaries." ISNAR, The Hague, 1984. Mimeo.

0027 Harvey, Nigel, ed. *Agricultural Research Centres: A World Directory of Organizations and Programmes.* Seventh Edition, Two Volumes. Harlow, U.K.: Longman, 1983.

0073 Oram, Peter A., and Vishva Bindlish. "Investment in Agricultural Research in Developing Countries: Progress, Problems, and the Determination of Priorities." IFPRI, Washington, D.C., January 1984. Mimeo.

0163 CGIAR. "National Agricultural Research." CGIAR, Washington, D.C., 1985. Mimeo.

0164 Association for the Advancement of Agricultural Sciences in Africa (AAASA). *Proceedings of the Workshop on Agricultural Research Administration, Nairobi, Kenya, 27–30 June 1977.* Proceedings Series PE-4. Addis Ababa, Ethiopia: AAASA and IDRC, August 1979.

0165 Evenson, R.E., and Y. Kislev. *Agricultural Research and Productivity.* New Haven: Yale University Press, 1975.

0175 Cooper, St.G.C. *Agricultural Research in Tropical Africa.* Kampala: East African Literature Bureau, 1970.

0238 World Bank. "Malawi National Rural Development Program Review." World Bank, Washington, D.C., 1982. Mimeo.

0240 Ministry of Agriculture – Department of Agricultural Research. "Staff and Expenditure – Department of Agricultural Research." Ministry of Agriculture, Malawi, 1981. Mimeo.

0242 Commonwealth Agricultural Bureaux (CAB). *List of Research Workers in Agricultural Sciences in the Commonwealth 1981.* Slough, England: CAB, 1981.

0243 Commonwealth Agricultural Bureaux (CAB). *List of Research Workers in the Agricultural Sciences in the Commonwealth and in the Republic of Ireland 1978.* Slough, England: CAB, 1978.

0244 Commonwealth Agricultural Bureaux (CAB). *List of Research Workers in the Agricultural Sciences in the Commonwealth and in the Republic of Ireland 1975.* Slough, England: CAB, 1975.

0266 UNESCO. *National Science Policies in Africa.* Science Policy Studies and Documents No. 31. Paris: UNESCO, 1974.

0279 Commonwealth Agricultural Bureaux (CAB). *List of Research Workers in the Agricultural Sciences in the Commonwealth and in the Republic of Ireland 1972.* Slough, England: CAB, 1972.

0285 Commonwealth Agricultural Bureaux (CAB). *List of Research Workers in the Agricultural Sciences in the Commonwealth and in the Republic of Ireland 1969.* Slough, England: CAB, 1969.

0286 Commonwealth Agricultural Bureaux (CAB). *List of Research Workers in Agriculture, Animal Health and Forestry in the Commonwealth and in the Republic of Ireland 1966.* Slough, England: CAB, 1966.

0287 Commonwealth Agricultural Bureaux (CAB). *List of Research Workers in Agriculture, Animal Health and Forestry in the British Commonwealth, the Republic of Sudan and the Republic of Ireland 1959.* Slough, England: CAB, 1959.

0360 Cooper, St.G.C. "Towards Trained Manpower for Agricultural Research in Africa." Paper presented at the Conference on Agricultural Research and Production in Africa, organized by the Association for the Advancement of Agricultural Sciences in Africa (AAASA), Addis Ababa, 29 August–4 September 1971.

0385 SADCC, and DEVRES, Inc. *SADCC Region Agricultural Research Resource Assessment – Data Base Management Information System: Diskettes and User's Guide.* Printouts. Washington, D.C.: SADCC, and DEVRES, Inc., n.d.

0400 UNESCO. *The Promotion of Scientific Activity in Tropical Africa.* Science Policy Studies and Documents No. 11. Paris: UNESCO, 1969.

0445 Swanson, Burton E., and Wade H. Reeves. "Agricultural Research Eastern and Southern Africa: Manpower and Training." World Bank, Washington, D.C., August 1986. Mimeo.

0446 Kyomo, M.L. "Agricultural Research in Eastern and Southern Africa: Issues and Priorities." Southern African Centre for Cooperation in Agricultural Research of SADCC, Gabarone, Botswana, 1986. Mimeo.

0467 DEVRES, Inc. *The Agricultural Research Resource Assessment Pilot Report for Botswana, Malawi and Swaziland.* Washington, D.C.: DEVRES, Inc., November 1983.

0532 UNESCO Field Science Office for Africa. *Survey on the Scientific and Technical Potential of the Countries of Africa.* Paris: UNESCO, 1970.

0589 Kassapu, Samuel. *Les Dépenses de Recherche Agricole dans 34 Pays d'Afrique Tropicale.* Paris: Centre de Développement de l'OCDE, 1976.

0653 Webster, B.N. *Index of Agricultural Research Institutions and Stations in Africa.* Rome: FAO, n.d.

0738 Mkandawire, N.A. "Study on Small Countries Research: Report on Malawi." IDRC, Ottawa, April 1986. Mimeo.

0740 Sands, Carolyn Marie. "The Theoretical and Empirical Basis for Analyzing Agricultural Technology Systems." PhD thesis, University of Illinois, Urbana-Champaign, Illinois, 1988.

0847 Kimura, J.H. "Financial and Administrative Management of Research Institutions in Eastern and Southern Africa: Report on Responses to a Questionnaire." In *Promotion of Technology Policy and Science Management in Africa,* edited by Karl Wolfgang Menck and Wolfgang Gmelin. Bonn: Deutsche Stiftung fur internationale Entwicklung (DSE), 1986.

0852 Evenson, Robert E., and Yoav Kislev. *Investment in Agricultural Research and Extension: A Survey of International Data.* Center Discussion Paper No. 124. New Haven, Connecticut: Economic Growth Center, Yale University, August 1971.

0872 Department of Agricultural Research Malawi. *Annual Report of the Department of Agricultural Research for the Years 1971/72, 1972/73, and 1973/74.* Zomba, Malawi: The Government Printer, various years.

0886 Evenson, Robert E., and Yoav Kislev. "Investment in Agricultural Research and Extension: A Survey of International Data." *Economic Development and Cultural Change* Vol. 23 (April 1975): 507–521.

Malaysia

Personnel

Period/Year	PhD	MSc	BSc	Subtotal	Expat	Total	Source
1960–64	17	22	66	105	27	132	715/716
1965–69	17	44	110	171	..	171	285
1970	..	..	..	..	..	..	
1971	..	..	..	..	..	223	279/295
1972	..	..	..	..	..	..	
1973	..	..	..	..	..	..	
1974	..	..	..	..	..	367	244/295
1975	..	..	..	..	..	..	
1976	..	..	..	..	..	..	
1977	..	..	..	..	..	..	
1978	..	..	..	..	..	..	
1979	..	..	..	..	..	..	
1980	..	..	..	..	..	..	
1981	..	..	..	..	..	741	794/276
1982	..	..	..	..	..	82	179/276/134
1983	171	358	318	847	15	862	17
1984	..	..	..	..	..	..	
1985	..	..	..	..	..	..	
1986	..	..	..	..	..	..	

1960–64: 1960   1965–69: 1968

Expenditure

LCU = Ringgits

Period/Year	Current LCU (millions)	Constant 1980 LCU (millions)	Constant 1980 US$ (millions) Atlas	Constant 1980 US$ (millions) PPP	Source
1960–64	9.675	17.867	8.061	14.878	715/716
1965–69	..	..	..	..	
1970	..	..	..	..	
1971	..	..	..	..	
1972	..	..	..	..	
1973	42.880	71.989	32.477	59.944	295
1974	57.480	85.592	38.614	71.271	295
1975	67.260	103.348	46.625	86.056	295
1976	53.870	73.412	33.119	61.129	295
1977	68.400	87.255	39.365	72.656	295
1978	76.490	92.783	41.859	77.259	295
1979	80.990	86.366	38.963	71.916	295
1980	108.652	108.652	49.018	90.473	303
1981	..	..	..	..	
1982	..	..	..	..	
1983	144.210	133.035	60.018	110.776	17
1984	..	..	..	..	
1985	..	..	..	..	
1986	..	..	..	..	

1960–64: 1960   1965–69:

Personnel Comments:

The Malaysian Agricultural Research and Development Institute (MARDI) was created in 1971. This institute took over agricultural research, which hitherto was executed by several divisions and departments of the (Federal) Ministry of Agriculture and by some semiautonomous research organizations. However, some organizations preserved their independent status, namely, the Rubber Research Institute (RRIM), the Forestry Research Institute, the Veterinary Research Institute, and the Fisheries Research Insitute.

In 1979 the Palm Oil Research Institute (PORIM) split from MARDI to become an independent institute. In addition to the research institutes already mentioned, some agricultural research is undertaken by the Federal and State Ministries of Agriculture and by the universities. For these last two categories no figures are included in the indicator series.

The table below gives a breakdown to institute level:

	1960	1965[b]	1968	1971	1974	1980	1981	1982	1983
MARDI[a]	36	13	71	73	155	404	411	472	447
PORIM	–	–	–	–	–	NA	37	41**	92
RRIM	68	37	66	110*	167*	NA	199	216*	211
FIRI	7	NA	NA	NA	NA	NA	43*	45*	54
FORI	14	8	14	13*	16*	48*	[51]	[55]	58
VRI	7	9	20	27*	29*	NA	NA	NA	NA
Total	132	67	171	223	367	NA	741	829	862
Sources	715	286	285	295	295	295	794	134	17
	716			279*	244*	242*	276*	276* 179**	

[ ] = estimated or constructed by authors.
a. Before 1971 the figures given under MARDI are a sum total of several departments and divisions of the Ministry of Agriculture and some other organizations, which more or less were taken over by MARDI.
b. Omitted from indicator series as institutional-level estimates appear abnormally low.

No information is available on the Veterinary Research Institute for the more recent years, while for the earlier years, information on the Fisheries Research Institute is missing.

In the table below, the 1983 figure has been broken down to institute level.

	PhD	MSc	BSc	Sub	Exp	Total
MARDI	55	240	152	447	0	447
PORIM	26	31	23	80	12	92
RRIM	84	70	55	209	2	211
FIRI	0	5	49	54	0	54
FORI	6	12	39	57	1	58
VRI	NA	NA	NA	NA	NA	NA
Total	171	358	318	847	15	862

One can divide Malaysia roughly into West Malaysia (the Peninsular) and East Malaysia (Sarawak and Sabah on Borneo). The organizations included in the indicator series are all based on the Peninsular. MARDI is required to serve the Peninsular as well as Sarawak and Sabah, but in general one can say that the indicators represent Peninsular Malaysia. Because of the federal-state structure and the fact that the individual states act rather independently, some sources report additional agricultural researchers in Sarawak and Sabah. The table below gives an overview of the number of agricultural researchers in these two states within the respective Ministries of Agriculture.

	1960	1965	1968	1971	1974	1980
Sarawak						
PhD	0	1	2	2	2	2
MSc	0	1	2	5	6	6
BSc	2	8	10	1	21	28
Total	2	10	14	18	29	36
Sabah						
PhD	0	0	1	0	2	
MSc	2	3	8	4	7	
BSc	12	5	11	9	14	
Total	14	8	20	13	23	
Source	715 716	286	285	279	244	242

Expenditure Comments:

The expenditure figure for 1960, given by sources 715 and 716, include the agricultural research expenditures of the Department of Agriculture, RRIM, FIRI, FORI, and VRI.

The expenditure figures from source 295 include the Malaysian Agricultural Research and Development Institute (MARDI), RRIM, FORI, and FIRI. The 1980 figure includes MARDI, RRIM, and the Palm Oil Research Institute of Malaysia (PORIM). PORIM was established in September 1979 and, according to source 295, now carries out all palm oil research previously carried out by MARDI.

The 1983 expenditure figure has been constructed as follows:

Institute	(million Ringgit)
MARDI	68.400
RRIM	36.820
PORIM	23.900
FIRI	7.180
FORI	7.910
VRI	NA
Total	144.210

Acronyms and Abbreviations:

FIRI: Fisheries Research Institute
FORI: Forestry Research Institute
MARDI: Malaysian Agricultural Research and Development Institute
PORIM: Palm Oil Research Institute Malaysia
RRIM : Rubber Research Institute Malaysia
VRI: Veterinary Research Institute

Sources:

0017 ISNAR, IFARD & AOAD. Survey of National Agricultural Research Systems: Unpublished Questionnaire Responses. ISNAR. The Hague, 1985.

0134 Malaysian Agricultural Research and Development Institute (MARDI). *Annual Report 1981*. Kuala Lumpur: MARDI, 1985.

0179 The Philippine Council for Agriculture and Resources Research and Development (PCARRD), and IDRC. *Agriculture and Resources Research Manpower Development in South and Southeast Asia.* Proceedings of the Workshop on Agriculture and Resources Research Manpower Development in South and Southeast Asia, 21–23 October 1981 and 4–6 February 1982, Singapore. Los Baños, Philippines: PCARRD, 1983.

0244 Commonwealth Agricultural Bureaux (CAB). *List of Research Workers in the Agricultural Sciences in the Commonwealth and in the Republic of Ireland 1975.* Slough, England: CAB, 1975.

0276 Ministry of Agriculture – Fisheries Division. *Annual Report for 1982, Fisheries Research Institute, Glugor, Penang.* Malaysia: Ministry of Agriculture, January 1984.

0279 Commonwealth Agricultural Bureaux (CAB). *List of Research Workers in the Agricultural Sciences in the Commonwealth and in the Republic of Ireland 1972.* Slough, England: CAB, 1972.

0285 Commonwealth Agricultural Bureaux (CAB). *List of Research Workers in the Agricultural Sciences in the Commonwealth and in the Republic of Ireland 1969.* Slough, England: CAB, 1969.

0286 Commonwealth Agricultural Bureaux (CAB). *List of Research Workers in Agriculture, Animal Health and Forestry in the Commonwealth and in the Republic of Ireland 1966.* Slough, England: CAB, 1966.

0295 Mustapha, Nik Ishak bin Nik. "The Agricultural Research Resource Allocation System in Peninsular Malaysia." In *Resource Allocation to Agricultural Research*. Proceedings of a workshop held in Singapore 8–10 June 1981, edited by Douglas Daniels and Barry Nestel, pp. 49–54, Ottawa: IDRC, 1981.

0715 Chang, C.W. *FAO Directory of Agricultural Research Institutes and Experiment Stations in Asia and the Far East*. Bangkok, Thailand: FAO – Regional Office for Asia and the Far East, September 1962.

0716 Chang, C.W. *FAO Directory of Agricultural Research Institutes and Experiment Stations in Asia and the Far East, Supplement*. Bangkok, Thailand: FAO – Regional Office for Asia and the Far East, May 1963.

0794 Yusof bin Hashim, Mohd, and Ahmad Shafri bin Man. "An Inventory of Research Manpower and Support Staff in Malaysia." In *Agriculture and Resources Research Manpower Development in South and Southeast Asia*, pp. 187–193. Los Baños, Philippines: Philippine Council for Agriculture and Resources Research and Development, 1983.

Additional References:

0010 Boyce, James K., and Robert E. Evenson. *National and International Agricultural Research and Extension Programs*. New York: Agricultural Development Council, Inc., 1975.

0014 Judd, M. Ann, James K. Boyce, and Robert E. Evenson. "Investing in Agricultural Supply." Economic Growth Center, Yale University, New Haven, Connecticut, 1983. Mimeo.

0016 Oram, Peter A., and Vishva Bindlish. *Resource Allocations to National Agricultural Research: Trends in the 1970s*. The Hague and Washington, D.C.: ISNAR and IFPRI, November 1981.

0018 Daniels, Douglas, and Barry Nestel, eds. *Resource Allocation to Agricultural Research*. Proceedings of a workshop held in Singapore 8–10 June 1981. Ottawa: IDRC, 1981.

0026 Oram, Peter A., and Mark Gieben. "Document Summaries." ISNAR, The Hague, 1984. Mimeo.

0027 Harvey, Nigel, ed. *Agricultural Research Centres: A World Directory of Organizations and Programmes*. Seventh Edition, Two Volumes. Harlow, U.K.: Longman, 1983.

0073 Oram, Peter A., and Vishva Bindlish. "Investment in Agricultural Research in Developing Countries: Progress, Problems, and the Determination of Priorities." IFPRI, Washington, D.C., January 1984. Mimeo.

0093 Bangladesh Agricultural Research Council (BARC), ISNAR, and Winrock International. *Management of Human Resources in Agricultural Research*. Report of the International Workshop on Management of Human Resources in Agricultural Research held March 3–5, 1986 in Dhaka, Bangladesh. Dhaka: BARC, ISNAR and Winrock International, 1986.

0095 FAO – CARIS. *Agricultural Research in Developing Countries – Volume 1: Research Institutions*. Rome: FAO – CARIS, 1978.

0165 Evenson, R.E., and Y. Kislev. *Agricultural Research and Productivity*. New Haven: Yale University Press, 1975.

0176 Moseman, Albert, ed. *National Agricultural Research Systems in Asia*. Report of the Regional Seminar held at the India International Centre, New Delhi, India, March 8–13, 1971. New York: ADC, Inc., 1971.

0242 Commonwealth Agricultural Bureaux (CAB). *List of Research Workers in Agricultural Sciences in the Commonwealth 1981*. Slough, England: CAB, 1981.

0243 Commonwealth Agricultural Bureaux (CAB). *List of Research Workers in the Agricultural Sciences in the Commonwealth and in the Republic of Ireland 1978*. Slough, England: CAB, 1978.

0287 Commonwealth Agricultural Bureaux (CAB). *List of Research Workers in Agriculture, Animal Health and Forestry in the British Commonwealth, the Republic of Sudan and the Republic of Ireland 1959*. Slough, England: CAB, 1959.

0304 Yusof bin Hashim, Mohd. "The Agricultural Research System in Malaysia: A Study of Resource Allocation." In *Resource Allocation to Agricultural Research*. Proceedings of a workshop held in Singapore 8–10 June 1981, edited by Douglas Daniels and Barry Nestel, pp. 145–149. Ottawa: IDRC, 1981.

0732 FAO – Regional Office for Asia and the Pacific. *Agricultural Research Systems in the Asia-Pacific Region*. Bangkok: FAO – Regional Office for Asia and the Pacific, 1986.

0744 *Agricultural Research Centres: A World Directory of Organizations and Programmes*. Eighth Edition, Two Volumes. Harlow, U.K.: Longman, 1986.

0797 Yusof bin Hashim, Mohd, and Ahmad Shafri bin Man. "The Manpower Development Program of the Malaysian Agricultural Research and Development Institute: Past, Present, and Future." In *Agriculture and Resources Research Manpower Development in South and Southeast Asia*, pp. 181–185. Los Baños, Philippines: Philippine Council for Agriculture and Resources Research and Development, 1983.

290

0852 Evenson, Robert E., and Yoav Kislev. *Investment in Agricultural Research and Extension: A Survey of International Data.* Center Discussion Paper No. 124. New Haven, Connecticut: Economic Growth Center, Yale University, August 1971.

0886 Evenson, Robert E., and Yoav Kislev. "Investment in Agricultural Research and Extension: A Survey of International Data." *Economic Development and Cultural Change* Vol. 23 (April 1975): 507–521.

Mali

Personnel

Period/Year	PhD	MSc	BSc	Subtotal	Expat	Total	Source
1960-64	..	..	..				
1965-69	..	..	..	..	..	9	653
1970	..	..	..	..	..	13	532
1971	..	..	..				
1972	..	..	..	..	..	24	431
1973	..	..	..				
1974	..	..	..				
1975	..	..	..				
1976	..	..	..				
1977	..	..	..				
1978	..	..	..				
1979	..	..	..				
1980	..	..	..				
1981	..	..	..				
1982	..	..	..				
1983	12	38	196	246	29	275	384/44
1984	..	..	..				
1985	..	..	..				
1986	..	..	..				

1960-64:   1965-69: 1966

Expenditure

LCU = CFA Francs (formerly Mali Francs)

Period/Year	Current LCU (millions)	Constant 1980 LCU (millions)	Constant 1980 US$ (millions)		Source
			Atlas	PPP	
1960–64	..	..	..	..	
1965–69	..	..	..	..	
1970	..	..	..	..	
1971	..	..	..	..	
1972	219.503	567.191	2.606	2.249	431
1973	160.946	374.293	1.720	1.484	589
1974	..	..	..	..	
1975	..	..	..	..	
1976	..	..	..	..	
1977	..	..	..	..	
1978	..	..	..	..	
1979	..	..	..	..	
1980	..	..	..	..	
1981	..	..	..	..	
1982	..	..	..	..	
1983	3897.620	3166.223	14.546	12.552	68/44
1984	..	..	..	..	
1985	..	..	..	..	
1986	..	..	..	..	

1960–64:   1965–69:

Personnel Comments:

Mali is a former colony of France. As in many former French colonies, France continued its overseas agricultural research program after Mali gained independence in the early sixties. The Mali government has only gradually assumed responsibility for agricultural research. As far as it has been possible to reconstruct the development of agricultural research in Mali, the French overseas research organizations, IRCT, IFAC, and IRAT, operated under their own names until 1975 or thereabouts. In the meantime the Mali government established IER within the Ministry of Production, now the Ministry of Agriculture. Initially IER had only a coordinating role in agricultural research, but over time new agricultural research stations were established and the French research stations were integrated into IER. In addition to IER, which is mainly responsible for agronomic research, several small research units on livestock, forestry, and hydrobiology developed over time within different ministries. In 1980 these research units were integrated into INRZFH. Only the Central Veterinarian Laboratory (LCV) and Operation N'Dama Yanfolila (ONDY) did not integrate into INRZFH but stayed within the Ministry of Rural Development.

Besides these national agricultural research institutes, Mali is host to an OICMA unit and an ILCA unit. Because of the international character of these institutes, they were not included in the indicator series.

The tables below give an overview of how the personnel indicators have been constructed.

	1966	1970	1972	1974
IRAT	3	5	11	
IRCT	5	5	7	14
IFAC	0	2	3	
IER/DRA	–	–	2	NA
MP/LHM	–	1	1	NA
MRE/CRZ	1	NA	NA	NA
Total	9	13	24	NA
Source	653	532	431	589

	1980	1983
IER	111	162*
INRZFH	NA	85
LCV	NA	23
ONDY	NA	5
Total	NA	275
Source	26	384/44*

Expenditure Comments:

The tables below give an overview of how the expenditure indicators have been constructed.

	1966	1971	1972	1973
IRAT	NA	NA	76.901	71.125
IRCT	132.720	162.465	109.119	56.000
IFAC	NA	NA	28.912	29.250
IER/DRA	–	–	2.752	[2.752]
MP/LHM	–	–	1.819	[1.819]
Total	NA	NA	219.503	160.946
Source	589	589	431	589

	1980	1982	1983
IER	1054.110	2260.098	2012.200*
INRZFH	NA	NA	1200.000
LCV	NA	NA	578.000
ONDY	NA	NA	107.420
Total	NA	NA	3897.620
Source	26	44	68*/44

[ ] = estimated or constructed by authors.

In 1979 the 'Division de Recherche sur les Systèmes de Production Rurale' was established within IER, in addition to the already existing 'Division de Recherche Agonomique' (DRA). In quite a few sources only DRA has been included in the post-1979 figures for IER. This may explain the conflicting figures given by these sources with the figures included in the indicator series.

Acronyms and Abbreviations:

IER/DRA: Institut d'Economie Rurale – Division de Recherche Agronomique

IFAC : Institut Français de Recherches Fruitières d'Outre-Mer

ILCA: International Livestock Center for Africa

INRZFH: Institut National de la Recherche Zootechnique, Forestière et Hydrobiologique

IRAT: Institut de Recherches Agronomiques Tropicales et des Cultures Vivrières

IRCT: Institut de Recherches du Coton et des Textiles Exotiques

LCV: Laboratoire Central Vétérinaire

MP/LHM: Ministère de la Production – Laboratoire Hydrobiologique de Mopti

MRE/CRZ: Ministry of Rural Economy – Centre de Recherches Zootechniques

OICMA: Organisation Internationale contre le Criquet Migrateur Africain

ONDY: Operation N'Dama Yanfolila

Sources:

0044 Institut du Sahel, and DEVRES, Inc. *Assessment of Agricultural Research Resources in the Sahel, Volume III: National Report Mali.* Washington, D.C.: DEVRES, Inc., August 1984.

0068 Conte, Stephan. "West Africa Agricultural Research Data Base Building Task: Termination Report." World Bank, Washington, D.C., 1985. Mimeo.

0384 Institut du Sahel, and DEVRES, Inc. *Sahel Region Agricultural Research Resource Assessment – Data Base Management Information System: Diskettes and User's Guide.* Printouts. Washington, D.C.: Institut du Sahel, and DEVRES, Inc., n.d.

0431 Current Agricultural Research Information System (CARIS). *Directory of Agricultural Research Institutions and Projects in West Africa – Pilot Project 1972–73.* Rome: FAO, 1973.

0532 UNESCO Field Science Office for Africa. *Survey on the Scientific and Technical Potential of the Countries of Africa.* Paris: UNESCO, 1970.

0589 Kassapu, Samuel. *Les Dépenses de Recherche Agricole dans 34 Pays d'Afrique Tropicale.* Paris: Centre de Développement de l'OCDE, 1976.

0653 Webster, B.N. *Index of Agricultural Research Institutions and Stations in Africa.* Rome: FAO, n.d.

Additional References:

0010 Boyce, James K., and Robert E. Evenson. *National and International Agricultural Research and Extension Programs.* New York: Agricultural Development Council, Inc., 1975.

0014 Judd, M. Ann, James K. Boyce, and Robert E. Evenson. "Investing in Agricultural Supply." Economic Growth Center, Yale University, New Haven, Connecticut, 1983. Mimeo.

0016 Oram, Peter A., and Vishva Bindlish. *Resource Allocations to National Agricultural Research: Trends in the 1970s.* The Hague and Washington, D.C.: ISNAR and IFPRI, November 1981.

0017 ISNAR, IFARD & AOAD. Survey of National Agricultural Research Systems: Unpublished Questionnaire Responses. ISNAR, The Hague, 1985.

0023 Bennell, Paul. *Agricultural Researchers in Sub-Saharan Africa: An Overview.* Working Paper No. 4. The Hague: ISNAR, October 1985.

0026 Oram, Peter A., and Mark Gieben. "Document Summaries." ISNAR, The Hague, 1984. Mimeo.

0027 Harvey, Nigel, ed. *Agricultural Research Centres: A World Directory of Organizations and Programmes.* Seventh Edition, Two Volumes. Harlow, U.K.: Longman, 1983.

0045 Institut du Sahel, and DEVRES, Inc. *Bilan des Ressources de la Recherche Agricole dans les Pays du Sahel, Volume I: Analyse et Stratégie Régionale.* Washington, D.C.: DEVRES, Inc., 1984.

0073 Oram, Peter A., and Vishva Bindlish. "Investment in Agricultural Research in Developing Countries: Progress, Problems, and the Determination of Priorities." IFPRI, Washington, D.C., January 1984. Mimeo.

0083 Institut du Sahel, and DEVRES, Inc. *Bilan des Ressources de la Recherche Agricole dans les Pays du Sahel, Volume II: Résumés des Rapports Nationaux.* Washington, D.C.: DEVRES, Inc., August 1984.

0086 World Bank. "West Africa Agricultural Research Review – Country Studies." World Bank, Washington, D.C., 1985. Mimeo.

0088 Swanson, Burton E., and Wade H. Reeves. "Agricultural Research West Africa: Manpower and Training." World Bank, Washington, D.C., November 1985. Mimeo.

0089 Oram, Peter. "Report on National Agricultural Research in West Africa." IFPRI, Washington, D.C., February 1986. Mimeo.

0095 FAO – CARIS. *Agricultural Research in Developing Countries – Volume 1: Research Institutions.* Rome: FAO – CARIS, 1978.

0144 Ministère de l'Agriculture. *Rapport Annuel de la Division de la Recherche Agronomique.* Mali: Ministère de l'Agriculture, 1983.

0145 AID. *Country Development Strategy Statement: Mali.* Washington, D.C.: AID, January 1983.

0163 CGIAR. "National Agricultural Research." CGIAR, Washington, D.C., 1985. Mimeo.

0165 Evenson, R.E., and Y. Kislev. *Agricultural Research and Productivity*. New Haven: Yale University Press, 1975.

0175 Cooper, St.G.C. *Agricultural Research in Tropical Africa*. Kampala: East African Literature Bureau, 1970.

0228 CIRAD. *Coopération Française en Recherche Agronomique: Activités du CIRAD dans les Pays du Sahel*. Paris: CIRAD, March 1985.

0258 "Effort Français en Faveur de la Recherche Agronomique dans les Pays du Sahel 1984–1985." 1985. Mimeo.

0266 UNESCO. *National Science Policies in Africa*. Science Policy Studies and Documents No. 31. Paris: UNESCO, 1974.

0360 Cooper, St.G.C. "Towards Trained Manpower for Agricultural Research in Africa." Paper presented at the Conference on Agricultural Research and Production in Africa, organized by the Association for the Advancement of Agricultural Sciences in Africa (AAASA), Addis Ababa, 29 August–4 September 1971.

0372 Gabas, Jean-Jacques. *La Recherche Agricole dans les Pays Membres du CILSS*. Club du Sahel, September 1986.

0373 Groupement d'Etudes et de Recherches pour le Développement de l'Agronomie Tropicale (GERDAT). *Rapport Général d'Activité pour 1982*. Paris: GERDAT, 1983. 0400 UNESCO. *The Promotion of Scientific Activity in Tropical Africa*. Science Policy Studies and Documents No. 11. Paris: UNESCO, 1969.

0466 World Bank – West Africa Agricultural Research Review Team. "National Agricultural Research Systems (NARS): A Regional Appraisal and Selected Issues." World Bank, Washington, D.C., August 1986. Mimeo.

0741 Casas, Joseph. "Compte Rendu de Mission au Mali (19–30 Octobre 1987)." ISNAR, The Hague, December 1987. Mimeo.

0852 Evenson, Robert E., and Yoav Kislev. *Investment in Agricultural Research and Extension: A Survey of International Data*. Center Discussion Paper No. 124. New Haven, Connecticut: Economic Growth Center, Yale University, August 1971.

0886 Evenson, Robert E., and Yoav Kislev. "Investment in Agricultural Research and Extension: A Survey of International Data." *Economic Development and Cultural Change* Vol. 23 (April 1975): 507–521.

Martinique

Personnel

Period/Year	PhD	MSc	BSc	Subtotal	Expat	Total	Source
1960–64	..	..	..	..	..	..	
1965–69	..	..	..	..	..	..	
1970	..	..	..	..	..	..	
1971	..	..	..	..	..	..	
1972	..	..	..	..	..	..	
1973	..	..	..	..	..	..	
1974	..	..	..	..	..	..	
1975	..	..	..	..	..	..	
1976	..	..	..	..	..	..	
1977	..	..	..	..	..	..	
1978	..	..	..	..	..	..	
1979	..	..	..	..	..	..	
1980	..	..	..	..	..	..	
1981	..	..	..	..	..	..	
1982	..	..	..	..	..	7	27
1983	..	..	..	..	..	..	
1984	..	..	..	..	..	..	
1985	..	..	..	..	..	..	
1986	..	..	..	..	..	..	

1960–64:   1965–69:

Expenditure   *LCU = French Francs*

Period/Year	Current LCU (millions)	Constant 1980 LCU (millions)	Constant 1980 US$ (millions)		Source
			Atlas	PPP	
1960–64	..	..	..	..	
1965–69	..	..	..	..	
1970	..	..	..	..	
1971	..	..	..	..	
1972	..	..	..	..	
1973	..	..	..	..	
1974	..	..	..	..	
1975	..	..	..	..	
1976	..	..	..	..	
1977	..	..	..	..	
1978	..	..	..	..	
1979	..	..	..	..	
1980	..	..	..	..	
1981	..	..	..	..	
1982	..	..	..	..	
1983	..	..	..	..	
1984	..	..	..	..	
1985	..	..	..	..	
1986	..	..	..	..	

1960–64:   1965–69:

296

Personnel Comments:
The 1982 figure includes 'Institut de Recherche Agronomiques Tropicales' (IRAT), 1 researcher, and 'Institut de Recherches sur les Fruits et Agrumes' (IFRA), 6 researchers.

Sources:
0027 Harvey, Nigel, ed. *Agricultural Research Centres: A World Directory of Organizations and Programmes.* Seventh Edition, Two Volumes. Harlow, U.K.: Longman, 1983.

Mauritania

Personnel

Period/Year	PhD	MSc	BSc	Subtotal	Expat	Total	Source
1960–64	..	..	..	..	..	..	
1965–69	..	..	..	..	..	4	653
1970	..	..	..	..	..	8	532
1971	..	..	..	..	..		
1972	..	..	..	..	..	5	431/589
1973	..	..	..	..	..	..	
1974	..	..	..	..	..	..	
1975	..	..	..	..	..	..	
1976	..	..	..	..	..	..	
1977	..	..	..	..	..	..	
1978	..	..	..	..	..	..	
1979	..	..	..	..	..	..	
1980	..	..	..	..	..	..	
1981	..	..	..	..	..	..	
1982	..	..	..	..	..	..	
1983	7	4	1	12	..	12	384
1984	..	..	..	..	..	..	
1985	..	..	..	..	..	..	
1986	..	..	..	..	..	..	

1960–64:   1965–69: 1966

Expenditure

LCU = Ouguiyas

Period/Year	Current LCU (millions)	Constant 1980 LCU (millions)	Constant 1980 US$ (millions)		Source
			Atlas	PPP	
1960–64	..	..	..	..	
1965–69	..	..	..	..	
1970	..	..	..	..	
1971	..	..	..	..	
1972	24.868	53.849	1.185	1.493	431
1973	..	..	..	..	
1974	..	..	..	..	
1975	..	..	..	..	
1976	..	..	..	..	
1977	..	..	..	..	
1978	..	..	..	..	
1979	..	..	..	..	
1980	..	..	..	..	
1981	..	..	..	..	
1982	24.000	18.072	0.398	0.501	128/142
1983	..	..	..	..	
1984	..	..	..	..	
1985	..	..	..	..	
1986	..	..	..	..	

1960–64:   1965–69:

Personnel Comments:

The table below gives an overview of the few scattered observations which could be obtained from the available sources.

	1966	1970	1972	1974	1980	1983	1985
IFAC[a]	2	5	3	3	–	–	–
CNRADA(IRAT)[b]	2	3	2	NA	8	6	7
CNERV[c]	–	–	–	NA	NA	6	NA
Subtotal	4	8	5	NA	NA	12	NA
CNROP(LPN)[d]	NA	1	NA	NA	NA	12	NA
Source	653	532	431/589	589	14	384	128

a. IFAC ceased operations in the mid-seventies and merged with CNRADA.
b. CNRADA was established in 1973 and took over the IRAT facilities in Kaedi.
c. In the literature no mention is made of the existence of veterinary or livestock research in Mauritania before the creation of CNERV in 1973.
d. In 1978 LPN was transformed into CNROP.

Expenditure Comments:

The table below gives an overview of construction details for the figures included in the indicator series.

	1966	1971	1972	1973	1975	1978
IFAC[a]	6.497	12.083	9.067	11.580	–	–
CNRADA(IRAT)[b]	NA	NA	15.814	NA	NA	7.524
CNERV[c]	–	–	–	–	13.000	NA
Subtotal	NA	NA	24.881	NA	NA	NA
Source	589	589	431	589	142	14

	1979	1980	1981	1982	1983	1984	1985
IFAC[a]	–	–	–	–	–	–	–
CNRADA(IRAT)[b]	6.55	5.000	7.000	14.000	14.000	14.000	14.004
CNERV[c]	NA	NA	NA	10.000*	NA	NA	NA
Subtotal	NA	NA	NA	24.000	NA	NA	NA
Source	14	128	128	128 142*	128	128	128

For footnotes see the table under personnel comments.

Acronyms and Abbreviations:

CNERV: Centre National de l'Elevage et de Recherche Vétérinaire

CNRADA: Centre National de Recherche Agronomique et de Développement Agricole

CNROP: Centre National de la Recherche Océanographique et des Pêches

IFAC: Institut Français de Recherches Fruitières d'Outre-Mer

IRAT: Institut de Recherches Agronomiques Tropicales et des Cultures Vivrières

LPN: Laboratoire des Pêches Nouadibhou

Sources:

0128 FAO. *Rapport d'une Mission de Consultant sur la Recherche Agronomique en Republique Islamique de Mauritanie.* Rome: FAO, 1985.

0142 Dolle, V., et al. *Etat de l'Agriculture Mauritanienne et Objectifs d'une Recherche pour son Développement.* Paris: Ministère des Relations Extérieures, May 1984.

0384 Institut du Sahel, and DEVRES, Inc. *Sahel Region Agricultural Research Resource Assessment – Data Base Management Information System: Diskettes and User's Guide.* Printouts. Washington, D.C.: Institut du Sahel, and DEVRES, Inc., n.d.

0431 Current Agricultural Research Information System (CARIS). *Directory of Agricultural Research Institutions and Projects in West Africa – Pilot Project 1972–73.* Rome: FAO, 1973.

0532 UNESCO Field Science Office for Africa. *Survey on the Scientific and Technical Potential of the Countries of Africa.* Paris: UNESCO, 1970.

0589 Kassapu, Samuel. *Les Dépenses de Recherche Agricole dans 34 Pays d'Afrique Tropicale.* Paris: Centre de Développement de l'OCDE, 1976.

0653 Webster, B.N. *Index of Agricultural Research Institutions and Stations in Africa.* Rome: FAO, n.d.

Additional References:

0014 Judd, M. Ann, James K. Boyce, and Robert E. Evenson. "Investing in Agricultural Supply." Economic Growth Center, Yale University, New Haven, Connecticut, 1983. Mimeo.

0016 Oram, Peter A., and Vishva Bindlish. *Resource Allocations to National Agricultural Research: Trends in the 1970s.* The Hague and Washington, D.C.: ISNAR and IFPRI, November 1981.

0023 Bennell, Paul. *Agricultural Researchers in Sub-Saharan Africa: An Overview.* Working Paper No. 4. The Hague: ISNAR, October 1985.

0027 Harvey, Nigel, ed. *Agricultural Research Centres: A World Directory of Organizations and Programmes.* Seventh Edition, Two Volumes. Harlow, U.K.: Longman, 1983.

0045 Institut du Sahel, and DEVRES, Inc. *Bilan des Ressources de la Recherche Agricole dans les Pays du Sahel, Volume I: Analyse et Stratégie Régionale.* Washington, D.C.: DEVRES, Inc., 1984.

0068 Conte, Stephan. "West Africa Agricultural Research Data Base Building Task: Termination Report." World Bank, Washington, D.C., 1985. Mimeo.

0073 Oram, Peter A., and Vishva Bindlish. "Investment in Agricultural Research in Developing Countries: Progress, Problems, and the Determination of Priorities." IFPRI, Washington, D.C., January 1984. Mimeo.

0083 Institut du Sahel, and DEVRES, Inc. *Bilan des Ressources de la Recherche Agricole dans les Pays du Sahel, Volume II: Résumés des Rapports Nationaux.* Washington, D.C.: DEVRES, Inc., August 1984.

0086 World Bank. "West Africa Agricultural Research Review – Country Studies." World Bank, Washington, D.C., 1985. Mimeo.

0088 Swanson, Burton E., and Wade H. Reeves. "Agricultural Research West Africa: Manpower and Training." World Bank, Washington, D.C., November 1985. Mimeo.

0089 Oram, Peter. "Report on National Agricultural Research in West Africa." IFPRI, Washington, D.C., February 1986. Mimeo.

0143 Institut du Sahel, and DEVRES, Inc. *Assessment of Agricultural Research Resources in the Sahel, Volume III: National Report Mauritania.* Washington, D.C.: DEVRES, Inc., August 1984.

0163 CGIAR. "National Agricultural Research." CGIAR, Washington, D.C., 1985. Mimeo.

0175 Cooper, St.G.C. *Agricultural Research in Tropical Africa.* Kampala: East African Literature Bureau, 1970.

0228 CIRAD. *Cooperation Française en Recherche Agronomique: Activités du CIRAD dans les Pays du Sahel.* Paris: CIRAD, March 1985.

0258 "Effort Français en Faveur de la Recherche Agronomique dans les Pays du Sahel 1984–1985." 1985. Mimeo.

300

0266 UNESCO. *National Science Policies in Africa.* Science Policy Studies and Documents No. 31. Paris: UNESCO, 1974.

0360 Cooper, St.G.C. "Towards Trained Manpower for Agricultural Research in Africa." Paper presented at the Conference on Agricultural Research and Production in Africa, organized by the Association for the Advancement of Agricultural Sciences in Africa (AAASA), Addis Ababa, 29 August–4 September 1971.

0372 Gabas, Jean-Jacques. *La Recherche Agricole dans les Pays Membres du CILSS.* Club du Sahel, September 1986.

0373 Groupement d'Etudes et de Recherches pour le Développement de l'Agronomie Tropicale (GERDAT). *Rapport Général d'Activité pour 1982.* Paris: GERDAT, 1983.

0466 World Bank – West Africa Agricultural Research Review Team. "National Agricultural Research Systems (NARS): A Regional Appraisal and Selected Issues." World Bank, Washington, D.C., August 1986. Mimeo.

0852 Evenson, Robert E., and Yoav Kislev. *Investment in Agricultural Research and Extension: A Survey of International Data.* Center Discussion Paper No. 124. New Haven, Connecticut: Economic Growth Center, Yale University, August 1971.

Mauritius

Personnel

Period/Year	PhD	MSc	BSc	Subtotal	Expat	Total	Source
1960–64	..	..	..	..	..	28	10
1965–69	..	..	..	..	..	36	10
1970	..	..	..	..	..	44	532
1971	..	..	..	..	..	51	10
1972	..	..	..	..	..	52	10
1973	..	..	..	..	..	61	10
1974	..	..	..	..	..	..	
1975	..	..	..	..	..	..	
1976	..	..	..	..	..	89	95
1977	..	..	..	..	..	..	
1978	..	..	..	..	..	..	
1979	..	..	..	..	..	..	
1980	13	22	37	72	..	72	185/242
1981	..	..	..	..	..	94	185
1982	..	..	..	..	..	96	185/27
1983	..	..	..	..	..	95	432
1984	–30–		78	108	..	108	445
1985	..	..	..	..	..	109	801
1986	..	..	..	..	..	120	801

1960–64: 1961   1965–69: 1966

Expenditure   *LCU = Mauritian Rupees*

Period/Year	Current LCU (millions)	Constant 1980 LCU (millions)	Constant 1980 US$ (millions)		Source
			Atlas	PPP	
1960–64	..	..	..	..	
1965–69	5.000	27.174	3.607	4.438	175
1970	..	..	..	..	
1971	6.060	27.297	3.623	4.458	589
1972	..	..	..	..	
1973	..	..	..	..	
1974	..	..	..	..	
1975	..	..	..	..	
1976	13.660	26.890	3.569	4.391	17/95
1977	..	..	..	..	
1978	..	..	..	..	
1979	..	..	..	..	
1980	..	..	..	..	
1981	..	..	..	..	
1982					
1983	45.184	35.081	4.656	5.729	17
1984	46.080	32.985	4.378	5.387	801
1985	50.052	33.819	4.489	5.523	801
1986	57.963	..	..	..	801

1960–64:   1965–69: 1966

Personnel Comments:

Agricultural research in Mauritius is carried out by several divisions of the Ministry of Agriculture, Natural Resources and the Environment (MANRE) and by the Mauritius Sugar Industries Research Institute (MSIRI). In addition, some agricultural research is undertaken by the School of Agriculture, the Forestry Department (1983: 6 researchers, source 432), the Fisheries Department (1983: 5 researchers, source 432), and some parastatal bodies. The indicator series, however, include only the Agricultural Services or Department of Agriculture of MANRE and MSIRI. There are some significant classification problems with the personnel figures. In the case of the Department of Agriculture, all professional staff seem to be included in the figures, including those who do not hold a research position but who are more service oriented (although they hold a title as scientific officer). Sources 244 and 243 give considerably lower estimates of the number of researchers within the Department of Agriculture. For reasons of consistency over time, the series include figures which may overestimate the research personnel capacity of the Department of Agriculture. In the case of MSIRI there were inconsistencies in the way technical personnel with a university degree had been classified. In some cases they seem to be included as researchers, in others not.

The table below gives a breakdown of the number of researchers at the institute level.

	1970	1976	1977	1980	1981	1982	1983	1984	1985	1986
MANRE	24	55	14	56	72	69	73	83	81	86
MSIRI	20	34	25	16*	[22]	27*	22	25	28	34
Total	44	89	39a	72	94	96	95	108	109	120
Source	532	95	243	185 242*	185	185 27*	432	445	801	801

[ ] = estimated or constructed by authors.
a. Not included in the indicator series because the MANRE figure appears to be too low.

	PhD	MSc	BSc	Total	Source
1983	9	7	0	16	17
1984	10	6	2	18	801
1985	10	6	2	18	801
1986	10	7	1	18	801
1986	10	7	1	18	801

Sources 17 and 801 provide the staffing figures for the School of Agriculture of the University of Mauritius, as shown opposite.

No indication is given as to how much time university staff spend on research.

Expenditure Comments:

The table below gives an institutional breakdown of the expenditure figures in current LCUs.

	1966	1966	1971	1976	1983	1984	1985	1986
MANRE	NA	NA	3.400	5.110	25.072	26.000	27.000	30.000
MSIRI	2.136	NA	2.660	8.550*	20.112	20.080	23.052	27.963
Total	NA	5.000	6.060	13.660	45.184	46.080	50.052	57.963
Source	589	175	589	17/95*	17	801	801	801

Various sources (10 and 95) report higher expenditure figures for some of the years reported here. According to information contained in source 589, these may often represent total MANRE expenditures, inclusive of extension, animal health, and other nonresearch functions.

Research expenditures by the School of Agriculture are presented in the table below:

	1983	1984	1985	1986	1987
	3.867[a]	1.000	1.000	1.000	1.200
Source	17	801	801	801	801

a. This figure probably represents total expenditures by the School of Agriculture. When compared with the other years, it suggests that roughly 25% of the total School of Agriculture expenditures are research related.

Acronyms and Abbreviations:
MANRE : Ministry of Agriculture, Natural Resources and the Environment
MSIRI : Mauritius Sugar Industry Research Institute

Sources:
0010 Boyce, James K., and Robert E. Evenson. *National and International Agricultural Research and Extension Programs.* New York: Agricultural Development Council, Inc., 1975.

0017 ISNAR, IFARD & AOAD. Survey of National Agricultural Research Systems: Unpublished Questionnaire Responses. ISNAR, The Hague, 1985.

0027 Harvey, Nigel, ed. *Agricultural Research Centres: A World Directory of Organizations and Programmes.* Seventh Edition, Two Volumes. Harlow, U.K.: Longman, 1983.

0095 FAO – CARIS. *Agricultural Research in Developing Countries – Volume 1: Research Institutions.* Rome: FAO – CARIS, 1978.

0175 Cooper, St.G.C. *Agricultural Research in Tropical Africa.* Kampala: East African Literature Bureau, 1970.

0185 Ministry of Agriculture and Natural Resources and the Environment. *Annual Report of the Ministry of Agriculture and Natural Resources and the Environment for the Year 1980, 1981, and 1982.* Mauritius: Ministry of Agriculture and Natural Resources and the Environment, various years.

0242 Commonwealth Agricultural Bureaux (CAB). *List of Research Workers in Agricultural Sciences in the Commonwealth 1981.* Slough, England: CAB, 1981.

0432 FAO. "Trained Agricultural Manpower Assessment in Africa – Country Report Mauritius." FAO, Rome, January 1984. Mimeo.

0445 Swanson, Burton E., and Wade H. Reeves. "Agricultural Research Eastern and Southern Africa: Manpower and Training." World Bank, Washington, D.C., August 1986. Mimeo.

0532 UNESCO Field Science Office for Africa. *Survey on the Scientific and Technical Potential of the Countries of Africa.* Paris: UNESCO, 1970.

0801 Owadally, A.L. Personal communication. Agricultural Services – Ministry of Agriculture, Fisheries, and Natural Resources. Reduit, Mauritius, February 1988.

Additional References:
0014 Judd, M. Ann, James K. Boyce, and Robert E. Evenson. "Investing in Agricultural Supply." Economic Growth Center, Yale University, New Haven, Connecticut, 1983. Mimeo.

0023 Bennell, Paul. *Agricultural Researchers in Sub-Saharan Africa: An Overview.* Working Paper No. 4. The Hague: ISNAR, October 1985.

0165 Evenson, R.E., and Y. Kislev. *Agricultural Research and Productivity.* New Haven: Yale University Press, 1975.

0184 Mauritius Sugar Industry Research Institute (MSIRI). *Annual Report 1982.* Mauritius: MSIRI, 1983.

0243 Commonwealth Agricultural Bureaux (CAB). *List of Research Workers in the Agricultural Sciences in the Commonwealth and in the Republic of Ireland 1978.* Slough, England: CAB, 1978.

0244 Commonwealth Agricultural Bureaux (CAB). *List of Research Workers in the Agricultural Sciences in the Commonwealth and in the Republic of Ireland 1975.* Slough, England: CAB, 1975.

0266 UNESCO. *National Science Policies in Africa.* Science Policy Studies and Documents No. 31. Paris: UNESCO, 1974.

0279 Commonwealth Agricultural Bureaux (CAB). *List of Research Workers in the Agricultural Sciences in the Commonwealth and in the Republic of Ireland 1972.* Slough, England: CAB, 1972.

0286 Commonwealth Agricultural Bureaux (CAB). *List of Research Workers in Agriculture, Animal Health and Forestry in the Commonwealth and in the Republic of Ireland 1966.* Slough, England: CAB, 1966.

0287 Commonwealth Agricultural Bureaux (CAB). *List of Research Workers in Agriculture, Animal Health and Forestry in the British*

Commonwealth, the Republic of Sudan and the Republic of Ireland 1959. Slough, England: CAB, 1959.

0360 Cooper, St.G.C. "Towards Trained Manpower for Agricultural Research in Africa." Paper presented at the Conference on Agricultural Research and Production in Africa, organized by the Association for the Advancement of Agricultural Sciences in Africa (AAASA), Addis Ababa, 29 August–4 September 1971.

0589 Kassapu, Samuel. Les Dépenses de Recherche Agricole dans 34 Pays d'Afrique Tropicale. Paris: Centre de Développement de l'OCDE, 1976.

0653 Webster, B.N. Index of Agricultural Research Institutions and Stations in Africa. Rome: FAO, n.d.

0744 Agricultural Research Centres: A World Directory of Organizations and Programmes. Eighth Edition, Two Volumes. Harlow, U.K.: Longman, 1986.

0852 Evenson, Robert E., and Yoav Kislev. Investment in Agricultural Research and Extension: A Survey of International Data. Center Discussion Paper No. 124. New Haven, Connecticut: Economic Growth Center, Yale University, August 1971.

0866 University of Mauritius – School of Agriculture. Research Programme 1982–83. Reduit: University of Mauritius, May 1982.

0886 Evenson, Robert E., and Yoav Kislev. "Investment in Agricultural Research and Extension: A Survey of International Data." Economic Development and Cultural Change Vol. 23 (April 1975): 507–521.

Mexico

Personnel

Period/Year	PhD	MSc	BSc	Subtotal	Expat	Total	Source
1960–64	..	..	..	..	..	192	10
1965–69	..	..	..	..	..	242	10/214
1970	..	..	..	..	..	235	26
1971	..	..	..	..	..	359	10
1972	..	..	..	..	..	474	10
1973	..	..	..	..	..	..	
1974	..	..	..	..	..	..	
1975	..	..	..	..	..	499	49
1976	..	..	..	..	..	..	
1977	52	120	409	581	..	581	49
1978	57	156	521	734	..	734	49
1979	61	197	642	900	..	900	49
1980	72	210	692	974	..	974	49
1981	79	231	696	1006	..	1006	49
1982	76	235	799	1110	..	1110	49
1983	..	..	..	..	..	..	
1984	..	..	..	..	..	..	
1985	..	..	..	..	..	..	
1986	..	..	..	..	..	..	

1960–64: 1962 & 64   1965–69: 1966 & 68

Expenditure

LCU = Mexican Pesos

Period/Year	Current LCU (millions)	Constant 1980 LCU (millions)	Constant 1980 US$ (millions)		Source
			Atlas	PPP	
1960–64	24.400	171.379	6.448	12.050	49
1965–69	24.700	147.803	5.561	10.392	49
1970	51.000	258.883	9.740	18.202	49
1971	56.000	269.231	10.129	18.930	49
1972	79.000	357.466	13.449	25.133	49
1973	128.000	512.000	19.263	35.999	49
1974	192.000	625.407	23.530	43.972	49
1975	320.000	901.408	33.914	63.378	49
1976	350.000	823.529	30.984	57.902	49
1977	521.000	940.433	35.382	66.122	49
1978	612.000	947.368	35.643	66.609	49
1979	985.000	1267.696	47.695	89.132	49
1980	1796.000	1796.000	67.571	126.277	49
1981	2647.000	2079.340	78.232	146.198	49
1982	3181.000	1550.951	58.352	109.047	49
1983	..	..	..	..	
1984	..	..	..	..	
1985	..	..	..	..	
1986	..	..	..	..	

1960–64: 1961, 62, 63 & 64   1965–69: 1965, 66, 67, 68 & 69

Personnel Comments:

These figures refer to the 'Instituto Nacional de Investigaciones Agrícolas' (INIA) only. According to source 26, the following institutes also execute agricultural research: INIP (300 researchers in 1981), INIF (134 researchers in 1981), INMECAFE (23 researchers in 1981), Sugar Production Improvement Institute, Monterrey Institute of Technology, and National Institute of Biotic Resources. No figures are available for these institutes for years other than 1981, so they have not been included in the indicator series. In several sources (e.g., source 10) it is stated that INIA represents approximately two-thirds of the National Agricultural Research System.

Expenditure Comments:

The expenditure figures refer to 'Instituto Nacional de Investigaciones Agrícolas' (INIA) only. Agricultural research expenditures by institutes other than INIA are not negligible. For 1981, source 26 gives the following expenditure figures:

	(million US$)
INIA	85.808
INIP	26.360
INIF	22.590
INMECAFE	0.880

Acronyms and Abbreviations:

INIA: Instituto Nacional de Investigaciones Agrícolas
INIF: Instituto Nacional de Investigaciones Forestales
INIP: Instituto Nacional de Investigaciones Pecuarias
INMECAFE : Instituto Mexicano de Café

Sources:

0010 Boyce, James K., and Robert E. Evenson. *National and International Agricultural Research and Extension Programs.* New York: Agricultural Development Council, Inc., 1975.

0026 Oram, Peter A., and Mark Gieben. "Document Summaries." ISNAR, The Hague, 1984. Mimeo.

0049 Matus, Jaime G., et al. "A Review of the Collaboration between the National Agricultural Research System and the International Research Centers of the CGIAR."

Centro de Economia, Colegio de Postgraduados, Chapingo, Mexico, December 1984. Mimeo.

0214 Venezian, Eduardo L., and William K. Gamble. *The Agricultural Development of Mexico – Its Structure and Growth Since 1950.* New York: Praeger Publishers, 1969.

Additional References:

0001 Trigo, Eduardo J., and Martín E. Piñeiro. "Funding Agricultural Research." In *Report of a Conference: Selected Issues in Agricultural Research in Latin America,* edited by Barry Nestel and Eduardo J. Trigo. The Hague: ISNAR, March 1984.

0014 Judd, M. Ann, James K. Boyce, and Robert E. Evenson. "Investing in Agricultural Supply." Economic Growth Center, Yale University, New Haven, Connecticut, 1983. Mimeo.

0016 Oram, Peter A., and Vishva Bindlish. *Resource Allocations to National Agricultural Research: Trends in the 1970s.* The Hague and Washington, D.C.: ISNAR and IFPRI, November 1981.

0027 Harvey, Nigel, ed. *Agricultural Research Centres: A World Directory of Organizations and Programmes.* Seventh Edition, Two Volumes. Harlow, U.K.: Longman, 1983.

0053 Piñeiro, Martín, and Eduardo Trigo, eds. *Cambio Técnico en el Agro Latino Americano – Situación y Perspectivas en la Década de 1980.* San José: IICA, 1983.

0073 Oram, Peter A., and Vishva Bindlish. "Investment in Agricultural Research in Developing Countries: Progress, Problems, and the Determination of Priorities." IFPRI, Washington, D.C., January 1984. Mimeo.

0163 CGIAR. "National Agricultural Research." CGIAR, Washington, D.C., 1985. Mimeo.

0165 Evenson, R.E., and Y. Kislev. *Agricultural Research and Productivity.* New Haven: Yale University Press, 1975.

0172 Trigo, Eduardo. Background material. ISNAR, The Hague, n.d.

0213 Centro Internacional de Investigación para el Desarrollo. "Asignación de Recursos para Investigación Agrícola en América Latina." Proyecto ARIAL Brasil – Estudio de Caso. Unidad de Apoyo a la Programación, April 1979. Mimeo.

0740 Sands, Carolyn Marie. "The Theoretical and Empirical Basis for Analyzing Agricultural Technology Systems." PhD thesis, University of Illinois, Urbana-Champaign, Illinois, 1988.

0852 Evenson, Robert E., and Yoav Kislev. *Investment in Agricultural Research and Extension: A Survey of International Data.* Center Discussion Paper

No. 124. New Haven, Connecticut: Economic Growth Center, Yale University, August 1971.

0886 Evenson, Robert E., and Yoav Kislev. "Investment in Agricultural Research and Extension: A Survey of International Data." *Economic Development and Cultural Change* Vol. 23 (April 1975): 507–521.

Montserrat

Personnel

Period/Year	PhD	MSc	BSc	Subtotal	Expat	Total	Source
1960–64	..	..	..	..	..	..	
1965–69	..	..	..	..	..	..	
1970	..	..	..	..	..	..	
1971	..	..	..	..	..	..	
1972	..	..	..	..	..	..	
1973	..	..	..	..	..	..	
1974	..	..	..	..	..	..	
1975	..	..	..	..	..	..	
1976	..	..	..	..	..	..	
1977	..	..	..	..	..	..	
1978	..	..	..	..	..	..	
1979	..	..	..	..	..	..	
1980	0	0	3	3	..	3	343
1981	..	..	..	4	..	4	342
1982	0	1	2	3	..	3	355
1983	..	..	..	..	..	..	
1984	0	1	1	2	..	2	349
1985	..	..	..	..	..	..	
1986	..	..	..	..	..	..	

1960–64:   1965–69:

Expenditure   *LCU = East Caribbean Dollars*

Period/Year	Current LCU (millions)	Constant 1980 LCU (millions)	Constant 1980 US$ (millions)		Source
			Atlas	PPP	
1960–64	..	..	..	..	
1965–69	..	..	..	..	
1970	..	..	..	..	
1971	..	..	..	..	
1972	..	..	..	..	
1973	..	..	..	..	
1974	..	..	..	..	
1975	..	..	..	..	
1976	..	..	..	..	
1977	..	..	..	..	
1978	..	..	..	..	
1979	..	..	..	..	
1980	..	..	..	..	
1981	..	..	..	..	
1982	..	..	..	..	
1983	..	..	..	..	
1984	..	..	..	..	
1985	..	..	..	..	
1986	..	..	..	..	

1960–64:   1965–69:

Personnel Comments:
These figures represent the Caribbean Agricultural Research and Development Institute (CARDI) unit at Montserrat only. There are no other organizations at Montserrat executing agricultural research.

Expenditure Comments:
Source 115 reports a cash contribution by the Government of Montserrat to the operations of CARDI of EC$ 40,000 in 1983.

Sources:
0342 Caribbean Agricultural Research and Development Institute (CARDI). *CARDI 1976–1981: Report of the Chairman.* St. Augustine, Trinidad & Tobago: CARDI, 1982.
0343 Caribbean Agricultural Research and Development Institute (CARDI). *Annual Report 1980.* St. Augustine, Trinidad & Tobago: CARDI, 1981.
0349 Caribbean Agricultural Research and Development Institute (CARDI). *Annual Report 1983–1984 – Leeward Islands Unit.* CARDI, 1984.
0355 Caribbean Agricultural Research and Development Institute (CARDI). *Research and Development Summary 1982–83.* St. Augustine, Trinidad: CARDI, 1983.

Additional References:
0115 Economic Commission for Latin America and the Caribbean (ECLAC). *Agricultural Research Policy and Management – Volumes I and II.* Papers presented at the Workshop on Agricultural Research Policy and Management, 26–30 September 1983, Port of Spain, Trinidad. Port of Spain, Trinidad: ECLAC, November 1984.
0242 Commonwealth Agricultural Bureaux (CAB). *List of Research Workers in Agricultural Sciences in the Commonwealth 1981.* Slough, England: CAB, 1981.
0243 Commonwealth Agricultural Bureaux (CAB). *List of Research Workers in the Agricultural Sciences in the Commonwealth and in the Republic of Ireland 1978.* Slough, England: CAB, 1978.
0286 Commonwealth Agricultural Bureaux (CAB). *List of Research Workers in Agriculture, Animal Health and Forestry in the Commonwealth and in the Republic of Ireland 1966.* Slough, England: CAB, 1966.
0287 Commonwealth Agricultural Bureaux (CAB). *List of Research Workers in Agriculture, Animal Health and Forestry in the British Commonwealth, the Republic of Sudan and the Republic of Ireland 1959.* Slough, England: CAB, 1959.
0353 Caribbean Agricultural Research and Development Institute (CARDI). *Paper Presented at a Meeting of the Governing Body of CARDI on January 29, 1982, Basseterre, St. Kitts.* St. Augustine, Trinidad: CARDI, 1982.
0354 Caribbean Agricultural Research and Development Institute (CARDI). *Annual Report Research and Development 1983–84: Highlights.* St. Augustine, Trinidad: CARDI, 1984.
0356 Caribbean Agricultural Research and Development Institute (CARDI). *Research and Development Summary 1983–84.* St. Augustine, Trinidad: CARDI, 1984.

Morocco

Personnel

Period/Year	PhD	MSc	BSc	Subtotal	Expat	Total	Source
1960–64	..	..	..	14	113	127	52
1965–69	..	..	..	27	76	103	52
1970	..	..	..	42	82	124	52
1971	..	..	..	48	82	130	52
1972	..	..	..	52	71	123	52
1973	..	..	..	48	53	101	52
1974	..	..	..	58	40	98	52
1975	..	..	..	68	35	103	52
1976	..	..	..	84	53	137	52
1977	..	..	..	103	49	152	52
1978	..	..	..	..		..	
1979	..	..	..	..		..	
1980	–121–		143	264	..	264	838
1981	–109–		133	242	..	242	838
1982	..	..		198	14	212	52
1983	4	96	126	226	4	230	17
1984	–95–		117	212	..	212	838
1985	–88–		103	191	..	191	838
1986	–85–		93	178	..	178	838

1960–64: 1963 & 64   1965–69: 1965, 66, 67, 68 & 69

Expenditure

LCU = Moroccan Dirhams

Period/Year	Current LCU (millions)	Constant 1980 LCU (millions)	Constant 1980 US$ (millions) Atlas	Constant 1980 US$ (millions) PPP	Source
1960–64	..	..	..	..	
1965–69	..	..	..	..	
1970	..	..	..	..	
1971	..	..	..	..	
1972	..	..	..	..	
1973	..	..	..	..	
1974	35.575	51.335	12.893	16.881	26
1975	..	..	..	..	
1976	..	..	..	..	
1977	44.457	54.953	13.801	18.071	26
1978	..	..	..	..	
1979	..	..	..	..	
1980	..	..	..	..	
1981	67.249	60.859	15.285	20.013	26
1982	..	..	..	..	
1983	92.000	72.670	18.251	23.896	17
1984	134.000	96.472	24.229	31.724	52
1985	..	..	..	..	
1986	133.805	..	..	..	838

1960–64:   1965–69:

Personnel Comments:

These figures refer to 'Direction de la Recherche Agronomique' (DRA) of the Ministry of Agriculture, or to 'Institut National de la Recherche Agronomique' (INRA). INRA, for the first time established in 1962, was reestablished in 1981, after it had been transferred to DRA in 1966. According to source 884, in 1985 or thereabouts, approximately 60% of all agricultural research in Morocco was done by INRA, 30% by 'Institut Agronomique et Vétérinaire Hassan II' (1982: 200 graduate research staff, source 27; 1985: 175 researchers or thereabouts according to source 418), and 10% by the universities.

Source 532 reports 8 researchers at 'Station de Recherches Forestières' and 4 at 'Institut des Pêches Maritimes du Maroc' for the year 1970.

Source 27 reports 15 researchers at 'Station de Recherches Forestières' in 1982.

Expenditure Comments:

The expenditure figures refer to 'Direction de la Recherche Agronomique' (DRA) of the Ministry of Agriculture or to 'Institut National de la Recherche Agronomique' (INRA) only.

According to source 418, the operating budget of 'Institut Agronomique et Veterinaire Hassan II' was 27 million DH in 1985 or thereabouts, while INRA had an operating budget of 112 million DH.

Sources:

0017 ISNAR, IFARD & AOAD. Survey of National Agricultural Research Systems: Unpublished Questionnaire Responses. ISNAR, The Hague, 1985.

0026 Oram, Peter A., and Mark Gieben. "Document Summaries." ISNAR, The Hague, 1984. Mimeo.

0052 ISNAR. *L'Institut National de la Recherche Agronomique du Maroc.* The Hague: ISNAR, January 1984.

0838 Khettouch, Moha, Khadija Rajraji, and Abderraouf Elaouni. *Données sur les Moyens Humains de l'Institut National de la Recherche Agronomique.* Rabat, Morocco: Institut National de la Recherche Agronomique, 1986.

Additional References:

0010 Boyce, James K., and Robert E. Evenson. *National and International Agricultural Research and Extension Programs.* New York: Agricultural Development Council, Inc., 1975.

0014 Judd, M. Ann, James K. Boyce, and Robert E. Evenson. "Investing in Agricultural Supply." Economic Growth Center, Yale University, New Haven, Connecticut, 1983. Mimeo.

0016 Oram, Peter A., and Vishva Bindlish. *Resource Allocations to National Agricultural Research: Trends in the 1970s.* The Hague and Washington, D.C.: ISNAR and IFPRI, November 1981.

0027 Harvey, Nigel, ed. *Agricultural Research Centres: A World Directory of Organizations and Programmes.* Seventh Edition, Two Volumes. Harlow, U.K.: Longman, 1983.

0073 Oram, Peter A., and Vishva Bindlish. "Investment in Agricultural Research in Developing Countries: Progress, Problems, and the Determination of Priorities." IFPRI, Washington, D.C., January 1984. Mimeo.

0095 FAO – CARIS. *Agricultural Research in Developing Countries – Volume 1: Research Institutions.* Rome: FAO – CARIS, 1978.

0165 Evenson, R.E., and Y. Kislev. *Agricultural Research and Productivity.* New Haven: Yale University Press, 1975.

0175 Cooper, St.G.C. *Agricultural Research in Tropical Africa.* Kampala: East African Literature Bureau, 1970.

0266 UNESCO. *National Science Policies in Africa.* Science Policy Studies and Documents No. 31. Paris: UNESCO, 1974.

0339 Institut National de la Recherche Agronomique (INRA). "Eléments de Réponse aux Questions Demandées par l'ISNAR." INRA, n.d. Mimeo.

0340 ISNAR. *The National Institute for Agricultural Research of Morocco – Volumes I and II.* The Hague: ISNAR, April 1986.

0360 Cooper, St.G.C. "Towards Trained Manpower for Agricultural Research in Africa." Paper presented at the Conference on Agricultural Research and Production in Africa, organized by the Association for the Advancement of Agricultural Sciences in Africa (AAASA), Addis Ababa, 29 August–4 September 1971.

0407 Centre International de Hautes Etudes Agronomiques Méditerranéennes. "Séminaire sur l'Orientation et l'Organisation de la Recherche Agronomique dans les Pays du Bassin Méditerranéen, Istanbul, 1–3 December 1986." Centre International de Hautes Etudes Agronomiques Méditerranéennes, Paris, December 1986. Mimeo.

0418 Lazzaoui, Mohamed. "Le Système de Recherche Agronomique Marocain." Paper presented at the Séminaire sur l'Orientation et l'Organisation de la Recherche Agronomique dans les Pays du

312

Bassin Méditerranéen, Centre International de Hautes Etudes Agronomiques Méditerranéennes, Istanbul, 1–3 December 1986.

0422 Casas, Joseph. "Les Ressources Humaines et Financières dans les Pays du Bassin Méditerranéen." Preliminary table. ISNAR, The Hague, 1986. Mimeo.

0532 UNESCO Field Science Office for Africa. *Survey on the Scientific and Technical Potential of the Countries of Africa.* Paris: UNESCO, 1970.

0653 Webster, B.N. *Index of Agricultural Research Institutions and Stations in Africa.* Rome: FAO, n.d.

0654 Office de la Recherche Scientifique et Technique Outre-Mer (ORSTOM). *Rapport d'Activité 1967.* Paris: ORSTOM, 1969.

0852 Evenson, Robert E., and Yoav Kislev. *Investment in Agricultural Research and Extension: A Survey of International Data.* Center Discussion Paper No. 124. New Haven, Connecticut: Economic Growth Center, Yale University, August 1971.

0861 Institut National de la Recherche Agronomique (INRA). *Prévisions Budgetaires – Budget de Fonctionnement de l'Exercice 1984.* Rabat, Morocco: INRA, Octobre 1983.

0865 Oram, Peter. "Agricultural Research Objectives and Priorities: Constraints to the Development of Agricultural Research Institutions in Arab countries." Review paper prepared for the first meeting of the Council for Arab Agricultural Research, organized by the Arab Fund for Economic and Social Development, Kuwait, 23 March 1988.

0884 Devred, R. Personal communication. Senior Research Officer, ISNAR. The Hague, 1986.

0886 Evenson, Robert E., and Yoav Kislev. "Investment in Agricultural Research and Extension: A Survey of International Data." *Economic Development and Cultural Change* Vol. 23 (April 1975): 507–521.

Mozambique

Personnel

Period/Year	PhD	MSc	BSc	Subtotal	Expat	Total	Source
1960–64	..	..	..	..		..	
1965–69	..	..	..	..		42	852
1970	..	..	..	..		21	10
1971	..	..	..	..		26	10
1972	..	..	..	..		23	10
1973	..	..	..	..			
1974	..	..	..	..			
1975	..	..	..	..			
1976	..	..	..	..			
1977	..	..	..	..			
1978	..	..	..	..			
1979	..	..	..	..			
1980	..	..	..	..		..	
1981	..	..	..	..		..	
1982	..	..	..	..		..	
1983	0	0	13	13	63	76	23
1984	0	0	11	11	66	77	6
1985	0	0	14	14	63	77	445
1986	..	..	..	..		..	

1960–64:   1965–69: 1967

Expenditure   *LCU = Meticals*

Period/Year	Current LCU (millions)	Constant 1980 LCU (millions)	Constant 1980 US$ (millions)		Source
			Atlas	PPP	
1960–64	..	..	..	..	
1965–69	36.225	108.755	3.089	5.968	852
1970	49.000	144.256	4.097	7.916	852
1971	48.900	126.950	3.605	6.966	852
1972	53.300	126.545	3.594	6.944	10
1973	..	..	..	..	
1974	..	..	..	..	
1975	..	..	..	..	
1976	..	..	..	..	
1977	..	..	..	..	
1978	..	..	..	..	
1979	..	..	..	..	
1980	..	..	..	..	
1981	..	..	..	..	
1982	..	..	..	..	
1983	..	..	..	..	
1984	..	..	..	..	
1985	..	..	..	..	
1986	..	..	..	..	

1960–64:   1965–69: 1967, 68 & 69

314

Personnel Comments:

The 1984 figure includes the following organizations: National Agricultural Research Institute, Center for Forestry Research, Institute of Animal Breeding and Reproduction, Secretariat of State for Cashew, National Sugar Institute, National Institute for Veterinary Research, Secretariat of State for Cotton, Rural Development Centers, and Citrus Directorate. For other observations, no indication is given as to which organizations are included.

Expenditure Comments:

No information is given by source 10 as to which organizations are included in these expenditure figures.

Sources:

0006 SADCC, and DEVRES, Inc. *Agricultural Research Resource Assessment in the SADCC Countries, Volume II: Country Report Mozambique.* Washington, D.C.: DEVRES, Inc., January 1985.

0010 Boyce, James K., and Robert E. Evenson. *National and International Agricultural Research and Extension Programs.* New York: Agricultural Development Council, Inc., 1975.

0023 Bennell, Paul. *Agricultural Researchers in Sub-Saharan Africa: An Overview.*

Working Paper No. 4. The Hague: ISNAR, October 1985.

0445 Swanson, Burton E., and Wade H. Reeves. "Agricultural Research Eastern and Southern Africa: Manpower and Training." World Bank, Washington, D.C., August 1986. Mimeo.

0852 Evenson, Robert E., and Yoav Kislev. *Investment in Agricultural Research and Extension: A Survey of International Data.* Center Discussion Paper No. 124. New Haven, Connecticut: Economic Growth Center, Yale University, August 1971.

Additional References:

0002 SADCC, and DEVRES, Inc. *Agricultural Research Resource Assessment in the SADCC Countries, Volume I: Regional Analysis and Strategy.* Washington, D.C.: DEVRES, Inc., January 1985.

0165 Evenson, R.E., and Y. Kislev. *Agricultural Research and Productivity.* New Haven: Yale University Press, 1975.

0446 Kyomo, M.L. "Agricultural Research in Eastern and Southern Africa: Issues and Priorities." Southern African Centre for Cooperation in Agricultural Research of SADCC, Gabarone, Botswana, 1986. Mimeo.

0886 Evenson, Robert E., and Yoav Kislev. "Investment in Agricultural Research and Extension: A Survey of International Data." *Economic Development and Cultural Change* Vol. 23 (April 1975): 507–521.

Nepal

Personnel

Period/Year	PhD	MSc	BSc	Subtotal	Expat	Total	Source
1960–64	..	..	..	..	..	..	
1965–69	..	..	..	..	..	96	10
1970	..	..	..	..	..	..	
1971	..	..	..	..	..	..	
1972	..	..	..	..	..	..	
1973	..	..	..	..	..	..	
1974	..	..	..	..	..	..	
1975	..	..	..	..	..	..	
1976	..	..	..	..	..	..	
1977	9	92	119	220		220	26
1978	..	..	..	..	..	..	
1979	..	..	..	..	..	..	
1980	14	157	217	388		388	59
1981	..	..	..	..	..	401	17
1982	..	..	..	..	..	418	17
1983	16	216	264	496	16	512	17
1984	..	..	..	..	..	455	472
1985	..	..	..	..	..	..	
1986	..	..	..	..	..	..	

1960–64:   1965–69: 1968

Expenditure

LCU = Nepalese Rupees

Period/Year	Current LCU (millions)	Constant 1980 LCU (millions)	Constant 1980 US$ (millions) Atlas	PPP	Source
1960–64	..	..	..	..	
1965–69	4.226	10.512	0.872	3.229	10
1970	..	..	..	..	
1971	13.310	28.140	2.334	8.643	59
1972	14.960	28.068	2.328	8.621	59
1973	14.280	27.782	2.305	8.533	59
1974	19.290	31.013	2.573	9.525	59
1975	24.700	30.991	2.571	9.519	59
1976	28.960	36.155	2.999	11.104	59
1977	30.140	38.991	3.235	11.976	59
1978	32.700	38.698	3.210	11.886	59
1979	39.200	42.196	3.500	12.960	59
1980	37.420	37.420	3.104	11.493	59
1981	37.750	34.986	2.902	10.746	59
1982	..	..	..	..	
1983	..	..	..	..	
1984	..	..	..	..	
1985	..	..	..	..	
1986	..	..	..	..	

1960–64:   1965–69: 1968

Personnel Comments:

The figures in the indicator series after 1975 refer to the Ministry of Agriculture (MOA) only. Within the ministry the following departments or units have research duties: the Department of Agriculture, the Department of Livestock and Animal Health, the Department of Food and Agricultural Marketing Services, the Central Food Research Laboratory, and the Agricultural Project Service Center. At MOA there is a substantial difference between the number of established posts and the number of posts actually filled. In the different sources, the actual number of researchers and the number of established research posts are often not clearly identified, so it is not always possible to ascertain whether the number given is the actual number or not.

In addition to the Ministry of Agriculture, which has primary responsibility for agricultural research in the country, there are several additional ministries, universities, and other organizations which execute agricultural research. Two units which are often mentioned are the Forestry Survey and Research Office (21, 24, and 24 researchers in 1980, 1983, and 1984, respectively; sources 294, 387, and 427) and the Department of Medicinal Plants (20, 107, and 106 researchers in 1983, 1983, and 1984, respectively; sources 17, 387, and 427).

Expenditure Comments:

The expenditure figures given by source 59 cover the research expenditures of the Ministry of Agriculture only. Compared with sources such as 294 and 17, source 59 gives rather high expenditure figures.

Sources:

0010 Boyce, James K., and Robert E. Evenson. *National and International Agricultural Research and Extension Programs.* New York: Agricultural Development Council, Inc., 1975.

0017 ISNAR, IFARD & AOAD. Survey of National Agricultural Research Systems: Unpublished Questionnaire Responses. ISNAR, The Hague, 1985.

0026 Oram, Peter A., and Mark Gieben. "Document Summaries." ISNAR, The Hague, 1984. Mimeo.

0059 Sharma, Ramesh P., and Jock R. Anderson. *Nepal and the CGIAR Centers: A Study of Their Collaboration in Agricultural Research.* CGIAR Study Paper Number 7. Washington, D.C.: World Bank, 1984.

0472 Pradhanang, A.M. *Report of the Agriculture Research and Technology Development in Nepal.* FAO – Asia & Pacific Region Office, November 1984.

Additional References:

0014 Judd, M. Ann, James K. Boyce, and Robert E. Evenson. "Investing in Agricultural Supply." Economic Growth Center, Yale University, New Haven, Connecticut, 1983. Mimeo.

0016 Oram, Peter A., and Vishva Bindlish. *Resource Allocations to National Agricultural Research: Trends in the 1970s.* The Hague and Washington, D.C.: ISNAR and IFPRI, November 1981.

0018 Daniels, Douglas, and Barry Nestel, eds. *Resource Allocation to Agricultural Research.* Proceedings of a workshop held in Singapore 8–10 June 1981. Ottawa: IDRC, 1981.

0027 Harvey, Nigel, ed. *Agricultural Research Centres: A World Directory of Organizations and Programmes.* Seventh Edition, Two Volumes. Harlow, U.K.: Longman, 1983.

0073 Oram, Peter A., and Vishva Bindlish. "Investment in Agricultural Research in Developing Countries: Progress, Problems, and the Determination of Priorities." IFPRI, Washington, D.C., January 1984. Mimeo.

0093 Bangladesh Agricultural Research Council (BARC), ISNAR, and Winrock International. *Management of Human Resources in Agricultural Research.* Report of the International Workshop on Management of Human Resources in Agricultural Research held March 3–5, 1986 in Dhaka, Bangladesh. Dhaka: BARC, ISNAR, and Winrock International, 1986.

0095 FAO – CARIS. *Agricultural Research in Developing Countries – Volume 1: Research Institutions.* Rome: FAO – CARIS, 1978.

0163 CGIAR. "National Agricultural Research." CGIAR, Washington, D.C., 1985. Mimeo.

0165 Evenson, R.E., and Y. Kislev. *Agricultural Research and Productivity.* New Haven: Yale University Press, 1975.

0176 Moseman, Albert, ed. *National Agricultural Research Systems in Asia.* Report of the Regional Seminar held at the India International Centre, New Delhi, India, March 8–13, 1971. New York: ADC, Inc., 1971.

0294 Sharma, Ramesh P. "Agricultural Research Resource Allocation in Nepal." In *Resource Allocation to Agricultural Research.* Proceedings of a workshop held in Singapore 8–10 June 1981, edited by Douglas Daniels and Barry Nestel, pp. 42–48. Ottawa: IDRC, 1981.

0329 Pant, Yadab Deb. "Managing Human Resources for Agricultural Research in Nepal." In *Management of Human Resources in Agricultural Research*. Report of the International Workshop on Management of Human Resources in Agricultural Research held 3–5 March 1986, edited by Theodore Hutchcroft, pp. 105–109. Dhaka: Bangladesh Agricultural Research Council, ISNAR and Winrock International, 1986.

0387 FAO, and the World Bank. *Report of the Nepal Agricultural Research Review*. Rome: FAO, 1983.

0590 UNESCO. *Science and Technology in Countries of Asia and the Pacific*. Science Policy Studies and Documents No. 52. Paris: UNESCO, n.d.

0732 FAO – Regional Office for Asia and the Pacific.

Agricultural Research Systems in the Asia-Pacific Region. Bangkok: FAO – Regional Office for Asia and the Pacific, 1986.

0737 Yadav, Ram P. *Agricultural Research in Nepal: Resource Allocation, Structure, and Incentives*. IFPRI Research Report No. 62. Washington D.C.: IFPRI, September 1987.

0852 Evenson, Robert E., and Yoav Kislev. *Investment in Agricultural Research and Extension: A Survey of International Data*. Center Discussion Paper No. 124. New Haven, Connecticut: Economic Growth Center, Yale University, August 1971.

0886 Evenson, Robert E., and Yoav Kislev. "Investment in Agricultural Research and Extension: A Survey of International Data." *Economic Development and Cultural Change* Vol. 23 (April 1975): 507–521.

Netherlands

Personnel

Period/Year	PhD	MSc	BSc	Subtotal	Expat	Total	Source
1960–64	..	..	..	..	..	752	852/282
1965–69	..	..	..	..	..	894	852/556/564/580
1970	..	..	..	..	..	971	556/564/581
1971	..	..	..	..	..	..	
1972	..	..	..	..	..	..	
1973	..	..	..	..	..	..	
1974	..	..	..	..	..	..	
1975	..	..	..	..	..	1249	426/544/581
1976	..	..	..	..	..	1312	426/544/581
1977	..	..	..	..	..	1352	426/544/581/283
1978	..	..	..	..	..	1376	426/544/581
1979	..	..	..	..	..	1388	426/544/581
1980	..	..	..	..	..	1403	426/544/581
1981	..	..	..	..	..	1500	376
1982	..	..	..	..	..	1573	428/545
1983	..	..	..	..	..	..	
1984	..	..	..	..	..	1691	430/546
1985	..	..	..	..	..	1755	424/564
1986	..	..	..	..	..	..	

1960–64: 1960 & 61   1965–69: 1965, 66, 67, 68 & 69

Expenditure

LCU = Dutch Guilders

Period/Year	Current LCU (millions)	Constant 1980 LCU (millions)	Constant 1980 US$ (millions)		Source
			Atlas	PPP	
1960–64	46.000	156.463	79.040	59.412	282
1965–69	129.000	317.734	160.509	120.651	430
1970	..	..	..	..	
1971	196.000	379.110	191.514	143.957	430
1972	..	..	..	..	
1973	..	..	..	..	
1974	296.000	438.519	221.525	166.516	430
1975	..	..	..	..	
1976	..	..	..	..	
1977	436.000	504.046	254.628	191.398	430
1978	..	..	..	..	
1979	470.000	496.304	250.717	188.458	430
1980	..	..	..	..	
1981	507.000	480.569	242.768	182.483	430
1982	..	..	..	..	
1983	..	..	..	..	
1984	602.000	521.212	263.300	197.916	430
1985	..	..	..	..	
1986	..	..	..	..	

1960–64: 1961   1965–69: 1967

Personnel Comments:
The main institutions involved in agricultural research in the Netherlands are the Ministry of Agriculture (including several directorates, of which the Directorate of Agricultural Research is the most important), TNO (Applied Scientific Research Organization – some of the TNO institutes execute agriculturally-related research, especially in the field of food processing), the Agricultural University Wageningen, and the Faculty of Veterinary Science at the State University of Utrecht. In addition the following smaller organizations can be identified: RIJP (State Authority for the IJsselmeerpolders), NIZO (Netherlands Institute for Dairy Research), and IRS (Institute for Efficient Sugar Production). The National Council of Agricultural Research (NRLO/TNO) is in charge of coordinating these independent entities.

The table below gives an overview of how the personnel indicators have been constructed.

	1966	1967	1968	1969	1970	1975	1976	1977	1978	1979	1980	1982	1984	1985
MLVa	524	527	541	530	530	[690]	[715]	[736]	[756]	[778]	[790]	[878]	857	865
AUWb	[156]*	[169]*	[183]*	[196]*	[209]*	244*	280*	297*	300*	289*	292*	361*	453*	509*
FVMc	[47]	[51]	[56]	60**	100**	100**	100**	100**	100*	100*	100**	[100]	137	[137]
TNO	⊤	⊤	⊤	⊤	⊤	[160]	[160]	160***	[160]	[160]	[160]	[165]	170	[170]
RIJP	[132]	[132]	[132]	[132]	[132]	⊤	⊤	⊤	⊤	⊤	⊤	⊤	26	[26]
NIZO						[55]	[57]	[59]	[60]	[61]	[61]	[69]	42	[42]
IRS	⊥	⊥	⊥	⊥	⊥	⊥	⊥	⊥	⊥	⊥	⊥	⊥	6	[6]
Total	859	879	912	918	971	1249	1312	1352	1376	1388	1403	1573	1691	1755
Source	556	556	556	556	556	426	426	426	426	426	426	428	430	424
	564*	564*	564*	564*	564*	544*	544*	544*	544*	544*	545*	545*	546*	564*
				580**	581**	581**	581**	581**	581**	581**	581**			
								283***						

[ ] = estimated or constructed by authors.
a. The figures for the years 1975–1980 for MLV were prorated from the total number of personnel, on the basis that the share of academics within the total personnel number increased from about 18% in 1970 to 25% in 1984.
b. The number of researchers expressed in full-time equivalent units at the AUW for the years 1966–1970 are constructed on the basis of the following assumptions: 1) Academic staff spend one third of their time on research, and 2) The estimate based on core faculty was increased by 25% because of research financed by external sources.
c. The figures given by source 581 are a rather rough estimation.

Expenditure Comments:
The table on the next page gives an overview of construction details for the expenditure indicators.
No precise statistics are available on agricultural research expenditures. Nearly all the figures presented here are based on informed assumptions – for example, the proportion of the university budget which has been imputed to agricultural research. Given these assumptions, it was preferable to rely on one source, rather than jump from one to another.

	1967	1971	1974	1977	1979	1981	1984
MLV	90.0	137.0	191.0	277.0	303.0	330.0	351.0
AUW		22.0	36.0	60.0	70.0	77.0	132.0
FVM		8.0	13.0	25.0	19.0	18.0	27.0
TNO	[39.0]	17.0	33.0	43.0	42.0	44.0	49.0
RIJP				13.0	14.0	15.0	18.0
NIZO		[12.0]	[23.0]	15.0	18.0	19.0	20.0
IRS				3.0	4.0	4.0	5.0
Total	129.0	196.0	296.0	436.0	470.0	507.0	602.0
Source	430	430	430	430	430	430	430

[ ] = estimated or constructed by authors.

Acronyms and Abbreviations:
AUW: Agricultural University Wageningen
FVM: Faculty of Veterinary Medicine, Utrecht State University
IRS: (Institute for Efficient Sugar Production)
MLV: (Ministry of Agriculture and Fisheries)
NIZO: (Netherlands Institute for Dairy Research)
NRLO: (National Council of Agricultural Research)
RIJP: (State Authority for the IJsselmeerpolders)
TNO: (Applied Scientific Research Organization)

Sources:

0282 EEG. De Organisatie van het Landbouwkundig Onderzoek in de Landen van de EEG. EEG Studies, Serie Landbouw 9. Brussel: EEG, 1963.

0283 European Economic Community (EEC). "Monograph on the Organization of Agricultural Research." EEC, Brussels, 1982. Mimeo.

0376 FAO. *Fourteenth FAO Regional Conference for Europe, Reykjavik, Iceland, 17–21 September 1984: Research in Support of Agricultural Policies in Europe.* Rome: FAO, 1984.

0424 Ministerie van Landbouw en Visserij. *Aard en Omvang van het Landbouwkundig Onderzoek.* Tweede Kamer der Staten Generaal, Vergaderjaar 1985–1986, 19307, nrs. 1–2. The Hague: Ministerie van Landbouw en Visserij, November 1985.

0426 Ministerie van Landbouw en Visserij – Directie Landbouwkundig Onderzoek *Meerjarenvisie 1977–1981 voor het Landbouwkundig en het Visserij-Onderzoek.* The Hague: Ministerie van Landbouw en Visserij, 1977.

0428 Directie Landbouwkundig Onderzoek (DLO). "Monograph on the Organization of Agricultural Research in the Netherlands (draft)." DLO, Wageningen, The Netherlands, 1985. Mimeo.

0430 Berg, A. van den. Personal communication. Nationale Raad voor Landbouwkundig Onderzoek TNO. The Hague, September 1986.

0544 Landbouwhogeschool Wageningen. *Jaarverslag.* Academic years: 1975/1976, 1976/1977, 1977/78, and 1978/1979. Wageningen: Landbouwhogeschool, various years.

0545 Landbouwhogeschool Wageningen. *Jaarverslag.* Academic years: 1980–1982. Wageningen: Landbouwhogeschool, various years.

0546 Landbouwhogeschool Wageningen. *Jaarverslag – Wetenschappelijk Verslag.* Academic years: 1983 and 1984. Wageningen: Landbouwhogeschool, various years.

0556 OECD. *Reviews of National Science Policy: Netherlands.* Paris: OECD, 1973.

0564 Kunst, Jan. Personal communication. Medewerker Financieel Economische Zaken, Agricultural University Wageningen. Wageningen, March 1987.

0580 Netherlands Central Bureau of Statistics (CBS). *Research and Development in the Netherlands 1969.* The Hague: Staatsuitgeverij, 1971.

0581 Netherlands Central Bureau of Statistics (CBS). *Research and Development in the Netherlands.* Years: 1970–1983. The Hague: Staatsuitgeverij, various years.

0852 Evenson, Robert E., and Yoav Kislev. *Investment in Agricultural Research and Extension: A Survey of International Data.* Center Discussion Paper No. 124. New Haven, Connecticut: Economic Growth Center, Yale University, August 1971.

Additional References:

0010 Boyce, James K., and Robert E. Evenson. *National and International Agricultural Research and Extension Programs.* New York: Agricultural Development Council, Inc., 1975.

0014 Judd, M. Ann, James K. Boyce, and Robert E. Evenson. "Investing in Agricultural Supply." Economic Growth Center, Yale University, New Haven, Connecticut, 1983. Mimeo.

0021 OECD. *OECD Science and Technology Indicators.* Paris: OECD, 1984.

0027 Harvey, Nigel, ed. *Agricultural Research Centres: A World Directory of Organizations and Programmes.* Seventh Edition, Two Volumes. Harlow, U.K.: Longman, 1983.

0039 Landbouw Economisch Instituut (LEI). *Landbouwcijfers.* Yearly publication. The Hague: LEI, various years.

0165 Evenson, R.E., and Y. Kislev. *Agricultural Research and Productivity.* New Haven: Yale University Press, 1975.

0425 Verkaik, A.P. *Research Management Studies 3: Organisatiestructuur landbouwkundig onderzoek en achtergronden van haar totstandkoming.* The Hague: Nationale Raad voor Landbouwkundig Onderzoek TNO, 1972.

0427 Nationale Raad voor Landbouwkundig Onderzoek TNO. *Studierapport 10: Meerjarenvisie Landbouwkundig Onderzoek 1982–1986.* The Hague: Nationale Raad voor Landbouwkundig Onderzoek TNO, 1981.

0473 OECD. *Science and Technology Indicators: Basic Statistical Series – Volume A: The Objectives of Government R&D Funding 1974–1985.* Paris: OECD, May 1983.

0474 OECD. *Science and Technology Indicators: Basic Statistical Series – Volume A: The Objectives of Government R&D, 1969–1981.* Paris: OECD, July 1981.

0509 OECD. *Survey of the Resources Devoted to R&D by OECD Member Countries, International Statistical Year 1971 – Volume 5: Total Tables, Statistical Tables and Notes.* Paris: OECD, August 1974.

0510 OECD. *International Survey of the Resources Devoted to R&D in 1967 by OECD Member Countries, Statistical Tables and Notes – Volume 5: Total Tables.* Paris: OECD, August 1970.

0511 OECD. *International Survey of the Resources Devoted to R&D in 1969 by OECD Member Countries, Statistical Tables and Notes – Volume 5: Total Tables.* Paris: OECD, June 1973.

0512 OECD. *Survey of the Resources Devoted to R&D by OECD Member Countries, International Statistical Year 1973 – Volume 5: Total Tables, Statistical Tables and Notes.* Paris: OECD, September 1976.

0513 OECD. *International Survey of the Resources Devoted to R&D by OECD Member Countries, International Statistical Year 1975 – International Volume: Statistical Tables and Notes.* Paris: OECD, March 1979.

0514 OECD. *A Study of Resources Devoted to R&D in OECD Member Countries in 1963/64, International Statistical Year for Research and Development – Volume 2: Statistical Tables and Notes.* Paris: OECD, 1968.

0530 OECD. *OECD Science and Technology Indicators No. 2 – R&D, Invention and Competitiveness.* Paris: OECD, 1986.

0547 Landbouwhogeschool Wageningen. "Wetenschappelijk Verslag over 1979." Landbouwhogeschool, Wageningen, n.d. Mimeo.

0548 Landbouwhogeschool Wageningen – Vaste Commissie voor de Wetenschapsbeoefening. *Op Weg naar een Onderzoekbeleid van de Landbouwhogeschool.* Wageningen: Landbouwhogeschool, 1974.

0549 Landbouwhogeschool Wageningen. *Wetenschappelijk Verslag 1981–1982.* Wageningen: Landbouwhogeschool, various years.

0550 Landbouwhogeschool Wageningen. *Ontwikkelingsplan Landbouwhoge– school 1980 tot en met 1983.* Wageningen: Landbouwhogeschool, November 1977.

0551 Ministerie van Landbouw en Visserij – Directie Landbouwkundig Onderzoek. "Bijdrage tot een Meerjarenplan voor het Landbouwkundig Onderzoek 1972–1976." Ministerie van Landbouw en Visserij, The Hague, December 1972. Mimeo.

0555 OECD. *Country Reports on the Organisation of Scientific Research: Netherlands.* Paris: OECD, February 1964.

0574 OECD. *Intellectual Investment in Agriculture for Economic and Social Development.* Documentation in Food and Agriculture No. 60. Paris: OECD, 1962.

0577 Netherlands Central Bureau of Statistics (CBS). *Research and Developement in the Netherlands 1959.* Volumes 1 and 2. Zeist, The Netherlands: Uitgeversmaatschappij W. de Haan N.V., 1963.

0578 Netherlands Central Bureau of Statistics (CBS). *Research and Development Activities in the Netherlands 1964.* The Hague: Staatsuitgeverij, 1967.

0579 Netherlands Central Bureau of Statistics (CBS). *Research and Development Activities in the Netherlands 1967.* The Hague: Staatsuitgeverij, 1969.

0601 OECD – STIID Data Bank. "Public Funding of R&D by Socio-Economic Objective." Tables. OECD, Paris, April 1987. Mimeo.

0602 OECD – STIID Data Bank. "R&D Expenditure in the Higher Education and Private Non-Profit Sectors." Tables. OECD, Paris, April 1987. Mimeo.

0606 EUROSTAT. *Government Financing of Research and Development 1975–1984.* Luxembourg: Office des Publications Officielles des Communautés Européennes, February 1985.

0703 Martin, Ben R., and John Irvine. *An International Comparison of Government Funding of Academic and Academically Related Research.* Advisory Board for the Research Council (ABRC) Science Policy Study No. 2. Brighton: ABRC – Science Policy Research Unit, University of Sussex, October 1986.

0706 EUROSTAT. *Government Financing of Research and Development 1975–1985.* Luxembourg: Office for Official Publications of the European Communities, 1987.

0707 OECD – STIID Data Bank. "R&D Expenditure in the Higher Education and Private Non-Profit Sectors: Higher Education, Agricultural Sciences, General Universities Funds." Table. OECD, Paris, June 1987. Mimeo.

0708 OECD – STIID Data Bank. "Gross Domestic Expenditure on R&D: All Fields of Sciences – Agriculture, Forestry and Fishing." Tables. OECD, Paris, June 1987. Mimeo.

0886 Evenson, Robert E., and Yoav Kislev. "Investment in Agricultural Research and Extension: A Survey of International Data." *Economic Development and Cultural Change* Vol. 23 (April 1975): 507–521.

New Caledonia

Personnel

Period/Year	PhD	MSc	BSc	Subtotal	Expat	Total	Source
1960–64	..	..	..	..	..	..	
1965–69	..	..	..	..	..	..	
1970	..	..	..	..	..	..	
1971	..	..	..	..	..	..	
1972	..	..	..	..	..	..	
1973	..	..	..	..	..	..	
1974	..	..	..	..	..	..	
1975	..	..	..	..	..	..	
1976	..	..	..	..	..	..	
1977	..	..	..	..	..	..	
1978	..	..	..	..	..	..	
1979	..	..	..	..	..	..	
1980	..	..	..	..	..	..	
1981	..	..	..	..	..	14	38
1982	..	..	..	..	..	..	
1983	..	..	..	..	..	..	
1984	..	..	..	..	..	..	
1985	..	..	..	..	..	..	
1986	..	..	..	..	..	..	

1960–64:   1965–69:

Expenditure   *LCU = CFP Francs*

Period/Year	Current LCU (millions)	Constant 1980 LCU (millions)	Constant 1980 US$ (millions)		Source
			Atlas	PPP	
1960–64	..	..	..	..	
1965–69	..	..	..	..	
1970	..	..	..	..	
1971	..	..	..	..	
1972	..	..	..	..	
1973	..	..	..	..	
1974	..	..	..	..	
1975	..	..	..	..	
1976	..	..	..	..	
1977	..	..	..	..	
1978	..	..	..	..	
1979	..	..	..	..	
1980	..	..	..	..	
1981	..	..	..	..	
1982	..	..	..	..	
1983	..	..	..	..	
1984	..	..	..	..	
1985	..	..	..	..	
1986	..	..	..	..	

1960–64:   1965–69:

Personnel Comments:
The 1983 figure, from source 38, has been constructed as follows:

Organization	Number of Researchers
Department of Agriculture	3
IFCC	1
ORSTOM	10[a]
Total	14

a.Source 27 and 744 identify a total of 64 and 67 ORSTOM researchers, respectively, but many of them work in non-agricultural areas (e.g., oceanography, archaeology, etc.).

Acronyms and Abbreviations:
IFCC: Institut Français du Café, du Cacao et autres plantes stimulantes

ORSTOM: Office de la Recherche Scientifique et Technique d'Outre-Mer

Sources:
0038 Gamble, W.K., R.M. Bourke, and C.W. Brookson. *South Pacific Agricultural Research Study.* Consultants' Report to the Asian Development Bank (two volumes). Manila: Asian Development Bank, June 1981.

Additional References:
0027 Harvey, Nigel, ed. *Agricultural Research Centres: A World Directory of Organizations and Programmes.* Seventh Edition, Two Volumes. Harlow, U.K.: Longman, 1983.
0654 Office de la Recherche Scientifique et Technique Outre-Mer (ORSTOM). *Rapport d'Activité 1967.* Paris: ORSTOM, 1969.
0744 *Agricultural Research Centres: A World Directory of Organizations and Programmes.* Eighth Edition, Two Volumes. Harlow, U.K.: Longman, 1986.

New Zealand

Personnel

Period/Year	PhD	MSc	BSc	Subtotal	Expat	Total	Source
1960–64	..	..	..	..	..	..	
1965–69	200	174	189	563	..	563	286/285
1970	..	..	..	..	..	..	
1971	341	215	205	761	..	761	279
1972	..	..	..	..	..	..	
1973	..	..	..	..	..	..	
1974	433	187	170	790	..	790	244
1975	..	..	..	..	..	..	
1976	..	..	..	..	..	..	
1977	587	260	235	1082	..	1082	243
1978	..	..	..	..	..	..	
1979	..	..	..	..	..	..	
1980	656	277	261	1194	..	1194	242
1981	..	..	..	..	..	..	
1982	..	..	..	..	..	..	
1983	..	..	..	..	..	..	
1984	..	..	..	..	..	..	
1985	..	..	..	..	..	..	
1986	..	..	..	..	..	..	

1960–64:   1965–69: 1965 & 68

Expenditure

LCU = New Zealand Dollars

Period/Year	Current LCU (millions)	Constant 1980 LCU (millions)	Constant 1980 US$ (millions)		Source
			Atlas	PPP	
1960–64	6.150	29.421	28.576	27.949	480/479
1965–69	11.028	45.776	44.461	43.486	480/479
1970	15.473	52.809	51.292	50.166	480/479
1971	19.115	56.721	55.092	53.883	480/479
1972	20.590	55.499	53.904	52.722	480/479
1973	24.038	59.648	57.934	56.663	480/479
1974	31.214	73.272	71.167	69.606	480/479
1975	35.974	60.767	59.021	57.726	480/479
1976	39.023	55.747	54.146	52.958	480/479
1977	45.497	68.007	66.054	64.604	480/479
1978	55.741	73.537	71.424	69.857	480/479
1979	61.381	71.125	69.082	67.566	480/479
1980	77.233	77.233	75.014	73.368	480/479
1981	97.042	81.275	78.940	77.208	480/479
1982	107.035	82.525	80.154	78.396	480/479
1983	111.878	78.401	76.149	74.478	480/479
1984	..	..	..	..	
1985	..	..	..	..	
1986	..	..	..	..	

1960–64: 1960, 61, 62, 63 & 64   1965–69: 1965, 66, 67, 68 & 69

Personnel Comments:

These figures include MAF and DSIR agricultural researchers, relevant Department of Internal Affairs, New Zealand Meteorological Service, and Ministry of Works and Development personnel, plus agricultural researchers from 11 publicly sponsored research associations and 8 universities, colleges, and research institutes. University researchers were prorated to full-time equivalents using a 0.5 weight. These personnel figures are lower than the compatible figures recorded in source 479. However, the source 479 figures report science staff ceilings rather than positions filled and, according to source 480, may also be inclusive of support staff.

Expenditure Comments:

These figures represent the sum of total agricultural research expenditures taken from source 480 plus public sector forestry and fisheries research expenditures calculated from source 479. The source 480 figures were obtained from detailed calculations using data from sources 477, 478, and 479. They include agricultural research expenditures of the MAF and DSIR, other governmental units plus research associations (whose primary funding sources include the DSIR and contributions of producers via a commodity-specific cess, see source 481) and colleges and universities. The institutional coverage is extensive and compatible with the personnel figures.

Acronyms and Abbreviations:

DSIR: Department of Science and Industry Research
MAF: Ministry of Agriculture and Fisheries
NRAC: National Research Advisory Council

Sources:

0242 Commonwealth Agricultural Bureaux (CAB). *List of Research Workers in Agricultural Sciences in the Commonwealth 1981.* Slough, England: CAB, 1981.

0243 Commonwealth Agricultural Bureaux (CAB). *List of Research Workers in the Agricultural Sciences in the Commonwealth and in the Republic of Ireland 1978.* Slough, England: CAB, 1978.

0244 Commonwealth Agricultural Bureaux (CAB). *List of Research Workers in the Agricultural Sciences in the Commonwealth and in the Republic of Ireland 1975.* Slough, England: CAB, 1975.

0279 Commonwealth Agricultural Bureaux (CAB). *List of Research Workers in the Agricultural Sciences in the Commonwealth and in the Republic of Ireland 1972.* Slough, England: CAB, 1972.

0285 Commonwealth Agricultural Bureaux (CAB). *List of Research Workers in the Agricultural Sciences in the Commonwealth and in the Republic of Ireland 1969.* Slough, England: CAB, 1969.

0286 Commonwealth Agricultural Bureaux (CAB). *List of Research Workers in Agriculture, Animal Health and Forestry in the Commonwealth and in the Republic of Ireland 1966.* Slough, England: CAB, 1966.

0479 National Research Advisory Council. *Report of the National Research Advisory Council for the Year Ended 31 March 1986.* Presented to the House of Representatives Pursuant to Section 11 of the National Research Advisory Council Act 1963; Yearly Reports 1965–1986. Wellington, New Zealand: Government Printer, various years.

0480 Eveleens, W.M. Personal communication. Ruakura Agricultural Research Centre, Ministry of Agriculture and Fisheries. Hamilton, New Zealand, 1986.

Additional References:

0010 Boyce, James K., and Robert E. Evenson. *National and International Agricultural Research and Extension Programs.* New York: Agricultural Development Council, Inc., 1975.

0014 Judd, M. Ann, James K. Boyce, and Robert E. Evenson. "Investing in Agricultural Supply." Economic Growth Center, Yale University, New Haven, Connecticut, 1983. Mimeo.

0027 Harvey, Nigel, ed. *Agricultural Research Centres: A World Directory of Organizations and Programmes.* Seventh Edition, Two Volumes. Harlow, U.K.: Longman, 1983.

0165 Evenson, R.E., and Y. Kislev. *Agricultural Research and Productivity.* New Haven: Yale University Press, 1975.

0287 Commonwealth Agricultural Bureaux (CAB). *List of Research Workers in Agriculture, Animal Health and Forestry in the British Commonwealth, the Republic of Sudan and the Republic of Ireland 1959.* Slough, England: CAB, 1959.

0453 Ministry of Agriculture and Fisheries. *New Zealand Agriculture.* Wellington, New Zealand: Ministry of Agriculture and Fisheries, 1974.

0473 OECD. *Science and Technology Indicators: Basic Statistical Series – Volume A: The Objectives of*

Government R&D Funding 1974–1985. Paris: OECD, May 1983.

0474 OECD. *Science and Technology Indicators: Basic Statistical Series – Volume A: The Objectives of Government R&D, 1969–1981.* Paris: OECD, July 1981.

0477 McBride, Conor P., and Christine de Joux. *Scientific Research in New Zealand: Government Expenditure and Manpower, 1926–1966.* National Research Advisory Council Publication No. 1. New Zealand: Department of Scientific and Industrial Research, 1966.

0478 McBride, Conor P. *Scientific Research in New Zealand: Expenditure and Manpower 1953–1962.* Information Series No. 41. New Zealand: Department of Scientific and Industrial Research, 1964.

0481 Ulyatt, M.J., J.P. Kerr, and E.W. Hewett. "Research and Development Funding in Agriculture." *New Zealand Agricultural Science* Vol. 20, No. 3 (August 1986): 163–168.

0512 OECD. *Survey of the Resources Devoted to R&D by OECD Member Countries, International Statistical Year 1973 – Volume 5: Total Tables, Statistical Tables and Notes.* Paris: OECD, September 1976.

0513 OECD. *International Survey of the Resources Devoted to R&D by OECD Member Countries, International Statistical Year 1975 – International Volume: Statistical Tables and Notes.* Paris: OECD, March 1979.

0530 OECD. *OECD Science and Technology Indicators No. 2 – R&D, Invention and Competitiveness.* Paris: OECD, 1986.

0590 UNESCO. *Science and Technology in Countries of Asia and the Pacific.* Science Policy Studies and Documents No. 52. Paris: UNESCO, n.d.

0601 OECD – STIID Data Bank. "Public Funding of R&D by Socio-Economic Objective." Tables. OECD, Paris, April 1987. Mimeo.

0602 OECD – STIID Data Bank. "R&D Expenditure in the Higher Education and Private Non-Profit Sectors." Tables. OECD, Paris, April 1987. Mimeo.

0603 Scobie, G.M., and W.M. Eveleens. *Agricultural Research: What's it worth?* Discussion paper 1/86. Hamilton, New Zealand: Economics Division, Ministry of Agriculture and Fisheries, June 1986.

0604 Fordyce, Alison M. *Industry Levies and Research Funding.* Technical Report No. 24. Palmerston North, New Zealand: Plant Physiology Division – Department of Scientific and Industrial Research, February 1986.

0756 Scobie, G.M., and W.M. Eveleens. *The Return to Investment in Agricultural Research in New Zealand: 1926–27 to 1983–84.* Research Report 1/87. Hamilton, New Zealand: MAF Economics Division, October 1987.

0886 Evenson, Robert E., and Yoav Kislev. "Investment in Agricultural Research and Extension: A Survey of International Data." *Economic Development and Cultural Change* Vol. 23 (April 1975): 507–521.

Nicaragua

Personnel

Period/Year	PhD	MSc	BSc	Subtotal	Expat	Total	Source
1960–64	..	..	..	..	..	..	
1965–69	..	..	..	..	..	24	208
1970	..	..	..	..	..		
1971	..	..	..	..	..	29	10
1972	..	..	..	..	..		
1973	..	..	..	..	..		
1974	..	..	..	..	..		
1975	..	..	..	..	..		
1976	..	..	..	..	..		
1977	..	..	..	..	..	44	208
1978	..	..	..	..	..		
1979	..	..	..	..	..		
1980	–10–		47	57	..	57	215
1981	..	..	..	..	..		
1982	..	..	..	..	..		
1983	..	..	..	..	..		
1984	..	..	..	..	..		
1985	..	..	..	..	..		
1986	..	..	..	..	..		

1960–64:   1965–69: 1966

Expenditure

LCU = Cordobas

Period/Year	Current LCU (millions)	Constant 1980 LCU (millions)	Constant 1980 US$ (millions)		Source
			Atlas	PPP	
1960–64	1.995	8.599	0.798	2.206	10
1965–69	3.520	14.426	1.338	3.702	10
1970	..	..	..	..	
1971	..	..	..	..	
1972	..	..	..	..	
1973	..	..	..	..	
1974	..	..	..	..	
1975	..	..	..	..	
1976	..	..	..	..	
1977	10.502	20.920	1.940	5.368	208
1978	..	..	..	..	
1979	..	..	..	..	
1980	14.070	14.070	1.305	3.610	215
1981	..	..	..	..	
1982	..	..	..	..	
1983	..	..	..	..	
1984	..	..	..	..	
1985	..	..	..	..	
1986	..	..	..	..	

1960–64: 1962   1965–69: 1965

Personnel Comments:

Over the last 10 to 15 years, agricultural research under the 'Ministry of Agriculture' (or whatever other name it may have had) has been moved quite often within the changing organizational structure of the ministry. Until 1976 agricultural research had been a directorate within the Ministry of Agriculture. In that year, agricultural research was moved to a semiautonomous institute called 'Instituto Nacional de Tecnología Agropecuaria' (INTA). In 1980 INTA was abolished and agricultural research was brought under the 'Dirección General de Tecnología Agropecuaria' within the newly created 'Ministerio de Desarrollo Agropecuario y Reforma Agraria'. In 1984, a further reorganization took place and agricultural research was moved to 'Dirección General de Agricultura' (sources 208, 215, and 382). The personnel indicator series refers to agricultural researchers within the 'Ministry of Agriculture,' respectively, INTA only.

Expenditure Comments:

The expenditure indicator series is consistent with the personnel series.

Sources:

0010 Boyce, James K., and Robert E. Evenson. *National and International Agricultural Research and Extension Programs.* New York: Agricultural Development Council, Inc., 1975.

0208 Segura, M. "Diagnóstico de la Investigación Agropecuaria en el Istmo Centroamericano." IICA, Guatemala, 1979. Mimeo.

0215 Soikes, Raul. "Programa de Investigación Agropecuaria de Nicaragua." IICA – OEA, Costa Rica, December 1980. Mimeo.

Additional References:

0001 Trigo, Eduardo J., and Martín E. Piñeiro. "Funding Agricultural Research." In *Report of a Conference: Selected Issues in Agricultural Research in Latin America*, edited by Barry Nestel and Eduardo J. Trigo. The Hague: ISNAR, March 1984.

0014 Judd, M. Ann, James K. Boyce, and Robert E. Evenson. "Investing in Agricultural Supply." Economic Growth Center, Yale University, New Haven, Connecticut, 1983. Mimeo.

0016 Oram, Peter A., and Vishva Bindlish. *Resource Allocations to National Agricultural Research: Trends in the 1970s.* The Hague and Washington, D.C.: ISNAR and IFPRI, November 1981.

0026 Oram, Peter A., and Mark Gieben. "Document Summaries." ISNAR, The Hague, 1984. Mimeo.

0053 Piñeiro, Martín, and Eduardo Trigo, eds. *Cambio Técnico en el Agro Latino Americano – Situación y Perspectivas en la Década de 1980.* San José: IICA, 1983.

0073 Oram, Peter A., and Vishva Bindlish. "Investment in Agricultural Research in Developing Countries: Progress, Problems, and the Determination of Priorities." IFPRI, Washington, D.C., January 1984. Mimeo.

0165 Evenson, R.E., and Y. Kislev. *Agricultural Research and Productivity.* New Haven: Yale University Press, 1975.

0172 Trigo, Eduardo. Background material. ISNAR, The Hague, n.d.

0382 "La Administración de la Investigación Agrícola en Nicaragua." Paper presented at Curso-Taller sobre la Administración de la Investigación Agrícola, Panama City, 14–25 July 1986.

0852 Evenson, Robert E., and Yoav Kislev. *Investment in Agricultural Research and Extension: A Survey of International Data.* Center Discussion Paper No. 124. New Haven, Connecticut: Economic Growth Center, Yale University, August 1971.

0886 Evenson, Robert E., and Yoav Kislev. "Investment in Agricultural Research and Extension: A Survey of International Data." *Economic Development and Cultural Change* Vol. 23 (April 1975): 507–521.

Niger

Personnel

Period/Year	PhD	MSc	BSc	Subtotal	Expat	Total	Source
1960–64	..	..	..	..	..	..	
1965–69	..	..	..	..	..	..	
1970	..	..	..	0	9	9	532
1971	..	..	..	..	..	..	
1972	..	..	..	..	..	12	431
1973	..	..	..	..	..	..	
1974	..	..	..	..	..	..	
1975	..	..	..	..	..	..	
1976	..	..	..	..	..	16	16
1977	..	..	..	..	..	..	
1978	..	..	..	..	..	..	
1979	..	..	..	..	..	..	
1980	..	..	..	..	..	..	
1981	..	..	..	..	..	..	
1982	..	..	..	..	..	46	27
1983	..	..	..	22	26	48	68/83
1984	..	..	..	27	30	57	171
1985	..	..	..	30	47	77	86
1986	..	..	..	..	..	..	

1960–64:   1965–69:

Expenditure

LCU = CFA Francs

Period/Year	Current LCU (millions)	Constant 1980 LCU (millions)	Constant 1980 US$ (millions)		Source
			Atlas	PPP	
1960–64	..	..	..	..	
1965–69	130.840	443.525	2.013	2.046	589
1970	..	..	..	..	
1971	176.498	526.860	2.391	2.431	589
1972	218.052	586.161	2.660	2.704	431/589
1973	167.000	406.326	1.844	1.875	589
1974	..	..	..	..	
1975	175.528	401.666	1.823	1.853	16
1976	272.000	489.209	2.220	2.257	261
1977	304.000	451.709	2.050	2.084	261
1978	398.000	525.760	2.386	2.426	261
1979	384.000	459.330	2.085	2.119	261
1980	598.000	598.000	2.714	2.759	261
1981	450.000	446.667	1.891	1.922	261
1982	583.000	454.050	2.061	2.095	261
1983	534.450	400.037	1.815	1.846	782
1984	534.450	366.815	1.665	1.692	782
1985	534.450	357.253	1.621	1.648	782
1986	..	..	..	..	

1960–64:   1965–69: 1966

Personnel Comments:

As in most former French colonies, France continued to conduct agricultural research in Niger after independence. IRAT, IEMVT, and CTFT had their own permanent bases, while IFAC, IRCT, and possibly also other French research organizations occasionally sent missions to Niger. In 1975, however, agricultural research was more or less nationalized and concentrated in INRAN. The figures prior to 1975 comprise the French research organizations, while the post-1975 figures refer to INRAN only. For reasons of consistency, we have excluded 'Station Avicole et Caprine de Maradi' and 'Station Sahélienne Expérimentale de Toukounous' from the pre-1975 figures in the indicator series. When agricultural researchers working on development projects and at universities and regional and international institutes in Niger are included, the estimated number of researchers for 1984 is more than double the INRAN figure (source 171).

Expenditure Comments:

The table below gives an overview of how the pre-1975 expenditure indicators have been constructed.

	1966	1971	1972	1973
IRAT	82.000	113.500	95.942[a]	70.750
IEMVT	21.000	25.700	28.246	47.250
CTFT[b]	27.840	37.298	37.083[c]	38.000
IFAC			43.158	11.000
IRCT			13.623	
Total	130.840	176.498	218.052	167.000
Source	589	589	431/589	589

a. Expenditures of 47.971 million CFA Francs for 'Station Canne à Sucre de Tillabery' (US$ 131,000, 1 scientist, source 431) have been excluded. When compared with other institutional-level estimates, these expenditures appear extraordinarily high.

b. The CTFT station in Niger was in charge of both Niger and Burkina Faso.

c. To maintain consistency in currency units, the expenditure figure given by source 431 and 589 has been scaled down by a factor of 10.

The post-1975 expenditure figures refer to INRAN only. They represent government contributions to INRAN, a small amount of 'autofinancement,' and contributions from 'Fonds National d'Investissement' (FNI) which, however, ceased in 1982 (source 86). Source 86 gives an expenditure figure of 633 million FCFA for 1984. This figure is inclusive of donor contributions which, according to source 86, represent about 20% of total expenditures. For reasons of consistency over time, this figure was excluded from the indicator series. Using an extremely broad definition of NARS, which includes universities, development projects and regional and international institutes, total agricultural research expenditures (inclusive of foreign aid and salaries of expatriates) has been estimated to be as high as 5000 million FCFA in 1985/1986 (source 171).

Acronyms and Abbreviations:

CTFT: Centre Technique Forestier Tropical

IEMVT: Institut d'Elevage et de Médecine Vétérinaire des Pays Tropicaux

IFAC: Institut Français de Recherches Fruitières d'Outre-Mer

INRAN: Institut National de Recherches Agronomiques du Niger

IRAT: Institut de Recherches Agronomiques Tropicales et des Cultures Vivrières

IRCT: Institut de Recherche du Coton et des Textiles Exotiques

Sources:

0016 Oram, Peter A., and Vishva Bindlish. *Resource Allocations to National Agricultural Research: Trends in the 1970s.* The Hague and Washington, D.C.: ISNAR and IFPRI, November 1981.

0027 Harvey, Nigel, ed. *Agricultural Research Centres: A World Directory of Organizations and Programmes.* Seventh Edition, Two Volumes. Harlow, U.K.: Longman, 1983.

0068 Conte, Stephan. "West Africa Agricultural Research Data Base Building Task: Termination Report." World Bank, Washington, D.C., 1985. Mimeo.

0083 Institut du Sahel, and DEVRES, Inc. *Bilan des Ressources de la Recherche Agricole dans les Pays du Sahel, Volume II: Résumés des Rapports Nationaux.* Washington, D.C.: DEVRES, Inc., August 1984.

0086 World Bank. "West Africa Agricultural Research Review – Country Studies." World Bank, Washington, D.C., 1985. Mimeo.

0171 Casas, Joseph, and Willem Stoop. "Mission Report: Brief Presentation of the Agricultural Research System of Niger." ISNAR, The Hague, May 1986. Mimeo.

0261 Ecole Supérieure de Gestion des Entreprises (ESGE). *Etude- Diagnostic de l'Institut National de Recherches Agronomiques Niger (INRAN): Recommandations a l'Institut de Recherche*. Etude réalisée par l'ESGE pour le CILSS. Niamey, Niger: ESGE, October 1985.

0431 Current Agricultural Research Information System (CARIS). *Directory of Agricultural Research Institutions and Projects in West Africa – Pilot Project 1972–73*. Rome: FAO, 1973.

0532 UNESCO Field Science Office for Africa. *Survey on the Scientific and Technical Potential of the Countries of Africa*. Paris: UNESCO, 1970.

0589 Kassapu, Samuel. *Les Dépenses de Recherche Agricole dans 34 Pays d'Afrique Tropicale*. Paris: Centre de Développement de l'OCDE, 1976.

0782 Institut National de Recherches Agronomiques du Niger (INRAN). *Recherche pour le Développement*. Niamey, Niger: INRAN, n.d.

Additional References:

0010 Boyce, James K., and Robert E. Evenson. *National and International Agricultural Research and Extension Programs*. New York: Agricultural Development Council, Inc., 1975.

0023 Bennell, Paul. *Agricultural Researchers in Sub-Saharan Africa: An Overview*. Working Paper No. 4. The Hague: ISNAR, October 1985.

0045 Institut du Sahel, and DEVRES, Inc. *Bilan des Ressources de la Recherche Agricole dans les Pays du Sahel, Volume I: Analyse et Stratégie Régionale*. Washington, D.C.: DEVRES, Inc., 1984.

0088 Swanson, Burton E., and Wade H. Reeves. "Agricultural Research West Africa: Manpower and Training." World Bank, Washington, D.C., November 1985. Mimeo.

0089 Oram, Peter. "Report on National Agricultural Research in West Africa." IFPRI, Washington, D.C., February 1986. Mimeo.

0163 CGIAR. "National Agricultural Research." CGIAR, Washington, D.C., 1985. Mimeo.

0175 Cooper, St.G.C. *Agricultural Research in Tropical Africa*. Kampala: East African Literature Bureau, 1970.

0258 "Effort Français en Faveur de la Recherche Agronomique dans les Pays du Sahel 1984–1985." 1985. Mimeo.

0266 UNESCO. *National Science Policies in Africa*. Science Policy Studies and Documents No. 31. Paris: UNESCO, 1974.

0360 Cooper, St.G.C. "Towards Trained Manpower for Agricultural Research in Africa." Paper presented at the Conference on Agricultural Research and Production in Africa, organized by the Association for the Advancement of Agricultural Sciences in Africa (AAASA), Addis Ababa, 29 August–4 September 1971.

0372 Gabas, Jean-Jacques. *La Recherche Agricole dans les Pays Membres du CILSS*. Club du Sahel, September 1986.

0373 Groupement d'Etudes et de Recherches pour le Développement de l'Agronomie Tropicale (GERDAT). *Rapport Général d'Activité pour 1982*. Paris: GERDAT, 1983.

0384 Institut du Sahel, and DEVRES, Inc. *Sahel Region Agricultural Research Resource Assessment – Data Base Management Information System: Diskettes and User's Guide*. Printouts. Washington, D.C.: Institut du Sahel, and DEVRES, Inc., n.d.

0466 World Bank – West Africa Agricultural Research Review Team. "National Agricultural Research Systems (NARS): A Regional Appraisal and Selected Issues." World Bank, Washington, D.C., August 1986. Mimeo.

0653 Webster, B.N. *Index of Agricultural Research Institutions and Stations in Africa*. Rome: FAO, n.d.

0747 ISNAR. *International Workshop on Agricultural Research Management*. Report of a workshop. The Hague: ISNAR, 1987.

0750 Soumana, Idrissa. "Strategic Evolution of Planning at the National Institute of Agricultural Research in Niger." Paper presented at the International Workshop on Agricultural Research Management, ISNAR, The Hague, 7–11 September 1987.

0785 Fournier, A. "La Recherche Agronomique Nigérienne: Présentation, Synthèse des Recherches, et Suggestions." Mission USAID Niger, Niamey, Niger, Janvier 1986. Mimeo.

0852 Evenson, Robert E., and Yoav Kislev. *Investment in Agricultural Research and Extension: A Survey of International Data*. Center Discussion Paper No. 124. New Haven, Connecticut: Economic Growth Center, Yale University, August 1971.

Nigeria

Personnel

Period/Year	PhD	MSc	BSc	Subtotal	Expat	Total	Source
1960–64	..	..	..	..	..	172	852
1965–69	..	..	..	..	..	306	268
1970	..	..	..	..	..	..	
1971	..	..	..	..	..	..	
1972	..	..	..	..	..	327	154
1973	..	..	..	..	..	369	154
1974	..	..	..	..	..	..	
1975	..	..	..	..	..	..	
1976	..	..	..	..	..	..	
1977	..	..	..	..	..	..	
1978	..	..	..	..	..	791	155
1979	..	..	..	..	..	..	
1980	..	..	..	..	..	1014	617
1981	..	..	..	..	..	1024	162
1982	..	..	..	..	..	912	86
1983	..	..	..	..	..	1088	17
1984	..	..	..	..	..	986	86
1985	..	..	..	..	..	..	
1986	..	..	..	..	..	..	

1960–64: 1963   1965–69: 1966

Expenditure   *LCU = Naira*

Period/Year	Current LCU (millions)	Constant 1980 LCU (millions)	Constant 1980 US$ (millions) Atlas	Constant 1980 US$ (millions) PPP	Source
1960–64	..	..	..	..	
1965–69	4.530	27.678	44.123	41.492	10
1970	..	..	..	..	
1971	..	..	..	..	
1972	..	..	..	..	
1973	..	..	..	..	
1974	..	..	..	..	
1975	..	..	..	..	
1976	33.368	56.942	90.774	85.362	167
1977	55.344	86.072	137.211	129.030	167
1978	63.248	82.785	131.972	124.104	167
1979	54.978	61.635	98.254	92.936	167
1980	70.792	70.792	112.853	106.124	167
1981	71.500	69.620	110.985	104.368	167
1982	62.155	56.556	90.158	84.783	167
1983	55.855	49.561	79.007	74.297	167
1984	..	..	..	..	
1985	..	..	..	..	
1986	..	..	..	..	

1960–64:   1965–69: 1966, 67 & 68

Personnel Comments:

The table below gives a breakdown of the number of researchers by institute. Source 157 indicates there is a large discrepancy (sometimes 50% or more) between the number of established positions and the number of positions actually filled. This could be a possible explanation for the sometimes large differences between the number of researchers cited by different sources for the same year. Unfortunately, it is not possible to quantify the magnitude of this problem with any degree of precision. For a particular year, estimates of the total number of researchers were based principally on information from a single source, and an attempt was made, wherever possible, to include only positions filled.

	1978	1980	1981	1982	1983	1984
IAR	104	129	129	[129]	166	115
NCRI	159	127	127	77	127	77
NRCRI	30	81	81	75	81	62
NIHORT	[30]	34	34	[34]	41	42
IAR&T	37	44	44	[44]	39	63
CRIN	43	45	49	32	45	28
NIFOR	[48]	68	74	71	68	69
RRIN	24	30	30	31	30	31
FRIN	85	83	83	86	83	80
NVRI	57	120	120	82	120	95
NAPRI	30	47	47	[47]	47	61
NITR	23	40	40	35	40	44
LERIN	7	17	17	18	19	17
LCRI	20	20	20	31	20	30
KLRI	[20]	32	32	27	32	34
NIOMR	33	46	46	42	46	64
AERLS	[30]	33	33	[33]	33	39
NSPRI	11	18	18	[18]	51	35
Total	791	1014	1024	912	1088	986
Source	155	617	162	86	17	86

[ ] = estimated or constructed by authors.

Expenditure Comments:

Conflicting information from different sources is caused mainly by the fact that there is a large discrepancy between the approved budgets of the institutes and the actual funds released by the government. Moreover, the fact that capital investments and recurrent costs are kept seperate means that only the recurrent costs are often reported. Another problem is the large number of institutes for which expenditure figures need to be aggregated in order to construct a system-level figure. The expenditure figures given here for the years 1976 to 1983 seem the most reasonable estimates currently available. They include the research expenditures of the following institutes: IAR, NCRI, NRCRI, NIHORT, IAR&T, CRIN, NIFOR, RRIN, FRIN, NVRI, NAPRI, NITR, LERIN, LCRI, KLRI, NIOMR, AERLS, and NSPRI. In 1980 there was a shift of the budget year, which means that total expenditures for 1980 probably cover a period of 15 months.

Acronyms and Abbreviations:

AERLS: Agricultural Extension, Research Liaison
    Service, Zaira
CRIN: Cocoa Research Institute of Nigeria
FRIN: Forest Research Institute of Nigeria
IAR: Institute of Agricultural Research
IAR&T: Institute of Agricultural Research and Training
KLR: Kainji Lake Research Institute
LCRI: Lake Chad Research Institute
LERI: Leather Research Institute of Nigeria
NAPRI: National Animal Production Research
    Institute

NCRI: National Cereals Research Institute
NIFOR: National Institute for Oil Palm Research
NIHORT: National Institute for Horticultural Research
NIOMR: Nigerian Institute of Oceanography and Marine Research
NIT: National Institute for Trypanosomiasis Research
NRCRI: National Root Crops Research Institute
NSPRI: National Stored Products Research Institute
NVRI : National Veterinary Research Institute
RRIN: Rubber Research Institute of Nigeria

Sources:

0010 Boyce, James K., and Robert E. Evenson. *National and International Agricultural Research and Extension Programs.* New York: Agricultural Development Council, Inc., 1975.

0017 ISNAR, IFARD & AOAD. Survey of National Agricultural Research Systems: Unpublished Questionnaire Responses. ISNAR, The Hague, 1985.

0086 World Bank. "West Africa Agricultural Research Review – Country Studies." World Bank, Washington, D.C., 1985. Mimeo.

0154 Nigerian Council for Science and Technology (NCST). *National Policies and Priorities for Research in Science and Technology.* Lagos: NCST, December 1975.

0155 Idachaba, F.S. *Agricultural Research Policy in Nigeria.* IFPRI Research Report No. 17. Washington, D.C.: IFPRI, 1980.

0162 Bennell, Paul. "Conditions of Service for Agricultural Research Scientists in Nigeria: What Room for Manoeuvre?" ISNAR, The Hague, January 1984. Mimeo.

0167 Idachaba, F.S. "Agricultural Research Policy in Nigeria." Paper presented at Agricultural Research Policy Seminar, University of Minnesota, Minneapolis/St. Paul, 22 April 1986.

0268 Consortium for the Study of Nigerian Rural Development. *A Survey of Nigerian Agricultural Research.* East Lansing, Michigan: Michigan State University, September 1966.

0617 Research Institutes Review Panel – Green Revolution National Committee. *Report of Research Institutes Review Panel 1980/81.* Two volumes. Ibadan, Nigeria: National Committee on Green Revolution, January 1981.

0852 Evenson, Robert E., and Yoav Kislev. *Investment in Agricultural Research and Extension: A Survey of International Data.* Center Discussion Paper No. 124. New Haven, Connecticut: Economic Growth Center, Yale University, August 1971.

Additional References

0014 Judd, M. Ann, James K. Boyce, and Robert E. Evenson. "Investing in Agricultural Supply." Economic Growth Center, Yale University, New Haven, Connecticut, 1983. Mimeo.

0016 Oram, Peter A., and Vishva Bindlish. *Resource Allocations to National Agricultural Research: Trends in the 1970s.* The Hague and Washington, D.C.: ISNAR and IFPRI, November 1981.

0018 Daniels, Douglas, and Barry Nestel, eds. *Resource Allocation to Agricultural Research.* Proceedings of a workshop held in Singapore 8–10 June 1981. Ottawa: IDRC, 1981.

0023 Bennell, Paul. *Agricultural Researchers in Sub-Saharan Africa: An Overview.* Working Paper No. 4. The Hague: ISNAR, October 1985.

0026 Oram, Peter A., and Mark Gieben. "Document Summaries." ISNAR, The Hague, 1984. Mimeo.

0027 Harvey, Nigel, ed. *Agricultural Research Centres: A World Directory of Organizations and Programmes.* Seventh Edition, Two Volumes. Harlow, U.K.: Longman, 1983.

0061 Okoro, D.E., and J.N. Onuoha. "The Impact of the Collaboration between the International Agricultural Research System and the National Agricultural Research System in Nigeria." CGIAR, Washington, D.C., April 1985. Mimeo.

0068 Conte, Stephan. "West Africa Agricultural Research Data Base Building Task: Termination Report." World Bank, Washington, D.C., 1985. Mimeo.

0073 Oram, Peter A., and Vishva Bindlish. "Investment in Agricultural Research in Developing Countries: Progress, Problems, and the Determination of Priorities." IFPRI, Washington, D.C., January 1984. Mimeo.

0088 Swanson, Burton E., and Wade H. Reeves. "Agricultural Research West Africa: Manpower and Training." World Bank, Washington, D.C., November 1985. Mimeo.

0089 Oram, Peter. "Report on National Agricultural Research in West Africa." IFPRI, Washington, D.C., February 1986. Mimeo.

0095 FAO – CARIS. *Agricultural Research in Developing Countries – Volume 1: Research Institutions.* Rome: FAO – CARIS, 1978.

0135 FAO, and UNDP. *National Agricultural Research – Report of an Evaluation Study in Selected Countries.* Rome: FAO and UNDP, 1984.

0156 Idachaba, F.S. "Agricultural Research Resource Allocation Priorities: The Nigerian Experience." Invited paper delivered at the International Workshop on Resource Allocation in National

Agricultural Research Systems Co-sponsored by IFARD and IDRC, Singapore, 8–10 June 1981.

0157 Clark, Norman. "Organizational Aspects of Nigeria's Research System." *Research Policy* Vol. 9 (1980): 148–172.

0163 CGIAR. "National Agricultural Research." CGIAR, Washington, D.C., 1985. Mimeo.

0164 Association for the Advancement of Agricultural Sciences in Africa (AAASA). *Proceedings of the Workshop on Agricultural Research Administration, Nairobi, Kenya, 27–30 June 1977.* Proceedings Series PE-4. Addis Ababa, Ethiopia: AAASA and IDRC, August 1979.

0165 Evenson, R.E., and Y. Kislev. *Agricultural Research and Productivity.* New Haven: Yale University Press, 1975.

0168 Adeyemo, Remi. "Agricultural Research in Nigeria: Priorities and Effectiveness." *Agricultural Administration* Vol. 17 (1984): 81–91.

0175 Cooper, St.G.C. *Agricultural Research in Tropical Africa.* Kampala: East African Literature Bureau, 1970.

0242 Commonwealth Agricultural Bureaux (CAB). *List of Research Workers in Agricultural Sciences in the Commonwealth 1981.* Slough, England: CAB, 1981.

0243 Commonwealth Agricultural Bureaux (CAB). *List of Research Workers in the Agricultural Sciences in the Commonwealth and in the Republic of Ireland 1978.* Slough, England: CAB, 1978.

0244 Commonwealth Agricultural Bureaux (CAB). *List of Research Workers in the Agricultural Sciences in the Commonwealth and in the Republic of Ireland 1975.* Slough, England: CAB, 1975.

0266 UNESCO. *National Science Policies in Africa.* Science Policy Studies and Documents No. 31. Paris: UNESCO, 1974.

0279 Commonwealth Agricultural Bureaux (CAB). *List of Research Workers in the Agricultural Sciences in the Commonwealth and in the Republic of Ireland 1972.* Slough, England: CAB, 1972.

0285 Commonwealth Agricultural Bureaux (CAB). *List of Research Workers in the Agricultural Sciences in the Commonwealth and in the Republic of Ireland 1969.* Slough, England: CAB, 1969.

0286 Commonwealth Agricultural Bureaux (CAB). *List of Research Workers in Agriculture, Animal Health and Forestry in the Commonwealth and in the Republic of Ireland 1966.* Slough, England: CAB, 1966.

0287 Commonwealth Agricultural Bureaux (CAB). *List of Research Workers in Agriculture, Animal Health and Forestry in the British Commonwealth, the Republic of Sudan and the Republic of Ireland 1959.* Slough, England: CAB, 1959.

0300 Idachaba, F.S. "Agricultural Research Resource Allocation Priorities: The Nigerian Experience." In *Resource Allocation to Agricultural Research.* Proceedings of a workshop held in Singapore 8–10 June 1981, edited by Douglas Daniels and Barry Nestel, pp. 104–121. Ottawa: IDRC, 1981.

0360 Cooper, St.G.C. "Towards Trained Manpower for Agricultural Research in Africa." Paper presented at the Conference on Agricultural Research and Production in Africa, organized by the Association for the Advancement of Agricultural Sciences in Africa (AAASA), Addis Ababa, 29 August–4 September 1971.

0400 UNESCO. *The Promotion of Scientific Activity in Tropical Africa.* Science Policy Studies and Documents No. 11. Paris: UNESCO, 1969.

0431 Current Agricultural Research Information System (CARIS). *Directory of Agricultural Research Institutions and Projects in West Africa – Pilot Project 1972–73.* Rome: FAO, 1973.

0466 World Bank – West Africa Agricultural Research Review Team. "National Agricultural Research Systems (NARS): A Regional Appraisal and Selected Issues." World Bank, Washington, D.C., August 1986. Mimeo.

0532 UNESCO Field Science Office for Africa. *Survey on the Scientific and Technical Potential of the Countries of Africa.* Paris: UNESCO, 1970.

0534 Famoriyo, O.A. "Agricultural Research Administration in Nigeria." *Science and Public Policy* Vol. 15, No. 5 (October 1986): 295–300.

0589 Kassapu, Samuel. *Les Dépenses de Recherche Agricole dans 34 Pays d'Afrique Tropicale.* Paris: Centre de Développement de l'OCDE, 1976.

0634 Nigerian Stored Products Institute – Federal Ministry of Trade. *Annual Report 1968.* Lagos, Nigeria: Federal Ministry of Information, 1969.

0651 Ogor, E. "Problems of Research Organizations in Nigeria." In *Papers presented to the Conference on Agricultural Research and Production in Africa, September 1971 – Part 3,* supplement to volume 2 of the Journal of the Association for the Advancement of Agricultural Sciences in Africa. Addis Ababa, Ethiopia: Association for the Advancement of Agricultural Sciences in Africa, June 1975.

0653 Webster, B.N. *Index of Agricultural Research Institutions and Stations in Africa.* Rome: FAO, n.d.

0744 *Agricultural Research Centres: A World Directory of Organizations and Programmes.* Eighth Edition, Two Volumes. Harlow, U.K.: Longman, 1986.

0770 Ruttan, Vernon W. *Agricultural Research Policy and Development.* FAO Research and Technology Paper 2. Rome: FAO, 1987.

0886 Evenson, Robert E., and Yoav Kislev. "Investment in Agricultural Research and Extension: A Survey of International Data." *Economic Development and Cultural Change* Vol. 23 (April 1975): 507–521.

Norway

Personnel

Period/Year	PhD	MSc	BSc	Subtotal	Expat	Total	Source
1960–64	..	..	..	..	..	410	523
1965–69	..	..	..	..	..	447	523
1970	..	..	..	..	..	457	523
1971	..	..	..	..	..	464	523
1972	..	..	..	..	..	546	523
1973	..	..	..	..	..	558	523/859
1974	..	..	..	..	..	574	523
1975	..	..	..	..	..	614	523/859
1976	..	..	..	..	..	654	523
1977	..	..	..	..	..	618	523/859
1978	..	..	..	..	..	638	523
1979	..	..	..	..	..	684	523/859
1980	..	..	..	..	..	729	523
1981	..	..	..	..	..	736	523/859
1982	..	..	..	..	..	752	523
1983	..	..	..	..	..	769	523/859
1984	..	..	..	..	..	756	523
1985	..	..	..	..	..	780	523/859
1986	..	..	..	..	..	..	

1960–64: 1963   1965–69: 1966, 67, 68 & 69

Expenditure

LCU = Norwegian Kroner

Period/Year	Current LCU (millions)	Constant 1980 LCU (millions)	Constant 1980 US$ (millions)		Source
			Atlas	PPP	
1960–64	40.900	132.362	25.322	21.076	523/494
1965–69	73.225	195.760	37.451	31.170	523/494
1970	88.700	198.879	38.047	31.667	523
1971	92.700	194.748	37.257	31.009	523
1972	114.000	228.457	43.705	36.376	523
1973	128.100	235.046	44.966	37.426	523/859
1974	148.200	246.589	47.174	39.264	523
1975	179.300	271.256	51.893	43.191	523/859
1976	221.300	311.252	59.544	49.560	523
1977	240.500	312.338	59.752	49.732	523/859
1978	280.900	342.561	65.534	54.545	523
1979	297.800	340.732	65.184	54.254	523/859
1980	322.800	322.800	61.754	51.398	523
1981	380.000	333.333	63.769	53.076	523/859
1982	450.500	358.678	68.617	57.111	523
1983	502.100	376.952	72.113	60.021	523/859
1984	529.600	373.484	71.450	59.469	523
1985	556.400	372.922	71.342	59.379	523/859
1986	..	..	..	..	

1960–64: 1963   1965–69: 1966, 67, 68 & 69

Personnel Comments:

The Agricultural Research Council of Norway (NLVF), which is attached to the Ministry of Agriculture, was formed in 1949. According to source 499, it was charged with the task of coordinating all research in agriculture (including research in forestry and freshwater fisheries) as well as advising the Ministry on research policy orientations and priorities. However, sea fisheries research is excluded from the mandate of the Ministry of Agriculture and NLVF. It falls within the mandate of the Ministry of Fisheries and the Fisheries Research Council (NFFR, founded in 1972). Given this organizational structure, the figures given by source 523 exclude sea-fisheries research.

Sources 494 and 859, however, provide figures on sea-fisheries research. The data given by these sources are not as complete as the data given by source 523. Therefore some earlier personnel figures have been constructed on the basis of sea-fisheries research expenditures relative to other agricultural research expenditures for these years. Over the whole period, several sea-fisheries research personnel figures have been based on interpolation. Source 523 and 859 both give personnel figures in full-time equivalents, and both sources also include university researchers.

The table below gives an overview of how the personnel figures have been constructed.

	1963	1966	1967	1968	1969	1970	1971	1972	1973	1974
NLVF	342	360	364	376	386	369	372	450	458	459
NFFR[a]	[68]	[58]	[78]	[81]	[84]	[88]	[92]	[96]	100*	[115]
Total	410	418	442	457	470	457	464	546	558	574
Source	523	523	523	523	523	523	523	523	523 859*	523

	1975	1976	1977	1978	1979	1980	1981	1982	1983	1984	1985
NLVF	484	509	458	463	494	520	509	516	525	509	530
NFFR[a]	130*	[145]	160*	[175]	190*	[209]	227*	[236]	244*	[247]	250*
Total	614	654	618	638	684	729	736	752	769	756	78
Source	523 859*	523	523 859*	523	523 859*	523	523 859*	523	523 859*	523	523 859*

[ ] = estimated or constructed by authors.

a. It is assumed that the NFFR figures before 1972 represent the sea fisheries research institutes which were merged into NFFR in 1972.

Expenditure Comments:
The expenditure figures have been constructed as follows:

	1963	1966	1967	1968	1969	1970	1971	1972	1973	1974	1975
NLVF	34.1	50.7	56.1	66.3	69.6	70.7	72.7	92.0	104.3	117.5	141.8
NFFR[a]	6.8*	8.2*	12.0*	[14.0]	[16.0]	[18.0]	[20.0]	[22.0]	23.8*	[30.7]	37.5
Total	40.9	58.9	68.1	80.3	85.6	88.7	92.7	114.0	128.1	148.2	179.3
Source	523 494*	523 494*	523 494*	523	523	523	523	523	523 859*	523	523

	1976	1977	1978	1979	1980	1981	1982	1983	1984	1985
NLVF	175.8	187.1	218.0	225.4	230.3	267.5	324.8	363.2	379.2	394.5
NFFR[a]	[45.5]	53.4*	[62.9]	72.4*	[92.5]	112.5*	[125.7]	138.9*	[150.4]	161.9*
Total	221.3	240.5	280.9	297.8	322.8	380.0	450.5	502.1	529.6	556.4
Source	523	523 859*	523	523 859*	523	523 859*	523	523 859*	523	523 859*

[ ] = estimated or constructed by authors.

a. It is assumed that the NFFR figures before 1972 represent the sea fisheries research institutes which were merged into NFFR in 1972.

Acronyms and Abbreviations:
NLVF: (Agricultural Research Council of Norway)
NFFR: (Norwegian Fisheries Research Council)

Sources:
0494 OECD. *Reviews of National Science Policy: Norway.* Paris: OECD, 1970.
0523 Sundsbø, Svein, and Mary Villa. Personal communication. Agricultural Research Council of Norway (NLVF). Oslo, March 1987.
0859 Vaage, Roald, and Endre Bjørgum. Personal communication. Norwegian Fisheries Research Council. Trondheim, Norway, April 1988.

Additional References:
0010 Boyce, James K., and Robert E. Evenson. *National and International Agricultural Research and Extension Programs.* New York: Agricultural Development Council, Inc., 1975.
0021 OECD. *OECD Science and Technology Indicators.* Paris: OECD, 1984.

0022 UNESCO. *Statistical Yearbook 1983.* Paris: UNESCO, 1983.
0027 Harvey, Nigel, ed. *Agricultural Research Centres: A World Directory of Organizations and Programmes.* Seventh Edition, Two Volumes. Harlow, U.K.: Longman, 1983.
0165 Evenson, R.E., and Y. Kislev. *Agricultural Research and Productivity.* New Haven: Yale University Press, 1975.
0376 FAO. *Fourteenth FAO Regional Conference for Europe, Reykjavik, Iceland, 17–21 September 1984: Research in Support of Agricultural Policies in Europe.* Rome: FAO, 1984.
0461 Norges Landbruksvitenskapelige Forskningsråd (NLVF). *Langtidsplan for Norsk Landbruksforskning fram til 1990.* Oslo: NLVF, 1984.
0473 OECD. *Science and Technology Indicators: Basic Statistical Series – Volume A: The Objectives of Government R&D Funding 1974–1985.* Paris: OECD, May 1983.
0474 OECD. *Science and Technology Indicators: Basic Statistical Series – Volume A: The Objectives of*

Government R&D, 1969–1981. Paris: OECD, July 1981.

0499 OECD. Reviews of National Science Policy: Norway. Paris: OECD, 1985.

0501 OECD. Country Reports on the Organisation of Scientific Research: Norway. Paris: OECD, April 1965.

0502 OECD. Policies for Science and Education, Country Reviews: Norway. Paris: OECD, October 1962.

0509 OECD. Survey of the Resources Devoted to R&D by OECD Member Countries, International Statistical Year 1971 – Volume 5: Total Tables, Statistical Tables and Notes. Paris: OECD, August 1974.

0510 OECD. International Survey of the Resources Devoted to R&D in 1967 by OECD Member Countries, Statistical Tables and Notes – Volume 5: Total Tables. Paris: OECD, August 1970.

0511 OECD. International Survey of the Resources Devoted to R&D in 1969 by OECD Member Countries, Statistical Tables and Notes – Volume 5: Total Tables. Paris: OECD, June 1973.

0512 OECD. Survey of the Resources Devoted to R&D by OECD Member Countries, International Statistical Year 1973 – Volume 5: Total Tables, Statistical Tables and Notes. Paris: OECD, September 1976.

0513 OECD. International Survey of the Resources Devoted to R&D by OECD Member Countries, International Statistical Year 1975 – International Volume: Statistical Tables and Notes. Paris: OECD, March 1979.

0514 OECD. A Study of Resources Devoted to R&D in OECD Member Countries in 1963/64, International Statistical Year for Research and Development – Volume 2: Statistical Tables and Notes. Paris: OECD, 1968.

0530 OECD. OECD Science and Technology Indicators No. 2 – R&D, Invention and Competitiveness. Paris: OECD, 1986.

0601 OECD – STIID Data Bank. "Public Funding of R&D by Socio-Economic Objective." Tables. OECD, Paris, April 1987. Mimeo.

0602 OECD – STIID Data Bank. "R&D Expenditure in the Higher Education and Private Non-Profit Sectors." Tables. OECD, Paris, April 1987. Mimeo.

0621 NORDFORSK. Forsknings Virksomhet i Norden i 1967 – Utgifter og Personale. Nordisk Statistisk Skriftserie 18. Oslo: NORDFORSK, 1970.

0623 Norges Landbruksvitenskapelige Forskningsråd (NLVF). Landbruksforskning Personale og Kostnader i 1966. Oslo: NLVF, 1967.

0624 Norges Landbruksvitenskapelige Forskningsråd (NLVF). Landbruksforskning Personale og Kostnader i 1963. Oslo: NLVF, 1964.

0625 NAVF, NLVF, and NTNF. Norsk Forskningsvirksomhet Utgifter og Personale Innen Forskning og Utviklingsarbeid 1968. Oslo: NAVF, NLVF, and NTNF, June 1970.

0626 NAVF, NLVF, and NTNF. Norsk Forskningsvirksomhet Utgifter og Arsverk 1966. Oslo: NAVF, NLFV, and NTNF, September 1968.

0627 NAVF, NLVF, and NTNF. Utgifter og Arsverk i Norsk Forskning – En Statistik Oversikt for Aret 1963. Oslo: NAVF, NLVF, and NTNF, 1966.

0628 Norges Landbruksvitenskapelige Forskningråd (NLVF). Forlag om Utbygging av de Landbruksvitenskapelige Institusjoner 1961–1970. Oslo: NLVF, 1960.

0707 OECD – STIID Data Bank. "R&D Expenditure in the Higher Education and Private Non-Profit Sectors: Higher Education, Agricultural Sciences, General Universities Funds." Table. OECD, Paris, June 1987. Mimeo.

0708 OECD – STIID Data Bank. "Gross Domestic Expenditure on R&D: All Fields of Sciences – Agriculture, Forestry and Fishing." Tables. OECD, Paris, June 1987. Mimeo.

0848 Forskningsrådenes Statistikkutvalg (FSU). FoU-Statistikk 1985, Forsknings- og Utviklingsarbeid: Utgifter og Personale. Oslo: FSU, August 1987.

0849 Forskningsrådenes Statistikkutvalg (FSU). FoU-Statistikk 1985: Universitets- og Høgskolesektoren. Oslo: FSU, October 1987.

0850 Forskningsrådenes Statistikkutvalg (FSU). FoU-Statistikk 1985: Instituttsektoren. Oslo: FSU, August 1987.

0851 Norwegian Research Councils' Committee for R&D Statistics (FSU). R&D Statistics 1985–87 (Resources Devoted to R&D in Norway). Oslo: FSU, August 1987.

0852 Evenson, Robert E., and Yoav Kislev. Investment in Agricultural Research and Extension: A Survey of International Data. Center Discussion Paper No. 124. New Haven, Connecticut: Economic Growth Center, Yale University, August 1971.

0864 Norwegian Research Councils' Committee for R&D Statistics. R&D Statistics: Resources Devoted to R&D in Norway. Oslo: Norwegian Research Councils' Committee for R&D Statistics, March 1986.

0886 Evenson, Robert E., and Yoav Kislev. "Investment in Agricultural Research and Extension: A Survey of International Data." Economic Development and Cultural Change Vol. 23 (April 1975): 507–521.

Oman

Personnel

Period/Year	PhD	MSc	BSc	Subtotal	Expat	Total	Source
1960–64	..	..	..	..	..	..	
1965–69	..	..	..	..	..	..	
1970	..	..	..	..	..	..	
1971	..	..	..	..	..	..	
1972	..	..	..	..	..	..	
1973	..	..	..	..	..	..	
1974	..	..	..	..	..	..	
1975	..	..	..	..	..	..	
1976	..	..	..	..	..	10	95
1977	..	..	..	..	..	..	
1978	..	..	..	..	..	..	
1979	..	..	..	..	..	..	
1980	..	..	..	..	..	..	
1981	..	..	..	..	..	..	
1982	..	..	..	..	..	..	
1983	1	2	7	10	32	42	17
1984	..	..	..	..	..	..	
1985	..	..	..	..	..	..	
1986	..	..	..	..	..	..	

1960–64:   1965–69:

Expenditure   *LCU = Rials Omani*

Period/Year	Current LCU (millions)	Constant 1980 LCU (millions)	Constant 1980 US$ (millions)		Source
			Atlas	PPP	
1960–64	..	..	..	..	
1965–69	..	..	..	..	
1970	..	..	..	..	
1971	..	..	..	..	
1972	..	..	..	..	
1973	..	..	..	..	
1974	..	..	..	..	
1975	..	..	..	..	
1976	..	..	..	..	
1977	..	..	..	..	
1978	..	..	..	..	
1979	..	..	..	..	
1980	..	..	..	..	
1981	..	..	..	..	
1982	..	..	..	..	
1983	..	..	..	..	
1984	..	..	..	..	
1985	..	..	..	..	
1986	..	..	..	..	

1960–64:   1965–69:

Personnel Comments:

The 1976 figure includes crop research only. The 1983 figure has been constructed as follows:

	PhD	MSc	BSc	Expat	Total
Directorate of Agricultural Research	1	1	5	25	32
Directorate of Animal Wealth	0	1	2	7	10
Total	1	2	7	32	42

Expenditure Comments:

Total expenditures of the Directorate of Agricultural Research in 1983 were 896,121 Rials Omani (approximately US$ 2.6 million). No expenditure figures, however, were given in the ISNAR, IFARD & AOAD questionnaire for the research section of the Directorate of Animal Wealth.

Sources:

0017 ISNAR, IFARD & AOAD. Survey of National Agricultural Research Systems: Unpublished Questionnaire Responses. ISNAR, The Hague, 1985.

0095 FAO – CARIS. *Agricultural Research in Developing Countries – Volume 1: Research Institutions.* Rome: FAO – CARIS, 1978.

Additional References:

0094 FAO – Near East Regional Office. *Directory of Agricultural Research Institutions in the Near East Region.* Cairo: FAO – Near East Regional Office, 1979.

0831 FAO. *Report of an Expert Consultation on Agricultural Extension and Research Linkages in the Near East, Amman, Jordan, 22–26 July 1985.* Rome: FAO, 1985.

0865 Oram, Peter. "Agricultural Research Objectives and Priorities: Constraints to the Development of Agricultural Research Institutions in Arab countries." Review paper prepared for the first meeting of the Council for Arab Agricultural Research, organized by the Arab Fund for Economic and Social Development, Kuwait, 23 March 1988.

Pakistan

Personnel

Period/Year	PhD	MSc	BSc	Subtotal	Expat	Total	Source
1960–64	..	..	..	..	..	..	
1965–69	129	552	722	1403	..	1403	454
1970	..	..	..	..	..	..	
1971	..	..	..	..	..	..	
1972	..	..	..	..	..	..	
1973	..	..	..	..	..	..	
1974	..	..	..	..	..	..	
1975	..	..	..	..	..	..	
1976	..	..	..	..	..	..	
1977	..	..	..	..	..	..	
1978	233	1405	1196	2834	..	2834	296
1979	..	..	..	..	..	..	
1980	..	..	..	..	..	..	
1981	221	1538	754	2513	..	2513	587
1982	..	..	..	..	..	..	
1983	..	..	..	..	..	..	
1984	299	2181	951	3431	..	3431	587
1985	..	..	..	..	..	..	
1986	..	..	..	..	..	..	

1960–64:   1965–69: 1968

Expenditure   *LCU = Pakistan Rupees*

Period/Year	Current LCU (millions)	Constant 1980 LCU (millions)	Constant 1980 US$ (millions) Atlas	PPP	Source
1960–64	9.912	44.153	4.450	12.710	281
1965–69	14.994	52.582	5.299	15.136	281
1970	10.412	32.037	3.229	9.222	281
1971	13.493	39.569	3.988	11.390	281
1972	15.265	41.938	4.227	12.072	281
1973	17.333	41.171	4.149	11.851	281
1974	38.036	73.429	7.400	21.137	281
1975	77.953	122.954	12.392	35.394	281
1976	97.001	136.429	13.750	39.272	281
1977	99.783	126.789	12.778	36.498	281
1978	123.662	144.128	14.526	41.489	281
1979	146.817	162.229	16.350	46.699	281
1980	168.573	168.573	16.989	48.525	281
1981	..	..	..	..	
1982	..	..	..	..	
1983	..	..	..	..	
1984	..	..	..	..	
1985	..	..	..	..	
1986	..	..	..	..	

1960–65: 1960, 61, 62, 63 & 64   1965–69: 1965, 66, 67, 68 & 69

Personnel Comments:

The personnel indicators more or less include all agricultural researchers working at the federal and provincial agricultural research institutes, Pakistan Agricultural Research Council (PARC), and the agricultural universities. In the case of university researchers, no full-time equivalent (FTE) ratio has been applied. Source 296 indicates that 632 of the 2834 researchers were university researchers and held the following qualifications: 120 PhD's, 385 MSc's, and 127 BSc's. For 1984, source 587 reports 689 university researchers, of which 151 had PhD's, 477 MSc's, and 61 BSc's. Although not documented, it is understood that the time spent on research by these university researchers is rather low (10–30%). This would mean that the number of researchers, especially those with a graduate degree, if expressed in FTEs would be considerably less than the estimate presented in the indicator series.

Expenditure Comments:

The expenditure figures given by source 281 are constructed by summing the agricultural research expenditures of the federal and provincial governments, Pakistan Agricultural Research Council (from 1973–74 on), the major donor agencies (USAID and the Australian Government), and the Pakistan Central Cotton Committee.

No clear indication is given by source 281 whether expenditures for university research are included or not. Source 587 gives the following figures for 1978, 1981, and 1984: (respectively) 227.4, 398.5, and 786.4 million Rupees. They explicitly include federal and provincial institutes, PARC, and agricultural universities. These figures, however, are budget figures. In the case of PARC, source 587 indicates that real expenditures over the period 1978–1984 fluctuated between 54% and 92% of the budgeted figure.

Sources:

0281 Nagy, Joseph Gilbert. "The Pakistan Agricultural Development Model: An Economic Evaluation of Agricultural Research and Extension Expenditures." PhD diss., University of Minnesota, Minneapolis, March 1984.

0296 Ahmad, Malik Mushtaq. "Resource Allocation to Agricultural Research in Pakistan." In *Resource Allocation to Agricultural Research*. Proceedings of a workshop held in Singapore 8–10 June 1981, edited by Douglas Daniels and Barry Nestel, pp. 55–60. Ottawa: IDRC, 1981.

0454 Wahid, Abdul. "State of Current Agricultural Research and Education in Pakistan." *Agriculture Pakistan* Vol. 19, No. 3 (1968): 261–285.

0587 Pakistan Agricultural Research Council (PARC). *National Agricultural Research Plan*. Islamabad: PARC – Directorate of Publications, 1986.

Additional References:

0010 Boyce, James K., and Robert E. Evenson. *National and International Agricultural Research and Extension Programs*. New York: Agricultural Development Council, Inc., 1975.

0014 Judd, M. Ann, James K. Boyce, and Robert E. Evenson. "Investing in Agricultural Supply." Economic Growth Center, Yale University, New Haven, Connecticut, 1983. Mimeo.

0016 Oram, Peter A., and Vishva Bindlish. *Resource Allocations to National Agricultural Research: Trends in the 1970s*. The Hague and Washington, D.C.: ISNAR and IFPRI, November 1981.

0018 Daniels, Douglas, and Barry Nestel, eds. *Resource Allocation to Agricultural Research*. Proceedings of a workshop held in Singapore 8–10 June 1981. Ottawa: IDRC, 1981.

0022 UNESCO. *Statistical Yearbook 1983*. Paris: UNESCO, 1983.

0026 Oram, Peter A., and Mark Gieben. "Document Summaries." ISNAR, The Hague, 1984. Mimeo.

0027 Harvey, Nigel, ed. *Agricultural Research Centres: A World Directory of Organizations and Programmes*. Seventh Edition, Two Volumes. Harlow, U.K.: Longman, 1983.

0073 Oram, Peter A., and Vishva Bindlish. "Investment in Agricultural Research in Developing Countries: Progress, Problems, and the Determination of Priorities." IFPRI, Washington, D.C., January 1984. Mimeo.

0093 Bangladesh Agricultural Research Council (BARC), ISNAR, and Winrock International. *Management of Human Resources in Agricultural Research*. Report of the International Workshop on Management of Human Resources in Agricultural Research held March 3–5, 1986 in Dhaka, Bangladesh. Dhaka: BARC, ISNAR and Winrock International, 1986.

0094 FAO – Near East Regional Office. *Directory of Agricultural Research Institutions in the Near East Region*. Cairo: FAO – Near East Regional Office, 1979.

0095 FAO – CARIS. *Agricultural Research in Developing Countries – Volume 1: Research Institutions.* Rome: FAO – CARIS, 1978.

0135 FAO, and UNDP. *National Agricultural Research – Report of an Evaluation Study in Selected Countries.* Rome: FAO and UNDP, 1984.

0165 Evenson, R.E., and Y. Kislev. *Agricultural Research and Productivity.* New Haven: Yale University Press, 1975.

0173 FAO. *FAO Near East Regional Studies on Organization and Administration of Agricultural Research.* Rome: FAO, 1974.

0174 Watson, J.M. "Comparative Study of Agricultural Research Organisation and Administration in the Near East Region." Paper presented at the Workshop on Organization and Administration of Agricultural Services in the Arab States, Cairo, 2–15 March 1964.

0179 The Philippine Council for Agriculture and Resources Research and Development (PCARRD), and IDRC. *Agriculture and Resources Research Manpower Development in South and Southeast Asia.* Proceedings of the Workshop on Agriculture and Resources Research Manpower Development in South and Southeast Asia, 21–23 October 1981 and 4–6 February 1982, Singapore. Los Baños, Philippines: PCARRD, 1983.

0190 Pakistan Agricultural Research Council (PARC). *Directory of PARC Scientists and Technicians.* Islamabad, Pakistan: PARC, 1983.

0191 Pakistan Agricultural Research Council (PARC). *Annual Report 1985.* Islamabad, Pakistan: PARC, 1985.

0192 Wahid, Abdul. *Review of Agricultural Research System in Pakistan.* Islamabad, Pakistan: PARC, 1982.

0193 Joint Pakistan-American Agricultural Research Review Team. *Report of the Joint Pakistan-American Agricultural Research Review Team.* Dacca: Bangladesh Agricultural Research Council, April 1968.

0285 Commonwealth Agricultural Bureaux (CAB). *List of Research Workers in the Agricultural Sciences in the Commonwealth and in the Republic of Ireland 1969.* Slough, England: CAB, 1969.

0286 Commonwealth Agricultural Bureaux (CAB). *List of Research Workers in Agriculture, Animal Health and Forestry in the Commonwealth and in the Republic of Ireland 1966.* Slough, England: CAB, 1966.

0287 Commonwealth Agricultural Bureaux (CAB). *List of Research Workers in Agriculture, Animal Health and Forestry in the British Commonwealth, the Republic of Sudan and the Republic of Ireland 1959.* Slough, England: CAB, 1959.

0318 Muhammed, A. "Pakistan's Development Perspective: Status of Agricultural Research Set-Up and Its Role in National Development." In *Strengthening National Agricultural Research.* Report from a SAREC workshop, 10–17 September 1979, edited by Bo Bengtsson and Getachew Tedla, pp. 112–122. Sweden: SAREC, 1980.

0330 Rizvi, M.H. "Managing the Human Resources in Agricultural Research in Pakistan." In *Management of Human Resources in Agricultural Research.* Report of the International Workshop on Management of Human Resources in Agricultural Research held 3–5 March 1986, edited by Theodore Hutchcroft, pp. 110–123. Dhaka: Bangladesh Agricultural Research Council, ISNAR and Winrock International, 1986.

0396 Muhammed, Amir. "National Agricultural Research System of Pakistan." Paper presented at the 1st International Meeting of National Agricultural Research Systems and 2nd IFARD Global Convention, IFARD/EMBRAPA, Brasília, Brasil, 6–11 October 1986.

0590 UNESCO. *Science and Technology in Countries of Asia and the Pacific.* Science Policy Studies and Documents No. 52. Paris: UNESCO, n.d.

0726 Webster, Brian, Carlos Valverde, and Alan Fletcher, eds. *The Impact of Research on National Agricultural Development.* Report on the First International Meeting of National Agricultural Research Systems and the Second IFARD Global Convention. The Hague: ISNAR, July 1987.

0727 Agricultural Research Review Committee. *Review of Agricultural Research Institutions in Pakistan.* Islamabad: Agricultural Research Division – Government of Pakistan, November 1984.

0728 Johnson, R.W.M., S. Mahfooz Ali Shah, Ghulam Rasul, and Sohail Munawar. *Agricultural Research Manpower Survey.* Islamabad: Pakistan Agricultural Research Council – Directorate of Planning, n.d.

0732 FAO – Regional Office for Asia and the Pacific. *Agricultural Research Systems in the Asia-Pacific Region.* Bangkok: FAO – Regional Office for Asia and the Pacific, 1986.

0767 Muhammed, Amir. "The National Agricultural Research System of Pakistan." In *The Impact of Research on National Agricultural Development,* edited by Brian Webster, Carlos Valverde, and Alan Fletcher, pp. 183–191. The Hague: ISNAR, July 1987.

0796 Hasnain, H., and M.H. Rizvi. "An Inventory of Agricultural Research Manpower in Pakistan." In *Agriculture and Resources Research Manpower Development in South and Southeast Asia,* pp.

207–215. Los Baños, Philippines: Philippine Council for Agriculture and Resources Research and Development, 1983.

0831 FAO. *Report of an Expert Consultation on Agricultural Extension and Research Linkages in the Near East, Amman, Jordan, 22–26 July 1985.* Rome: FAO, 1985.

0852 Evenson, Robert E., and Yoav Kislev. *Investment in Agricultural Research and Extension: A Survey of International Data.* Center Discussion Paper No. 124. New Haven, Connecticut: Economic Growth Center, Yale University, August 1971.

0865 Oram, Peter. "Agricultural Research Objectives and Priorities: Constraints to the Development of Agricultural Research Institutions in Arab countries." Review paper prepared for the first meeting of the Council for Arab Agricultural Research, organized by the Arab Fund for Economic and Social Development, Kuwait, 23 March 1988.

0878 Pray, C.E. "The Economics of Agricultural Research in British Punjab and Pakistani Punjab, 1905–1975." PhD diss., University of Pennsylvania, Pennsylvania, 1978.

0886 Evenson, Robert E., and Yoav Kislev. "Investment in Agricultural Research and Extension: A Survey of International Data." *Economic Development and Cultural Change* Vol. 23 (April 1975): 507–521.

Panama

Personnel

Period/Year	PhD	MSc	BSc	Subtotal	Expat	Total	Source
1960–64	..	..	..	..	..	..7	10
1965–69	..	..	..	..	..	..	
1970	..	..	..	..	..	..	
1971	..	..	..	..	..	..	
1972	..	..	..	..	..	..	
1973	..	..	..	..	..		
1974					..		
1975	2	8	13	23	..	23	218
1976	2	11	17	30	..	30	218
1977	3	12	18	33	..	33	218
1978	3	14	21	38	..	38	218
1979	5	16	30	51	..	51	218
1980	5	17	42	64	..	64	218
1981	5	19	45	69	..	69	218
1982	8	28	86	122	..	122	218
1983	8	32	89	129	6	135	17
1984	9	33	91	133	..	133	218
1985	..	..	..		..		
1986	9	41	90	140	..	140	833

1960–64: 1961   1965–69:

Expenditure

LCU = Balboas

Period/Year	Current LCU (millions)	Constant 1980 LCU (millions)	Constant 1980 US$ (millions)		Source
			Atlas	PPP	
1960–64	0.263	0.625	0.618	0.905	10
1965–69	..	..	..	..	
1970	..	..	..	..	
1971	..	..	..	..	
1972	..	..	..	..	
1973	..	..	..	..	
1974	..	..	..	..	
1975	0.492	0.701	0.693	1.016	208
1976	0.884	1.204	1.191	1.745	216
1977	1.080	1.406	1.390	2.038	216
1978	1.146	1.382	1.367	2.003	216
1979	2.597	2.866	2.834	4.154	216
1980	3.553	3.553	3.512	5.149	216
1981	3.429	3.278	3.241	4.750	218
1982	5.438	4.971	4.914	7.203	218
1983	4.125	3.703	3.661	5.366	218
1984	5.910	5.060	5.002	7.332	218
1985	4.740	3.953	3.908	5.729	338
1986	..	..	..	..	

1960–64: 1961   1965–69:

Personnel Comments:

These figures refer to the 'Instituto de Investigación Agropecuaria Panamá' (IDIAP) only. Source 218 reports only national researchers. The ISNAR, IFARD & AOAD questionnaire (source 17) gives additional information on other organizations which conduct agricultural research in Panama, namely, United Fruit Company (1983: 7 researchers), Compañía Panameña de Alimentos (NESTLE), Cítricos de Chirique, and the Faculty of Agriculture of the University of Panama (1983: 32 researchers).

Expenditure Comments:

These expenditure figures refer to the 'Instituto de Investigación Agropecuaria Panamá' (IDIAP) only. According to source 17, other organizations which conduct agricultural research are United Fruit Company (1983: B$ 0.5 million research expenditure), Compañía Panameña de Alimentos (NESTLE), Cítricos de Chirique, and the Faculty of Agriculture of the University of Panama (1983: B$ 0.6 million research expenditure).

Sources:

0010 Boyce, James K., and Robert E. Evenson. *National and International Agricultural Research and Extension Programs.* New York: Agricultural Development Council, Inc., 1975.

0017 ISNAR, IFARD & AOAD. Survey of National Agricultural Research Systems: Unpublished Questionnaire Responses. ISNAR, The Hague, 1985.

0208 Segura, M. "Diagnóstico de la Investigación Agropecuaria en el Istmo Centroamericano." IICA, Guatemala, 1979. Mimeo.

0216 Rosales, Franklin E. "Situación del Sistema Nacional de Investigación Agronómica en Panamá." IICA – OEA, San José, December 1980. Mimeo.

0218 FAO. *Informe Sobre el Instituto de Investigación Agropecuaria (IDIAP) de Panamá.* Rome: FAO, June 1985.

0338 Cuellar, Miguel, Damaris Chea, Jorge Jonas, and Hermel López. "Estudio del Manejo y Organización de la Investigación en Fincas – Caso: Panamá." Paper prepared for the ISNAR Study on the Organization and Management of On-Farm Research. ISNAR, The Hague, September 1986. Mimeo.

0833 Noriega, Tomás. Personal communication. Director General, Instituto de Investigación Agropecuaria de Panamá. Panamá, March 1988.

Additional References:

0001 Trigo, Eduardo J., and Martín E. Piñeiro. "Funding Agricultural Research." In *Report of a Conference: Selected Issues in Agricultural Research in Latin America*, edited by Barry Nestel and Eduardo J. Trigo. The Hague: ISNAR, March 1984.

0014 Judd, M. Ann, James K. Boyce, and Robert E. Evenson. "Investing in Agricultural Supply." Economic Growth Center, Yale University, New Haven, Connecticut, 1983. Mimeo.

0016 Oram, Peter A., and Vishva Bindlish. *Resource Allocations to National Agricultural Research: Trends in the 1970s.* The Hague and Washington, D.C.: ISNAR and IFPRI, November 1981.

0053 Piñeiro, Martín, and Eduardo Trigo, eds. *Cambio Técnico en el Agro Latino Americano – Situación y Perspectivas en la Década de 1980.* San José: IICA, 1983.

0073 Oram, Peter A., and Vishva Bindlish. "Investment in Agricultural Research in Developing Countries: Progress, Problems, and the Determination of Priorities." IFPRI, Washington, D.C., January 1984. Mimeo.

0095 FAO – CARIS. *Agricultural Research in Developing Countries – Volume 1: Research Institutions.* Rome: FAO – CARIS, 1978.

0165 Evenson, R.E., and Y. Kislev. *Agricultural Research and Productivity.* New Haven: Yale University Press, 1975.

0172 Trigo, Eduardo. Background material. ISNAR, The Hague, n.d.

0217 Elliott, Howard, Reed Hertford, Judith Snow, and Eduardo Trigo. *Identificación de Oportunidades para el Mejoramiento de los Sistemas de Gestión de Tecnología Agropecuaria en Latino América: Una Metodología y Caso de Ensayo.* New Jersey and The Hague: Rutgers University and ISNAR, 1985.

0464 Instituto de Investigación Agropecuaria de Panamá (IDIAP). *Memoria Anual 1982.* Panama: IDIAP, 1982.

0852 Evenson, Robert E., and Yoav Kislev. *Investment in Agricultural Research and Extension: A Survey of International Data.* Center Discussion Paper No. 124. New Haven, Connecticut: Economic Growth Center, Yale University, August 1971.

0886 Evenson, Robert E., and Yoav Kislev. "Investment in Agricultural Research and Extension: A Survey of International Data." *Economic Development and Cultural Change* Vol. 23 (April 1975): 507–521.

Papua New Guinea

Personnel

Period/Year	PhD	MSc	BSc	Subtotal	Expat	Total	Source
1960–64	..	..	..	..	..	..	
1965–69	4	4	53	61	..	61	286/285
1970	..	..	..	..	..		
1971	12	16	79	107	..	107	279
1972	..	..	..	..	..	..	
1973	..	..	..	..	..	106	470
1974	..	..	..	..			
1975	..	..	..	..			
1976	..	..	..	..	..	73	95
1977	..	..	..	..			
1978	..	..	..	..			
1979	..	..	..	..			
1980	..	..	..	42	65	107	470
1981	..	..	..	..			
1982	..	..	..	..	..		
1983	..	..	..	..			
1984	..	..	..	..			
1985	..	..	..	..			
1986	..	..	..	..	..		

1960–64:   1965–69: 1965 & 68

Expenditure

LCU = Kina

Period/Year	Current LCU (millions)	Constant 1980 LCU (millions)	Constant 1980 US$ (millions)		Source
			Atlas	PPP	
1960–64	..	..	..	..	
1965–69	..	..	..	..	
1970	..	..	..	..	
1971	..	..	..	..	
1972	..	..	..	..	
1973	..	..	..	..	
1974	..	..	..	..	
1975	..	..	..	..	
1976	..	..	..	..	
1977	..	..	..	..	
1978	..	..	..	..	
1979	..	..	..	..	
1980	6.689	6.689	9.518	18.044	468/482
1981	7.072	7.220	10.273	19.477	469/482
1982	8.118	7.955	11.319	21.458	469/482
1983	..	..	..	..	
1984	..	..	..	..	
1985	..	..	..	..	
1986	..	..	..	..	

1960–64:   1965–69:

Personnel Comments:

The figures prior to 1975 represent the number of researchers in crop, livestock, fisheries, and forestry research working in the Department of Agriculture, Stock and Fisheries, and the Department of Forests. Following independence in 1975, these two departments were merged into the newly established Department of Primary Industries (DPI). The figures after 1975 represent the number of researchers working at DPI in crop, livestock, fisheries, and forestry research.

With independence in 1975, a large number of expatriate researchers left the country, so the decline in the number of researchers which the 1976 observation suggests appears plausible. The fairly rapid recovery in the total number of researchers in the years to follow masks a substantial decline in the average level of formal qualifications and years of working experience within the research system (sources 35 and 470).

During the seventies, agricultural research has ceased to be the sole preserve of DPI. Industry organizations such as the Oil Palm Research Association and the Coffee Industry Board now have substantial involvement, as do some provincial integrated rural development projects such as the Southern Highlands Rural Development Project. There are numerous other small research projects being conducted by the University of Papua New Guinea and the Institute of Technology and several nongovernment organizations and enterprises (source 35).

Source 38 identifies about 45 agricultural researchers working outside DPI in 1981 which for reasons of consistency are not included in the indicator series.

	1965	1968	1971	1973
Department of Agriculture, Stock and Fisheries				
Division of Animal Industry	10	8	11	14
Division of Research and Surveys			65	68
Division of Plant Industry	34	33		
Fisheries Research	[7]	[7]	(9)	(10)
Department of Forests	8	14	31	24
Total	59	62	107	106
Source	286	285	279	470

	1976	1980	1981	1982
Department of Primary Industries				
Livestock Branch	18	22	NA	NA
Agricultural Branch	38	56	63	69
Fisheries Branch	[12]	14	NA	NA
Forestry Office	5	15	NA	NA
Total	73	107	NA	NA
Source	95	470	38	35

[ ] = estimated or constructed by authors.
( ) = already included in the Division of Research and Surveys figure.

Expenditure Comments:

The expenditure figures for the years 1980, 1981, and 1982 have been constructed on the basis of estimates of revenue and expenditure of the Department of Primary Industries, as published by sources 468 and 469. Separating research expenditures from total research and development expenditures required several (informed) assumptions. In the case of crop research expenditures, it was assumed, on the basis of information provided by source 35, that 65% of the 'Crop Production, Research and Services' expenditures (function 3, activity 7, sources 468 and 469) were dedicated to crop research. In the case of livestock research, it was assumed, on the basis of information provided by source 482, that 50% of the 'Livestock Production Research and Health Services' expenditures (function 3, activity 6, sources 468 and 469) were dedicated to livestock research. Forestry research is the sum of 'Forest Products Research,' 'Botanical Research,' and 'Forest Management Research' expenditures (function 3, activities 1, 2, and 3, sources 468 and 469). In the case of Fisheries Research expenditures, it was assumed, on the basis of information provided by source 482, that 30% of the 'Fisheries Administration: Production Research and Inspection Services' (function 6, activity 1) expenditures were research related.

	1980	1981	1982
Crop Research	3.046	3.596	4.314
Livestock Research	1.690	1.650	1.870
Forestry Research	1.024	1.047	0.959
Fisheries Research	0.929	0.779	0.975
Total	6.689	7.072	8.118
Sources	468/482	469/482	469/482

Sources:

0095 FAO – CARIS. *Agricultural Research in Developing Countries – Volume 1: Research Institutions.* Rome: FAO – CARIS, 1978.

0279 Commonwealth Agricultural Bureaux (CAB). *List of Research Workers in the Agricultural Sciences in the Commonwealth and in the Republic of Ireland 1972.* Slough, England: CAB, 1972.

0285 Commonwealth Agricultural Bureaux (CAB). *List of Research Workers in the Agricultural Sciences in the Commonwealth and in the Republic of Ireland 1969.* Slough, England: CAB, 1969.

0286 Commonwealth Agricultural Bureaux (CAB). *List of Research Workers in Agriculture, Animal Health and Forestry in the Commonwealth and in the Republic of Ireland 1966.* Slough, England: CAB, 1966.

0468 Government of Papua New Guinea. *Estimates of Revenue and Expenditure for the Year Ending 31st December, 1981.* Report presented by Mr. John Kaputin, M.P., Minister of the National Parliament for Finance on the Occasion of the Budget, 1981. Papua New Guinea: Government of Papua New Guinea, 1981.

0469 Government of Papua New Guinea. *Estimates of Revenue and Expenditure for the Year Ending 31st December, 1982.* Report presented by Mr John Kaputin, M.P., Minister of the National Parliament for Finance on the Occasion of the Budget, 1982. Papua New Guinea: Government of Papua New Guinea, 1982.

0470 Doery, R.C., and A.E. Charles, eds. *Research Programme Survey 1980.* Papua New Guinea: Department of Primary Industries, 1980.

0482 Dagg, Matthew. Personal communication. Senior Research Officer, ISNAR. The Hague, March 1987.

Additional References:

0016 Oram, Peter A., and Vishva Bindlish. *Resource Allocations to National Agricultural Research: Trends in the 1970s.* The Hague and Washington, D.C.: ISNAR and IFPRI, November 1981.

0026 Oram, Peter A., and Mark Gieben. "Document Summaries." ISNAR, The Hague, 1984. Mimeo.

0027 Harvey, Nigel, ed. *Agricultural Research Centres: A World Directory of Organizations and Programmes.* Seventh Edition, Two Volumes. Harlow, U.K.: Longman, 1983.

0035 ISNAR. *Report to the Government of Papua New Guinea: Review of the Program and Organization for Crops Research in Papua New Guinea.* The Hague: ISNAR, June 1982.

0038 Gamble, W.K., R.M. Bourke, and C.W. Brookson. *South Pacific Agricultural Research Study.* Consultants' Report to the Asian Development Bank (two volumes). Manila: Asian Development Bank, June 1981.

0073 Oram, Peter A., and Vishva Bindlish. "Investment in Agricultural Research in Developing Countries: Progress, Problems, and the Determination of Priorities." IFPRI, Washington, D.C., January 1984. Mimeo.

0242 Commonwealth Agricultural Bureaux (CAB). *List of Research Workers in Agricultural Sciences in the Commonwealth 1981.* Slough, England: CAB, 1981.

0471 Bomo, Nelson. "Agricultural Research and Extension Programme in Papua New Guinea." *Alafua Agricultural Bulletin* Vol. 5, No. 1 (January–March 1980): 8–13.

0475 Jarrett, F.G. *Report on Agricultural Research and Extension, Department of Agriculture Stock and Fisheries (DASF) – Papua New Guinea.* Adelaide, Australia: The University of Adelaide, March 1972.

0590 UNESCO. *Science and Technology in Countries of*

Asia and the Pacific. Science Policy Studies and Documents No. 52. Paris: UNESCO, n.d.

0732 FAO – Regional Office for Asia and the Pacific. *Agricultural Research Systems in the Asia-Pacific Region.* Bangkok: FAO – Regional Office for Asia and the Pacific, 1986.

0759 Kula, G.R. Personal communication. Deputy Director, Crop Research, Department of Agriculture and Livestock. Konedobu, Papua New Guinea, December 1987.

0771 Fernando, L.H. "ISNAR/IRETA Workshop on Planning and Management of Agricultural Research in the South Pacific: Survey of Agricultural Research Resources." ISNAR, The Hague, n.d. Mimeo.

Paraguay

Personnel

Period/Year	PhD	MSc	BSc	Subtotal	Expat	Total	Source
1960–64	..	..	..	..	..	10	10
1965–69	0	5	8	13	..	13	267
1970	..	..	..	..	..	..	
1971	..	..	..	..	..	26	717
1972	..	..	..	..	..	..	
1973	..	..	..	..	..	..	
1974	..	..	..	..	..	..	
1975	..	..	..	..	..	..	
1976	2	12	19	33	..	33	26
1977	2	15	19	36	..	36	26
1978	3	16	22	41	..	41	26
1979	2	18	24	44	..	44	26
1980	3	20	30	53	..	53	26
1981	..	..	..	..	..	..	
1982	..	..	..	..	..	..	
1983	3	25	52	80	6	86	17
1984	..	..	..	..	..	..	
1985	..	..	..	..	..	..	
1986	2	36	74	112	..	112	267

1960–64: 1963   1965–69: 1967

Expenditure

LCU = Guaranies

Period/Year	Current LCU (millions)	Constant 1980 LCU (millions)	Constant 1980 US$ (millions)		Source
			Atlas	PPP	
1960–64	..	..	..	..	
1965–69	..	..	..	..	
1970	..	..	..	..	
1971	..	..	..	..	
1972	..	..	..	..	
1973	..	..	..	..	
1974	..	..	..	..	
1975	..	..	..	..	
1976	75.575	128.311	0.941	1.351	26
1977	89.599	139.345	1.022	1.467	26
1978	104.630	147.366	1.081	1.552	26
1979	142.355	166.303	1.219	1.751	26
1980	390.550	390.550	2.864	4.113	26
1981	..	..	..	..	
1982	..	..	..	..	
1983	1346.015	962.815	7.060	10.139	17
1984	..	..	..	..	
1985	..	..	..	..	
1986	..	..	..	..	

1960–64:   1965–69:

Personnel Comments:

These figures represent researchers working at 'Dirección de Investigación y Extensión Agropecuaria y Forestal' (DIEAF) only. The total number of graduate-level staff within DIEAF is considerably higher than the number of researchers, because many of them work within the extension section. Comparing across all available sources suggests that some prior compilations mistakenly considered all graduate-level staff working within DIEAF to be researchers (e.g., source 32). In source 17 (ISNAR, IFARD & AOAD questionnaire) no figure is given for the number of researchers with a BSc or equivalent. The 1983 BSc figure was therefore approximated on the basis of the 1980 and 1986 observations.

Expenditure Comments:

In Paraguay, 'Dirección de Investigación y Extensión Agropecuaria y Forestal' (DIEAF) is the main institution which undertakes both agricultural research and extension. This mixture of functions makes it rather difficult to divide total expenditures into research and extension expenditures. Although source 26 gives research expenditure only, most other sources give total expenditures of DIEAF (i.e., inclusive of extension). Although the ISNAR, IFARD & AOAD questionnaire explicitly sought research expenditures only, it is possible that the 1983 figure includes nonresearch expenditures.

Sources:

0010 Boyce, James K., and Robert E. Evenson. *National and International Agricultural Research and Extension Programs.* New York: Agricultural Development Council, Inc., 1975.

0017 ISNAR, IFARD & AOAD. Survey of National Agricultural Research Systems: Unpublished Questionnaire Responses. ISNAR, The Hague, 1985.

0026 Oram, Peter A., and Mark Gieben. "Document Summaries." ISNAR, The Hague, 1984. Mimeo.

0267 Alvarez, Luis Alberto. *Una Introspectiva del Sistema de Investigación Agropecuaria del Ministerio de Agricultura y Ganadería (MAG).* Publicacion Miscelánea No. 17. Asunción,

Paraguay: MAG – Dirección de Investigación y Extensión Agropecuaria y Forestal, 1986.

0717 Centro de Documentación sobre Investigación y Enseñanza Superior Agropecuaria de la Zona Sur. *Instituciones de Investigación Agrícola de la Zona Sur: Argentina, Chile, Paraguay y Uruguay.* Serie: Informaciones No. 2. Buenos Aires: Facultad de Agronomía y Veterinaria – Biblioteca Central, IICA – Zona Sur, 1972.

Additional References:

0001 Trigo, Eduardo J., and Martín E. Piñeiro. "Funding Agricultural Research." In *Report of a Conference: Selected Issues in Agricultural Research in Latin America,* edited by Barry Nestel and Eduardo J. Trigo. The Hague: ISNAR, March 1984.

0014 Judd, M. Ann, James K. Boyce, and Robert E. Evenson. "Investing in Agricultural Supply." Economic Growth Center, Yale University, New Haven, Connecticut, 1983. Mimeo.

0016 Oram, Peter A., and Vishva Bindlish. *Resource Allocations to National Agricultural Research: Trends in the 1970s.* The Hague and Washington, D.C.: ISNAR and IFPRI, November 1981.

0032 Castronovo, Alfonso J.P. "Diagnóstico Descriptivo de la Situación Actual del Sistema Nacional de Investigación y Extensión en Argentina, Chile, Paraguay y Uruguay." Buenos Aires, December 1980. Mimeo.

0053 Piñeiro, Martín, and Eduardo Trigo, eds. *Cambio Técnico en el Agro Latino Americano – Situación y Perspectivas en la Década de 1980.* San José: IICA, 1983.

0073 Oram, Peter A., and Vishva Bindlish. "Investment in Agricultural Research in Developing Countries: Progress, Problems, and the Determination of Priorities." IFPRI, Washington, D.C., January 1984. Mimeo.

0165 Evenson, R.E., and Y. Kislev. *Agricultural Research and Productivity.* New Haven: Yale University Press, 1975.

0172 Trigo, Eduardo. Background material. ISNAR, The Hague, n.d.

0852 Evenson, Robert E., and Yoav Kislev. *Investment in Agricultural Research and Extension: A Survey of International Data.* Center Discussion Paper No. 124. New Haven, Connecticut: Economic Growth Center, Yale University, August 1971.

0886 Evenson, Robert E., and Yoav Kislev. "Investment in Agricultural Research and Extension: A Survey of International Data." *Economic Development and Cultural Change* Vol. 23 (April 1975): 507–521.

Peru

Personnel

Period/Year	PhD	MSc	BSc	Subtotal	Expat	Total	Source
1960–64	..	..	..	..	..	117	852/10
1965–69	..	..	..	..	..	131	852
1970	..	..	..	..	..	174	335
1971	..	..	..	..	..	179	14
1972	..	..	..	..	..	..	
1973	..	..	..	..	..	..	
1974	..	..	..	..	..	..	
1975	..	..	..	..	..	254	14
1976	..	..	..	..	..	223	335
1977	..	..	..	..	..	269	14
1978	6	33	218	257	..	257	202
1979	..	..	..	..	..	..	
1980	4	30	239	273	..	273	205
1981	..	..	..	..	..	..	
1982	..	..	..	..	..	..	
1983	..	..	..	..	..	250	335
1984	..	..	..	..	..	..	
1985	..	..	..	..	..	273	203
1986	..	..	..	..	..	..	

1960–64: 1963 & 64   1965–69: 1965

Expenditure

LCU = Soles (1980)

Period/Year	Current LCU (millions)	Constant 1980 LCU (millions)	Constant 1980 US$ (millions)		Source
			Atlas	PPP	
1960–64	25.977	798.665	2.384	5.812	10/852
1965–69	40.230	1005.750	3.003	7.319	852
1970	193.500	2845.588	8.495	20.707	852
1971	..	..	..	..	
1972	..	..	..	..	
1973	..	..	..	..	
1974	..	..	..	..	
1975	..	..	..	..	
1976	471.326	2909.420	8.686	21.172	205
1977	348.565	1556.094	4.646	11.324	205
1978	370.672	1023.956	3.057	7.451	205
1979	787.218	1218.604	3.638	8.868	205
1980	1133.248	1133.248	3.383	8.247	205
1981	5391.000	3230.078	9.643	23.505	72
1982	6925.000	2539.421	7.581	18.479	72
1983	14208.000	2517.810	7.517	18.322	72
1984	40869.000	3402.348	10.157	24.759	72
1985	..	..	..	..	
1986	..	..	..	..	

1960–64: 1962, 63 & 64   1965–69: 1965

Personnel Comments:
Over the last 25 years the Peruvian agricultural research system has undergone numerous reorganizations, particularly that part of the agricultural research system which is under the responsibility of the Ministry of Agriculture and/or Food.

During the sixties, agricultural research under the responsibility of the Ministry of Agriculture, was conducted by the semiautonomous institute 'Servicio de Investigación y Promoción Agraria' (SIPA). In 1970 SIPA was abolished as a semiautonomous institute. The research component was placed under the Ministry of Agriculture, and the extension service disappeared. In 1975/1976 the Ministry of Agriculture was split into the Ministry of Food and the Ministry of Agriculture. The Ministry of Food took over all food crop/commodity research, leaving research on industrial crops, natural resources, and irrigation to the Ministry of Agriculture. In 1978 the ministries were again merged.

In 1979 'Instituto Nacional de Investigación Agraria' (INIA) was created and given responsibility for all research formerly executed by 'Dirección General de Investigación,' 'Dirección de Investigación Forestal y Fauna,' 'Sub Dirección de Investigación de Aguas y Suelos,' and 'Instituto de Investigaciones Agroindustriales' (source 205). In 1981 the extension services merged with INIA, whose name was changed to 'Instituto Nacional de Investigacion y Promocion Agraria' (INIPA). At the same time, research in the areas of forestry and fauna, water and soils, and agroindustry were moved out of INIPA. According to source 203, these three research divisions were transferred to the following institutes: INFOR, INAF, and INDDA, respectively.

In 1987, INIPA was broken up and the research component of INIPA merged with INDDA to form 'Instituto Nacional de Investigacion Agraria y Agroindustrial' (INIA).

In addition to these research agencies under the Ministry of Agriculture and Food, a considerable amount of research is also conducted by the National Agricultural University (UNA), and to a lesser extent by 15 regional universities. Total academic staff of UNA was 308 in 1969, 342 in 1976, and 406 in 1980 (source 203). Some nonpublicly funded institutes such as 'Instituto Central de Investigaciones Azucareras,' 'Hoja Peruana de Tabaco,' and 'Instituto de Estudios Andinos' also conduct some agricultural research.

For reasons of consistency, the indicator series only includes the principal agricultural research agencies – either the main semiautonomous agricultural research institutes (SIPA, INIA, INIPA) or the 'Direccion General de Investigacion' – under the Ministry of Agriculture and/or Food. For 1980, researchers working on forestry and fauna research and agroindustry research were excluded from the total number of researchers working at INIA in order to obtain an observation which is consistent with earlier and later observations.

Expenditure Comments:
The expenditure figures are generally consistent with the personnel indicators. Only the 1979 and 1980 figures include some minor expenditures on forestry and fauna research and on water and soils research, and may therefore be somewhat out of line. The post-1980 figures given by source 72, are constructed on the basis of expenditures on research, plus a proportion of administrative expenditures.

Source 72 indicates that in the early 1980s Peru undertook a revitalization of its agricultural research and extension system. Substantial financial assistance was provided by external donors, including USAID, World Bank, and Inter-American Development Bank, along with increased counterpart funds from local sources. Thus the rapidly increased levels of support afforded agricultural research in the early 1980s, following a decline in real support during the later half of the 1970s, which these figures suggest, are plausible. Nevertheless, unavoidable inconsistencies in the series reported by source 205 and 72 may overstate the rate of increase in real expenditures to some degree.

In particular, the figures obtained from source 205 represent proposed or budgeted expenditures and are exclusive of capital investments, while those obtained from source

72 include capital investments and are actual expenditures, not simply budgeted expenditures. The hyperinflation experienced by Peru during this period resulted in actual expenditures over the course of a year exceeding budgeted or proposed expenditures at the beginning of the fiscal year as extra funds were made available to compensate public agencies such as INIPA for the declining purchasing power of their budget allocations.

The table below gives some additional observations of research expenditures in millions of current LCUs not included, for reasons of consistency, in the indicator series.

	1976	1977	1978	1979	1980
IIA	27.500	33.000	43.077	67.456	98.760
DIFF	6.211	6.754	44.004		
SDIAS			3.879		
Source	205	205	205	205	205

Acronyms and Abbreviations:
DIFF: Dirección de Investigación Forestal y Fauna
IIA: Instituto de Investigaciones Agro-industriales
INAF: Instituto Nacional de Ampliación de la Frontera Agrícola
INDDA: Instituto Nacional de Desarrollo Agroindustrial
INFOR: Instituto Nacional de Forestal
INIA: Instituto Nacional de Investigación Agraria
INIAA: Instituto Nacional de Investigación Agraria y Agroindustrial
INIPA: Instituto Nacional de Investigación y Promoción Agraria
SDIAS: Sub Dirección de Investigación de Aguas y Suelos
SIPA: Servicio de Investigación y Promoción Agraria
UNA: Universidad Nacional Agraria

Sources:
0010 Boyce, James K., and Robert E. Evenson. *National and International Agricultural Research and Extension Programs*. New York: Agricultural Development Council, Inc., 1975.
0014 Judd, M. Ann, James K. Boyce, and Robert E. Evenson. "Investing in Agricultural Supply." Economic Growth Center, Yale University, New Haven, Connecticut, 1983. Mimeo.

0072 Norton, George W., and Víctor G. Ganoza. "The Benefits of Agricultural Research and Extension in Peru." Virginia Polytechnic Institute and North Carolina State University, USA, June 1985. Mimeo.
0202 Instituto Nacional de Investigacion Agraria (INIA) – Ministry of Agriculture and Food. *Baseline Study of the Peruvian Agricultural Research, Education and Extension System*. Peru: INIA – Ministry of Agriculture and Food, December 1979.
0203 Coutu, A.J., and K. Raven. "Background Paper on the National Agricultural Research, Extension and Educational System of Peru." North Carolina State University and National Agrarian University, North Carolina and Lima, August 1985. Mimeo.
0205 Blasco, Mario, Pedro González A., and Luis Scarneo W. "Perú: Análisis de la Cooperación Entre los Centros Nacionales y los Centros Internacionales de Investigación Agropecuaria." IICA-OEA: Oficina en Colombia, and PROTAAL, Bogotá, December 1980. Mimeo.
0335 Paz Silva, Luis J. *Peru and the CGIAR Centers – A Study of Their Collaboration in Agricultural Research*. CGIAR Study Paper No. 12. Washington, D.C.: World Bank, 1986.
0852 Evenson, Robert E., and Yoav Kislev. *Investment in Agricultural Research and Extension: A Survey of International Data*. Center Discussion Paper No. 124. New Haven, Connecticut: Economic Growth Center, Yale University, August 1971.

Additional References:
0001 Trigo, Eduardo J., and Martín E. Piñeiro. "Funding Agricultural Research." In *Report of a Conference: Selected Issues in Agricultural Research in Latin America*, edited by Barry Nestel and Eduardo J. Trigo. The Hague: ISNAR, March 1984.
0016 Oram, Peter A., and Vishva Bindlish. *Resource Allocations to National Agricultural Research: Trends in the 1970s*. The Hague and Washington, D.C.: ISNAR and IFPRI, November 1981.
0026 Oram, Peter A., and Mark Gieben. "Document Summaries." ISNAR, The Hague, 1984. Mimeo.
0027 Harvey, Nigel, ed. *Agricultural Research Centres: A World Directory of Organizations and Programmes*. Seventh Edition, Two Volumes. Harlow, U.K.: Longman, 1983.
0053 Piñeiro, Martín, and Eduardo Trigo, eds. *Cambio Técnico en el Agro Latino Americano – Situación y Perspectivas en la Década de 1980*. San José: IICA, 1983.

0073 Oram, Peter A., and Vishva Bindlish. "Investment in Agricultural Research in Developing Countries: Progress, Problems, and the Determination of Priorities." IFPRI, Washington, D.C., January 1984. Mimeo.

0095 FAO – CARIS. *Agricultural Research in Developing Countries – Volume 1: Research Institutions.* Rome: FAO – CARIS, 1978.

0135 FAO, and UNDP. *National Agricultural Research – Report of an Evaluation Study in Selected Countries.* Rome: FAO and UNDP, 1984.

0163 CGIAR. "National Agricultural Research." CGIAR, Washington, D.C., 1985. Mimeo.

0165 Evenson, R.E., and Y. Kislev. *Agricultural Research and Productivity.* New Haven: Yale University Press, 1975.

0172 Trigo, Eduardo. Background material. ISNAR, The Hague, n.d.

0204 AID. "Peru Project Paper: Agricultural Research, Extension and Education." AID, Washington, D.C., 1980. Mimeo.

0305 Ardila, Jorge, Eduardo Trigo, and Martín Piñeiro. "Human Resources in Agricultural Research: Three Cases in Latin America." In *Resource Allocation to Agricultural Research.* Proceedings of a workshop held in Singapore 8–10 June 1981, edited by Douglas Daniels and Barry Nestel, pp. 151–167. Ottawa: IDRC, 1981.

0839 Pelaez Bardales, Mario. Personal communication. Instituto Nacional de Investigación Agraria y Agroindustrial (INIAA). Lima, Peru, Marzo 1988.

0886 Evenson, Robert E., and Yoav Kislev. "Investment in Agricultural Research and Extension: A Survey of International Data." *Economic Development and Cultural Change* Vol. 23 (April 1975): 507–521.

Philippines

Personnel

Period/Year	PhD	MSc	BSc	Subtotal	Expat	Total	Source
1960–64	..	..	..	..	..	219	852
1965–69	..	..	..	..	..	..	
1970	..	..	..	..	..	..	
1971	38	106	673	817	..	817	788
1972	..	..	..	..	..	..	
1973	..	..	..	..	..	..	
1974	90	211	827	1128	..	1128	786
1975	..	..	..	..	..	..	
1976	..	..	..	..	..	..	
1977	76	218	1096	1390	..	1390	786
1978	..	..	..	..	..	..	
1979	..	..	..	..	..	..	
1980	..	..	..	..	..	..	
1981	136	448	908	1492	..	1492	732
1982	..	..	..	..	..	..	
1983	..	..	..	..	..	..	
1984	..	..	..	..	..	..	
1985	213	655	1570	2438	..	2438	802
1986	..	..	..	..	..	..	

1960–64: 1960   1965–69:

Expenditure

LCU = Philippine Pesos

Period/Year	Current LCU (millions)	Constant 1980 LCU (millions)	Constant 1980 US$ (millions)		Source
			Atlas	PPP	
1960–64	10.480	58.476	7.436	16.472	62
1965–69	19.600	83.761	10.651	23.595	62
1970	37.647	128.928	16.394	36.318	62
1971	41.100	125.305	15.934	35.297	62
1972	50.277	143.649	18.266	40.464	62
1973	76.603	185.479	23.585	52.248	62
1974	89.749	165.895	21.095	46.731	62
1975	73.402	125.259	15.928	35.284	62
1976	96.751	151.173	19.223	42.584	62
1977	92.800	135.080	17.177	38.051	62
1978	103.100	137.284	17.457	38.671	62
1979	90.200	104.277	13.260	29.374	62
1980	101.900	101.900	12.957	28.704	62
1981	161.800	145.766	18.535	41.061	62
1982	136.900	113.799	14.471	32.056	62
1983	117.400	87.416	11.116	24.624	62
1984	161.221	80.090	10.184	22.561	732
1985	194.843	82.247	10.458	23.168	732
1986	..	..	..	..	

1960–64: 1960, 61, 62, 63 & 64   1965–69: 1965, 66, 67, 68 & 69

Personnel Comments:

The Philippine agricultural research system consists of many different organizations, which – from 1972 on – are coordinated by the Philippine Council for Agriculture and Resources Research and Development (PCARRD). For the years 1971 and 1974, sources 786 and 788 report both total and full-time equivalent (FTE) researchers. The indicator series includes the preferred FTE estimates. The 1974 institution-specific ratios of FTEs to total researchers were used to prorate the later years. Over the whole period, the total researcher count was approximately double the FTE figures. The table below gives construction details for the years 1971, 1974, and 1977.

	1971				1974				1977			
	PhD	MSc	BSc	Total	PhD	MSc	BSc	Total	PhD	MSc	BSc	Total
MOA					2	29	293	324	2	23	289	314
MNR[a]	2	27	315	344	0	6	117	123	1	16	265	282
CI	0	3	72	75	1	2	77	80				
NSDB	5	31	184	220	12	19	116	147	3	17	97	117
OGA					1	17	54	72	3	13	163	179
CAU	1	45	102	178	74	138	170	382	67	149	282	498
Total	38	106	673	817	90	211	827	1128	76	218	1096	1390

a. In 1971 these figures refer to the former Department of Agriculture and Natural Resources.

Explicitly excluded from the indicator series, although mentioned by sources 788 and 786, is research by private-sector and international agencies (mainly IRRI).

Expenditure Comments:

The source 62 and 732 figures include only direct expenditures of the Philippine government on agricultural research and do not include grants or loans from international sources.

Source 732 indicates that a significant proportion of these external funds where spent on manpower and research facilities development. To the extent that the research facilities component represented infrastructural (i.e., capital) investments, the indicator series underestimates total research expenditures in the Philippines.

Acronyms and Abbreviations:
CAU: Colleges and Universities
CI: Commodity Institutes
IRRI: International Rice Research Institute
MNR: Ministry of Natural Resources
MOA: Ministry of Agriculture
NSDB: National Science Development Board
OGA: Other Government Agencies

PCARRD: Philippine Council for Agriculture and Resources Research and Development

Sources:
0062 Gomez, Arturo S. *Philippines and the CGIAR Centers: A Study of Their Collaboration in Agricultural Research.* CGIAR Study Paper Number 15. Washington, D.C.: World Bank, 1986.
0732 FAO – Regional Office for Asia and the Pacific. *Agricultural Research Systems in the Asia-Pacific Region.* Bangkok: FAO – Regional Office for Asia and the Pacific, 1986.
0786 Librero, Aida R. "An Inventory of Research Manpower in Agriculture and Natural Resources in the Philippines." In *Agriculture and Resources Research Manpower Development in South and Southeast Asia*, pp. 131–141. Los Baños, Philippines: Philippine Council for Agriculture and Resources Research and Development, 1983.
0788 National Agricultural Research System Survey Technical Panel (NARSSTP). *The Philippine Agricultural Research System – Evaluation and Recommendations.* Volume 1. Manila, Philippines: NARSSTP, December 1971.
0802 Gapasin, Dely P. Personal communication. Deputy Executive Director for Research and Development, Philippine Council for Agriculture and Resources Research and Development. Los Baños, Philippines, February 1988.

362

Additional References:

0010 Boyce, James K., and Robert E. Evenson. *National and International Agricultural Research and Extension Programs*. New York: Agricultural Development Council, Inc., 1975.

0014 Judd, M. Ann, James K. Boyce, and Robert E. Evenson. "Investing in Agricultural Supply." Economic Growth Center, Yale University, New Haven, Connecticut, 1983. Mimeo.

0016 Oram, Peter A., and Vishva Bindlish. *Resource Allocations to National Agricultural Research: Trends in the 1970s*. The Hague and Washington, D.C.: ISNAR and IFPRI, November 1981.

0017 ISNAR, IFARD & AOAD. Survey of National Agricultural Research Systems: Unpublished Questionnaire Responses. ISNAR, The Hague, 1985.

0018 Daniels, Douglas, and Barry Nestel, eds. *Resource Allocation to Agricultural Research*. Proceedings of a workshop held in Singapore 8–10 June 1981. Ottawa: IDRC, 1981.

0022 UNESCO. *Statistical Yearbook 1983*. Paris: UNESCO, 1983.

0026 Oram, Peter A., and Mark Gieben. "Document Summaries." ISNAR, The Hague, 1984. Mimeo.

0027 Harvey, Nigel, ed. *Agricultural Research Centres: A World Directory of Organizations and Programmes*. Seventh Edition, Two Volumes. Harlow, U.K.: Longman, 1983.

0073 Oram, Peter A., and Vishva Bindlish. "Investment in Agricultural Research in Developing Countries: Progress, Problems, and the Determination of Priorities." IFPRI, Washington, D.C., January 1984. Mimeo.

0095 FAO – CARIS. *Agricultural Research in Developing Countries – Volume 1: Research Institutions*. Rome: FAO – CARIS, 1978.

0163 CGIAR. "National Agricultural Research." CGIAR, Washington, D.C., 1985. Mimeo.

0165 Evenson, R.E., and Y. Kislev. *Agricultural Research and Productivity*. New Haven: Yale University Press, 1975.

0176 Moseman, Albert, ed. *National Agricultural Research Systems in Asia*. Report of the Regional Seminar held at the India International Centre, New Delhi, India, March 8–13, 1971. New York: ADC, Inc., 1971.

0179 Philippine Council for Agriculture and Resources Research and Development (PCARRD), and IDRC. *Agriculture and Resources Research Manpower Development in South and Southeast Asia*. Proceedings of the Workshop on Agriculture and Resources Research Manpower Development in South and Southeast Asia, 21–23

October 1981 and 4–6 February 1982, Singapore. Los Baños, Philippines: PCARRD, 1983.

0284 ISNAR. *Review of the Research System of the Ministry of Agriculture and Food in the Philippines*. The Hague: ISNAR, January 1986.

0299 Drilon, J.D., and Aida R. Librero. "Defining Research Priorities for Agriculture and Natural Resources in the Philippines." In *Resource Allocation to Agricultural Research*. Proceedings of a workshop held in Singapore 8–10 June 1981, edited by Douglas Daniels and Barry Nestel, pp. 97–103. Ottawa: IDRC, 1981.

0434 Anh, Do. "Agricultural Research Management Systems in the Philippines, India, and Bangladesh." In *Proceedings of the Research Extension Linkage Workshop Held in Hanoi, Viet Nam, 9–13 June 1986*. Rome: FAO, 1986.

0590 UNESCO. *Science and Technology in Countries of Asia and the Pacific*. Science Policy Studies and Documents No. 52. Paris: UNESCO, n.d.

0787 Philippine Council for Agriculture and Resources Research and Development (PCARRD). *The PCARRD Corplan 1984–1988*. Philippines: PCARRD, 1983.

0789 International Agricultural Development Service (IADS). *Report of the Review Team Philippine Council for Agriculture and Resources Research*. New York: IADS, November 1980.

0790 Pray, Carl E. *Agricultural Research and Technology Transfer by the Private Sector in the Philippines*. St. Paul: University of Minnesota – Economic Development Center, October 1986.

0791 Valmayor, Ramon V., and Nora M. Valera. "Agriculture and Resources Research Manpower Development Program in the Philippines: Past, Present, and Future." In *Agriculture and Resources Research Manpower Development in South and Southeast Asia*, pp. 107–117. Los Baños, Philippines: PCARRD, 1983.

0852 Evenson, Robert E., and Yoav Kislev. *Investment in Agricultural Research and Extension: A Survey of International Data*. Center Discussion Paper No. 124. New Haven, Connecticut: Economic Growth Center, Yale University, August 1971.

0869 Boyce, James K. "Agricultural Research in Indonesia, The Philippines, Bangladesh, South Korea and India: A Documentary History." University of Minnesota Asian Agricultural Research Review Project, July 1980. Mimeo.

0886 Evenson, Robert E., and Yoav Kislev. "Investment in Agricultural Research and Extension: A Survey of International Data." *Economic Development and Cultural Change* Vol. 23 (April 1975): 507–521

Portugal

Personnel

Period/Year	PhD	MSc	BSc	Subtotal	Expat	Total	Source
1960–64	..	..	..	..	..	306	574/514
1965–69	..	..	..	..	..	..	
1970							
1971	..	..	..	..	..	..	
1972	..	..	..	..	..	376	663
1973	..	..	..	..	..	..	
1974	..	..	..	..	..	..	
1975	..	..	..	..	..	..	
1976	..	..	..	..	..	351	665
1977	..	..	..	..	..	..	
1978	..	..	..	..	..	..	
1979	..	..	..	..	..	..	
1980	..	..	..	..	..	..	
1981	..	..	..	..	..	..	
1982	..	..	..	..	..	438	521
1983	..	..	..	..	..	..	
1984	..	..	..	..	..	..	
1985	..	..	..	..	..	460	422
1986	..	..	..	..	..		

1960–64: 1960 & 64   1965–69:

Expenditure   *LCU = Portuguese Escudos*

Period/Year	Current LCU (millions)	Constant 1980 LCU (millions)	Constant 1980 US$ (millions)		Source
			Atlas	PPP	
1960–64	55.154	297.735	5.731	8.943	574/514
1965–69	..	..	..	..	
1970	..	..	..	..	
1971	174.417	711.906	13.703	21.384	662
1972	146.585	555.246	10.688	16.678	663
1973	..	..	..	..	
1974	..	..	..	..	
1975	..	..	..	..	
1976	347.514	753.826	14.510	22.643	665
1977	..	..	..	..	
1978	565.377	801.953	15.437	24.089	664
1979	..	..	..	..	
1980	..	..	..	..	
1981	..	..	..	..	
1982	..	..	..	..	
1983	..	..	..	..	
1984	1642.200	760.278	14.634	22.837	666/667
1985	..	..	..	..	
1986	..	..	..	..	

1960–64: 1960 & 64   1965–69:

Personnel Comments:

Both the personnel and expenditure indicators are based on a classification of scientific disciplines, rather than the preferred socioeconomic objective. Several sources also give figures by socioeconomic objective, but they are, in general, less consistent than those by scientific discipline.

The personnel indicators have been constructed as shown below.

Consistency in institutional coverage between observations is difficult to verify, particularly for the earlier years.

	1960	1964	1972	1976	1982	1985
Government	NA	271	355	310	330	350
PNP	NA	19	1			
Universities	NA	8	20	41	108	110
Total	314	298	376	351	438	460
Source	574	514	663	665	521	422

Expenditure Comments:

The expenditure indicators have been constructed as follows:

	1960	1964	1971	1972	1976	1978	1984
Government	NA	62.590	167.718	141.433	336.138	475.793	1411.200
PNP	NA	4.899		0.679		0.431	
Universities	NA	0.819	6.699	4.473	11.376	89.153	231.000*
Total	42.000	68.308	174.417	146.585	347.514	565.377	1642.200
Source	574	514	662	663	665	664	666 667*

Acronyms and Abbreviations:

PNP: Private Nonprofit sector

Sources:

0422 Casas, Joseph. "Les Ressources Humaines et Financières dans les Pays du Bassin Méditerranéen." Preliminary table. ISNAR, The Hague, 1986. Mimeo.

0514 OECD. *A Study of Resources Devoted to R&D in OECD Member Countries in 1963/64, International Statistical Year for Research and Development – Volume 2: Statistical Tables and Notes.* Paris: OECD, 1968.

0521 OECD. *Reviews of National Science and Technology Policy: Portugal.* Paris: OECD, 1986.

0574 OECD. *Intellectual Investment in Agriculture for Economic and Social Development.* Documentation in Food and Agriculture No. 60. Paris: OECD, 1962.

0662 Junta Nacional de Investigação Científica e Tecnológica. *Resursos em Ciência e Tecnologia – Inventário de 1971.* Lisboa: Junta Nacional de Investigação Científica e Tecnológia, n.d.

0663 Junta Nacional de Investigação Científica e Tecnológica. *Recursos Inventário em Ciência e Tecnologia 1972.* Lisboa: Junta Nacional de Investigação Científica e Tecnológica, n.d.

0664 Junta Nacional de Investigação Científica e Tecnológica. *Recursos em Ciência e Tecnologia Portugal 1978.* Lisboa: Junta Nacional de Investigação Científica e Tecnológica, 1981.

0665 Junta Nacional de Investigação Científica e Tecnológica. *Investigação & Desenvolvimento Portugal 1976.* Lisboa: Junta Nacional de Investigação Científica e Tecnológica, n.d.

0666 Junta Nacional de Investigação Científica e Tecnológica. *Investigação & Desenvolvimento Experimental – Inquérito ao Potencial Científico e Tecnológico Nacional Actualização a 84.12.31: Sector Estado (Dados Provisórios).* Lisboa: Junta Nacional de Investigação Científica e Tecnológica, June 1986.

0667 Junta Nacional de Investigação Científica e Tecnológica. *Investigação & Desenvolvimento Experimental – Inquérito ao Potencial Científico e Tecnológico Nacional Actualização a 84.12.31: Sector Ensino Superior (Dados Provisórios).* Lisboa: Junta Nacional de Investigação Científica e Tecnologica, July 1986.

Additional References:

0010 Boyce, James K., and Robert E. Evenson. *National and International Agricultural Research and Extension Programs.* New York: Agricultural Development Council, Inc., 1975.

0021 OECD. *OECD Science and Technology Indicators.* Paris: OECD, 1984.

0022 UNESCO. *Statistical Yearbook 1983.* Paris: UNESCO, 1983.

0027 Harvey, Nigel, ed. *Agricultural Research Centres: A World Directory of Organizations and Programmes.* Seventh Edition, Two Volumes. Harlow, U.K.: Longman, 1983.

0165 Evenson, R.E., and Y. Kislev. *Agricultural Research and Productivity.* New Haven: Yale University Press, 1975.

0473 OECD. *Science and Technology Indicators: Basic Statistical Series – Volume A: The Objectives of Government R&D Funding 1974–1985.* Paris: OECD, May 1983.

0474 OECD. *Science and Technology Indicators: Basic Statistical Series – Volume A: The Objectives of Government R&D, 1969–1981.* Paris: OECD, July 1981.

0509 OECD. *Survey of the Resources Devoted to R&D by OECD Member Countries, International Statistical Year 1971 – Volume 5: Total Tables, Statistical Tables and Notes.* Paris: OECD, August 1974.

0512 OECD. *Survey of the Resources Devoted to R&D by OECD Member Countries, International Statistical Year 1973 – Volume 5: Total Tables, Statistical Tables and Notes.* Paris: OECD, September 1976.

0513 OECD. *International Survey of the Resources Devoted to R&D by OECD Member Countries, International Statistical Year 1975 – International Volume: Statistical Tables and Notes.* Paris: OECD, March 1979.

0522 OECD. *Country Reports on the Organisation of Scientific Research: Portugal.* Paris: OECD, December 1963.

0530 OECD. *OECD Science and Technology Indicators No. 2 – R&D, Invention and Competitiveness.* Paris: OECD, 1986.

0601 OECD – STIID Data Bank. "Public Funding of R&D by Socio-Economic Objective." Tables. OECD, Paris, April 1987. Mimeo.

0602 OECD – STIID Data Bank. "R&D Expenditure in the Higher Education and Private Non-Profit Sectors." Tables. OECD, Paris, April 1987. Mimeo.

0708 OECD – STIID Data Bank. "Gross Domestic Expenditure on R&D: All Fields of Sciences – Agriculture, Forestry and Fishing." Tables. OECD, Paris, June 1987. Mimeo.

0852 Evenson, Robert E., and Yoav Kislev. *Investment in Agricultural Research and Extension: A Survey of International Data.* Center Discussion Paper No. 124. New Haven, Connecticut: Economic Growth Center, Yale University, August 1971.

0886 Evenson, Robert E., and Yoav Kislev. "Investment in Agricultural Research and Extension: A Survey of International Data." *Economic Development and Cultural Change* Vol. 23 (April 1975): 507–521.

Puerto Rico

Personnel

Period/Year	PhD	MSc	BSc	Subtotal	Expat	Total	Source
1960–64	..	..	..	..	..	133	492
1965–69	..	..	..	..	..	131	492
1970	..	..	..	..	..	118	488
1971	..	..	..	..	..	122	488
1972	..	..	..	..	..	111	488
1973	..	..	..	..	..	101	488
1974	..	..	..	..	..	96	488
1975	..	..	..	..	..	96	488
1976	..	..	..	..	..	106	488
1977	..	..	..	..	..	103	488
1978	..	..	..	..	..	112	488
1979	..	..	..	..	..	102	488
1980	..	..	..	..	..	96	488
1981	..	..	..	..	..	91	488
1982	..	..	..	..	..	81	488
1983	..	..	..	..	..	68	488
1984	..	..	..	..	..	76	488
1985	..	..	..	..	..	71	488
1986	..	..	..	..	..	..	

1960–64: 1964   1965–69: 1965, 66, 67, 68 & 69

Expenditure   LCU = U.S. Dollars

Period/Year	Current LCU (millions)	Constant 1980 LCU (millions)	Constant 1980 US$ (millions)		Source
			Atlas	PPP	
					489/490/
1960–64	2.814	7.554	7.510	8.457	491/492
1965–69	4.405	10.339	10.279	11.574	492
1970	5.402	11.024	10.961	12.342	492
1971	5.981	11.546	11.480	12.926	492
1972	5.827	10.751	10.689	12.036	492
1973	6.421	11.109	11.045	12.437	492
1974	7.176	11.397	11.331	12.752	493
1975	7.727	11.166	11.102	12.501	488
1976	7.353	9.990	9.933	11.184	488
1977	7.902	10.066	10.008	11.269	488
1978	8.695	10.314	10.255	11.547	488
1979	9.084	9.906	9.849	11.090	488
1980	9.002	9.002	8.950	10.078	488
1981	9.850	8.987	8.936	10.061	488
1982	8.675	7.434	7.391	8.322	488
1983	8.402	6.932	6.893	7.761	488
1984	9.035	7.176	7.176	8.034	488
1985	9.232	7.096	7.096	7.944	488
1986	..	..	..	..	

1960–64: 1960, 61, 62, 63 & 64   1965–69: 1965, 66, 67, 68 & 69

Personnel Comments:

The personnel figures are measured in scientist years (SYs). The 1972–85 figures represent an aggregation of project-level data obtained directly from the USDA's Current Research Information System (source 488). The 1964–69 figures represent adjusted full-time equivalent (FTE) counts from source 492. The adjustments ensured the higher FTE figures were in line with the somewhat lower SY estimates obtained for the years 1970–1975 in which the two series overlapped.

Expenditure Comments:

The 1960–74 figures are research expenditures, not simply funds available, while the 1975–85 figures from source 488 represent obligations.

Sources:

0488 US Department of Agriculture – Cooperative State Research Service. *Inventory of Agricultural Research – Fiscal Years: 1970–1985*. Washington, D.C.: US Government Printing Office, various years.

0489 US Department of Agriculture – Agricultural Research Service. *Funds for Research at State Agricultural Experiment Stations, 1960*. Washington, D.C.: US Government Printing Office, April 1961.

0490 US Department of Agriculture – Cooperative State Experiment Station Service. *Funds for Research at State Agricultural Experiment Stations*. Annual issues 1961 and 1962. Washington, D.C.: US Government Printing Office, various years.

0491 US Department of Agriculture – Cooperative State Research Service. *Funds for Research at State Agricultural Experiment Stations, 1963*. Washington, D.C.: Government Printing Office, March 1964.

0492 US Department of Agriculture – Cooperative State Research Service. *Funds for Research at State Agricultural Experiment Stations and Other State Institutions*. Annual issues 1964 to 1974. Washington, D.C.: US Government Printing Office, various years.

0493 US Department of Agriculture – Cooperative State Research Service. *Funds for Research at State Agricultural Experiment Stations and Other Cooperating Institutions*. Annual issues 1974 and 1975. Washington, D.C.: US Government Printing Office, various years.

Additional References:

0027 Harvey, Nigel, ed. *Agricultural Research Centres: A World Directory of Organizations and Programmes*. Seventh Edition, Two Volumes. Harlow, U.K.: Longman, 1983.

0095 FAO – CARIS. *Agricultural Research in Developing Countries – Volume 1: Research Institutions*. Rome: FAO – CARIS, 1978.

Qatar

Personnel

Period/Year	PhD	MSc	BSc	Subtotal	Expat	Total	Source
1960–64	..	..	..	..	..	..	
1965–69	..	..	..	..	..	..	
1970	..	..	..	..	..	..	
1971	..	..	..	..	..	..	
1972	..	..	..	..	..	..	
1973	..	..	..	..	..	..	
1974	..	..	..	..	..	..	
1975	..	..	..	..	..	..	
1976	..	..	..	..	..	..	
1977	..	..	..	..	..	..	
1978	..	..	..	..	..	..	
1979	..	..	..	..	..	..	
1980	..	..	..	..	..	..	
1981	..	..	..	..	..	..	
1982	..	..	..	..	..	..	
1983	0	0	6	6	1	7	17
1984	..	..	..	..	..	..	
1985	..	..	..	..	..	..	
1986	..	..	..	..	..	..	

1960–64:   1965–69:

Expenditure   *LCU = Qatar Riyals*

Period/Year	Current LCU (millions)	Constant 1980 LCU (millions)	Constant 1980 US$ (millions)		Source
			Atlas	PPP	
1960–64	..	..	..	..	
1965–69	..	..	..	..	
1970	..	..	..	..	
1971	..	..	..	..	
1972	..	..	..	..	
1973	..	..	..	..	
1974	..	..	..	..	
1975	..	..	..	..	
1976	..	..	..	..	
1977	..	..	..	..	
1978	..	..	..	..	
1979	..	..	..	..	
1980	..	..	..	..	
1981	..	..	..	..	
1982	..	..	..	..	
1983	8.847	7.504	1.746	1.541	17
1984	..	..	..	..	
1985	..	..	..	..	
1986	..	..	..	..	

1960–64:   1965–69:

Personnel Comments:
The 1983 figure includes researchers in the Department of Agriculture, the Department of Fisheries, and the Department of Agricultural and Water Research which all fall under the Ministry of Industry and Agriculture.

Sources:
0017 ISNAR, IFARD & AOAD. Survey of National Agricultural Research Systems: Unpublished Questionnaire Responses. ISNAR, The Hague, 1985.

Additional References:
0094 FAO – Near East Regional Office. *Directory of Agricultural Research Institutions in the Near East Region.* Cairo: FAO – Near East Regional Office, 1979.

0865 Oram, Peter. "Agricultural Research Objectives and Priorities: Constraints to the Development of Agricultural Research Institutions in Arab Countries." Review paper prepared for the first meeting of the Council for Arab Agricultural Research, organized by the Arab Fund for Economic and Social Development, Kuwait, 23 March 1988.

Rwanda

Personnel

Period/Year	PhD	MSc	BSc	Subtotal	Expat	Total	Source
1960–64	..	..	..	..	..	9	10/742
1965–69	..	..	..	..	..	7	10
1970	..	..	..	0	10	10	175/532
1971	..	..	..	1	15	16	589
1972	..	..	..	..	..	..	
1973	..	..	..	1	15	16	742
1974	..	..	..	..	..	..	
1975	..	..	..	..	..	..	
1976	..	..	..	..	..	23	95
1977	..	..	..	..	..	..	
1978	..	..	..	..	..	..	
1979	..	..	..	..	..	..	
1980	..	..	..	..	..	..	
1981	1	–14–		15	10	25	235
1982	..	..	..	26	4	30	23
1983	2	–22–		24	7	31	17
1984	..	..	..	..	..	41	230
1985	..	..	..	26	17	43	445
1986	..	..	..	..	..	..	

1960–64: 1961 & 63   1965–69: 1966 & 67

Expenditure   *LCU = Rwanda Francs*

Period/Year	Current LCU (millions)	Constant 1980 LCU (millions)	Constant 1980 US$ (millions)		Source
			Atlas	PPP	
1960–64	..	..	..	..	
1965–69	124.700	166.043	0.747	1.192	589
1970	..	..	..	..	
1971	33.700	72.473	0.820	1.308	589
1972	..	..	..	..	
1973	..	..	..	..	
1974	..	..	..	..	
1975	..	..	..	..	
1976	57.000	81.779	0.925	1.476	95
1977	..	..	..	..	
1978	..	..	..	..	
1979	..	..	..	..	
1980	..	..	..	..	
1981	..	..	..	..	
1982	120.000	106.385	1.203	1.923	13
1983	136.400	115.495	1.306	2.084	17
1984	..	..	..	..	
1985	..	..	..	..	
1986	..	..	..	..	

1960–64:   1965–69: 1966

Personnel Comments:

The figures refer to ISAR, formerly INEAC, only. ISAR was created in June 1962. Prior to that date, agricultural research in Rwanda was performed by INEAC.

For Rwanda, the 'ingénieur agronome' degree does not quite correspond to the BSc or MSc categories used in this series. It is somewhere in between. For 1981 and 1983, 14 and 22, respectively, 'ingénieur agronome' were reported.

Expenditure Comments:

The research expenditure figures refer to ISAR only.

Acronyms and Abbreviations:

INEAC: Institut National pour l'Etude Agronomique au Congo
ISAR: Institut des Sciences Agronomiques du Rwanda

Sources:

0010 Boyce, James K., and Robert E. Evenson. *National and International Agricultural Research and Extension Programs.* New York: Agricultural Development Council, Inc., 1975.

0013 ISNAR. *Report to the Government of Rwanda: The National Agricultural Research System of Rwanda.* The Hague: ISNAR, February 1983.

0017 ISNAR, IFARD & AOAD. Survey of National Agricultural Research Systems: Unpublished Questionnaire Responses. ISNAR, The Hague, 1985.

0023 Bennell, Paul. *Agricultural Researchers in Sub-Saharan Africa: An Overview.* Working Paper No. 4. The Hague: ISNAR, October 1985.

0095 FAO – CARIS. *Agricultural Research in Developing Countries – Volume 1: Research Institutions.* Rome: FAO – CARIS, 1978.

0175 Cooper, St.G.C. *Agricultural Research in Tropical Africa.* Kampala: East African Literature Bureau, 1970.

0230 Institut des Sciences Agronomiques du Rwanda (ISAR). *Rapport Annuel 1984.* Rwanda: ISAR, 1985.

0235 FAO. "Trained Agricultural Manpower Assessment (TAMA): Original Answer Sheets." FAO, Rome, 1982. Mimeo.

0445 Swanson, Burton E., and Wade H. Reeves. "Agricultural Research Eastern and Southern Africa: Manpower and Training." World Bank, Washington, D.C., August 1986. Mimeo.

0532 UNESCO Field Science Office for Africa. *Survey on the Scientific and Technical Potential of the Countries of Africa.* Paris: UNESCO, 1970.

0589 Kassapu, Samuel. *Les Dépenses de Recherche Agricole dans 34 Pays d'Afrique Tropicale.* Paris: Centre de Développement de l'OCDE, 1976.

0742 Institut des Sciences Agronomiques du Rwanda (ISAR). *Institut des Sciences Agronomiques du Rwanda: 1962–1987 Point de la Recherche.* Butare, Rwanda: ISAR, Novembre 1987.

Additional References:

0016 Oram, Peter A., and Vishva Bindlish. *Resource Allocations to National Agricultural Research: Trends in the 1970s.* The Hague and Washington, D.C.: ISNAR and IFPRI, November 1981.

0026 Oram, Peter A., and Mark Gieben. "Document Summaries." ISNAR, The Hague, 1984. Mimeo.

0073 Oram, Peter A., and Vishva Bindlish. "Investment in Agricultural Research in Developing Countries: Progress, Problems, and the Determination of Priorities." IFPRI, Washington, D.C., January 1984. Mimeo.

0090 Institut de Recherche Agronomique et Zootechnique – Communauté Economique des Pays des Grands Lacs. *Répertoire des Recherches Agronomiques en Cours; Burundi-Rwanda-Zaire: 1984.* Gitega, Burundi: Institut de Recherche Agronomique et Zootechnique, 1984.

0165 Evenson, R.E., and Y. Kislev. *Agricultural Research and Productivity.* New Haven: Yale University Press, 1975.

0266 UNESCO. *National Science Policies in Africa.* Science Policy Studies and Documents No. 31. Paris: UNESCO, 1974.

0360 Cooper, St.G.C. "Towards Trained Manpower for Agricultural Research in Africa." Paper presented at the Conference on Agricultural Research and Production in Africa, organized by the Association for the Advancement of Agricultural Sciences in Africa (AAASA), Addis Ababa, 29 August–4 September 1971.

0400 UNESCO. *The Promotion of Scientific Activity in Tropical Africa.* Science Policy Studies and Documents No. 11. Paris: UNESCO, 1969.

0858 ISNAR. "A Review of Agricultural Systems Research in Rwanda." ISNAR, The Hague, January 1988. Mimeo.

0886 Evenson, Robert E., and Yoav Kislev. "Investment in Agricultural Research and Extension: A Survey of International Data." *Economic Development and Cultural Change* Vol. 23 (April 1975): 507–521.

Sao Tome & Principe

Personnel

Period/Year	PhD	MSc	BSc	Subtotal	Expat	Total	Source
1960–64	..	..	..	..	..	..	
1965–69	..	..	..	..	..	..	
1970	..	..	..	..	..	..	
1971	..	..	..	..	..	..	
1972	..	..	..	..	..	..	
1973	..	..	..	..	..	..	
1974	..	..	..	..	..	..	
1975	..	..	..	..	..	..	
1976	..	..	..	..	..	..	
1977	..	..	..	..	..	..	
1978	..	..	..	..	..	..	
1979	..	..	..	..	..	..	
1980	..	..	..	..	..	..	
1981	..	..	..	..	..	..	
1982	..	..	..	..	..	4	27
1983	..	..	..	..	..	..	
1984	..	..	..	..	..	2	130
1985	..	..	..	..	..	3	68
1986	..	..	..	..	..	..	

1960–64:   1965–69:

Expenditure

LCU = Dobras

Period/Year	Current LCU (millions)	Constant 1980 LCU (millions)	Constant 1980 US$ (millions)		Source
			Atlas	PPP	
1960–64	..	..	..	..	
1965–69	..	..	..	..	
1970	..	..	..	..	
1971	..	..	..	..	
1972	..	..	..	..	
1973	..	..	..	..	
1974	..	..	..	..	
1975	..	..	..	..	
1976	..	..	..	..	
1977	..	..	..	..	
1978	..	..	..	..	
1979	..	..	..	..	
1980	..	..	..	..	
1981	..	..	..	..	
1982	..	..	..	..	
1983	..	..	..	..	
1984	4.146	3.030	0.087	0.164	130
1985	..	..	..	..	
1986	..	..	..	..	

1960–64:   1965–69:

Personnel Comments:
Agricultural research at Sao Tomé & Principe is conducted within the Ministry of Agriculture and Livestock by the 'Direction du Développement Technique' (DDT).

Expenditure Comments:
The 1984 expenditure figure is based on the budget figure for 1984 of 'Direction du Développement Technique' (DDT).

Sources:
0027 Harvey, Nigel, ed. *Agricultural Research Centres: A World Directory of Organizations and Programmes.* Seventh Edition, Two Volumes. Harlow, U.K.: Longman, 1983.
0068 Conte, Stephan. "West Africa Agricultural Research Data Base Building Task: Termination Report." World Bank, Washington, D.C., 1985. Mimeo.
0130 FAO. *Rapport de Mission: La Recherche Agronomique à Sao Tomé & Principe.* Rome: FAO, April 1985.

Additional References:
0088 Swanson, Burton E., and Wade H. Reeves. "Agricultural Research West Africa: Manpower and Training." World Bank, Washington, D.C., November 1985. Mimeo.
0095 FAO – CARIS. *Agricultural Research in Developing Countries – Volume 1: Research Institutions.* Rome: FAO – CARIS, 1978.
0466 World Bank – West Africa Agricultural Research Review Team. "National Agricultural Research Systems (NARS): A Regional Appraisal and Selected Issues." World Bank, Washington, D.C., August 1986. Mimeo.

Saudi Arabia

Personnel

Period/Year	PhD	MSc	BSc	Subtotal	Expat	Total	Source
1960–64	..	..	..	..	..	..	
1965–69	..	..	..	..	..	..	
1970	..	..	..	..	..	..	
1971	..	..	..	..	..	..	
1972	..	..	..	..	..	..	
1973	..	..	..	..	..	..	
1974	..	..	..	..	..	..	
1975	..	..	..	..	..	..	
1976	..	..	..	..	..	..	
1977	..	..	..	..	..	..	
1978	..	..	..	..	..	..	
1979	..	..	..	..	..	..	
1980	..	..	..	..	..	..	
1981	..	..	..	..	..	..	
1982	..	..	..	..	..	..	
1983	2	12	80	94	77	171	17
1984	..	..	..	..	..	..	
1985	..	..	..	..	..	..	
1986	..	..	..	..	..	..	

1960–64:   1965–69:

Expenditure   *LCU = Saudi Arabian Riyals*

Period/Year	Current LCU (millions)	Constant 1980 LCU (millions)	Constant 1980 US$ (millions)		Source
			Atlas	PPP	
1960–64	..	..	..	..	
1965–69	..	..	..	..	
1970	..	..	..	..	
1971	..	..	..	..	
1972	..	..	..	..	
1973	..	..	..	..	
1974	..	..	..	..	
1975	..	..	..	..	
1976	..	..	..	..	
1977	..	..	..	..	
1978	..	..	..	..	
1979	..	..	..	..	
1980	..	..	..	..	
1981	..	..	..	..	
1982	..	..	..	..	
1983	..	..	..	..	
1984	..	..	..	..	
1985	..	..	..	..	
1986	..	..	..	..	

1960–64:   1965–69:

Personnel Comments:

The 1983 figure has been constructed on the basis of source 17 as follows:

	PhD	MSc	BSc	Expat	Total
National Centre For Horticultural Research and Development	0	1	4	14	19
Gassim Agriculture Research Centre	0	1	15	15	31
Regional Agriculture and Water Research Center	1	8	24	23	56
Range and Animal Development Research Center	0	1	12	13	26
Wadi Gizan Development Project	0	0	7	12	19
Fisheries Research Center	1	1	18	0	20
Total	2	12	80	77	171

Expenditure Comments:

It was not possible to construct a reliable and complete expenditure figure from the ISNAR, IFARD & AOAD questionnaire response (source 17).

Sources:

0017 ISNAR, IFARD & AOAD. Survey of National Agricultural Research Systems: Unpublished Questionnaire Responses. ISNAR, The Hague, 1985.

Additional References:

0094 FAO – Near East Regional Office. *Directory of Agricultural Research Institutions in the Near East Region.* Cairo: FAO – Near East Regional Office, 1979.

0173 FAO. *FAO Near East Regional Studies on Organization and Administration of Agricultural Research.* Rome: FAO, 1974.

0865 Oram, Peter. "Agricultural Research Objectives and Priorities: Constraints to the Development of Agricultural Research Institutions in Arab countries." Review paper prepared for the first meeting of the Council for Arab Agricultural Research, organized by the Arab Fund for Economic and Social Development, Kuwait, 23 March 1988.

Senegal

Personnel

Period/Year	PhD	MSc	BSc	Subtotal	Expat	Total	Source
1960-64	..	..	..	..	..	60	852
1965-69	..	..	..	..	..	51	10
1970	..	..	..	..	..	60	532
1971	..	..	..	..	..	..	
1972	..	..	..	..	..	63	431
1973	..	..	..		..	..	
1974	..	..	..	10	53	63	589
1975	..	..	..	..	..	..	
1976	..	..	..	..	..	83	95
1977	..	..	..	..		..	
1978	..	..	..	37	62	99	259
1979	..	..	..	..	..	126	26
1980	..	..	..	..	..	..	
1981	..	..	..	..	..	..	
1982	..	..	..	..	..	..	
1983	..	..	..	..	..	165	15
1984	..	..	..	129	54	183	815
1985	..	..	..	..	..	..	
1986	..	..	..	..	..	..	

1960–64: 1963   1965–69: 1966

Expenditure   *LCU = CFA Francs*

Period/Year	Current LCU (millions)	Constant 1980 LCU (millions)	Constant 1980 US$ (millions)		Source
			Atlas	PPP	
1960–64	..	..	..	..	
1965–69	750.990	1816.386	8.302	11.980	589
1970	..	..	..	..	
1971	1085.774	2323.681	10.620	15.326	589
1972	1010.983	2079.790	9.505	13.718	431
1973	941.488	1797.533	8.215	11.856	589
1974	..	..	..	..	
1975	..	..	..	..	
1976	1349.000	1906.381	8.713	12.574	26
1977	1498.000	1955.820	8.939	12.900	26
1978	1613.000	1929.716	8.820	12.728	26
1979	1739.000	1938.985	8.862	12.789	26
1980	1579.000	1579.000	7.217	10.415	26
1981	2254.000	2094.342	9.572	13.814	26
1982	..	..	..	..	
1983	3024.000	2333.398	10.665	15.390	15
1984	..	..	..	..	
1985	..	..	..	..	
1986	5099.578	..	..	..	814

1960–64:   1965–69: 1966

Personnel Comments:
Until 1975 agricultural research in Senegal was dominated by French overseas agricultural research institutes. In 1975 these institutes were nationalized and integrated into ISRA. In some cases, like forestry and fisheries research, only a name change took place, while in the case of crop research, a reorganization was initiated. The table below gives an overview of how the indicator series has been constructed. The post-1975 figures include all institutes integrated into ISRA, while the pre-1975 figures include the predecessors of these institutes.

	1966	1970	1972	1974		1976	1978	1979	1983	1984
IRAT	23	27	24		ISRA Headquarters	[3]	5	5	[7]	8
IFAC	NA	2	3		CNRA Bambey	30	36	42	39	29
IRHO	5	6	[6]		CNRA Djibelor	4	5	10	17	19
IRCT			2		CNRA Kaolack	3	8	11	8	11
					CR Richard Toll	4	8	13	12	17
					CDH[c]				13	14
IEMVT	NA	16	17		LNERV	23	18	11	25	25
CRZ de Dahra	NA	2	2		CRZ de Dahra	2	1	2	7	9
CRZ de Kolda[a]	–	–	–		CRZ de Kolda	[2]	2	2	5	6
CRODT[b]	NA	5	[6]		CRODT	9	15	14	17	24
CTFT	NA	2	3		CNRF	3	1	16	15	21
Total	NA	60	63	63		83	99	126	165	183
Source	653	532	431	589		95	259	26	15	815

[.] = estimated or constructed by authors.
a. Established in 1972.
b. Originally established by ORSTOM.
c. Integrated into ISRA in 1979.

Other organizations which, it is understood, conduct agricultural research but for various reasons are not included in the indicator series, are ITA (1976: 13 researchers, source 95; 1982: 25 researchers, source 27), Faculty of Science, University of Dakar (some research, source 68), Ecole Nationale Supérieure Universitaire de Technologie (some research, source 68), and ORSTOM (1972: 39 researchers, source 10; 1976: 55 researchers, source 95; 1985: 106 researchers, source 258). ORSTOM-Senegal is one of the biggest ORSTOM centers in Africa and has a regional, not specifically Senegalese, mandate. In addition, ORSTOM does not exclusively conduct agricultural research.

Expenditure Comments:
The expenditure figures included in the indicator series generally correspond with the personnel indicators.

The table on the next page gives an overview of construction details for the expenditure indicators.

	1966	1971	1972	1973		1978	1983
IRAT	379.790	598.740	521.637	481.027	ISRA Headquarters	80.050	NA
IFAC	9.800	34.000	24.985	33.250	CNRA Bambey	572.559	707.000
IRHO	57.500	62.906	[65.653]	68.400	CNRA Djbelor	112.287	255.000
IRCT			26.300	6.000	CNRA Kaolack	198.789	313.000
					CR Richard Toll	85.209	244.000
					CDH[a]	–	286.000
IEMVT	198.900	275.393	[243.000]	210.625	LNERV	308.791	528.000
CRZ de Dahta	44.000	44.000	[49.247]	[54.494]	CRZ de Dahra	80.728	185.000
CRZ de Kolda[b]	–	–	–	NA	CRZ de Kolda	49.422	71.000
CRODT[c]	32.000	35.800	[42.621]	[49.442]	CRODT	83.546	263.000
CTFT	29.000	34.935	37.540	38.250	CNRF	41.342	172.000
Total	750.990	1085.774	1010.983	941.488		612.723	3024.000
Source	589	589	431	589		16/259	15

[ ] = estimated or constructed by authors.
a. Integrated into ISRA in 1979.
b. Established in 1972.
c. Originally established by ORSTOM.

Acronyms and Abbreviations:
CDH: Centre pour le Développement de l'Horticulture
CNRA: Centre National de Recherches Agronomiques
CNRF: Centre National de Recherche Forestière
CR: Centre de Recherches
CRODT: Centre de Recherches Océanographiques Dakar-Tiaroye
CRZ : Centre de Recherches Zootechniques
CTFT: Centre Technique Forestier Tropical
IEMVT : Institut d'Elevage et de Médecine Vétérinaire des Pays Tropicaux
IFAC: Institut Français de Recherches Fruitières d'Outre mer
IRAT: Institut de Recherches Agronomiques Tropicales et des Cultures Vivrières
IRCT: Institut de Recherches du Coton et des Textiles
IRFA : Institut de Recherches sur les Fruits et Agrumes
IRHO: Institut de Recherches pour les Huiles et Oléagineux
ISRA: Institut Sénégalais pour la Recherche Agricole
ITA: Institut de Technologie Alimentaire
LNERV : Laboratoire National de l'Elevage et de la Recherche Vétérinaire
ORSTOM: Organization de Recherche Scientifique et Technique Outre-Mer

Sources:
0010 Boyce, James K., and Robert E. Evenson. *National and International Agricultural Research and Extension Programs*. New York: Agricultural Development Council, Inc., 1975.

0015 Institut du Sahel, and DEVRES, Inc. *Assessment of Agricultural Research Resources in the Sahel, Volume III: National Report Senegal.* Washington, D.C.: DEVRES, Inc., August 1984.

0026 Oram, Peter A., and Mark Gieben. "Document Summaries." ISNAR, The Hague, 1984. Mimeo.

0095 FAO – CARIS. *Agricultural Research in Developing Countries – Volume 1: Research Institutions.* Rome: FAO – CARIS, 1978.

0259 Senegalese-IADS Study Team. "Senegal Agricultural Research Review." Draft of a Senegalese-IADS Study Team Report to the General Delegation for Scientific and Technical Research. Dakar, Senegal, December 1978. Mimeo.

0431 Current Agricultural Research Information System (CARIS). *Directory of Agricultural Research Institutions and Projects in West Africa – Pilot Project 1972–73.* Rome: FAO, 1973.

0532 UNESCO Field Science Office for Africa. *Survey on the Scientific and Technical Potential of the Countries of Africa.* Paris: UNESCO, 1970.

0589 Kassapu, Samuel. *Les Dépenses de Recherche Agricole dans 34 Pays d'Afrique Tropicale.* Paris: Centre de Développement de l'OCDE, 1976.

0814 Institut Sénégalais de Recherches Agricoles (ISRA). "Compte Prévisionnel 1987 Volet Gestion." ISRA, Dakar, Sénégal. Mimeo.

0815 ISNAR. "Développement et Gestion des Ressources Humaines à l'Institut Sénégalais de Recherches Agricoles." Une Etude Préliminaire. ISNAR, The Hague, February 1988. Mimeo.

0852 Evenson, Robert E., and Yoav Kislev. *Investment in Agricultural Research and Extension: A Survey of International Data.* Center Discussion Paper No. 124. New Haven, Connecticut: Economic Growth Center, Yale University, August 1971.

Additional References:

0016 Oram, Peter A., and Vishva Bindlish. *Resource Allocations to National Agricultural Research: Trends in the 1970s.* The Hague and Washington, D.C.: ISNAR and IFPRI, November 1981.

0023 Bennell, Paul. *Agricultural Researchers in Sub-Saharan Africa: An Overview.* Working Paper No. 4. The Hague: ISNAR, October 1985.

0027 Harvey, Nigel, ed. *Agricultural Research Centres: A World Directory of Organizations and Programmes.* Seventh Edition, Two Volumes. Harlow, U.K.: Longman, 1983.

0045 Institut du Sahel, and DEVRES, Inc. *Bilan des Ressources de la Recherche Agricole dans les Pays du Sahel, Volume I: Analyse et Stratégie Régionale.* Washington, D.C.: DEVRES, Inc., 1984.

0068 Conte, Stephan. "West Africa Agricultural Research Data Base Building Task: Termination Report." World Bank, Washington, D.C., 1985. Mimeo.

0073 Oram, Peter A., and Vishva Bindlish. "Investment in Agricultural Research in Developing Countries: Progress, Problems, and the Determination of Priorities." IFPRI, Washington, D.C., January 1984. Mimeo.

0083 Institut du Sahel, and DEVRES, Inc. *Bilan des Ressources de la Recherche Agricole dans les Pays du Sahel, Volume II: Résumés des Rapports Nationaux.* Washington, D.C.: DEVRES, Inc., August 1984.

0088 Swanson, Burton E., and Wade H. Reeves. "Agricultural Research West Africa: Manpower and Training." World Bank, Washington, D.C., November 1985. Mimeo.

0089 Oram, Peter. "Report on National Agricultural Research in West Africa." IFPRI, Washington, D.C., February 1986. Mimeo.

0135 FAO, and UNDP. *National Agricultural Research – Report of an Evaluation Study in Selected Countries.* Rome: FAO and UNDP, 1984.

0141 IAR&T Consultancy Group – University of Ife. *A Consultancy Report on Agricultural Research Delivery Systems in West Africa.* Rome: FAO, December 1979.

0152 IRAT, and Centre National de Recherches Agronomiques de Bambey (CNRA). *Rapport Annuel 1974 IRAT-Senegal.* Bambey: IRAT, and CNRA, 1975.

0153 Institut Sénégalais de Recherches Agricoles (ISRA). *Le Centre National de Recherches Agronomiques de Bambey.* Bambey: ISRA, November 1976.

0163 CGIAR. "National Agricultural Research." CGIAR, Washington, D.C., 1985. Mimeo.

0165 Evenson, R.E., and Y. Kislev. *Agricultural Research and Productivity.* New Haven: Yale University Press, 1975.

0175 Cooper, St.G.C. *Agricultural Research in Tropical Africa.* Kampala: East African Literature Bureau, 1970.

0258 "Effort Français en Faveur de la Recherche Agronomique dans les Pays du Sahel 1984–1985." 1985. Mimeo.

0260 Centre Africain d'Etudes Supérieures en Gestion (CESAG). *Rapport Introductif de la Commission No 1: Organisation de Système de Recherche au Sénégal.* Etude réalisée par le CESAG pour le CILSS. Senegal: CESAG, June 1986.

0262 Pages, P.D., and J.D. Drilon, Jr. "Organizational Trends in National Research Management: A Global Overview." n.d. Mimeo.

0266 UNESCO. *National Science Policies in Africa.* Science Policy Studies and Documents No. 31. Paris: UNESCO, 1974.

0360 Cooper, St. G.C. "Towards Trained Manpower for Agricultural Research in Africa." Paper presented at the Conference on Agricultural Research and Production in Africa, organized by the Association for the Advancement of Agricultural Sciences in Africa (AAASA), Addis Ababa, 29 August–4 September 1971.

0372 Gabas, Jean-Jacques. *La Recherche Agricole dans les Pays Membres du CILSS.* Club du Sahel, September 1986.

0373 Groupement d'Etudes et de Recherches pour le Développement de l'Agronomie Tropicale (GERDAT). *Rapport Général d'Activité pour 1982.* Paris: GERDAT, 1983.

0384 Institut du Sahel, and DEVRES, Inc. *Sahel Region Agricultural Research Resource Assessment - Data Base Management Information System: Diskettes and User's Guide.* Printouts. Washington, D.C.: Institut du Sahel, and DEVRES, Inc, n.d.

0400 UNESCO. *The Promotion of Scientific Activity in Tropical Africa.* Science Policy Studies and Documents No. 11. Paris: UNESCO, 1969.

0456 Bingen, R. James, and Jacques Faye. "Agricultural Research and Extension in Francophone West

Africa: The Senegal Experience." Paper presented at the Farming Systems Research and Extension Symposium: Management and Methodology, Kansas State University, Manhattan, Kansas, 13–16 October 1985.

0466 World Bank – West Africa Agricultural Research Review Team. "National Agricultural Research Systems (NARS): A Regional Appraisal and Selected Issues." World Bank, Washington, D.C., August 1986. Mimeo.

0653 Webster, B.N. *Index of Agricultural Research Institutions and Stations in Africa*. Rome: FAO, n.d.

0654 Office de la Recherche Scientifique et Technique Outre-Mer (ORSTOM). *Rapport d'Activité 1967*. Paris: ORSTOM, 1969.

0886 Evenson, Robert E., and Yoav Kislev. "Investment in Agricultural Research and Extension: A Survey of International Data." *Economic Development and Cultural Change* Vol. 23 (April 1975): 507–521.

Seychelles

Personnel

Period/Year	PhD	MSc	BSc	Subtotal	Expat	Total	Source
1960–64	..	..	..	..	..	..	
1965–69	..	..	..	..	..	..	
1970	..	..	..	..	..	..	
1971	..	..	..	..	..	..	
1972	..	..	..	..	..	..	
1973	..	..	..	..	..	..	
1974	..	..	..	..	..	..	
1975	..	..	..	..	..	..	
1976	..	..	..	..	..	..	
1977	..	..	..	..	..	..	
1978	..	..	..	..	..	..	
1979	..	..	..	..	..	..	
1980	..	..	..	..	..	..	
1981	..	..	..	..	..	..	
1982	..	..	..	..	..	..	
1983	..	..	..	3	2	5	23
1984	..	..	..	..	..	..	
1985	0	0	5	5	3	8	445
1986	..	..	..	..	..	..	

1960–64:   1965–69:

Expenditure

LCU = Seychelle Rupees

Period/Year	Current LCU (millions)	Constant 1980 LCU (millions)	Constant 1980 US$ (millions)		Source
			Atlas	PPP	
1960–64	..	..	..	..	
1965–69	..	..	..	..	
1970	..	..	..	..	
1971	..	..	..	..	
1972	..	..	..	..	
1973	..	..	..	..	
1974	..	..	..	..	
1975	..	..	..	..	
1976	..	..	..	..	
1977	..	..	..	..	
1978	..	..	..	..	
1979	..	..	..	..	
1980	..	..	..	..	
1981	..	..	..	..	
1982	..	..	..	..	
1983	..	..	..	..	
1984	..	..	..	..	
1985	..	..	..	..	
1986	..	..	..	..	

1960–64:   1965–69:

Personnel Comments:

The 1983 figure includes only crop research, while the 1985 figure can be broken down as follows:

	National	Expatriate
Ministry of National Development		
Agricultural Department		
Agricultural Promotion Division	3	0
Fisheries Division	2	1
Private Tea Research	0	2
Total	5	3

Sources:

0023 Bennell, Paul. *Agricultural Researchers in Sub-Saharan Africa: An Overview.* Working Paper No. 4. The Hague: ISNAR, October 1985.

0445 Swanson, Burton E., and Wade H. Reeves. "Agricultural Research Eastern and Southern Africa: Manpower and Training." World Bank, Washington, D.C., August 1986. Mimeo.

Additional References:

0244 Commonwealth Agricultural Bureaux (CAB). *List of Research Workers in the Agricultural Sciences in the Commonwealth and in the Republic of Ireland 1975.* Slough, England: CAB, 1975.

0279 Commonwealth Agricultural Bureaux (CAB). *List of Research Workers in the Agricultural Sciences in the Commonwealth and in the Republic of Ireland 1972.* Slough, England: CAB, 1972.

0285 Commonwealth Agricultural Bureaux (CAB). *List of Research Workers in the Agricultural Sciences in the Commonwealth and in the Republic of Ireland 1969.* Slough, England: CAB, 1969.

0286 Commonwealth Agricultural Bureaux (CAB). *List of Research Workers in Agriculture, Animal Health and Forestry in the Commonwealth and in the Republic of Ireland 1966.* Slough, England: CAB, 1966.

0287 Commonwealth Agricultural Bureaux (CAB). *List of Research Workers in Agriculture, Animal Health and Forestry in the British Commonwealth, the Republic of Sudan and the Republic of Ireland 1959.* Slough, England: CAB, 1959.

Sierra Leone

Personnel

Period/Year	PhD	MSc	BSc	Subtotal	Expat	Total	Source
1960–64	..	..	..	..	..	..	
1965–69	..	..	..	..	..	..	
1970	..	..	..	..	..	..	
1971	..	..	..	..	..	..	
1972	..	..	..	..	..	..	
1973	..	..	..	..	..	..	
1974	..	..	..	..	..	..	
1975	..	..	..	..	..	..	
1976	..	..	..	..	..	..	
1977	..	..	..	..	..	..	
1978	..	..	..	..	..	..	
1979	..	..	..	..	..	..	
1980	..	..	..	..	..	..	
1981	..	..	..	..	..	..	
1982	..	..	..	..	..	..	
1983	..	..	..	..	..	..	
1984	..	..	..	..	..	..	
1985	..	..	..	..	..	46	86
1986	..	..	..	..	..		

1960–64:   1965–69:

Expenditure

LCU = Leones

Period/Year	Current LCU (millions)	Constant 1980 LCU (millions)	Constant 1980 US$ (millions)		Source
			Atlas	PPP	
1960–64	..	..	..	..	
1965–69	..	..	..	..	
1970	..	..	..	..	
1971	..	..	..	..	
1972	..	..	..	..	
1973	..	..	..	..	
1974	..	..	..	..	
1975	..	..	..	..	
1976	..	..	..	..	
1977	..	..	..	..	
1978	..	..	..	..	
1979	..	..	..	..	
1980	..	..	..	..	
1981	..	..	..	..	
1982	..	..	..	..	
1983	..	..	..	..	
1984	3.142	0.906	0.841	1.183	86
1985	2.800	0.543	0.504	0.709	466
1986	..	..	..	..	

1960–64:   1965–69:

Personnel Comments:
The 1985 figure includes the Rokpur Rice Research Station, the Adaptive Crop Research and Extension Project, and the Land and Water Development Division. It is consistent with the 1984 and 1985 expenditure figure.

	1965	1968	1970	1971	1972	1974	1977	1980	1981	1982	1985
RRRS		3	3	2	3						14
VRL	3		1		2						
AHS				4							
FR&TC	2		2							2	
MOA/FD[a]	4		3								
IMB&O			3								
ACREP[b]									6		6
LWDD[c]											26
FBC/FPAS			40		37						
NUC/FA	24	29	17		42	46	29	38			
Source	286	285	532	279	431	244	243	242	68	68	86

a.According to source 68 the research capacity of the Fisheries Division has been transferred to Fourah Bay College. Year not given.
b.Initiated in 1979 under external financial and technical assistance from USAID and in collaboration with a consortium of the Southern University, Louisiana State University, and the Njala University College.
c.Grew out of the FAO/UNDP Land Resources Survey Project which terminated in the late 1970s.

Given only scattered observations, as presented above, it has not been possible to construct a time series. It appears the universities have played an important role in agricultural research in Sierra Leone. For instance, the Rokpur Rice Research Station was developed from the Njala University College, and the Institute of Marine Biology and Oceanography from the Fourah Bay College.

Expenditure Comments:
The 1984 expenditure figure includes the Rokpur Rice Research Station, the Land and Water Development Division, and the Adaptive Crop Research and Extension Project.

Acronyms and Abbreviations:
ACREP: Adaptive Crop Research and Extension Project
AHS: Animal Husbandry Station
FBC/FPAS: Fourah Bay College – Faculty of Pure and Applied Science
FR & TC: Forestry Research & Training Centre
IMB & O: Institute of Marine Biology and Oceanography
LWDD : Land and Water Development Division
MOA/FD: Ministry of Agriculture – Fisheries Division
NUC/FA : Njala University College – Faculty of Agriculture
RRRS : Rokpur Rice Research Station
VRL : Veterinary Research Laboratory

Sources:
0086 World Bank. "West Africa Agricultural Research Review – Country Studies." World Bank, Washington, D.C., 1985. Mimeo.
0466 World Bank – West Africa Agricultural Research Review Team. "National Agricultural Research Systems (NARS): A Regional Appraisal and Selected Issues." World Bank, Washington, D.C., August 1986. Mimeo.

Additional References:
0010 Boyce, James K., and Robert E. Evenson. *National and International Agricultural Research and Extension Programs.* New York: Agricultural Development Council, Inc., 1975.
0014 Judd, M. Ann, James K. Boyce, and Robert E. Evenson. "Investing in Agricultural Supply." Economic Growth Center, Yale University, New Haven, Connecticut, 1983. Mimeo.
0016 Oram, Peter A., and Vishva Bindlish. *Resource Allocations to National Agricultural Research: Trends in the 1970s.* The Hague and Washington, D.C.: ISNAR and IFPRI, November 1981.
0017 ISNAR, IFARD & AOAD. Survey of National Agricultural Research Systems: Unpublished Questionnaire Responses. ISNAR, The Hague, 1985.
0023 Bennell, Paul. *Agricultural Researchers in Sub-Saharan Africa: An Overview.* Working Paper No. 4. The Hague: ISNAR, October 1985.

0068 Conte, Stephan. "West Africa Agricultural Research Data Base Building Task: Termination Report." World Bank, Washington, D.C., 1985. Mimeo.

0073 Oram, Peter A., and Vishva Bindlish. "Investment in Agricultural Research in Developing Countries: Progress, Problems, and the Determination of Priorities." IFPRI, Washington, D.C, January 1984. Mimeo.

0088 Swanson, Burton E., and Wade H. Reeves. "Agricultural Research West Africa: Manpower and Training." World Bank, Washington, D.C., November 1985. Mimeo.

0089 Oram, Peter. "Report on National Agricultural Research in West Africa." IFPRI, Washington, D.C., February 1986. Mimeo.

0164 Association for the Advancement of Agricultural Sciences in Africa (AAASA). *Proceedings of the Workshop on Agricultural Research Administration, Nairobi, Kenya, 27–30 June 1977.* Proceedings Series PE-4. Addis Ababa, Ethiopia: AAASA and IDRC, August 1979.

0165 Evenson, R.E., and Y. Kislev. *Agricultural Research and Productivity.* New Haven: Yale University Press, 1975.

0175 Cooper, St.G.C. *Agricultural Research in Tropical Africa.* Kampala: East African Literature Bureau, 1970.

0242 Commonwealth Agricultural Bureaux (CAB). *List of Research Workers in Agricultural Sciences in the Commonwealth 1981.* Slough, England: CAB, 1981.

0243 Commonwealth Agricultural Bureaux (CAB). *List of Research Workers in the Agricultural Sciences in the Commonwealth and in the Republic of Ireland 1978.* Slough, England: CAB, 1978.

0244 Commonwealth Agricultural Bureaux (CAB). *List of Research Workers in the Agricultural Sciences in the Commonwealth and in the Republic of Ireland 1975.* Slough, England: CAB, 1975.

0266 UNESCO. *National Science Policies in Africa.* Science Policy Studies and Documents No. 31. Paris: UNESCO, 1974.

0279 Commonwealth Agricultural Bureaux (CAB). *List of Research Workers in the Agricultural Sciences in the Commonwealth and in the Republic of Ireland 1972.* Slough, England: CAB, 1972.

0285 Commonwealth Agricultural Bureaux (CAB). *List of Research Workers in the Agricultural Sciences in the Commonwealth and in the Republic of Ireland 1969.* Slough, England: CAB, 1969.

0286 Commonwealth Agricultural Bureaux (CAB). *List of Research Workers in Agriculture, Animal Health and Forestry in the Commonwealth and in the Republic of Ireland 1966.* Slough, England: CAB, 1966.

0287 Commonwealth Agricultural Bureaux (CAB). *List of Research Workers in Agriculture, Animal Health and Forestry in the British Commonwealth, the Republic of Sudan and the Republic of Ireland 1959.* Slough, England: CAB, 1959.

0360 Cooper, St.G.C. "Towards Trained Manpower for Agricultural Research in Africa." Paper presented at the Conference on Agricultural Research and Production in Africa, organized by the Association for the Advancement of Agricultural Sciences in Africa (AAASA), Addis Ababa, 29 August–4 September 1971.

0431 Current Agricultural Research Information System (CARIS). *Directory of Agricultural Research Institutions and Projects in West Africa – Pilot Project 1972–73.* Rome: FAO, 1973.

0532 UNESCO Field Science Office for Africa. *Survey on the Scientific and Technical Potential of the Countries of Africa.* Paris: UNESCO, 1970.

0589 Kassapu, Samuel. *Les Dépenses de Recherche Agricole dans 34 Pays d'Afrique Tropicale.* Paris: Centre de Développement de l'OCDE, 1976.

0653 Webster, B.N. *Index of Agricultural Research Institutions and Stations in Africa.* Rome: FAO, n.d.

0852 Evenson, Robert E., and Yoav Kislev. *Investment in Agricultural Research and Extension: A Survey of International Data.* Center Discussion Paper No. 124. New Haven, Connecticut: Economic Growth Center, Yale University, August 1971.

0886 Evenson, Robert E., and Yoav Kislev. "Investment in Agricultural Research and Extension: A Survey of International Data." *Economic Development and Cultural Change* Vol. 23 (April 1975): 507–521.

Singapore

Personnel

Period/Year	PhD	MSc	BSc	Subtotal	Expat	Total	Source
1960–64	1	4	3	8	..	8	715
1965–69	..	..	..	..	..	8	10
1970	..	..	..	..	..	..	
1971	..	..	..	..	..	10	10
1972	..	..	..	..	..	..	
1973	..	..	..	..	..	..	
1974	..	..	..	..	..	..	
1975	..	..	..	..	..	..	
1976	..	..	..	..	..	..	
1977	..	..	..	..	..	..	
1978	..	..	..	..	..	..	
1979	..	..	..	..	..	..	
1980	..	..	..	..	..	..	
1981	..	..	..	..	..	..	
1982	..	..	..	..	..	..	
1983	..	..	..	..	..	..	
1984	..	..	..	..	..	..	
1985	..	..	..	..	..	..	
1986	..	..	..	..	..	..	

1960–64: 1960   1965–69: 1968

Expenditure

LCU = Singapore Dollars

Period/Year	Current LCU (millions)	Constant 1980 LCU (millions)	Constant 1980 US$ (millions) Atlas	Constant 1980 US$ (millions) PPP	Source
1960–64	0.540	1.087	0.499	0.628	715
1965–69	3.673	6.891	3.166	3.984	10
1970	..	..	..	..	
1971	..	..	..	..	
1972	..	..	..	..	
1973	..	..	..	..	
1974	..	..	..	..	
1975	..	..	..	..	
1976	..	..	..	..	
1977	..	..	..	..	
1978	..	..	..	..	
1979	..	..	..	..	
1980	..	..	..	..	
1981	..	..	..	..	
1982	..	..	..	..	
1983	..	..	..	..	
1984	..	..	..	..	
1985	..	..	..	..	
1986	..	..	..	..	

1960–64: 1960   1965–69: 1968

Personnel Comments:

The 1960 figure refers to the Central Experiment Station of the Department of Primary Products. According to source 10, the figures include a great deal of fisheries research. It probably accounts for most of the agricultural research carried out in the country.

Sources:

0010 Boyce, James K., and Robert E. Evenson. *National and International Agricultural Research and Extension Programs.* New York: Agricultural Development Council, Inc., 1975.

0715 Chang, C.W. *FAO Directory of Agricultural Research Institutes and Experiment Stations in Asia and the Far East.* Bangkok, Thailand: FAO – Regional Office for Asia and the Far East, September 1962.

Additional References:

0027 Harvey, Nigel, ed. *Agricultural Research Centres: A World Directory of Organizations and Programmes.* Seventh Edition, Two Volumes. Harlow, U.K.: Longman, 1983.

0716 Chang, C.W. *FAO Directory of Agricultural Research Institutes and Experiment Stations in Asia and the Far East, Supplement.* Bangkok, Thailand: FAO – Regional Office for Asia and the Far East, May 1963.

0852 Evenson, Robert E., and Yoav Kislev. *Investment in Agricultural Research and Extension: A Survey of International Data.* Center Discussion Paper No. 124. New Haven, Connecticut: Economic Growth Center, Yale University, August 1971.

Solomon Islands

Personnel

Period/Year	PhD	MSc	BSc	Subtotal	Expat	Total	Source
1960–64	..	..	..	..	..	..	
1965–69	..	..	..	..	..	7	723
1970	..	..	..	..	..	6	723
1971	..	..	..	..	..	10	723
1972	..	..	..	..	..	7	723
1973	..	..	..	..	..	7	723
1974	..	..	..	..	..	6	723
1975	..	..	..	..	..	9	723
1976	..	..	..	..	..	10	723
1977	..	..	..	..	..	10	723
1978	..	..	..	..	..	11	723
1979	..	..	..	..	..	11	723
1980	..	..	..	..	..	11	723
1981	..	..	..	..	..	11	723
1982	..	..	..	..	..	12	723
1983	..	..	..	..	..	8	723
1984	..	..	..	..	..	13	723
1985	..	..	..	..	..	13	723
1986	0	1	3	4	9	13	723/771

1960–64:   1965–69: 1966, 67, 68 & 69

Expenditure

LCU = Solomon Islands Dollars

Period/Year	Current LCU (millions)	Constant 1980 LCU (millions)	Constant 1980 US$ (millions) Atlas	PPP	Source
1960–64	..	..	..	..	
1965–69	0.093	0.248	0.276	0.437	723
1970	0.101	0.230	0.257	0.407	723
1971	0.089	0.197	0.219	0.347	723
1972	0.336	0.690	0.770	1.218	723
1973	0.276	0.549	0.612	0.969	723
1974	0.104	0.175	0.195	0.309	723
1975	0.160	0.265	0.296	0.468	723
1976	0.177	0.280	0.312	0.495	723
1977	0.295	0.410	0.457	0.724	723
1978	0.250	0.318	0.355	0.562	723
1979	0.451	0.525	0.585	0.927	723
1980	0.314	0.314	0.350	0.555	723
1981	0.322	0.281	0.313	0.496	723
1982	0.211	0.168	0.188	0.297	723
1983	0.443	0.328	0.366	0.579	723
1984	0.326	..	..	..	723
1985	0.775	..	..	..	723
1986	1.205	..	..	..	723

1960–64:   1965–69: 1965, 66, 67, 68 & 69

Personnel Comments:
The personnel indicators include the Department of Research of the Ministry of Agriculture and Lands only (source 808). Not included are the Fisheries Division (1983: 3 researchers, source 17) and the Forestry Division (1982: 1 researcher, source 27) of the Ministry of Natural Resources. Also not included are three private companies which undertake agricultural research, namely, Solomon Rice Company Limited (1983: 1 researcher, source 17; ceased operations in 1986, source 808), Solomon Islands Plantation Limited (1983: 2 researchers, source 17; 1987: 1 researcher, source 808), and Lever Solomons Limited (1987: 1 researcher, source 808).

Expenditure Comments:
The expenditure indicators include the Department of Research of the Ministry of Agriculture and Lands only. The table below provides additional information concerning forestry and fisheries research expenditures in thousands of LCUs (source 723).

	1965	1966	1967	1968	1969	1970	1972
Forestry research	25.044	54.534	35.190	2.977	19.521	8.500	11.172
Fisheries research							0.250

	1973	1974	1975	1976	1977	1978	1979
Forestry research	34.750	103.550	43.600	42.822	8.467	5.218	6.000
Fisheries research	2.100	1.000	6.500	19.000	35.500	64.000	26.000

	1980	1981	1982	1983	1984	1985	1986
Forestry research	6.000	2.600	3.000	306.084	243.850	77.535	86.465
Fisheries research	108.000	75.000	46.500	104.500	103.000	1253.500	1297.500

Sources:
0723 Fernando, L.H. "South Pacific NARS Resource Commitment Survey." ADB and ISNAR, The Hague, 1987. Mimeo.
0771 Fernando, L.H. "ISNAR/IRETA Workshop on Planning and Management of Agricultural Research in the South Pacific: Survey of Agricultural Research Resources." ISNAR, The Hague, n.d. Mimeo.

Additional References:
0016 Oram, Peter A., and Vishva Bindlish. *Resource Allocations to National Agricultural Research: Trends in the 1970s.* The Hague and Washington, D.C.: ISNAR and IFPRI, November 1981.
0017 ISNAR, IFARD & AOAD. Survey of National Agricultural Research Systems: Unpublished Questionnaire Responses. ISNAR, The Hague, 1985.
0026 Oram, Peter A., and Mark Gieben. "Document Summaries." ISNAR, The Hague, 1984. Mimeo.
0027 Harvey, Nigel, ed. *Agricultural Research Centres: A World Directory of Organizations and Programmes.* Seventh Edition, Two Volumes. Harlow, U.K.: Longman, 1983.
0038 Gamble, W.K., R.M. Bourke, and C.W. Brookson. *South Pacific Agricultural Research Study.* Consultants' Report to the Asian Development Bank (two volumes). Manila: Asian Development Bank, June 1981.
0041 ISNAR. *Solomon Islands: Agricultural Research, Extension, and Support Facilities Project.* The Hague: ISNAR, December 1982.
0163 CGIAR. "National Agricultural Research." CGIAR, Washington, D.C., 1985. Mimeo.

390

0243 Commonwealth Agricultural Bureaux (CAB). *List of Research Workers in the Agricultural Sciences in the Commonwealth and in the Republic of Ireland 1978.* Slough, England: CAB, 1978.

0244 Commonwealth Agricultural Bureaux (CAB). *List of Research Workers in the Agricultural Sciences in the Commonwealth and in the Republic of Ireland 1975.* Slough, England: CAB, 1975.

0279 Commonwealth Agricultural Bureaux (CAB). *List of Research Workers in the Agricultural Sciences in the Commonwealth and in the Republic of Ireland 1972.* Slough, England: CAB, 1972.

0285 Commonwealth Agricultural Bureaux (CAB). *List of Research Workers in the Agricultural Sciences in the Commonwealth and in the Republic of Ireland 1969.* Slough, England: CAB, 1969.

0286 Commonwealth Agricultural Bureaux (CAB). *List of Research Workers in Agriculture, Animal Health and Forestry in the Commonwealth and in the Republic of Ireland 1966.* Slough, England: CAB, 1966.

0287 Commonwealth Agricultural Bureaux (CAB). *List of Research Workers in Agriculture, Animal Health and Forestry in the British Commonwealth, the Republic of Sudan and the Republic of Ireland 1959.* Slough, England: CAB, 1959.

0808 Abington, J.B. Personal communication. Chief Research Officer, Ministry of Agriculture & Lands. Honiara, Solomon Islands, February 1988.

Somalia

Personnel

Period/Year	PhD	MSc	BSc	Subtotal	Expat	Total	Source
1960–64	..	..	..	..	..	8	852
1965–69	..	..	..	..	..	12	175
1970	..	..	..	..	..	9	532
1971	..	..	..	..	..	..	
1972	..	..	..	..	..	..	
1973	..	..	..	..	..	..	
1974	..	..	..	..	..	..	
1975	..	..	..	..	..	9	16
1976	..	..	..	..	..	24	95
1977	..	..	..	..	..	22	16
1978	..	..	..	..	..	..	
1979	..	..	..	..	..	..	
1980	0	1	9	10	..	10	870
1981	..	..	..	..	..	25	40
1982	1	2	30	33	..	33	761
1983	..	..	..	30	4	34	40
1984	..	..	..	..	..	..	
1985	..	..	..	..	..	..	
1986	1	10	24	35	..	35	873

1960–64: 1963   1965–69: 1967 & 69

Expenditure

LCU = Somalia Shillings

Period/Year	Current LCU (millions)	Constant 1980 LCU (millions)	Constant 1980 US$ (millions)		Source
			Atlas	PPP	
1960–64	..	..	..	..	
1965–69	2.500	10.279	1.097	2.309	589/175
1970	..	..	..	..	
1971	0.350	1.276	0.136	0.287	589
1972	..	..	..	..	
1973	..	..	..	..	
1974	..	..	..	..	
1975	..	..	..	..	
1976	..	..	..	..	
1977	..	..	..	..	
1978	..	..	..	..	
1979	..	..	..	..	
1980	..	..	..	..	
1981	..	..	..	..	
1982	..	..	..	..	
1983	3.560	1.434	0.153	0.322	40
1984	..	..	..	..	
1985	..	..	..	..	
1986	..	..	..	..	

1960–64:   1965–69: 1966 & 67

Personnel Comments:

These figures refer to the Agricultural Research Institute (ARI) of the Ministry of Agriculture only. ARI was established in 1965. During the early seventies ARI steadily declined for lack of consistent support. A rejuvenation came about in 1975 with the UNDP/FAO/ARI project to strengthen agricultural research (source 40). In 1980, 17 staff members were reported to be on training abroad. They have not been included in the 1980 figure. Figures for the later years may well include some staff on training.

Livestock and range research is not undertaken by ARI, but by the Department of Animal Health, the Department of Animal Production, the Trypanosomiasis Control Project, and the National Range Agency. In total, about 8 researchers were identified within these organizations in 1983 by source 40. They are not included in the indicator series. In addition 21 teacher-researchers at the Faculty of Agriculture and 15 at the Faculty of Veterinary and Animal Science were reported by source 761 for the early eighties.

Expenditure Comments:

The expenditure figures refer to the Agricultural Research Institute (ARI) only. The 1966 and 1967 figure consists mainly of foreign donor money. Given that ARI was established in 1965, it is likely that much of this donor money supported infrastructural development. Sources 589 and 40 indicate that the 1983 figure represents expenditures from domestic sources only.

Sources:

0016 Oram, Peter A., and Vishva Bindlish. *Resource Allocations to National Agricultural Research: Trends in the 1970s.* The Hague and Washington, D.C.: ISNAR and IFPRI, November 1981.

0040 ISNAR. *Report to the Government of the Democratic Republic of Somalia: Development of an Agricultural Research System.* The Hague: ISNAR, November 1983.

0095 FAO – CARIS. *Agricultural Research in Developing Countries – Volume 1: Research Institutions.* Rome: FAO – CARIS, 1978.

0175 Cooper, St.G.C. *Agricultural Research in Tropical Africa.* Kampala: East African Literature Bureau, 1970.

0532 UNESCO Field Science Office for Africa. *Survey on the Scientific and Technical Potential of the Countries of Africa.* Paris: UNESCO, 1970.

0589 Kassapu, Samuel. *Les Dépenses de Recherche Agricole dans 34 Pays d'Afrique Tropicale.* Paris: Centre de Développement de l'OCDE, 1976.

0761 Noor, M.A., M. Abdullahi, and A.N. Alio. "Status of Agricultural Research in Somalia." In *Proceedings of the Eastern Africa-ACIAR Consultation on Agricultural Research*, pp. 16–23. Australia: ACIAR, 1984.

0852 Evenson, Robert E., and Yoav Kislev. *Investment in Agricultural Research and Extension: A Survey of International Data.* Center Discussion Paper No. 124. New Haven, Connecticut: Economic Growth Center, Yale University, August 1971.

0870 Kira, M. Taher. *Report on Agricultural Research Manpower Assessment in People's Democratic Republic of Yemen and Democratic Republic of Somalia.* Cairo: FAO Near East Regional Office, January 1980.

0873 Abdalla, Abdalla Ahmed. "Agricultural Research in the IGADD Sub-Region and Related Manpower Training." Inter-Governmental Authority on Drought and Development (IGADD) and FAO, 1987. Mimeo.

Additional References:

0010 Boyce, James K., and Robert E. Evenson. *National and International Agricultural Research and Extension Programs.* New York: Agricultural Development Council, Inc., 1975.

0023 Bennell, Paul. *Agricultural Researchers in Sub-Saharan Africa: An Overview.* Working Paper No. 4. The Hague: ISNAR, October 1985.

0026 Oram, Peter A., and Mark Gieben. "Document Summaries." ISNAR, The Hague, 1984. Mimeo.

0073 Oram, Peter A., and Vishva Bindlish. "Investment in Agricultural Research in Developing Countries: Progress, Problems, and the Determination of Priorities." IFPRI, Washington, D.C., January 1984. Mimeo.

0094 FAO – Near East Regional Office. *Directory of Agricultural Research Institutions in the Near East Region.* Cairo: FAO – Near East Regional Office, 1979.

0163 CGIAR. "National Agricultural Research." CGIAR, Washington, D.C., 1985. Mimeo.

0165 Evenson, R.E., and Y. Kislev. *Agricultural Research and Productivity.* New Haven: Yale University Press, 1975.

0173 FAO. *FAO Near East Regional Studies on Organization and Administration of Agricultural Research.* Rome: FAO, 1974.

0227 Australian Centre for International Agricultural Research (ACIAR). *Proceedings of the Eastern*

Africa-ACIAR Consultation on Agricultural Research, 18–22 July 1983, Nairobi, Kenya. Australia: ACIAR, 1984.

0266 UNESCO. *National Science Policies in Africa.* Science Policy Studies and Documents No. 31. Paris: UNESCO, 1974.

0287 Commonwealth Agricultural Bureaux (CAB). *List of Research Workers in Agriculture, Animal Health and Forestry in the British Commonwealth, the Republic of Sudan and the Republic of Ireland 1959.* Slough, England: CAB, 1959.

0360 Cooper, St.G.C. "Towards Trained Manpower for Agricultural Research in Africa." Paper presented at the Conference on Agricultural Research and Production in Africa, organized by the Association for the Advancement of Agricultural Sciences in Africa (AAASA), Addis Ababa, 29 August–4 September 1971.

0445 Swanson, Burton E., and Wade H. Reeves. "Agricultural Research Eastern and Southern Africa: Manpower and Training." World Bank, Washington, D.C., August 1986. Mimeo.

0846 Alio, A.N. "Agricultural Research System in Somalia." Paper presented at the Second General Conference of AARINENA, Nicosia, Cyprus, 15–17 December 1987.

0886 Evenson, Robert E., and Yoav Kislev. "Investment in Agricultural Research and Extension: A Survey of International Data." *Economic Development and Cultural Change* Vol. 23 (April 1975): 507–521.

South Africa

Personnel

Period/Year	PhD	MSc	BSc	Subtotal	Expat	Total	Source
1960–64	..	..	..	..	..	..	
1965–69	..	..	..	..	..	..	
1970	..	..	..	..	..	..	
1971	..	..	..	..	..	..	
1972	..	..	..	..	..	..	
1973	..	..	..	..	..	..	
1974	..	..	..	..	..	..	
1975	..	..	..	..	..	..	
1976	..	..	..	..	..	..	
1977	..	..	..	..	..	..	
1978	..	..	..	..	..	..	
1979	..	..	..	..	..	..	
1980	..	..	..	..	..	..	
1981	..	..	..	..	..	..	
1982	..	..	..	..	..	..	
1983	..	..	..	..	..	..	
1984	..	..	..	..	..	..	
1985	..	..	..	..	..	1647	719/744
1986	..	..	..	..	..	..	

1960–64:   1965–69:

Expenditure

LCU = South African Rands

Period/Year	Current LCU (millions)	Constant 1980 LCU (millions)	Constant 1980 US$ (millions) Atlas	Constant 1980 US$ (millions) PPP	Source
1960–64	..	..	..	..	
1965–69	..	..	..	..	
1970	..	..	..	..	
1971	..	..	..	..	
1972	..	..	..	..	
1973	..	..	..	..	
1974	..	..	..	..	
1975	..	..	..	..	
1976	..	..	..	..	
1977	..	..	..	..	
1978	..	..	..	..	
1979	..	..	..	..	
1980	..	..	..	..	
1981	..	..	..	..	
1982	..	..	..	..	
1983	..	..	..	..	
1984	102.300	64.421	68.898	126.109	783
1985	..	..	..	..	
1986	..	..	..	..	

1960–64:   1965–69:

Personnel Comments:

The 1985 personnel indicator has been constructed as follows:

	PhD	MSc	BSc	Total
Department of Agriculture and Water Supply				
Research Institutes				
Animal and Dairy Science	24	30	50	104
Botanical	13	17	10	40
Citrus and Subtropical	2	9	18	39
Fruit and Fruit Technology	14	21	19	54
Grain Crops	20	17	52	89
Horticultural	5	13	28	46
Plant Protection	36	31	28	95
Soil and Irrigation	9	8	26	53
Tobacco and Cotton	4	11	8	23
Veterinary	19	13	52	84
Viticultural and Oenological	6	7	5	48
Subtotal	162	197	316	675
Regional Agricultural Research Organizations				
Eastern Cape	2	8	4	14
Free State	2	8	17	27
Highveld	4	8	25	37
Karoo	3	7	9	19
Natal	2	11	13	26
Transvaal	2	6	13	21
Winter Rainfall	1	12	17	30
Subtotal	16	60	98	174
Directorates				
Agricultural Engineering and Water Supply	2	9	76	87
Agricultural Production Economics	0	5	41	46
Biometric and Datametric Services	4	5	24	33
Subtotal	6	19	141	166
Total	184	276	555	1015
Department of Agricultural Economics and Marketing				
Directorates				
Agricultural Economic Trends	0	2	19	21
Investigations	0	1	12	13
Subtotal	0	3	31	34
Department of Environmental Affairs[a]				
Research Institutes				
Sea Fisheries				78
South African Forestry				45
Subtotal				123

(continued on next page)

	PhD	MSc	BSc	Total
Council for Scientific and Industrial Research[a]				
Research Institutes				
National Food Research Institute				77
National Institute for Water Research				113
National Research Institute for Oceanology				75
National Timber Research Institute				34
South African Wool and Textile Research Institute				22
Sugar Melting Research Institute				20
Subtotal				341
Total				1513
Faculties of Agriculture and Veterinary Science[b]	58	30	36	124
Grand Total				1637

a. The figures for the research institutes affiliated to the Department of Environmental Affairs and the Council for Scientific and Industrial Research are taken from source 744. All other figures are extracted from source 719.
b. It is assumed that faculty members spent one-third of their time on research.

Expenditure Comments:

On the basis of information given by source 783, total 'State' expenditures on agricultural research (both commodity-specific and related to agricultural resources) was estimated to be 102.3 million Rands in 1984.

Sources:

0719 Republic of South Africa – Department of Agriculture and Water Supply. *Official List of Professional and Research Workers, Lecturing Staff, Extension and Other Workers in the Agricultural Fields, April 1985.* Pretoria: Department of Agriculture and Water Supply, 1985.

0744 Professional and Information Publishing Division Longman House. *Agricultural Research Centres: A World Directory of Organizations and Programmes.* Eighth Edition, Two Volumes. Harlow, U.K.: Longman, 1986.

0783 Donovan, P.A., and M.G. Lynas. "The Political, Structural and Managerial Advantages of Commodity-Controlled Agricultural Research and Development in South Africa." *Agricultural Administration and Extension* No. 28 (January 1988): 19–28.

Additional References:

0010 Boyce, James K., and Robert E. Evenson. *National and International Agricultural Research and Extension Programs.* New York: Agricultural Development Council, Inc., 1975.

0027 Harvey, Nigel, ed. *Agricultural Research Centres: A World Directory of Organizations and Programmes.* Seventh Edition, Two Volumes. Harlow, U.K.: Longman, 1983.

0165 Evenson, R.E., and Y. Kislev. *Agricultural Research and Productivity.* New Haven: Yale University Press, 1975.

0287 Commonwealth Agricultural Bureaux (CAB). *List of Research Workers in Agriculture, Animal Health and Forestry in the British Commonwealth, the Republic of Sudan and the Republic of Ireland 1959.* Slough, England: CAB, 1959.

0637 Department of Agricultural Technical Services (DATS). *Agricultural Research 1968: Part II – 4 & 5. Registered Projects, and Research Workers.* Republic of South Africa: DATS, n.d.

0638 Department of Agricultural Technical Services (DATS). *Agricultural Research 1968: Part I – 1, 2 & 3. Interim and Final Reports on Research Projects, Summaries of Theses, and Scientific Publications.* Republic of South Africa: DATS, n.d.

0852 Evenson, Robert E., and Yoav Kislev. *Investment in Agricultural Research and Extension: A Survey of International Data.* Center Discussion Paper No. 124. New Haven, Connecticut: Economic Growth Center, Yale University, August 1971.

0886 Evenson, Robert E., and Yoav Kislev. "Investment in Agricultural Research and Extension: A Survey of International Data." *Economic Development and Cultural Change* Vol. 23 (April 1975): 507–521.

Spain

Personnel

Period/Year	PhD	MSc	BSc	Subtotal	Expat	Total	Source
1960–64	..	..	..	..	..	589	514
1965–69	..	..	..	..	..	511	661/659
1970	..	..	..	..	..	484	659
1971	..	..	..	..	..	717	658
1972	..	..	..	..	..	717	658
1973	..	..	..	..	..	520	660
1974	..	..	..	..	..	592	660
1975	..	..	..	..	..		
1976	..	..	..	..	..		
1977	..	..	..	..	..		
1978	..	..	..	..	..		
1979	..	..	..	..	..		
1980	..	..	..	..	..		
1981	..	..	..	..	..		
1982	..	..	..	..	..	1273	657
1983	..	..	..	..	..	1224	657
1984	..	..	..	..	..		
1985	..	..	..	..	..		
1986	..	..	..	..	..		

1960–64: 1963   1965–66: 1967 & 69

Expenditure   *LCU = Spanish Pesetas*

Period/Year	Current LCU (millions)	Constant 1980 LCU (millions)	Constant 1980 US$ (millions)		Source
			Atlas	PPP	
1960–64	221.544	1549.259	20.366	23.384	514
1965–69	301.732	1403.021	18.443	21.177	661/659
1970	230.945	942.633	12.391	14.228	659
1971	420.929	1594.428	20.960	24.066	658
1972	493.227	1718.561	22.591	25.940	658
1973	1068.480	3328.598	43.756	50.241	660
1974	1269.507	3394.404	44.621	51.235	660
1975	..	..	..	..	
1976	..	..	..	..	
1977	..	..	..	..	
1978	..	..	..	..	
1979	..	..	..	..	
1980	..	..	..	..	
1981	..	..	..	..	
1982	7469.856	5863.309	77.076	88.500	657
1983	8328.975	5857.226	76.996	88.408	657
1984	..	..	..	..	
1985	..	..	..	..	
1986	..	..	..	..	

1960–64: 1963   1965–66: 1967 & 69

Personnel Comments:
The table below gives a breakdown of the total number of researchers into 'Administración Pública' (the non-university sector), and 'Enseñanza Superior' (the university sector). The fluctuations observed in the earlier years could well be due to changes in coverage from one source to another.

	1963	1967	1969	1970	1971	1972	1973	1974	1982	1983
Administración Pública	537	549	355	393	565	533	444	506	972	889
Enseñanza Superior	52[a]	41	76	91	152	184	76	86	301	335
Total	589	590	431	484	717	717	520	592	1273	1224
Source	514	661	659	659	658	658	660	660	657	657

a. To express the original figure in full-time equivalents, it was assumed that university researchers spent, on average, one-third of their time on research.

Expenditure Comments:
The table below gives an overview of how the expenditure indicators have been constructed.

	1963	1967	1969	1970	1971
Administración Pública	213.740	401.600	176.794	206.920	370.080
Enseñanza Superior	7.804	7.200	17.870	24.025	50.849
Total	221.544	408.800	194.664	230.945	420.929
Source	514	661	659	659	658

	1972	1973	1974	1982	1983
Administración Pública	459.353	1023.180	1225.547	6697.820	7380.380
Enseñanza Superior	33.874	45.300	43.960	772.036	948.595
Total	493.227	1068.480	1269.507	7469.856	8328.975
Source	658	660	660	657	657

Sources:

0514 OECD. *A Study of Resources Devoted to R&D in OECD Member Countries in 1963/64, International Statistical Year for Research and Development – Volume 2: Statistical Tables and Notes.* Paris: OECD, 1968.

0657 Instituto Nacional de Estadística (INE). *Estadística sobre las Actividades en Investigación Científica y Desarrollo Tecnológico: Años 1982 y 1983.* Madrid: INE, 1986.

0658 Ministerio de Planificación del Desarrollo – Instituto Nacional de Estadística (INE). *Estadística sobre las Actividades en Investigación Científica y Desarrollo Tecnológico: Años 1971–72.* Madrid: INE, n.d.

0659 Instituto Nacional de Estadística (INE). *Estadística sobre las Actividades en*

Investigación Científica y Desarrollo Tecnológico: Años 1969 y 1970. Madrid: INE, 1973.

0660 Ministerio de Economía – Instituto Nacional de Estadística (INE). *Estadística sobre las Actividades en Investigación Científica y Desarrollo Tecnológico: Años 1973–974*. Madrid: INE, 1978.

0661 Patronato 'Juan de la Cierva' – Consejo Superior de Investigaciones Científicas. *Encuesta sobre Actividades de Investigación Científica y Técnica en España en 1967*. Madrid: Consejo Superior de Investigaciones Científicas, 1970.

Additional References:

0010 Boyce, James K., and Robert E. Evenson. *National and International Agricultural Research and Extension Programs*. New York: Agricultural Development Council, Inc., 1975.

0021 OECD. *OECD Science and Technology Indicators*. Paris: OECD, 1984.

0022 UNESCO. *Statistical Yearbook 1983*. Paris: UNESCO, 1983.

0027 Harvey, Nigel, ed. *Agricultural Research Centres: A World Directory of Organizations and Programmes*. Seventh Edition, Two Volumes. Harlow, U.K.: Longman, 1983.

0165 Evenson, R.E., and Y. Kislev. *Agricultural Research and Productivity*. New Haven: Yale University Press, 1975.

0407 Centre International de Hautes Etudes Agronomiques Méditerranéennes. "Séminaire sur l'Orientation et l'Organisation de la Recherche Agronomique dans les Pays du Bassin Méditerranéen, Istanbul, 1–3 December 1986." Centre International de Hautes Etudes Agronomiques Méditerranéennes, Paris, December 1986. Mimeo.

0410 Moreno, Jesús. "La Recherche Agricole en Espagne." Paper presented at the Séminaire sur l'Orientation et l'Organisation de la Recherche Agronomique dans les Pays du Bassin Méditerranéen, Centre International de Hautes Etudes Agronomiques Méditerranéennes, Istanbul, 1–3 December 1986.

0422 Casas, Joseph. "Les Ressources Humaines et Financières dans les Pays du Bassin Méditerranéen." Preliminary table. ISNAR, The Hague, 1986. Mimeo.

0473 OECD. *Science and Technology Indicators: Basic Statistical Series – Volume A: The Objectives of Government R&D Funding 1974–1985*. Paris: OECD, May 1983.

0474 OECD. *Science and Technology Indicators: Basic Statistical Series – Volume A: The Objectives of Government R&D, 1969–1981*. Paris: OECD, July 1981.

0509 OECD. *Survey of the Resources Devoted to R&D by OECD Member Countries, International Statistical Year 1971 –Volume 5: Total Tables, Statistical Tables and Notes*. Paris: OECD, August 1974.

0510 OECD. *International Survey of the Resources Devoted to R&D in 1967 by OECD Member Countries, Statistical Tables and Notes – Volume 5: Total Tables*. Paris: OECD, August 1970.

0511 OECD. *International Survey of the Resources Devoted to R&D in 1969 by OECD Member Countries, Statistical Tables and Notes – Volume 5: Total Tables*. Paris: OECD, June 1973.

0512 OECD. *Survey of the Resources Devoted to R&D by OECD Member Countries, International Statistical Year 1973 – Volume 5: Total Tables, Statistical Tables and Notes*. Paris: OECD, September 1976.

0513 OECD. *International Survey of the Resources Devoted to R&D by OECD Member Countries, International Statistical Year 1975 - International Volume: Statistical Tables and Notes*. Paris: OECD, March 1979.

0530 OECD. *OECD Science and Technology Indicators No. 2 – R&D, Invention and Competitiveness*. Paris: OECD, 1986.

0538 OECD. *Country Reports on the Organisation of Scientific Research: Spain*. Paris: OECD, June 1964.

0539 OECD. *Reviews of National Science Policy: Spain*. Paris: OECD, 1971.

0540 Ministerio de Agricultura y Pesca – Instituto Nacional de Investigaciones Agrarias. *Agricultural Research*. Madrid: Ministerio de Agricultura y Pesca, 1981.

0574 OECD. *Intellectual Investment in Agriculture for Economic and Social Development*. Documentation in Food and Agriculture No. 60. Paris: OECD, 1962.

0601 OECD – STIID Data Bank. "Public Funding of R&D by Socio-Economic Objective." Tables. OECD, Paris, April 1987. Mimeo.

0602 OECD – STIID Data Bank. "R&D Expenditure in the Higher Education and Private Non-Profit Sectors." Tables. OECD, Paris, April 1987. Mimeo.

0707 OECD – STIID Data Bank. "R&D Expenditure in the Higher Education and Private Non-Profit Sectors: Higher Education, Agricultural Sciences, General Universities Funds." Table. OECD, Paris, June 1987. Mimeo.

0708 OECD – STIID Data Bank. "Gross Domestic Expenditure on R&D: All Fields of Sciences – Agriculture, Forestry and Fishing." Tables. OECD, Paris, June 1987. Mimeo.

0784 Castells, Manuel, Antonio Barrera, et al. *Nuevas Tecnologias, Economia y Sociedad en Espana.* Volumen 2. Madrid: Alianza Editorial, 1986.

0852 Evenson, Robert E., and Yoav Kislev. *Investment in Agricultural Research and Extension: A Survey of International Data.* Center Discussion Paper No. 124. New Haven, Connecticut: Economic Growth Center, Yale University, August 1971.

0862 "Science in Iberia: A Renaissance in the Making." *Nature* Vol. 324 (27 November 1986): 313–328.

0886 Evenson, Robert E., and Yoav Kislev. "Investment in Agricultural Research and Extension: A Survey of International Data." *Economic Development and Cultural Change* Vol. 23 (April 1975): 507–521.

Sri Lanka

Personnel

Period/Year	PhD	MSc	BSc	Subtotal	Expat	Total	Source
1960–64	..	..	..	..	..	51	287/286
1965–69	40	20	30	90	..	90	286/285
1970	..	..	..	..	..	..	
1971	53	31	63	147	..	147	279
1972	..	..	..	..	..	..	
1973	..	..	..	..	..	..	
1974	53	32	64	149	..	149	244
1975	..	..	..	..	..	..	
1976	..	..	..	..	..	..	
1977	68	58	121	247	..	247	243
1978	..	..	..	..	..	..	
1979	..	..	..	..	..	..	
1980	68	63	110	241	..	241	242
1981	..	..	..	..	..	..	
1982	..	..	..	..	..	..	
1983	67	116	208	391	..	391	34
1984	..	..	..	..	..	..	
1985	..	..	..	..	..	..	
1986	..	..	..		..	..	

1960–64: 1960   1965–69: 1965 & 68

Expenditure   *LCU = Sri Lanka Rupees*

Period/Year	Current LCU (millions)	Constant 1980 LCU (millions)	Constant 1980 US$ (millions)		Source
			Atlas	PPP	
1960–64	..	..	..	..	
1965–69	13.188	43.801	2.533	11.325	877
1970	18.551	56.905	3.291	14.713	877
1971	18.120	54.090	3.128	13.985	877
1972	..	..	..	..	
1973	19.634	49.959	2.889	12.917	877
1974	21.824	44.721	2.586	11.563	877
1975	24.610	47.146	2.726	12.190	265/297/877
1976	28.850	50.792	2.937	13.133	265/297
1977	43.760	66.404	3.840	17.169	265/297
1978	52.680	72.964	4.219	18.865	265/297
1979	64.540	77.479	4.480	20.033	265/297
1980	75.400	75.400	4.360	19.495	265/297
1981	..	..	..	..	
1982	..	..	..	..	
1983	130.458	87.263	5.046	22.562	34
1984	..	..	..	..	
1985	..	..	..	..	
1986	..	..	..	..	

1960–64: 1960   1965–69: 1966, 67, 68 & 69

Personnel Comments:

Agricultural research in Sri Lanka is executed by numerous agricultural research institutes related to many different ministries. Various reorganizations over time, including the creation of new institutes, make it difficult to present a consistent time series. The table below gives a breakdown of the total number of research scientists at the institutional level. The subtotal figure represents the indicator series.

	1958	1960	1965	1968	1971	1974	1977	1980	1983
MADR/DOA	13	[15]	21	43	47	57	129	130	258
MPI/TRI	6	[10]	20	18	25	25	29	31	31
MPI/RRI	7	[10]	16	11	34	34	26	20	31
MCI/CRI	9	[9]	9	9	8	8	20	2	37
MLLD/FD/FRI	4	[4]	3	3	5	5	4	3	4
MRID/DAPH/VRI[a]	2	[3]	5	21	28	20	39	3	30
Subtotal	41	[51]	74	105	147	49	247	24	391
MLLD/ID/LUD							4	7	11
MF/NARA									33
MADR/SRI									11
MADR/DMEC									23
MADR/ARTI									37
Source	287	287 286	286	285	279	244	243	242	34

[ ] = estimated or constructed by authors.
a. Until 1977 DAPH/VRI was reported to be under the jurisdiction of DOA.

In addition, another 151 teacher-researchers within the university sector, involved in part-time agricultural research, are reported by sources 34 and 732 for the year 1983.

Expenditure Comments:

The expenditure figures have been constructed as shown on the next page.

Acronyms and Abbreviations:

ARTI: Agricultural Research and Training Institute
CC : Cashew Corporation
CRI : Coconut Research Institute
DAPH: Department of Animal Production and Health
DMEC: Department of Minor Export Crops
DOA : Department of Agriculture
FRI : Forestry Research Institute
ID : Irrigation Department
LUD : Land Use Division
MADR: Ministry of Agricultural Development and Research
MCI : Ministry of Coconut Industries
MF : Ministry of Fisheries
MLLD: Ministry of Lands and Land Development
MPI : Ministry of Plantation Industries

MRID: Ministry of Rural Industrial Development
NARA: National Aquatic Resources Agency
RRI : Rubber Research Institute
SRI : Sugar Research Institute
TC : Tobacco Company
TRI : Tea Research Institute
VRI : Veterinary Research Institute

Sources:

0034 Joint Review Group: Agricultural Research Group, Sri Lanka, and ISNAR. *Report to the Government of Sri Lanka: The Agricultural Research System in Sri Lanka*. The Hague: ISNAR, June 1984.

0242 Commonwealth Agricultural Bureaux (CAB). *List of Research Workers in Agricultural Sciences in the Commonwealth 1981. Slough, England: CAB, 1981.*

0243 Commonwealth Agricultural Bureaux (CAB). *List of Research Workers in the Agricultural Sciences in the Commonwealth and in the Republic of Ireland 1978. Slough, England: CAB, 1978.*

0244 Commonwealth Agricultural Bureaux (CAB). *List of Research Workers in the Agricultural Sciences in the Commonwealth and in the*

Research Expenditures

	1966	1967	1968	1969	1970	1971	1973	1974
MADR/DOA	1.634	2.607	4.055	5.553	6.391	5.949	6.326	6.814
MPI/TRI	4.920	4.739	3.551	3.148	4.004	4.157	2.368	3.842
MPI/RRI	1.670	1.903	1.771	1.975	2.905	2.763	3.704	3.385
MCI/CRI	1.359	1.262	1.489	1.771	1.984	1.657	3.296	4.014
MLLD/FD/ FRI	0.038	0.041	0.044	0.073	0.057	0.143	0.140	0.161
MRID/DAPH/ VRI	[1.394]	[2.061]	[2.727]	[2.968]	[3.210]	[3.451]	[3.800]	[3.968]
Subtotal	11.015	12.613	13.637	15.488	18.551	18.120	19.634	21.824
MLLD/ID/ LUD	0.568	0.800	0.874	0.930	1.000	1.033	1.711	1.645
MF/NARA	0.110	0.110	0.098	0.141	0.159	0.313	1.097	3.106
MADR/SRI				0.336	0.269	0.284	0.474	0.762
MADR/DMEC							0.047	0.483
MADR/ARIT							0.731	1.395
CC								
TC						0.020	0.252	0.263
Source	877	877	877	877	877	877	877	877

	1975	1976	1977	1978	1979	1980	1983
MADR/DOA	9.050	9.620	23.610	28.310	36.180	35.310	48.870
MPI/TRI	5.150*	5.140*	5.060	6.840	9.870	10.360	24.865
MPI/RRI	3.160*	3.590*	4.420	6.910	4.720	7.200	12.927
MCI/CRI	12.670*	2.850*	3.050	4.050	5.960	7.090	33.325
MLLD/FD/ FRI	0.440*	0.470*	0.650	0.640	0.690	0.800	1.724
MRID/DAPH/ VRI	4.140	7.180	6.970	5.930	7.120	14.640	8.747
Subtotal	24.610	28.850	43.760	52.680	64.540	75.400	130.458
MLLD/ID/ LUD	2.592**						4.250
MF/NARA	3.079**						9.000
MADR/SRI	2.250*	2.810*	2.620*	2.970*	2.940*	2.930*	12.500
MADR/DMEC	2.790	3.220	2.280	2.900	5.340	9.680	8.950
MADR/ARIT	2.520	2.880	1.610	2.030	3.160	4.480	12.640
CC	0.000**	0.000*	0.040*	0.080*	0.110*	0.140*	
TC	0.483**						
Source	297	297	297	297	297	297	34
	265*	265*	265*	265*	265*	265*	
	877**						

[ ] = estimated or constructed by authors.

404

Republic of Ireland 1975. Slough, England: CAB, 1975.

0265 Herath, H.M.G., and Y.D.A. Senanayake. "A Study of the Financial and Human Resources in Plantation Crops Research in Sri Lanka." *Journal of Agrarian Studies* Vol. 3, No. 2 (1982): 96–109.

0279 Commonwealth Agricultural Bureaux (CAB). *List of Research Workers in the Agricultural Sciences in the Commonwealth and in the Republic of Ireland 1972.* Slough, England: CAB, 1972.

0285 Commonwealth Agricultural Bureaux (CAB). *List of Research Workers in the Agricultural Sciences in the Commonwealth and in the Republic of Ireland 1969.* Slough, England: CAB, 1969.

0286 Commonwealth Agricultural Bureaux (CAB). *List of Research Workers in Agriculture, Animal Health and Forestry in the Commonwealth and in the Republic of Ireland 1966.* Slough, England: CAB, 1966.

0297 Senanayake, Y.D.A., and H.M.G. Herath. "Resource Allocation to Agricultural Research in Sri Lanka." In *Resource Allocation to Agricultural Research.* Proceedings of a workshop held in Singapore 8–10 June 1981, edited by Douglas Daniels and Barry Nestel, pp. 60–67. Ottawa: IDRC, 1981.

0319 Bengtsson, B. "Agricultural Research in Sri Lanka." In *Strengthening National Agricultural Research.* Report from a SAREC workshop, 10–17 September 1979, edited by Bo Bengtsson and Getachew Tedla, pp. 123–137. Sweden: SAREC, 1980.

0877 Liyanage, Shantha, T. Wijesinghe, N. Anbalagan, and S. Peiris. *A Survey of the Expenditure on Research and Experimental Development in Sri Lanka 1966–1975.* Colombo: National Science Council of Sri Lanka, September 1977.

Additional References:

0010 Boyce, James K., and Robert E. Evenson. *National and International Agricultural Research and Extension Programs.* New York: Agricultural Development Council, Inc., 1975.

0014 Judd, M. Ann, James K. Boyce, and Robert E. Evenson. "Investing in Agricultural Supply." Economic Growth Center, Yale University, New Haven, Connecticut, 1983. Mimeo.

0016 Oram, Peter A., and Vishva Bindlish. *Resource Allocations to National Agricultural Research: Trends in the 1970s.* The Hague and Washington, D.C.: ISNAR and IFPRI, November 1981.

0017 ISNAR, IFARD & AOAD. Survey of National Agricultural Research Systems: Unpublished

Questionnaire Responses. ISNAR, The Hague, 1985.

0018 Daniels, Douglas, and Barry Nestel, eds. *Resource Allocation to Agricultural Research.* Proceedings of a workshop held in Singapore 8–10 June 1981. Ottawa: IDRC, 1981.

0027 Harvey, Nigel, ed. *Agricultural Research Centres: A World Directory of Organizations and Programmes.* Seventh Edition, Two Volumes. Harlow, U.K.: Longman, 1983.

0073 Oram, Peter A., and Vishva Bindlish. "Investment in Agricultural Research in Developing Countries: Progress, Problems, and the Determination of Priorities." IFPRI, Washington, D.C., January 1984. Mimeo.

0077 ISNAR. *Human Resources Planning for Agricultural Research Personnel in Sri Lanka: The Research Scientist Cadre.* The Hague: ISNAR, August 1985.

0095 FAO – CARIS. *Agricultural Research in Developing Countries – Volume 1: Research Institutions.* Rome: FAO – CARIS, 1978.

0165 Evenson, R.E., and Y. Kislev. *Agricultural Research and Productivity.* New Haven: Yale University Press, 1975.

0176 Moseman, Albert, ed. *National Agricultural Research Systems in Asia.* Report of the Regional Seminar held at the India International Centre, New Delhi, India, March 8–13, 1971. New York: ADC, Inc., 1971.

0287 Commonwealth Agricultural Bureaux (CAB). *List of Research Workers in Agriculture, Animal Health and Forestry in the British Commonwealth, the Republic of Sudan and the Republic of Ireland 1959.* Slough, England: CAB, 1959.

0436 Patel, R.K. "Agricultural Research in Sri Lanka." In *Proceedings of the Research Extension Linkage Workshop Held in Hanoi, Viet Nam, 9–13 June 1986.* Rome: FAO, 1986.

0457 Joint Review Group: Agricultural Research Group, Sri Lanka, and ISNAR. *The Agricultural Research System in Sri Lanka.* The Hague: ISNAR, October 1986.

0715 Chang, C.W. *FAO Directory of Agricultural Research Institutes and Experiment Stations in Asia and the Far East.* Bangkok, Thailand: FAO – Regional Office for Asia and the Far East, September 1962.

0716 Chang, C.W. *FAO Directory of Agricultural Research Institutes and Experiment Stations in Asia and the Far East, Supplement.* Bangkok, Thailand: FAO – Regional Office for Asia and the Far East, May 1963.

0732 FAO – Regional Office for Asia and the Pacific. *Agricultural Research Systems in Regional Office for Asia and the Pacific, 1986.*

0852 Evenson, Robert E., and Yoav Kislev. *Investment in Agricultural Research and Extension: A Survey of International Data*. Center Discussion Paper No. 124. New Haven, Connecticut: Economic Growth Center, Yale University, August 1971.

0886 Evenson, Robert E., and Yoav Kislev. "Investment in Agricultural Research and Extension: A Survey of International Data." *Economic Development and Cultural Change* Vol. 23 (April 1975): 507–521.

St Kitts-Nevis

Personnel

Period/Year	PhD	MSc	BSc	Subtotal	Expat	Total	Source
1960–64	..	..	..	..	..	..	
1965–69	0	1	2	3	..	3	285
1970	..	..	..	..	..	..	
1971	..	..	..	..	..	..	
1972	..	..	..	..	..	..	
1973	..	..	..	..	..	..	
1974	0	1	0	1	..	1	244
1975	..	..	..	..	..	..	
1976	..	..	..	..	..	..	
1977	..	..	..	..	..	..	
1978	0	1	0	1	..	1	341
1979	..	..	..	..	..	..	
1980	1	0	2	3	..	3	343
1981	..	..	..	5	..	5	342
1982	..	..	..	..	..	..	
1983	0	2	3	5	..	5	115
1984	1	3	2	6	..	6	349/354/356
1985	..	..	..	..	..	..	
1986	..	..	..	..	..	..	

1960–64:   1965–69: 1968

Expenditure

LCU = East Caribbean Dollars

Period/Year	Current LCU (millions)	Constant 1980 LCU (millions)	Constant 1980 US$ (millions)		Source
			Atlas	PPP	
1960–64	..	..	..	..	
1965–69	..	..	..	..	
1970	..	..	..	..	
1971	..	..	..	..	
1972	..	..	..	..	
1973	..	..	..	..	
1974	..	..	..	..	
1975	..	..	..	..	
1976	0.108	0.150	0.055	0.079	115
1977	0.108	0.139	0.051	0.073	115
1978	0.108	0.130	0.047	0.068	115
1979	0.108	0.122	0.045	0.064	115
1980	0.108	0.108	0.039	0.057	115
1981	0.130	0.117	0.043	0.061	115
1982	..	..	..	..	
1983	..	..	..	..	
1984	..	..	..	..	
1985	..	..	..	..	
1986	..	..	..	..	

1960–64:   1965–69:

Personnel Comments:

The figures before 1975 refer to the Ministry of Agriculture, and the figures after 1975 to the Caribbean Agricultural Research and Development Institute (CARDI). Currently CARDI appears to be the only organization at St Kitts-Nevis that executes agricultural research (source 115).

Expenditure Comments:

These figures include the government contribution to the Caribbean Agricultural Research and Development Institute (CARDI) only. Source 115 indicates that the United States Agency for International Development (USAID) and the European Development Fund (EDF) also contribute to the system, but no figures are available.

Sources:

0115 Economic Commission for Latin America and the Caribbean (ECLAC). *Agricultural Research Policy and Management – Volumes I and II.* Papers presented at the Workshop on Agricultural Research Policy and Management, 26–30 September 1983, Port of Spain, Trinidad. Port of Spain, Trinidad: ECLAC, November 1984.

0244 Commonwealth Agricultural Bureaux (CAB). *List of Research Workers in the Agricultural Sciences in the Commonwealth and in the Republic of Ireland 1975.* Slough, England: CAB, 1975.

0285 Commonwealth Agricultural Bureaux (CAB). *List of Research Workers in the Agricultural Sciences in the Commonwealth and in the Republic of Ireland 1969.* Slough, England: GAB, 1969.

0341 Caribbean Agricultural Research and Development Institute (CARDI). *Annual Report 1977–1978.* St. Augustine, Trinidad & Tobago: CARDI, November 1978.

0342 Caribbean Agricultural Research and Development Institute (CARDI). *CARDI 1976–1981: Report of the Chairman.* St. Augustine, Trinidad & Tobago: CARDI, 1982.

0343 Caribbean Agricultural Research and Development Institute (CARDI). *Annual Report 1980.* St. Augustine, Trinidad & Tobago: CARDI, 1981.

0349 Caribbean Agricultural Research and Development Institute (CARDI). *Annual Report 1983–1984 – Leeward Islands Unit.* CARDI, 1984.

0354 Caribbean Agricultural Research and Development Institute (CARDI). *Annual Report Research and Development 1983–84: Highlights.* St. Augustine, Trinidad: CARDI, 1984.

0356 Caribbean Agricultural Research and Development Institute (CARDI). *Research and Development Summary 1983–84.* St. Augustine, Trinidad: CARDI, 1984.

Additional References:

0279 Commonwealth Agricultural Bureaux (CAB). *List of Research Workers in the Agricultural Sciences in the Commonwealth and in the Republic of Ireland 1972.* Slough, England: CAB, 1972.

0287 Commonwealth Agricultural Bureaux (CAB). *List of Research Workers in Agriculture, Animal Health and Forestry in the British Commonwealth, the Republic of Sudan and the Republic of Ireland 1959.* Slough, England: CAB, 1959.

0353 Caribbean Agricultural Research and Development Institute (CARDI). *Paper Presented at a Meeting of the Governing Body of CARDI on January 29, 1982, Basseterre, St. Kitts.* St. Augustine, Trinidad: CARDI, 1982.

0355 Carribbean Agricultural Research and Development Institute (CARDI). *Research and Development Summary 1982–83.* St. Augustine, Trinidad: CARDI, 1983.

St Lucia

Personnel

Period/Year	PhD	MSc	BSc	Subtotal	Expat	Total	Source
1960–64	..	..	..	..	..	..	
1965–69	1	3	1	5	..	5	286/285
1970	..	..	..	..	..	..	
1971	1	4	6	11	..	11	279
1972	..	..	..	..	..	..	
1973	..	..	..	..	..	..	
1974	2	4	9	15	..	15	244
1975	..	..	..	..	..	..	
1976	..	..	..	..	..	..	
1977	..	..	..	..	..	..	
1978	..	..	..	..	..	..	
1979	..	..	..	..	..	..	
1980	..	..	..	..	..	..	
1981	..	..	..	..	..	..	
1982	..	..	..	..	..	..	
1983	6	5	10	21	..	21	115
1984	..	..	..	..	..	..	
1985	..	..	..	..	..	..	
1986	..	..	..	..	..	..	

1960–64:   1965–69: 1965 & 68

Expenditure   *LCU = East Caribbean Dollars*

Period/Year	Current LCU (millions)	Constant 1980 LCU (millions)	Constant 1980 US$ (millions)		Source
			Atlas	PPP	
1960–64	..	..	..	..	
1965–69	..	..	..	..	
1970	..	..	..	..	
1971	..	..	..	..	
1972	..	..	..	..	
1973	..	..	..	..	
1974	..	..	..	..	
1975	..	..	..	..	
1976	..	..	..	..	
1977	..	..	..	..	
1978	..	..	..	..	
1979	..	..	..	..	
1980	..	..	..	..	
1981	..	..	..	..	
1982	..	..	..	..	
1983	4.000	3.418	1.185	1.791	115
1984	..	..	..	..	
1985	..	..	..	..	
1986	..	..	..	..	

1960–64:   1965–69:

Personnel Comments:

The figures for the years 1965, 1968, 1971, and 1974 include agricultural researchers at the Ministry of Agriculture and at the Winward Islands Banana Growers Association (WINBAN) which is headquartered at St Lucia.

According to source 115, the 1983 figure includes researchers at the Ministry of Agriculture, WINBAN, the Caribbean Agricultural Research and Development Institute (CARDI), and a French mission. In addition, other sources (27 and 355, respectively) noted 9 researchers at WINBAN in 1982 and 5 researchers at the St. Lucia unit of CARDI in 1983.

Expenditure Comments:

The 1983 expenditure figure includes the Ministry of Agriculture, WINBAN, CARDI, and a French mission.

Sources:

0115 Economic Commission for Latin America and the Caribbean (ECLAC). *Agricultural Research Policy and Management – Volumes I and II.* Papers presented at the Workshop on Agricultural Research Policy and Management, 26–30 September 1983, Port of Spain, Trinidad. Port of Spain, Trinidad: ECLAC, November 1984.

0244 Commonwealth Agricultural Bureaux (CAB). *List of Research Workers in the Agricultural Sciences in the Commonwealth and in the Republic of Ireland 1975.* Slough, England: CAB, 1975.

0279 Commonwealth Agricultural Bureaux (CAB). *List of Research Workers in the Agricultural Sciences in the Commonwealth and in the Republic of Ireland 1972.* Slough, England: CAB, 1972.

0285 Commonwealth Agricultural Bureaux (CAB). *List of Research Workers in the Agricultural Sciences in the Commonwealth and in the Republic of Ireland 1969.* Slough, England: CAB, 1969.

0286 Commonwealth Agricultural Bureaux (CAB). *List of Research Workers in Agriculture, Animal Health and Forestry in the Commonwealth and in the Republic of Ireland 1966.* Slough, England: CAB, 1966.

Additional References:

0027 Harvey, Nigel, ed. *Agricultural Research Centres: A World Directory of Organizations and Programmes.* Seventh Edition, Two Volumes. Harlow, U.K.: Longman, 1983.

0063 ISNAR. *Report to the Board of Directors of CARDI: Analysis, Evaluation and Proposals to Strengthen CARDI's Regional Capacity.* The Hague: ISNAR, 1985.

0287 Commonwealth Agricultural Bureaux (CAB). *List of Research Workers in Agriculture, Animal Health and Forestry in the British Commonwealth, the Republic of Sudan and the Republic of Ireland 1959.* Slough, England: CAB, 1959.

0341 Caribbean Agricultural Research and Development Institute (CARDI). *Annual Report 1977–1978.* St. Augustine, Trinidad & Tobago: CARDI, November 1978.

0342 Caribbean Agricultural Research and Development Institute (CARDI). *CARDI 1976–1981: Report of the Chairman.* St. Augustine, Trinidad & Tobago: CARDI, 1982.

0343 Caribbean Agricultural Research and Development Institute (CARDI). *Annual Report 1980.* St. Augustine, Trinidad & Tobago: CARDI, 1981.

0353 Caribbean Agricultural Research and Development Institute (CARDI). *Paper Presented at a Meeting of the Governing Body of CARDI on January 29, 1982, Basseterre, St. Kitts.* St. Augustine, Trinidad: CARDI, 1982.

0354 Caribbean Agricultural Research and Development Institute (CARDI). *Annual Report Research and Development 1983–84: Highlights.* St. Augustine, Trinidad: CARDI, 1984.

0355 Caribbean Agricultural Research and Development Institute (CARDI). *Research and Development Summary 1982–83.* St. Augustine, Trinidad: CARDI, 1983.

0356 Caribbean Agricultural Research and Development Institute (CARDI). *Research and Development Summary 1983–84.* St. Augustine, Trinidad: CARDI, 1984.

St Vincent

Personnel

Period/Year	PhD	MSc	BSc	Subtotal	Expat	Total	Source
1960–64	..	..	..	..	..		
1965–69	1	1	2	4	..	4	286/285
1970	..	..	..	..	..		
1971	1	1	3	5	..	5	279
1972	..	..	..	..	..		
1973					..		
1974	0	0	1	1	..	1	244
1975	..	..	..	..	..		
1976	..	..	..	..	..		
1977	..	..	..	..	..		
1978	..	..	..	..	..		
1979	..	..	..	..	..		
1980	0	1	4	5	..	5	343
1981	..	..	..	..	..	5	342
1982			..	..	..		
1983	0	1	3	4	..	4	115/355
1984	0	4	2	6	..	6	354
1985				..	..		
1986	0	2	2	4	1	5	810

1960–64:   1965–69: 1965 & 68

Expenditure   *LCU = East Carribean Dollars*

Period/Year	Current LCU (millions)	Constant 1980 LCU (millions)	Constant 1980 US$ (millions)		Source
			Atlas	PPP	
1960–64	..	..	..	..	
1965–69	..	..	..	..	
1970	..	..	..	..	
1971	..	..	..	..	
1972	..	..	..	..	
1973	..	..	..	..	
1974	..	..	..	..	
1975	..	..	..	..	
1976	..	..	..	..	
1977	..	..	..	..	
1978	..	..	..	..	
1979	..	..	..	..	
1980	..	..	..	..	
1981	..	..	..	..	
1982	..	..	..	..	
1983	..	..	..	..	
1984	..	..	..	..	
1985	..	..	..	..	
1986	..	..	..	..	

1960–64:   1965–69:

Personnel Comments:
According to source 810, St Vincent (and the Grenadines) does not have a full-fledged agricultural research institute. Some agricultural research, however, has been conducted over the last 25 years or so by the Department of Agriculture in collaboration with the Regional Research Centre and, after 1975, with CARDI (Caribbean Agricultural Research and Development Institute). The Regional Research Centre which merged with the University of the West Indies in 1966/67 was detached from the university and transformed into CARDI in 1975.

CARDI was given a decentralized structure and maintains research units in each of the member countries.

Another organization involved in agricultural research on St Vincent is WINBAN (Winward Islands Banana Growers Association). According to source 810, however, WINBAN, which is headquartered on St Lucia, only conducts trials at St Vincent and has no researchers stationed at St Vincent.

The table below gives an overview of construction details for the indicator series:

	1965	1968	1971	1974	1980	1981	1983	1984	1986
Department of Agriculture	3	4	5	1	[1]	[1]	1	[1]	3
CARDI	–	–	–	–	4	4	3*	5	2
Total	3	4	5	1	5	5	4	6	5
Source	286	285	279	244	343	342	115 355*	354	810

[ ] = estimated or constructed by authors.

Sources:

0115 Economic Commission for Latin America and the Caribbean (ECLAC). *Agricultural Research Policy and Management – Volumes I and II.* Papers presented at the Workshop on Agricultural Research Policy and Management, 26–30 September 1983, Port of Spain, Trinidad. Port of Spain, Trinidad: ECLAC, November 1984.

0244 Commonwealth Agricultural Bureaux (CAB). *List of Research Workers in the Agricultural Sciences in the Commonwealth and in the Republic of Ireland 1975.* Slough, England: CAB, 1975.

0279 Commonwealth Agricultural Bureaux (CAB). *List of Research Workers in the Agricultural Sciences in the Commonwealth and in the Republic of Ireland 1972.* Slough, England: CAB, 1972.

0285 Commonwealth Agricultural Bureaux (CAB). *List of Research Workers in the Agricultural Sciences in the Commonwealth and in the Republic of Ireland 1969.* Slough, England: CAB, 1969.

0286 Commonwealth Agricultural Bureaux (CAB). *List of Research Workers in Agriculture, Animal Health and Forestry in the Commonwealth and in the Republic of Ireland 1966.* Slough, England: CAB, 1966.

0342 Caribbean Agricultural Research and Development Institute (CARDI). *CARDI 1976–1981: Report of the Chairman.* St. Augustine, Trinidad & Tobago: CARDI, 1982.

0343 Caribbean Agricultural Research and Development Institute (CARDI). *Annual Report 1980.* St. Augustine, Trinidad & Tobago: CARDI, 1981.

0354 Caribbean Agricultural Research and Development Institute (CARDI). *Annual Report Research and Development 1983–84: Highlights.* St. Augustine, Trinidad: CARDI, 1984.

0355 Caribbean Agricultural Research and Development Institute (CARDI). *Research and Development Summary 1982–83.* St. Augustine, Trinidad: CARDI, 1983.

0810 Gunsam, Charles. Personal communication. Agricultural Research Officer, Ministry of Trade, Industry and Agriculture. Saint Vincent & the Grenadines, February 1988.

412

Additional References:

0063 ISNAR. *Report to the Board of Directors of CARDI: Analysis, Evaluation and Proposals o Strengthen CARDI's Regional Capacity.* The Hague: ISNAR, 1985.

0287 Commonwealth Agricultural Bureaux (CAB). *List of Research Workers in Agriculture, Animal Health and Forestry in the British Commonwealth, the Republic of Sudan and the Republic of Ireland 1959.* Slough, England: CAB, 1959.

0353 Caribbean Agricultural Research and Development Institute (CARDI). *Paper Presented at a Meeting of the Governing Body of CARDI on January 29, 1982, Basseterre, St. Kitts.* St. Augustine, Trinidad: CARDI, 1982.

0356 Caribbean Agricultural Research and Development Institute (CARDI). *Research and Development Summary 1983–84.* St. Augustine, Trinidad: CARDI, 1984.

Sudan

Personnel

Period/Year	PhD	MSc	BSc	Subtotal	Expat	Total	Source
1960–64	..	..	..	..	..	45	852
1965–69	..	..	..	..	..	52	10
1970	..	..	..	79	3	82	175
1971	..	..	..	..	..	..	
1972	..	..	..	74	2	76	266
1973	..	..	..	..	..	..	
1974	..	..	..	..	..	..	
1975	..	..	..	..	..	..	
1976	..	..	..	..	..	..	
1977	–112–		54	166	..	166	252
1978	..	..	..	..	..	..	
1979	–123–		38	161	..	161	251
1980	..	..	..	..	..	164	73
1981	104	60	54	218	..	218	251
1982	–150–		50	200	..	200	250
1983	122	51	29	202	..	202	76
1984	..	..	..	..	..	..	
1985	145	35	24	204	..	204	17
1986	129	68	51	248	..	248	710

1960–64: 1963   1965–69: 1967

Expenditure

LCU = Sudanese pounds

Period/Year	Current LCU (millions)	Constant 1980 LCU (millions)	Constant 1980 US$ (millions)		Source
			Atlas	PPP	
1960–64	0.384	2.018	4.260	6.148	10
1965–69	0.897	4.283	9.042	13.049	589
1970	..	..	..	..	
1971	1.131	3.989	8.422	12.154	252
1972	1.419	4.480	9.459	13.651	252
1973	1.337	3.610	7.621	10.999	252
1974	1.519	3.250	6.861	9.901	252
1975	1.586	3.152	6.654	9.604	252
1976	2.200	4.232	8.936	12.896	252
1977	2.697	4.946	10.441	15.069	252
1978	2.938	4.610	9.733	14.047	252
1979	3.262	4.074	8.600	12.412	445
1980	5.947	5.947	12.556	18.120	445
1981	5.930	4.733	9.992	14.420	445
1982	6.314	4.360	9.206	13.286	445
1983	7.170	3.738	7.892	11.390	445
1984	8.452	3.675	7.758	11.197	445
1985	..	..	..	..	
1986	..	..	..	..	

1960–64: 1961   1965–69: 1966

Personnel Comments:

These figures refer to the Agricultural Research Corporation (ARC) only. ARC was created in 1967 out of the Research Division of the Ministry of Agriculture to investigate the scientific basis of crop production in the Sudan (source 709). In 1975 the research functions of the Food Processing Centre, Fisheries and Marine Biology, Forestry, Range and Pastures, and Game and Wildlife were transferred to ARC. This caused a jump in the indicator series in the mid-seventies. From 1981 on, the indicators have also included the Western Sudan Agricultural Research Project, which falls under ARC (1983: 17 researchers, source 76). Although nearly all agricultural research has been integrated into ARC, two research organizations have remained outside ARC, namely, 1) Animal Production Research Administration (1977: 29 researchers, source 252) and 2) Veterinary Research Administration (1970: 17 researchers, source 532; 1972: 16 researchers, source 448; 1976: 73 researchers, source 95; 1977: 35 researchers, source 252; 1983: 69 researchers, source 847). For reasons of consistency, these organizations are not included in the indicator series.

Expenditure Comments:

These expenditure indicators refer to ARC only. For the period 1971 to 1984, sources 252 and 445 indicate that the figures represent the budgeted, not realized, government contribution to ARC. Contributions by foreign donors are not included. From 1980 on, ARC has often received only a small fraction of its budgeted allocation (source 445). Currently, donor funds appear to be substituting for domestic support of research.

Sources:

0010 Boyce, James K., and Robert E. Evenson. *National and International Agricultural Research and Extension Programs.* New York: Agricultural Development Council, Inc., 1975.

0017 ISNAR, IFARD & AOAD. Survey of National Agricultural Research Systems: Unpublished Questionnaire Responses. ISNAR, The Hague, 1985.

0073 Oram, Peter A., and Vishva Bindlish. "Investment in Agricultural Research in Developing Countries: Progress, Problems, and the Determination of Priorities." IFPRI, Washington, D.C., January 1984. Mimeo.

0076 Agricultural Research Corporation (ARC). *List of Research Scientists and Senior Administrators as on July 1983.* Sudan: ARC, July 1983.

0175 Cooper, St.G.C. *Agricultural Research in Tropical Africa.* Kampala: East African Literature Bureau, 1970.

0250 Bunting, A.H., F. Haworth, and J.D. Robinson. "Report on a Visit to the Sudan, 27 March to 8 April 1982." ISNAR, The Hague, 1982. Mimeo.

0251 Lacy, W.B., L. Busch, and P. Marcotte. *The Sudan Agricultural Research Corporation: Organization, Practices, and Policy Recommendations.* Lexington, Kentucky: University of Kentucky, October 1983.

0252 International Agricultural Development Service (IADS). *Sudan Agricultural Research Capabilities.* New York: IADS, November 1977.

0266 UNESCO. *National Science Policies in Africa.* Science Policy Studies and Documents No. 31. Paris: UNESCO, 1974.

0445 Swanson, Burton E., and Wade H. Reeves. "Agricultural Research Eastern and Southern Africa: Manpower and Training." World Bank, Washington, D.C., August 1986. Mimeo.

0589 Kassapu, Samuel. *Les Dépenses de Recherche Agricole dans 34 Pays d'Afrique Tropicale.* Paris: Centre de Développement de l'OCDE, 1976.

0710 Ministry of Agriculture and Natural Resources – Agricultural Research Corporation (ARC). *List of Scientific Staff and Senior Administrators.* Wad Medani, Sudan: Ministry of Agriculture and Natural Resources – ARC, December 1986.

0852 Evenson, Robert E., and Yoav Kislev. *Investment in Agricultural Research and Extension: A Survey of International Data.* Center Discussion Paper No. 124. New Haven, Connecticut: Economic Growth Center, Yale University, August 1971.

Additional References:

0014 Judd, M. Ann, James K. Boyce, and Robert E. Evenson. "Investing in Agricultural Supply." Economic Growth Center, Yale University, New Haven, Connecticut, 1983. Mimeo.

0016 Oram, Peter A., and Vishva Bindlish. *Resource Allocations to National Agricultural Research: Trends in the 1970s.* The Hague and Washington, D.C.: ISNAR and IFPRI, November 1981.

0022 UNESCO. *Statistical Yearbook 1983.* Paris: UNESCO, 1983.

0023 Bennell, Paul. *Agricultural Researchers in Sub-Saharan Africa: An Overview.* Working Paper No. 4. The Hague: ISNAR, October 1985.

0026 Oram, Peter A., and Mark Gieben. "Document Summaries." ISNAR, The Hague, 1984. Mimeo.

0027 Harvey, Nigel, ed. *Agricultural Research Centres: A World Directory of Organizations and Programmes.* Seventh Edition, Two Volumes. Harlow, U.K.: Longman, 1983.

0094 FAO – Near East Regional Office. *Directory of Agricultural Research Institutions in the Near East Region.* Cairo: FAO – Near East Regional Office, 1979.

0095 FAO – CARIS. *Agricultural Research in Developing Countries – Volume 1: Research Institutions.* Rome: FAO – CARIS, 1978.

0158 ISNAR. *Improving the Management of Agricultural Research: Report and Recommendations for the Agricultural Research Corporation of Sudan.* The Hague: ISNAR, February 1984.

0163 CGIAR. "National Agricultural Research." CGIAR, Washington, D.C., 1985. Mimeo.

0165 Evenson, R.E., and Y. Kislev. *Agricultural Research and Productivity.* New Haven: Yale University Press, 1975.

0174 Watson, J.M. "Comparative Study of Agricultural Research Organisation and Administration in the Near East Region." Paper presented at the Workshop on Organization and Administration of Agricultural Services in the Arab States, Cairo, 2–15 March 1964.

0227 Australian Centre for International Agricultural Research (ACIAR). *Proceedings of the Eastern Africa-ACIAR Consultation on Agricultural Research, 18–22 July 1983, Nairobi, Kenya.* Australia: ACIAR, 1984.

0253 Agricultural Research Corporation (ARC). "ARC Research Scientists Employed at Major Research Stations." ARC, Sudan, July 1983. Mimeo.

0287 Commonwealth Agricultural Bureaux (CAB). *List of Research Workers in Agriculture, Animal Health and Forestry in the British Commonwealth, the Republic of Sudan and the Republic of Ireland 1959.* Slough, England: CAB, 1959.

0360 Cooper, St.G.C. "Towards Trained Manpower for Agricultural Research in Africa." Paper presented at the Conference on Agricultural Research and Production in Africa, organized by the Association for the Advancement of Agricultural Sciences in Africa (AAASA), Addis Ababa, 29 August–4 September 1971.

0532 UNESCO Field Science Office for Africa. *Survey on the Scientific and Technical Potential of the Countries of Africa.* Paris: UNESCO, 1970.

0653 Webster, B.N. *Index of Agricultural Research Institutions and Stations in Africa.* Rome: FAO, n.d.

0709 Ministry of Agriculture and Natural Resources – Agricultural Research Corporation (ARC). *A Note on the Agricultural Research Corporation.* Wad Medani, Sudan: ARC, January 1987.

0762 Musnad, H.A., H.H.M. Faki, and A.H. El Jack. "Agricultural Research Priorities in the Sudan." In *Proceedings of the Eastern Africa-ACIAR Consultation on Agricultural Research*, pp. 24–30. Australia: ACIAR, 1984.

0831 FAO. *Report of an Expert Consultation on Agricultural Extension and Research Linkages in the Near East, Amman, Jordan, 22–26 July 1985.* Rome: FAO, 1985.

0834 Saeed, Hassan Mohamed, Arif Jamal Mohamed Ahmed, and Mohamed El Hussein El Tahir. "Technology Transfer and Feedback: Agricultural Research Extension." Part IV, chapter 19 in *The Agricultural Sector of Sudan: Policy and Systems Studies*, edited by A.B. Zahlan and W.Y. Magar, pp. 381–408. London: Ithaca Press, 1986.

0847 Kimura, J.H. "Financial and Administrative Management of Research Institutions in Eastern and Southern Africa: Report on Responses to a Questionnaire." In *Promotion of Technology Policy and Science Management in Africa*, edited by Karl Wolfgang Menck and Wolfgang Gmelin. Bonn: Deutsche Stiftung für internationale Entwicklung (DSE), 1986.

0873 Abdalla, Abdalla Ahmed. "Agricultural Research in the IGADD Sub-Region and Related Manpower Training." Inter-Governmental Authority on Drought and Development (IGADD) and FAO, 1987. Mimeo.

0886 Evenson, Robert E., and Yoav Kislev. "Investment in Agricultural Research and Extension: A Survey of International Data." *Economic Development and Cultural Change* Vol. 23 (April 1975): 507–521.

Suriname

Personnel

Period/Year	PhD	MSc	BSc	Subtotal	Expat	Total	Source
1960–64	..	..	..	..	..	14	451
1965–69	..	..	..	..	..	12	451/583
1970	..	..	..	..	..	16	583
1971	2	10	0	12	..	12	583
1972	3	16	0	19	..	19	583
1973	3	17	0	20	..	20	583
1974	2	17	0	19	..	19	583
1975	3	20	2	25	..	25	583
1976	2	16	2	20	..	20	583
1977	..	..	..	..	..	17	451
1978	..	..	..	..	..	15	451
1979	..	..	..	..	..	..	
1980	..	..	..	..	..	..	
1981	..	..	..	..	..	..	
1982	..	..	..	..	..	..	
1983	..	..	..	..	..	..	
1984	..	..	..	..	..	..	
1985	..	..	..	..	..	..	
1986	..	..	..	..	..	..	

1960–64: 1960, 61, 62, 63 & 64   1965–69: 1965, 66, 67, 68 & 69

Expenditure

LCU = Suriname Guilders

Period/Year	Current LCU (millions)	Constant 1980 LCU (millions)	Constant 1980 US$ (millions)		Source
			Atlas	PPP	
1960–64	..	..	..	..	
1965–69	..	..	..	..	
1970	..	..	..	..	
1971	..	..	..	..	
1972	..	..	..	..	
1973	..	..	..	..	
1974	..	..	..	..	
1975	..	..	..	..	
1976	..	..	..	..	
1977	..	..	..	..	
1978	..	..	..	..	
1979	..	..	..	..	
1980	..	..	..	..	
1981	..	..	..	..	
1982	..	..	..	..	
1983	..	..	..	..	
1984	..	..	..	..	
1985	..	..	..	..	
1986	..	..	..	..	

1960–64:   1965–69:

417

Personnel Comments:
The figures include the Agricultural Experiment Station ('Landbouwproefstation') only. On the basis of source 840 and source 115, it is understood that in addition to the Agricultural Experiment Station, there are some foundations which execute agricultural research on specific crops. Data on the number of researchers working at these foundations are not available.

Sources:

0451 Landbouwproefstation. *Agricultural Experiment Station 75th Anniversary.* Parimaribo, Suriname: Landbouwproefstation, 1978.

0583 Ehrencron, V.K.R., ed. *Jaarverslag.* Years: 1966–1970. Suriname: Landbouwproefstation, various years.

Additional References:

0027 Harvey, Nigel, ed. *Agricultural Research Centres: A World Directory of Organizations and Programmes.* Seventh Edition, Two Volumes. Harlow, U.K.: Longman, 1983.

0053 Pineiro, Martin, and Eduardo Trigo, eds. *Cambio Tecnico en el Agro Latino Americano – Situacion y Perspectivas en la Decada de 1980.* San José: IICA, 1983.

0115 Economic Commission for Latin America and the Caribbean (ECLAC). *Agricultural Research Policy and Management – Volumes I and II.* Papers presented at the Workshop on Agricultural Research Policy and Management, 26–30 September 1983, Port of Spain, Trinidad. Port of Spain, Trinidad: ECLAC, November 1984.

0582 Samson, J.A., ed. *Jaarverslag.* Years: 1962 and 1965. Suriname: Landbouwproefstation, various years.

0584 Landbouwproefstation Suriname. *Jaarverslag.* Years: 1971–1976. Suriname: Landbouwproefstation, various years.

0840 Sital, J.T. Personal communication. Department of Development Economics, Wageningen Agricultural University. Wageningen, The Netherlands, Spring 1987.

Swaziland

Personnel

Period/Year	PhD	MSc	BSc	Subtotal	Expat	Total	Source
1960–64	..	..	..	..	..	6	10
1965–69	..	..	..	..	..	10	10/175
1970	..	..	..	..	..	9	532
1971	..	..	..	..	..	12	17
1972	..	..	..	..	..	12	17
1973	..	..	..	..	..	..	
1974	..	..	..	..	..	..	
1975	..	..	..	..	..	..	
1976	..	..	..	..	..	12	17
1977	..	..	..	..	..	12	17
1978	..	..	..	..	..	12	17
1979	..	..	..	..	..	4	17
1980	..	..	..	..	..	2	17
1981	..	..	..	..	..	3	17
1982	..	2	..	..	..	18	17
1983	0	2	10	12	6	18	17
1984	..	..	..	..	..	15	8
1985	..	..	..	..	..	..	
1986	..	..	..	..	..	..	

1960–64: 1961   1965–69: 1966 & 67

Expenditure

LCU = Emulangeni/South African Rands

Period/Year	Current LCU (millions)	Constant 1980 LCU (millions)	Constant 1980 US$ (millions)		Source
			Atlas	PPP	
1960–64	0.053	0.210	0.233	0.309	10
1965–69	0.133	0.470	0.520	0.689	589/175
1970	0.140	0.442	0.489	0.648	17
1971	0.220	0.641	0.710	0.941	17/589
1972	0.660	1.838	2.036	2.698	17
1973	..	..	..	..	
1974	..	..	..	..	
1975	..	..	..	..	
1976	0.100	0.165	0.182	0.242	17
1977	0.270	0.402	0.446	0.591	17
1978	0.400	0.550	0.609	0.808	17
1979	0.500	0.595	0.658	0.873	17
1980	0.620	0.620	0.687	0.910	17
1981	1.220	1.082	1.198	1.588	17
1982	2.900	2.400	2.657	3.523	17
1983	3.549	2.635	2.917	3.867	17
1984	..	..	..	..	
1985	..	..	..	..	
1986	..	..	..	..	

1960–64: 1961   1965–69: 1966 & 67

Personnel Comments:

These figures refer to the Agricultural Research Division of the Ministry of Agriculture and Cooperatives. Source 17 indicates that in 1972 the Agricultural Research Division was transferred to the then University of Botswana, Lesotho, and Swaziland. In 1978 the Research Division was transferred back to the Government of Swaziland. Just after 1978 the Research Division lost most of its research officers because of delays by the Government in the creation of research posts. It was not until 1982 that these posts were created and researchers could be attracted to fill them. Source 532 indicates that the 1970 figure includes the Malkerns Research Station only.

Expenditure Comments:

The expenditure figures refer to the Agricultural Research Division.

Sources:

0008 SADCC, and DEVRES, Inc. *Agricultural Research Resource Assessment in the SADCC Countries, Volume II: Country Report Swaziland.* Washington, D.C.: DEVRES, Inc., January 1985.

0010 Boyce, James K., and Robert E. Evenson. *National and International Agricultural Research and Extension Programs.* New York: Agricultural Development Council, Inc., 1975.

0017 ISNAR, IFARD & AOAD. Survey of National Agricultural Research Systems: Unpublished Questionnaire Responses. ISNAR, The Hague, 1985.

0175 Cooper, St.G.C. *Agricultural Research in Tropical Africa.* Kampala: East African Literature Bureau, 1970.

0532 UNESCO Field Science Office for Africa. *Survey on the Scientific and Technical Potential of the Countries of Africa.* Paris: UNESCO, 1970.

0589 Kassapu, Samuel. *Les Dépenses de Recherche Agricole dans 34 Pays d'Afrique Tropicale.* Paris: Centre de Développement de l'OCDE, 1976.

Additional References:

0002 SADCC, and DEVRES, Inc. *Agricultural Research Resource Assessment in the SADCC Countries, Volume I: Regional Analysis and Strategy.* Washington, D.C.: DEVRES, Inc., January 1985.

0023 Bennell, Paul. *Agricultural Researchers in Sub-Saharan Africa: An Overview.* Working Paper No. 4. The Hague: ISNAR, October 1985.

0027 Harvey, Nigel, ed. *Agricultural Research Centres: A World Directory of Organizations and Programmes.* Seventh Edition, Two Volumes. Harlow, U.K.: Longman, 1983.

0095 FAO – CARIS. *Agricultural Research in Developing Countries – Volume 1: Research Institutions.* Rome: FAO – CARIS, 1978.

0165 Evenson, R.E., and Y. Kislev. *Agricultural Research and Productivity.* New Haven: Yale University Press, 1975.

0244 Commonwealth Agricultural Bureaux (CAB). *List of Research Workers in the Agricultural Sciences in the Commonwealth and in the Republic of Ireland 1975.* Slough, England: CAB, 1975.

0279 Commonwealth Agricultural Bureaux (CAB). *List of Research Workers in the Agricultural Sciences in the Commonwealth and in the Republic of Ireland 1972.* Slough, England: CAB, 1972.

0286 Commonwealth Agricultural Bureaux (CAB). *List of Research Workers in Agriculture, Animal Health and Forestry in the Commonwealth and in the Republic of Ireland 1966.* Slough, England: CAB, 1966.

0287 Commonwealth Agricultural Bureaux (CAB). *List of Research Workers in Agriculture, Animal Health and Forestry in the British Commonwealth, the Republic of Sudan and the Republic of Ireland 1959.* Slough, England: CAB, 1959.

0360 Cooper, St.G.C. "Towards Trained Manpower for Agricultural Research in Africa." Paper presented at the Conference on Agricultural Research and Production in Africa, organized by the Association for the Advancement of Agricultural Sciences in Africa (AAASA), Addis Ababa, 29 August–4 September 1971.

0385 SADCC, and DEVRES, Inc. *SADCC Region Agricultural Research Resource Assessment – Data Base Management Information System: Diskettes and User's Guide.* Printouts. Washington, D.C.: SADCC, and DEVRES, Inc., n.d.

0445 Swanson, Burton E., and Wade H. Reeves. "Agricultural Research Eastern and Southern Africa: Manpower and Training." World Bank, Washington, D.C., August 1986. Mimeo.

0446 Kyomo, M.L. "Agricultural Research in Eastern and Southern Africa: Issues and Priorities." Southern African Centre for Cooperation in Agricultural Research of SADCC, Gabarone, Botswana, 1986. Mimeo.

0467 DEVRES, Inc. *The Agricultural Research Resource Assessment Pilot Report for Botswana, Malawi and Swaziland.* Washington, D.C.: DEVRES, Inc., November 1983.

420

0653 Webster, B.N. *Index of Agricultural Research Institutions and Stations in Africa.* Rome: FAO, n.d.

0852 Evenson, Robert E., and Yoav Kislev. *Investment in Agricultural Research and Extension: A Survey of International Data.* Center Discussion Paper No. 124. New Haven, Connecticut: Economic Growth Center, Yale University, August 1971.

0886 Evenson, Robert E., and Yoav Kislev. "Investment in Agricultural Research and Extension: A Survey of International Data." *Economic Development and Cultural Change* Vol. 23 (April 1975): 507–521.

Sweden

Personnel

Period/Year	PhD	MSc	BSc	Subtotal	Expat	Total	Source
1960–64	..	..	..	..		407	613/885
1965–69	..	..	..	..		503	613/885
1970	..	..	..	..		601	613/885
1971	..	..	..	..		585	613/885
1972	..	..	..	..		612	613/885
1973	..	..	..	..		640	613/885
1974	..	..	..	..		697	613/885
1975	..	..	..	..		717	613/885
1976	..	..	..	..		758	613/885
1977	..	..	..	..		827	613/885
1978	..	..	..	..		817	613/885
1979	..	..	..	..		813	613/885
1980	..	..	..	..		837	613/885
1981	..	..	..	..		909	613/885
1982	..	..	..	..		930	613/885
1983	..	..	..	..		1002	613/885
1984	..	..	..	..		1068	613/885
1985	..	..	..	..		1158	613/885
1986	..					..	

1960–64: 1961, 62, 63 & 64   1965–69: 1965, 66, 67, 68 & 69

Expenditure   *LCU = Swedish Kroner*

Period/Year	Current LCU (millions)	Constant 1980 LCU (millions)	Constant 1980 US$ (millions)		Source
			Atlas	PPP	
1960–64	43.625	151.574	34.418	21.261	613/885
1965–69	82.979	235.178	53.402	32.988	613/885
1970	129.287	324.842	73.763	45.565	613/885
1971	135.257	317.504	72.096	44.535	613/885
1972	146.592	321.475	72.998	45.092	613/885
1973	168.107	344.482	78.222	48.319	613/885
1974	182.308	341.401	77.522	47.887	613/885
1975	202.453	330.805	75.117	46.401	613/885
1976	212.799	310.655	70.541	43.575	613/885
1977	240.797	318.094	72.230	44.618	613/885
1978	249.953	301.511	68.465	42.292	613/885
1979	295.774	330.474	75.041	46.355	613/885
1980	364.738	364.738	82.822	51.161	613/885
1981	391.010	357.087	81.084	50.088	613/885
1982	453.053	381.037	86.523	53.447	613/885
1983	489.809	375.332	85.227	52.647	613/885
1984	571.183	406.536	92.313	57.024	613/885
1985	622.142	414.209	94.055	58.100	613/885
1986	..	..	..	..	

1960–64: 1961, 62, 63 & 64   1965–69: 1965, 66, 67, 68 & 69

Personnel Comments:

These figures are measured in full-time equivalents and include university and an estimate of non-university research personnel – exclusive of support staff. In Sweden there is only one university of agricultural sciences, the Swedish University of Agricultural Sciences (SUAS). It belongs to the Department of Agriculture and was created in 1977 in a merger of the formerly independent Colleges of Agriculture, Forestry and Veterinary Medicine. According to source 613, research on fish production was introduced at the university in recent years but this constitutes a relatively small proportion of the national commitment to fisheries research. During 1961–65 university-based research personnel averaged 82% of total researchers, while during 1981–85 this had increased slightly to 87%.

Expenditure Comments:

These figures represent an estimate of university plus non-university research expenditures, not simply funds available. Total university expenditures were scaled down by 20% – an estimate of the nonresearch (teaching plus information services) component of total university spending. Both university and non-university expenditures exclude capital costs.

Although the expenditure and personnel series are commensurable, they do imply a decline in real spending per scientist during the later part of the 1970s and into the 1980s. Source 613 suggests this may be due to a decline in real (university) salaries over this period, coupled with a change in the research personnel structure within the university system. In 1971–72, approximately 30% of total SAUS research personnel were at the higher level (roughly equivalent to a full professor in the US system), but by 1985 this had dropped to approximately 17%.

Sources:

0613 Croon, Ingemar. Personal communication. Department of Economics and Statistics, Swedish University of Agricultural Sciences. Uppsala, Sweden, July 1988.

0885 Croon, Ingemar. "Resursinsatser i Offentlig Jordbruksforskning i Sverige Under Perioden 1945–1985." Draft. Swedish University of Agricultural Science, Uppsala, Sweden, July 1988. [Mimeo]

Additional References:

0010 Boyce, James K., and Robert E. Evenson. *National and International Agricultural Research and Extension Programs.* New York: Agricultural Development Council, Inc., 1975.

0021 OECD. *OECD Science and Technology Indicators.* Paris: OECD, 1984.

0027 Harvey, Nigel, ed. *Agricultural Research Centres: A World Directory of Organizations and Programmes.* Seventh Edition, Two Volumes. Harlow, U.K.: Longman, 1983.

0165 Evenson, R.E., and Y. Kislev. *Agricultural Research and Productivity.* New Haven: Yale University Press, 1975.

0376 FAO. *Fourteenth FAO Regional Conference for Europe, Reykjavik, Iceland, 17–21 September 1984: Research in Support of Agricultural Policies in Europe.* Rome: FAO, 1984.

0473 OECD. *Science and Technology Indicators: Basic Statistical Series – Volume A: The Objectives of Government R&D Funding 1974–1985.* Paris: OECD, May 1983.

0474 OECD. *Science and Technology Indicators: Basic Statistical Series – Volume A: The Objectives of Government R&D, 1969–1981.* Paris: OECD, July 1981.

0509 OECD. *Survey of the Resources Devoted to R&D by OECD Member Countries, International Statistical Year 1971 – Volume 5: Total Tables, Statistical Tables and Notes.* Paris: OECD, August 1974.

0510 OECD. *International Survey of the Resources Devoted to R&D in 1967 by OECD Member Countries, Statistical Tables and Notes – Volume 5: Total Tables.* Paris: OECD, August 1970.

0511 OECD. *International Survey of the Resources Devoted to R&D in 1969 by OECD Member Countries, Statistical Tables and Notes – Volume 5: Total Tables.* Paris: OECD, June 1973.

0512 OECD. *Survey of the Resources Devoted to R&D by OECD Member Countries, International Statistical Year 1973 – Volume 5: Total Tables, Statistical Tables and Notes.* Paris: OECD, September 1976.

0513 OECD. *International Survey of the Resources Devoted to R&D by OECD Member Countries, International Statistical Year 1975 – International Volume: Statistical Tables and Notes.* Paris: OECD, March 1979.

0514 ΘECD. *A Study of Resources Devoted to R&D in OECD Member Countries in 1963/64, International Statistical Year for Research and Development – Volume 2: Statistical Tables and Notes.* Paris: OECD, 1968.

0530 OECD. *OECD Science and Technology Indicators No. 2 – R&D, Invention and Competitiveness.* Paris: OECD, 1986.

0574 OECD. *Intellectual Investment in Agriculture for Economic and Social Development.* Documentation in Food and Agriculture No. 60. Paris: OECD, 1962.

0601 OECD – STIID Data Bank. "Public Funding of R&D by Socio-Economic Objective." Tables. OECD, Paris, April 1987. Mimeo.

0602 OECD – STIID Data Bank. "R&D Expenditure in the Higher Education and Private Non-Profit Sectors." Tables. OECD, Paris, April 1987. Mimeo.

0614 OECD. *OECD Policies for Science and Education, Country Reviews: Sweden.* Paris: OECD, October 1962.

0615 OECD. *Country Reports on the Organisation of Scientific Research: Sweden.* Paris: OECD, November 1963.

0616 OECD. *Reviews of National Policies for Science and Education – Scientific Policy: Sweden.* Paris: OECD, January 1964.

0621 NORDFORSK. *Forsknings Virksomhet i Norden i 1967 – Utgifter og Personale.* Nordisk Statistisk Skriftserie 18. Oslo: NORDFORSK, 1970.

0707 OECD – STIID Data Bank. "R&D Expenditure in the Higher Education and Private Non-Profit Sectors: Higher Education, Agricultural Sciences, General Universities Funds." Table. OECD, Paris, June 1987. Mimeo.

0852 Evenson, Robert E., and Yoav Kislev. *Investment in Agricultural Research and Extension: A Survey of International Data.* Center Discussion Paper No. 124. New Haven, Connecticut: Economic Growth Center, Yale University, August 1971.

0886 Evenson, Robert E., and Yoav Kislev. "Investment in Agricultural Research and Extension: A Survey of International Data." *Economic Development and Cultural Change* Vol. 23 (April 1975): 507–521.

Switzerland

Personnel

Period/Year	PhD	MSc	BSc	Subtotal	Expat	Total	Source
1960–64	..	..	..	..	..	112	620
1965–69	..	..	..	..	..	143	620
1970	..	..	..	..	..	178	620
1971	..	..	..	..	..	186	620
1972	..	..	..	..	..	192	620
1973	..	..	..	..	..	..	
1974	..	..	..	..	..	..	
1975	..	..	..	..	..	..	
1976	..	..	..	..	..	..	
1977	..	..	..	..	..	..	
1978	..	..	..	..	..	..	
1979	..	..	..	..	..	..	
1980	..	..	..	..	..	..	
1981	..	..	..	..	..	..	
1982	..	..	..	..	..	..	
1983	..	..	..	..	..	..	
1984	..	..	..	..	..	..	
1985	..	..	..	..	..	..	
1986	..	..	..	..	..	..	

1960–64: 1960   1965–69: 1965, 67, 68 & 69

Expenditure   *LCU= Swiss Francs*

Period/Year	Current LCU (millions)	Constant 1980 LCU (millions)	Constant 1980 US$ (millions)		Source
			Atlas	PPP	
1960–64	7.400	18.782	11.685	7.050	620
1965–69	23.600	44.101	27.437	16.553	620
1970	37.500	62.189	38.690	23.342	620
1971	32.600	49.544	30.823	18.596	601
1972	41.100	56.925	35.415	21.366	601
1973	50.000	64.020	39.829	24.029	601
1974	59.000	70.659	43.959	26.521	601
1975	38.600	43.128	26.832	16.188	601
1976	45.000	48.966	30.463	18.379	601
1977	31.660	34.354	21.373	12.903	668
1978	..	..	..	..	
1979	35.560	36.509	22.714	13.703	668
1980	..	..	..	..	
1981	47.950	44.855	27.906	16.836	668
1982	..	..	..	..	
1983	66.000	55.696	34.650	20.905	867
1984	..	..	..	..	
1985	..	..	..	..	
1986	77.000	59.140	36.793	22.198	867

1960–64:1960   1965–69: 1965, 67, 68 & 69

Personnel Comments:

The indicator series includes only those researchers working at the seven federal agricultural research stations which are located at various sites around the country. In addition to these seven federal stations, several universities undertake agricultural research (108 researchers in 1970, source 10; 157 researchers in 1978, source 669), as does the Federal Institute of Forestry Research (27 researchers in 1967, source 619).

Expenditure Comments:

The pre–1970 figures represent net expenditures (exclusive receipts from the sale of produce) plus investments of the seven federal agricultural research stations only. The post–1970 figures represent public funding of agriculture as defined by OECD. In this case, only federal-level contributions to agricultural research are included, while university research expenditures are excluded. Post–1976 estimates from source 601 are in fact obtained from the agency who compiles the statistics reported in source 668. Source 669 indicates that the university sector spent 20.52 million Swiss Francs on agricultural (including veterinary) research in 1978. For the sake of consistency this figure is omitted from the indicator series.

According to source 867, the 1986 figure includes a 16 million Swiss Franc component of 'development aid,' some of which (according to source 868) is spent on research within Switzerland but most of which is spent externally. It is unclear whether this component was included in earlier figures, but it is probable that it was, and for reasons of consistency, it is retained in the present series. Source 868 also indicates that 'central costs' (e.g., computing facilities and the like) incurred by federal research agencies have been excluded, at least from the later figures.

In qualitative terms the decline in agricultural research spending during the last half of the 1970s, and subsequent growth in the 1980s, which these figures demonstrate, is plausible, according to source 868. The federal research agencies responsible for agricultural research are also responsible for monitoring and evaluating environmental quality. The staff ceilings imposed on federal agencies in 1974 by the Swiss Confederation (and still in place), when combined with the growing concern over environmental issues during the latter half of the 1970s, meant that resources were diverted from research to environmental management issues during this period.

Sources:

0601 OECD – STIID Data Bank. "Public Funding of R&D by Socio-Economic Objective." Tables. OECD, Paris, April 1987. Mimeo.

0620 Pellanda, Sanzio. *Sur la Rentabilité de la Recherche Agronomique du Secteur Public en Suisse – Etude Econométrique.* Fribourg, Switzerland: Université de Fribourg, June 1972.

0668 Bundesamt für Statistik. *Forschung und Entwicklung des Bundes 1978–1981.* Beitrage zur Schweizerischen Statistik – Heft 104. Bern: Bundesamt für Statistik, 1983.

0867 Buri, Markus, Raul Suarez de Miguel, and Bruno Walder. "Forschung und Entwicklung 1986 in der Schweiz." Bericht des BFS für den Basisbericht "OECD-Examen 1988 der schweizerischen Forschungs- und Technologiepolitik". Bundesamt für Statistik, Bern, April 1988. Mimeo.

Additional References:

0010 Boyce, James K., and Robert E. Evenson. *National and International Agricultural Research and Extension Programs.* New York: Agricultural Development Council, Inc., 1975.

0021 OECD. *OECD Science and Technology Indicators.* Paris: OECD, 1984.

0022 UNESCO. *Statistical Yearbook 1983.* Paris: UNESCO, 1983.

0027 Harvey, Nigel, ed. *Agricultural Research Centres: A World Directory of Organizations and Programmes.* Seventh Edition, Two Volumes. Harlow, U.K.: Longman, 1983.

0165 Evenson, R.E., and Y. Kislev. *Agricultural Research and Productivity.* New Haven: Yale University Press, 1975.

0473 OECD. *Science and Technology Indicators: Basic Statistical Series – Volume A: The Objectives of Government R&D Funding 1974–1985.* Paris: OECD, May 1983.

0474 OECD. *Science and Technology Indicators: Basic Statistical Series – Volume A: The Objectives of Government R&D, 1969–1981.* Paris: OECD, July 1981.

0509 OECD. *Survey of the Resources Devoted to R&D by OECD Member Countries, International Statistical Year 1971 – Volume 5: Total Tables,*

426

Statistical Tables and Notes. Paris: OECD, August 1974.

0510 OECD. *International Survey of the Resources Devoted to R&D in 1967 by OECD Member Countries, Statistical Tables and Notes – Volume 5: Total Tables.* Paris: OECD, August 1970.

0511 OECD. *International Survey of the Resources Devoted to R&D in 1969 by OECD Member Countries, Statistical Tables and Notes – Volume 5: Total Tables.* Paris: OECD, June 1973.

0513 OECD. *International Survey of the Resources Devoted to R&D by OECD Member Countries, International Statistical Year 1975 – International Volume: Statistical Tables and Notes.* Paris: OECD, March 1979.

0524 OECD. *Reviews of National Science Policy: Switzerland.* Paris: OECD, 1971.

0530 OECD. *OECD Science and Technology Indicators No. 2 – R&D, Invention and Competitiveness.* Paris: OECD, 1986.

0574 OECD. *Intellectual Investment in Agriculture for Economic and Social Development.* Documentation in Food and Agriculture No. 60. Paris: OECD, 1962.

0602 OECD – STIID Data Bank. "R&D Expenditure in the Higher Education and Private Non-Profit Sectors." Tables. OECD, Paris, April 1987. Mimeo.

0619 Sekretariat des Schweizerischen Wissenschaftsrates. *Wissenschafts- Politik.* Jahrgang 4, Nr. 6/1970. Bern: Sekretariat des Schweizerischen Wissenschaftsrates, 1970.

0669 Bundesamt für Statistik. *Forschung und Entwicklung an den schweizerischen Hochschulen.* Beitrage zur schweizerischen Statistik – Heft 98. Bern: Bundesamt für Statistik, 1982.

0708 OECD – STIID Data Bank. "Gross Domestic Expenditure on R&D: All Fields of Sciences – Agriculture, Forestry and Fishing." Tables. OECD, Paris, June 1987. Mimeo.

0852 Evenson, Robert E., and Yoav Kislev. *Investment in Agricultural Research and Extension: A Survey of International Data.* Center Discussion Paper No. 124. New Haven, Connecticut: Economic Growth Center, Yale University, August 1971.

0868 Buri, Markus. Personal communication. Bundesamt für Statistik. Bern, May 1988.

0886 Evenson, Robert E., and Yoav Kislev. "Investment in Agricultural Research and Extension: A Survey of International Data." *Economic Development and Cultural Change* Vol. 23 (April 1975): 507–521.

Syrian Arab Republic

Personnel

Period/Year	PhD	MSc	BSc	Subtotal	Expat	Total	Source
1960–64	..	..	..	..	..		
1965–69	..	..	..	..	..	15	886
1970	..	..	..	..	..	69	852
1971	..	..	..	..	..	..	
1972	..	..	..	..	..	..	
1973	..	..	..	..	..	..	
1974	..	..	..	..	..	..	
1975	..	..	..	..	..	..	
1976	..	..	..	..	..	153	95
1977	..	..	..	..	..	..	
1978	..	..	..	..	..	..	
1979	..	..	..	..	..	..	
1980	..	..	..	..	..	..	
1981	15	6	160	181	..	181	256
1982	..	..	..	..	..	..	
1983	24	9	217	250	3	253	56/17
1984	..	..	..	..	..	..	
1985	..	..	..	..	..	..	
1986	..	..	..	..	..	..	

1960–64: 1965–69: 1965

Expenditure   *LCU = Syrian Pounds (Lira)*

Period/Year	Current LCU (millions)	Constant 1980 LCU (millions)	Constant 1980 US$ (millions) Atlas	PPP	Source
1960–64	..	..	..	..	
1965–69	1.681	5.797	1.451	2.990	886
1970	3.820	11.646	2.916	6.008	852
1971	..	..	..	..	
1972	..	..	..	..	
1973	..	..	..	..	
1974	..	..	..	..	
1975	..	..	..	..	
1976	..	..	..	..	
1977	..	..	..	..	
1978	..	..	..	..	
1979	..	..	..	..	
1980	..	..	..	..	
1981	..	..	..	..	
1982	..	..	..	..	
1983	15.910	12.779	3.199	6.592	17
1984	..	..	..	..	
1985	..	..	..	..	
1986	..	..	..	..	

Personnel Comments:

The personnel figures refer to the Scientific Agricultural Research Department (SARD) of the Ministry of Agriculture and Agricultural Reform (MAAR) only. Although SARD represents the major part of the Syrian agricultural research system, several other organizations also conduct agricultural research. The table below gives some scattered personnel observations for these other organizations.

	1976	1982	1983
MAAR/SD			169
MAAR/CB	15	15	34
GADEB	31		6*
GOS			10*
GOT	5		19*
UOA/FA		120[a]	
Source	95	27	17
			56*

a. Total staff with teaching and research duties.

Given the regional character of the Arab Centre for Studies of Arid Zones and Dry Land (ASCAD) and the international focus of ICARDA, they have been excluded from this series. ASCAD was established in 1971 as a technical agency of the Arab League and headquartered at Douma near Damascus. In 1983 ASCAD had 78 researchers, of whom 33 held a PhD degree (source 254).

Expenditure Comments:

The expenditure figures are commensurable with the personnel figures. Some scattered expenditure figures for organizations not included in the indicator series are given in the table below.

	1976	1983
MAAR/SD		7.632
MAAR/CB	0.771	3.194
GADEB	2.000	
GOT	0.482	
Source	95	17

Acronyms and Abbreviations:

CB: Cotton Bureau
GADEB: General Administration for the Development of the Euphrates Basin
GOS: General Organization for Sugar
GOT: General Organization for Tobacco
MAAR: Ministry of Agriculture & Agricultural Reform
SARD: Scientific Agricultural Research Department
SD: Soils Department
UOA/FA: University of Aleppo – Faculty of Agriculture

Sources:

0017 ISNAR, IFARD & AOAD. Survey of National Agricultural Research Systems: Unpublished Questionnaire Responses. ISNAR, The Hague, 1985.

0056 El–Akhrass, Hisham. *Syria and the CGIAR Centers: A Study of Their Collaboration in Agricultural Research.* CGIAR Study Paper Number 13. Washington, D.C.: World Bank, 1986.

0095 FAO – CARIS. *Agricultural Research in Developing Countries – Volume 1: Research Institutions.* Rome: FAO – CARIS, 1978.

0256 Carmen, M.L. "Reorganization of Agricultural Research in the Syrian Arab Republic." FAO, Rome, December 1981. Mimeo.

0852 Evenson, Robert E., and Yoav Kislev. *Investment in Agricultural Research and Extension: A Survey of International Data.* Center Discussion Paper No. 124. New Haven, Connecticut: Economic Growth Center, Yale University, August 1971.

0886 Evenson, Robert E., and Yoav Kislev. "Investment in Agricultural Research and Extension: A Survey of International Data." *Economic Development and Cultural Change* Vol. 23 (April 1975): 507–521.

Additional References:

0010 Boyce, James K., and Robert E. Evenson. *National and International Agricultural Research and Extension Programs.* New York: Agricultural Development Council, Inc., 1975.

0016 Oram, Peter A., and Vishva Bindlish. *Resource Allocations to National Agricultural Research: Trends in the 1970s.* The Hague and Washington, D.C.: ISNAR and IFPRI, November 1981.

0026 Oram, Peter A., and Mark Gieben. "Document Summaries." ISNAR, The Hague, 1984. Mimeo.

0027 Harvey, Nigel, ed. *Agricultural Research Centres: A World Directory of Organizations and Programmes.* Seventh Edition, Two Volumes. Harlow, U.K.: Longman, 1983.

0073 Oram, Peter A., and Vishva Bindlish. "Investment in Agricultural Research in Developing Countries: Progress, Problems, and the Determination of Priorities." IFPRI, Washington, D.C., January 1984. Mimeo.

0094 FAO – Near East Regional Office. *Directory of Agricultural Research Institutions in the Near East Region*. Cairo: FAO – Near East Regional Office, 1979.

0163 CGIAR. "National Agricultural Research." CGIAR, Washington, D.C., 1985. Mimeo.

0165 Evenson, R.E., and Y. Kislev. *Agricultural Research and Productivity*. New Haven: Yale University Press, 1975.

0174 Watson, J.M. "Comparative Study of Agricultural Research Organisation and Administration in the Near East Region." Paper presented at the Workshop on Organization and Administration of Agricultural Services in the Arab States, Cairo, 2–15 March 1964.

0254 UN Economic Commission for Western Asia (ECWA), and FAO. *Agriculture and Development in Western Asia*. Number 7. Baghdad: ECWA, November 1984.

0831 FAO. *Report of an Expert Consultation on Agricultural Extension and Research Linkages in the Near East, Amman, Jordan, 22–26 July 1985*. Rome: FAO, 1985.

0865 Oram, Peter. "Agricultural Research Objectives and Priorities: Constraints to the Development of Agricultural Research Institutions in Arab countries." Review paper prepared for the first meeting of the Council for Arab Agricultural Research, organized by the Arab Fund for Economic and Social Development, Kuwait, 23 March 1988.

Taiwan

Personnel

Period/Year	PhD	MSc	BSc	Subtotal	Expat	Total	Source
1960–64	..	..	..	..	..		
1965–69	..	..	..	..	..		
1970	..	..	..	..	..		
1971	..	..	..	..	..		
1972	..	..	..	..	..		
1973	..	..	..	..	..		
1974	..	..	..	..	..		
1975	..	..	..	..	..		
1976	..	..	..	..	..		
1977	..	..	..	..	..		
1978	..	..	..	..	..	..	
1979	..	..	..	..	..	..	
1980	..	..	..	..	..	1142	773
1981	..	..	..	..	..	1417	773
1982	..	..	..	..	..	1581	773
1983	..	..	..	..	..	1446	773
1984	286	569	889	1744	..	1744	773
1985	262	577	1009	1848	..	1848	773
1986	..	..	..	..	..	..	

1960–64:   1965–69:

Expenditure   *LCU = New Taiwanese Dollars*

Period/Year	Current LCU (millions)	Constant 1980 LCU (millions)	Constant 1980 US$ (millions)		Source
			Atlas	PPP	
1960–64	..	..	..	..	
1965–69	..	..	..	..	
1970	..	..	..	..	
1971	..	..	..	..	
1972	..	..	..	..	
1973	..	..	..	..	
1974	..	..	..	..	
1975	..	..	..	..	
1976	..	..	..	..	
1977	..	..	..	..	
1978	1200.830	1634.791	41.947	58.931	773
1979	1450.220	1682.828	43.179	60.663	773
1980	1436.350	1436.350	36.855	51.778	773
1981	1956.330	1744.963	44.774	62.903	773
1982	2732.300	2356.080	60.454	84.933	773
1983	2354.020	1991.594	51.102	71.794	773
1984	2178.380	1827.796	46.899	65.889	773
1985	2467.980	2066.424	53.022	74.491	773
1986	..	..	..	..	

1960–64:   1965–69:

Personnel Comments:

The personnel figures given by source 773 are based on the National Survey on Science and Technology which has been conducted annually by the National Science Council since 1978.

When compared with figures from source 857, the coverage of this nationwide survey appears to be fairly complete and also seems to include the university component of the NARS.

The more detailed information which is available for 1984 and 1985 suggests that 24% of the researchers who are included in the total figures are qualified below the BSc level. For this reason, the 1980–83 total research figure from source 773 was scaled down by 24%.

Expenditure Comments:

According to source 773, the expenditure figures include agricultural research expenditures by governmental agricultural research institutes, universities, and agribusiness. It is not clear if the agribusiness component includes private, nonprofit or private, for-profit institutions. Source 857 also mentions two government enterprise system institutes and three autonomous research organizations. It may well be that they are also included in the agribusiness component.

Sources:

0773   Lin, J.Y. Personal communication. Director, Division of Life Sciences, National Science Council. Taipei, January 1988.

Additional References:

0010   Boyce, James K., and Robert E. Evenson. *National and International Agricultural Research and Extension Programs.* New York: Agricultural Development Council, Inc., 1975.

0014   Judd, M. Ann, James K. Boyce, and Robert E. Evenson. "Investing in Agricultural Supply." Economic Growth Center, Yale University, New Haven, Connecticut, 1983. Mimeo.

0027   Harvey, Nigel, ed. *Agricultural Research Centres: A World Directory of Organizations and Programmes.* Seventh Edition, Two Volumes. Harlow, U.K.: Longman, 1983.

0165   Evenson, R.E., and Y. Kislev. *Agricultural Research and Productivity.* New Haven: Yale University Press, 1975.

0176   Moseman, Albert, ed. *National Agricultural Research Systems in Asia.* Report of the Regional Seminar held at the India International Centre, New Delhi, India, March 8–13, 1971. New York: ADC, Inc., 1971.

0715   Chang, C.W. *FAO Directory of Agricultural Research Institutes and Experiment Stations in Asia and the Far East.* Bangkok, Thailand: FAO – Regional Office for Asia and the Far East, September 1962.

0716   Chang, C.W. *FAO Directory of Agricultural Research Institutes and Experiment Stations in Asia and the Far East, Supplement.* Bangkok, Thailand: FAO – Regional Office for Asia and the Far East, May 1963.

0740   Sands, Carolyn Marie. "The Theoretical and Empirical Basis for Analyzing Agricultural Technology Systems." PhD diss., University of Illinois, Urbana-Champaign, Illinois, 1988.

0769   Bureau of Statistics, Directorate-General of Budget, Accounting & Statistics, Executive Yuan, ed. *Statistical Yearbook of the Republic of China 1986.* Taipei, Taiwan: Directorate-General of Budget, Accounting & Statistics, Executive Yuan, November 1986.

0852   Evenson, Robert E., and Yoav Kislev. *Investment in Agricultural Research and Extension: A Survey of International Data.* Center Discussion Paper No. 124. New Haven, Connecticut: Economic Growth Center, Yale University, August 1971.

0857   Johnson, Eldon L., Frederick C. Fliegel, John L. Woods, and Mel C. Chu. *The Agricultural Technology System of Taiwan.* Urbana-Champaign, Illinois: INTERPAKS, Office of International Agriculture, University of Illinois, December 1987.

0886   Evenson, Robert E., and Yoav Kislev. "Investment in Agricultural Research and Extension: A Survey of International Data." *Economic Development and Cultural Change* Vol. 23 (April 1975): 507–521.

Tanzania

Personnel

Period/Year	PhD	MSc	BSc	Subtotal	Expat	Total	Source
1960–64	..	..	..	..	..	107	518
1965–69	11	19	57	87	..	87	285
1970	..	..	..	..	..	110	532
1971	24	46	57	127	..	127	279
1972	..	..	..	..	..	..	
1973	..	..	..	..	..	..	
1974	..	..	..	..	..	..	
1975	..	..	..	..	..	..	
1976	..	..	..	..	..	..	
1977	..	..	..	..	..	210	243
1978	..	..	..	..	..	..	
1979	..	..	..	..	..	..	
1980	..	..	..	..	..	..	
1981	..	..	..	..	..	..	
1982	..	..	..	..	..	247	27
1983	..	..	..	..	..	..	
1984	19	98	120	237	68	305	232
1985	..	..	..	..	..	..	
1986	..	..	..	..	..	..	

1960–64: 1964   1965–69: 1968

Expenditure

LCU = Tanzania Shillings

Period/Year	Current LCU (millions)	Constant 1980 LCU (millions)	Constant 1980 US$ (millions) Atlas	PPP	Source
1960–64	11.989	36.177	4.434	5.891	518
1965–69	..	..	..	..	
1970	31.159	88.269	10.819	14.375	58
1971	35.987	99.412	12.185	16.189	58
1972	35.671	92.412	11.327	15.050	58
1973	54.389	123.893	15.186	20.176	58
1974	58.488	111.832	13.707	18.212	58
1975	..	..	..	..	
1976	..	..	..	..	
1977	..	..	..	..	
1978	..	..	..	..	
1979	..	..	..	..	
1980	..	..	..	..	
1981	..	..	..	..	
1982	..	..	..	..	
1983	246.000	149.362	18.307	24.324	58
1984	197.800	101.384	12.427	16.511	232
1985	..	..	..	..	
1986	..	..	..	..	

1960–64: 1964   1965–69:

Personnel Comments:

In 1980 the following institutions were created: TARO, TALIRO, TAFORI, and TAFIRI. Each of these institutes took over the research facilities and programs in their respective fields, namely, agriculture (mainly crops), livestock, forestry, and fisheries, respectively. This research was formerly carried out by different Ministries and Divisions. In addition to these four newly created institutes, TPRI and UAC continued their activities, and some research was left at the Directorate of Agricultural Research within the Ministry of Agriculture. In the table below, an attempt has been made to classify the historical data according to the newly created institutes by adding a separate entry for TPRI, including UAC and the Directorate of Agricultural Research under TARO, and where possible, an identifiable portion was also allocated to TALIRO.

	1964	1965	1968	1970	1971	1974
Agricultural Research[a]/TARO[b]	61	25	45	59	73	101
Livestock Research/TALIRO[d]	19	13	14	23	28	NA
Forestry Research/TAFORI	5	4	5	6	[7]	NA
Fisheries Research/TAFIRI	4	[5]	6	5	7	7
Pesticides Research/TPRI[e]	18	12	17	17	12	NA
Total	107	59[c]	87	110	127	NA
Source	518	286	85	532	279	244

	1977	1979	1980	1982	1984	1985
Agricultural Research[a]/TARO[b]	136	147	167	172	204	189
Livestock Research/TALIRO[d]	42	NA	NA	27	56	NA
Forestry Research/TAFORI	[10]	NA	NA	20	12	20
Fisheries Research/TAFIRI	[8]	NA	NA	[11]	12	NA
Pesticides Research/TPRI[e]	[14]	NA	NA	17	21	35
Total	210	NA	NA	247	305	NA
Source	243	313	242	27	232	744

[ ] = estimated or constructed by authors.

a. Mainly crop research.

b. Inclusive Uyole Agricultural Center and Directorate of Agricultural Research.

c. Not included in the indicator series. The number of agricultural researchers seems to be underreported.

d. In those instances where it was possible to identify the livestock research component of UAC, it was included in this category. In all other cases, UAC was included in the agricultural research category.

e. The pre–1977 figures represent the East African Pesticides Research Institute. Although located in Tanzania, it had a mandate covering the whole East African community (Kenya, Tanzania, and Uganda). After the collapse of the community in 1977, its research institutes were nationalized by the respective countries. In this case, the procedure here was to include the pre–1977 data for "East African" institutes in the indicator series for the country where the institute was located.

Expenditure Comments:

The 1964 expenditure figure includes Crop, Livestock, Forestry, and Fisheries research expenditures and TPRI research expenditures. In certain cases the expenditure figures reported by source 518 only partially cover total expenditures, e.g., they exclude capital investments or salaries. Therefore, the reported figure underestimates total agricultural research expenditures.

434

Source 58 indicates that the agricultural research expenditures which it reports for the years 1970–74 include crop, livestock, and training aspects. In the appendix of source 58, a time series for the period 1975–1983 is given. In the copy available to this project, however, this appendix is not readable.

The 1983 expenditure figure includes the following research organizations: TARO, TALIRO, UAC, and TPRI. In addition, the 1984 expenditure figure also includes TAFORI and TAFIRI.

Acronyms and Abbreviations:
TAFIRI : Tanzania Fisheries Research Institute
TAFORI : Tanzania Forestry Research Institute
TALIRO : Tanzania Livestock Research Organization
TARO : Tanzania Agricultural Research Organization
TPRI : Tropical Pesticides Research Institute (formerly East African)
UAC : Uyole Agricultural Center

Sources:
0027 Harvey, Nigel, ed. *Agricultural Research Centres: A World Directory of Organizations and Programmes.* Seventh Edition, Two Volumes. Harlow, U.K.: Longman, 1983.
0058 Ndunguru, B. "The Impact of the Collaboration between the International Agricultural Research System and the National Agricultural Research System in Tanzania." CGIAR, Washington, D.C., September 1984. Mimeo.
0232 Malima, V.F., F.M. Shao, F. Turchi, and D. Sorrenti. *Agricultural Research Resource Assessment in the SADCC Countries, Volume II: Country Report Tanzania.* Washington, D.C.: SADCC and DEVRES, Inc., May 1985.
0243 Commonwealth Agricultural Bureaux (CAB). *List of Research Workers in the Agricultural Sciences in the Commonwealth and in the Republic of Ireland 1978.* Slough, England: CAB, 1978.
0279 Commonwealth Agricultural Bureaux (CAB). *List of Research Workers in the Agricultural Sciences in the Commonwealth and in the Republic of Ireland 1972.* Slough, England: CAB, 1972.
0285 Commonwealth Agricultural Bureaux (CAB). *List of Research Workers in the Agricultural Sciences in the Commonwealth and in the Republic of Ireland 1969.* Slough, England: CAB, 1969.
0518 East African Academy. *Research Services in East Africa.* Nairobi: East African Publishing House, 1966.
0532 UNESCO Field Science Office for Africa. *Survey on the Scientific and Technical Potential of the Countries of Africa.* Paris: UNESCO, 1970.

Additional References:
0002 SADCC, and DEVRES, Inc. *Agricultural Research Resource Assessment in the SADCC Countries, Volume I: Regional Analysis and Strategy.* Washington, D.C.: DEVRES, Inc., January 1985.
0010 Boyce, James K., and Robert E. Evenson. *National and International Agricultural Research and Extension Programs.* New York: Agricultural Development Council, Inc., 1975.
0014 Judd, M. Ann, James K. Boyce, and Robert E. Evenson. "Investing in Agricultural Supply." Economic Growth Center, Yale University, New Haven, Connecticut, 1983. Mimeo.
0016 Oram, Peter A., and Vishva Bindlish. *Resource Allocations to National Agricultural Research: Trends in the 1970s.* The Hague and Washington, D.C.: ISNAR and IFPRI, November 1981.
0019 Bengtsson, Bo, and Tedla Getachew, eds. *Strengthening National Agricultural Research.* Report from a SAREC workshop, 10–17 September 1979. Part 1: Background Documents; Part II: Summary and Conclusions. Sweden: SAREC, 1980.
0023 Bennell, Paul. *Agricultural Researchers in Sub-Saharan Africa: An Overview.* Working Paper No. 4. The Hague: ISNAR, October 1985.
0026 Oram, Peter A., and Mark Gieben. "Document Summaries." ISNAR, The Hague, 1984. Mimeo.
0073 Oram, Peter A., and Vishva Bindlish. "Investment in Agricultural Research in Developing Countries: Progress, Problems, and the Determination of Priorities." IFPRI, Washington, D.C., January 1984. Mimeo.
0095 FAO – CARIS. *Agricultural Research in Developing Countries – Volume 1: Research Institutions.* Rome: FAO – CARIS, 1978.
0135 FAO, and UNDP. *National Agricultural Research – Report of an Evaluation Study in Selected Countries.* Rome: FAO and UNDP, 1984.
0163 CGIAR. "National Agricultural Research." CGIAR, Washington, D.C., 1985. Mimeo.
0164 Association for the Advancement of Agricultural Sciences in Africa (AAASA). *Proceedings of the Workshop on Agricultural Research Administration, Nairobi, Kenya, 27–30 June 1977.* Proceedings Series PE-4. Addis Ababa, Ethiopia: AAASA and IDRC, August 1979.

0165 Evenson, R.E., and Y. Kislev. *Agricultural Research and Productivity*. New Haven: Yale University Press, 1975.

0175 Cooper, St.G.C. *Agricultural Research in Tropical Africa*. Kampala: East African Literature Bureau, 1970.

0227 Australian Centre for International Agricultural Research (ACIAR). *Proceedings of the Eastern Africa-ACIAR Consultation on Agricultural Research, 18–22 July 1983, Nairobi, Kenya*. Australia: ACIAR, 1984.

0231 Foote, R.J. and D. Rosjo. *A Survey of Five Agricultural Research Institutes in Tanzania, with Emphasis on Ways to Improve Management of the Research System*. Rural Economy Research Paper No. 7. Dar Es Salaam, and Mzumbe, Tanzania: University of Dar Es Salaam, and Institute of Development Management, May 1978.

0242 Commonwealth Agricultural Bureaux (CAB). *List of Research Workers in Agricultural Sciences in the Commonwealth 1981*. Slough, England: CAB, 1981.

0244 Commonwealth Agricultural Bureaux (CAB). *List of Research Workers in the Agricultural Sciences in the Commonwealth and in the Republic of Ireland 1975*. Slough, England: CAB, 1975.

0266 UNESCO. *National Science Policies in Africa*. Science Policy Studies and Documents No. 31. Paris: UNESCO, 1974.

0286 Commonwealth Agricultural Bureaux (CAB). *List of Research Workers in Agriculture, Animal Health and Forestry in the Commonwealth and in the Republic of Ireland 1966*. Slough, England: CAB, 1966.

0287 Commonwealth Agricultural Bureaux (CAB). *List of Research Workers in Agriculture, Animal Health and Forestry in the British Commonwealth, the Republic of Sudan and the Republic of Ireland 1959*. Slough, England: CAB, 1959.

0313 Liwenga, J.M. "Agricultural Research in Tanzania." In *Strengthening National Agricultural Research*. Report from a SAREC workshop, 10–17 September 1979, edited by Bo Bengtsson and Getachew Tedla, pp. 55–61. Sweden: SAREC, 1980.

0360 Cooper, St.G.C. "Towards Trained Manpower for Agricultural Research in Africa." Paper presented at the Conference on Agricultural Research and Production in Africa, organized by the Association for the Advancement of Agricultural Sciences in Africa (AAASA), Addis Ababa, 29 August–4 September 1971.

0385 SADCC, and DEVRES, Inc. *SADCC Region Agricultural Research Resource Assessment – Data Base Management Information System:*

Diskettes and User's Guide. Printouts. Washington, D.C.: SADCC, and DEVRES, Inc., n.d.

0445 Swanson, Burton E., and Wade H. Reeves. "Agricultural Research Eastern and Southern Africa: Manpower and Training." World Bank, Washington, D.C., August 1986. Mimeo.

0446 Kyomo, M.L. "Agricultural Research in Eastern and Southern Africa: Issues and Priorities." Southern African Centre for Cooperation in Agricultural Research of SADCC, Gabarone, Botswana, 1986. Mimeo.

0589 Kassapu, Samuel. *Les Dépenses de Recherche Agricole dans 34 Pays d'Afrique Tropicale*. Paris: Centre de Développement de l'OCDE, 1976.

0640 East African Agriculture and Forestry Research Organization. *Record of Research – Annual Report 1969*. Nairobi: East African Community, 1970.

0641 East African Veterinary Research Organization. *Record of Research – Annual Report 1968*. Kenya: East African Community, 1969.

0642 Keen, B.A. *The East African Agriculture and Forestry Research Organization – Its Origin and Objects*. Nairobi: East African Community, n.d.

0643 Binns, H.R. *The East African Veterinary Research Organisation – Its Development, Objects and Scientific Activities*. Nairobi: East African Community, n.d.

0653 Webster, B.N. *Index of Agricultural Research Institutions and Stations in Africa*. Rome: FAO, n.d.

0744 *Agricultural Research Centres: A World Directory of Organizations and Programmes*. Eighth Edition, Two Volumes. Harlow, U.K.: Longman, 1986.

0747 ISNAR. *International Workshop on Agricultural Research Management*. Report of a workshop. The Hague: ISNAR, 1987.

0749 Macha, Augustine M. "Priority Setting in the Tanzania Livestock Research Organization." Paper presented at the International Workshop on Agricultural Research Management, ISNAR, The Hague, 7–11 September 1987.

0764 Kasembe, J.R.N., A.M. Macha, and A.N. Mphuru. "Agricultural Research in Tanzania." In *Proceedings of the Eastern Africa-ACIAR Consultation on Agricultural Research*, pp. 44–56. Australia: ACIAR, 1984.

0770 Ruttan, Vernon W. *Agricultural Research Policy and Development*. FAO Research and Technology Paper 2. Rome: FAO, 1987.

0847 Kimura, J.H. "Financial and Administrative Management of Research Institutions in Eastern and Southern Africa: Report on Responses to a Questionnaire." In *Promotion of Technology Policy and Science Management in Africa*, edited by Karl Wolfgang Menck and Wolfgang Gmelin. Bonn: Deutsche Stiftung für internationale Entwicklung (DSE), 1986.

0852 Evenson, Robert E., and Yoav Kislev. *Investment in Agricultural Research and Extension: A Survey of International Data.* Center Discussion Paper No. 124. New Haven, Connecticut: Economic Growth Center, Yale University, August 1971.

0886 Evenson, Robert E., and Yoav Kislev. "Investment in Agricultural Research and Extension: A Survey of International Data." *Economic Development and Cultural Change* Vol. 23 (April 1975): 507–521.

Thailand

Personnel

Period/Year	PhD	MSc	BSc	Subtotal	Expat	Total	Source
1960–64	..	..	..	..	..	..	
1965–69	..	..	..	..	..	..	
1970	..	..	..	..	..	..	
1971	..	..	..	..	..	..	
1972	..	..	..	..	..	..	
1973	..	..	..	..	..	..	
1974	9	32	533	574	..	574	293
1975	9	33	554	596	..	596	293
1976	50	196	1034	1280	..	1280	293
1977	48	187	1085	1320	..	1320	293
1978	47	193	1101	1341	..	1341	293
1979	50	192	1187	1429	..	1429	293
1980	..	..	..	..	..	..	
1981	..	..	..	..	..	..	
1982	..	..	..	..	..	..	
1983	..	..	..	..	..	..	
1984	..	..	..	..	..	..	
1985	..	..	..	..	..	..	
1986	..	..	..	..	..	..	

1960–64:   1965–69:

Expenditure

LCU = Baht

Period/Year	Current LCU (millions)	Constant 1980 LCU (millions)	Constant 1980 US$ (millions)		Source
			Atlas	PPP	
1960–64	..	..	..	..	
1965–69	229.300	587.949	27.253	67.550	718
1970	281.900	726.546	33.677	83.473	718
1971	285.200	723.858	33.552	83.165	718
1972	247.900	579.206	26.847	66.545	718
1973	226.900	441.440	20.462	50.717	718
1974	202.700	331.751	15.377	38.115	293
1975	307.000	488.854	22.659	56.165	293
1976	313.500	480.092	22.253	55.158	293
1977	384.000	541.608	25.105	62.226	293
1978	397.700	516.494	23.940	59.340	293
1979	392.700	457.159	21.190	52.523	293
1980	555.700	555.700	25.758	63.845	332
1981	674.600	624.630	28.953	71.764	332
1982	767.500	687.724	31.877	79.013	332
1983	850.900	739.270	34.267	84.935	332
1984	..	..	..	..	
1985	..	..	..	..	
1986	..	..	..	..	

1960–64:   1965–69: 1969

Personnel Comments:

The collective statistical picture of agricultural research in Thailand, presented by the sources and references cited below, is exceptionally confusing. The indicator series attempts to report person-year (research) equivalents of professional personnel (i.e., BSc or higher) working in the Ministry of Agriculture and Cooperatives (MOAC) and the universities.

These appear to be the most comprehensive and accurate system-level figures available. Other references which do not use standard person-year equivalents include sources 14, 54, and 179. Furthermore, source 179 includes only researchers at the Department of Agriculture within MOAC and excludes researchers at the other departments within MOAC, i.e., the Department of Forests, the Department of Livestock, the Department of Fisheries, the Office of Agricultural Economics, and the Land Development Department. The system-level figures given in source 14 are virtually identical with the professional personnel figures given in source 179 (pp. 170–171) for the years in which there is a relevant overlap (1976 and 1980). Unfortunately, the expenditure figures reported in source 14 appear to be internally inconsistent with its personnel figures as the expenditure figures closely match the 1975 and 1977–80 MOAC figures reported in source 293, when both are converted to comparable constant LCUs. The figures reported in a more recent study (source 54) were not included here. Although source 54 gives agricultural research personnel figures for all the departments within MOAC, the universities, and several other ministries, they appear to substantially overestimate the level of scientific years (SYs) dedicated to agricultural research as, according to the source itself (p. 32), they include all government officials and permanent and temporary employees engaged under research projects.

Expenditure Comments:

The indicator series includes agricultural research expenditures by the Ministry of Agriculture and Cooperatives (MOAC), the universities, the Ministry of Industry, and the Ministry of Finance. The expenditures by the latter two ministries are rather small (5.2 million Baht in 1979). The personnel and expenditure indicators are therefore reasonably compatible.

The more recent figures given by source 54 were not reported. Like the personnel figures, they appear to be substantially biased upward.

They report commitments rather than actual expenditures; include wages and salaries of all researchers, some of whom may only work part-time on research; and do not account for the (end of fiscal year) transfer of funds to nonresearch activities.

Sources:

0293 Isarangkura, Rungruang. "Inventory of Agricultural Research Expenditure and Manpower in Thailand." In *Resource Allocation to Agricultural Research*. Proceedings of a workshop held in Singapore 8–10 June 1981, edited by Douglas Daniels and Barry Nestel, pp. 32–41. Ottawa: IDRC, 1981.

0332 Adulavidhaya, Kamphol, Rungruang Isarangkura, Preeyanuch Apibunyopas, and Nittaya Dulyasatit. "Evaluation in National Agricultural Research Systems in Thailand." Paper presented at the International Workshop on The Role of Evaluation in National Agricultural Research Systems, IFARD, and IDRC, Singapore, 7–9 July 1986.

0718 Adulavidhaya, Kamphol, Rungruang Isarangkura, Preeyanuch Apibunyopas, and Nittaya Dulyasatit. "Evaluation of Agriculutral Research in Thailand." In *Evaluation in National Agricultural Research*. Proceedings of a workshop held in Singapore, 7–9 July 1986, edited by Douglas Daniels, pp. 51–61. Canada: IDRC, 1987.

Additional References:

0010 Boyce, James K., and Robert E. Evenson. *National and International Agricultural Research and Extension Programs*. New York: Agricultural Development Council, Inc., 1975.

0014 Judd, M. Ann, James K. Boyce, and Robert E. Evenson. "Investing in Agricultural Supply." Economic Growth Center, Yale University, New Haven, Connecticut, 1983. Mimeo.

0016 Oram, Peter A., and Vishva Bindlish. *Resource Allocations to National Agricultural Research: Trends in the 1970s*. The Hague and Washington, D.C.: ISNAR and IFPRI, November 1981.

0017 ISNAR, IFARD & AOAD. Survey of National Agricultural Research Systems: Unpublished

Questionnaire Responses. ISNAR, The Hague, 1985.

0018 Daniels, Douglas, and Barry Nestel, eds. *Resource Allocation to Agricultural Research.* Proceedings of a workshop held in Singapore 8–10 June 1981. Ottawa: IDRC, 1981.

0026 Oram, Peter A., and Mark Gieben. "Document Summaries." ISNAR, The Hague, 1984. Mimeo.

0027 Harvey, Nigel, ed. *Agricultural Research Centres: A World Directory of Organizations and Programmes.* Seventh Edition, Two Volumes. Harlow, U.K.: Longman, 1983.

0054 Isarangkura, Rungruang. *Thailand and the CGIAR Centers: A Study of Their Collaboration in Agricultural Research.* CGIAR Study Paper Number 16. Washington, D.C.: World Bank, 1986.

0073 Oram, Peter A., and Vishva Bindlish. "Investment in Agricultural Research in Developing Countries: Progress, Problems, and the Determination of Priorities." IFPRI, Washington, D.C., January 1984. Mimeo.

0093 Bangladesh Agricultural Research Council (BARC), ISNAR, and Winrock International. *Management of Human Resources in Agricultural Research.* Report of the International Workshop on Management of Human Resources in Agricultural Research held March 3–5, 1986 in Dhaka, Bangladesh. Dhaka: BARC, ISNAR, and Winrock International, 1986.

0095 FAO – CARIS. *Agricultural Research in Developing Countries – Volume 1: Research Institutions.* Rome: FAO – CARIS, 1978.

0163 CGIAR. "National Agricultural Research." CGIAR, Washington, D.C., 1985. Mimeo.

0165 Evenson, R.E., and Y. Kislev. *Agricultural Research and Productivity.* New Haven: Yale University Press, 1975.

0179 The Philippine Council for Agriculture and Resources Research and Development (PCARRD), and IDRC. *Agriculture and Resources Research Manpower Development in South and Southeast Asia.* Proceedings of the Workshop on Agriculture and Resources Research Manpower Development in South and Southeast Asia, 21–23 October 1981 and 4–6 February 1982, Singapore. Los Baños, Philippines: PCARRD, 1983.

0272 Wongsiri, Tanongchit. "Human Resource Planning and Management." Paper presented at the Workshop on Improving Agricultural Research Organizations and Management: Progess and Issues for the Future, ISNAR, The Hague, 8–12 September 1986.

0320 Sabhasri, S. "Agricultural Research in Thailand." In *Strengthening National*

Agricultural Research. Report from a SAREC workshop, 10–17 September 1979, edited by Bo Bengtsson and Getachew Tedla, pp. 138–143. Sweden: SAREC, 1980.

0331 Vattanatangum, Anan. "Current Status and Activities in Human Resources Management of the Department of Agriculture, Thailand." In *Management of Human Resources in Agricultural Research.* Report of the International Workshop on Management of Human Resources in Agricultural Research held 3–5 March 1986, edited by Theodore Hutchcroft, pp. 124–131. Dhaka: Bangladesh Agricultural Research Council, ISNAR, and Winrock International, 1986.

0455 World Bank. "Thailand National Agricultural Research Project - Appraisal Project File." World Bank, Washington, D.C., November 1980. Mimeo.

0590 UNESCO. *Science and Technology in Countries of Asia and the Pacific.* Science Policy Studies and Documents No. 52. Paris: UNESCO, n.d.

0732 FAO – Regional Office for Asia and the Pacific. *Agricultural Research Systems in the Asia-Pacific Region.* Bangkok: FAO – Regional Office for Asia and the Pacific, 1986.

0792 Isarangkura, Rungruang. "Agricultural Research Manpower in Thailand." In *Agriculture and Resources Research Manpower Development in South and Southeast Asia,* pp. 149–164. Los Baños, Philippines: Philippine Council for Agriculture and Resources Research and Development, 1983.

0793 Tongpan, Sopin. "Agricultural Research Resources in Thailand." In *Agriculture and Resources Research Manpower Development in South and Southeast Asia,* pp. 165–180. Los Baños, Philippines: Philippine Council for Agriculture and Resources Research and Development, 1983.

0852 Evenson, Robert E., and Yoav Kislev. *Investment in Agricultural Research and Extension: A Survey of International Data.* Center Discussion Paper No. 124. New Haven, Connecticut: Economic Growth Center, Yale University, August 1971.

0855 ISNAR. "Working Papers prepared for the Committee on Conditions of Service in the Department of Agriculture (Royal Thai Government)." ISNAR, The Hague, n.d. Mimeo.

0886 Evenson, Robert E., and Yoav Kislev. "Investment in Agricultural Research and Extension: A Survey of International Data." *Economic Development and Cultural Change* Vol. 23 (April 1975): 507–521.

Togo

Personnel

Period/Year	PhD	MSc	BSc	Subtotal	Expat	Total	Source
1960–64	..	..	..	..	..	..	
1965–69	..	..	..	..	..	5	653
1970	..	..	..	..	..	13	532
1971	..	..	..	..	..	..	
1972	..	..	..	..	..	18	431
1973	..	..	..	..	..	..	
1974	..	..	..	..	..	..	
1975	..	..	..	..	..	..	
1976	..	..	..	..	..	..	
1977	..	..	..	..	..	..	
1978	..	..	..	..	..	..	
1979	..	..	..	..	..	..	
1980	..	..	..	30	9	39	86/275
1981	..	..	..	..	..	..	
1982	..	..	..	..	..	..	
1983	..	..	..	..	..	..	
1984	..	..	..	44	14	58	86/275
1985	..	..	..	..	..	..	
1986	..	..	..	..	..	..	

1960–64:   1965–69: 1966

Expenditure

LCU = CFA Francs

Period/Year	Current LCU (millions)	Constant 1980 LCU (millions)	Constant 1980 US$ (millions)		Source
			Atlas	PPP	
1960–64	..	..	..	..	
1965–69	..	..	..	..	
1970	..	..	..	..	
1971	..	..	..	..	
1972	141.040	351.721	1.662	2.372	431
1973	..	..	..	..	
1974	..	..	..	..	
1975	..	..	..	..	
1976	..	..	..	..	
1977	..	..	..	..	
1978	..	..	..	..	
1979	..	..	..	..	
1980	1049.290	1049.290	4.960	7.075	86
1981	..	..	..	..	
1982	..	..	..	..	
1983	1183.160	869.971	4.112	5.866	86
1984	1166.860	827.560	3.912	5.580	86
1985	..	..	..	..	
1986	..	..	..	..	

1960–64:   1965–69:

Personnel Comments:

The table below gives a breakdown of researchers at the institutional level.

	1966	1970	1972	1980	1984
IRCT	3	2	3	6	6
IRCC	–	2	2	2	6
IRAT	2	1	2	3	2
CREAT	–	–	–	6	7
DPV	–	4	[4]	3	8
DRA	–	0	[0]	6	11
INPT	–	–	–	NA	NA
DNTA	–	–	–	4	4
DEPEG	–	4	7	9	14
Total	5	13	18	39	58
Source	653	532	431	86	86

[ ] = estimated or constructed by authors.

'Office de la Recherche Scientifique et Technique Outre-Mer' (ORSTOM), 'Bureau d'Etudes Forestières' (BEF), and 'Direction de la Météreologie Nationale' (DMN) are not included. Other sources, that also exclude ORSTOM, give a considerably higher number of researchers. This difference is probably caused by a more comprehensive definition of researcher than was used to construct this table.

Expenditure Comments:

Total agricultural research expenditures (current LCUs) were constructed as follows:

	1971	1972	1980	1983	1984
IRCT	NA	52.074	90.500	170.800	112.100
IRCC	16.030	40.975	199.170	219.840	230.360
IRAT	NA	14.991	299.840	136.300	91.830
CREAT	–	–	126.250[a]	161.220[a]	187.850[a]
DPV	NA	NA	52.000[a]	[93.690][a]	107.580[a]
DRA	NA	NA	110.250	[183.840]	208.370
INPT	–	–	29.870	43.020	43.320
DNTA	–	–	42.470	[52.900]	56.370
DEPEG	33.000	[33.000]	98.940	[121.550]	129.080
Total	NA	141.040	1049.290	1183.160	1166.860
Source	589	431	86	86	86

[ ] = estimated or constructed by authors.
a. Salaries of expatriates not included.

Acronyms and Abbreviations:

BEF: Bureau d'Etudes Forestières
CREAT: Centre de Recherche en Elevage d'Avetonou
DEPEG: Direction des Etudes Pédologiques et de l'Ecologie Générale
DMN: Direction de la Météreologie Nationale

DNTA : Direction Nationale de Technologie Alimentaire
DPV: Direction de la Protection des Végétaux
DRA: Direction de la Recherche Agronomique
INPT: Institut des Plantes à Tubercules
IRAT: Institut de Recherche en Agronomie Tropicale

IRCC : Institut de Recherche sur le Café et Cacao
IRCT: Institut de Recherche sur le Coton et les Fibres Textiles
ORSTOM : Office de la Recherche Scientifique et Technique Outre-Mer

Sources:

0086 World Bank. "West Africa Agricultural Research Review – Country Studies." World Bank, Washington, D.C., 1985. Mimeo.

0275 Aithnard, T., and C.T. Nguyen-Vu. "La Recherche Agricole au Togo: Bilan et Perspectives." Paper presented at the Workshop on Improving Agricultural Research Organizations and Management: Progress and Issues for the Future, ISNAR, The Hague, 8–12 September 1986.

0431 Current Agricultural Research Information System (CARIS). *Directory of Agricultural Research Institutions and Projects in West Africa – Pilot Project 1972–73.* Rome: FAO, 1973.

0532 UNESCO Field Science Office for Africa. *Survey on the Scientific and Technical Potential of the Countries of Africa.* Paris: UNESCO, 1970.

0653 Webster, B.N. *Index of Agricultural Research Institutions and Stations in Africa.* Rome: FAO, n.d.

Additional References:

0010 Boyce, James K., and Robert E. Evenson. *National and International Agricultural Research and Extension Programs.* New York: Agricultural Development Council, Inc., 1975.

0016 Oram, Peter A., and Vishva Bindlish. *Resource Allocations to National Agricultural Research: Trends in the 1970s.* The Hague and Washington, D.C.: ISNAR and IFPRI, November 1981.

0022 UNESCO. *Statistical Yearbook 1983.* Paris: UNESCO, 1983.

0023 Bennell, Paul. *Agricultural Researchers in Sub-Saharan Africa: An Overview.* Working Paper No. 4. The Hague: ISNAR, October 1985.

0027 Harvey, Nigel, ed. *Agricultural Research Centres: A World Directory of Organizations and Programmes.* Seventh Edition, Two Volumes. Harlow, U.K.: Longman, 1983.

0068 Conte, Stephan. "West Africa Agricultural Research Data Base Building Task: Termination Report." World Bank, Washington, D.C., 1985. Mimeo.

0073 Oram, Peter A., and Vishva Bindlish. "Investment in Agricultural Research in Developing Countries: Progress, Problems, and the Determination of Priorities." IFPRI, Washington, D.C., January 1984. Mimeo.

0088 Swanson, Burton E., and Wade H. Reeves. "Agricultural Research West Africa: Manpower and Training." World Bank, Washington, D.C., November 1985. Mimeo.

0089 Oram, Peter. "Report on National Agricultural Research in West Africa." IFPRI, Washington, D.C., February 1986. Mimeo.

0163 CGIAR. "National Agricultural Research." CGIAR, Washington, D.C., 1985. Mimeo.

0175 Cooper, St.G.C. *Agricultural Research in Tropical Africa.* Kampala: East African Literature Bureau, 1970.

0266 UNESCO. *National Science Policies in Africa.* Science Policy Studies and Documents No. 31. Paris: UNESCO, 1974.

0360 Cooper, St.G.C. "Towards Trained Manpower for Agricultural Research in Africa." Paper presented at the Conference on Agricultural Research and Production in Africa, organized by the Association for the Advancement of Agricultural Sciences in Africa (AAASA), Addis Ababa, 29 August–4 September 1971.

0373 Groupement d'Etudes et de Recherches pour le Développement de l'Agronomie Tropicale (GERDAT). *Rapport Général d'Activité pour 1982.* Paris: GERDAT, 1983.

0400 UNESCO. *The Promotion of Scientific Activity in Tropical Africa.* Science Policy Studies and Documents No. 11. Paris: UNESCO, 1969.

0466 World Bank – West Africa Agricultural Research Review Team. "National Agricultural Research Systems (NARS): A Regional Appraisal and Selected Issues." World Bank, Washington, D.C., August 1986. Mimeo.

0589 Kassapu, Samuel. *Les Dépenses de Recherche Agricole dans 34 Pays d'Afrique Tropicale.* Paris: Centre de Développement de l'OCDE, 1976.

0654 Office de la Recherche Scientifique et Technique Outre-Mer (ORSTOM). *Rapport d'Activité 1967.* Paris: ORSTOM, 1969.

0852 Evenson, Robert E., and Yoav Kislev. *Investment in Agricultural Research and Extension: A Survey of International Data.* Center Discussion Paper No. 124. New Haven, Connecticut: Economic Growth Center, Yale University, August 1971.

Tonga

Personnel

Period/Year	PhD	MSc	BSc	Subtotal	Expat	Total	Source
1960–64	..	..	..	..	..	..	
1965–69	..	..	..	..	..	..	
1970	..	..	..	..	..	1	723
1971	..	..	..	..	..	2	723
1972	..	..	..	1	4	5	723
1973	..	..	..	..	..	6	723
1974	..	..	..	..	..	4	723
1975	..	..	..	..	..	5	723
1976	..	..	..	..	..	5	723
1977	..	..	..	..	..	6	723
1978	..	..	..	..	..	5	723
1979	..	..	..	..	..	5	723
1980	..	..	..	..	..	9	723
1981	..	..	..	4	4	8	38
1982	..	..	..	..	..	12	723
1983	0	1	4	5	8	13	723
1984	..	..	..	..	..	15	723
1985	..	..	..	..	..	17	723
1986	0	3	6	9	12	21	723

1960–64:   1965–69:

Expenditure   *LCU = Tongan Pa'anga*

Period/Year	Current LCU (millions)	Constant 1980 LCU (millions)	Constant 1980 US$ (millions)		Source
			Atlas	PPP	
1960–64	..	..	..	..	
1965–69	..	..	..	..	
1970	..	..	..	..	
1971	..	..	..	..	
1972	..	..	..	..	
1973	0.028	0.059	0.065	0.097	723
1974	0.033	0.061	0.067	0.099	723
1975	0.040	0.067	0.073	0.109	723
1976	0.054	0.092	0.101	0.150	723
1977	0.073	0.104	0.114	0.170	723
1978	0.200	0.246	0.270	0.402	723
1979	0.340	0.388	0.425	0.634	723
1980	0.280	0.280	0.306	0.457	723
1981	0.280	0.254	0.278	0.414	723
1982	0.220	0.191	0.209	0.311	723
1983	0.315	0.267	0.292	0.436	723
1984	0.281	..	..	..	723
1985	0.317	..	..	..	723
1986	0.422	..	..	..	723

1960–64:   1965–69:

Personnel Comments:

The indicator series refers to the Research Division of the Ministry of Agriculture, Fisheries and Forests. According to source 723, this division only undertakes crop and livestock research.

Expenditure Comments:

The expenditure indicators are commensurable with the personnel series. Source 723 indicates that the salaries of expatriates are not included in the expenditure figures, but capital investments paid with foreign donor money are.

Sources:

0038 Gamble, W.K., R.M. Bourke, and C.W. Brookson. *South Pacific Agricultural Research Study.* Consultants' Report to the Asian Development Bank (two volumes). Manila: Asian Development Bank, June 1981.

0723 Fernando, L.H. "South Pacific NARS Resource Commitment Survey." ADB and ISNAR, The Hague, 1987. Mimeo.

Additional References:

0016 Oram, Peter A., and Vishva Bindlish. *Resource Allocations to National Agricultural Research: Trends in the 1970s.* The Hague and Washington, D.C.: ISNAR and IFPRI, November 1981.

0163 CGIAR. "National Agricultural Research." CGIAR, Washington, D.C., 1985. Mimeo.

0771 Fernando, L.H. "ISNAR/IRETA Workshop on Planning and Management of Agricultural Research in the South Pacific: Survey of Agricultural Research Resources." ISNAR, The Hague, n.d. Mimeo.

Trinidad & Tobago

Personnel

Period/Year	PhD	MSc	BSc	Subtotal	Expat	Total	Source
1960–64	..	..	..		..	..	
1965–69	6	14	4	24	..	24	286/285
1970	..	..	..		..	..	
1971	6	19	9	34	..	34	279
1972	..	..	..	..	..		
1973	..	..	..	..	..		
1974	4	15	15	34	..	34	244
1975	..	..	..	..	..		
1976	..	..	..	..	..		
1977	6	16	23	45	..	45	243
1978	..	..	..	..	..		
1979	..	..	..	..	..		
1980	7	21	20	48	..	48	242
1981	..	..	..	..	..		
1982	..	..	..	..	..		
1983	6	20	32	58	..	58	115
1984	..	..	..	..	..		
1985	..	..	..	..	..		
1986	..	..	..	..	..		

1960–64:   1965–69: 1965 & 68

Expenditure   *LCU = Trinidad & Tobago Dollars*

Period/Year	Current LCU (millions)	Constant 1980 LCU (millions)	Constant 1980 US$ (millions)		Source
			Atlas	PPP	
1960–64	..	..	..		
1965–69	..	..	..		
1970	..	..	..		
1971	..	..	..		
1972	..	..	..		
1973	..	..	..	..	
1974	..	..	..	..	
1975	..	..	..	..	
1976	..	..	..	..	
1977	4.800	7.488	2.724	3.694	219
1978	..	..	..	..	
1979	..	..	..	..	
1980	..	..	..	..	
1981	..	..	..	..	
1982	..	..	..	..	
1983	..	..	..	..	
1984	..	..	..	..	
1985	..	..	..	..	
1986	..	..	..	..	

1960–64:   1965–69:

Personnel Comments:

These figures refer to agricultural research under the aegis of the Ministry of Agriculture only. However, two other important agricultural research organizations can be identified. First is the University of West Indies (UWI). This university has a strong Faculty of Agriculture with a higher number of PhD's than the Ministry of Agriculture (e.g., 33 of the 44 academic staff reported by source 363 in 1980 held a PhD degree). The other important agricultural research organization is CARDI, which has its headquarters at the campus of UWI. CARDI was created in 1975 out of the Regional Research

	1965	1968	1971	1974	1975	1977
Ministry of Agriculture	27					
Crop Research Division		13	22	21		23
Animal Health Division		1	3	2		2
Animal Prod.& Res.Div.		2	4	7		6
Forestry Division		1	1	3		6
Fisheries Division			3			7
Other		4	1	1		1
Subtotal	27	21	34	34	NA	45
Caroni Research Inst.			5			
UWI	49	50	58			
CARDI					18	
CARIRI			7			
IMA						
Source	286	285	279	244	342	243

	1978	1980	1981	1982	1983	1984
Ministry of Agriculture	40					
Crop Research Division		25		28	41**	
Animal Health Division		5				
Animal Prod.& Res.Div.		6			9**	
Forestry Division		8		1*		
Fisheries Division						
Other		4				
Subtotal	40	48	NA	NA	58*	NA
Caroni Research Inst.	7	10			10*	
UWI	48	44**	46*	40*		
CARDI	16*	18*	21		20	16
CARIRI		4				
IMA		3				
Source	325	242	342	361	355	354
	341*	343*	362*	27*	115*	
		363**			224**	

Centre, which until then was an integral part of the UWI. At the same time a decision was made to decentralize CARDI and accordingly units have been established in Belize, Jamaica, St Kitts/Nevis, St Lucia, Barbados, and Guyana in addition to Trinidad and Tobago.

In addition to these major agricultural research organizations, the following small institutes can be identified: the Caroni Research Station (10 researchers in 1983, source 115) and the Sugar Feed Centre – both under the Ministry of State Enterprises – the Commonwealth Institute of Biological Control, CARIRI (4 agricultural researchers in 1980, source 242), and IMA (3 researchers in 1980, source 242).

The ISNAR, IFARD & AOAD questionnaire response (source 17), which includes the Ministry of Agriculture, CARDI, UWI, and most of the smaller institutes, gives a total of 103 researchers of which 31 held a PhD, 35 an MSc, and 17 a BSc, plus 20 expatriates for 1983.

Expenditure Comments:
The 1977 expenditure figure is, according to source 219, a conservative estimate of the expenditures on agricultural research by the Ministry of Agriculture. For 1983, the ISNAR, IFARD & AOAD questionnaire response (source 17) gives a total of TT$ 37,550,000, which includes the Ministry of Agriculture, the University of the West Indies, CARDI, and some smaller research institutes. On the basis of the information provided by source 17, it was impossible to isolate the Ministry of Agriculture.

Acronyms and Abbreviations:
CARDI: Caribbean Agricultural Research and Development Institute
CARIRI: Caribbean Industrial Research Institute
IMA: Institute of Marine Affairs
UWI: University of West Indies

Sources:
0115 Economic Commission for Latin America and the Caribbean (ECLAC). *Agricultural Research Policy and Management – Volumes I and II.* Papers presented at the Workshop on Agricultural Research Policy and Management, 26–30 September 1983, Port of Spain, Trinidad. Port of Spain, Trinidad: ECLAC, November 1984.

0219 Department of Agriculture, Natural Resources and Rural Development (DARNDR), and IICA. "Réunion Technique Régionale sur les Systèmes de Recherche Agricole dans les Antilles." DARNDR, and IICA, Port-au-Prince, Haiti, November/December 1977. Mimeo.
0242 Commonwealth Agricultural Bureaux (CAB). *List of Research Workers in Agricultural Sciences in the Commonwealth 1981.* Slough, England: CAB, 1981.
0243 Commonwealth Agricultural Bureaux (CAB). *List of Research Workers in the Agricultural Sciences in the Commonwealth and in the Republic of Ireland 1978.* Slough, England: CAB, 1978.
0244 Commonwealth Agricultural Bureaux (CAB). *List of Research Workers in the Agricultural Sciences in the Commonwealth and in the Republic of Ireland 1975.* Slough, England: CAB, 1975.
0279 Commonwealth Agricultural Bureaux (CAB). *List of Research Workers in the Agricultural Sciences in the Commonwealth and in the Republic of Ireland 1972.* Slough, England: CAB, 1972.
0285 Commonwealth Agricultural Bureaux (CAB). *List of Research Workers in the Agricultural Sciences in the Commonwealth and in the Republic of Ireland 1969.* Slough, England: CAB, 1969.
0286 Commonwealth Agricultural Bureaux (CAB). *List of Research Workers in Agriculture, Animal Health and Forestry in the Commonwealth and in the Republic of Ireland 1966.* Slough, England: CAB, 1966.

Additional References:
0010 Boyce, James K., and Robert E. Evenson. *National and International Agricultural Research and Extension Programs.* New York: Agricultural Development Council, Inc., 1975.
0014 Judd, M. Ann, James K. Boyce, and Robert E. Evenson. "Investing in Agricultural Supply." Economic Growth Center, Yale University, New Haven, Connecticut, 1983. Mimeo.
0016 Oram, Peter A., and Vishva Bindlish. *Resource Allocations to National Agricultural Research: Trends in the 1970s.* The Hague and Washington, D.C.: ISNAR and IFPRI, November 1981.
0027 Harvey, Nigel, ed. *Agricultural Research Centres: A World Directory of Organizations and Programmes.* Seventh Edition, Two Volumes. Harlow, U.K.: Longman, 1983.
0053 Piñeiro, Martín, and Eduardo Trigo, eds. *Cambio Técnico en el Agro Latino Americano – Situación y*

Perspectivas en la Década de 1980. San José: IICA, 1983.

0063   ISNAR. *Report to the Board of Directors of CARDI: Analysis, Evaluation and Proposals to Strengthen CARDI's Regional Capacity*. The Hague: ISNAR, 1985.

0163   CGIAR. "National Agricultural Research." CGIAR, Washington, D.C., 1985. Mimeo.

0172   Trigo, Eduardo. Background material. ISNAR, The Hague, n.d.

0224   Ministry of Agriculture, Livestock, and Food Production. *Research Division – Annual Report 1983*. Trinidad and Tobago: Ministry of Agriculture, Livestock, and Food Production, 1983.

0287   Commonwealth Agricultural Bureaux (CAB). *List of Research Workers in Agriculture, Animal Health and Forestry in the British Commonwealth, the Republic of Sudan and the Republic of Ireland 1959*. Slough, England: CAB, 1959.

0325   Ahmad, N. "Agricultural Research in Trinidad and Tobago." In *Strengthening National Agricultural Research*. Report from a SAREC workshop, 10–17 September 1979, edited by Bo Bengtsson and Getachew Tedla, pp. 182–197. Sweden: SAREC, 1980.

0341   Caribbean Agricultural Research and Development Institute (CARDI). *Annual Report 1977–1978*. St. Augustine, Trinidad & Tobago: CARDI, November 1978.

0342   Caribbean Agricultural Research and Development Institute (CARDI). *CARDI 1976–1981: Report of the Chairman*. St. Augustine, Trinidad & Tobago: CARDI, 1982.

0343   Caribbean Agricultural Research and Development Institute (CARDI). *Annual Report 1980*. St. Augustine, Trinidad & Tobago: CARDI, 1981.

0353   Caribbean Agricultural Research and Development Institute (CARDI). *Paper Presented at a Meeting of the Governing Body of CARDI on January 29, 1982, Basseterre, St. Kitts*. St. Augustine, Trinidad: CARDI, 1982.

0354   Caribbean Agricultural Research and Development Institute (CARDI). *Annual Report Research and Development 1983–84: Highlights*. St. Augustine, Trinidad: CARDI, 1984.

0355   Caribbean Agricultural Research and Development Institute (CARDI). *Research and Development Summary 1982–83*. St. Augustine, Trinidad: CARDI, 1983.

0356   Caribbean Agricultural Research and Development Institute (CARDI). *Research and Development Summary 1983–84*. St. Augustine, Trinidad: CARDI, 1984.

0361   Central Experiment Station. *Research Division – Crop Research: Annual Report 1982 (summary)*. Trinidad & Tobago: Central Experiment Station, 1982.

0362   University of the West Indies – Faculty of Agriculture. *Information Booklet for Academic Year 1981–1982*. St. Augustine, Trinidad & Tobago: University of the West Indies – Faculty of Agriculture, September 1981.

0363   University of the West Indies – Faculty of Agriculture. *A Compendium of Research*. St. Augustine, Trinidad & Tobago: University of the West Indies – Faculty of Agriculture, January 1980.

0364   Caribbean Industrial Research Institute (CARIRI). *CARIRI 1983 Annual Report*. Trinidad & Tobago: CARIRI, 1984.

0744   *Agricultural Research Centres: A World Directory of Organizations and Programmes*. Eighth Edition, Two Volumes. Harlow, U.K.: Longman, 1986.

Tunisia

Personnel

Period/Year	PhD	MSc	BSc	Subtotal	Expat	Total	Source
1960–64	..	..	..	..	..	39	852
1965–69	..	..	..	..	..	..	
1970	..	..	..	..	..	101	532
1971	..	..	..	56	56	112	647
1972	..	..	..	..	..	114	266
1973	..	..	..	..	..	..	
1974	..	..	..	..	..	..	
1975	..	..	..	..	..	..	
1976	..	..	..	71	17	88	647
1977	..	..	..	..	..	..	
1978	..	..	..	..	..	..	
1979	..	..	..	..	..	..	
1980	..	..	..	91	2	93	647
1981	..	..	..	..	..	94	419
1982	..	..	..	..	..	106	419
1983	..	..	..	..	..	120	419
1984	..	..	..	..	..	141	419
1985	..	..	..	129	13	142	647
1986	..	..	..	..	..	..	

1960–64: 1963   1965–69:

Expenditure

LCU = Tunisian Dinars

Period/Year	Current LCU (millions)	Constant 1980 LCU (millions)	Constant 1980 US$ (millions)		Source
			Atlas	PPP	
1960–64	..	..	..	..	
1965–69	..	..	..	..	
1970	..	..	..	..	
1971	..	..	..	..	
1972	1.403	3.037	7.213	10.136	266
1973	..	..	..	..	
1974	..	..	..	..	
1975	..	..	..	..	
1976	..	..	..	..	
1977	..	..	..	..	
1978	..	..	..	..	
1979	..	..	..	..	
1980	..	..	..	..	
1981	..	..	..	..	
1982	..	..	..	..	
1983	..	..	..	..	
1984	..	..	..	..	
1985	7.000	4.324	10.720	14.431	647
1986	..	..	..	..	

1960–64:   1965–69:

Personnel Comments:

The table below gives an overview of how the indicator series has been constructed. Quite a number of institutes have not been included in the series because of limited statistical data. Some of the institutes have also been excluded because they are not exclusively research oriented. Many also have education or service roles, which makes their data on the number of researchers rather difficult to interpret.

	1970	1971	1976	1980	1981	1982	1983	1984	1985
INRAT	65	64	55	48	49	59	63	76	72
CRGR[a]	19	20	9	11	11	16	17	16	18
INRF	17	28	24	16	16	14	13	14	15
IRA[b]	–	–	–	18	18	17	17	20	22
IO[c]	–	–	–	–	–	–	10	15	15
Subtotal	101	112	88	93	94	106	120	141	142
INSTOP	6		12*				20*		
IRVT								30*	27
DS	19		11			20*	18*		24
BIRH/DRE	10		10*				31*		
INAT	25						54*		
ORSTOM	7		13*				11*		
Source	532	647	647	647	419	419	419	419	647
			95*			20*	255*	648*	

a. The 1970 figures include CRUESI, and the 1971 figures include CRUESI and CATID. According to source 647, these institutes were attached to CRGR in 1972.
b. Created in 1976.
c. Created in 1982.

Expenditure Comments:

The 1972 and 1985 expenditure figures correspond with the respective personnel figures. The 1985 figure is an estimate of the total research expenditures of INRAT, CRGR, INRF, IO, and IRA. When other institutes like IRVT, DS, and universities are included, total agricultural research expenditures in 1985 are estimated (by source 647) at 12.1 – 13.0 million Dinars.

Acronyms and Abbreviations:

BIRH/DRE : Bureau de l'Inventaire et des Recherches Hydrologiques – Division des Ressources en Eau
CATID: Centre d'Amélioration des Techniques d'Irrigation et de Drainage
CRGR: Centre de Recherche du Génie Rural
CRUESI: Centre de Recherches pour l'Utilisation des Eaux Salées en Irrigation
DS: Division des Sols
INAT: Institut National Agronomique de Tunis (AgriculturalUniversity)

INRAT: Institut National de la Recherche Agronomique de Tunisie
INRF: Institut National de Recherches Forestières
INSTOP: Institut National Scientifique et Technique d'Océanographie et de Pêche
IO: Institut de l'Olivier
IRA: Institut des Régions Arides
IRVT: Institut de la Recherche Vétérinaire de Tunisie
ORSTOM: Office de la Recherche Scientifique et Technique d'Outre-Mer

Sources:

0266 UNESCO. *National Science Policies in Africa.* Science Policy Studies and Documents No. 31. Paris: UNESCO, 1974.

0419 Lasram, Mustapha. "Rapport sur le Système National de la Recherche Agricole en Tunisie." Paper presented at the Séminaire sur l'Orientation et l'Organisation de la Recherche Agronomique dans les Pays du Bassin Méditerranéen, Centre International de Hautes Etudes Agronomiques Méditerranéennes, Istanbul, 1–3 December 1986.

0532 UNESCO Field Science Office for Africa. *Survey on the Scientific and Technical Potential of the Countries of Africa.* Paris: UNESCO, 1970.

0647 ISNAR. *Programme de Développement de la Recherche Agricole en Tunisie.* Three volumes. The Hague: ISNAR, May 1987.

0852 Evenson, Robert E., and Yoav Kislev. *Investment in Agricultural Research and Extension: A Survey of International Data.* Center Discussion Paper No. 124. New Haven, Connecticut: Economic Growth Center, Yale University, August 1971.

Additional References:

0010 Boyce, James K., and Robert E. Evenson. *National and International Agricultural Research and Extension Programs.* New York: Agricultural Development Council, Inc., 1975.

0014 Judd, M. Ann, James K. Boyce, and Robert E. Evenson. "Investing in Agricultural Supply." Economic Growth Center, Yale University, New Haven, Connecticut, 1983. Mimeo.

0016 Oram, Peter A., and Vishva Bindlish. *Resource Allocations to National Agricultural Research: Trends in the 1970s.* The Hague and Washington, D.C.: ISNAR and IFPRI, November 1981.

0026 Oram, Peter A., and Mark Gieben. "Document Summaries." ISNAR, The Hague, 1984. Mimeo.

0027 Harvey, Nigel, ed. *Agricultural Research Centres: A World Directory of Organizations and Programmes.* Seventh Edition, Two Volumes. Harlow, U.K.: Longman, 1983.

0073 Oram, Peter A., and Vishva Bindlish. "Investment in Agricultural Research in Developing Countries: Progress, Problems, and the Determination of Priorities." IFPRI, Washington, D.C, January 1984. Mimeo.

0094 FAO – Near East Regional Office. *Directory of Agricultural Research Institutions in the Near East Region.* Cairo: FAO – Near East Regional Office, 1979.

0095 FAO – CARIS. *Agricultural Research in Developing Countries – Volume 1: Research Institutions.* Rome: FAO – CARIS, 1978.

0163 CGIAR. "National Agricultural Research." CGIAR, Washington, D.C., 1985. Mimeo.

0175 Cooper, St.G.C. *Agricultural Research in Tropical Africa.* Kampala: East African Literature Bureau, 1970.

0255 Centre National de Documentation Agricole (CNDA). *La Recherche en Cours.* Tunis: CNDA, September 1983.

0360 Cooper, St.G.C. "Towards Trained Manpower for Agricultural Research in Africa." Paper presented at the Conference on Agricultural Research and Production in Africa, organized by the Association for the Advancement of Agricultural Sciences in Africa (AAASA), Addis Ababa, 29 August–4 September 1971.

0407 Centre International de Hautes Etudes Agronomiques Méditerranéennes. "Séminaire sur l'Orientation et l'Organisation de la Recherche Agronomique dans les Pays du Bassin Méditerranéen, Istanbul, 1–3 December 1986." Centre International de Hautes Etudes Agronomiques Méditerranéennes, Paris, December 1986. Mimeo.

0422 Casas, Joseph. "Les Ressources Humaines et Financières dans les Pays du Bassin Méditerranéen." Preliminary table. ISNAR, The Hague, 1986. Mimeo.

0586 UNESCO – Division of Statistics on Science and Technology, Office of Statistics. *Human and Financial Resources for Research and Experimental Development in Agriculture.* Paris: UNESCO, July 1983.

0648 Institut de la Recherche Vétérinaire de Tunisie. "Rapport sur l'Activite de l'Institut de la Recherche Vétérinaire de Tunisie pour l'Annee 1984." Institut de la Recherche Vétérinaire de Tunisie, La Rabta, Tunis, n.d. Mimeo.

0653 Webster, B.N. *Index of Agricultural Research Institutions and Stations in Africa.* Rome: FAO, n.d.

0654 Office de la Recherche Scientifique et Technique Outre-Mer (ORSTOM). *Rapport d'Activité 1967.* Paris: ORSTOM, 1969.

0865 Oram, Peter. "Agricultural Research Objectives and Priorities: Constraints to the Development of Agricultural Research Institutions in Arab countries." Review paper prepared for the first meeting of the Council for Arab Agricultural Research, organized by the Arab Fund for Economic and Social Development, Kuwait, 23 March 1988.

Turkey

Personnel

Period/Year	PhD	MSc	BSc	Subtotal	Expat	Total	Source
1960–64	..	..	..	..	..	..	
1965–69	..	..	..	..	..	397	886
1970	..	..	..	..	..	479	852
1971	..	..	..	..	..	..	
1972	..	..	..	..	..	..	
1973	..	..	..	..	..	..	
1974	..	..	..	..	..	..	
1975	..	..	..	..	..	..	
1976	..	..	..	..	..	..	
1977	..	..	..	..	..	..	
1978	..	..	..	..	..	..	
1979	..	..	..	..	..	..	
1980	..	..	..	..	..	..	
1981	..	..	..	..	..	..	
1982	..	..	..	..	..	1424	27
1983	..	..	..	..	..	..	
1984	..	..	..	..	..	..	
1985	..	..	..	..	..	1800	422
1986	..	..	..	..	..	..	

1960–64:   1965–69: 1965

Expenditure   *LCU = Turkey Liras*

Period/Year	Current LCU (millions)	Constant 1980 LCU (millions)	Constant 1980 US$ (millions)		Source
			Atlas	PPP	
1960–64	..	..	..	..	
1965–69	31.500	2788.466	11.584	18.758	886
1970	111.815	2021.623	29.701	48.095	852
1971	127.370	1953.436	28.699	46.473	10
1972	..	..	..	..	
1973	204.000	2203.615	32.375	52.425	601
1974	318.000	2688.119	39.493	63.951	601
1975	394.800	2869.872	42.163	68.275	601
1976	470.100	2903.213	42.653	69.069	601
1977	970.800	4813.408	70.717	114.513	601
1978	1029.000	3551.472	52.177	84.491	601
1979	1398.800	2828.785	41.559	67.298	601
1980	1930.800	1930.800	28.367	45.934	601
1981	3140.600	2211.690	32.493	52.617	26
1982	..	..	..	..	
1983	..	..	..	..	
1984	..	..	..	..	
1985	..	..	..	..	
1986	..	..	..	..	

1960–64:   1965–69: 1965

Personnel Comments:

The organizational structure of agricultural research in Turkey is quite complex. In 1981 (according to source 26) agricultural research was undertaken by four major entities, namely, 1) Ministry of Agriculture and Forestry (for livestock, nutrition, forestry, and crops), 2) Ministry of Village Affairs (soil, water, and irrigation), 3) Ministry of Sugar, and 4) Faculties of Agriculture in 4 major universities (especially Ankara and Ismir). Within the Ministry of Agriculture and Forestry, research falls under 8 General Directorates (GD) and is carried out in more than 50 regional and subregional institutes and stations.

The 1982 personnel figure given by source 27 excludes researchers within the higher education sector and the forestry research institute – which was reported to have 689 graduate staff, but which seems to be out of proportion when compared with the other institutes. The 1985 estimate given by source 422 also excludes researchers within the higher education sector.

Expenditure Comments:

The post–1973 expenditure figures explicitly exclude university research expenditures.

Sources:

0010 Boyce, James K., and Robert E. Evenson. *National and International Agricultural Research and Extension Programs.* New York: Agricultural Development Council, Inc., 1975.

0026 Oram, Peter A., and Mark Gieben. "Document Summaries." ISNAR, The Hague, 1984. Mimeo.

0027 Harvey, Nigel, ed. *Agricultural Research Centres: A World Directory of Organizations and Programmes.* Seventh Edition, Two Volumes. Harlow, U.K.: Longman, 1983.

0422 Casas, Joseph. "Les Ressources Humaines et Financieres dans les Pays du Bassin Méditerranéen." Preliminary table. ISNAR, The Hague, 1986. Mimeo.

0601 OECD – STIID Data Bank. "Public Funding of R&D by Socio-Economic Objective." Tables. OECD, Paris, April 1987. Mimeo.

0852 Evenson, Robert E., and Yoav Kislev. *Investment in Agricultural Research and Extension: A Survey of International Data.* Center Discussion Paper No. 124. New Haven, Connecticut: Economic Growth Center, Yale University, August 1971.

0886 Evenson, Robert E., and Yoav Kislev. "Investment in Agricultural Research and Extension: A Survey of International Data." *Economic Development and Cultural Change* Vol. 23 (April 1975): 507–521.

Additional References:

0022 UNESCO. *Statistical Yearbook 1983.* Paris: UNESCO, 1983.

0073 Oram, Peter A., and Vishva Bindlish. "Investment in Agricultural Research in Developing Countries: Progress, Problems, and the Determination of Priorities." IFPRI, Washington, D.C., January 1984. Mimeo.

0095 FAO – CARIS. *Agricultural Research in Developing Countries – Volume 1: Research Institutions.* Rome: FAO – CARIS, 1978.

0165 Evenson, R.E., and Y. Kislev. *Agricultural Research and Productivity.* New Haven: Yale University Press, 1975.

0407 Centre International de Hautes Etudes Agronomiques Méditerranéennes. "Séminaire sur l'Orientation et l'Organisation de la Recherche Agronomique dans les Pays du Bassin Méditerranéen, Istanbul, 1–3 December 1986." Centre International de Hautes Etudes Agronomiques Méditerranéennes, Paris, December 1986. Mimeo.

0420 Ergun, Yusuf. "Recherche Agricole en Turquie." Paper presented at the Séminaire sur l'Orientation et l'Organisation de la Recherche Agronomique dans les Pays du Bassin Méditerranéen, Centre International de Hautes Etudes Agronomiques Méditerranéennes, Istanbul, 1–3 December 1986.

0473 OECD. *Science and Technology Indicators: Basic Statistical Series – Volume A: The Objectives of Government R&D Funding 1974–1985.* Paris: OECD, May 1983.

0474 OECD. *Science and Technology Indicators: Basic Statistical Series – Volume A: The Objectives of Government R&D, 1969–1981.* Paris: OECD, July 1981.

0513 OECD. *International Survey of the Resources Devoted to R&D by OECD Member Countries, International Statistical Year 1975 – International Volume: Statistical Tables and Notes.* Paris: OECD, March 1979.

0516 Aresvik, Oddvar. *The Agricultural Development of Turkey.* Praeger Special Studies in International Economics and Development. New York: Praeger Publishers, 1975.

454

0530 OECD. *OECD Science and Technology Indicators No. 2 – R&D, Invention and Competitiveness.* Paris: OECD, 1986.

0560 OECD. *Country Reports on the Organisation of Scientific Research: Turkey.* Paris: OECD, January 1964.

0602 OECD – STIID Data Bank. "R&D Expenditure in the Higher Education and Private Non-Profit Sectors." Tables. OECD, Paris, April 1987. Mimeo.

0744 *Agricultural Research Centres: A World Directory of Organizations and Programmes.* Eighth Edition, Two Volumes. Harlow, U.K.: Longman, 1986.

0831 FAO. *Report of an Expert Consultation on Agricultural Extension and Research Linkages in the Near East, Amman, Jordan, 22–26 July 1985.* Rome: FAO, 1985.

0865 Oram, Peter. "Agricultural Research Objectives and Priorities: Constraints to the Development of Agricultural Research Institutions in Arab countries." Review paper prepared for the first meeting of the Council for Arab Agricultural Research, organized by the Arab Fund for Economic and Social Development, Kuwait, 23 March 1988.

Tuvalu

Personnel

Period/Year	PhD	MSc	BSc	Subtotal	Expat	Total	Source
1960–64	..	..	..	..	..	..	
1965–69	..	..	..	..	..	..	
1970	..	..	..	..	..	..	
1971	..	..	..	..	..	..	
1972	..	..	..	..	..	..	
1973	..	..	..	..	..	..	
1974	..	..	..	..	..	..	
1975	..	..	..	..	..	..	
1976	..	..	..	..	..	..	
1977	..	..	..	..	..	..	
1978	..	..	..	..	..	..	
1979	..	..	..	..	..	..	
1980	..	..	..	..	..	..	
1981	..	..	..	..	..	..	
1982	..	..	..	..	..	..	
1983	..	..	..	0	1	1	723
1984	..	..	..	..	..	..	
1985	..	..	..	..	..	..	
1986	..	..	..	0	1	1	723

1960–64:   1965–69:

Expenditure   LCU = *Tuvalu Dollars (Australian Dollars)*

Period/Year	Current LCU (millions)	Constant 1980 LCU (millions)	Constant 1980 US$ (millions)		Source
			Atlas	PPP	
1960–64	..	..	..	..	
1965–69	..	..	..	..	
1970	..	..	..	..	
1971	..	..	..	..	
1972	..	..	..	..	
1973	..	..	..	..	
1974	..	..	..	..	
1975	..	..	..	..	
1976	..	..	..	..	
1977	..	..	..	..	
1978	..	..	..	..	
1979	..	..	..	..	
1980	..	..	..	..	
1981	..	..	..	..	
1982	..	..	..	..	
1983	0.060	0.045	0.051	0.076	723
1984	..	..	..	..	
1985	..	..	..	..	
1986	0.040	0.025	0.028	0.042	723

1960–64:   1965–69:

Personnel Comments:

Some very limited agricultural research is executed through the Division of Agriculture of the Ministry of Commerce and Natural Resources by an expatriate researcher who works part-time at Tuvalu.

Expenditure Comments:

The expenditure figures correspond with the personnel indicators.

Sources:

0723 Fernando, L.H. "South Pacific NARS Resource Commitment Survey." ADB and ISNAR, The Hague, 1987. Mimeo.

Additional References:

0038 Gamble, W.K., R.M. Bourke, and C.W. Brookson. *South Pacific Agricultural Research Study.* Consultants Report to the Asian Development Bank (two volumes). Manila: Asian Development Bank, June 1981.

0243 Commonwealth Agricultural Bureaux (CAB). *List of Research Workers in the Agricultural Sciences in the Commonwealth and in the Republic of Ireland 1978.* Slough, England: CAB, 1978.

0771 Fernando, L.H. "ISNAR/IRETA Workshop on Planning and Management of Agricultural Research in the South Pacific: Survey of Agricultural Research Resources." ISNAR, The Hague, n.d. Mimeo.

Uganda

Personnel

Period/Year	PhD	MSc	BSc	Subtotal	Expat	Total	Source
1960–64	..	..	..	..	..	67	518
1965–69	..	..	..	..	..	89	286/285
1970	..	..	..	..	..	125	532
1971	..	..	..	..	..	..	
1972	..	..	..	..	..	..	
1973	..	..	..	..	..	..	
1974	..	..	..	..	..	..	
1975	..	..	..	..	..	..	
1976	..	..	..	..	..	..	
1977	..	..	..	..	..	..	
1978	..	..	..	..	..	..	
1979	..	..	..	..	..	..	
1980	..	..	..	..	..	..	
1981	..	..	..	..	..	..	
1982	..	..	..	..	..	185	27
1983	..	..	..	..	..	..	
1984	..	..	..	..	..	..	
1985	..	..	..	..	..	..	
1986	..	..	..	..	..	..	

1960–64: 1964   1965–69: 1965 & 68

Expenditure

LCU = Uganda Shillings

Period/Year	Current LCU (millions)	Constant 1980 LCU (millions)	Constant 1980 US$ (millions)		Source
			Atlas	PPP	
1960–64	..	..			
1965–69	..	..			
1970	..	..	..		
1971	..	..	..		
1972	..	..	..		
1973	..	..	..		
1974	..	..	..		
1975	..	..	..		
1976	..	..	..		
1977	..	..	..		
1978	..	..	..		
1979	..	..	..		
1980	..	..	..		
1981	..	..	..		
1982	..	..	..		
1983	..	..	..		
1984	..	..	..		
1985	..	..	..		
1986	..	..	..		

1960–64:   1965–69: 1967

Personnel Comments:

The table below gives an overview of how the indicators have been constructed:

	1964	1965	1966	1968	1970	1971	1974	1980	1981	1982	1983	1984
DOA/RD	18	22	17	30	45	55	54	114	119	91	86	142
DOVS & AI	13	12	10	12	19	NA	12	NA	NA	20	NA	NA
DOF/RD	2	[2]	NA	[9]	13	NA	NA	NA	NA	NA	NA	NA
FD/RD	3	6	NA	8	6	NA	6	NA	NA	20	NA	NA
UFFRO[a]	6	[6]	NA	12	8	11	NA	NA	NA	20	30*	NA
UTRO[a]	8	10	5	12	12	13*	NA	NA	NA	[17]	17*	NA
CGC/CRC[b]	12	11	11	[9]	8	–	–	–	–	–	–	–
MA/FOA&F[c]	5	5	NA	11	14	16	NA	NA	18	17	NA	NA
Total	67	74	NA	103	125	NA	NA	NA	NA	185	NA	NA
Source	518	286	653	285	532 589*	279	244	242	79	27	160 847*	227

[ ] = estimated or constructed by authors.
a. The pre–1977 figures represent the East African Freshwater Fisheries Research Organization and the East African Trypanosomiasis Research Organization. Both organizations had mandates covering the whole community until the abolishment of the East African Community (Kenya, Tanzania, and Uganda) in 1977. In this case the procedure used here was to include the pre–1977 data for these 'East African' organizations in the indicator series for the country where the institute is located.
b. Merged with the Department of Agriculture in 1971.
c. It is assumed that faculty staff spend one-third of their time on research.

Given the political instability over the last 15 years or so, only very limited statistical information of doubtful quality is available.

Expenditure Comments:

Very limited quantifiable information concerning agricultural research expenditures is available for Uganda. It was not possible to construct an expenditure figure that is compatible with the personnel figures.

Acronyms and Abbreviations:

CGC/CRC: (Empire) Cotton Growing Corporation – Cotton Research Center
DOA/RD: Department of Agriculture – Research Division
DOVS & AI: Department of Veterinary Services and Animal Industry
DOF/RD: Department of Fisheries – Research Division
FD/RD: Forestry Department – Research Division
UFFRO: Uganda Freshwater Fisheries Research Organization (formerly East African)
UTRO: Uganda Trypanosomiasis Research Organization (formerly East African)
MA/FOA&F: Makarere University – Faculty of Agriculture and Forestry

Sources:

0027 Harvey, Nigel, ed. *Agricultural Research Centres: A World Directory of Organizations and Programmes.* Seventh Edition, Two Volumes. Harlow, U.K.: Longman, 1983.

0285 Commonwealth Agricultural Bureaux (CAB). *List of Research Workers in the Agricultural Sciences in the Commonwealth and in the Republic of Ireland 1969.* Slough, England: CAB, 1969.

0286 Commonwealth Agricultural Bureaux (CAB). *List of Research Workers in Agriculture, Animal Health and Forestry in the Commonwealth and in the Republic of Ireland 1966.* Slough, England: CAB, 1966.

0518 East African Academy. *Research Services in East Africa.* Nairobi: East African Publishing House, 1966.

0532 UNESCO Field Science Office for Africa. *Survey on the Scientific and Technical Potential of the Countries of Africa.* Paris: UNESCO, 1970.

Additional References:

0010 Boyce, James K., and Robert E. Evenson. *National and International Agricultural Research and Extension Programs.* New York: Agricultural Development Council, Inc., 1975.

459

0016 Oram, Peter A., and Vishva Bindlish. *Resource Allocations to National Agricultural Research: Trends in the 1970s*. The Hague and Washington, D.C.: ISNAR and IFPRI, November 1981.

0023 Bennell, Paul. *Agricultural Researchers in Sub-Saharan Africa: An Overview*. Working Paper No. 4. The Hague: ISNAR, October 1985.

0073 Oram, Peter A., and Vishva Bindlish. "Investment in Agricultural Research in Developing Countries: Progress, Problems, and the Determination of Priorities." IFPRI, Washington, D.C., January 1984. Mimeo.

0079 International Development Research Center (IDRC). *Agricultural Research in Uganda: A Program for Rehabilitation*. Ottawa: IDRC, April 1982.

0095 FAO – CARIS. *Agricultural Research in Developing Countries – Volume 1: Research Institutions*. Rome: FAO – CARIS, 1978.

0160 FAO. *Agricultural Research Review Mission: Uganda*. Rome: FAO, 1984.

0163 CGIAR. "National Agricultural Research." CGIAR, Washington, D.C., 1985. Mimeo.

0164 Association for the Advancement of Agricultural Sciences in Africa (AAASA). *Proceedings of the Workshop on Agricultural Research Administration, Nairobi, Kenya, 27–30 June 1977*. Proceedings Series PE–4. Addis Ababa, Ethiopia: AAASA and IDRC, August 1979.

0165 Evenson, R.E., and Y. Kislev. *Agricultural Research and Productivity*. New Haven: Yale University Press, 1975.

0166 Opio-Odongo, J.M.A. *Agricultural Research Policy in Uganda*. Kampala: Makerere University, March 1986.

0175 Cooper, St.G.C. *Agricultural Research in Tropical Africa*. Kampala: East African Literature Bureau, 1970.

0227 Australian Centre for International Agricultural Research (ACIAR). *Proceedings of the Eastern Africa-ACIAR Consultation on Agricultural Research, 18–22 July 1983, Nairobi, Kenya*. Australia: ACIAR, 1984.

0233 Opio-Odongo, J.M.A. "The Contextual Imperatives of Agricultural Research Planning and Management in Uganda." Paper presented for the Agricultural Research Planning and Management Short Course organised by the Overseas Development Group, University of East Anglia, Norwich, England, 28 September–9 November 1983.

0242 Commonwealth Agricultural Bureaux (CAB). *List of Research Workers in Agricultural Sciences in the Commonwealth 1981*. Slough, England: CAB, 1981.

0244 Commonwealth Agricultural Bureaux (CAB). *List of Research Workers in the Agricultural Sciences in the Commonwealth and in the Republic of Ireland 1975*. Slough, England: CAB, 1975.

0266 UNESCO. *National Science Policies in Africa*. Science Policy Studies and Documents No. 31. Paris: UNESCO, 1974.

0279 Commonwealth Agricultural Bureaux (CAB). *List of Research Workers in the Agricultural Sciences in the Commonwealth and in the Republic of Ireland 1972*. Slough, England: CAB, 1972.

0287 Commonwealth Agricultural Bureaux (CAB). *List of Research Workers in Agriculture, Animal Health and Forestry in the British Commonwealth, the Republic of Sudan and the Republic of Ireland 1959*. Slough, England: CAB, 1959.

0360 Cooper, St.G.C. "Towards Trained Manpower for Agricultural Research in Africa." Paper presented at the Conference on Agricultural Research and Production in Africa, organized by the Association for the Advancement of Agricultural Sciences in Africa (AAASA), Addis Ababa, 29 August–4 September 1971.

0445 Swanson, Burton E., and Wade H. Reeves. "Agricultural Research Eastern and Southern Africa: Manpower and Training." World Bank, Washington, D.C., August 1986. Mimeo.

0589 Kassapu, Samuel. *Les Dépenses de Recherche Agricole dans 34 Pays d'Afrique Tropicale*. Paris: Centre de Développement de l'OCDE, 1976.

0622 King, H.E., ed. *Progress Reports from Experiment Stations Season 1969–70: Uganda*. London: Cotton Research Corporation, 1970.

0639 Serere Research Station. "Annual Report April 1969–March 1970, Part I." Serere Research Station, Entebbe, Uganda, 1970. Mimeo.

0640 East African Agriculture and Forestry Research Organization. *Record of Research – Annual Report 1969*. Nairobi: East African Community, 1970.

0641 East African Veterinary Research Organization. *Record of Research – Annual Report 1968*. Kenya: East African Community, 1969.

0642 Keen, B.A. *The East African Agriculture and Forestry Research Organization – Its Origin and Objects*. Nairobi: East African Community, n.d.

0643 Binns, H.R. *The East African Veterinary Research Organisation – Its Development, Objects and Scientific Activities*. Nairobi: East African Community, n.d.

0653 Webster, B.N. *Index of Agricultural Research Institutions and Stations in Africa*. Rome: FAO, n.d.

0765 Nsubuga, H.S.K., W.A. Sakira, and P. Karani. "Agricultural Research in Uganda." In *Proceedings of the Eastern Africa-ACIAR*

460

Consultation on Agricultural Research, pp. 57–65. Australia: ACIAR, 1984.

0847 Kimura, J.H. "Financial and Administrative Management of Research Institutions in Eastern and Southern Africa: Report on Responses to a Questionnaire." In *Promotion of Technology Policy and Science Management in Africa*, edited by Karl Wolfgang Menck and Wolfgang Gmelin. Bonn: Deutsche Stiftung für internationale Entwicklung (DSE), 1986.

0852 Evenson, Robert E., and Yoav Kislev. *Investment in Agricultural Research and Extension: A Survey of International Data*. Center Discussion Paper No. 124. New Haven, Connecticut: Economic Growth Center, Yale University, August 1971.

0873 Abdalla, Abdalla Ahmed. "Agricultural Research in the IGADD Sub-Region and Related Manpower Training." Inter-Governmental Authority on Drought and Development (IGADD) and FAO, 1987. Mimeo.

0886 Evenson, Robert E., and Yoav Kislev. "Investment in Agricultural Research and Extension: A Survey of International Data." *Economic Development and Cultural Change* Vol. 23 (April 1975): 507–521.

United Arab Emirates

Personnel

Period/Year	PhD	MSc	BSc	Subtotal	Expat	Total	Source
1960–64	..	..	..	..			
1965–69	..	..	..	..			
1970	..	..	..	..			
1971	..	..	..	..			
1972	..	..	..	..			
1973	..	..	..	..			
1974	..	..	..	..			
1975	..	..	..	..			
1976	..	..	..	..			
1977	..	..	..	..			
1978	..	..	..	..			
1979	..	..	..	..			
1980	..	..	..	..	..		
1981	..	..	..	..	..		
1982	..	..	..	..	..	..	
1983	0	0	2	2	10	12	17
1984	..	..	..	..	..	..	
1985	..	..	..	..	..		
1986	..	..	..	..	..		

1960–64:   1965–69:

Expenditure   *LCU = Dirhams*

Period/Year	Current LCU (millions)	Constant 1980 LCU (millions)	Constant 1980 US$ (millions)		Source
			Atlas	PPP	
1960–64	..	..	..	..	
1965–69	..	..	..	..	
1970	..	..	..	..	
1971	..	..	..	..	
1972	..	..	..	..	
1973	..	..	..	..	
1974	..	..	..	..	
1975	..	..	..	..	
1976	..	..	..	..	
1977	..	..	..	..	
1978	..	..	..	..	
1979	..	..	..	..	
1980	..	..	..	..	
1981	..	..	..	..	
1982	..	..	..	..	
1983	5.500	5.248	1.406	1.244	17
1984	..	..	..	..	
1985	..	..	..	..	
1986	..	..	..	..	

1960–64:   1965–69:

Personnel Comments:

The 1983 figure includes the following organizations: the Water and Soil Investigation Project for Agricultural Development and Al Ain Central Laboratory. Source 27 gives 4 researchers in the Al Ain Central Laboratory for 1982.

Expenditure Comments:

The 1983 expenditure figure includes the Water and Soil Investigation Project for Agricultural Development and Al Ain Central Laboratory.

Sources:

0017   ISNAR, IFARD & AOAD. Survey of National Agricultural Research Systems: Unpublished Questionnaire Responses. ISNAR, The Hague, 1985.

Additional References:

0027   Harvey, Nigel, ed. *Agricultural Research Centres: A World Directory of Organizations and Programmes.* Seventh Edition, Two Volumes. Harlow, U.K.: Longman, 1983.

0094   FAO – Near East Regional Office. *Directory of Agricultural Research Institutions in the Near East Region.* Cairo: FAO – Near East Regional Office, 1979.

0165   Evenson, R.E., and Y. Kislev. *Agricultural Research and Productivity.* New Haven: Yale University Press, 1975.

0865   Oram, Peter. "Agricultural Research Objectives and Priorities: Constraints to the Development of Agricultural Research Institutions in Arab countries." Review paper prepared for the first meeting of the Council for Arab Agricultural Research, organized by the Arab Fund for Economic and Social Development, Kuwait, 23 March 1988.

United Kingdom

Personnel

Period/Year	PhD	MSc	BSc	Subtotal	Expat	Total	Source
1960–64	..	..	..	..	..	..	
1965–69	1313	402	680	2395	..	2395	286/287
1970	..	..	..	..	..	..	
1971	1731	442	782	2955	..	2955	279
1972	..	..	..	..	..	..	
1973	..	..	..	..	..	..	
1974	1972	474	826	3272	..	3272	244
1975	..	..	..	..	..	..	
1976	..	..	..	..	..	..	
1977	2169	460	828	3457	..	3457	243
1978	..	..	..	..	..	..	
1979	..	..	..	..	..	..	
1980	2345	492	840	3677	..	3677	242
1981	..	..	..	..	..	..	
1982	..	..	..	..	..	..	
1983	..	..	..	..	..	..	
1984	..	..	..	..	..	..	
1985	..	..	..	..	..	..	
1986	..	..	..	..	..	..	

1960–64:   1965–69: 1965 & 68

Expenditure

LCU = Pounds Sterling

Period/Year	Current LCU (millions)	Constant 1980 LCU (millions)	Constant 1980 US$ (millions)		Source
			Atlas	PPP	
1960–64	12.453	64.957	124.256	129.047	542
1965–69	21.798	93.351	178.572	185.457	542/694/691
1970	33.669	124.018	237.235	246.822	542/692
1971	38.686	130.696	250.008	259.648	542/692
1972	41.891	130.502	249.637	259.262	542/692
1973	47.143	137.443	262.915	273.052	542/692
1974	57.896	146.572	280.378	291.189	542/702
1975	75.155	149.711	286.383	297.425	542/702
1976	87.172	151.340	289.499	300.661	542/702
1977	96.607	147.267	281.708	292.570	542/701
1978	107.582	147.575	282.296	293.181	542/700
1979	125.877	150.751	288.372	299.490	542/688/283
1980	160.060	160.060	306.179	317.984	542
1981	190.550	170.897	326.909	339.514	542/688
1982	211.228	175.877	336.435	349.407	542/688
1983	219.720	174.105	333.045	345.886	542/688
1984	233.641	177.809	340.131	353.246	542/688
1985	..	..	..	..	
1986	..	..	..	..	

1960–64: 1960, 61, 62, 63 & 64   1965–69: 1965, 66, 67, 68 & 69

Personnel Comments:

These figures were constructed by an aggregation of detailed research personnel figures compiled at the institute level. For example, in 1980 the listing included researchers from 53 tertiary institutes (universities plus colleges and polytechnics) and 33 noneducational institutes. Personnel from tertiary institutes were prorated at the 50% rate.

Expenditure Comments:

The two major organizations in agricultural research in the UK are the Agricultural and Food Research Council (AFRC, previously ARC) and the Ministry of Agriculture, Fisheries, and Food (MAFF). In addition, the following organizations fund or execute smaller programs in agricultural research: the Department of Agriculture and Fisheries for Scotland (DAFS); the Department of Agriculture, Northern Ireland (DANI); the Forestry Commission; the Department of the Environment (DOE); the Natural Environment Research Council (NERC); and the universities.

Although MAFF is an important executing agency for agricultural research, it commissions a great deal of agricultural research to other organizations, especially AFRC. AFRC receives about half its budget from MAFF and the other half is provided by the Science Budget of the Department of Education and Science. In addition, some industries and other bodies provide a small – but increasing (source 557) – proportion of the AFRC budget. AFRC and MAFF only serve England and Wales. Scotland and Northern Ireland have their own agricultural research organizations. In the case of Scotland, however, there is an organizational link between AFRC and DAFS, in that AFRC advises DAFS on research in the Scottish Agricultural Research Institutes. Together these two organizations form the Agricultural and Food Research Service (AFRS).

To construct the expenditure indicators, the following procedure was adopted. In sources 688 to 702, total agricultural research expenditures are presented in such a way that AFRC expenditures (internally and externally) can be identified. The remaining non-AFRC expenditures include the agricultural research expenditures of MAFF, DAFS, DANI, the Forestry Commission, DOE, and NERC. Source 542 gives a complete series of AFRC expenditures for the period 1960–1984. In most cases these AFRC expenditure figures were more or less the same as those identified in the total expenditure figures. In some cases, however, they differ. It seems this was caused by omitting expenditures for the central administration of AFRC, particularly by source 688. It was therefore decided to replace, for all years, the AFRC expenditure figures from sources 688 to 702 with those of source 542, and add the non-AFRC expenditures to them. The table on the next page gives an overview of how the indicators have been constructed.

The U.K. financial year runs from April 1 to March 31, so according to the standard practice in this series, a 1980 figure, for example, refers to the April 1980 to March 1981 period.

Acronyms and Abbreviations:

AFRC: Agricultural and Food Research Council
AFRS: Agricultural and Food Research Service
ARC: Agricultural Research Council
DAFS: Department of Agriculture and Fisheries for Scotland
DANI: Department of Agriculture, Northern Ireland
DOE: Department of the Environment
MAFF: Ministry of Agriculture, Fisheries and Food
NERC: Natural Environment Research Council
SARI: Scottish Agricultural Research Institutes

Sources:

0242 Commonwealth Agricultural Bureaux (CAB). *List of Research Workers in Agricultural Sciences in the Commonwealth 1981.* Slough, England: CAB, 1981.

0243 Commonwealth Agricultural Bureaux (CAB). *List of Research Workers in the Agricultural Sciences in the Commonwealth and in the Republic of Ireland 1978.* Slough, England: CAB, 1978.

0244 Commonwealth Agricultural Bureaux (CAB). *List of Research Workers in the Agricultural Sciences in the Commonwealth and in the Republic of Ireland 1975.* Slough, England: CAB, 1975.

0279 Commonwealth Agricultural Bureaux (CAB). *List of Research Workers in the Agricultural Sciences in the Commonwealth and in the Republic of Ireland 1972.* Slough, England: CAB, 1972.

	1960	1961	1962	1963	1964	1965	1966	1967	1968
AFRC[a] non-AFRC[b,c] correction[d]	5.860 [4.497] –	6.456 [4.954] –	6.814 [5.229] –	7.566 [5.806] –	8.551 [6.532] –	9.132 6.520* [0.695]	10.437 6.876* [0.695]	12.432 7.854* [0.608]	13.910 9.167* [0.846]
Total	10.357	11.410	12.043	13.372	15.083	16.347	18.008	20.894	23.923
Source	542	542	542	542	542	542 694*	542 691*	542 691*	542 691*

	1969	1970	1971	1972	1973	1974	1975	1976
AFRC[a] non-AFRC[b,c] correction[d]	15.754 13.191* [0.871]	17.854 14.781* [1.034]	20.827 16.628* [1.231]	22.983 17.462* [1.446]	24.836 20.659* [1.648]	30.336 25.400* [2.160]	36.375 35.300 [3.480]	43.732 39.600* [3.840]
Total	29.816	33.669	38.686	41.891	47.143	57.896	75.155	87.172
Source	542 691*	542 692*	542 692*	542 692*	542 692*	542 702*	542 702*	542 702*

	1977	1978	1979	1980	1981	1982	1983	1984
AFRC[a] non-AFRC[b,c] correction[d]	48.207 44.200* [4.200]	55.522 47.500* [4.560]	63.067 59.900* 2.910**	78.200 [77.800] [3.780]	90.200 95.700* [4.650]	97.038 108.900* [5.290]	101.580 112.200* [5.940]	104.141 123.500 [6.000]
Total	96.607	107.582	125.877	160.060	190.550	211.228	219.720	233.641
Source	542 701*	542 700*	542 688* 283**	542	542 688*	542 688*	542 688*	542 543

[ ] = estimated or constructed by authors.

a. Gross expenditure figures inclusive of funds derived from the sale of produce, fees, and miscellaneous sources.

b. In several instances, NERC expenditures were classified as agricultural research but footnoted as mainly marine biology research. In most cases, however, this type of research expenditure was not included. They were therefore omitted from the series in the few cases they were included by the original source.

c. For the years 1960–64 non-AFRC expenditures have been estimated on the basis of a fixed percentage (1965–69 average) of AFRC expenditures. Both AFRC and non-AFRC expenditures include some funding for agricultural research executed by universities. However, the expenditures of agricultural research executed by universities and financed through General University Funds (GUF or core budget of the universities) are not included in the indicator series. According to source 703, these GUF expenditures on agricultural research amounted to 6.8 million, 14.6 million, and 19.2 million pounds, respectively, in 1975, 1980, and 1982. On average these GUF agricultural research expenditures amount to around 9% of the expenditure indicators in the respective years. One may increase the whole indicator series by 9% to include an estimate of the GUF component.

d. For the years 1960–1978 no breakdown is given of Northern Ireland Government research. On the basis of later observations, it was estimated that 60% of Northern Ireland Government research expenditures are agriculturally related. For the years 1966–1978, 60% of the Northern Ireland Government research expenditures were added to non-AFRC agricultural research expenditures. Post–1978 non-AFRC agricultural research expenditures excluded forestry research. Source 283 indicates that in 1979, 2.91 million pounds was spent on forestry research in the U.K. For the later years, a forestry research expenditure estimate has been constructed on the basis of the 1979 figure and the trend of non-AFRC expenditures.

0283 European Economic Community (EEC). "Monograph on the Organization of Agricultural Research." EEC, Brussels, 1982. Mimeo.

0285 Commonwealth Agricultural Bureaux (CAB). *List of Research Workers in the Agricultural Sciences in the Commonwealth and in the Republic of Ireland 1969*. Slough, England: CAB, 1969.

0286 Commonwealth Agricultural Bureaux (CAB). *List of Research Workers in Agriculture, Animal Health and Forestry in the Commonwealth and in the Republic of Ireland 1966*. Slough, England: CAB, 1966.

0542 Beck, Harvey. Personal communication. Department of Agricultural Economics, University of Reading. Reading, England, December 1986.

0688 Ministry of Agriculture, Fisheries and Food. *Report on Research and Development 1979–80, 1981–82, 1982–83, and 1983–84*. Yearly reports. London: Her Majesty's Stationery Office, various years.

0691 Central Statistical Office. *Research and Development Expenditure*. London: Her Majesty's Stationery Office, 1973.

0692 Central Statistical Office. *Research and Development – Expenditure and Employment*. London: Her Majesty's Stationery Office, 1976.

0694 Department of Education and Science – Ministry of Technology *Statistics of Science and Technology*. London: Her Majesty's Stationery Office, 1967.

0700 Bowles, J.R. "Research and Development: Expenditure and Employment in the Seventies." *Economic Trends* No. 334 (August 1981): 94–111.

0701 Arrundale, A. "Research and Development: Expenditure and Employment, 1978." *Economic Trends* No. 321 (July 1980): 99–109.

0702 Arrundale, R. "Research and Development: Expenditure and Employment." *Economic Trends* No. 309 (July 1979): 100–124.

Additional References:

0010 Boyce, James K., and Robert E. Evenson. *National and International Agricultural Research and Extension Programs*. New York: Agricultural Development Council, Inc., 1975.

0014 Judd, M. Ann, James K. Boyce, and Robert E. Evenson. "Investing in Agricultural Supply." Economic Growth Center, Yale University, New Haven, Connecticut, 1983. Mimeo.

0020 Cabinet Office. *Annual Review of Government Funded R&D 1983, 1984, 1985, and 1986*. London: Her Majesty's Stationery Office, various years.

0021 OECD. *OECD Science and Technology Indicators*. Paris: OECD, 1984.

0027 Harvey, Nigel, ed. *Agricultural Research Centres: A World Directory of Organizations and Programmes*. Seventh Edition, Two Volumes. Harlow, U.K.: Longman, 1983.

0165 Evenson, R.E., and Y. Kislev. *Agricultural Research and Productivity*. New Haven: Yale University Press, 1975.

0287 Commonwealth Agricultural Bureaux (CAB). *List of Research Workers in Agriculture, Animal Health and Forestry in the British Commonwealth, the Republic of Sudan and the Republic of Ireland 1959*. Slough, England: CAB, 1959.

0376 FAO. *Fourteenth FAO Regional Conference for Europe, Reykjavik, Iceland, 17–21 September 1984: Research in Support of Agricultural Policies in Europe*. Rome: FAO, 1984.

0394 Agricultural and Food Research Council (AFRC). *Report of the Agricultural and Food Research Council for the year 1985–86*. England: AFRC, November 1986.

0473 OECD. *Science and Technology Indicators: Basic Statistical Series – Volume A: The Objectives of Government R&D Funding 1974–1985*. Paris: OECD, May 1983.

0474 OECD. *Science and Technology Indicators: Basic Statistical Series – Volume A: The Objectives of Government R&D, 1969–1981*. Paris: OECD, July 1981.

0509 OECD. *Survey of the Resources Devoted to R&D by OECD Member Countries, International Statistical Year 1971 – Volume 5: Total Tables, Statistical Tables and Notes*. Paris: OECD, August 1974.

0510 OECD. *International Survey of the Resources Devoted to R&D in 1967 by OECD Member Countries, Statistical Tables and Notes – Volume 5: Total Tables*. Paris: OECD, August 1970.

0511 OECD. *International Survey of the Resources Devoted to R&D in 1969 by OECD Member Countries, Statistical Tables and Notes – Volume 5: Total Tables*. Paris: OECD, June 1973.

0512 OECD. *Survey of the Resources Devoted to R&D by OECD Member Countries, International Statistical Year 1973 – Volume 5: Total Tables, Statistical Tables and Notes*. Paris: OECD, September 1976.

0513 OECD. *International Survey of the Resources Devoted to R&D by OECD Member Countries, International Statistical Year 1975 – International Volume: Statistical Tables and Notes*. Paris: OECD, March 1979.

0514 OECD. *A Study of Resources Devoted to R&D in OECD Member Countries in 1963/64, International Statistical Year for Research and Development – Volume 2: Statistical Tables and Notes*. Paris: OECD, 1968.

0530 OECD. *OECD Science and Technology Indicators No. 2 – R&D, Invention and Competitiveness.* Paris: OECD, 1986.

0541 OECD. *Country Reports on the Organisation of Scientific Research: United Kingdom.* Paris: OECD, February 1966.

0543 Chief Scientists Group – Ministry of Agriculture, Fisheries & Food. *Report on Research and Development 1984–85.* London: Ministry of Agriculture, Fisheries & Food, n.d.

0557 Corbett, D.C.M. Personal communication. Agricultural and Food Research Council. London, July 1986.

0574 OECD. *Intellectual Investment in Agriculture for Economic and Social Development.* Documentation in Food and Agriculture No. 60. Paris: OECD, 1962.

0592 Cooke, G.W., ed. *Agricultural Research 1931–1981.* London: Agricultural Research Council, 1981.

0594 Agricultural Research Institute of Northern Ireland. *Jubilee Report 1926–1976.* Hillsborough, Co. Down, Northern Ireland: Agricultural Research Institute of Northern Ireland, 1976.

0596 Agricultural and Food Research Council (AFRC). *Corporate Plan 1987–1992.* London: AFRC, March 1987.

0601 OECD – STIID Data Bank. "Public Funding of R&D by Socio-Economic Objective." Tables. OECD, Paris, April 1987. Mimeo.

0602 OECD – STIID Data Bank. "R&D Expenditure in the Higher Education and Private Non-Profit Sectors." Tables. OECD, Paris, April 1987. Mimeo.

0606 EUROSTAT. *Government Financing of Research and Development 1975–1984.* Luxembourg: Office des Publications Officielles des Communautés Européennes, February 1985.

0635 Agricultural Research Council (ARC). *Index of Agricultural Research 1969.* London: ARC, n.d.

0636 Agricultural Research Council (ARC). *The Agricultural Research Service.* London: ARC, 1969.

0689 Ministry of Agriculture, Fisheries and Food. *Report on Research and Development 1976, 1977, and 1978.* Yearly reports. London: Her Majesty's Stationery Office, various years.

0690 Ministry of Agriculture, Fisheries and Food & Department of Agriculture and Fisheries for Scotland. *Fisheries Research and Development Board Third Report 1976–1977.* London: Her Majesty's Stationery Office, n.d.

0693 Department of Education and Science – Ministry of Technology. *Statistics of Science and Technology.* London: Her Majesty's Stationery Office, 1968.

0695 Bowles J.R. "Central Government Expenditure on Research and Development in 1984." *Economic Trends* No. 394 (August 1986): 82–89.

0696 Bowles J.R. "Research and Development in the United Kingdom in 1983." *Economic Trends* No. 382 (August 1985): 82–93.

0697 Bowles J.R. "Research and Development in the United Kingdom in 1981." *Economic Trends* No. 370 (August 1984): 81–89.

0698 Bowles J.R. "Research and Development: Preliminary Estimates of Expenditure in the United Kingdom in 1981." *Economic Trends* No. 359 (1983): 108–121.

0699 Bowles J.R. "Central Government Expenditure on Research and Development." *Economic Trends* No. 346 (August 1982): 82–85.

0703 Martin, Ben R., and John Irvine. *An International Comparison of Government Funding of Academic and Academically Related Research.* Advisory Board for the Research Council (ABRC) Science Policy Study No. 2. Brighton: ABRC – Science Policy Research Unit, University of Sussex, October 1986.

0706 EUROSTAT. *Government Financing of Research and Development 1975–1985.* Luxembourg: Office for Official Publications of the European Communities, 1987.

0770 Ruttan, Vernon W. *Agricultural Research Policy and Development.* FAO Research and Technology Paper 2. Rome: FAO, 1987.

0852 Evenson, Robert E., and Yoav Kislev. *Investment in Agricultural Research and Extension: A Survey of International Data.* Center Discussion Paper No. 124. New Haven, Connecticut: Economic Growth Center, Yale University, August 1971.

0886 Evenson, Robert E., and Yoav Kislev. "Investment in Agricultural Research and Extension: A Survey of International Data." *Economic Development and Cultural Change* Vol. 23 (April 1975): 507–521.

United States

Personnel

Period/Year	PhD	MSc	BSc	Subtotal	Expat	Total	Source
1960–64	..	..	..	..	..	12738	500/489/490/491/492
1965–69	..	..	..	..	..	13296	500/492
1970	..	..	..	..	..	12952	500/492
1971	..	..	..	..	..	13203	500/492
1972	..	..	..	..	..	13226	500/492
1973	..	..	..	..	..	13242	500/492
1974	..	..	..	..	..	12585	500/493
1975	..	..	..	..	..	12328	500/488
1976	..	..	..	..	..	13223	500/488
1977	..	..	..	..	..	13512	500/488
1978	..	..	..	..	..	13679	500/488
1979	..	..	..	..	..	13709	500/488
1980	..	..	..	..	..	13645	500/488
1981	..	..	..	..	..	14079	500/488
1982	..	..	..	..	..	14133	500/488
1983	..	..	..	..	..	14454	500/488
1984	..	..	..	..	..	..	
1985	..	..	..	..	..	..	
1986	..	..	..	..	..	..	

1960–64: 1960, 61, 62, 63 & 64   1965–69: 1965, 66, 67, 68 & 69

Expenditure   *LCU = U.S. Dollars*

Period/Year	Current LCU (millions)	Constant 1980 LCU (millions)	Constant 1980 US$ (millions)		Source
			Atlas	PPP	
1960–64	292.192	783.816	783.816	783.816	500/489/490/491/492
1965–69	445.707	1047.584	1047.584	1047.584	500/492
1970	535.741	1093.349	1093.349	1093.349	500/492
1971	579.708	1119.127	1119.127	1119.127	500/492
1972	621.493	1146.666	1146.666	1146.666	500/492
1973	665.415	1151.237	1151.237	1151.237	500/492
1974	706.004	1120.641	1120.641	1120.641	500/493
1975	807.416	1166.786	1166.786	1166.786	500/488
1976	896.019	1217.417	1217.417	1217.417	500/488
1977	1005.220	1280.535	1280.535	1280.535	500/488
1978	1110.485	1317.301	1317.301	1317.301	500/488
1979	1233.649	1345.310	1345.310	1345.310	500/488
1980	1344.868	1344.868	1344.868	1344.868	500/488
1981	1500.989	1369.516	1369.516	1369.516	500/488
1982	1637.750	1403.385	1403.385	1403.385	500/488
1983	1714.489	1414.595	1414.595	1414.595	500/488
1984	..	..	..	..	
1985	..	..	..	..	
1986	..	..	..	..	

1960–64: 1960, 61, 62, 63 & 64   1965–69: 1965, 66, 67, 68 & 69

Personnel Comments:

These personnel figures are measured in full-time equivalent (FTE) units and include the 50 State Agricultural Experiment Stations (SAES) plus the US Department of Agriculture. The total SAES FTE figures were derived by prorating state-level statistics on the number of researchers holding full- and part-time (i.e., jointly with teaching and/or extension) research appointments. The USDA figures were estimated by applying a (preliminary) average SAES ratio of noncapital cost per researcher to noncapital USDA obligations. The series excludes observations for the 1976 transition quarter when the reporting period shifted from a year ending June 30 to a year ending September 30.

Expenditure Comments:

These figures represent the sum of a 50–state total State Agricultural Experiment Station (SAES) expenditure figure (not simply funds available) plus United States Department of Agriculture (USDA) agricultural research obligations. In 1960 the SAES component accounted for 57% of the total, gradually rising to 63% by 1983. As with the personnel figures, the 1976 transition quarter was omitted.

Sources:

0488 US Department of Agriculture – Cooperative State Research Service. *Inventory of Agricultural Research – Fiscal Years: 1970–1985.* Washington, D.C.: US Government Printing Office, various years.

0489 US Department of Agriculture – Agricultural Research Service. *Funds for Research at State Agricultural Experiment Stations, 1960.* Washington, D.C.: US Government Printing Office, April 1961.

0490 US Department of Agriculture – Cooperative State Experiment Station Service. *Funds for Research at State Agricultural Experiment Stations.* Annual issues 1961 and 1962. Washington, D.C.: US Government Printing Office, various years.

0491 US Department of Agriculture – Cooperative State Research Service. *Funds for Research at State Agricultural Experiment Stations, 1963.* Washington, D.C.: Government Printing Office, March 1964.

0492 US Department of Agriculture – Cooperative State Research Service. *Funds for Research at State Agricultural Experiment Stations and other State Institutions.* Annual issues 1964 to 1974. Washington, D.C.: US Government Printing Office, various years.

0493 US Department of Agriculture -- Cooperative State Research Service. *Funds for Research at State Agricultural Experiment Stations and Other Cooperating Institutions.* Annual issues 1974 and 1975. Washington, D.C.: US Government Printing Office, various years.

0500 Hallaway, Michelle L. "The Statistical Evolution of US Agricultural Research: A Study in Institutional Development." MSc thesis, University of Minnesota, St. Paul, Minnesota, forthcoming.

Additional References:

0010 Boyce, James K., and Robert E. Evenson. *National and International Agricultural Research and Extension Programs.* New York: Agricultural Development Council, Inc., 1975.

0014 Judd, M. Ann, James K. Boyce, and Robert E. Evenson. "Investing in Agricultural Supply." Economic Growth Center, Yale University, New Haven, Connecticut, 1983. Mimeo.

0021 OECD. *OECD Science and Technology Indicators.* Paris: OECD, 1984.

0027 Harvey, Nigel, ed. *Agricultural Research Centres: A World Directory of Organizations and Programmes.* Seventh Edition, Two Volumes. Harlow, U.K.: Longman, 1983.

0165 Evenson, R.E., and Y. Kislev. *Agricultural Research and Productivity.* New Haven: Yale University Press, 1975.

0473 OECD. *Science and Technology Indicators: Basic Statistical Series – Volume A: The Objectives of Government R&D Funding 1974–1985.* Paris: OECD, May 1983.

0474 OECD. *Science and Technology Indicators: Basic Statistical Series – Volume A: The Objectives of Government R&D, 1969–1981.* Paris: OECD, July 1981.

0495 Pardey, Philip G., and Barbara Craig. "Causal Relationships between Public Sector Agricultural Research Expenditures and Output." *American Journal of Agricultural Economics* Vol. 71, no. 1 (February 1989): 9–19.

0496 Davis, J.S. "Stability of the Research Production Function Coefficient for US Agriculture." PhD diss., University of Minnesota, St. Paul, Minnesota, 1979.

0497 Pardey, Philip G., Barbara Craig, and Michelle Hallaway. "US Agricultural Research Deflators: 1890–1985." *Research Policy* Vol. 18 (forthcoming, 1989).

0498 Huffman, Wallace E. "The Institutional Development of the Public Agricultural Experiment Station System: Scientists and Departments." Iowa State University, Ames, Iowa, May 1985. Mimeo.

0509 OECD. *Survey of the Resources Devoted to R&D by OECD Member Countries, International Statistical Year 1971 – Volume 5: Total Tables, Statistical Tables and Notes.* Paris: OECD, August 1974.

0510 OECD. *International Survey of the Resources Devoted to R&D in 1967 by OECD Member Countries, Statistical Tables and Notes – Volume 5: Total Tables.* Paris: OECD, August 1970.

0511 OECD. *International Survey of the Resources Devoted to R&D in 1969 by OECD Member Countries, Statistical Tables and Notes – Volume 5: Total Tables.* Paris: OECD, June 1973.

0512 OECD. *Survey of the Resources Devoted to R&D by OECD Member Countries, International Statistical Year 1973 – Volume 5: Total Tables, Statistical Tables and Notes.* Paris: OECD, September 1976.

0513 OECD. *International Survey of the Resources Devoted to R&D by OECD Member Countries, International Statistical Year 1975 – International Volume: Statistical Tables and Notes.* Paris: OECD, March 1979.

0514 OECD. *A Study of Resources Devoted to R&D in OECD Member Countries in 1963/64, International Statistical Year for Research and Development – Volume 2: Statistical Tables and Notes.* Paris: OECD, 1968.

0530 OECD. *OECD Science and Technology Indicators No. 2 – R&D, Invention and Competitiveness.* Paris: OECD, 1986.

0563 OECD. *Country Reports on the Organisation of Scientific Research: United States.* Paris: OECD, June 1963.

0601 OECD – STIID Data Bank. "Public Funding of R&D by Socio-Economic Objective." Tables. OECD, Paris, April 1987. Mimeo.

0602 OECD – STIID Data Bank. "R&D Expenditure in the Higher Education and Private Non-Profit Sectors." Tables. OECD, Paris, April 1987. Mimeo.

0703 Martin, Ben R., and John Irvine. *An International Comparison of Government Funding of Academic and Academically Related Research.* Advisory Board for the Research Council (ABRC) Science Policy Study No. 2. Brighton: ABRC – Science Policy Research Unit, University of Sussex, October 1986.

0770 Ruttan, Vernon W. *Agricultural Research Policy and Development.* FAO Research and Technology Paper 2. Rome: FAO, 1987.

0852 Evenson, Robert E., and Yoav Kislev. *Investment in Agricultural Research and Extension: A Survey of International Data.* Center Discussion Paper No. 124. New Haven, Connecticut: Economic Growth Center, Yale University, August 1971.

0886 Evenson, Robert E., and Yoav Kislev. "Investment in Agricultural Research and Extension: A Survey of International Data." *Economic Development and Cultural Change* Vol. 23 (April 1975): 507–521.

Uruguay

Personnel

Period/Year	PhD	MSc	BSc	Subtotal	Expat	Total	Source
1960–64	..	..	..	..	..	20	10
1965–69	..	..	..	..	..	47	10/458/645
1970	2	18	42	62	..	62	645
1971	2	18	51	71	..	71	645
1972	3	19	42	64	..	64	645
1973	2	27	56	85	..	85	645
1974	1	22	56	79	..	79	645
1975	0	24	52	76	..	76	645
1976	0	25	51	76	..	76	645
1977	0	28	48	76	..	76	645
1978	0	27	40	67	..	67	645
1979	0	22	48	70	..	70	645
1980	0	18	54	72	..	72	645
1981	0	14	65	79	..	79	645
1982	0	15	64	79	..	79	645
1983	0	18	62	80	..	80	645
1984	0	19	59	78	..	78	645
1985	0	18	51	69	..	69	645
1986	..	..	..	..	..	..	

1960–64: 1960 & 63   1965–69: 1965, 66 & 69

Expenditure   *LCU = New Uraguayan Pesos*

Period/Year	Current LCU (millions)	Constant 1980 LCU (millions)	Constant 1980 US$ (millions)		Source
			Atlas	PPP	
1960–64	..	..	..	..	
1965–69	..	..	..	..	
1970	..	..	..	..	
1971	..	..	..	..	
1972	0.284	15.438	1.381	2.191	755
1973	1.100	29.114	2.605	4.132	26
1974	2.114	32.499	2.907	4.612	26
1975	4.051	36.703	3.284	5.209	26
1976	5.873	35.751	3.198	5.074	26
1977	7.753	30.302	2.711	4.300	26
1978	12.692	33.620	3.008	4.771	26
1979	17.038	25.719	2.301	3.650	26
1980	32.853	32.853	2.939	4.662	26
1981	48.333	37.094	3.319	5.264	26
1982	51.396	33.992	3.041	4.824	26
1983	58.860	25.036	2.240	3.553	17
1984	73.923	19.818	1.773	2.813	26
1985	..	..	..	..	
1986	387.582	34.421	3.079	4.885	755

1960–64:   1965–69:

Personnel Comments:

These figures represent 'Centro de Investigaciones Agrícolas "Alberto Boerger"' (CIAAB) only. Other organizations that execute agricultural research in Uruguay, but which are not included in the indicator series, are 'Centro de Investigaciones en Fruticultura y Horticultura' (1971: 21 researchers, source 717; in later reports it seems to have disappeared completely), 'Centro de Investigaciones Veterinarias "Miguel C. Rubino"' (1971: 36 researchers, source 717; 1980: 37 researchers, source 32), 'Facultad de Agronomía' (1980: 151 part-time researchers, source 32), 'Facultad de Medecina Veterinaria,' and 'Instituto Nacional de Pesca' (1982: 4 researchers, source 27).

Expenditure Comments:

These expenditure figures represent those of CIAAB only.

Sources:

0010 Boyce, James K., and Robert E. Evenson. *National and International Agricultural Research and Extension Programs.* New York: Agricultural Development Council, Inc., 1975.

0017 ISNAR, IFARD & AOAD. Survey of National Agricultural Research Systems: Unpublished Questionnaire Responses. ISNAR, The Hague, 1985.

0026 Oram, Peter A., and Mark Gieben. "Document Summaries." ISNAR, The Hague, 1984. Mimeo.

0458 Brannon, Russell H. *The Agricultural Development of Uruguay.* New York: Praeger Publishers, 1967.

0645 Grierson, John. Personal communication. Centro de Investigaciones Agricolas 'Alberto Boerger'. Montevideo, 1987.

0755 Rabufetti, Armando. "Institutional Reorganization of Agricultural Research in Uruguay." Paper presented at the International Workshop on Agricultural Research Management, ISNAR, The Hague, 7–11 September 1987.

Additional References:

0001 Trigo, Eduardo J., and Martín E. Piñeiro. "Funding Agricultural Research." In *Report of a Conference: Selected Issues in Agricultural Research in Latin America,* edited by Barry Nestel and Eduardo J. Trigo. The Hague: ISNAR, March 1984.

0014 Judd, M. Ann, James K. Boyce, and Robert E. Evenson. "Investing in Agricultural Supply." Economic Growth Center, Yale University, New Haven, Connecticut, 1983. Mimeo.

0016 Oram, Peter A., and Vishva Bindlish. *Resource Allocations to National Agricultural Research: Trends in the 1970s.* The Hague and Washington, D.C.: ISNAR and IFPRI, November 1981.

0027 Harvey, Nigel, ed. *Agricultural Research Centres: A World Directory of Organizations and Programmes.* Seventh Edition, Two Volumes. Harlow, U.K.: Longman, 1983.

0032 Castronovo, Alfonso J.P. "Diagnóstico Descriptivo de la Situación Actual del Sistema Nacional de Investigación y Extensión en Argentina, Chile, Paraguay y Uruguay." Buenos Aires, December 1980. Mimeo.

0053 Piñeiro, Martín, and Eduardo Trigo, eds. *Cambio Técnico en el Agro Latino Americano – Situación y Perspectivas en la Década de 1980.* San José: IICA, 1983.

0073 Oram, Peter A., and Vishva Bindlish. "Investment in Agricultural Research in Developing Countries: Progress, Problems, and the Determination of Priorities." IFPRI, Washington, D.C., January 1984. Mimeo.

0165 Evenson, R.E., and Y. Kislev. *Agricultural Research and Productivity.* New Haven: Yale University Press, 1975.

0172 Trigo, Eduardo. Background material. ISNAR, The Hague, n.d.

0515 UNESCO. *Statistical Yearbook 1972.* Paris: UNESCO, 1973.

0717 Centro de Documentación sobre Investigación y Enseñanza Superior Agropecuaria de la Zona Sur. *Instituciones de Investigación Agrícola de la Zona Sur: Argentina, Chile, Paraguay y Uruguay.* Serie: Informaciones No. 2. Buenos Aires: Facultad de Agronomía y Veterinaria – Biblioteca Central, IICA – Zona Sur, 1972.

0747 ISNAR. *International Workshop on Agricultural Research Management.* Report of a workshop. The Hague: ISNAR, 1987.

0852 Evenson, Robert E., and Yoav Kislev. *Investment in Agricultural Research and Extension: A Survey of International Data.* Center Discussion Paper No. 124. New Haven, Connecticut: Economic Growth Center, Yale University, August 1971.

0886 Evenson, Robert E., and Yoav Kislev. "Investment in Agricultural Research and Extension: A Survey of International Data." *Economic Development and Cultural Change* Vol. 23 (April 1975): 507–521.

Vanuatu

Personnel

Period/Year	PhD	MSc	BSc	Subtotal	Expat	Total	Source
1960–64	..	..	..	..	..		
1965–69	..	..	..	..	..		
1970	..	..	..	..	..		
1971	..	..	..	..	..		
1972	..	..	..	..	..		
1973	..	..	..	..	..		
1974	..	..	..	..	..		
1975	..	..	..	..	..	3	22
1976	..	..	..	..	..		
1977	..	..	..	..	..		
1978	..	..	..	..	..		
1979	..	..	..	..	..		
1980	..	..	..	..	..	..	
1981	..	..	..	..	..	3	38
1982	..	..	..	..	..		
1983	..	..	..	..	..		
1984	..	..	..	..	..		
1985	..	..	..	0	12	12	771
1986	..	..	..	..	..		

1960–64:   1965–69:

Expenditure

LCU = Vatu

Period/Year	Current LCU (millions)	Constant 1980 LCU (millions)	Constant 1980 US$ (millions)		Source
			Atlas	PPP	
1960–64	..	..	..	..	
1965–69	..	..	..	..	
1970	..	..	..	..	
1971	..	..	..	..	
1972	..	..	..	..	
1973	..	..	..	..	
1974	..	..	..	..	
1975	21.603	..	..	..	22
1976	..	..	..	..	
1977	..	..	..	..	
1978	..	..	..	..	
1979	..	..	..	..	
1980	..	..	..	..	
1981	..	..	..	..	
1982	..	..	..	..	
1983	..	..	..	..	
1984	..	..	..	..	
1985	..	..	..	..	
1986	..	..	..	..	

1960–64:   1965–69:

Personnel Comments:

The number of researchers refers to IRHO only. Source 38 mentions some additional researchers working at Togale Experiment Station without specifying the exact number. ORSTOM, which is stationed at New Caledonia, undertakes soil surveys in Vanuatu.

Expenditure Comments:

The expenditure figure corresponds with the personnel figure.

Acronyms and Abbreviations:

IRHO: Institut de Recherches pour les Huiles et les Oleagineux

ORSTOM: Office de la Recherche Scientifique et Technique d'Outre-Mer

Sources:

0022 UNESCO. *Statistical Yearbook 1983.* Paris: UNESCO, 1983.

0038 Gamble, W.K., R.M. Bourke, and C.W. Brookson. *South Pacific Agricultural Research Study.* Consultants' Report to the Asian Development Bank (two volumes). Manila: Asian Development Bank, June 1981.

0771 Fernando, L.H. "ISNAR/IRETA Workshop on Planning and Management of Agricultural Research in the South Pacific: Survey of Agricultural Research Resources." ISNAR, The Hague, n.d. Mimeo.

Venezuela

Personnel

Period/Year	PhD	MSc	BSc	Subtotal	Expat	Total	Source
1960–64	..	..	..	..	..	124	10/206
1965–69	7	41	127	175	..	175	206
1970	9	61	138	208	..	208	206
1971	9	66	151	226	..	226	206
1972	11	72	202	285	..	285	206
1973	12	76	260	348	..	348	206
1974	12	84	258	354	..	354	206
1975	12	103	250	365	..	365	206
1976	..	..	..	..	..	..	
1977	..	..	..	..	..	392	22
1978	..	..	..	..	..	..	
1979	..	..	..	..	..	..	
1980	..	..	..	..	..	..	
1981	..	..	..	..	..	..	
1982	..	..	..	..	..	..	
1983	18	135	230	383	..	383	17
1984	..	..	..	..	..	..	
1985	..	..	..	..	..	..	
1986	..	..	..	..	..	..	

1960–64: 1960, 62, 63 & 64   1965–69: 1965, 66, 67, 68 & 69

Expenditure

LCU = Bolivares

Period/Year	Current LCU (millions)	Constant 1980 LCU (millions)	Constant 1980 US$ (millions)		Source
			Atlas	PPP	
1960–64	14.775	95.501	19.516	25.970	10
1965–69	26.892	162.067	33.119	44.071	10
1970	..	..	..	..	
1971	..	..	..	..	
1972	..	..	..	..	
1973	..	..	..	..	
1974	..	..	..	..	
1975	82.290	168.627	34.460	45.855	206
1976	92.460	175.114	35.785	47.619	206
1977	112.380	188.241	38.468	51.189	206
1978	123.460	186.778	38.169	50.791	206
1979	118.420	147.656	30.174	40.152	206
1980	168.440	168.440	34.422	45.804	206
1981	..	..	..	..	
1982	..	..	..	..	
1983	164.700	128.471	26.254	34.935	17
1984	..	..	..	..	
1985	..	..	..	..	
1986	..	..	..	..	

1960–64: 1962 & 64   1965–69: 1965, 66, 67, 68 & 69

Personnel Comments:

These figures refer to 'Fondo Nacional de Investigaciones Agropecuarias' (FONAIAP) only. Other organizations conducting agricultural research in Venezuela are FUSAGRI, FONOCOPAL, FONDEFRU, FONALI, 'Fondo del Café,' 'Fondo del Cacao,' FDA, and FUDECO. Some of these organizations conduct their own research, while some only finance research done by FONAIAP or the universities. For two of these organizations source 27 gives the number of graduate research staff for the year 1982, namely, FUSAGRI, 70, and FUDECO, 69. Given only limited information on these organizations, they were excluded from the indicator series.

For 1983, source 17 reported 70 researchers whose degree status was not identified. For reasons of consistency, they were included here as BSc–equivalent researchers.

Expenditure Comments:

These expenditure figures refer to FONAIAP only. Agricultural research carried out by organizations other than FONAIAP is not negligible but because of lack of information, it is excluded from the expenditure series.

Acronyms and Abbreviations:

FDA: Fondo de Desarrollo Algodonero
FONAIAP: Fondo Nacional de Investigaciones Agropecuarias
FONALI: Fondo para el Desarrollo del Ajonjolí
FONDEFRU: Fondo de Desarrollo Frutícola
FONOCOPAL: Fondo de Coco y Palma
FUDECO: Fundación para el Desarrollo de la Región Centro Occidental de Venezuela
FUSAGRI: Fundación para Servicio al Agricultor

Sources:

0010 Boyce, James K., and Robert E. Evenson. *National and International Agricultural Research and Extension Programs.* New York: Agricultural Development Council, Inc., 1975.

0017 ISNAR, IFARD & AOAD. Survey of National Agricultural Research Systems: Unpublished Questionnaire Responses. ISNAR, The Hague, 1985.

0022 UNESCO. *Statistical Yearbook 1983.* Paris: UNESCO, 1983.

0206 Ardila V. Jorge, and Hernán Chaverra G. *Diagnóstico Sobre la Investigación Agropecuaria a Venezuela.* Bogotá: IICA – OEA: Oficina en Colombia, and PROTAAL, December 1980.

Additional References:

0001 Trigo, Eduardo J., and Martín E. Piñeiro. "Funding Agricultural Research." In *Report of a Conference: Selected Issues in Agricultural Research in Latin America*, edited by Barry Nestel and Eduardo J. Trigo. The Hague: ISNAR, March 1984.

0014 Judd, M. Ann, James K. Boyce, and Robert E. Evenson. "Investing in Agricultural Supply." Economic Growth Center, Yale University, New Haven, Connecticut, 1983. Mimeo.

0016 Oram, Peter A., and Vishva Bindlish. *Resource Allocations to National Agricultural Research: Trends in the 1970s.* The Hague and Washington, D.C.: ISNAR and IFPRI, November 1981.

0026 Oram, Peter A., and Mark Gieben. "Document Summaries." ISNAR, The Hague, 1984. Mimeo.

0027 Harvey, Nigel, ed. *Agricultural Research Centres: A World Directory of Organizations and Programmes.* Seventh Edition, Two Volumes. Harlow, U.K.: Longman, 1983.

0053 Piñeiro, Martín, and Eduardo Trigo, eds. *Cambio Técnico en el Agro Latino Americano – Situación y Perspectivas en la Década de 1980.* San José: IICA, 1983.

0073 Oram, Peter A., and Vishva Bindlish. "Investment in Agricultural Research in Developing Countries: Progress, Problems, and the Determination of Priorities." IFPRI, Washington, D.C., January 1984. Mimeo.

0095 FAO – CARIS. *Agricultural Research in Developing Countries – Volume 1: Research Institutions.* Rome: FAO – CARIS, 1978.

0165 Evenson, R.E., and Y. Kislev. *Agricultural Research and Productivity.* New Haven: Yale University Press, 1975.

0172 Trigo, Eduardo. Background material. ISNAR, The Hague, n.d.

0852 Evenson, Robert E., and Yoav Kislev. *Investment in Agricultural Research and Extension: A Survey of International Data.* Center Discussion Paper No. 124. New Haven, Connecticut: Economic Growth Center, Yale University, August 1971.

0886 Evenson, Robert E., and Yoav Kislev. "Investment in Agricultural Research and Extension: A Survey of International Data." *Economic Development and Cultural Change* Vol. 23 (April 1975): 507–521.

Virgin Islands (US)

Personnel

Period/Year	PhD	MSc	BSc	Subtotal	Expat	Total	Source
1960–64	..	..	..	..	..	..	
1965–69	..	..	..	..	..	..	
1970	..	..	..	..	..	..	
1971	..	..	..	..	..	..	
1972	..	..	..	..	..	..	
1973	..	..	..	..	..	1	488
1974	..	..	..	..	..	2	488
1975	..	..	..	..	..	3	488
1976	..	..	..	..	..	2	488
1977	..	..	..	..	..	5	488
1978	..	..	..	..	..	4	488
1979	..	..	..	..	..	6	488
1980	..	..	..	..	..	5	488
1981	..	..	..	..	..	6	488
1982	..	..	..	..	..	6	488
1983	..	..	..	..	..	6	488
1984	..	..	..	..	..	6	488
1985	..	..	..	..	..	5	488
1986	..	..	..	..	..	..	

1960–64:   1965–69:

Expenditure   *LCU = U.S. Dollars*

Period/Year	Current LCU (millions)	Constant 1980 LCU (millions)	Constant 1980 US$ (millions) Atlas	PPP	Source
1960–64	..	..	..	..	
1965–69	..	..	..	..	
1970	..	..	..	..	
1971	..	..	..	..	
1972	..	..	..	..	
1973	0.043	0.074	0.074	0.065	488
1974	0.125	0.198	0.198	0.173	488
1975	0.175	0.253	0.253	0.221	488
1976	0.136	0.185	0.185	0.161	488
1977	0.193	0.246	0.246	0.215	488
1978	0.245	0.291	0.290	0.254	488
1979	0.276	0.301	0.301	0.263	488
1980	0.313	0.313	0.313	0.273	488
1981	0.500	0.456	0.456	0.398	488
1982	0.567	0.486	0.486	0.424	488
1983	0.621	0.512	0.512	0.447	488
1984	0.777	0.617	0.617	0.539	488
1985	0.809	0.622	0.622	0.543	488
1986	..	..	..	..	

1960–64:   1965–69:

478

Personnel Comments:

The personnel figures from source 488 are measured in scientific year units and represent an aggregation of project-level data obtained from the USDA's Current Research Information System.

Expenditure Comments:

The expenditure figures from source 488 represent annual obligations incurred by the local research system.

Sources:

0488 US Department of Agriculture – Cooperative State Research Service. *Inventory of Agricultural Research – Fiscal Years: 1970–1985*. Washington, D.C.: US Government Printing Office, various years.

Additional References:

0027 Harvey, Nigel, ed. *Agricultural Research Centres: A World Directory of Organizations and Programmes*. Seventh Edition, Two Volumes. Harlow, U.K.: Longman, 1983.

0095 FAO – CARIS. *Agricultural Research in Developing Countries – Volume 1: Research Institutions*. Rome: FAO – CARIS, 1978.

Western Samoa

Personnel

Period/Year	PhD	MSc	BSc	Subtotal	Expat	Total	Source
1960–64	..	..	..	..	..	..	
1965–69	..	..	..	..	..	..	
1970	..	..	..	..	..	..	
1971	..	..	..	..	..	..	
1972	..	..	..	..	..	..	
1973	..	..	..	..	..	..	
1974	..	..	..	..	..	..	
1975	..	..	..	..	..	..	
1976	..	..	..	..	..	..	
1977	..	..	..	..	..	..	
1978	..	..	..	..	..	..	
1979	..	..	..	..	..	..	
1980	..	..	..	..	..	8	16
1981	..	..	..	..	..	..	
1982	..	..	..	..	..	..	
1983	0	1	3	4	5	9	723
1984	..	..	..	..	..	..	
1985	..	..	..	..	..	..	
1986	0	2	3	5	5	10	723

1960–64:   1965–69:

Expenditure   *LCU = Tala*

Period/Year	Current LCU (millions)	Constant 1980 LCU (millions)	Constant 1980 US$ (millions)		Source
			Atlas	PPP	
1960–64	..	..	..	..	
1965–69	..	..	..	..	
1970	..	..	..	..	
1971	..	..	..	..	
1972	..	..	..	..	
1973	..	..	..	..	
1974	..	..	..	..	
1975	..	..	..	..	
1976	..	..	..	..	
1977	0.094	0.142	0.150	0.218	22
1978	..	..	..	..	
1979	..	..	..	..	
1980	0.151	0.151	0.160	0.232	16
1981	0.288	0.239	0.253	0.368	38
1982	..	..	..	..	
1983	..	..	..	..	
1984	0.560	0.302	0.319	0.464	723
1985	0.850	0.420	0.444	0.646	723
1986	1.130	..	..	..	723

1960–64:   1965–69:

Personnel Comments:

The personnel figures refer to the number of researchers at the Western Samoan Ministry of Agriculture. For 1980, source 36 reports a total number of 29 graduate staff at the ministry "having some research duties." Compared with other available observations, this figure seems too large, and for that reason is not included in the indicator series.

Expenditure Comments:

The expenditure indicators correspond to the personnel series.

Sources:

0016 Oram, Peter A., and Vishva Bindlish. *Resource Allocations to National Agricultural Research: Trends in the 1970s.* The Hague and Washington, D.C.: ISNAR and IFPRI, November 1981.

0022 UNESCO. *Statistical Yearbook 1983.* Paris: UNESCO, 1983.

0038 Gamble, W.K., R.M. Bourke, and C.W. Brookson. *South Pacific Agricultural Research Study.* Consultants' Report to the Asian Development Bank (two volumes). Manila: Asian Development Bank, June 1981.

0723 Fernando, L.H. "South Pacific NARS Resource Commitment Survey." ADB and ISNAR, The Hague, 1987. Mimeo.

Additional References:

0026 Oram, Peter A., and Mark Gieben. "Document Summaries." ISNAR, The Hague, 1984. Mimeo.

0036 ISNAR. *Report to the Government of Western Samoa: The Agricultural Research System in Western Samoa.* The Hague: ISNAR, December 1983.

0263 Jain, H.K. "Update on Western Samoa – A Back to Office Report." ISNAR, The Hague, July 1986. Mimeo.

0732 FAO – Regional Office for Asia and the Pacific. *Agricultural Research Systems in the Asia-Pacific Region.* Bangkok: FAO – Regional Office for Asia and the Pacific, 1986.

0771 Fernando, L.H. "ISNAR/IRETA Workshop on Planning and Management of Agricultural Research in the South Pacific: Survey of Agricultural Research Resources." ISNAR, The Hague, n.d. Mimeo.

Yemen Arab Republic

Personnel

Period/Year	PhD	MSc	BSc	Subtotal	Expat	Total	Source
1960–64	..	..	..	..	..	25	
1965–69	..	..	..	..	..	..	
1970	..	..	..	..	..	..	
1971	..	..	..	..	..	..	
1972	..	..	..	..	..	..	
1973	..	..	..	..	..	..	
1974	..	..	..	..	..	..	
1975	..	..	..	..	..	..	
1976	..	..	..	..	..	..	
1977	..	..	..	..	..	..	
1978	..	..	..	..	..	..	
1979	..	..	..	..	..	..	
1980	0	0	12	12	13	25	26
1981	..	..	..	..	..	..	
1982	..	..	..	..	..	..	
1983	1	17	25	43	13	56	812
1984	3	17	37	57	13	70	812
1985	6	21	37	64	16	80	812
1986	11	24	30	65	11	76	812

1960–64:   1965–69:

Expenditure

LCU = Yemen Rials

Period/Year	Current LCU (millions)	Constant 1980 LCU (millions)	Constant 1980 US$ (millions)		Source
			Atlas	PPP	
1960–64	..	..	..	..	
1965–69	..	..	..	..	
1970	..	..	..	..	
1971	..	..	..	..	
1972	..	..	..	..	
1973	..	..	..	..	
1974	..	..	..	..	
1975	..	..	..	..	
1976	..	..	..	..	
1977	..	..	..	..	
1978	..	..	..	..	
1979	..	..	..	..	
1980	..	..	..	..	
1981	..	..	..	..	
1982	..	..	..	..	
1983	6.435	5.746	1.229	2.949	812
1984	38.138	32.320	6.914	16.588	812
1985	25.037	..	..	..	812
1986	34.357	..	..	..	812

1960–64:   1965–69:

Personnel Comments:

According to source 812, the Agricultural Research Authority (ARA) was established in 1983 as a semiautonomous institution within the Ministry of Agriculture and Fisheries, and given full responsibility for all agricultural research in the country.

In the 10 years preceding the creation of ARA, several agricultural research stations were established and nationals were sent for training abroad with assistance from FAO, IDA, and UNDP.

Precise personnel figures for this period are difficult to construct. The 1980 figure, given by source 26, represents all researchers within the Central Agricultural Research Services Project.

From 1983 on, the indicators represent total ARA research staff. These figures, however, include nationals on training.

Expenditure Comments:

Over the last 15 or so years, three major international assistance projects have contributed to a strengthening of the national agricultural research system in the Arab Republic of Yemen. For the first project, the FAO/UNDP Project 1973–1978, expenditure data were not available.

For the second project, the IDA/UNDP Project 1979–1984 – also called the Central Agricultural Research Services Project – it is apparent from source 812 that the total approved budget for 5 years was US$ 6.5 million, with a government contribution in kind of YR 39.3 million. It appears that this project was replaced after three years by a third project, the IDA/IFAD/Italy/FAO Project 1983–1988, with the objective of creating ARA.

The table below gives a breakdown of the total agricultural research expenditures of ARA, by source of funding.

	1983	1984	1985	1986	1987
Government	6.435	14.316	18.310	19.447	19.800
External Loan/Assistance	–	23.822	6.727	14.810	NA
Total	6.435	38.138	25.037	34.257	NA
Source	812	812	812	812	841

Sources:

0026 Oram, Peter A., and Mark Gieben. "Document Summaries." ISNAR, The Hague, 1984. Mimeo.

0812 Sallam, Abdul Rahman. Personal communication. Director General, Agricultural Research Authority, Ministry of Agriculture and Fisheries. Taiz, Yemen Arab Republic, February 1988.

Additional References:

0094 FAO – Near East Regional Office. *Directory of Agricultural Research Institutions in the Near East Region.* Cairo: FAO – Near East Regional Office, 1979.

0813 Agricultural Research Authority (ARA). *Agricultural Research Authority Annual Report 1984.* Taiz, Yemen Arab Republic: ARA – Ministry of Agriculture and Fisheries, 1985.

0831 FAO. *Report of an Expert Consultation on Agricultural Extension and Research Linkages in the Near East, Amman, Jordan, 22–26 July 1985.* Rome: FAO, 1985.

0841 Hariri, G. "Agricultural Research Authority (Yemen)/ISNAR Potential Collaboration: Back-to-Office Report." ISNAR, The Hague, March 1988. Mimeo.

0865 Oram, Peter. "Agricultural Research Objectives and Priorities: Constraints to the Development of Agricultural Research Institutions in Arab countries." Review paper prepared for the first meeting of the Council for Arab Agricultural Research, organized by the Arab Fund for Economic and Social Development, Kuwait, 23 March 1988.

Yemen, P.D.R.

Personnel

Period/Year	PhD	MSc	BSc	Subtotal	Expat	Total	Source
1960–64	..	..	..	..	..	..	
1965–69	..	..	..	..	..	..	
1970	..	..	..	..	..	..	
1971	..	..	..	..	..	..	
1972	..	..	..	..	..	..	
1973	..	..	..	..	..	..	
1974	..	..	..	..	..	..	
1975	..	..	..	..	..	..	
1976	..	..	..	..	..	..	
1977	..	..	..	..	..	..	
1978	..	..	..	..	..	..	
1979	..	..	..	..	..	..	
1980	..	..	..	..	..	..	
1981	0	12	34	46	..	46	870
1982	..	..	..	..	..	..	
1983	8	37	25	70	..	70	17
1984	..	..	..	..	..	..	
1985	9	43	31	83	..	83	845
1986	..	..	..	..	..	..	

1960–64:   1965–69:

Expenditure   *LCU = Yemen Dinars*

Period/Year	Current LCU (millions)	Constant 1980 LCU (millions)	Constant 1980 US$ (millions)		Source
			Atlas	PPP	
1960–64	..	..	..	..	
1965–69	..	..	..	..	
1970	..	..	..	..	
1971	..	..	..	..	
1972	..	..	..	..	
1973	..	..	..	..	
1974	..	..	..	..	
1975	..	..	..	..	
1976	..	..	..	..	
1977	..	..	..	..	
1978	..	..	..	..	
1979	..	..	..	..	
1980	..	..	..	..	
1981	..	..	..	..	
1982	..	..	..	..	
1983	0.550	0.435	1.203	2.364	17
1984	..	..	..	..	
1985	..	..	..	..	
1986	..	..	..	..	

1960–64:   1965–69:

484

Personnel Comments:
The personnel figures include the agricultural research centers at El-Kod and Seiyum. According to source 844, agricultural research was restructured in 1985 and merged with extension into the Department of Research and Extension.

Expenditure Comments:
The 1983 expenditure figure is consistent with the personnel figure.

Sources:
0017 ISNAR, IFARD & AOAD. Survey of National Agricultural Research Systems: Unpublished Questionnaire Responses. ISNAR, The Hague, 1985.
0845 "Information on Agricultural Research in Peoples' Democratic Republic of Yemen." Handout submitted at the Rainfed Agricultural Information Network Workshop, Amman, Jordan, 17–20 March 1985.
0870 Kira, M. Taher. *Report on Agricultural Research Manpower Assessment in People's Democratic Republic of Yemen and Democratic Republic of Somalia.* Cairo: FAO Near East Regional Office, January 1980.

Additional References:
0094 FAO – Near East Regional Office. *Directory of Agricultural Research Institutions in the Near East Region.* Cairo: FAO – Near East Regional Office, 1979.
0173 FAO. *FAO Near East Regional Studies on Organization and Administration of Agricultural Research.* Rome: FAO, 1974.
0831 FAO. *Report of an Expert Consultation on Agricultural Extension and Research Linkages in the Near East, Amman, Jordan, 22–26 July 1985.* Rome: FAO, 1985.
0844 Bamakharama, Hussain Salem. "The Current Agricultural Research System in PDR Yemen." Paper presented at the Second AARINENA Conference, Nicosia, Cyprus, 15–17 December 1987.
0865 Oram, Peter. "Agricultural Research Objectives and Priorities: Constraints to the Development of Agricultural Research Institutions in Arab countries." Review paper prepared for the first meeting of the Council for Arab Agricultural Research, organized by the Arab Fund for Economic and Social Development, Kuwait, 23 March 1988.

Zaire

Personnel

Period/Year	PhD	MSc	BSc	Subtotal	Expat	Total	Source
1960–64	..	..	..		..	..	
1965–69	..	..	..	0	37	37	653
1970	..	..	..	5	16	21	532/175
1971	..	..	..	21	28	49	589
1972	..	..	..	30	37	67	589
1973	..	..	..	29	47	76	589
1974	..	..	..	..	..	..	
1975	..	..	..	..	..	..	
1976	..	..	..	..	..	37	95
1977	..	..	..	..	..	..	
1978	..	..	..	..	..	..	
1979	..	..	..	..	..	..	
1980	..	..	..	..	..	..	
1981	..	..	..	..	..	..	
1982	..	..	..	..	..	..	
1983	2	8	33	43	..	43	151
1984	..	..	..	..		..	
1985	..	..	..	..		..	
1986	..	..	..	..		..	

1960–64: 1965–69: 1966

Expenditure LCU = *Francs Congolais/Zaires*

Period/Year	Current LCU (millions)	Constant 1980 LCU (millions)	Constant 1980 US$ (millions)		Source
			Atlas	PPP	
1960–64	0.208	22.373	9.129	8.432	589
1965–69	0.428	13.081	5.338	4.930	589
1970	0.941	16.538	6.748	6.233	589
1971	1.213	20.207	8.245	7.615	589
1972	1.190	17.983	7.338	6.777	589
1973	1.757	22.523	9.191	8.488	589
1974	..	..	..	..	
1975	..	..	..	..	
1976	..	..	..	..	
1977	..	..	..	..	
1978	..	..	..	..	
1979	..	..	..	..	
1980	..	..	..	..	
1981	..	..	..	..	
1982	21.811	12.137	4.953	4.574	466
1983	35.817	10.641	4.342	4.010	466
1984	50.175	7.634	3.115	2.877	466
1985	..	..	..	..	
1986	..	..	..	..	

1960–64: 1960, 61, 62, 63 & 64 1965–69: 1965, 66, 67, 68 & 69

Personnel Comments:

The indicator series includes INERA (formerly INEAC) only. According to source 653, in 1966 or so an attempt was made through international and bilateral technical assistance to revive, in part, the former excellent research organization developed by INEAC. Five of the 7 major research and experiment stations were entrusted to research teams from Italy, West Germany, France, and Belgium, involving a total of 37 expatriate researchers. For the year 1983 or so, source 151 identifies about 150 researchers within organizations other than INERA, which are in some way involved in agricultural or agriculturally related research. These organizations are PN (17 researchers), CREN-K (10), CRAAL (6), CRSN (18), IRSS (3), MAB (2), SCZ (4), LVK (8), development projects (18), and universities (70). For reasons of consistency, they are not included in the indicator series.

Expenditure Comments:

The expenditure figures refer to INERA (formerly INEAC) only.

Acronyms and Abbreviations:

CRAAL: Centre de Recherche Agro-Alimentaire de Lubumbashi
CREN-K: Centre Régional d'Etudes Nucléaires de Kinshasa
CRSN: Centre de Recherche au Sciences Naturelles
INEAC: Institut National pour l'Etude Agronomique du Congo
INERA: Institut National pour l'Etude et la Recherche Agronomiques
IRSS: Institut de Recherche des Sciences de la Santé
LVK: Laboratoire Vétérinaire de Kinshasa
MAB: (Man and Biosphere)
PN: Programmes Nationaux
SCZ: Société des Cultures du Zaïre

Sources:

0095 FAO – CARIS. *Agricultural Research in Developing Countries – Volume 1: Research Institutions.* Rome: FAO – CARIS, 1978.
0151 ISNAR, and Groupe d'Etude de la Reorganisation du Systeme National de Recherche Agronomique. *Etude de la Reorganisation du Systeme National de Recherche Agronomique du Zaire.* The Hague: ISNAR, February 1985.
0175 Cooper, St. G.C. *Agricultural Research in Tropical Africa.* Kampala: East African Literature Bureau, 1970.
0466 World Bank – West Africa Agricultural Research Review Team. "National Agricultural Research Systems (NARS): A Regional Appraisal and Selected Issues." World Bank, Washington, D.C., August 1986. Mimeo.
0532 UNESCO Field Science Office for Africa. *Survey on the Scientific and Technical Potential of the Countries of Africa.* Paris: UNESCO, 1970.
0589 Kassapu, Samuel. *Les Dépenses de Recherche Agricole dans 34 Pays d'Afrique Tropicale.* Paris: Centre de Développement de l'OCDE, 1976.
0653 Webster, B.N. *Index of Agricultural Research Institutions and Stations in Africa.* Rome: FAO, n.d.

Additional References:

0010 Boyce, James K., and Robert E. Evenson. *National and International Agricultural Research and Extension Programs.* New York: Agricultural Development Council, Inc., 1975.
0014 Judd, M. Ann, James K. Boyce, and Robert E. Evenson. "Investing in Agricultural Supply." Economic Growth Center, Yale University, New Haven, Connecticut, 1983. Mimeo.
0016 Oram, Peter A., and Vishva Bindlish. *Resource Allocations to National Agricultural Research: Trends in the 1970s.* The Hague and Washington, D.C.: ISNAR and IFPRI, November 1981.
0023 Bennell, Paul. *Agricultural Researchers in Sub-Saharan Africa: An Overview.* Working Paper No. 4. The Hague: ISNAR, October 1985.
0026 ORAM, Peter A., and Mark Gieben. "Document Summaries." ISNAR, The Hague, 1984. Mimeo.
0027 Harvey, Nigel, ed. *Agricultural Research Centres: A World Directory of Organizations and Programmes.* Seventh Edition, Two Volumes. Harlow, U.K.: Longman, 1983.
0073 Oram, Peter A., and Vishva Bindlish. "Investment in Agricultural Research in Developing Countries: Progress, Problems, and the Determination of Priorities." IFPRI, Washington, D.C., January 1984. Mimeo.
0086 World Bank. "West Africa Agricultural Research Review – Country Studies." World Bank, Washington, D.C., 1985. Mimeo.
0089 Oram, Peter. "Report on National Agricultural Research in West Africa." IFPRI, Washington, D.C., February 1986. Mimeo.
0090 Institut de Recherche Agronomique et Zootechnique – Communauté Economique des Pays des Grands Lacs. *Répertoire des Recherches Agronomiques en Cours; Burundi-Rwanda-Zaire:*

1984. Gitega, Burundi: Institut de Recherche Agronomique et Zootechnique, 1984.

0163 CGIAR. "National Agricultural Research." CGIAR, Washington, D.C., 1985. Mimeo.

0266 UNESCO. *National Science Policies in Africa.* Science Policy Studies and Documents No. 31. Paris: UNESCO, 1974.

0360 Cooper, St.G.C. "Towards Trained Manpower for Agricultural Research in Africa." Paper presented at the Conference on Agricultural Research and Production in Africa, organized by the Association for the Advancement of Agricultural Sciences in Africa (AAASA), Addis Ababa, 29 August–4 September 1971.

0445 Swanson, Burton E., and Wade H. Reeves. "Agricultural Research Eastern and Southern Africa: Manpower and Training." World Bank, Washington, D.C., August 1986. Mimeo.

0852 Evenson, Robert E., and Yoav Kislev. *Investment in Agricultural Research and Extension: A Survey of International Data.* Center Discussion Paper No. 124. New Haven, Connecticut: Economic Growth Center, Yale University, August 1971.

Zambia

Personnel

Period/Year	PhD	MSc	BSc	Subtotal	Expat	Total	Source
1960–64	0	0	0	0	21	21	852/806
1965–69	0	1	0	1	42	43	806
1970	0	1	1	2	51	53	806
1971	0	2	1	3	69	72	806
1972	0	2	3	5	74	79	806
1973	0	1	4	5	61	66	806
1974	0	1	7	8	77	85	806
1975	0	2	7	9	67	76	806
1976	0	3	8	11	74	85	806
1977	0	3	10	13	48	61	806
1978	1	2	11	14	48	62	806
1979	1	2	12	15	41	56	806
1980	1	9	29	39	39	78	806
1981	1	9	26	36	45	81	806
1982	1	9	32	42	48	90	806
1983	1	9	44	54	62	116	806
1984	2	13	45	60	63	123	806
1985	1	13	67	81	57	138	806
1986	4	24	57	85	68	153	806

1960–64: 1963 & 64   1965–69: 1965, 66, 67, 68 & 69

Expenditure   *LCU = Kwacha*

Period/Year	Current LCU (millions)	Constant 1980 LCU (millions)	Constant 1980 US$ (millions) Atlas	PPP	Source
1960–64	..	..	..	..	
1965–69	1.075	1.839	2.191	2.497	175
1970	0.939	1.710	2.038	2.323	806
1971	1.482	2.795	3.331	3.796	806
1972	1.375	2.603	3.102	3.536	806
1973	1.387	2.357	2.809	3.201	806
1974	1.402	2.332	2.779	3.167	806
1975	1.636	3.172	3.781	4.309	806
1976	1.839	3.146	3.749	4.273	806
1977	2.022	3.104	3.699	4.216	806
1978	2.258	3.077	3.667	4.179	806
1979	2.537	2.836	3.379	3.851	806
1980	3.301	3.301	3.934	4.483	806
1981	2.807	2.619	3.122	3.558	806
1982	5.117	4.499	5.361	6.110	806
1983	2.868	1.453	1.732	1.974	806
1984	5.087	2.269	2.704	3.081	806
1985	5.267	1.656	1.973	2.249	806
1986	13.732	..	..	..	806

1960–64:   1965–69: 1968

Personnel Comments:

The main agricultural research organization in Zambia is the Research Branch of the Department of Agriculture (DOA) within the Ministry of Agriculture and Water Development (MOA&WD). Other departments which also conduct some agricultural research are the Department of Fisheries (DOF) and the Department of Veterinary and Tsetse Control (DOV&TC) – both within MOA&WD – and the Forestry Department (FD) within the Ministry of Land and Natural Resources (MOL&NR). In addition to agricultural research within the Ministries, there are several agricultural or agriculturally related research institutes under the responsibility of the National Council for Scientific Research (NCSR) and formerly the ARC. The indicator series includes only the Research Branch of the Department of Agriculture (DOA/RB), given only limited data – especially financial data – on the other research units. The table below gives an overview of some observations for additional research institutions.

	1965	1968	1969	1970	1971	1972	1973	1974	1975	1976
MOA&WD/DOF				26*						
MOA&WD /DOV&TC	7	10		7*	6*					
MOL&NR/FD	11	13		14*	26	28	24	21	21	18
NCSR/ARC		36	36	36	37	44	60	64	71	
Source	286	285	806	806	806 532*	806 279*	806	806	806	806

	1977	1978	1979	1980	1981	1982	1983	1984	1985
MOA&WD/DOF				11*		8*	11*	12*	8*
MOA&WD /DOV&TC	12*			16*		20*	25*	14*	
MOL&NR/FD	13	21	27	32	26	27	32	32	32
NCSR/ARC	69	83	77	75	81	86	86	94	
Source	806 243*	806	806	806 242*	806	806 27*	806 17*	806 385*	80 744*

Expenditure Comments:

The expenditure figures given by source 806 refer to the Research Branch of the Department of Agriculture only. The expenditure figure given by source 175 explicitly excludes university research and forestry, fisheries, and veterinary research. It is very likely that this strict definition only includes the Research Branch.

Acronyms and Abbreviations:

APRU: Animal Production Research Unit
ARC: Agricultural Research Council
DOA/RB : Department of Agriculture – Research Branch
DOF: Department of Fisheries
DOV&TC : Department of Veterinary and Tsetse Control
FD: Forestry Department
MOA&WD: Ministry of Agriculture and Water Development
MOL&NR: Ministry of Land and Natural Resources
NCSR/ARC: National Council for Scientific Research – Agricultural Research Council

Sources:

0175 Cooper, St.G.C. *Agricultural Research in Tropical Africa.* Kampala: East African Literature Bureau, 1970.
0806 Patel, B.K. Personal communication. Assistant Director, Department of Agriculture. Lusaka, February 1988.

0852 Evenson, Robert E., and Yoav Kislev. *Investment in Agricultural Research and Extension: A Survey of International Data.* Center Discussion Paper No. 124. New Haven, Connecticut: Economic Growth Center, Yale University, August 1971.

Additional References:

0002 SADCC, and DEVRES, Inc. *Agricultural Research Resource Assessment in the SADCC Countries, Volume I: Regional Analysis and Strategy.* Washington, D.C.: DEVRES, Inc., January 1985.

0009 SADCC, and DEVRES, Inc. *Agricultural Research Resource Assessment in the SADCC Countries, Volume II: Country Report Zambia.* Washington, D.C.: DEVRES, Inc., January 1985.

0010 Boyce, James K., and Robert E. Evenson. *National and International Agricultural Research and Extension Programs.* New York: Agricultural Development Council, Inc., 1975.

0014 Judd, M. Ann, James K. Boyce, and Robert E. Evenson. "Investing in Agricultural Supply." Economic Growth Center, Yale University, New Haven, Connecticut, 1983. Mimeo.

0016 Oram, Peter A., and Vishva Bindlish. *Resource Allocations to National Agricultural Research: Trends in the 1970s.* The Hague and Washington, D.C.: ISNAR and IFPRI, November 1981.

0017 ISNAR, IFARD & AOAD. Survey of National Agricultural Research Systems: Unpublished Questionnaire Responses. ISNAR, The Hague, 1985.

0019 Bengtsson, Bo, and Tedla Getachew, eds. *Strengthening National Agricultural Research.* Report from a SAREC workshop, 10–17 September 1979. Part 1: Background Documents; Part II: Summary and Conclusions. Sweden: SAREC, 1980.

0022 UNESCO. *Statistical Yearbook 1983.* Paris: UNESCO, 1983.

0023 Bennell, Paul. *Agricultural Researchers in Sub-Saharan Africa: An Overview.* Working Paper No. 4. The Hague: ISNAR, October 1985.

0026 Oram, Peter A., and Mark Gieben. "Document Summaries." ISNAR, The Hague, 1984. Mimeo.

0027 Harvey, Nigel, ed. *Agricultural Research Centres: A World Directory of Organizations and Programmes.* Seventh Edition, Two Volumes. Harlow, U.K.: Longman, 1983.

0095 FAO – CARIS. *Agricultural Research in Developing Countries – Volume 1: Research Institutions.* Rome: FAO – CARIS, 1978.

0163 CGIAR. "National Agricultural Research." CGIAR, Washington, D.C., 1985. Mimeo.

0164 Association for the Advancement of Agricultural Sciences in Africa (AAASA). *Proceedings of the Workshop on Agricultural Research Administration, Nairobi, Kenya, 27–30 June 1977.* Proceedings Series PE–4. Addis Ababa, Ethiopia: AAASA and IDRC, August 1979.

0242 Commonwealth Agricultural Bureaux (CAB). *List of Research Workers in Agricultural Sciences in the Commonwealth 1981.* Slough, England: CAB, 1981.

0243 Commonwealth Agricultural Bureaux (CAB). *List of Research Workers in the Agricultural Sciences in the Commonwealth and in the Republic of Ireland 1978.* Slough, England: CAB, 1978.

0244 Commonwealth Agricultural Bureaux (CAB). *List of Research Workers in the Agricultural Sciences in the Commonwealth and in the Republic of Ireland 1975.* Slough, England: CAB, 1975.

0245 Chibasa, W.M. "Agricultural Research Efforts in Zambia." Paper presented at the inaugural meetings of the consultative technical meetings under SADCC Food Security Project One, November 1982.

0266 UNESCO. *National Science Policies in Africa.* Science Policy Studies and Documents No. 31. Paris: UNESCO, 1974.

0279 Commonwealth Agricultural Bureaux (CAB). *List of Research Workers in the Agricultural Sciences in the Commonwealth and in the Republic of Ireland 1972.* Slough, England: CAB, 1972.

0285 Commonwealth Agricultural Bureaux (CAB). *List of Research Workers in the Agricultural Sciences in the Commonwealth and in the Republic of Ireland 1969.* Slough, England: CAB, 1969.

0286 Commonwealth Agricultural Bureaux (CAB). *List of Research Workers in Agriculture, Animal Health and Forestry in the Commonwealth and in the Republic of Ireland 1966.* Slough, England: CAB, 1966.

0314 Nsereko, J. "Agricultural Research in Zambia and Its Relationship with the International Agricultural Research System." In *Strengthening National Agricultural Research.* Report from a SAREC workshop, 10–17 September 1979, edited by Bo Bengtsson and Getachew Tedla, pp. 70–76. Sweden: SAREC, 1980.

0360 Cooper, St.G.C. "Towards Trained Manpower for Agricultural Research in Africa." Paper presented at the Conference on Agricultural Research and Production in Africa, organized by the Association for the Advancement of Agricultural Sciences in Africa (AAASA), Addis Ababa, 29 August–4 September 1971.

0385 SADCC, and DEVRES, Inc. *SADCC Region Agricultural Research Resource Assessment –*

Data Base Management Information System: Diskettes and User's Guide. Printouts. Washington, D.C.: SADCC, and DEVRES, Inc., n.d.

0388 Kean, Stewart. "Research Branch Expenditure (1976–1985)." Preliminary table made for the Zambian case study of the ISNAR Study on the Organization and Management of On-Farm Research. ISNAR, The Hague, 1986. Mimeo.

0400 UNESCO. *The Promotion of Scientific Activity in Tropical Africa.* Science Policy Studies and Documents No. 11. Paris: UNESCO, 1969.

0445 Swanson, Burton E., and Wade H. Reeves. "Agricultural Research Eastern and Southern Africa: Manpower and Training." World Bank, Washington, D.C., August 1986. Mimeo.

0446 Kyomo, M.L. "Agricultural Research in Eastern and Southern Africa: Issues and Priorities."

Southern African Centre for Cooperation in Agricultural Research of SADCC, Gabarone, Botswana, 1986. Mimeo.

0532 UNESCO Field Science Office for Africa. *Survey on the Scientific and Technical Potential of the Countries of Africa.* Paris: UNESCO, 1970.

0589 Kassapu, Samuel. *Les Dépenses de Recherche Agricole dans 34 Pays d'Afrique Tropicale.* Paris: Centre de Développement de l'OCDE, 1976.

0847 Kimura, J.H. "Financial and Administrative Management of Research Institutions in Eastern and Southern Africa: Report on Responses to a Questionnaire." In *Promotion of Technology Policy and Science Management in Africa*, edited by Karl Wolfgang Menck and Wolfgang Gmelin. Bonn: Deutsche Stiftung für internationale Entwicklung (DSE), 1986.

Zimbabwe

Personnel

Period/Year	PhD	MSc	BSc	Subtotal	Expat	Total	Source
1960–64	..	..	..	..	..	..	
1965–69	16	20	78	114	..	114	286/285
1970	..	..	..	..	..	112	43
1971	..	..	..	..	..	..	
1972	..	..	..	..	..	..	
1973	..	..	..	..	..	..	
1974	..	..	..	..	..	133	43
1975	..	..	..	..	..	..	
1976	..	..	..	..	..	..	
1977	..	..	..	..	..	124	43
1978	..	..	..	..	..	..	
1979	..	..	..	..	..	..	
1980	22	29	63	114	..	114	242
1981	..	..	..	..	..	..	
1982	..	..	..	..	..	..	
1983	16	51	84	151	..	151	17
1984	..	..	..	153	..	153	7
1985	..	..	..	..	..	193	744
1986	..	..	..	..	..	..	

1960–64:   1965–69: 1965 & 68

Expenditure   *LCU = Zimbabwe Dollars*

Period/Year	Current LCU (millions)	Constant 1980 LCU (millions)	Constant 1980 US$ (millions)		Source
			Atlas	PPP	
1960–64	..	..	..	..	
1965–69	2.182	4.949	7.064	9.097	67/805
1970	2.846	6.164	8.798	11.330	67/805
1971	3.180	6.510	9.292	11.966	67/805
1972	3.591	7.007	10.002	12.880	67/805
1973	4.084	7.397	10.558	13.596	67/805
1974	4.324	7.245	10.340	13.316	67/805
1975	5.047	7.625	10.884	14.015	67/805
1976	6.608	9.142	13.049	16.805	67/805
1977	6.906	8.639	12.332	15.880	67/805
1978	7.494	9.316	13.297	17.124	67/805
1979	8.529	9.301	13.276	17.097	67/805
1980	9.428	9.428	13.457	17.330	67/805
1981	11.573	10.419	14.871	19.150	67/805
1982	11.484	9.085	12.968	16.700	67/805
1983	12.554	8.702	12.121	15.996	67/805
1984	14.038	8.766	12.512	16.113	67/805
1985	16.963	10.553	15.063	19.398	67/805
1986	..	..	..	..	

1960–64:   1965–69: 1965, 66, 67, 68 & 69

Personnel Comments:
The personnel indicator series comprises the Department of Research and Specialist Services (DRSS) within the Ministry of Agriculture, the Tobacco Research Board, and the Pig Industry Board. The following organizations are also involved in agricultural research: the Department of Veterinary Services and the Branch of Agricultural Engineering within the Ministry of Agriculture, the Forestry Commission, the Department of National Parks and Wildlife (forestry and fisheries research), the Faculty of Agriculture – Zimbabwe University, and some private organizations like the Zimbabwe Sugar Association.

The table below gives an institutional breakdown of the personnel indicators, plus some scattered observations on organizations which are not included in the series.

	1965	1968	1970	1974	1977	1980	1982	1983	1984	1985	1986
DRSS	78	105	97	118	108	98	NA	128	128	165	163
TRB	29	15	[15]	[15]	[16]	16	23	23	23	25	NA
PIB	0	0	[0]	[0]	[0]	0	0	[0]	2	3	NA
Subtotal	107	120	112	133	124	114	NA	151	153	193	NA
DVS	20	21				14	19		10	28	
FC	5	4				4	8	4	4	8	
BAE									7		
DNPW	7	15				12			21		
FA	22	19				19		19	19		
ZSA						4	9		5	9	
RARS									3		
Source	286	285	43	43	43	242	27	17	7	744	809

[ ] = estimated or constructed by authors.

Expenditure Comments:
The indicator series includes the expenditures of DRSS, TRB, PIB, plus all agricultural research expenditures funded by the Agricultural Research Council (ARC). Some of the ARC expenditures are not necessarily spent by the organizations mentioned above.

Acronyms and Abbreviations:
ARC: Agricultural Research Council
BAE : Branch of Agricultural Engineering
DNPW: Department of National Parks and Wildlife
DRSS: Department of Research and Specialist Services
DVS: Department of Veterinary Services
FA: Faculty of Agriculture
FC: Forestry Commission
PIB: Pig Industry Board
RARS: Rattray Arnold Research Station
TRB : Tobacco Research Board
ZSA: Zimbabwe Sugar Association

Sources:
0007 SADCC, and DEVRES, Inc. *Agricultural Research Resource Assessment in the SADCC Countries, Volume II: Country Report Zimbabwe.* Washington, D.C.: DEVRES, Inc., January 1985.
0017 ISNAR, IFARD & AOAD. Survey of National Agricultural Research Systems: Unpublished Questionnaire Responses. ISNAR, The Hague, 1985.
0043 Department of Research and Specialist Services – Ministry of Agriculture, and ISNAR. *A Training Plan for the Department of Research and Specialist Services, Zimbabwe, 1985 to 1988.* The Hague: ISNAR, February 1985.
0067 Billing, K.J. *Zimbabwe and the CGIAR Centers: A Study of Their Collaboration in Agricultural Research.* CGIAR Study Paper Number 6. Washington, D.C.: World Bank, 1985.
0242 Commonwealth Agricultural Bureaux (CAB). *List of Research Workers in Agricultural Sciences in the Commonwealth 1981.* Slough, England: CAB, 1981.

0285 Commonwealth Agricultural Bureaux (CAB). *List of Research Workers in the Agricultural Sciences in the Commonwealth and in the Republic of Ireland 1969.* Slough, England: CAB, 1969.

0286 Commonwealth Agricultural Bureaux (CAB). *List of Research Workers in Agriculture, Animal Health and Forestry in the Commonwealth and in the Republic of Ireland 1966.* Slough, England: CAB, 1966.

0744 *Agricultural Research Centres: A World Directory of Organizations and Programmes.* Eighth Edition, Two Volumes. Harlow, U.K.: Longman, 1986.

0805 Mudimu, Godfrey D. "Planning and Evaluating Agricultural Research in Zimbabwe: A Preliminary Assessment and Research Proposal." MSc thesis, Department of Agricultural Economics, Michigan State University, Michigan, 1986.

Additional References:

0002 SADCC, and DEVRES, Inc. *Agricultural Research Resource Assessment in the SADCC Countries, Volume I: Regional Analysis and Strategy.* Washington, D.C.: DEVRES, Inc., January 1985.

0010 Boyce, James K., and Robert E. Evenson. *National and International Agricultural Research and Extension Programs.* New York: Agricultural Development Council, Inc., 1975.

0014 Judd, M. Ann, James K. Boyce, and Robert E. Evenson. "Investing in Agricultural Supply." Economic Growth Center, Yale University, New Haven, Connecticut, 1983. Mimeo.

0016 Oram, Peter A., and Vishva Bindlish. *Resource Allocations to National Agricultural Research: Trends in the 1970s.* The Hague and Washington, D.C.: ISNAR and IFPRI, November 1981.

0017 ISNAR, IFARD & AOAD. Survey of National Agricultural Research Systems: Unpublished Questionnaire Responses. ISNAR, The Hague, 1985.

0023 Bennell, Paul. *Agricultural Researchers in Sub-Saharan Africa: An Overview.* Working Paper No. 4. The Hague: ISNAR, October 1985.

0026 Oram, Peter A., and Mark Gieben. "Document Summaries." ISNAR, The Hague, 1984. Mimeo.

0027 Harvey, Nigel, ed. *Agricultural Research Centres: A World Directory of Organizations andProgrammes.* Seventh Edition, Two Volumes. Harlow, U.K.: Longman, 1983.

0073 Oram, Peter A., and Vishva Bindlish. "Investment in Agricultural Research in Developing Countries: Progress, Problems, and the Determination of Priorities." IFPRI, Washington, D.C., January 1984. Mimeo.

0163 CGIAR. "National Agricultural Research." CGIAR, Washington, D.C., 1985. Mimeo.

0165 Evenson, R.E., and Y. Kislev. *Agricultural Research and Productivity.* New Haven: Yale University Press, 1975.

0246 Ministry of Agriculture – Department of Research and Specialist Services (DRSS). "Establishment Tables 1981–1982." Ministry of Agriculture, Zimbabwe, 1982. Mimeo.

0247 Ministry of Agriculture – Department of Research and Specialist Services (DRSS). "Establishment Tables 1983–1984." Ministry of Agriculture, Zimbabwe, 1984. Mimeo.

0287 Commonwealth Agricultural Bureaux (CAB). *List of Research Workers in Agriculture, Animal Health and Forestry in the British Commonwealth, the Republic of Sudan and the Republic of Ireland 1959.* Slough, England: CAB, 1959.

0385 SADCC, and DEVRES, Inc. *SADCC Region Agricultural Research Resource Assessment – Data Base Management Information System: Diskettes and User's Guide.* Printouts. Washington, D.C.: SADCC, and DEVRES, Inc., n.d.

0445 Swanson, Burton E., and Wade H. Reeves. "Agricultural Research Eastern and Southern Africa: Manpower and Training." World Bank, Washington, D.C., August 1986. Mimeo.

0446 Kyomo, M.L. "Agricultural Research in Eastern and Southern Africa: Issues and Priorities." Southern African Centre for Cooperation in Agricultural Research of SADCC, Gabarone, Botswana, 1986. Mimeo.

0653 Webster, B.N. *Index of Agricultural Research Institutions and Stations in Africa.* Rome: FAO, n.d.

0809 Fenner, R.J. Personal communication. Director, Department of Research and Specialist Services. Harare, Zimbabwe, February 1988.

0832 Mugabe, N.R. *Project for the Strengthening of Agricultural Research in SADCC Countries: National Report for the Zimbabwe Agricultural Research Resource Assessment.* Harare: Ministry of Agriculture, August 1984.

0852 Evenson, Robert E., and Yoav Kislev. *Investment in Agricultural Research and Extension: A Survey of International Data.* Center Discussion Paper No. 124. New Haven, Connecticut: Economic Growth Center, Yale University, August 1971.

0886 Evenson, Robert E., and Yoav Kislev. "Investment in Agricultural Research and Extension: A Survey of International Data." *Economic Development and Cultural Change* Vol. 23 (April 1975): 507–521.

References for Part II

Abdalla, Abdalla Ahmed. "Agricultural Research in the IGADD Sub-Region and Related Manpower Training." Inter-Governmental Authority on Drought and Development (IGADD) and FAO, 1987. Mimeo. [0873]

Abebe, M., M. Mekuria, and T. Gebremeskel. "Status of Agricultural Research in Ethiopia." In *Proceedings of the Eastern Africa-ACIAR Consultation on Agricultural Research*, pp. 9–15. Australia: ACIAR, 1984. [0760]

Abington, J.B. Personal communication. Chief Research Officer, Ministry of Agriculture & Lands. Honiara, Solomon Islands, February 1988. [0808]

Academy of Finland. *Kertomus Suomen Akatemian Toiminnasta 1984.* Helsinki: The Academy of Finland, 1985. [0605]

Acharya, R.M. Personal communication. Deputy Director General, Indian Council of Agricultural Research. New Delhi, January 1988. [0777]

Adeyemo, Remi. "Agricultural Research in Nigeria: Priorities and Effectiveness." *Agricultural Administration* Vol. 17 (1984): 81–91. [0168]

Adulavidhaya, Kamphol, Rungruang Isarangkura, Preeyanuch Apibunyopas, and Nittaya Dulyasatit. "Evaluation in National Agricultural Research Systems in Thailand." Paper presented at the International Workshop on The Role of Evaluation in National Agricultural Research Systems, IFARD, and IDRC, Singapore, 7–9 July 1986. [0332]

Adulavidhaya, Kamphol, Rungruang Isarangkura, Preeyanuch Apibunyopas, and Nittaya Dulyasatit. "Evaluation of Agriculutral Research in Thailand." In *Evaluation in National Agricultural Research*. Proceedings of a workshop held in Singapore, 7–9 July 1986, edited by Douglas Daniels, pp. 51–61. Canada: IDRC, 1987. [0718]

Agble, W.K. "Agricultural Research in Ghana." In *Strengthening National Agricultural Research*. Report from a SAREC workshop, 10–17 September 1979, edited by Bo Bengtsson and Getachew Tedla, pp. 36–43. Sweden: SAREC, 1980. [0310]

Agency for Agricultural Research and Development (AARD). *5 Years (1976–1980) of Agricultural Research and Development for Indonesia.* Indonesia: AARD, n.d. [0447]

Agency for Agricultural Research and Development (AARD). *This is AARD.* Indonesia: AARD, 1985. [0182]

Agency for Agricultural Research and Development (AARD). *An Evaluation of the Industrial Crop Research Program of AARD.* Two volumes. Indonesia: AARD, August 1985. [0159]

Agency for Agricultural Research and Development (AARD). *AARD Reviews 1984–1987: An Analysis of their Recommendations and their Implementation.* Jakarta, Indonesia: AARD, August 1988. [0811]

Agricultural and Food Research Council (AFRC). *Report of the Agricultural and Food Research Council for the year 1985–86.* England: AFRC, November 1986. [0394]

Agricultural and Food Research Council (AFRC). *Corporate Plan 1987–1992.* London: AFRC, March 1987. [0596]

Agricultural Research Authority (ARA). *Agricultural Research Authority Annual Report 1984.* Taiz, Yemen Arab Republic: ARA – Ministry of Agriculture and Fisheries, 1985. [0813]

Agricultural Research Centres: A World Directory of Organizations and Programmes. Eighth Edition, Two Volumes. Harlow, U.K.: Longman, 1986. [0744]

Agricultural Research Corporation (ARC). *List of Research Scientists and Senior Administrators as of July 1983.* Sudan: ARC, July 1983. [0076]

Agricultural Research Corporation (ARC). "ARC Research Scientists Employed at Major Research Stations." ARC, Sudan, July 1983. Mimeo. [0253]

Agricultural Research Council (ARC). *Index of Agricultural Research 1969.* London: ARC, n.d. [0635]

Agricultural Research Council (ARC). *The Agricultural Research Service.* London: ARC, 1969. [0636]

Agricultural Research Institute of Northern Ireland. *Jubilee Report 1926–1976.* Hillsborough, Co. Down, Northern Ireland: Agricultural Research Institute of Northern Ireland, 1976. [0594]

Agricultural Research Review Committee. *Review of Agricultural Research Institutions in Pakistan.* Islamabad: Agricultural Research Division – Government of Pakistan, November 1984. [0727]

Agriculture Affairs and Fish Resources Authority (AAFRA). Handout submitted during the Rainfed Agricultural Information Network Workshop, Amman, Jordan, 17–20 March 1985. [0843]

Ahmad, Malik Mushtaq. "Resource Allocation to Agricultural Research in Pakistan." In *Resource Allocation to Agricultural Research*. Proceedings of a workshop held in Singapore 8–10 June 1981, edited by Douglas Daniels and Barry Nestel, pp. 55–60. Ottawa: IDRC, 1981. [0296]

Ahmad, N. "Agricultural Research in Trinidad and Tobago." In *Strengthening National Agricultural Research*. Report from a SAREC workshop, 10–17 September 1979, edited by Bo Bengtsson and Getachew Tedla, pp. 182–197. Sweden: SAREC, 1980. [0325]

Ahsan, Ekramul. *Bangladesh: Resources Allocation in Agricultural Research.* Bangladesh: Bangladesh Agricultural Research Council, n.d. [0101]

Ahsan, Ekramul. "Resource Allocation to Agricultural Research in Bangladesh." In *Resource Allocation to Agricultural Research*. Proceedings of a workshop held in Singapore 8–10 June 1981, edited by Douglas Daniels and Barry Nestel, pp. 129–136. Ottawa: IDRC, 1981. [0302]

Ahsan, Ekramul. "Upgrading Manpower Resources for Agricultural Research in Bangladesh." In *Agriculture and Resources Research Manpower Development in South and Southeast Asia*, pp. 201–206. Los Baños, Philippines: Philippine Council for Agriculture and Resources Research and Development, 1983. [0795]

Ahsan, Ekramul. "Keynote Address: Management of Human Resources in Agricultural Research." In *Management of Human Resources in Agricultural Research*. Report of the International Workshop on Management of Human Resources in Agricultural Research held 3–5 March 1986, edited by Theodore Hutchcroft, pp. 27–31. Dhaka: Bangladesh Agricultural Research Council, ISNAR, and Winrock International, 1986. [0326]

AID. "Peru Project Paper: Agricultural Research, Extension and Education." AID, Washington, D.C., 1980. Mimeo. [0204]

AID. *Country Development Strategy Statement: Mali.* Washington, D.C.: AID, January 1983. [0145]

Aithnard, T., and C.T. Nguyen-Vu. "La Recherche Agricole au Togo: Bilan et Perspectives." Paper presented at the Workshop on Improving Agricultural Research Organizations and Management: Progress and Issues for the Future, ISNAR, The Hague, 8–12 September 1986. [0275]

Alarcón Millan, Enrique. *El Modelo Institucional para la Investigación Agropecuaria: Problemática y Planteamiento para su Cambio.* Bogotá: Instituto Colombiano Agropecuario – Subgerencia de Investigación y Transferencia, May 1986. [0734]

Alassane Salif, N'Diaye. Personal communication. Le Ministre de la Recherche Scientifique, Ministère de la Recherche Scientifique. Abidjan, Côte d'Ivoire, January 1988. [0774]

Alfaro, R. "La Investigación Agrícola en el Ministerio de Agricultura y Ganadería de Costa Rica."

Paper presented at Curso-Taller sobre la Administración de la Investigación Agrícola, Panamá City, 14–25 July 1986. [0380]

Alio, A.N. "Agricultural Research System in Somalia." Paper presented at the Second General Conference of AARINENA, Nicosia, Cyprus, 15–17 December 1987. [0846]

Alvarez, Luis Alberto. *Una Introspectiva del Sistema de Investigación Agropecuaria del Ministerio de Agricultura y Ganadería (MAG).* Publicación Miscelánea No. 17. Asunción, Paraguay: MAG – Dirección de Investigación y Extensión Agropecuaria y Forestal, 1986. [0267]

Alves, Eliseu, and Elisio Contini. "A Modernização da Agricultura Brasileira." EMBRAPA, Brazil, 13 March 1987. Mimeo. [0725]

Alves, Eliseu Roberto de Andrade, et al. *Pesquisa Agropecuária: Perspectiva Histórica e Desenvolvimento Institucional.* Brasília: Empresa Brasileira de Pesquisa Agropecuária, 1985. [0108].

An Foras Taluntais. *Annual Report.* Years: 1964/65 – 1969/70. Dublin: An Foras Taluntais, various years. [0631]

Anh, Do. "Agricultural Research Management Systems in the Philippines, India, and Bangladesh." In *Proceedings of the Research Extension Linkage Workshop Held in Hanoi, Viet Nam, 9–13 June 1986.* Rome: FAO, 1986. [0434]

Anh, Do. "Research-Extension Linkage in Viet Nam: The State of the Art." In *Proceedings of the Research Extension Linkage Workshop Held in Hanoi, Viet Nam, 9–13 June 1986.* Rome: FAO, 1986. [0438]

Ardila V., Jorge. "Colombia: Análisis de la Cooperación entre los Centros Nacionales y los Centros Internacionales de Investigación Agropecuaria." BID, OWA & IICA, 1980. Mimeo. [0030]

Ardila V., Jorge. "Organización Institucional para la Investigación Agropecuaria en Colombia, en Relación al Plan Nacional de Investigaciones Agropecuarias." ICA, Cali, April 1982. Mimeo. [0121]

Ardila V., Jorge, and Hernán Chaverra G. *Diagnóstico Sobre la Investigación Agropecuaria a Venezuela.* Bogota: IICA and OEA: Oficina en Colombia, and PROTAAL, December 1980. [0206]

Ardila V., Jorge, N. Reichart, and A. Rincón. *Sistemas Nacionales de Investigación Agropecuaria en América Latina: Análisis Comparativo de los Recursos Humanos en Países Seleccionados – El Caso del INTA en Argentina.* Documento PROTAAL 48. Costa Rica: IICA, 1980. [0097]

Ardila V., Jorge, and Eduardo Trigo, et al. *Sistemas Nacionales de Investigación Agropecuaria en América Latina: Análisis Comparativo de los*

Recursos Humanos en Países Seleccionados. Colombia: IICA – Oficina en Colombia, February 1980. [0194]

Ardila V., Jorge, Eduardo Trigo, and Martín Piñeiro. "Human Resources in Agricultural Research: Three Cases in Latin America." In *Resource Allocation to Agricultural Research.* Proceedings of a workshop held in Singapore 8–10 June 1981, edited by Douglas Daniels and Barry Nestel, pp. 151–167. Ottawa: IDRC, 1981. [0305]

Ardila V., Jorge, Eduardo Trigo, and Martín Piñeiro. "Human Resources in Agricultural Research: Three Cases in Latin America." IICA – PROTAAL, San José, March 1981. Mimeo. [0096]

Aresvik, Oddvar. *The Agricultural Development of Turkey.* Praeger Special Studies in International Economics and Development. New York: Praeger Publishers, 1975. [0516]

Aresvik, Oddvar. *The Agricultural Development of Iran.* Praeger Special Studies in International Economics and Development. New York: Praeger Publishers, 1976. [0517]

Arrundale, R. "Research and Development: Expenditure and Employment." *Economic Trends* No. 309 (July 1979): 100–124. [0702]

Arrundale, R. "Research and Development: Expenditure and Employment, 1978." *Economic Trends* No. 321 (July 1980): 99–109. [0701]

Association for the Advancement of Agricultural Sciences in Africa (AAASA). *Proceedings of the Workshop on Agricultural Research Administration, Nairobi, Kenya, 27–30 June 1977.* Proceedings Series PE–4. Addis Ababa, Ethiopia: AAASA and IDRC, August 1979. [0164]

Australian Centre for International Agricultural Research (ACIAR). *Proceedings of the Eastern Africa-ACIAR Consultation on Agricultural Research, 18–22 July 1983, Nairobi, Kenya.* Australia: ACIAR, 1984. [0227]

Ayuk-Takem, J.A. Personal communication. Interim Director of the Institute of Agronomic Research. Yaounde, Cameroon, March 1988. [0837]

Badruddoza, K.M. "Agricultural Research System in Indonesia." In *Proceedings of the Research Extension Linkage Workshop Held in Hanoi, Viet Nam, 9–13 June 1986.* Rome: FAO, 1986. [0435]

Bamakharama, Hussain Salem. "The Current Agricultural Research System in PDR Yemen." Paper presented at the Second AARINENA Conference, Nicosia, Cyprus, 15–17 December 1987. [0844]

Bangladesh Agricultural Research Council (BARC). *Agricultural Research in Bangladesh.* Dhaka: BARC, 1983. [0114]

Bangladesh Agricultural Research Council (BARC). *National Agricultural Research Plan 1984–1989.* Bangladesh: BARC, May 1984. [0099]

Bangladesh Agricultural Research Council (BARC). "Progress Report Manpower Planning in Agricultural Research." BARC, Dhaka, July 1986. Mimeo. [0804]

Bangladesh Agricultural Research Council (BARC). *Manpower Planning and Development in Agricultural Research.* Dhaka: BARC, September 1987. [0854]

Bangladesh Agricultural Research Council (BARC), ISNAR, and Winrock International. *Management of Human Resources in Agricultural Research.* Report of the International Workshop on Management of Human Resources in Agricultural Research held March 3–5, 1986 in Dhaka, Bangladesh. Dhaka: BARC, ISNAR, and Winrock International, 1986. [0093]

Barker, G., L.A. Bell, and A. Wahab. "Agricultural Research in Jamaica." In *Strengthening National Agricultural Research.* Report from a SAREC workshop, 10–17 September 1979, edited by Bo Bengtsson and Getachew Tedla, pp. 175–181. Sweden: SAREC, 1980. [0324]

Barquero, Vargas A. "Costa Rica – Some Facts and Figures about Agriculture and Agricultural Research." In *Strengthening National Agricultural Research.* Report from a SAREC workshop, 10–17 September 1979, edited by Bo Bengtsson and Getachew Tedla, pp. 155–166. Sweden: SAREC, 1980. [0322]

Beck, Harvey. Personal communication. Department of Agricultural Economics, University of Reading. Reading, England, December 1986. [0542]

Bengtsson, B. "Agricultural Research in Sri Lanka." In *Strengthening National Agricultural Research.* Report from a SAREC workshop, 10–17 September 1979, edited by Bo Bengtsson and Getachew Tedla, pp. 123–137. Sweden: SAREC, 1980. [0319]

Bengtsson, Bo, and Tedla Getachew, eds. *Strengthening National Agricultural Research.* Report from a SAREC workshop, 10–17 September 1979. Part 1: Background Documents; Part II: Summary and Conclusions. Sweden: SAREC, 1980. [0019]

Bennell, Paul. "Conditions of Service for Agricultural Research Scientists in Nigeria: What Room for Manoeuvre?" ISNAR, The Hague, January 1984. Mimeo. [0162]

Bennell, Paul. *Agricultural Researchers in Sub-Saharan Africa: An Overview.* Working Paper No. 4. The Hague: ISNAR, October 1985. [0023]

Berg, A. van den. Personal communication. Nationale Raad voor Landbouwkundig Onderzoek TNO. The Hague, September 1986. [0430]

Bernal C., Fernando. "Análisis de la Estructura de

Generación y Tranferencia de Tecnología: La Subgerencia de Investigación y Transferencia del ICA." Bogotá, October 1987. Mimeo. [0733]

Berner, Boel. *The Organization and Economy of Pest Control in China.* Discussion Paper Series 128. Sweden: Research Policy Institute, University of Lund, July 1979. [0277]

Berner, Boel. *The Organization and Planning of Scientific Research in China Today.* Discussion Paper Series 134. Sweden: Research Policy Institute, University of Lund, November 1979. [0278]

Bilbeisi, Usama. "The Jordanian Experiment in Agricultural Research and Extension Linkage: An Analytic Overview." Paper presented at the International Workshop on Agricultural Research Management, ISNAR, The Hague, 7–11 September 1987. [0752]

Billgren, Boel, and Jon Sigurdson. *An Estimate of Research and Development Expenditures in the People's Republic of China in 1973.* Industry and Technology Occasional Paper No. 16. Paris: OECD Development Centre, July 1977. [0554]

Billing, K.J. *Zimbabwe and the CGIAR Centers: A Study of Their Collaboration in Agricultural Research.* CGIAR Study Paper Number 6. Washington, D.C.: World Bank, 1985. [0067]

Bingen, R. James, and Jacques Faye. "Agricultural Research and Extension in Francophone West Africa: The Senegal Experience." Paper presented at the Farming Systems Research and Extension Symposium: Management and Methodology, Kansas State University, Manhattan, Kansas, 13–16 October 1985. [0456]

Binns, H.R. *The East African Veterinary Research Organisation – Its Development, Objects and Scientific Activities.* Nairobi: East African Community, n.d. [0643]

Birowo, A.T. "Policy Issues for Agricultural Research in Indonesia." In *Strengthening National Agricultural Research.* Report from a SAREC workshop, 10–17 September 1979, edited by Bo Bengtsson and Getachew Tedla, pp. 98–111. Sweden: SAREC, 1980. [0317]

Blasco, Mario, Pedro González A., and Luis Scarneo W. "Perú: Análisis de la Cooperación Entre los Centros Nacionales y los Centros Internacionales de Investigación Agropecuaria." IICA – OEA: Oficina en Colombia, and PROTAAL, Bogotá, December 1980. Mimeo. [0205]

Bomo, Nelson. "Agricultural Research and Extension Programme in Papua New Guinea." *Alafua Agricultural Bulletin* Vol. 5, No. 1 (January– March 1980): 8–13. [0471]

Bonilla, Sergio. "Proceso de Formación y Evolución del INIA en Chile." In *Organización y Administración de la Generación y Transferencia de Tecnología Agropecuaria*, Serie: Ponencias, Resultados y Recomendaciones de Eventos Técnicos No. A4/UY-86-001, edited by Horacio H. Stagno and Mario Allegri. Montevideo, Uruguay: IICA, and Centro de Investigaciones Agrícolas 'Alberto Boerger', 1986. [0652]

Bouchet, Frederic C. "An Analysis of Sources of Growth in French Agriculture 1960–1984." PhD diss., Virginia Polytechnic Institute and State University, Blacksburg, Virginia, August 1987. [0875]

Bowles, J.R. "Research and Development: Expenditure and Employment in the Seventies." *Economic Trends* No. 334 (August 1981): 94–111. [0700]

Bowles J.R. "Central Government Expenditure on Research and Development." *Economic Trends* No. 346 (August 1982): 82–85. [0699]

Bowles J.R. "Research and Development: Preliminary Estimates of Expenditure in the United Kingdom in 1981." *Economic Trends* No. 359 (1983): 108–121. [0698]

Bowles J.R. "Research and Development in the United Kingdom in 1981." *Economic Trends* No. 370 (August 1984): 81–89. [0697]

Bowles J.R. "Research and Development in the United Kingdom in 1983." *Economic Trends* No. 382 (August 1985): 82–93. [0696]

Bowles J.R. "Central Government Expenditure on Research and Development in 1984." *Economic Trends* No. 394 (August 1986): 82–89. [0695]

Boyce, James K. "Agricultural Research in Indonesia, The Philippines, Bangladesh, South Korea and India: A Documentary History." University of Minnesota Asian Agricultural Research Review Project, July 1980. Mimeo. [0869]

Boyce, James K., and Robert E. Evenson. *National and International Agricultural Research and Extension Programs.* New York: Agricultural Development Council, Inc., 1975. [0010]

Branco Marcelino, Fernando. Personal communication. Director, Instituto de Investigação Agronomica. Chianga, Angola, February 1988. [0803]

Brannon, Russell H. *The Agricultural Development of Uruguay.* New York: Praeger Publishers, 1967. [0458]

Brazilian Agricultural Research Corporation (EMBRAPA). *Formation of Human Capital and Returns on Investment in Manpower Training by EMBRAPA.* Brasilia: EMBRAPA, 1984. [0070]

Brazilian Agricultural Research Corporation (EMBRAPA). *The Socio-Economic Impact of Investments in Research by EMBRAPA: Results Obtained, Profitability and Future Prospects.* Brasilia: EMBRAPA, 1985. [0071]

Brinkman, George L. *An Analysis of Sources of Multifactor Productivity Growth in Canadian Agriculture, 1961 to 1980, with Projections to 2000.*

Report prepared for the Development Policy Directorate of the Regional Development Branch of Agriculture Canada. Ontario: School of Agricultural Economics and Extension Education, University of Guelph, December 1984. [0607]

Britos Dos Santos, F.L. "Some Information Regarding Agricultural Investigations in Guinea Bissau." In *Strengthening National Agricultural Research*. Report from a SAREC workshop, 10–17 September 1979, edited by Bo Bengtsson and Getachew Tedla, pp. 44–47. Sweden: SAREC, 1980. [0311]

Bundesamt für Statistik. *Forschung und Entwicklung an den schweizerischen Hochschulen*. Beitrage zur schweizerischen Statistik – Heft 98. Bern: Bundesamt für Statistik, 1982. [0669]

Bundesamt für Statistik. *Forschung und Entwicklung des Bundes 1978–1981*. Beitrage zur schweizerischen Statistik – Heft 104. Bern: Bundesamt für Statistik, 1983. [0668]

Bundesministerium für Land- und Forstwirtschaft. *Das Forschungs- und Versuchswesen im Bereich des Bundesministeriums für Land- und Forstwirtschaft*. Wien: Bundesministerium für Land- und Forstwirtschaft, March 1980. [0678]

Bundesministerium für Wissenschaft und Forschung. *Bericht 1972 der Bundesregierung an den Nationalrat*. Wien: Bundesministerium für Wissenschaft und Forschung, April 1972. [0680]

Bundesministerium für Wissenschaft und Forschung. *Bericht 1977 der Bundesregierung an den Nationalrat*. Wien: Bundesministerium für Wissenschaft und Forschung, 1977. [0674]

Bundesministerium für Wissenschaft und Forschung. *Bericht 1980 der Bundesregierung an den Nationalrat*. Wien: Bundesministerium für Wissenschaft und Forschung, 1980. [0675]

Bundesministerium für Wissenschaft und Forschung. *Bericht 1985 der Bundesregierung an den Nationalrat*. Wien: Bundesministerium für Wissenschaft und Forschung, 1985. [0676]

Bundesministerium für Wissenschaft und Forschung. *Bericht 1986 der Bundesregierung an den Nationalrat*. Wien: Bundesministerium für Wissenschaft und Forschung, 1986. [0677]

Bunting, A.H., F. Haworth, and J.D. Robinson. "Report on a Visit to the Sudan, 27 March to 8 April 1982." ISNAR, The Hague, 1982. Mimeo. [0250]

Bureau of Statistics, Directorate-General of Budget, Accounting & Statistics, Executive Yuan, ed. *Statistical Yearbook of the Republic of China 1986*. Taipei, Taiwan: Directorate-General of Budget, Accounting & Statistics, Executive Yuan, November 1986. [0769]

Buri, Markus, Raul Suarez de Miguel, and Bruno Walder. "Forschung und Entwicklung 1986 in der Schweiz." Bericht des BFS für den Basisbericht "OECD-Examen 1988 der schweizerischen Forschungs- und Technologiepolitik". Bundesamt für Statistik, Bern, April 1988. Mimeo. [0867]

Buri, Markus. Personal communication. Bundesamt für Statistik. Bern, May 1988. [0868]

Cabinet Office. *Annual Review of Government Funded R&D 1983, 1984, 1985, and 1986*. London: Her Majesty's Stationery Office, various years. [0020]

Cameron, R.J. *Research and Experimental Development, Higher Education Organisations, Australia, 1981*. Canberra: Australian Bureau of Statistics, October 1983. [0611]

Cameron, R.J. *Research and Experimental Development – General Government Organisations, Australia, 1981–82*. Catalogue No. 8109.0. Canberra: Australian Bureau of Statistics, January 1984. [0610]

Canada Grains Council. *Agricultural Research in Canada*. Manitoba, Canada: Canada Grains Council, 1986. [0599]

Caribbean Agricultural Research and Development Institute (CARDI) – Barbados Unit. *Annual Report 1985*. St. Michael, Barbados: CARDI, 1985. [0129]

Caribbean Agricultural Research and Development Institute (CARDI). *Annual Report 1977–1978*. St. Augustine, Trinidad & Tobago: CARDI, November 1978. [0341]

Caribbean Agricultural Research and Development Institute (CARDI). *Annual Report 1980*. St. Augustine, Trinidad & Tobago: CARDI, 1981. [0343]

Caribbean Agricultural Research and Development Institute (CARDI). *Annual Report 1980 – Barbados Unit*. Barbados: CARDI, 1982. [0344]

Caribbean Agricultural Research and Development Institute (CARDI). *Annual Report 1981 – Barbados Unit*. Barbados: CARDI, 1982. [0345]

Caribbean Agricultural Research and Development Institute (CARDI). *Paper Presented at a Meeting of the Governing Body of CARDI on January 29, 1982, Basseterre, St. Kitts*. St. Augustine, Trinidad: CARDI, 1982. [0353]

Caribbean Agricultural Research and Development Institute (CARDI). *CARDI 1976–1981: Report of the Chairman*. St. Augustine, Trinidad & Tobago: CARDI, 1982. [0342]

Caribbean Agricultural Research and Development Institute (CARDI). *Annual Report 1983 – Jamaica Unit*. Kingston, Jamaica: CARDI, 1983. [0346]

Caribbean Agricultural Research and Development Institute (CARDI). *Research and Development Summary 1982–83*. St. Augustine, Trinidad: CARDI, 1983. [0355]

Caribbean Agricultural Research and Development Institute (CARDI). *Annual Report 1983–1984 –*

Belize Unit. St. Augustine, Trinidad: CARDI, 1984. [0352]

Caribbean Agricultural Research and Development Institute (CARDI). *Annual Report 1983–1984 – Leeward Islands Unit*. CARDI, 1984. [0349]

Caribbean Agricultural Research and Development Institute (CARDI). *Annual Report 1983 – Antigua Unit*. St. John's, Antigua: CARDI, 1984. [0347]

Caribbean Agricultural Research and Development Institute (CARDI). *Annual Report 1984 – Barbados Unit*. Barbados: CARDI, 1984. [0350]

Caribbean Agricultural Research and Development Institute (CARDI). *Research and Development Summary 1983–84*. St. Augustine, Trinidad: CARDI, 1984. [0356]

Caribbean Agricultural Research and Development Institute (CARDI). *Annual Report Research and Development 1983–84: Highlights*. St. Augustine, Trinidad: CARDI, 1984. [0354]

Caribbean Agricultural Research and Development Institute (CARDI). *Annual Report 1985 – Barbados Unit*. Barbados: CARDI, 1985. [0348]

Caribbean Agricultural Research and Development Institute (CARDI). *Annual Report 1984 – Jamaica Unit*. St. Augustine, Trinidad: CARDI, 1985. [0351]

Caribbean Industrial Research Institute (CARIRI). *CARIRI 1983 Annual Report*. Trinidad & Tobago: CARIRI, 1984. [0364]

Carmen, M.L. "Reorganization of Agricultural Research in the Syrian Arab Republic." FAO, Rome, December 1981. Mimeo. [0256]

Carranza, Germán. "Sector Agropecuario de Costa Rica." Ministerio de Agricultura y Ganadería, San José, Costa Rica, March 1987. Mimeo. [0713]

Casas, Joseph, and Alfred Conesa. "Bref Aperçu du Système Francais de Recherche Agricole et Agro-Alimentaire." Centre International des Hautes Etudes Agronomiques Méditerranéennes (CIHEAM), Paris, September 1987. Mimeo. [0853]

Casas, Joseph, and Willem Stoop. "Mission Report: Brief Presentation of the Agricultural Research System of Niger." ISNAR, The Hague, May 1986. Mimeo. [0171]

Casas, Joseph, Pierre Carrière, and Philippe Lacombe. *La Recherche Agronomique et la Diffusion du Progrès Technique en Union Soviétique*. Series Notes et Documents No. 33. Montpellier: Institut National de Recherche Agronomique, June 1979. [0366]

Casas, Joseph. "Cuba: A Small Country, a Large Agricultural Research Potential." Paper presented at the Workshop on Agricultural Research Policy, Minneapolis, Minnesota, April 1985. [0065]

Casas, Joseph. "Les Ressources Humaines et Financières dans les Pays du Bassin Méditerranéen."

Preliminary table. ISNAR, The Hague, 1986. Mimeo. [0422]

Casas, Joseph. "Dispositif de Recherche-Développement du Secteur Agricole – Les Budgets des Organismes." ISNAR, The Hague, 1986. Mimeo. [0423]

Casas, Joseph. "Compte Rendu de Mission au Mali (19–30 Octobre 1987)." ISNAR, The Hague, December 1987. Mimeo. [0741]

Castells, Manuel, Antonio Barrera, et al. *Nuevas Tecnologías, Economía y Sociedad en España*. Volumen 2. Madrid: Alianza Editorial, 1986. [0784]

Castronovo, Alfonso J.P. "Diagnóstico Descriptivo de la Situación Actual del Sistema Nacional de Investigación y Extensión en Argentina, Chile, Paraguay y Uruguay." Buenos Aires, December 1980. Mimeo. [0032]

Central Experiment Station. *Research Division – Crop Research: Annual Report 1982 (summary)*. Trinidad & Tobago: Central Experiment Station, 1982. [0361]

Central Statistical Office of Finland. *Research Activity*. Official Statistics of Finland XXXVIII:1–7, bi-annually 1971–1983. Helsinki: Central Statistical Office of Finland, various years. [0609]

Central Statistical Office. *Research and Development Expenditure*. London: Her Majesty's Stationery Office, 1973. [0691]

Central Statistical Office. *Research and Development – Expenditure and Employment*. London: Her Majesty's Stationery Office, 1976. [0692]

Centre Africain d'Etudes Supérieures en Gestion (CESAG). *Rapport Introductif de la Commission No 1: Organisation de Système de Recherche au Senegal*. Etude réalisée par le CESAG pour le CILSS. Senegal: CESAG, June 1986. [0260]

Centre International de Hautes Etudes Agronomiques Méditerranéennes. "Séminaire sur l'Orientation et l'Organisation de la Recherche Agronomique dans les Pays du Bassin Méditerranéen, Istanbul, 1–3 December 1986." Centre International de Hautes Etudes Agronomiques Méditerranéennes, Paris, December 1986. Mimeo. [0407]

Centre National de Documentation Agricole (CNDA). *La Recherche en Cours*. Tunis: CNDA, September 1983. [0255]

Centre National du Machinisme Agricole, du Génie Rural, des Eaux et des Forêts (CEMAGREF). *Programmation des Actions CEMAGREF 1986–1988*. Antony, France: CEMAGREF, October 1986. [0537]

Centro de Documentación sobre Investigación y Enseñanza Superior Agropecuaria de la Zona Sur. *Instituciones de Investigación Agrícola de la Zona Sur; Argentina, Chile, Paraguay y Uruguay*.

Serie: Informaciones No. 2. Buenos Aires: Facultad de Agronomía y Veterinaria – Biblioteca Central, IICA – Zona Sur, 1972. [0717]

Centro Internacional de Investigación para el Desarrallo. "Asignación de Recursos para Investigación Agrícola en América Latina." Proyecto ARIAL Brasil – Estudio de Caso. Unidad de Apoyo a la Programación, April 1979. Mimeo. [0213]

Centro Nacional de Tecnología Agropecuaria – Ministerio de Agricultura y Ganadería El Salvador. "Informes Sobre Actividades de Investigación y Extensión Agropecuarias realizadas por Instituciones del Sector Público Agropecuario." Documento para el Proyecto BID-IICA-Iowa State University. IICA, San Andrés, Colombia, 1980. Mimeo. [0209]

CGIAR. "National Agricultural Research." CGIAR, Washington D.C., 1985. Mimeo. [0163]

Chan, T.K. Personal communication. Agriculture and Fisheries Department. Hong Kong, January 1988. [0779]

Chang, C.W. *FAO Directory of Agricultural Research Institutes and Experiment Stations in Asia and the Far East.* Bangkok, Thailand: FAO – Regional Office for Asia and the Far East, September 1962. [0715]

Chang, C.W. *FAO Directory of Agricultural Research Institutes and Experiment Stations in Asia and the Far East, Supplement.* Bangkok, Thailand: FAO – Regional Office for Asia and the Far East, May 1963. [0716]

Chaparro, Fernando, et al. "Research Priorities and Resource Allocation in Agriculture: The Case of Colombia." Paper presented at the International Workshop on Resources Allocation in National Agricultural Research Systems, IDRC-IFARD, Singapore, 8–10 June 1981. [0120]

Chaparro, Fernando, et al. "Research Priorities and Resource Allocation in Agriculture: The Case of Colombia." In *Resource Allocation to Agricultural Research.* Proceedings of a workshop held in Singapore 8–10 June 1981, edited by Douglas Daniels and Barry Nestel, pp. 68–96. Ottawa: IDRC, 1981. [0298]

Chesney, H.A.D. "Agricultural Research in Guyana." In *Strengthening National Agricultural Research.* Report from a SAREC workshop, 10–17 September 1979, edited by Bo Bengtsson and Getachew Tedla, pp. 167–174. Sweden: SAREC, 1980. [0323]

Chibasa, W.M. "Agricultural Research Efforts in Zambia." Paper presented at the inaugural meetings of the consultative technical meetings under SADCC Food Security Project One, November 1982. [0245]

Chief Scientists Group – Ministry of Agriculture, Fisheries & Food. *Report on Research and Development 1984–85.* London: Ministry of Agriculture, Fisheries & Food, n.d. [0543]

Christie, Leonard A. *Agricultural Research in Canada: Priorities, Funding and Manpower.* Ottawa: Library of Parliament – Science and Technology Division, Research Branch, April 1980. [0600]

CIRAD. *Coopération Francaise en Recherche Agronomique: Activités du CIRAD dans les Pays du Sahel.* Paris: CIRAD, March 1985. [0228]

Clark, Norman. "Organisational Aspects of Nigeria's Research System." *Research Policy Vol. 9* (1980): 148–172. [0157]

Comitato per la Ricerca Scientifica e Tecnica (CREST). *Il Finanziamento Pubblico della Ricerca e dello Sviluppo nei Paesi della Comunita: Analisi per Categorie di Obiettivi 1969–1973.* Luxembourg: Comunita Europee, n.d. [0826]

Comitato per la Ricerca Scientifica e Tecnica (CREST). *Il Finanziamento Pubblico della Ricerca e dello Sviluppo nei Paesi della Comunita: Analisi Sommaria per Obiettivi Principali di Ricerca e Sviluppo 1974.* Luxembourg: Comunita Europee, April 1975. [0829]

Commissariat Général du Plan d'Equipement et de la Productivité. *5e Plan de Développement Economique et Social – La Recherche Scientifique et Technique, Tome 1.* Paris: Commissariat Général du Plan d'Equipement et de la Productivité, 1970. [0535]

Committee on Scientific and Technical Research (CREST). *Public Expenditure on Research and Development in the Community Countries: Summary Report 1974–1976.* Luxembourg: European Communities, 1976. [0827]

Commonwealth Agricultural Bureaux (CAB). *List of Research Workers in Agriculture, Animal Health and Forestry in the British Commonwealth, the Republic of Sudan and the Republic of Ireland 1959.* Slough, England: CAB, 1959. [0287]

Commonwealth Agricultural Bureaux (CAB). *List of Research Workers in Agriculture, Animal Health and Forestry in the Commonwealth and in the Republic of Ireland 1966.* Slough, England: CAB, 1966. [0286]

Commonwealth Agricultural Bureaux (CAB). *List of Research Workers in the Agricultural Sciences in the Commonwealth and in the Republic of Ireland 1969.* Slough, England: CAB, 1969. [0285]

Commonwealth Agricultural Bureaux (CAB). *List of Research Workers in the Agricultural Sciences in the Commonwealth and in the Republic of Ireland 1972.* Slough, England: CAB, 1972. [0279]

Commonwealth Agricultural Bureaux (CAB). *List of Research Workers in the Agricultural Sciences in*

502

the Commonwealth and in the Republic of Ireland 1975. Slough, England: CAB, 1975. [0244]

Commonwealth Agricultural Bureaux (CAB). List of Research Workers in the Agricultural Sciences in the Commonwealth and in the Republic of Ireland 1978. Slough, England: CAB, 1978. [0243]

Commonwealth Agricultural Bureaux (CAB). List of Research Workers in Agricultural Sciences in the Commonwealth 1981. Slough, England: CAB, 1981. [0242]

Commonwealth Scientific & Industrial Research Organization (CSIRO). Twelfth Annual Report 1959–60. Australia: CSIRO, n.d. [0629]

Commonwealth Scientific and Industrial Research Organization (CSIRO). Seventeenth Annual Report 1964–65. Australia: CSIRO, n.d. [0630]

Conesa, A.P. "Quelques Aspects du Dispositif de Recherche Agricole et Agro-Alimentaire en France." Paper presented at the Séminaire sur l'Orientation et l'Organisation de la Recherche Agronomique dans les Pays du Bassin Méditerranéen, Centre International de Hautes Etudes Agronomiques Méditerranéennes, Istanbul, 1–3 December 1986. [0411]

Consejo Nacional de Investigaciones Científicas y Tecnológicas (CNICT). Un Analisis del Desarrollo Científico Tecnológico del Sector Agropecuario de Costa Rica. Volumes 1, 2 and 3. San José: CNICT, 1980. [0124]

Consiglio Nazionale delle Ricerche (CNR). Il Finanziamento Pubblico della Ricerca e dello Sviluppo nei Paesi della Comunita: Analisi per Obiettivi 1967–1970. Roma: CNR, 1970. [0828]

Consortium for the Study of Nigerian Rural Development. A Survey of Nigerian Agricultural Research. East Lansing, Michigan: Michigan State University, September 1966. [0268]

Conte, Stephan. "West Africa Agricultural Research Data Base Building Task: Termination Report." World Bank, Washington, D.C., 1985. Mimeo. [0068]

Cook, Theodore W., Gaylord L. Walker, and Carl. J. Metzger. Final Report Agricultural Research Project Jamaica. Arlington, USA: Multinational Agribusiness Systems Incorporated, September 1983. [0359]

Cooke, G.W., ed. Agricultural Research 1931–1981. London: Agricultural Research Council, 1981. [0592]

Cooper, St.G.C. Agricultural Research in Tropical Africa. Kampala: East African Literature Bureau, 1970. [0175]

Cooper, St.G.C. "Towards Trained Manpower for Agricultural Research in Africa." Paper presented at the Conference on Agricultural Research and Production in Africa, organized by the Association for the Advancement of

Agricultural Sciences in Africa (AAASA), Addis Ababa, 29 August–4 September 1971. [0360]

Corbett, D.C.M. Personal communication. Agricultural and Food Research Council. London, July 1986. [0557]

Coutu, A.J., and K. Raven. "Background Paper on the National Agricultural Research, Extension and Educational System of Peru." North Carolina State University and National Agrarian University, North Carolina and Lima, August 1985. Mimeo. [0203]

Croon, Ingemar. Personal communication. Department of Economics and Statistics, Swedish University of Agricultural Sciences. Uppsala, Sweden, July 1988. [0613]

Croon, Ingemar. "Resursinsatser i offentlig jordbruksforskning i Sverige under perioden 1945–1985." Draft. Swedish University of Agricultural Sciences, Uppsala, Sweden, July 1988. Mimeo. [0885]

Cuellar, Miguel, Damaris Chea, Jorge Jonas, and Hermel López. "Estudio del Manejo y Organización de la Investigación en Fincas – Caso: Panamá." Paper prepared for the ISNAR Study on the Organization and Management of On-Farm Research. ISNAR, The Hague, September 1986. Mimeo. [0338]

Current Agricultural Research Information System (CARIS). Directory of Agricultural Research Institutions and Projects in West Africa – Pilot Project 1972–73. Rome: FAO, 1973. [0431]

Cushing, R.L. Master Plan for Bangladesh Agricultural Research Council Phase II. Dhaka: Bangladesh Agricultural Research Council, November 1982. [0807]

Cyprus Agricultural Research Institute (CARI). Cyprus Agricultural Research Institute, Annual Report for 1983. Nicosia, Cyprus: Ministry of Agriculture and Natural Resources, March 1984. [0187]

Dagg, Matthew. Personal communication. Senior Research Officer, ISNAR. The Hague, March 1987. [0482]

Daniels, Douglas, and Barry Nestel, eds. Resource Allocation to Agricultural Research. Proceedings of a workshop held in Singapore 8–10 June 1981. Ottawa: IDRC, 1981. [0018]

Davis, J.S. "Stability of the Research Production Function Coefficient for US Agriculture." PhD diss., University of Minnesota, St. Paul, Minnesota, 1979. [0496]

Debela, Seme. Personal communication. General Manager, Institute of Agricultural Research. Addis Ababa, January 1988. [0772]

Departamento de Economía Agraria, Universidad Católica de Chile. "La Investigación Agropecuaria en Chile." In Panorama Económico de la Agricultura, edited by Paul Aldunate V., et al., pp. 2–8. Santiago, Chile: Departamento de Economía Agraria – Universidad Católica de Chile, March 1985. [0585]

Department of Agricultural Research – Ministry of Agriculture. *Malawi Agricultural Research Strategy Plan.* Malawi: Ministry of Agriculture, September 1983. [0241]

Department of Agricultural Research Malawi. *Annual Report of the Department of Agricultural Research for the Years 1971/72, 1972/73, and 1973/74.* Zomba, Malawi: The Government Printer, various years. [0872]

Department of Agricultural Technical Services (DATS). *Agricultural Research 1968: Part I – 1, 2 & 3.* Interim and Final Reports on Research Projects, Summaries of Theses, and Scientific Publications. Republic of South Africa: DATS, n.d. [0638]

Department of Agricultural Technical Services (DATS). *Agricultural Research 1968: Part II – 4 & 5.* Registered Projects and Research Workers. Republic of South Africa: DATS, n.d. [0637]

Department of Agriculture and Fisheries. *Annual Report of the Veterinary Research Laboratory 1968–69.* Dublin: Stationary Office, n.d. [0633]

Department of Agriculture, Natural Resources and Rural Development (DARNDR), and IICA. "Réunion Technique Régionale sur les Systèmes de Recherche Agricole dans les Antilles." DARNDR and IICA, Port-au-Prince, Haiti, November/December 1977. Mimeo. [0219]

Department of Agriculture, Natural Resources and Rural Development (DARNDR). "Diagnosis of the Situation of the Agricultural Research and Extension System in Haiti (A Summary)." Paper presented at the Caribbean Workshop on the Organization and Administration of Agricultural Research, Bridgetown, Barbados, February 1981. [0222]

Department of Agriculture Nyasaland/Malawi. *Annual Report of the Department of Agriculture for the Years 1959/60, 1960/61, 1963/64, and 1967/68.* Part II. Zomba, Nyasaland/Malawi: The Government Printer, various years. [0871]

Department of Education and Science – Ministry of Technology. *Statistics of Science and Technology.* London: Her Majesty's Stationery Office, 1967. [0694]

Department of Education and Science – Ministry of Technology. *Statistics of Science and Technology.* London: Her Majesty's Stationery Office, 1968. [0693]

Department of Research and Specialist Services – Ministry of Agriculture, and ISNAR. *A Training Plan for the Department of Research and Specialist Services, Zimbabwe, 1985 to 1988.* The Hague: ISNAR, February 1985. [0043]

Department of Science. *Project SCORE Report 5: Summary of All Expenditures by Australia on Research and Development during 1968–69.* Canberra: Australian Government Publishing Service, 1973. [0685]

Department of Science. *Project SCORE Research and Development in Australia 1973–74 – Volume 2: All-Sector Results, Business Enterprise Sector, and Higher Education Sector.* Canberra: Australian Government Publishing Service, 1977. [0686]

Department of Science and the Environment. *Project SCORE Research and Development in Australia 1976–77 – All-Sector Results.* Canberra: Australian Government Publishing Service, 1980. [0687]

Der Bundesminister für Forschung und Technologie. *Fünfter Forschungsbericht der Bundesregierung.* Bonn: Der Bundesminister für Forschung und Technologie, 1975. [0670]

Der Bundesminister für Forschung und Technologie. *Bundesbericht Forschung VI.* Reihe: Berichte und Dokumentationen, Band 4. Bonn: Der Bundesminister für Forschung und Technologie, 1979. [0671]

Der Bundesminister für Forschung und Technologie. *Faktenbericht 1981 zum Bundesbericht Forschung.* Reihe: Berichte und Dokumentationen. Bonn: Der Bundesminister für Forschung und Technologie, 1982. [0672]

Der Bundesminister für Forschung und Technologie. *Faktenbericht 1986 zum Bundesbericht Forschung.* Bonn: Der Bundesminister für Forschung und Technologie, May 1986. [0673]

Deutsche Akademie der Landwirtschaftswissenschaften. *10 Jahre Deutsche Akademie der Landwirtschaftswissenschaften zu Berlin.* Berlin: Deutsche Akademie der Landwirtschaftswissenschaften, 1962. [0525]

Development Alternatives, Inc. *Agricultural Sector Assistance Strategy for Upper Volta.* Washington, D.C.: Development Alternatives, Inc., March 1982. [0110]

"Development and Administration of Agricultural Research and Its Contribution to Agricultural Development in Malawi." n.d. Mimeo. [0239]

Devred, R. Personal communication. Senior Research Officer, ISNAR. The Hague, 1986. [0884]

DEVRES, Inc. *The Agricultural Research Resource Assessment Pilot Report for Botswana, Malawi and Swaziland.* Washington, D.C.: DEVRES, Inc., November 1983. [0467]

Dikoumé, Cosme, Oscar Cordeiro, and Mathieu Gracia. *Affectation des Ressources à la Recherche Agricole au Cameroun.* Douala, Cameroon:

504

Institut Panafricain pour le Développement, May 1984. [0069]

Directie Landbouwkundig Onderzoek (DLO). "Monograph on the Organization of Agricultural Research in the Netherlands (draft)." DLO, Wageningen, The Netherlands, 1985. Mimeo. [0428]

Doery, R.C., and A.E. Charles, eds. *Research Programme Survey 1980.* Papua New Guinea: Department of Primary Industries, 1980. [0470]

Dolle, V., et al. *Etat de l'Agriculture Mauritanienne et Objectifs d'une Recherche pour son Développement.* Paris: Ministère des Relations Exterieures, May 1984. [0142]

Donovan, P.A., and M.G. Lynas. "The Political, Structural and Managerial Advantages of Commodity-Controlled Agricultural Research and Development in South Africa." *Agricultural Administration and Extension* No. 28 (January 1988): 19–28. [0783]

Doulou, V. Personal communication. Directeur des Activités Scientifiques et Technologiques, Direction Générale de la Recherche Scientifique et Technique, Ministère de la Recherche Scientifique et de l'Environnement. Brazzaville, Congo, April 1988. [0856]

Downer. "Proposal for a National Agricultural Research Service in Guyana." 1979. Mimeo. [0200]

Drilon, J.D., and Aida R. Librero. "Defining Research Priorities for Agriculture and Natural Resources in the Philippines." In *Resource Allocation to Agricultural Research.* Proceedings of a workshop held in Singapore 8–10 June 1981, edited by Douglas Daniels and Barry Nestel, pp. 97–103. Ottawa: IDRC, 1981. [0299]

East African Academy. *Research Services in East Africa.* Nairobi: East African Publishing House, 1966. [0518]

East African Agriculture and Forestry Research Organization. *Record of Research – Annual Report 1969.* Nairobi: East African Community, 1970. [0640]

East African Veterinary Research Organization. *Record of Research – Annual Report 1968.* Kenya: East African Community, 1969. [0641]

Ecole Superiéure de Gestion des Entreprises (ESGE). *Etude-Diagnostic de l'Institut National de Recherches Agronomiques Niger (INRAN): Recommandations a l'Institut de Recherche.* Etude réalisée par l'ESGE pour le CILSS. Niamey, Niger: ESGE, October 1985. [0261]

Economic Commission for Latin America and the Caribbean (ECLAC). *Agricultural Research Policy and Management – Volumes I and II.* Papers presented at the Workshop on Agricultural Research Policy and Management, 26–30 September 1983, Port of Spain, Trinidad. Port of Spain, Trinidad: ECLAC, November 1984. [0115]

Edwards, D.T. "An Economic View of Jamaican Agricultural Research." *Social and Economic Studies* Vol. 10, No. 3 (September 1961): 306–339. [0880]

EEG. *De Organisatie van het Landbouwkundig Onderzoek in de Landen van de EEG.* EEG Studies, Serie Landbouw 9. Brussel: EEG, 1963. [0282]

"Effort Francais en Faveur de la Recherche Agronomique dans les Pays du Sahel 1984–1985." 1985. Mimeo. [0258]

Ehrencron, V.K.R., ed. *Jaarverslag.* Years: 1966–1970. Suriname: Landbouwproefstation, various years. [0583]

El-Akhrass, Hisham. *Syria and the CGIAR Centers: A Study of Their Collaboration in Agricultural Research.* CGIAR Study Paper Number 13. Washington, D.C.: World Bank, 1986. [0056]

El-Sharkawy, Ahmed. "Les Systèmes de Recherche Agricole en Egypte." Paper presented at the Séminaire sur l'Orientation et l'Organisation de la Recherche Agronomique dans les Pays du Bassin Méditerranéen, Centre International de Hautes Etudes Agronomiques Méditerranéennes, Istanbul, 1–3 December 1986. [0409]

Elias, S.M. "Manpower Requirements and Developments for Agricultural Research in Bangladesh." Bangladesh Agricultural Research Council, Bangladesh, 1980. Mimeo. [0112]

Elias, S.M. "Manpower Developments for Agricultural Research in Bangladesh." In *Resource Allocation to Agricultural Research.* Proceedings of a workshop held in Singapore 8–10 June 1981, edited by Douglas Daniels and Barry Nestel, pp. 168–170. Ottawa: IDRC, 1981. [0306]

Ellinger, K.R. "Ivory Coast Agricultural Research." World Bank – Regional Mission in Western Africa, Abidjan, 1982. Mimeo. [0147]

Elliott, Howard, Reed Hertford, Judith Snow, and Eduardo Trigo. *Identificación de Oportunidades para el Mejoramiento de los Sistemas de Gestión de Tecnología Agropecuaria en Latino América: Una Metodología y Caso de Ensayo.* New Jersey and The Hague: Rutgers University and ISNAR, 1985. [0217]

Empresa Brasileira de Pesquisa Agropecuária (EMBRAPA). *EMBRAPA Ano 11 – Destaque dos Resultados da Pesquisa de 1983.* Brasilia: EMBRAPA, 1984. [0264]

Ergun, Yusuf. "Recherche Agricole en Turquie." Paper presented at the Séminaire sur l'Orientation et l'Organisation de la Recherche Agronomique dans les Pays du Bassin Méditerranéen, Centre International de Hautes Etudes Agronomiques Méditerranéennes, Istanbul, 1–3 December 1986. [0420]

European Economic Community (EEC). "Monograph on the Organization of Agricultural Research." EEC, Brussels, 1982. Mimeo. [0283]

EUROSTAT. *Government Financing of Research and Development 1970–1977*. Luxembourg: Office for Official Publications of the European Communities, 1977. [0819]

EUROSTAT. *Government Financing of Research and Development 1970–1978*. Luxembourg: Office for Official Publications of the European Communities, 1978. [0818]

EUROSTAT. *Government Financing of Research and Development 1970–1979*. Luxembourg: Office for Official Publications of the European Communities, 1980. [0820]

EUROSTAT. *Government Financing of Research and Development 1970–1980*. Luxembourg: Office for Official Publications of the European Communities, 1981. [0821]

EUROSTAT. *Government Financing of Research and Development 1975–1981*. Luxembourg: Office for Official Publications of the European Communities, 1982. [0822]

EUROSTAT. *Government Financing of Research and Development 1975–1982*. Luxembourg: Office for Official Publications of the European Communities, 1983. [0823]

EUROSTAT. *Government Financing of Research and Development 1975–1983*. Luxembourg: Office for Official Publications of the European Communities, 1984. [0824]

EUROSTAT. *Government Financing of Research and Development 1975–1984*. Luxembourg: Office des Publications Officielles des Communautés Européennes, February 1985. [0606]

EUROSTAT. *Government Financing of Research and Development 1975–1985*. Luxembourg: Office for Official Publications of the European Communities, 1987. [0706]

Eveleens, W.M. Personal communication. Ruakura Agricultural Research Centre, Ministry of Agriculture and Fisheries. Hamilton, New Zealand, 1986. [0480]

Evenson, R.E., and Y. Kislev. *Investment in Agricultural Research and Extension: A Survey of International Data*. Center Discussion Paper No. 124. New Haven, Connecticut: Economic Growth Center, Yale University, August 1971. [0852]

Evenson, R.E., and Y. Kislev. *Agricultural Research and Productivity*. New Haven: Yale University Press, 1975. [0165]

Evenson, R.E., and Y. Kislev. "Investment in Agricultural Research and Extension: A Survey of International Data." *Economic Development and Cultural Change* Vol. 23 (April 1975): 507–521.

Famoriyo, O.A. "Agricultural Research Administration in Nigeria." *Science and Public Policy* Vol. 15, No. 5 (October 1986): 295–300. [0534]

FAO – CARIS. *Agricultural Research in Developing Countries – Volume 1: Research Institutions*. Rome: FAO – CARIS, 1978. [0095]

FAO – Near East Regional Office. *Directory of Agricultural Research Institutions in the Near East Region*. Cairo: FAO – Near East Regional Office, 1979. [0094]

FAO – Regional Office for Asia and the Pacific. *Agricultural Research Systems in the Asia-Pacific Region*. Bangkok: FAO – Regional Office for Asia and the Pacific, 1986. [0732]

FAO, and the World Bank. *Report of the Nepal Agricultural Research Review*. Rome: FAO, 1983. [0387]

FAO, and UNDP. *National Agricultural Research – Report of an Evaluation Study in Selected Countries*. Rome: FAO and UNDP, 1984. [0135]

FAO. *Report to the Government of Iraq: Organization and Administration of Agricultural Research*. Rome: FAO, 1964. [0189]

FAO. *FAO Near East Regional Studies on Organization and Administration of Agricultural Research*. Rome: FAO, 1974. [0173]

FAO. *Institute of Agricultural Research Ethiopia – Project Findings and Recommendations*. Rome: FAO, 1982. [0745]

FAO. "Trained Agricultural Manpower Assessment (TAMA): Original Answer Sheets." FAO, Rome, 1982. Mimeo. [0235]

FAO. *Agricultural Research Review Mission: Uganda*. Rome: FAO, 1984. [0160]

FAO. *Assistance à l'Institut des Sciences Agronomiques Burundi*. Rome: FAO, 1984. [0087]

FAO. *Fourteenth FAO Regional Conference for Europe, Reykjavik, Iceland. 17–21 September 1984: Research in Support of Agricultural Policies in Europe*. Rome: FAO, 1984. [0376]

FAO. "Trained Agricultural Manpower Assessment in Africa – Country Report Mauritius." FAO, Rome, January 1984. Mimeo. [0432]

FAO. *Report of an FAO Mission: Organization and Administration of Applied Agricultural Research in Iraq*. Rome: FAO, February 1984. [0188]

FAO. *Report of a Review Mission: Agricultural Research in Liberia*. Rome: FAO, June 1984. [0133]

FAO. *Rapport d'une Mission de Consultation: La Recherche Agronomique au Tchad*. Rome: FAO, December 1984. [0131]

FAO. *Rapport d'une Mission de Consultant sur la Recherche Agronomique en Republique Islamique de Mauritanie*. Rome: FAO, 1985. [0128]

FAO. *Report of an Expert Consultation on Agricultural Extension and Research Linkages in the Near*

506

East, Amman, Jordan, 22–26 July 1985. Rome: FAO, 1985. [0831]

FAO. *Review of the Organization of Agricultural and Fisheries Research, Technology and Development in Egypt*. Rome: FAO, 1985. [0248]

FAO. *Rapport de Mission: La Recherche Agronomique à Sao Tomé & Principe*. Rome: FAO, April 1985. [0130]

FAO. *Informe Sobre el Instituto de Investigación Agropecuaria (IDIAP) de Panamá*. Rome: FAO, June 1985. [0218]

FAO. *Assistance à la Réorganisation et au Renforcement de la Recherche Agricole Guinée*. Rome: FAO, 1986. [0132]

Federación Nacional de Cafeteros de Colombia. *Informe del Gerente General al XLI Congreso Nacional de Cafeteros*. Anexo 2. Bogotá: Federación Nacional de Cafeteros de Colombia, November 1982. [0729]

Federación Nacional de Cafeteros de Colombia. *Informe del Gerente General al XLIV Congreso Nacional de Cafeteros*. Anexo. Bogotá: Federación Nacional de Cafeteros de Colombia, November 1986. [0730]

Fenner, R.J. Personal communication. Director, Department of Research and Specialist Services. Harare, Zimbabwe, February 1988. [0809]

Fernando, L.H. "South Pacific NARS Resource Commitment Survey." ADB and ISNAR, The Hague, 1987. Mimeo. [0723]

Fernando, L.H. "ISNAR/IRETA Workshop on Planning and Management of Agricultural Research in the South Pacific: Survey of Agricultural Research Resources." ISNAR, The Hague, n.d. Mimeo. [0771]

Fideghelli, C. "La Recherche Agricole en Italie." Paper presented at the Séminaire sur l'Orientation et l'Organisation de la Recherche Agronomique dans les Pays du Bassin Méditerranéen, Centre International de Hautes Etudes Agronomiques Méditerranéennes, Istanbul, 1–3 December 1986. [0416]

Fletcher, R., et al. "Agricultural Research in Guyana." March 1978. Mimeo. [0197]

Focan, A., and E. Sebatutisi. "La Recherche Agronomique au Burundi et ses Applications." In *Papers presented at the Conference on Agricultural Research and Production in Africa, September 1971 – Part 3*, supplement to volume 2 of the Journal of the Association for the Advancement of Agricultural Sciences in Africa. Addis Ababa, Ethiopia: Association for the Advancement of Agricultural Sciences in Africa, June 1975. [0650]

Foote, R.J. and D. Rosjo. *A Survey of Five Agricultural Research Institutes in Tanzania, with Emphasis on Ways to Improve Management of the Research System*. Rural Economy Research Paper No. 7. Dar Es Salaam, and Mzumbe, Tanzania: University of Dar Es Salaam, and Institute of Development Management, May 1978. [0231]

Fordyce, Alison M. *Industry Levies and Research Funding*. Technical Report No. 24. Palmerston North, New Zealand: Plant Physiology Division – Department of Scientific and Industrial Research, February 1986. [0604]

Forskningsrådenes Statistikkutvalg (FSU). *FoU-Statistikk 1985, Forsknings- og Utviklingsarbeid: Utgifter og Personale*. Oslo: FSU, August 1987. [0848]

Forskningsrådenes Statistikkutvalg (FSU). *FoU-Statistikk 1985: Instituttsektoren*. Oslo: FSU, August 1987. [0850]

Forskningsrådenes Statistikkutvalg (FSU). *FoU-Statistikk 1985: Universitets- og Høgskolesektoren*. Oslo: FSU, October 1987. [0849]

Forskningssekretariatet. *Forskningsstatistik 1982. Ressourceforbruget ved forskning og udviklingsarbejde i den offentlige sektor*. København: Forskningssekretariatet, 1986. [0505]

Fournier, A. "La Recherche Agronomique Nigérienne: Présentation, Synthèse des Recherches, et Suggestions." Mission USAID Niger, Niamey, Niger, January 1986. Mimeo. [0785]

Freitas Rivaldo, Ormuz. "Strategies for Strengthening the Brazilian Agricultural Research System." In *The Impact of Research on National Agricultural Development*, edited by Brian Webster, Carlos Valverde, and Alan Fletcher, pp. 161–181. The Hague: ISNAR, July 1987. [0766]

Gabas, Jean-Jacques. *La Recherche Agricole dans les Pays Membres du CILSS*. Club du Sahel, September 1986. [0372]

Galante, Ennio, and Cesare Sala. Personal communication. Istituto Biosintesi Vegetali, Consiglio Nazionale delle Ricerche. Milano, Italy, July 1988. [0881]

Gamble, W.K., R.M. Bourke, and C.W. Brookson. *South Pacific Agricultural Research Study*. Consultants' Report to the Asian Development Bank (two volumes). Manila: Asian Development Bank, June 1981. [0038]

Gapasin, Dely P. Personal communication. Deputy Executive Director for Research and Development, Philippine Council for Agriculture and Resources Research and Development. Los Baños, Philippines, February 1988. [0802]

García Berríos, Carlos Mario, and Víctor Manuel Rodríguez A. "Estructura, Logros y Objetivos." Centro de Tecnología Agrícola – División de

Investigación Agrícola, San Andrés, El Salvador, December 1987. Mimeo. [0836]

García Berríos, Carlos Mario. Personal communication. Centro de Tecnología Agrícola. San Andrés, El Salvador, March 1988. [0835]

Gebre, H. "Agricultural Research in Ethiopia." In *Strengthening National Agricultural Research.* Report from a SAREC workshop, 10–17 September 1979, edited by Bo Bengtsson and Getachew Tedla, pp. 25-35. Sweden: SAREC, 1980. [0309]

Giacich, Walter. "Document de Projet Révisé UNDP/FAO/Lao/82/011." Vientiane, Laos, June 1987. Mimeo. [0817]

Gill, Gerard J. *Operational Funding Constraints on Agricultural Research in Bangladesh.* BARC Agricultural Economics and Rural Social Science Papers No. 9.. Dacca: Bangladesh Agricultural Research Council (BARC), April 1981. [0476]

Gollifer, D.E. Personal communication. Director of Research, Department of Agricultural Research. Gabarone, Botswana, October 1986. [0720]

Gómez Moncayo, Fernando. *Instituto Colombiano Agropecuario – Informe de Gerencia 1985.* Bogotá: Instituto Colombiano Agropecuario, n.d. [0735]

Gómez, Arturo S. *Philippines and the CGIAR Centers: A Study of Their Collaboration in Agricultural Research.* CGIAR Study Paper Number 15. Washington, D.C.: World Bank, 1986. [0062]

Government of Papua New Guinea. *Estimates of Revenue and Expenditure for the Year Ending 31st December, 1981.* Report presented by Mr John Kaputin, M.P., Minister of the National Parliament for Finance on the Occasion of the Budget, 1981. Papua New Guinea: Government of Papua New Guinea, 1981. [0468]

Government of Papua New Guinea. *Estimates of Revenue and Expenditure for the Year Ending 31st December, 1982.* Report presented by Mr John Kaputin, M.P., Minister of the National Parliament for Finance on the Occasion of the Budget, 1982. Papua New Guinea: Government of Papua New Guinea, 1982. [0469]

Grierson, John. Personal communication. Centro de Investigaciones Agrícolas 'Alberto Boerger'. Montevideo, 1987. [0645]

Groupement d'Etudes et de Recherches pour le Développement de l'Agronomie Tropicale (GERDAT). *Rapport Général d'Activité pour 1982.* Paris: GERDAT, 1983. [0373]

Gunsam, Charles. Personal communication. Agricultural Research Officer, Ministry of Trade, Industry and Agriculture. Saint Vincent & the Grenadines, February 1988. [0810]

Haidar, Wael. "La Recherche Agronomique au Liban." Paper presented at the Séminaire sur l'Orientation et l'Organisation de la Recherche Agronomique dans les Pays du Bassin Méditerranéen, Centre International de Hautes Etudes Agronomiques Méditerranéennes, Istanbul, 1–3 December 1986. [0417]

Hallaway, Michelle L. "The Statistical Evolution of US Agricultural Research: A Study in Institutional Development." MSc thesis, University of Minnesota, St. Paul, Minnesota, forthcoming. [0500]

Hamilton, D.G. *Evaluation of Research and Development in Agriculture and Food in Canada.* Ontario: Canadian Agricultural Research Council, January 1980. [0597]

Hariri, G. "Agricultural Research Authority (Yemen)/ISNAR Potential Collaboration: Back-to-Office Report." ISNAR, The Hague, March 1988. Mimeo. [0841]

Harvey, Nigel, ed. *Agricultural Research Centres: A World Directory of Organizations and Programmes.* Seventh Edition, Two Volumes. Harlow, U.K.: Longman, 1983. [0027]

Hasnain, H., and M.H. Rizvi. "An Inventory of Agricultural Research Manpower in Pakistan." In *Agriculture and Resources Research Manpower Development in South and Southeast Asia,* pp. 207–215. Los Baños, Philippines: Philippine Council for Agriculture and Resources Research and Development, 1983. [0796]

Hastings, Trevor. "The Impact of Scientific Research on Australian Rural Productivity." *Australian Journal of Agricultural Economics* Vol. 25, No. 1 (April 1981): 48–59. [0571]

Hayami, Yujiro, and Masakatsu Akino. "Organization and Productivity of Agricultural Research Systems in Japan." In *Resource Allocation and Productivity in National and International Agricultural Research,* edited by Thomas M. Arndt, Dana G. Dalrymple, and Vernon W. Ruttan, pp. 29–43. Minneapolis, Minnesota: University of Minnesota Press, 1977. [0333]

Herath, H.M.G., and Y.D.A. Senanayake. "A Study of the Financial and Human Resources in Plantation Crops Research in Sri Lanka." *Journal of Agrarian Studies* Vol. 3, No. 2 (1982): 96–109. [0265]

Hernández López, Jesús. Personal communication. Director General, Investigación y Extensión Agrícola, Ministerio de Agricultura y Ganadería. San José, Costa Rica, January 1988. [0775]

Herskovic, Shlomo. Personal communication. Head, Department of R&D Economics, National Council for Research and Development. Jerusalem, 10 July 1987. [0724]

Hill, Jeffery Max. "Agricultural Research Useability Bias toward Small or Large Scale Farmers: An Examination of Crop Production Research in

Kenya." MSc thesis, University of California, Davis, California, 1981. [0274]

Hoan, Nguyen Trong. "Agricultural Research Planning, Monitoring and Evaluation in Viet Nam." Paper presented at the International Workshop on Agricultural Research Management, ISNAR, The Hague, 7–11 September 1987. [0748]

Hocombe, S.D. "FAO/World Bank Cooperative Programme – Project Brief Cote d'Ivoire: National Agricultural Research Project." Report No.: 4/86 CP-IVC 12 PB. FAO and the World Bank, Rome/Washington, D.C., January 1986. Mimeo. [0146]

Homem de Melo, Fernando. *Brazil and the CGIAR Centers: A Study of Their Collaboration in Agricultural Research.* CGIAR Study Paper Number 9. Washington, D.C.: World Bank, 1986. [0082]

Huffman, Wallace E. "The Institutional Development of the Public Agricultural Experiment Station System: Scientists and Departments." Iowa State University, Ames, Iowa, May 1985. Mimeo. [0498]

IAR&T Consultancy Group – University of Ife. *A Consultancy Report on Agricultural Research Delivery Systems in West Africa.* Rome: FAO, December 1979. [0141]

IBRD – Western Africa Projects Department Agriculture Division 4. "The Gambia Agricultural Development Project II." IBRD, Washington, D.C., April 1984. Mimeo. [0137]

ICRISAT. "Botswana." ICRISAT, 1980. Mimeo. [0105]

Idachaba, F.S. *Agricultural Research Policy in Nigeria.* IFPRI Research Report No. 17. Washington, D.C.: IFPRI, 1980. [0155]

Idachaba, F.S. "Agricultural Research Resource Allocation Priorities: The Nigerian Experience." Invited Paper delivered at the International Workshop on Resource Allocation in National Agricultural Research Systems Co-sponsored by IFARD and IDRC, Singapore, 8–10 June 1981. [0156]

Idachaba, F.S. "Agricultural Research Resource Allocation Priorities: The Nigerian Experience." In *Resource Allocation to Agricultural Research.* Proceedings of a workshop held in Singapore 8–10 June 1981, edited by Douglas Daniels and Barry Nestel, pp. 104–121. Ottawa: IDRC, 1981. [0300]

Idachaba, F.S. "Agricultural Research Policy in Nigeria." Paper presented at Agricultural Research Policy Seminar, University of Minnesota, Minneapolis/St. Paul, 22 April 1986. [0167]

IICA. "Strengthening the Agricultural Research Capability of the Ministry of Agriculture of Guyana." IICA, n.d. Mimeo. [0198]

IICA. "Reunión de Programación de Investigación Agrícola del Istmo Centro-Americano." IICA, Guatemala, 1979. Mimeo. [0126]

Indian Council of Agricultural Research (ICAR). *Report of the Review Committee on Agricultural Universities.* New Delhi: ICAR, October 1978. [0781]

Industries Assistance Commission. *Financing Rural Research.* Canberra: Australian Government Publishing Service, June 1976. [0565]

"Information on Agricultural Research in People's Democratic Republic of Yemen." Handout submitted at the Rainfed Agricultural Information Network Workshop, Amman, Jordan, 17–20 March 1985. [0845]

Institut d'Etudes et de Recherches Agricoles. "Lignes Directrices d'Organisation et d'Administration des Recherches Agricoles." MESRS, CNRST, and INERA, Ouagadougou, n.d. Mimeo. [0391]

Institut de la Recherche Agronomique (IRA). *Rapport d'Activités Techniques 1981.* Yaounde: IRA/Delegation Générale à la Recherche Scientifique et Technique (DGRST), 1981. [0085]

Institut de la Recherche Agronomique (IRA). "Note Succincte sur l'Institut de la Recherche Agronomique." IRA, Cameroon, April 1985. Mimeo. [0074]

Institut de la Recherche Vétérinaire de Tunisie. "Rapport sur l'Activité de l'Institut de la Recherche Vétérinaire de Tunisie pour l'Année 1984." Institut de la Recherche Vétérinaire de Tunisie, La Rabta, Tunis, n.d. Mimeo. [0648]

Institut de Recherche Agronomique et Zootechnique – Communauté Economique des Pays des Grands Lacs. *Répertoire des Recherches Agronomiques en Cours; Burundi-Rwanda-Zaire: 1984.* Gitega, Burundi: Institut de Recherche Agronomique et Zootechnique, 1984. [0090]

Institut des Recherches Zootechniques (IRZ). *Annual Report, 1981–1982.* Yaounde: IRZ, 1982. [0170]

Institut des Recherches Zootechniques (IRZ). *Annual Report, 1982–1983.* Yaounde, Cameroon: IRZ, 1983. [0406]

Institut des Sciences Agronomiques du Rwanda (ISAR). *Rapport Annuel 1984.* Rwanda: ISAR, 1985. [0230]

Institut des Sciences Agronomiques du Rwanda (ISAR). *Institut des Sciences Agronomiques du Rwanda: 1962–1987 Point de la Recherche.* Butare, Rwanda: ISAR, November 1987. [0742]

Institut du Sahel, and DEVRES Inc. *Sahel Region Agricultural Research Resource Assessment – Data Base Management Information System: Diskettes and User's Guide.* Printouts. Washington, D.C.: Institut du Sahel, and DEVRES, Inc., n.d. [0384]

Institut du Sahel, and DEVRES, Inc. *Bilan des Ressources de la Recherche Agricole dans les Pays du Sahel, Volume I: Analyse et Stratégie*

Régionale. Washington, D.C.: DEVRES, Inc., 1984. [0045]

Institut du Sahel, and DEVRES, Inc. *Bilan des Ressources de la Recherche Agricole dans les Pays du Sahel, Volume II: Résumés des Rapports Nationaux*. Washington, D.C.: DEVRES, Inc., August 1984. [0083]

Institut du Sahel, and DEVRES, Inc. *Assessment of Agricultural Research Resources in the Sahel, Volume III: National Report Cape Verde*. Washington, D.C.: DEVRES, Inc., August 1984. [0046]

Institut du Sahel, and DEVRES, Inc. *Assessment of Agricultural Research Resources in the Sahel, Volume III: National Report Chad*. Washington, D.C.: DEVRES, Inc., August 1984. [0092]

Institut du Sahel, and DEVRES, Inc. *Assessment of Agricultural Research Resources in the Sahel, Volume III: National Report The Gambia*. Washington, D.C.: DEVRES, Inc., August 1984. [0139]

Institut du Sahel, and DEVRES, Inc. *Assessment of Agricultural Research Resources in the Sahel, Volume III: National Report Mali*. Washington, D.C.: DEVRES, Inc., August 1984. [0044]

Institut du Sahel, and DEVRES, Inc. *Assessment of Agricultural Research Resources in the Sahel, Volume III: National Report Mauritania*. Washington, D.C.: DEVRES, Inc., August 1984. [0143]

Institut du Sahel, and DEVRES, Inc. *Assessment of Agricultural Research Resources in the Sahel, Volume III: National Report Senegal*. Washington, D.C.: DEVRES, Inc., August 1984. [0015]

Institut National de la Recherche Agronomique (INRA). "Eléments de Réponse aux Questions Demandées par l'ISNAR." INRA, n.d. Mimeo. [0339]

Institut National de la Recherche Agronomique (INRA). *L'Institut National de la Recherche Agronomique, Edition du 20e Anniversaire 1946–1966*. Paris: INRA, 1966. [0533]

Institut National de la Recherche Agronomique (INRA). *Prévisions Budgétaires – Budget de Fonctionnement de l'Exercice 1984*. Rabat, Morocco: INRA, Octobre 1983. [0861]

Institut National de la Recherche Agronomique (INRA). *Rapport d'Activité 1985*. Paris: INRA, 1985. [0403]

Institut National de la Recherche Agronomique (INRA). *INRA 1986*. Paris: INRA, 1986. [0405]

Institut National de la Recherche Agronomique (INRA). *Analyse Rétrospective des Moyens de l'INRA de 1970 à 1985*. Paris: INRA, 1986. [0404]

Institut National de Recherches Agronomiques du Niger (INRAN). *Recherche pour le Développement*. Niamey, Niger: INRAN, n.d. [0782]

Institut Sénégalais de Recherches Agricoles (ISRA). *Le Centre National de Recherches Agronomiques de Bambey*. Bambey: ISRA, November 1976. [0153]

Institut Sénégalais de Recherches Agricoles (ISRA). "Compte Prévisionnel 1987 Volet Gestion." ISRA, Dakar, Sénégal. Mimeo. [0814]

Institute of Agricultural Research (IAR). *Report for the Period April 1969 to March 1970*. Addis Ababa: IAR, 1970. [0618]

Instituto Centrale di Statistica (ICS). Personal communication. Director General, ICS. Rome, April 1988. [0860]

Instituto Colombiano Agropecuario (ICA). "Informe de Gerencia 1982." ICA, Colombia, n.d. Mimeo. [0195]

Instituto Colombiano Agropecuario (ICA). *Plan Nacional de Investigación Agropecuaria del ICA*. Tomo 1. Colombia: ICA, January 1981. [0091]

Instituto de Ciencia y Tecnología Agrícolas (ICTA). *Objectivos, Programas de Acción, Realizaciones, Metas*. Guatemala: ICTA, August 1983. [0078]

Instituto de Investigación Agropecuaria de Panamá (IDIAP). *Memoria Anual 1982*. Panama: IDIAP, 1982. [0464]

Instituto de Investigaciones Agropecuarias (INIA). *Memoria del INIA*. Years: 1964/1965 - 1976/1977. Santiago, Chile: INIA, various years. [0576]

Instituto de Investigaciones Agropecuarias (INIA). *Memoria del INIA, 1980, 1981, 1982, 1983, 1984, 1985*. Santiago: INIA, various years. [0064]

Instituto de Investigaciones Agropecuarias (INIA). *Investigación Agropecuaria*. Santiago, Chile: INIA, 1970. [0646]

Instituto de Investigaciones Agropecuarias (INIA). *Decimocuarta Memoria del INIA 1977/1979*. Santiago, Chile: INIA, 1981. [0575]

Instituto Nacional de Estadística (INE). *Estadística sobre las Actividades en Investigación Científica y Desarrollo Tecnológico: Años 1969 y 1970*. Madrid: INE, 1973. [0659]

Instituto Nacional de Estadística (INE). *Estadística sobre las Actividades en Investigación Científica y Desarrollo Tecnológico: Años 1982 y 1983*. Madrid: INE, 1986. [0657]

Instituto Nacional de Investigación Agraria (INIA) – Ministry of Agriculture and Food. *Baseline Study of the Peruvian Agricultural Research, Education and Extension System*. Peru: INIA – Ministry of Agriculture and Food, December 1979. [0202]

Instituto Nacional de Tecnología Agropecuaria (INTA). "Cuadro de Erogaciones de Ejecución del Presupuesto y Cuadro de Recursos Presupuestarios, 1956–1986 (En Valores Históricos y en Valores Actualizados – Año Base 1986)." INTA, Buenos Aires, n.d. Mimeo. [0711]

510

Instituto Nacional de Tecnología Agropecuaria (INTA). *Memoria y Balance Anual 1973.* Buenos Aires: INTA, 1974. [0081]

Instituto Nacional Tecnología Agropecuaria (INTA). *Memoria y Balance Anual 1978.* Buenos Aires: INTA, 1979. [0288]

Instituto Nacional Tecnología Agropecuaria (INTA). *Memoria y Balance Anual 1979.* Buenos Aires: INTA, 1980. [0289]

Instituto Nacional Tecnología Agropecuaria (INTA). *Memoria y Balance Anual 1980.* Buenos Aires: INTA, 1981. [0290]

Instituto Nacional Tecnología Agropecuaria (INTA). *Memoria y Balance Anual 1983.* Buenos Aires: INTA, 1984. [0291]

Instituto Nacional de Tecnología Agropecuaria (INTA). "Elementos para Priorización Institucional." Documento Preliminar para Discusión. INTA, Buenos Aires, Argentina, August 1987. Mimeo. [0374]

Instituut tot Aanmoediging van het Wetenschappelijk Onderzoek in Nijverheid en Landbouw (IWONL). *De Centra voor Landbouwkundig Onderzoek gesubsidieerd door het IWONL.* Belgium: IWONL, n.d. [0463]

International Agricultural Development Service (IADS). *Sudan Agricultural Research Capabilities.* New York: IADS, November 1977. [0252]

International Agricultural Development Service (IADS). *Report of the Review Team Philippine Council for Agriculture and Resources Research.* New York: IADS, November 1980. [0789]

International Development Research Center (IDRC). *Agricultural Research in Uganda: A Program for Rehabilitation.* Ottawa: IDRC, April 1982. [0079]

IRAT, and Centre National de Recherches Agronomiques de Bambey (CNRA). *Rapport Annuel 1974 IRAT-Senegal.* Bambey: IRAT, and CNRA, 1975. [0152]

IRRI. *Rice Research and Production in China: An IRRI Team's View.* Los Baños, Philippines: IRRI, 1978. [0178]

IRRI. *Education for Agriculture.* Proceedings of the Symposium on Education for Agriculture, 12–16 November 1984. Manila: IRRI, 1985. [0181]

Isarangkura, Rungruang. "Inventory of Agricultural Research Expenditure and Manpower in Thailand." In *Resource Allocation to Agricultural Research.* Proceedings of a workshop held in Singapore 8–10 June 1981, edited by Douglas Daniels and Barry Nestel, pp. 32–41. Ottawa: IDRC, 1981. [0293]

Isarangkura, Rungruang. "Agricultural Research Manpower in Thailand." In *Agriculture and Resources Research Manpower Development in South and Southeast Asia,* pp. 149–164. Los Baños, Philippines: Philippine Council for Agriculture and Resources Research and Development, 1983. [0792]

Isarangkura, Rungruang. *Thailand and the CGIAR Centers: A Study of Their Collaboration in Agricultural Research.* CGIAR Study Paper Number 16. Washington, D.C.: World Bank, 1986. [0054]

Islam, Mohammad Zahirul. *Directory of Agricultural Scientists and Technologists in Bangladesh.* Dhaka: Bangladesh Agricultural Research Council, 1985. [0830]

ISNAR. "Working Papers prepared for the Committee on Conditions of Service in the Department of Agriculture (Royal Thai Government)." ISNAR, The Hague, n.d. Mimeo. [0855]

ISNAR. *Informe al Gobierno de Costa Rica: El Sistema de Investigación Agropecuaria y Transferencia de Tecnología en Costa Rica.* The Hague: ISNAR, June 1981. [0031]

ISNAR. *A Report to the Government of Kenya: Kenya's National Agricultural Research System.* The Hague: ISNAR, September 1981. [0051]

ISNAR. *Agrotechnical Research in Ivory Coast.* The Hague: ISNAR, 1982. [0169]

ISNAR. *Guyana: The Agricultural Research System.* The Hague: ISNAR, March 1982. [0201]

ISNAR. *Report to the Government of Papua New Guinea: Review of the Program and Organization for Crops Research in Papua New Guinea.* The Hague: ISNAR, June 1982. [0035]

ISNAR. *Report to the Government of Malawi: A Review of the Agricultural Research System of Malawi.* The Hague: ISNAR, August 1982. [0011]

ISNAR. *Report to the Ministry of Agriculture and Fisheries of Fiji: A Review of the Agricultural Research Division.* The Hague: ISNAR, September 1982. [0042]

ISNAR. *Solomon Islands: Agricultural Research, Extension, and Support Facilities Project.* The Hague: ISNAR, December 1982. [0041]

ISNAR. *Report to the Government of Rwanda: The National Agricultural Research System of Rwanda.* The Hague: ISNAR, February 1983. [0013]

ISNAR. *Rapport au Gouvernement de la Repoblika Demokratika Malagasy: La Recherche Agricole a Madagascar – Bilan et Perspectives du FOFIFA.* The Hague: ISNAR, August 1983. [0037]

ISNAR. *Informe al Gobierno de la República Dominicana; El Sistema de Investigación Agropecuaria en la República Dominicana.* The Hague: ISNAR, August 1983. [0028]

ISNAR. *Report to the Government of the Democratic Republic of Somalia: Development of an*

511

Agricultural Research System. The Hague: ISNAR, November 1983. [0040]

ISNAR. *Report to the Government of Western Samoa: The Agricultural Research System in Western Samoa.* The Hague: ISNAR, December 1983. [0036]

ISNAR. *L'Institut National de la Recherche Agronomique du Maroc.* The Hague: ISNAR, January 1984. [0052]

ISNAR. *Improving the Management of Agricultural Research: Report and Recommendations for the Agricultural Research Corporation of Sudan.* The Hague: ISNAR, February 1984. [0158]

ISNAR. *Report to the Government of Ivory Coast: Agricultural Research in Ivory Coast.* The Hague: ISNAR, December 1984. [0012]

ISNAR. *Report to the Board of Directors of CARDI: Analysis, Evaluation and Proposals to Strengthen CARDI's Regional Capacity.* The Hague: ISNAR, 1985. [0063]

ISNAR. *Agricultural Research Plan Fiji.* The Hague: ISNAR, June 1985. [0186]

ISNAR. *Human Resources Planning for Agricultural Research Personnel in Sri Lanka: The Research Scientist Cadre.* The Hague: ISNAR, August 1985. [0077]

ISNAR. *Kenya Agricultural Research Strategy and Plan, Volumes 1 & 2.* The Hague: ISNAR, December 1985. [0280]

ISNAR. "Ethiopia: Agricultural Sector Study – Agricultural Research." ISNAR, The Hague, 1986. Mimeo. [0465]

ISNAR. "INERA – Masterplan." ISNAR, The Hague, 1986. Mimeo. [0392]

ISNAR. "Tables prepared by the Ethiopia Review Team." ISNAR, The Hague, 1986. Mimeo. [0383]

ISNAR. *Review of the Research System of the Ministry of Agriculture and Food in the Philippines.* The Hague: ISNAR, January 1986. [0284]

ISNAR. *The National Institute for Agricultural Research of Morocco – Volumes I and II.* The Hague: ISNAR, April 1986. [0340]

ISNAR. *International Workshop on Agricultural Research Management.* Report of a workshop. The Hague: ISNAR, 1987. [0747]

ISNAR. *Review of Research Program Management and Manpower Planning at the Institute of Agricultural Research in Ethiopia.* The Hague: ISNAR, March 1987. [0712]

ISNAR. *Programme de Développement de la Recherche Agricole en Tunisie.* Three volumes. The Hague: ISNAR, May 1987. [0647]

ISNAR. *An Analysis of Structure and Management of the Institute of Agricultural Research (IRA) and the Institute of Animal Research (IRZ) of Cameroon.* Report to the Ministry of Higher Education and Scientific Research of the Republic of Cameroon. The Hague: ISNAR, December 1987. [0757]

ISNAR. "A Review of Agricultural Systems Research in Rwanda." ISNAR, The Hague, January 1988. Mimeo. [0858]

ISNAR. "Développement et Gestion des Ressources Humaines à l'Institut Sénégalais de Recherches Agricoles." Une Etude Préliminaire. ISNAR, The Hague, February 1988. Mimeo. [0815]

ISNAR, and Groupe d'Etude de la Réorganisation du Système National de Recherche Agronomique. *Etude de la Réorganisation du Système National de Recherche Agronomique du Zaïre.* The Hague: ISNAR, February 1985. [0151]

ISNAR, and Institut Panafricain pour le Développement – Afrique Centrale (IPD/AC). *Rapport au Ministère de l'Enseignement Supérieur et de la Recherche Scientifique du Cameroun: L'Amélioration de la Gestion de la Recherche Agricole au Cameroun.* The Hague: ISNAR, June 1984. [0117]

ISNAR, IFARD & AOAD. Survey of National Agricultural Research Systems: Unpublished Questionnaire Responses. ISNAR, The Hague, 1985. [0017]

Istituto Centrale di Statistica (ICS). *Statistiche sulla Ricerca Scientifica Anni 1967–1974.* Estratto da: Bollettino Mensile di Statistica 1973:10, 1975:1, 1976:5, 1977:1. Roma: ICS, various years. [0683]

Istituto Centrale di Statistica (ICS). *Indagine Statistica sulla Ricerca Scientifica 1975–1983.* Supplemento al Bollettino Mensile di Statistica 1978:1&22, 1980:1, 1981:13, 1982:19, 1983:17&27, 1984:11, 1985:19. Roma: ICS, various years. [0681]

Istituto Centrale di Statistica (ICS). *Indagine Statistica sulla Ricerca Scientifica – Consuntivo 1984 / Previsione 1985 e 1986.* Collana d'Informazione Anno 1987, No. 1. Roma: ICS, 1987. [0682]

Jain, H.K. "Reorganization and Intensification of Agricultural Research in India." In *Strengthening National Agricultural Research.* Report from a SAREC workshop, 10–17 September 1979, edited by Bo Bengtsson and Getachew Tedla, pp. 89–97. Sweden: SAREC, 1980. [0316]

Jain, H.K. "Update on Western Samoa – A Back to Office Report." ISNAR, The Hague, July 1986. Mimeo. [0263]

Jamieson, Barbara M. "Resource Allocation to Agricultural Research in Kenya from 1963 to 1978." PhD diss., University of Toronto, Canada, 1981. [0229]

Jarrett, F.G. *Report on Agricultural Research and Extension, Department of Agriculture Stock and Fisheries (DASF) – Papua New Guinea.*

Adelaide, Australia: The University of Adelaide, March 1972. [0475]

Jarrett, F.G., and R.K. Lindner. "Rural Research in Australia." In *Agriculture in the Australian Economy* (second edition), edited by D.B. Williams, pp. 94–98. Sydney: Sydney University Press, 1980. [0568]

Jeffers, J.P.W. "Agricultural Research Profile, Barbados." Ministry of Agriculture, Food and Consumer Affairs, Barbados, 1977. Mimeo. [0103]

Jeffers, J.P.W. Personal communication. Deputy Chief Agricultural Officer (Research), Ministry of Agriculture, Food and Consumer Affairs. St. George, Barbados, February 1988. [0800]

Johnson, Eldon L., Frederick C. Fliegel, John L. Woods, and Mel C. Chu. *The Agricultural Technology System of Taiwan.* Urbana-Champaign, Illinois: INTERPAKS, Office of International Agriculture, University of Illinois, December 1987. [0857]

Johnson, R.W.M., S. Mahfooz Ali Shah, Ghulam Rasul, and Sohail Munawar. *Agricultural Research Manpower Survey.* Islamabad: Pakistan Agricultural Research Council – Directorate of Planning, n.d. [0728]

Joint Agriculture Research Survey Team. *Report of Survey and Recommendations on Organizations, Systems, and Requirements for Research in Agriculture and Related Industries in Indonesia.* Djakarta, Indonesia: Ministry of Agriculture, July 1969. [0441]

Joint Pakistan-American Agricultural Research Review Team. *Report of the Joint Pakistan-American Agricultural Research Review Team.* Dacca: Bangladesh Agricultural Research Council, April 1968. [0193]

Joint Review Group: Agricultural Research Group, Sri Lanka, and ISNAR. *Report to the Government of Sri Lanka: The Agricultural Research System in Sri Lanka.* The Hague: ISNAR, June 1984. [0034]

Joint Review Group: Agricultural Research Group, Sri Lanka, and ISNAR. *The Agricultural Research System in Sri Lanka.* The Hague: ISNAR, October 1986. [0457]

Judd, M. Ann, James K. Boyce, and Robert E. Evenson. "Investing in Agricultural Supply." Economic Growth Center, Yale University, New Haven, Connecticut, 1983. Mimeo. [0014]

Junta Nacional de Investigação Científica e Tecnológica. *Resursos em Ciênciae Tecnologia – Inventário de 1971.* Lisboa: Junta Nacional de Investigação Científica e Tecnológica, n.d. [0662]

Junta Nacional de Investigação Científica e Tecnológica. *Récursos Inventário em Ciência e Tecnologia 1972.* Lisboa: Junta Nacional de Investigação Científica e Tecnológica, n.d. [0663]

Junta Nacional de Investigação Científica e Tecnológica. *Investigação & Desenvolvimento Portugal 1976.* Lisboa: Junta Nacional de Investigação Científica e Tecnológica, n.d. [0665]

Junta Nacional de Investigação Científica e Tecnológica. *Recursos em Ciência e Tecnologia Portugal 1978.* Lisboa: Junta Nacional de Investigação Científica e Tecnológica, 1981. [0664]

Junta Nacional de Investigação Científica e Tecnológica. *Investigação & Desenvolvimento Experimental – Inquérito ao Potencial Científico e Tecnológico Nacional Actualização a 84.12.31: Sector Estado (Dados Provisórios).* Lisboa: Junta Nacional de Investigação Científica e Tecnológica, June 1986. [0666]

Junta Nacional de Investigação Científica e Tecnológica. *Investigação & Desenvolvimento Experimental – Inquérito ao Potencial Científico e Tecnológico Nacional Actualização a 84.12.31: Sector Ensino Superior (Dados Provisórios).* Lisboa: Junta Nacional de Investigação Científica e Tecnológica, July 1986. [0667]

Karama, A. Syarifuddin. "Planning, Programming and Evaluation of Research and Development Indonesia Experiences." Paper presented at the Workshop on Improving Agricultural Research Organizations and Management: Progress and Issues for the Future, ISNAR, The Hague, 8–12 September 1986. [0270]

Kasembe, J.R.N., A.M. Macha, and A.N. Mphuru. "Agricultural Research inTanzania." In *Proceedings of the Eastern Africa-ACIAR Consultation on Agricultural Research*, pp. 44–56. Australia: ACIAR, 1984. [0764]

Kassapu, Samuel. *Les Dépenses de Recherche Agricole dans 34 Pays d'Afrique Tropicale.* Paris: Centre de Développement de l'OCDE, 1976. [0589]

Kean, Stewart. "Research Branch Expenditure (1976–1985)." Preliminary table made for the Zambian case study of the ISNAR Study on the Organization and Management of On-Farm Research. ISNAR, The Hague, 1986. Mimeo. [0388]

Keen, B.A. *The East African Agriculture and Forestry Research Organization – Its Origin and Objects.* Nairobi: East African Community, n.d. [0642]

Kellou, R. "La Recherche Agricole en Algérie." Paper presented at the Séminaire sur l'Orientation et l'Organisation de la Recherche Agronomique dans les Pays du Bassin Méditerranéen, Centre International de Hautes Etudes Agronomiques Méditerranéennes, Istanbul, 1–3 December 1986. [0408]

Kellou, R. "Agricultural Research in Algeria and Its Human Resources." Paper presented at the International Workshop on Agricultural

Research Management, ISNAR, The Hague, 7–11 September 1987. [0753]

Khettouch, Moha, Khadija Rajraji, and Abderraouf Elaouni. *Données sur les Moyens Humains de l'Institut National de la Recherche Agronomique.* Rabat, Morocco: Institut National de la Recherche Agronomique, 1986. [0838]

Kim, Young Sang. "The Progress and Contribution of Research Projects for Agricultural Development in Korea." Paper presented at the 1st International Meeting of National Agricultural Research Systems and 2nd IFARD Global Convention, IFARD/EMBRAPA, Brasilia, Brasil, 6–11 October 1986. [0393]

Kim, Young Sang. "The Progress and Contribution of Research Projects for Agricultural Development in Korea." In *The Impact of Research on National Agricultural Development*, edited by Brian Webster, Carlos Valverde and Alan Fletcher, pp. 211–224. The Hague: ISNAR, July 1987. [0768]

Kimura, J.H. "Financial and Administrative Management of Research Institutions in Eastern and Southern Africa: Report on Responses to a Questionnaire." In *Promotion of Technology Policy and Science Management in Africa*, edited by Karl Wolfgang Menck and Wolfgang Gmelin. Bonn: Deutsche Stiftung für internationale Entwicklung (DSE), 1986. [0847]

King, H.E., ed. *Progress Reports from Experiment Stations Season 1969–70: Uganda.* London: Cotton Research Corporation, 1970. [0622]

Kira, M. Taher. *Report on Agricultural Research Manpower Assessment in People's Democratic Republic of Yemen and Democratic Republic of Somalia.* Cairo: FAO Near East Regional Office, January 1980. [0870]

Klein, K.K., and W.H. Furtan, eds. *Economics of Agricultural Research in Canada.* Papers presented at a conference held September 1983. Calgary, Alberta, Canada: The University of Calgary Press, 1985. [0874]

Knudsen, P.H., ed. *Agriculture in Denmark.* Copenhagen: Agricultural Council of Denmark, 1977. [0452]

Kong, Guasong. Personal communication. Second Secretary of Science and Technology of the Embassy of the People's Republic of China in the Netherlands. The Netherlands, April 1987. [0588]

Kong, Guasong. Personal communication. Second Secretary of Science and Technology of the Embassy of the People's Republic of China in the Netherlands. The Hague, March 1988. [0883]

Kula, G.R. Personal communication. Deputy Director, Crop Research, Department of Agriculture and Livestock. Konedobu, Papua New Guinea, December 1987. [0759]

Kunst, Jan. Personal communication. Medewerker Financieel Economische Zaken, Agricultural University Wageningen. Wageningen, March 1987. [0564]

Kyomo, M.L. "Agricultural Research in Eastern and Southern Africa: Issues and Priorities." Southern African Centre for Cooperation in Agricultural Research of SADCC, Gaborone, Botswana, 1986. Mimeo. [0446]

"La Administración de la Investigación Agrícola en Nicaragua." Paper presented at Curso-Taller sobre la Administración de la Investigación Agrícola, Panama City, 14–25 July 1986. [0382]

Lacy, W.B., L. Busch, and P. Marcotte. *The Sudan Agricultural Research Corporation: Organization, Practices, and Policy Recommendations.* Lexington, Kentucky: University of Kentucky, October 1983. [0251]

Landbouw Economisch Instituut (LEI). *Landbouwcijfers.* Yearly publication. The Hague: LEI, various years. [0039]

Landbouwhogeschool Wageningen. *Jaarverslag.* Academic years: 1980–1982. Wageningen: Landbouwhogeschool, various years. [0545]

Landbouwhogeschool Wageningen. *Jaarverslag.* Academic years: 1975/1976, 1976/1977, 1977/1978, and 1978/1979. Wageningen: Landbouwhogeschool, various years. [0544]

Landbouwhogeschool Wageningen. *Jaarverslag – Wetenschappelijk Verslag.* Academic years: 1983 and 1984. Wageningen: Landbouwhogeschool, various years. [0546]

Landbouwhogeschool Wageningen. "Wetenschappelijk Verslag over 1979." Landbouwhogeschool, Wageningen, n.d. Mimeo. [0547]

Landbouwhogeschool Wageningen. *Wetenschappelijk Verslag 1981–1982.* Wageningen: Landbouwhogeschool, various years. [0549]

Landbouwhogeschool Wageningen. *Ontwikkelingsplan Landbouwhogeschool 1980 tot en met 1983.* Wageningen: Landbouwhogeschool, November 1977. [0550]

Landbouwproefstation Suriname. *Jaarverslag.* Years: 1971–1976. Suriname: Landbouwproefstation, various years. [0584]

Landbouwproefstation. *Agricultural Experiment Station 75th Anniversary.* Parimaribo, Suriname: Landbouwproefstation, 1978. [0451]

Landbouwhogeschool Wageningen – Vaste Commissie voor de Wetenschapsbeoefening. *Op Weg naar een Onderzoekbeleid van de Landbouwhogeschool.* Wageningen: Landbouwhogeschool, 1974. [0548]

Larrea Herrera, Pablo. Personal communication. Director General, Instituto Nacional de Investigaciones Agropecuarias. Quito, Ecuador, January 1988. [0758]

Lasram, Mustapha. "Rapport sur le Système National de la Recherche Agricole en Tunisie." Paper presented at the Séminaire sur l'Orientation et l'Organisation de la Recherche Agronomique dans les Pays du Bassin Méditerranéen, Centre International de Hautes Etudes Agronomiques Méditerranéennes, Istanbul, 1–3 December 1986. [0419]

Lazzaoui, Mohamed. "Le Système de Recherche Agronomique Marocain." Paper presented at the Séminaire sur l'Orientation et l'Organisation de la Recherche Agronomique dans les Pays du Bassin Méditerranéen, Centre International de Hautes Etudes Agronomiques Méditerranéennes, Istanbul, 1–3 December 1986. [0418]

Liacos, Leonidas. "La Recherche Vétérinaire en Grèce." Paper presented at the Séminaire sur l'Orientation et l'Organisation de la Recherche Agronomique dans les Pays du Bassin Méditerranéen, Centre International de Hautes Etudes Agronomiques Méditerranéennes, Istanbul, 1–3 December 1986. [0412]

Librero, Aida R. "An Inventory of Research Manpower in Agriculture and Natural Resources in the Philippines." In *Agriculture and Resources Research Manpower Development in South and Southeast Asia*, pp. 131–141. Los Baños, Philippines: Philippine Council for Agriculture and Resources Research and Development, 1983. [0786]

Lin, J.Y. Personal communication. Director, Division of Life Sciences, National Science Council. Taipei, January 1988. [0773]

Lindarte, Eduardo. "Proyecto Asignación de Recursos para Investigación Agrícola en América Latina (ARIAL) – Colombia: Estudio de Caso." Unidad de Apoyo a la Programación – PSU, March 1979. Mimeo. [0863]

Lindner, R.K. "Accountability in Research through Industry Funding." In *Evaluation of Agricultural Research*, Miscellaneous publication 8–1981, edited by G.W. Norton, et al., pp. 177–181. St. Paul, Minnesota: Minnesota Agricultural Experiment Station, University of Minnesota, April 1981. [0569]

Liwenga, J.M. "Agricultural Research in Tanzania." In *Strengthening National Agricultural Research*. Report from a SAREC workshop, 10–17 September 1979, edited by Bo Bengtsson and Getachew Tedla, pp. 55–61. Sweden: SAREC, 1980. [0313]

Liyanage, Shantha, T. Wijesinghe, N. Anbalagan, and S. Peiris. *A Survey of the Expenditure on Research and Experimental Development in Sri Lanka 1966–1975*. Colombo: National Science Council of Sri Lanka, September 1977. [0877]

Lopes, J.R.B. "Agricultural Research in Brazil." In *Strengthening National Agricultural Research*. Report from a SAREC workshop, 10–17 September 1979, edited by Bo Bengtsson and Getachew Tedla, pp. 144–154. Sweden: SAREC, 1980. [0321]

Lyonga, S.N., and E.T. Pamo. "The Impact of the Collaboration between the International Agricultural Research System and the National Agricultural Research System in Cameroon." CGIAR, Washington, D.C., March 1985. Mimeo. [0060]

Macha, Augustine M. "Priority Setting in the Tanzania Livestock Research Organization." Paper presented at the International Workshop on Agricultural Research Management, ISNAR, The Hague, 7–11 September 1987. [0749]

Malaysian Agricultural Research and Development Institute (MARDI). *Annual Report 1981*. Kuala Lumpur: MARDI, 1985. [0134]

Malima, V.F., F.M. Shao, F. Turchi, and D. Sorrenti. *Agricultural Research Resource Assessment in the SADCC Countries, Volume II: Country Report Tanzania*. Washington, D.C.: SADCC and DEVRES, Inc., May 1985. [0232]

"Manpower Analysis Gambia." n.d. Mimeo. [0386]

Marroquín, Víctor René. "La Investigación Agrícola Salvadoreña." Paper presented at Curso–Taller sobre la Administración de la Investigación Agrícola, Panama City, 14–25 July 1986. [0381]

Martin, Ben R., and John Irvine. *An International Comparison of Government Funding of Academic and Academically Related Research*. Advisory Board for the Research Council (ABRC) Science Policy Study No. 2. Brighton: ABRC – Science Policy Research Unit, University of Sussex, October 1986. [0703]

Marzocca, Angel. "Proceso de Formación y Evolución del INTA, en Argentina." In *Organización y Administración de la Generación y Transferencia de Tecnología Agropecuaria*, Serie: Ponencias, Resultados y Recomendaciones de Eventos Técnicos No. A4/UY-86–001, edited by Horacio H. Stagno and Mario Allegri. Montevideo, Uruguay: IICA and Centro de Investigaciones Agrícolas 'Alberto Boerger', 1986. [0655]

Maticic, Brane. "L'Orientation et l'Organisation de la Recherche Agronomique en Yougoslavie." Paper presented at the Séminaire sur l'Orientation et l'Organisation de la Recherche Agronomique dans les Pays du Bassin Méditerranéen, Centre International de Hautes Etudes Agronomiques Méditerranéennes, Istanbul, 1–3 December 1986. [0421]

Matus, Jaime G., et al. "A Review of the Collaboration between the National Agricultural Research System and the International Research Centers of the CGIAR." Centro de Economia, Colegio de Postgraduados, Chapingo, Mexico, December 1984. Mimeo. [0049]

Mauritius Sugar Industry Research Institute (MSIRI). *Annual Report 1982.* Mauritius: MSIRI, 1983. [0184]

McBride, Conor P. *Scientific Research in New Zealand: Expenditure and Manpower 1953–1962.* Information Series No. 41. New Zealand: Department of Scientific and Industrial Research, 1964. [0478]

McBride, Conor P., and Christine de Joux. *Scientific Research in New Zealand: Government Expenditure and Manpower, 1926–1966.* National Research Advisory Council Publication No. 1. New Zealand: Department of Scientific and Industrial Research, 1966. [0477]

McDermott, J.K., and David Bathrick. *Guatemala: Development of the Institute of Agricultural Science and Technology (ICTA) and Its Impact on Agricultural and Farm Productivity.* AID Project Impact Evaluation No. 30. Washington, D.C.: AID, February 1982. [0211]

McGlaughlin, Glen R. "A Study Regarding Agricultural Research in Canada." Research Division, Saskatchewan Wheat Pool, January 1977. Mimeo. [0598]

McLean, Diana. "Agricultural Research Analysis." AID, Banjul, The Gambia, September 1984. Mimeo. [0138]

Messmann, Karl. "Forschung und Experimentelle Entwicklung in Österreich 1966/67 bis 1985." *Statistische Nachrichten* Vol. 40, No. 5 (1985): 291–300. [0679]

Ministère de l'Agriculture. "Bilan Diagnostic du Secteur Agricole de 1960 à 1980." Ministère de l'Agriculture, Yaounde, 1980. Mimeo. [0118]

Ministère de l'Agriculture. *Rapport Annuel de la Division de la Recherche Agronomique.* Mali: Ministère de l'Agriculture, 1983. [0144]

Ministère de l'Agriculture et de l'Elevage. *Institut des Sciences Agronomiques du Burundi (ISABU) – Rapport des Activités de Recherches 1983.* Bujumbura, Burundi: Ministère de l'Agriculture et de l'Elevage, 1984. [0225]

Ministère de la Coopération et du Développement de France. "Procès-Verbal de la IXeme Commission Mixte Franco-Ivoirienne de Recherche." Ministère de la Coopération et du Développement de France, Paris, January 1982. Mimeo. [0148]

Ministère de l'Enseignement Superiéur et de la Recherche Scientifique (MESRS). "Organigramme du Ministère de l'Enseignement Superiéur et de la Recherche Scientifique." MESRS, n.d. Mimeo. [0389]

Ministère du Plan. *Plan Quinquennial de Développement Economique, Social et Culturel (1981–1985): Politique Sectionnelle.* Abidjan: Ministère du Plan, 1981. [0149]

Ministerie van Landbouw en Visserij. *Aard en Omvang van het Landbouwkundig Onderzoek.* Tweede Kamer der Staten Generaal, Vergaderjaar 1985– 1986, 19307, nrs. 1–2. The Hague: Ministerie van Landbouw en Visserij, November 1985. [0424]

Ministerie van Landbouw en Visserij – Directie Akkerbouw en Tuinbouw. "Verslag van een Studiereis naar de Bondsrepubliek Duitsland in de Periode 11 t/m 15 Augustus 1975." Ministerie van Landbouw en Visserij, The Hague, 1975. Mimeo. [0449]

Ministerie van Landbouw en Visserij – Directie Akkerbouw en Tuinbouw. "Indrukken van een Orientatiereis naar Belgie 25 t/m 29 Juli 1977." Ministerie van Landbouw en Visserij, The Hague, March 1978. Mimeo. [0450]

Ministerie van Landbouw en Visserij – Directie Landbouwkundig Onderzoek. "Bijdrage tot een Meerjarenplan voor het Landbouwkundig Onderzoek 1972–1976." Ministerie van Landbouw en Visserij, The Hague, December 1972. Mimeo. [0551]

Ministerie van Landbouw en Visserij – Directie Landbouwkundig Onderzoek *Meerjarenvisie 1977–1981 voor het Landbouwkundig en het Visserij-Onderzoek.* The Hague: Ministerie van Landbouw en Visserij, 1977. [0426]

Ministerio de Agricultura y Ganadería – Dirección General de Investigaciones Agrícolas. *Memoria Anual de la Investigación Agrícola 1983.* San José: Ministerio de Agricultura y Ganadería, 1984. [0125]

Ministerio de Agricultura y Pesca – Instituto Nacional de Investigaciones Agrarias. *Agricultural Research.* Madrid: Ministerio de Agricultura y Pesca, 1981. [0540]

Ministerio de Asuntos Campesinos y Agropecuarios. "Gastos para la Investigación en Agricultura, Forestal y Pesca en Bolivia." Table. Bolivia, n.d. Mimeo. [0721]

Ministerio de Economía – Instituto Nacional de Estadística (INE). *Estadística sobre las Actividades en Investigación Científica y Desarrollo Tecnológico: Años 1973–1974.* Madrid: INE, 1978. [0660]

Ministerio de Planificación del Desarrollo – Instituto Nacional de Estadística (INE). *Estadística sobre las Actividades en Investigación Científica y Desarrollo Tecnológico: Años 1971–1972.* Madrid: INE, n.d.. [0658]

Ministry of Agriculture – Animal Production Research Unit. *Ten Years of Animal Production and Range*

516

Research in Botswana. Gaborone: Ministry of Agriculture, 1980. [0336]

Ministry of Agriculture – Department of Agricultural Research. "Staff andExpenditure – Department of Agricultural Research." Ministry of Agriculture, Malawi, 1981. Mimeo. [0240]

Ministry of Agriculture – Department of Research and Specialist Services (DRSS). "Establishment Tables 1983–1984." Ministry of Agriculture, Zimbabwe, 1984. Mimeo. [0247]

Ministry of Agriculture – Department of Research and Specialist Services (DRSS). "Establishment Tables 1981–1982." Ministry of Agriculture, Zimbabwe, 1982. Mimeo. [0246]

Ministry of Agriculture – Fisheries Division. *Annual Report for 1982, Fisheries Research Institute, Glugor, Penang*. Malaysia: Ministry of Agriculture, January 1984. [0276]

Ministry of Agriculture and Fisheries. *New Zealand Agriculture*. Wellington, New Zealand: Ministry of Agriculture and Fisheries, 1974. [0453]

Ministry of Agriculture and Natural Resources – Agricultural Research Corporation (ARC). *List of Scientific Staff and Senior Administrators*. Wad Medani, Sudan: Ministry of Agriculture and Natural Resources – ARC, December 1986. [0710]

Ministry of Agriculture and Natural Resources – Agricultural Research Corporation (ARC). *A Note on the Agricultural Research Corporation*. Wad Medani, Sudan: ARC, January 1987. [0709]

Ministry of Agriculture and Natural Resources and the Environment. *Annual Report of the Ministry of Agriculture and Natural Resources and the Environment for the Year 1980, 1981, and 1982*. Mauritius: Ministry of Agriculture and Natural Resources and the Environment, various years. [0185]

Ministry of Agriculture Egypt. *Allocation of Resources in Agricultural Research*. Cairo: Ministry of Agriculture, November 1983. [0257]

Ministry of Agriculture, Fisheries and Food. *Report on Research and Development 1976, 1977, and 1978*. Yearly reports. London: Her Majesty's Stationery Office, various years. [0689]

Ministry of Agriculture, Fisheries and Food. *Report on Research and Development 1979–80, 1981–82, 1982–83, and 1983–84*. Yearly reports. London: Her Majesty's Stationery Office, various years. [0688]

Ministry of Agriculture, Fisheries and Food & Department of Agriculture and Fisheries for Scotland. *Fisheries Research and Development Board Third Report 1976–1977*. London: Her Majesty's Stationery Office, n.d.. [0690]

Ministry of Agriculture, Forestry and Fisheries. *Present Situation of Research Relating to Agriculture, Forestry and Fisheries 1984*. Tokyo, Japan: Ministry of Agriculture, Forestry and Fisheries, 1984. [0375]

Ministry of Agriculture, Livestock, and Food Production. *Research Division – Annual Report 1983*. Trinidad and Tobago: Ministry of Agriculture, Livestock, and Food Production, 1983. [0224]

Mkandawire, N.A. "Study on Small Countries Research: Report on Malawi." IDRC, Ottawa, April 1986. Mimeo. [0738]

Mohan, Rakesh, D. Jha, and Robert Evenson. "The Indian Agricultural Research System." *Economic and Political Weekly* Vol. VIII, No. 13 (March 1973): 223–228. [0876]

Montes Llamas, Gabriel. *Instituto Colombiano Agropecuario – Informe de Gerencia 1986*. Bogotá: Instituto Colombiano Agropecuario, n.d. [0736]

Morehouse, Ward, and Brijen Gupta. "India: Success and Failure." Chapter 8 in *Learning by Doing – Science and Technology in the Developing World*, edited by Aaron Segal, pp. 189–212. Boulder: Westview Press, 1987. [0684]

Moreno, Jesus. "La Recherche Agricole en Espagne." Paper presented at the Séminaire sur l'Orientation et l'Organisation de la Recherche Agronomique dans les Pays du Bassin Méditerranéen, Centre International de Hautes Etudes Agronomiques Méditerranéennes, Istanbul, 1–3 December 1986. [0410]

Morras, Héctor José María. Personal communication. Director (Int.) de Desarrollo de Recursos Humanos, Instituto Nacional de Tecnología Agropecuaria. Buenos Aires, Argentina, March 1988. [0029]

Moseman, A.H. "Agricultural Research in Bangladesh." FAO/DANIDA, Rome/Dacca, n.d. Mimeo. [0113]

Moseman, A.H., et al. *Bangladesh Agricultural Research System*. Bangladesh: Ministry of Agriculture and Fisheries, July 1980. [0111]

Moseman, Albert, ed. *National Agricultural Research Systems in Asia*. Report of the Regional Seminar held at the India International Centre, New Delhi, India, March 8–13, 1971. New York: ADC, Inc., 1971. [0176]

Mudimu, Godfrey D. "Planning and Evaluating Agricultural Research in Zimbabwe: A Preliminary Assessment and Research Proposal." MSc thesis, Department of Agricultural Economics, Michigan State University, Michigan, 1986. [0805]

Mugabe, N.R. *Project for the Strengthening of Agricultural Research in SADCC Countries: National Report for the Zimbabwe Agricultural Research Resource Assessment*. Harare: Ministry of Agriculture, August 1984. [0832]

Muhammed, A. "Pakistan's Development Perspective: Status of Agricultural Research Set-Up and Its Role in National Development." In *Strengthening National Agricultural Research.* Report from a SAREC workshop, 10–17 September 1979, edited by Bo Bengtsson and Getachew Tedla, pp. 112–122. Sweden: SAREC, 1980. [0318]

Muhammed, Amir. "National Agricultural Research System of Pakistan." Paper presented at the 1st International Meeting of National Agricultural Research Systems and 2nd IFARD Global Convention, IFARD/EMBRAPA, Brasilia, Brasil, 6–11 October 1986. [0396]

Muhammed, Amir. "The National Agricultural Research System of Pakistan." In *The Impact of Research on National Agricultural Development,* edited by Brian Webster, Carlos Valverde and Alan Fletcher, pp. 183–191. The Hague: ISNAR, July 1987. [0767]

Murphy, Diarmuid. *Research and Development in Ireland 1967.* Dublin: National Science Council, October 1969. [0632]

Musnad, H.A., H.H.M. Faki, and A.H. El Jack. "Agricultural Research Priorities in the Sudan." In *Proceedings of the Eastern Africa-ACIAR Consultation on Agricultural Research,* pp. 24–30. Australia: ACIAR, 1984. [0762]

Mustapha, Nik Ishak bin Nik. "The Agricultural Research Resource Allocation System in Peninsular Malaysia." In *Resource Allocation to Agricultural Research.* Proceedings of a workshop held in Singapore 8–10 June 1981, edited by Douglas Daniels and Barry Nestel, pp. 49–54. Ottawa: IDRC, 1981. [0295]

Muturi, S.N. "The System of Resource Allocation to Agricultural Research in Kenya." In *Resource Allocation to Agricultural Research.* Proceedings of a workshop held in Singapore 8–10 June 1981, edited by Douglas Daniels and Barry Nestel, pp. 123–128. Ottawa: IDRC, 1981. [0301]

Muturi, S.N. Personal communication. Director of Research Development, Ministry of Research, Science and Technology. Nairobi, January 1988. [0780]

Muturi, Stachys N. *Training in the CGIAR System: The Case of Kenya.* Rome: CGIAR – TAC, October 1984. [0798]

Nagy, Joseph Gilbert. "The Pakistan Agricultural Development Model: An Economic Evaluation of Agricultural Research and Extension Expenditures." PhD diss., University of Minnesota, Minneapolis, March 1984. [0281]

National Agricultural Research System Survey Technical Panel (NARSSTP). *The Philippine Agricultural Research System – Evaluation and Recommendations.* Volume 1. Manila, Philippines: NARSSTP, December 1971. [0788]

National Council for Science and Technology (NCST), and ISNAR. *A Manpower and Training Plan for the Agricultural Research System in Kenya 1983–1987.* The Hague: ISNAR, November 1982. [0033]

National Research Advisory Council. *Report of the National Research Advisory Council for the Year Ended 31 March 1986.* Presented to the House of Representatives Pursuant to Section 11 of the National Research Advisory Council Act 1963; Yearly Reports 1965–1986. Wellington, New Zealand: Government Printer, various years. [0479]

Nationale Raad voor Landbouwkundig Onderzoek TNO. *Studierapport 10: Meerjarenvisie Landbouwkundig Onderzoek 1982–1986.* The Hague: Nationale Raad voor Landbouwkundig Onderzoek TNO, 1981. [0427]

NAVF, NLVF, and NTNF. *Utgifter og Arsverk i Norsk Forskning – En Statistik Oversikt for Aret 1963.* Oslo: NAVF, NLVF, and NTNF, 1966. [0627]

NAVF, NLVF, and NTNF. *Norsk Forskningsvirksomhet Utgifter og Arsverk 1966.* Oslo: NAVF, NLFV, and NTNF, September 1968. [0626]

NAVF, NLVF, and NTNF. *Norsk Forskningsvirksomhet Utgifter og Personale Innen Forskning og Utviklingsarbeid 1968.* Oslo: NAVF, NLVF, and NTNF, June 1970. [0625]

Ndunguru, B. "The Impact of the Collaboration between the International Agricultural Research System and the National Agricultural Research System in Tanzania." CGIAR, Washington, D.C., September 1984. Mimeo. [0058]

Nestel, Barry. *Indonesia and the CGIAR Centers: A Study of Their Collaboration in Agricultural Research.* CGIAR Study Paper Number 10. Washington, D.C.: World Bank, 1985. [0047]

Netherlands Central Bureau of Statistics (CBS). *Research and Development in the Netherlands.* Years: 1970–1983. The Hague: Staatsuitgeverij, various years. [0581]

Netherlands Central Bureau of Statistics (CBS). *Research and Developement in the Netherlands 1959.* Volumes 1 and 2. Zeist, The Netherlands: Uitgeversmaatschappij W. de Haan N.V., 1963. [0577]

Netherlands Central Bureau of Statistics (CBS). *Research and Development Activities in the Netherlands 1964.* The Hague: Staatsuitgeverij, 1967. [0578]

Netherlands Central Bureau of Statistics (CBS). *Research and Development Activities in the Netherlands 1967.* The Hague: Staatsuitgeverij, 1969. [0579]

518

Netherlands Central Bureau of Statistics (CBS). *Research and Development in the Netherlands 1969*. The Hague: Staatsuitgeverij, 1971. [0580]

Ngatchou, Nya J. *Evolution de la Recherche Scientifique et Technique au Cameroun*. Yaounde: Délégation Générale à la Recherche Scientifique et Technique (DGRST), March 1982. [0084]

Nidup, Namgey. "Management of Human Resources in Agricultural Research in Bhutan." In *Management of Human Resources in Agricultural Research*. Report of the International Workshop on Management of Human Resources held 3–5 March 1986, edited by Theodore Hutchcroft, pp. 83–86. Dhaka: Bangladesh Agricultural Research Council, ISNAR, and Winrock International, 1986. [0327]

Nigerian Council for Science and Technology (NCST). *National Policies and Priorities for Research in Science and Technology*. Lagos: NCST, December 1975. [0154]

Nigerian Stored Products Institute – Federal Ministry of Trade. *Annual Report 1968*. Lagos, Nigeria: Federal Ministry of Information, 1969. [0634]

Noor, M.A., M. Abdullahi, and A.N. Alio. "Status of Agricultural Research in Somalia." In *Proceedings of the Eastern Africa-ACIAR Consultation on Agricultural Research*, pp. 16–23. Australia: ACIAR, 1984. [0761]

NORDFORSK. *Forsknings Virksomhet i Norden i 1967 – Utgifter og Personale*. Nordisk Statistisk Skriftserie 18. Oslo: NORDFORSK, 1970. [0621]

Norges Landbruksvitenskapelige Forskningråd (NLVF). *Forlag om Utbygging av de Landbruksvitenskapelige Institusjoner 1961–1970*. Oslo: NLVF, 1960. [0628]

Norges Landbruksvitenskapelige Forskningsråd (NLVF). *Landbruksforskning Personale og Kostnader i 1963*. Oslo: NLVF, 1964. [0624]

Norges Landbruksvitenskapelige Forskningsråd (NLVF). *Landbruksforskning Personale og Kostnader i 1966*. Oslo: NLVF, 1967. [0623]

Norges Landbruksvitenskapelige Forskningsråd (NLVF). *Langtidsplan for Norsk Landbruksforskning fram til 1990*. Oslo: NLVF, 1984. [0461]

Noriega, Tomás. Personal communication. Director General, Instituto de Investigación Agropecuaria de Panamá. Panamá, March 1988. [0833]

Norton, George W., and Víctor G. Ganoza. "The Benefits of Agricultural Research and Extension in Peru." Virginia Polytechnic Institute and North Carolina State University, USA, June 1985. Mimeo. [0072]

Norwegian Research Councils' Committee for R&D Statistics. *R&D Statistics: Resources Devoted to R&D in Norway*. Oslo: Norwegian Research Councils' Committee for R&D Statistics, March 1986. [0864]

Norwegian Research Councils' Committee for R&D Statistics (FSU). *R&D Statistics 1985–87 (Resources Devoted to R&D in Norway)*. Oslo: FSU, August 1987. [0851]

Nsereko, J. "Agricultural Research in Zambia and Its Relationship with the International Agricultural Research System." In *Strengthening National Agricultural Research*. Report from a SAREC workshop, 10–17 September 1979, edited by Bo Bengtsson and Getachew Tedla, pp. 70–76. Sweden: SAREC, 1980. [0314]

Nsubuga, H.S.K., W.A. Sakira, and P.Karani. "Agricultural Research in Uganda." In *Proceedings of the Eastern Africa-ACIAR Consultation on Agricultural Research*, pp. 57–65. Australia: ACIAR, 1984. [0765]

OECD. *Intellectual Investment in Agriculture for Economic and Social Development*. Documentation in Food and Agriculture No. 60. Paris: OECD, 1962. [0574]

OECD. *Policies for Science and Education, Country Reviews: Norway*. Paris: OECD, October 1962. [0502]

OECD. *OECD Policies for Science and Education, Country Reviews: Sweden*. Paris: OECD, October 1962. [0614]

OECD. *Country Reports on the Organisation of Scientific Research: United States*. Paris: OECD, June 1963. [0563]

OECD. *Country Reports on the Organisation of Scientific Research: Iceland*. Paris: OECD, July 1963. [0504]

OECD. *Rapports par Pays sur l'Organisation de la Recherche Scientifique: Luxembourg*. Paris: OECD, July 1963. [0561]

OECD. *Country Reports on the Organisation of Scientific Research: Belgium*. Paris: OECD, October 1963. [0558]

OECD. *Country Reports on the Organisation of Scientific Research: Germany*. Paris: OECD, November 1963. [0562]

OECD. *Country Reports on the Organisation of Scientific Research: Greece*. Paris: OECD, November 1963. [0528]

OECD. *Country Reports on the Organisation of Scientific Research: Sweden*. Paris: OECD, November 1963. [0615]

OECD. *Country Reports on the Organisation of Scientific Research: Portugal*. Paris: OECD, December 1963. [0522]

OECD. *Country Reports on the Organisation of Scientific Research: Austria*. Paris: OECD, January 1964. [0526]

OECD. *Country Reports on the Organisation of Scientific Research: Turkey.* Paris: OECD, January 1964. [0560]

OECD. *Reviews of National Policies for Science and Education – Scientific Policy: Sweden.* Paris: OECD, January 1964. [0616]

OECD. *Country Reports on the Organisation of Scientific Research: Netherlands.* Paris: OECD, February 1964. [0555]

OECD. *Country Reports on the Organisation of Scientific Research: France.* Paris: OECD, May 1964. [0552]

OECD. *Country Reports on the Organisation of Scientific Research: Spain.* Paris: OECD, June 1964. [0538]

OECD. *Country Reports on the Organisation of Scientific Research: Denmark.* Paris: OECD, July 1964. [0506]

OECD. *Country Reports on the Organisation of Scientific Research: Italy.* Paris: OECD, November 1964. [0519]

OECD. *Country Reports on the Organisation of Scientific Research: Norway.* Paris: OECD, April 1965. [0501]

OECD. *Reviews of National Science Policy: Greece.* Paris: OECD, September 1965. [0529]

OECD. *Reviews of National Science Policy: Belgium.* Paris: OECD, 1966. [0559]

OECD. *Reviews of National Science Policy: France.* Paris: OECD, 1966. [0553]

OECD. *Country Reports on the Organisation of Scientific Research: United Kingdom.* Paris: OECD, February 1966. [0541]

OECD. *Reviews of National Science Policy: Japan.* Paris: OECD, 1967. [0595]

OECD. *A Study of Resources Devoted to R&D in OECD Member Countries in 1963/64, International Statistical Year for Research and Development – Volume 2: Statistical Tables and Notes.* Paris: OECD, 1968. [0514]

OECD. *Reviews of National Science Policy: Canada.* Paris: OECD, 1969. [0649]

OECD. *Reviews of National Science Policy: Italy.* Paris: OECD, 1969. [0520]

OECD. *Reviews of National Science Policy: Norway.* Paris: OECD, 1970. [0494]

OECD. *International Survey of the Resources Devoted to R&D in 1967 by OECD Member Countries, Statistical Tables and Notes – Volume 5: Total Tables.* Paris: OECD, August 1970. [0510]

OECD. *Reviews of National Science Policy: Austria.* Paris: OECD, 1971. [0527]

OECD. *Reviews of National Science Policy: Spain.* Paris: OECD, 1971. [0539]

OECD. *Reviews of National Science Policy: Switzerland.* Paris: OECD, 1971. [0524]

OECD. *Reviews of National Science Policy: Iceland.* Paris: OECD, 1972. [0503]

OECD. *Reviews of National Science Policy: Netherlands.* Paris: OECD, 1973. [0556]

OECD. *International Survey of the Resources Devoted to R&D in 1969 by OECD Member Countries, Statistical Tables and Notes – Volume 5: Total Tables.* Paris: OECD, June 1973. [0511]

OECD. *Survey of the Resources Devoted to R&D by OECD Member Countries, International Statistical Year 1971 – Volume 5: Total Tables, Statistical Tables and Notes.* Paris: OECD, August 1974. [0509]

OECD. *Survey of the Resources Devoted to R&D by OECD Member Countries, International Statistical Year 1973 – Volume 5: Total Tables, Statistical Tables and Notes.* Paris: OECD, September 1976. [0512]

OECD. *Reviews of National Science Policy: Australia.* Paris: OECD, 1977. [0572]

OECD. *Science and Technology in the People's Republic of China.* Paris: OECD, 1977. [0487]

OECD. *International Survey of the Resources Devoted to R&D by OECD Member Countries, International Statistical Year 1975 – International Volume: Statistical Tables and Notes.* Paris: OECD, March 1979. [0513]

OECD. *Science and Technology Indicators: Basic Statistical Series – Volume A: The Objectives of Government R&D, 1969–1981.* Paris: OECD, July 1981. [0474]

OECD. *Science and Technology Indicators: Basic Statistical Series – Volume A: The Objectives of Government R&D Funding 1974–1985.* Paris: OECD, May 1983. [0473]

OECD. *OECD Science and Technology Indicators.* Paris: OECD, 1984. [0021]

OECD. *Reviews of National Science Policy: Norway.* Paris: OECD, 1985. [0499]

OECD. *OECD Science and Technology Indicators No. 2 – R&D, Invention and Competitiveness.* Paris: OECD, 1986. [0530]

OECD. *Reviews of National Science and Technology Policy: Australia.* Paris: OECD, 1986. [0573]

OECD. *Reviews of National Science and Technology Policy: Portugal.* Paris: OECD, 1986. [0521]

OECD. Personal communication. Background information from the OECD Science and Technology Indicators Division. Paris, Summer 1987. [0714]

OECD – Directorate for Science, Technology and Industry. "Public Funding of R&D by Socio-Economic Objectives in Austria." Table 0.1, version 2. OECD, Paris, January 1985. Mimeo. [0704]

OECD – Directorate for Science, Technology and Industry. "Public R&D Funding by Socio-Economic

Objective in Austria." Table 0.1, version 2. OECD, Paris, January 1986. Mimeo. [0705]

OECD – STIID Data Bank. "Public Funding of R&D by Socio-Economic Objective." Tables. OECD, Paris, April 1987. Mimeo. [0601]

OECD – STIID Data Bank. "R&D Expenditure in the Higher Education and Private Non-Profit Sectors." Tables. OECD, Paris, April 1987. Mimeo. [0602]

OECD – STIID Data Bank. "Gross Domestic Expenditure on R&D: All Fields of Sciences – Agriculture, Forestry and Fishing." Tables. OECD, Paris, June 1987. Mimeo. [0708]

OECD – STIID Data Bank. "R&D Expenditure in the Higher Education and Private Non-Profit Sectors: Higher Education, Agricultural Sciences, General Universities Funds." Table. OECD, Paris, June 1987. Mimeo. [0707]

Office de la Recherche Scientifique et Technique Outre-Mer (ORSTOM). *Rapport d'Activité 1985.* Paris: ORSTOM, n.d. [0656]

Office de la Recherche Scientifique et Technique Outre-Mer (ORSTOM). *Rapport d'Activité 1967.* Paris: ORSTOM, 1969. [0654]

Office de la Recherche Scientifique et Technique Outre-Mer. (ORSTOM). *Activités de l'ORSTOM en Côte d'Ivoire.* Abidjan: ORSTOM, 1980. [0150]

Office of Rural Development. *Annual Research Report for 1982.* South Korea: Office of Rural Development, 1983. [0177]

Ogor, E. "Problems of Research Organizations in Nigeria." In *Papers presented to the Conference on Agricultural Research and Production in Africa, September 1971 – Part 3,* supplement to volume 2 of the Journal of the Association for the Advancement of Agricultural Sciences in Africa. Addis Ababa, Ethiopia: Association for the Advancement of Agricultural Sciences in Africa, June 1975. [0651]

Ohashi, Tetsuro. Personal communication. Director, International Research Division – Agriculture, Forestry and Fisheries Research Council. Tokyo, April 1987. [0612]

Okoro, D.E., and J.N. Onuoha. "The Impact of the Collaboration between the International Agricultural Research System and the National Agricultural Research System in Nigeria." CGIAR, Washington, D.C., April 1985. Mimeo. [0061]

Oland, K. "Agricultural Research in Botswana." In *Strengthening National Agricultural Research.* Report from a SAREC workshop, 10–17 September 1979, edited by Bo Bengtsson and Getachew Tedla, pp. 7–19. Sweden: SAREC, 1980. [0307]

Oland, Kristian. "Agricultural Research in Botswana." Gaborone, August 1980. Mimeo. [0104]

Opio-Odongo, J.M.A. "The Contextual Imperatives of Agricultural Research Planning and Management in Uganda." Paper presented for the Agricultural Research Planning and Management Short Course organised by the Overseas Development Group, University of East Anglia, Norwich, England, 28 September–9 November 1983. [0233]

Opio-Odongo, J.M.A. *Agricultural Research Policy in Uganda.* Kampala: Makerere University, March 1986. [0166]

Oram, Peter. "Report on National Agricultural Research in West Africa." IFPRI, Washington, D.C., February 1986. Mimeo. [0089]

Oram, Peter. "Agricultural Research Objectives and Priorities: Constraints to the Development of Agricultural Research Institutions in Arab Countries." Review paper prepared for the first meeting of the Council for Arab Agricultural Research, organized by the Arab Fund for Economic and Social Development, Kuwait, 23 March 1988. [0865]

Oram, Peter A., and Vishva Bindlish. *Resource Allocations to National Agricultural Research: Trends in the 1970s.* The Hague and Washington, D.C.: ISNAR and IFPRI, November 1981. [0016]

Oram, Peter A., and Vishva Bindlish. "Investment in Agricultural Research in Developing Countries: Progress, Problems, and the Determination of Priorities." IFPRI, Washington, D.C., January 1984. Mimeo. [0073]

Oram, Peter A., and Mark Gieben. "Document Summaries." ISNAR, The Hague, 1984. Mimeo. [0026]

Orleans, Leo A. "Research and Development in Communist China." *Science* Vol. 157 (28 July 1967): 392–400. [0367]

Orleans, Leo A. *Science and Technology in the People's Republic of China.* Paris: OECD, 1977. [0483]

Orleans, Leo A. Personal communication. Library of Congress, Science and Technology Division. Washington D.C., March 1987. [0882]

Oteifa, B.A. "Status of Agricultural Research in Egypt." In *Strengthening National Agricultural Research.* Report from a SAREC workshop, 10–17 September 1979, edited by Bo Bengtsson and Getachew Tedla, pp. 20–24. Sweden: SAREC, 1980. [0308]

Oteifa, Bakir A. "Role of Universities and Private Sector Organizations in the National Agricultural Research System of Egypt." Paper presented at the 1st International Meeting of National Agricultural Research Systems and 2nd IFARD Global Convention, IFARD/EMBRAPA, Brasilia, Brasil, 6–11 October 1986. [0395]

Ouali, Ibrahim Firmin. *Burkina Faso and the CGIAR Centers: A Study of Their Collaboration in Agricultural Research.* CGIAR Study Paper

Number 23. Washington D.C.: World Bank, 1987. [0746]

Owadally, A.L. Personal communication. Agricultural Services – Ministry of Agriculture, Fisheries, and Natural Resources. Reduit, Mauritius, February 1988. [0801]

Pages, P.D., and J.D. Drilon, Jr. "Organizational Trends in National Research Management: A Global Overview." n.d. Mimeo. [0262]

Pakistan Agricultural Research Council (PARC). *Directory of PARC Scientists and Technicians.* Islamabad, Pakistan: PARC, 1983. [0190]

Pakistan Agricultural Research Council (PARC). *Annual Report 1985.* Islamabad, Pakistan: PARC, 1985. [0191]

Pakistan Agricultural Research Council (PARC). *National Agricultural Research Plan.* Islamabad: PARC – Directorate of Publications, 1986. [0587]

Pant, Yadab Deb. "Managing Human Resources for Agricultural Research in Nepal." In *Management of Human Resources in Agricultural Research.* Report of the International Workshop on Management of Human Resources in Agricultural Research held 3–5 March 1986, edited by Theodore Hutchcroft, pp. 105–109. Dhaka: Bangladesh Agricultural Research Council, ISNAR, and Winrock International, 1986. [0329]

Papanastasis, Vasilios. "La Recherche Forestière en Grèce." Paper presented at the Séminaire sur l'Orientation et l'Organisation de la Recherche Agronomique dans les Pays du Bassin Méditerranéen, Centre International de Hautes Etudes Agronomiques Méditerranéennes, Istanbul, 1–3 December 1986. [0413]

Papanastasis, Vasilios. "Quelques Détails sur la Recherche Forestière." Paper presented at the Séminaire sur l'Orientation et l'Organisation de la Recherche Agronomique dans les Pays du Bassin Méditerranéen, Centre International de Hautes Etudes Agronomiques Méditerranéennes, Istanbul, 1–3 December 1986. [0414]

Pardey, Philip G., and Barbara Craig. "Causal Relationships between Public Sector Agricultural Research Expenditures and Output." *American Journal of Agricultural Economics* Vol. 71, no. 1 (February 1989): 1–19. [0495]

Pardey, Philip G., Barbara Craig, and Michelle Hallaway. "US Agricultural Research Deflators: 1890–1985." *Research Policy* Vol. 18 (forthcoming, 1989). [0497]

Patel, B.K. Personal communication. Assistant Director, Department of Agriculture. Lusaka, February 1988. [0806]

Patel, R.K. "Agricultural Research in Sri Lanka." In *Proceedings of the Research Extension Linkage Workshop Held in Hanoi, Viet Nam, 9–13 June 1986.* Rome: FAO, 1986. [0436]

Patronato 'Juan de la Cierva' – Consejo Superior de Investigaciones Científicas. *Encuesta sobre Actividades de Investigación Científica y Técnica en España en 1967.* Madrid: Consejo Superior de Investigaciones Científicas, 1970. [0661]

Payne, Anthony, and Paul Sutton, eds. *Dependence under Challenge – The Political Economy of the Commonwealth Caribbean.* Manchester, England: Manchester University Press, 1984. [0357]

Paz Silva, Luis J. *Peru and the CGIAR Centers: A Study of Their Collaboration in Agricultural Research.* CGIAR Study Paper No. 12. Washington, D.C.: World Bank, 1986. [0335]

Pelaez Bardales, Mario. Personal communication. Instituto Nacional de Investigacion Agraria y Agroindustrial (INIAA). Lima, Peru, Marzo 1988. [0839]

Pellanda, Sanzio. *Sur la Rentabilité de la Recherche Agronomique du Secteur Public en Suisse – Etude Econométrique.* Fribourg, Switzerland: Université de Fribourg, June 1972. [0620]

Persaud, H.B., and M. Granger. "Resources for Agricultural Research in Guyana – Central Agricultural Station." Ministry of Agriculture, Georgetown, 1981. Mimeo. [0358]

Philippine Council for Agriculture and Resources Research and Development (PCARRD). *The PCARRD Corplan 1984–1988.* Philippines: PCARRD, 1983. [0787]

Philippine Council for Agriculture and Resources Research and Development (PCARRD), and IDRC. *Agriculture and Resources Research Manpower Development in South and Southeast Asia.* Proceedings of the Workshop on Agriculture and Resources Research Manpower Development in South and Southeast Asia, 21–23 October 1981 and 4–6 February 1982, Singapore. Los Baños, Philippines: PCARRD, 1983. [0179]

Pinchinat, Antonio. *Barbados Agricultural Research System – A Condensed Report.* Santo Domingo: IICA, December 1980. [0102]

Piñeiro, Martín, and Eduardo Trigo, eds. *Cambio Técnico en el Agro Latino Americano – Situación y Perspectivas en la Década de 1980.* San José: IICA, 1983. [0053]

Plessix, C.J. du, et al. "La Programmation de la Recherche Agronomique en Côte d'Ivoire." Extrait du Compte rendu du 10e Colloque de l'Institut International de la Potasse, Abidjan, Côte d'Ivoire, December 1973. [0161]

Posada Torres, Rafael. *Ecuador and the CGIAR Centers: A Study of Their Collaboration in Agricultural Research.* CGIAR Study Paper Number 11. Washington, D.C.: World Bank, 1986. [0368]

Poutiainen, Esko. Personal communication. Director General of the Agricultural Research Centre. Jokioinen, Finland, January 1987. [0508]

Pradhanang, A.M. *Report of the Agriculture Research and Technology Development in Nepal.* FAO – Asia & Pacific Region Office, November 1984. [0472]

Pray, C.E. "The Economics of Agricultural Research in British Punjab and Pakistani Punjab, 1905–1975." PhD diss., University of Pennsylvania, Pennsylvania, 1978. [0878]

Pray, C.E. "The Economics of Agricultural Research in Bangladesh." *Bangladesh Journal of Agricultural Economics* Vol. 2 (December 1979): 1–36. [0879]

Pray, Carl E. *Agricultural Research and Technology Transfer by the Private Sector in the Philippines.* St. Paul: University of Minnesota – Economic Development Center, October 1986. [0790]

Pray, Carl E., and Jock R. Anderson. *Bangladesh and the CGIAR Centers: A Study of Their Collaboration in Agricultural Research.* CGIAR Study Paper Number 8. Washington, D.C.: World Bank, 1985. [0050]

Qasem, Subhi. "Study on Small Country Research: Research in Jordan." IDRC, Ottawa, n.d. Mimeo. [0739]

Qasem, Subhi. "Technology Transfer and Feedback: Agricultural Research and Extension." Part IV, Chapter 11 in *The Agricultural Sector of Jordan: Policy and Systems Studies*, edited by A.B. Zahlan and Subhi Qasem, pp. 343–375. London: Ithaca Press, 1985. [0075]

Rabufetti, Armando. "Institutional Reorganization of Agricultural Research in Uruguay." Paper presented at the International Workshop on Agricultural Research Management, ISNAR, The Hague, 7–11 September 1987. [0755]

Raharinosy, V., and C.P. Ravohitrarivo. "Planification et Programmation de la Recherche à Madagascar (La Recherche Agricole)." Paper presented at the Workshop on Improving Agricultural Research Organizations and Management: Progress and Issues for the Future, ISNAR, The Hague, 8–12 September 1986. [0271]

Rahman, M.M. "Agricultural Research in Bangladesh." In *Strengthening National Agricultural Research.* Report from a SAREC workshop, 10–17 September 1979, edited by Bo Bengtsson and Getachew Tedla, pp. 77–88. Sweden: SAREC, 1980. [0315]

Rajki, Sandor, ed. *The First Twenty Years of Martonvasar.* Martonvasar, Hungary: Agricultural Research Institute of the Hungarian Academy of Sciences, 1971. [0593]

Raman, K.V. "Management of Human Resources in Agricultural Research: Indian Efforts and Experiences." In *Management of Human Resources in Agricultural Research.* Report of the International Workshop on Management of Human Resources in Agricultural Research held 3–5 March 1986, edited by Theodore Hutchcroft, pp. 87–96. Dhaka: Bangladesh Agricultural Research Council, ISNAR, and Winrock International, 1986. [0328]

Raman, K.V., and C. Prasad. "Trends in Agricultural Research and Extension in India." Paper presented at the regional seminar organised by the Commonwealth Association of Scientific Agricultural Societies and the Agricultural Institute of Malaysia, Kuala Lumpur, November 1986. [0433]

Randhawa, N.S. *Agricultural Research in India: An Overview of Its Organization, Management and Operations.* FAO Research and Technology Paper 3. Rome: FAO, 1987. [0778]

Randhawa, N.S., K.V. Raman, and M. Rajagopalan. *National Agricultural Research System in India and Its Impact on Agricultural Production and Productivity.* New Delhi: Indian Council of Agricultural Research, October 1986. [0337]

Rasmussen, Ole Kjeldsen. Personal communication. Landbrugets Samråd for Forskning og Forsøg. Copenhagen, January – March 1987. [0536]

Republic of South Africa – Department of Agriculture and Water Supply. *Official List of Professional and Research Workers, Lecturing Staff, Extension and Other Workers in the Agricultural Fields, April 1985.* Pretoria: Department of Agriculture and Water Supply, 1985. [0719]

Research Institutes Review Panel – Green Revolution National Committee. *Report of Research Institutes Review Panel 1980/81.* Two volumes. Ibadan, Nigeria: National Committee on Green Revolution, January 1981. [0617]

Retzlaff, Ralph H. "Human Resources: Lessons from AARD in Indonesia." In *Management of Human Resources in Agricultural Research.* Report of the International Workshop on Management of Human Resources in Agricultural Research held 3–5 March 1986, edited by Theodore Hutchcroft, pp. 58–65. Dhaka: Bangladesh Agricultural Research Council, ISNAR, and Winrock International, 1986. [0180]

Rizvi, M.H. "Managing the Human Resources in Agricultural Research in Pakistan." In *Management of Human Resources in Agricultural Research.* Report of the International Workshop on Management of Human Resources in Agricultural Research held 3–5 March 1986, edited by Theodore Hutchcroft, pp. 110–123. Dhaka: Bangladesh Agricultural Research Council, ISNAR, and Winrock International, 1986. [0330]

Rodríguez, P. "Importancia de la Investigación Agrícola en la República de Cuba." Paper presented at Curso–Taller sobre la Administración de la Investigación Agrícola, Panama City, 14–25 July 1986. [0378]

Rosales, Franklin E. "Situación del Sistema Nacional de Investigación Agronómica en Costa Rica." IICA – OEA, San José, December 1980. Mimeo. [0207]

Rosales, Franklin E. "Situación del Sistema Nacional de Investigación Agronómica en Honduras." IICA – OEA, San José, December 1980. Mimeo. [0212]

Rosales, Franklin E. "Situación del Sistema Nacional de Investigación Agronómica en Panamá." IICA – OEA, San José, December 1980. Mimeo. [0216]

Rost, E. Personal communication. Bundesministerium für Forschung und Technologie. Bonn, February 1988. [0799]

Ruttan, Vernon W. Agricultural Research Policy and Development. FAO Research and Technology Paper 2. Rome: FAO, 1987. [0770]

Sabhasri, S. "Agricultural Research in Thailand." In Strengthening National Agricultural Research. Report from a SAREC workshop, 10–17 September 1979, edited by Bo Bengtsson and Getachew Tedla, pp. 138–143. Sweden: SAREC, 1980. [0320]

SADCC. SADCC Agricultural Research Conference Gabarone Botswana. Sebele, Botswana: SADCC, April 1984. [0234]

SADCC, and DEVRES, Inc. Agricultural Research Resource Assessment in the SADCC Countries, Volume I: Regional Analysis and Strategy. Washington, D.C.: DEVRES, Inc., January 1985. [0002]

SADCC, and DEVRES, Inc. Agricultural Research Resource Assessment in the SADCC Countries, Volume II: Country Report Botswana. Washington, D.C.: DEVRES, Inc., January 1985. [0003]

SADCC, and DEVRES, Inc. Agricultural Research Resource Assessment in the SADCC Countries, Volume II: Country Report Lesotho. Washington, D.C.: DEVRES, Inc., January 1985. [0004]

SADCC, and DEVRES, Inc. Agricultural Research Resource Assessment in the SADCC Countries, Volume II: Country Report Malawi. Washington, D.C.: DEVRES, Inc., January 1985. [0005]

SADCC, and DEVRES, Inc. Agricultural Research Resource Assessment in the SADCC Countries, Volume II: Country Report Mozambique. Washington, D.C.: DEVRES, Inc., January 1985. [0006]

SADCC, and DEVRES, Inc. Agricultural Research Resource Assessment in the SADCC Countries, Volume II: Country Report Swaziland.

Washington, D.C.: DEVRES, Inc., January 1985. [0008]

SADCC, and DEVRES, Inc. Agricultural Research Resource Assessment in the SADCC Countries, Volume II: Country Report Zambia. Washington, D.C.: DEVRES, Inc., January 1985. [0009]

SADCC, and DEVRES, Inc. Agricultural Research Resource Assessment in the SADCC Countries, Volume II: Country Report Zimbabwe. Washington, D.C.: DEVRES, Inc., January 1985. [0007]

SADCC, and DEVRES, Inc. SADCC Region Agricultural Research Resource Assessment – Data Base Management Information System: Diskettes and User's Guide. Printouts. Washington, D.C.: SADCC, and DEVRES, Inc, n.d. [0385]

Saeed, Hassan Mohamed, Arif Jamal Mohamed Ahmed, and Mohamed El Hussein El Tahir. "Technology Transfer and Feedback: Agricultural Research Extension." Part IV, chapter 19 in The Agricultural Sector of Sudan: Policy and Systems Studies, edited by A.B. Zahlan and W.Y. Magar, pp. 381–408. London: Ithaca Press, 1986. [0834]

Saich, Tony. "Linking Research to the Productive Sector: Reforms of the Civilian Science and Technology System in Post-Mao China." Development and Change Vol. 17 (1986): 3–33. [0486]

Saint-Clair, P.M. "Réflexions sur l'Organisation et le Fonctionnement Actuels de la Recherche au FOFIFA." ISNAR, Madagascar, April 1987. Mimeo. [0816]

Sallam, Abdul Rahman. Personal communication. Director General, Agricultural Research Authority, Ministry of Agriculture and Fisheries. Taiz, Yemen Arab Republic, February 1988. [0812]

Salmon, David Christopher. "An Evaluation of Investment in Agricultural Research in Indonesia: 1965–1977." PhD diss., University of Minnesota, Minneapolis, April 1984. [0731]

Samson, J.A., ed. Jaarverslag. Years: 1962 and 1965. Suriname: Landbouwproefstation, various years. [0582]

Sanches da Fonseca, Maria A., and José Roberto Mendonça de Barros. "A Preliminary Attempt to Evaluate the Agricultural Research System in Brazil." In Resource Allocation to Agricultural Research. Proceedings of a workshop held in Singapore 8–10 June 1981, edited by Douglas Daniels and Barry Nestel, pp. 137–144. Ottawa: IDRC, 1981. [0303]

Sanches da Fonseca, Maria A., and Roberto Mendonça de Barros. "A Preliminary Attempt to Evaluate the Agricultural Research System in Brazil." Brazil, n.d. Mimeo. [0106]

Sánchez, Pedro A., and Grant M. Scobie. Cuba and the CGIAR Centers: A Study of Their Collaboration *in Agricultural Research*. CGIAR Study Paper No. 14. Washington, D.C.: World Bank, 1986. [0334]

Sands, Carolyn Marie. "The Theoretical and Empirical Basis for Analyzing Agricultural Technology Systems." PhD diss., University of Illinois, Urbana-Champaign, Illinois, 1988. [0740]

Scarda, Anna Maria. Personal communication. Consiglio Nazionale delle Ricerche – Istituto di Studi sulla Ricerca e Documentazione Scientifica. Roma, February 1988. [0825]

"Science in Iberia: A Renaissance in the Making." *Nature* Vol. 324 (27 November 1986): 313–328. [0862]

Scobie, G.M., and W.M. Eveleens. *Agricultural Research: What's It Worth?* Discussion paper 1/86. Hamilton, New Zealand: Economics Division, Ministry of Agriculture and Fisheries, June 1986. [0603]

Scobie, G.M., and W.M. Eveleens. *The Return to Investment in Agricultural Research in New Zealand: 1926–27 to 1983–84.* Research Report 1/87. Hamilton, New Zealand: MAF Economics Division, October 1987. [0756]

Secretaria Ejecutiva de Planificación Sectorial Agropecuaria (SEPSA). "Proyecto de Investigación y Transferencia de Tecnología: Información Preparada para la Misión FAO-Banco Mundial." SEPSA, San José, 1981. Mimeo. [0127]

Secretariado Técnico de la Presidencia – Departamento de Ciencia y Tecnología. *Estudio de Base del Sector Agropecuario y Forestal.* Santo Domingo: Secretariado Técnico de la Presidencia, August 1982. [0220]

Segura, M. "Diagnóstico de la Investigación Agropecuaria en el Istmo Centroamericano." IICA, Guatemala, 1979. Mimeo. [0208]

Sekretariat des Schweizerischen Wissenschaftsrates. *Wissenschafts-Politik.* Jahrgang 4, Nr. 6/1970. Bern: Sekretariat des Schweizerischen Wissenschaftsrates, 1970. [0619]

Senanayake, Y.D.A., and H.M.G. Herath. "Resource Allocation to Agricultural Research in Sri Lanka." In *Resource Allocation to Agricultural Research.* Proceedings of a workshop held in Singapore 8–10 June 1981, edited by Douglas Daniels and Barry Nestel, pp. 60–67. Ottawa: IDRC, 1981. [0297]

Senegalese-IADS Study Team. "Senegal Agricultural Research Review." Draft of a Senegalese-IADS Study Team Report to the General Delegation for Scientific and Technical Research. Dakar, Senegal, December 1978. Mimeo. [0259]

Serere Research Station. "Annual Report April 1969–March 1970, Part I." Serere Research Station, Entebbe, Uganda, 1970. Mimeo. [0639]

Setai, Bethuel. "Assessment of Future Trained Manpower Requirements in the Agricultural Sector of Lesotho." FAO, Rome, 1983. Mimeo. [0236]

Sficas, A.G. "L'Organisation et l'Orientation de la Recherche." Paper presented at the Séminaire sur l'Orientation et l'Organisation de la Recherche Agronomique dans les Pays du Bassin Méditerranéen, Centre International de Hautes Etudes Agronomiques Méditerranéennes, Istanbul, 1–3 December 1986. [0415]

Sharma, Ramesh P. "Agricultural Research Resource Allocation in Nepal." In *Resource Allocation to Agricultural Research.* Proceedings of a workshop held in Singapore 8–10 June 1981, edited by Douglas Daniels and Barry Nestel, pp. 42–48. Ottawa: IDRC, 1981. [0294]

Sharma, Ramesh P., and Jock R. Anderson. *Nepal and the CGIAR Centers: A Study of Their Collaboration in Agricultural Research.* CGIAR Study Paper Number 7. Washington, D.C.: World Bank, 1984. [0059]

Sharma, S.K. "Research Management System in Republic of Korea." In *Proceedings of the Research Extension Linkage Workshop Held in Hanoi, Viet Nam, 9–13 June 1986.* Rome: FAO, 1986. [0437]

Shawel, Hailu, and Akalu Negewo. "The Impact of the Collaboration between the International Agricultural Research System and the National Agricultural Research System in Ethiopia." CGIAR, Washington, D.C., December 1984. Mimeo. [0048]

Shehata, A.H. "Agricultural Research System in Egypt." Paper presented at the International Workshop on Agricultural Research Management, ISNAR, The Hague, August 1987. [0751]

Shin, Joseph. Personal communication. Agriculture and Fisheries Department. Hong Kong, May 1987. [0644]

Sigurdson, Jon. *Vietnam's Science and Technology – A Tentative Description of Structure and Planning.* Lund, Sweden: Research Policy Institute, University of Lund, 1982. [0743]

Sigurdson, Jon, and Zhang Wei. "Technological Transformation in the People's Republic of China – Achievements and Problems Faced." Paper prepared for the World Institute for Development Economics Research (WIDER), Helsinki, Finland. University of Lund and Tsinghua University, Sweden and China, August 1986. Mimeo. [0370]

Sital, J.T. Personal communication. Department of Development Economics, Wageningen Agricultural University. Wageningen, The Netherlands, Spring 1987. [0840]

Sivan, P. Personal communication. Director of Research, Ministry of Primary Industries. Nausori, Fiji, January 1988. [0776]

Snyder, Monteze M. *The Organizational Context and Manpower Implications for the Ministry of Agriculture The Gambia: Part II*. Banjul, The Gambia: AID, November 1984. [0136]

Soikes, Raul. "Programa de Investigación Agropecuaria de Nicaragua." IICA-OEA, Costa Rica, December 1980. Mimeo. [0215]

Soikes, Raul. "Situación del Sistema Nacional de Investigación y Extensión de Guatemala." IICA, Costa Rica, 1980. Mimeo. [0210]

Soil Research Institute. *Annual Report of the Soil Research Institute January 1965–March 1966*. Accra, Ghana: Ghana Academy of Sciences, 1966. [0462]

Solis, Rómulo V. "Documento Sobre Investigación Agropecuaria, Extensión Agrícola y Rural y Principales Programas de Investigación que Reciben." IICA, Quito, 1980. Mimeo. [0196]

Sompo-Ceesay, M.S. "Organization and Structure of Agricultural Research in The Gambia: Past, Present and Future." Paper presented at the Workshop on Improving Agricultural Research Organization and Management: Progress and Issues for the Future, ISNAR, The Hague, 8–12 September 1986. [0273]

Soumana, Idrissa. "Strategic Evolution of Planning at the National Institute of Agricultural Research in Niger." Paper presented at the International Workshop on Agricultural Research Management, ISNAR, The Hague, 7–11 September 1987. [0750]

South-East Consortium for International Development, and Tuskegee Institute. *Baseline Study of Agricultural Research, Education, and Extension in Guyana*. Chapel Hill, North Carolina: The South-East Consortium for International Development, June 1981. [0199]

Stagno, H., and R. Pineda. *Situación Actual del Sistema Nacional de Investigación y Extensión, y Descripción de los Programas Principales en Cultivos Alimenticios*. Informe para el Proyecto BID-IICA-Iowa State University. Santo Domingo: IICA, 1980. [0221]

Staicu, Irimie. *Scientific Research in Romanian Agriculture*. Bucharest: Meridiane Publishing House, 1966. [0440]

State Science and Technology Commission of the People's Republic of China. *Statistical Data on Science and Technology of 1985*. Beijing, China: The State Science and Technology Commission of the People's Republic of China, 1986. [0371]

Statistical Yearbook of China 1984. Beijing, China, 1984. [0484]

Stavis, Benedict. "Agricultural Research and Extension Services in China." *World Development* Vol. 6 (1978): 631–645. [0485]

Stewart, Rigoberto. *Costa Rica and the CGIAR Centers: A Study of Their Collaboration in Agricultural Research*. CGIAR Study Paper Number 4. Washington, D.C.: World Bank, 1985. [0098]

Stewart, Rigoberto. *Guatemala and the CGIAR Centers: A Study of Their Collaboration in Agricultural Research*. CGIAR Study Paper Number 5. Washington, D.C.: World Bank, 1985. [0066]

Sumelius, John. Personal communication. Agricultural Economics Research Institute. Helsinki, Finland, December 1986. [0507]

Sumelius, John. "The Returns to Investments in Agricultural Research in Finland 1950–1984." *Journal of Agricultural Science in Finland* Vol. 59, No. 4 (1987): 251–354. [0429]

Sundsbø, Svein, and Mary Villa. Personal communication. Agricultural Research Council of Norway (NLVF). Oslo, March 1987. [0523]

Sussman, Jean. "Agricultural Research in Cuba." University of Minnesota, St. Paul, July 1982. Mimeo. [0116]

Swanson, Burton E., and Wade H. Reeves. "Agricultural Research West Africa: Manpower and Training." World Bank, Washington, D.C., November 1985. Mimeo. [0088]

Swanson, Burton E., and Wade H. Reeves. "Agricultural Research Eastern and Southern Africa: Manpower and Training." World Bank, Washington, D.C., August 1986. Mimeo. [0445]

Swanson, Burton E., et al. "An INTERPAKS Case Study of the Agricultural Technology System in Malawi." INTERPAKS – University of Illinois, Urbana, Illinois, March 1986. Mimeo. [0439]

Tejada, S. "La Planificación de las Investigaciones Agropecuarias en la República Dominicana." Paper presented at Curso-Taller sobre la Administración de la Investigación Agrícola, Panama City, 14–25 July 1986. [0379]

Thairu, D.M. "Agricultural Research System in Kenya." In *Strengthening National Agricultural Research*. Report from a SAREC workshop, 10–17 September 1979, edited by Bo Bengtsson and Getachew Tedla, pp. 48–54. Sweden: SAREC, 1980. [0312]

Tisdell, C.A. "Research and Development in Australia – A Summary of Expenditure and the Allocation of Funds." In *Science and Industry Forum Report No. 5: The Influence of Research and Development on Economic Growth*, edited by the Australian Academy of Science, pp. 28–34. Canberra: Australian Academy of Science, June 1972. [0566]

526

Tongpan, Sopin. "Agricultural Research Resources in Thailand." In *Agriculture and Resources Research Manpower Development in South and Southeast Asia*, pp. 165–180. Los Baños, Philippines: Philippine Council for Agriculture and Resources Research and Development, 1983. [0793]

Trigo, Eduardo. Background material. ISNAR, The Hague, n.d. [0172]

Trigo, Eduardo, and Guillermo E. Gálvez. "Trip Report of the CIAT and IICA Mission to the Caribbean English-Speaking Countries for UNDP." 1979. Mimeo. [0365]

Trigo, Eduardo J., and Martín E. Piñeiro. "Funding Agricultural Research." In *Report of a Conference: Selected Issues in Agricultural Research in Latin America*, edited by Barry Nestel and Eduardo J. Trigo. The Hague: ISNAR, March 1984. [0001]

Trigo, Eduardo, Martín Piñeiro, and Jorge Ardila. *Organización de la Investigación Agropecuaria en América Latina*. San José: IICA, 1982. [0080]

Trouchaud, Jean-Pierre. "Rapport de Mission d'Appui à l'Organisation de la Recherche Scientifique et Technologique en Haute-Volta." 1979. Mimeo. [0390]

US Department of Agriculture – Agricultural Research Service. *Funds for Research at State Agricultural Experiment Stations, 1960*. Washington, D.C.: US Government Printing Office, April 1961. [0489]

US Department of Agriculture – Cooperative State Experiment Station Service. *Funds for Research at State Agricultural Experiment Stations*. Annual issues 1961 and 1962. Washington, D.C.: US Government Printing Office, various years. [0490]

US Department of Agriculture – Cooperative State Research Service. *Funds for Research at State Agricultural Experiment Stations and Other State Institutions*. Annual issues 1964 to 1974. Washington, D.C.: US Government Printing Office, various years. [0492]

US Department of Agriculture – Cooperative State Research Service. *Funds for Research at State Agricultural Experiment Stations and other Cooperating Institutions*. Annual issues 1974 and 1975. Washington, D.C.: US Government Printing Office, various years. [0493]

US Department of Agriculture – Cooperative State Research Service. *Inventory of Agricultural Research – Fiscal Years: 1970–1985*. Washington, D.C.: US Government Printing Office, various years. [0488]

US Department of Agriculture – Cooperative State Research Service. *Funds for Research at State Agricultural Experiment Stations, 1963*.

Washington, D.C.: US Government Printing Office, March 1964. [0491]

Ulyatt, M.J., J.P. Kerr, and E.W. Hewett. "Research and Development Funding in Agriculture." *New Zealand Agricultural Science* Vol. 20, No. 3 (August 1986): 163–168. [0481]

UN Economic Commission for Western Asia (ECWA), and FAO. *Agriculture and Development in Western Asia*. Number 7. Baghdad: ECWA, November 1984. [0254]

Underwood, E.J. "Agricultural Research in Australia: A Critical Appraisal." *Search* Vol. 4, No. 5 (May 1973): 155–160. [0570]

Underwood, E.J. "The Organisation of Agricultural Research in Australia." *Agricultural Administration* Vol. 1, No. 1 (1974): 73–81. [0567]

UNDP, and FAO. *Agricultural Research Institute, Nicosia, Cyprus – Final Report*. Rome: FAO, 1968. [0442]

UNDP, and FAO. *Report to the Government of Liberia: Agricultural Research Organization*. Project Report No. TA 3299. Rome: FAO, 1974. [0443]

UNESCO. *Science and Technology in Countries of Asia and the Pacific*. Science Policy Studies and Documents No. 52. Paris: UNESCO, n.d. [0590]

UNESCO. *La Politique Scientifique et l'Organisation de la Recherche Scientifique en Belgique*. Etudes et Documents de Politique Scientifique No. 1. Paris: UNESCO, 1965. [0398]

UNESCO. *Science Policy and the Organization of Scientific Research in the Socialist Federal Republic of Yugoslavia*. Science Policy Studies and Documents No. 9. Paris: UNESCO, 1968. [0399]

UNESCO. *The Promotion of Scientific Activity in Tropical Africa*. Science Policy Studies and Documents No. 11. Paris: UNESCO, 1969. [0400]

UNESCO. *National Science Policy and Organization of Research in Poland*. Science Policy Studies and Documents No. 21. Paris: UNESCO, 1970. [0401]

UNESCO. *La Politique Scientifique et l'Organisation de la Recherche en France*. Etudes et Documents de Politique Scientifique No. 24. Paris: UNESCO, 1971. [0444]

UNESCO. *La Politique Scientifique et l'Organisation de la Recherche Scientifique en Hongrie*. Etudes et Documents de Politique Scientifique No. 23. Paris: UNESCO, 1971. [0402]

UNESCO. *Science Policy and the European States*. Science Policy Studies and Documents No. 25. Paris: UNESCO, 1971. [0531]

UNESCO. *Statistical Yearbook 1972*. Paris: UNESCO, 1973. [0515]

UNESCO. *La Politique Scientifique et l'Organisation de la Recherche Scientifique dans la République Populaire de Bulgarie*. Etudes et Documents de

Politique Scientifique No. 32. Paris: UNESCO, 1974. [0397]

UNESCO. *National Science Policies in Africa.* Science Policy Studies and Documents No. 31. Paris: UNESCO, 1974. [0448]

UNESCO. *National Science Policies in Africa.* Science Policy Studies and Documents No. 31. Paris: UNESCO, 1974. [0266]

UNESCO. *Statistical Yearbook 1983.* Paris: UNESCO, 1983. [0022]

UNESCO – Division of Statistics on Science and Technology, Office of Statistics. *Human and Financial Resources for Research and Experimental Development in Agriculture.* Paris: UNESCO, July 1983. [0586]

UNESCO Field Science Office for Africa. *Survey on the Scientific and Technical Potential of the Countries of Africa.* Paris: UNESCO, 1970. [0532]

UNESCO, and the Academy of Sciences of the GDR. *Science and Technology Policy and the Organization of Research in the German Democratic Republic.* Science Policy Studies and Documents No. 57. Paris: UNESCO, n.d. [0591]

United States Team of Consultants to the Ministry of Agriculture of the Arab Republic of Egypt. "Increasing Egyptian Agricultural Production through Strengthened Research and Extension Programs." Ministry of Agriculture, Cairo, September 1983. Mimeo. [0249]

University of Mauritius – School of Agriculture. *Research Programme 1982–83.* Reduit: University of Mauritius, May 1982. [0866]

University of the West Indies – Faculty of Agriculture. *A Compendium of Research.* St. Augustine, Trinidad & Tobago: University of the West Indies – Faculty of Agriculture, January 1980. [0363]

University of the West Indies – Faculty of Agriculture. *Information Booklet for Academic Year 1981–1982.* St. Augustine, Trinidad & Tobago: University of the West Indies – Faculty of Agriculture, September 1981. [0362]

University of Wisconsin. "Gambia Agricultural Research and Diversification Project." Discussion paper. University of Wisconsin, Madison, 1985. Mimeo. [0140]

USAID – Highlands Agricultural Development Project. Title not known. Jordan, n.d. Mimeo. [0722]

Vaage, Roald, and Endre Bjørgum. Personal communication. Norwegian Fisheries Research Council. Trondheim, Norway, April 1988. [0859]

Valmayor, Ramon V., and Nora M. Valera. "Agriculture and Resources Research Manpower Development Program in the Philippines: Past, Present, and Future." In *Agriculture and Resources Research Manpower Development in South and Southeast Asia,* pp. 107–117. Los Baños, Philippines: PCARRD, 1983. [0791]

Vattanatangum, Anan. "Current Status and Activities in Human Resources Management of the Department of Agriculture, Thailand." In *Management of Human Resources in Agricultural Research.* Report of the International Workshop on Management of Human Resources in Agricultural Research held 3–5 March 1986, edited by Theodore Hutchcroft, pp. 124–131. Dhaka: Bangladesh Agricultural Research Council, ISNAR, and Winrock International, 1986. [0331]

Venezian, Eduardo. *Chile and the CGIAR Centers: A Study of Their Collaboration in Agricultural Research.* CGIAR Study Paper Number 20. Washington, D.C.: World Bank, April 1987. [0025]

Venezian, Eduardo L., and William K. Gamble. *The Agricultural Development of Mexico – Its Structure and Growth Since 1950.* New York: Praeger Publishers, 1969. [0214]

Verkaik, A.P. *Research Management Studies 3: Organisatiestructuur landbouwkundig onderzoek en achtergronden van haar totstandkoming.* The Hague: Nationale Raad voor Landbouwkundig Onderzoek TNO, 1972. [0425]

Wahab, Abdul H., and Noel Singh. *Agricultural Research in Jamaica.* Miscellaneous Publication No. 274. Jamaica: IICA, 1981. [0223]

Wahid, Abdul. "State of Current Agricultural Research and Education in Pakistan." *Agriculture Pakistan* Vol. 19, No. 3 (1968): 261–285. [0454]

Wahid, Abdul. *Review of Agricultural Research System in Pakistan.* Islamabad, Pakistan: PARC, 1982. [0192]

Wang'ati, F.J. "Allocation of Resources to Agricultural Research: An Inventory of the Current Situation in Kenya." In *Resource Allocation to Agricultural Research.* Proceedings of a workshop held in Singapore 8–10 June 1981, edited by Douglas Daniels and Barry Nestel, pp. 27–31. Ottawa: IDRC, 1981. [0292]

Wang'ati, F.J. "Agricultural Research in Kenya." In *Proceedings of the Eastern Africa–ACIAR Consultation on Agricultural Research,* pp. 31–43. Australia: ACIAR, 1984. [0763]

Wapakala, William W. "Improving Agricultural Research Organisation and Management – A Historical Perspective of the Kenyan Situation." Paper presented at the Workshop on Improving Agricultural Research Organisation and Management: Progress and Issues for the Future, ISNAR, The Hague, 8–12 September 1986. [0269]

Watson, J.M. "Comparative Study of Agricultural Research Organisation and Administration in the Near East Region." Paper presented at the

Workshop on Organization and Administration of Agricultural Services in the Arab States, Cairo, 2–15 March 1964. [0174]

Webster, B.N. *Index of Agricultural Research Institutions and Stations in Africa.* Rome: FAO, n.d. [0653]

Webster, Brian, Carlos Valverde, and Alan Fletcher, eds. *The Impact of Research on National Agricultural Development.* Report on the First International Meeting of National Agricultural Research Systems and the Second IFARD Global Convention. The Hague: ISNAR, July 1987. [0726]

Weersma-Haworth, Teresa. "Colombian Case Study." CGIAR Impact Study. CGIAR, Washington, D.C., October 1984. Mimeo. [0055]

Wennergren, Boyd E., Charles H. Antholt, and Morris D. Whitaker. *Agricultural Development in Bangladesh.* Boulder, Colorado: Westview Press, 1984. [0100]

Wongsiri, Tanongchit. "Human Resource Planning and Management." Paper presented at the Workshop on Improving Agricultural Research Organizations and Management: Progess and Issues for the Future, ISNAR, The Hague, 8–12 September 1986. [0272]

World Bank. "Agricultural Research and Farmer Advisory Services in Central America and Panama." World Bank, Washington, D.C., 1979. Mimeo. [0123]

World Bank. "Madagascar Agricultural Research Subsector Review." Report No. 3188-MAG. World Bank, Washington, D.C., October 1980. Mimeo. [0237]

World Bank. "Thailand National Agricultural Research Project – Appraisal Project File." World Bank, Washington, D.C., November 1980. Mimeo. [0455]

World Bank. *Brazil: Agricultural Research II Project.* Washington, D.C.: World Bank, 1981. [0109]

World Bank. "Malawi National Rural Development Program Review." World Bank, Washington, D.C., 1982. Mimeo. [0238]

World Bank. "Ethiopia Agricultural Research Project." World Bank, Washington, D.C., 1984. Mimeo. [0226]

World Bank. "West Africa Agricultural Research Review – Country Studies." World Bank, Washington, D.C., 1985. Mimeo. [0086]

World Bank. *India Staff Appraisal Report of the National Agricultural Research Project II.* Report No. 5469-IN. Washington, D.C.: World Bank, September 1985. [0377]

World Bank – East Asia and Pacific Projects Department. *Appraisal of Agricultural Research and Extension Project Indonesia.* Washington, D.C.: World Bank, April 1975. [0459]

World Bank – East Asia and the Pacific Regional Office. *Indonesia National Agricultural Research Project: Staff Appraisal Report.* Washington, D.C.: World Bank, April 1980. [0183]

World Bank – Eastern Africa Regional Office. *The Comoros: Problems and Prospects of a Small, Island Economy.* Washington, D.C.: World Bank, July 1979. [0122]

World Bank, FAO, and ISNAR. *Agricultural and Livestock Research: Upper Volta.* Report prepared for the Government of the Republic of Upper Volta on the joint mission of World Bank-FAO-ISNAR. The Hague: ISNAR, March 1983. [0024]

World Bank – Projects Department, East Asia and Pacific Regional Office. *Indonesia: Technical and Professional Manpower in Agriculture.* Two volumes. Washington, D.C.: World Bank, March 1981. [0460]

World Bank – West Africa Agricultural Research Review Team. "National Agricultural Research Systems (NARS): A Regional Appraisal and Selected Issues." World Bank, Washington, D.C., August 1986. Mimeo. [0466]

World Bank – Western Africa Projects Department. "Staff Appraisal Report Cameroon – National Agricultural Research Project." World Bank, Washington, D.C., 1984. Mimeo. [0119]

Wu, Yuan-li, and Robert B. Sheeks. *The Organization and Support of Scientific Research and Development in Mainland China.* New York: Praeger Publishers Inc., 1970. [0369]

Yadav, Ram P. *Agricultural Research in Nepal: Resource Allocation, Structure, and Incentives.* IFPRI Research Report No. 62. Washington D.C.: IFPRI, September 1987. [0737]

Yeganiantz, Levon, ed. *Brazilian Agriculture and Agricultural Research.* Brasilia: Brazilian Agricultural Research Corporation – EMBRAPA, 1984. [0107]

Yun Cho, Chae. Personal communication. Director General, International Technical Cooperation Center – Rural Development Administration. Suweon, Republic of Korea, March 1988. [0842]

Yusof bin Hashim, Mohd. "The Agricultural Research System in Malaysia: A Study of Resource Allocation." In *Resource Allocation to Agricultural Research.* Proceedings of a workshop held in Singapore 8–10 June 1981, edited by Douglas Daniels and Barry Nestel, pp. 145–149. Ottawa: IDRC, 1981. [0304].

Yusof bin Hashim, Mohd, and Ahmad Shafri bin Man. "An Inventory of Research Manpower and Support Staff in Malaysia." In *Agriculture and Resources Research Manpower Development in South and Southeast Asia,* pp. 187–193. Los

Baños, Philippines: Philippine Council for Agriculture and Resources Research and Development, 1983. [0794]

Yusof bin Hashim, Mohd, and Ahmad Shafri bin Man. "The Manpower Development Program of the Malaysian Agricultural Research and Development Institute: Past, Present, and Future." In *Agriculture and Resources Research Manpower Development in South and Southeast Asia*, pp. 181–185. Los Baños, Philippines: Philippine Council for Agriculture and Resources Research and Development, 1983. [0797]

Zentner, Robert Paul. "An Economic Evaluation of Public Wheat Research Expenditures in Canada." PhD diss., University of Minnesota, Minneapolis, January 1982. [0608]

Zhou, Fang. "Organization and Structure of the National Agricultural Research System in China." Paper presented at the International Workshop on Agricultural Research Management, ISNAR, The Hague, 7–11 September 1987. [0754]

Zin, Kyaw. *Burma and the CGIAR Centers: A Study of Their Collaboration in Agricultural Research*. CGIAR Study Paper Number 19. Washington, D.C.: World Bank, November 1986. [0057]

Appendix A

Table A.1: 1980 Atlas, Purchasing Power Parity (PPP), and Average Annual Exchange Rates (AAER) (1 US$ = ...LCU)

COUNTRY	Base Year (b = 1980) Exchange Rates		
	Atlas (e_b^*)	PPP	AAER(e_b)
AFGHANISTAN	41.293	15.630	44.129
ALGERIA	4.276	4.334	3.838
ANGOLA	33.699	35.842	29.918
ANTIGUA	2.801	1.908^f	2.700
ARGENTINA	5049.622^e	2305.245	1837.000
AUSTRALIA	0.891	1.067	0.878
AUSTRIA	13.043	16.035	12.938
BAHAMAS	1.004	0.893^f	1.000
BANGLADESH	15.846	4.883	15.454
BARBADOS	2.157	1.560	2.011
BELGIUM	28.861	37.702	29.243
BELIZE	1.985	1.414^f	2.000
BENIN	220.970	127.513	211.280
BERMUDA	0.938	1.332^f	1.000
BOLIVIA	42.886^d	15.113	24.510
BOTSWANA	0.918	0.578	0.777
BRAZIL	49.083^c	32.340	52.710
BRUNEI	3.294	2.864^f	2.151
BURKINA FASO	215.804	122.237	21.280
BURMA	6.311	2.390	6.594
BURUNDI	100.410	58.436	90.000
CAMEROON	217.395	223.377	211.280
CANADA	1.188	1.112	1.169
CAPE VERDE	36.219	21.389^f	40.175
CENTRAL AFRICAN REPUBLIC	229.409	151.375	211.280
CHAD	215.589	134.353	211.280
CHILE	44.576	22.673	39.000
CHINA	1.501	0.798^f	1.498
COLOMBIA	49.728	23.895	47.280
COMOROS	223.144	112.495^f	211.302
CONGO	237.527	228.985	211.280
COOK ISLANDS	1.030	0.726^f	1.027
COSTA RICA	9.100	6.086	8.570
COTE D'IVOIRE	199.733	240.717	211.280
CUBA	..	..	..
CYPRUS	0.378	0.276	0.353
DENMARK	5.435	7.602	5.636
DOMINICA	2.926	1.908^f	2.700
DOMINICAN REPUBLIC	1.041	0.638	1.000
ECUADOR	27.211	14.820	25.000
EGYPT	0.775	0.429	0.700
EL SALVADOR	2.607	1.319	2.500
ETHIOPIA	1.950	0.722	2.070
FIJI	0.874	0.505	0.818
FINLAND	3.906	4.655	3.730
FRANCE	4.440	5.314	4.226
FRENCH POLYNESIA	79.801	87.989^f	76.829

Table A.1 (contd.)

COUNTRY	Base Year (b = 1980) Exchange Rates		
	Atlas (e_b*)	PPP	AAER(e_b)
GABON	262.480[a]	277.639	211.280
GAMBIA	1.863	1.153	1.719
GERMANY, FED.REP.OF	1.800	2.456	1.818
GHANA	9.840	7.652	2.750
GREECE	41.993	40.476	42.617
GRENADA	2.841[a]	1.844[f]	2.700
GUADELOUPE	4.501	3.775[f]	4.226
GUAM	1.066	0.893[f]	1.000
GUATEMALA	1.005	0.556	1.000
GUINEA	18.599	13.757	18.969
GUINEA-BISSAU	46.756	18.001[f]	33.811
GUYANA	2.572	1.173	2.550
HAITI	5.302	1.990	5.000
HONDURAS	2.009	1.255	2.000
HONG KONG	5.220	3.741	4.976
ICELAND	5.262	6.538[g]	4.798
INDIA	8.349	3.020	7.863
INDONESIA	673.055	291.938	626.990
IRAN	87.970	55.149	70.614
IRAQ	0.295[b]	0.190	0.295
IRELAND	0.533	0.529	0.486
ISRAEL	5.836	4.633	5.124
ITALY	925.191	838.120	856.450
JAMAICA	1.818	1.174	1.781
JAPAN	209.200	248.872	226.740
JORDAN	0.318	0.181	0.298
KENYA	7.541	4.779	7.420
KOREA, REPUBLIC OF	595.088	391.758	607.430
KUWAIT	0.309	0.279	0.270
LAOS, P.D.R.	14.000[b]	7.453[f]	14.000
LEBANON	3.623[a]	2.590[f]	3.436
LESOTHO	0.867	0.346	0.779
LIBERIA	1.002	0.720	1.000
LIBYAN ARAB REPUBLIC	0.354	0.339[f]	0.296
MADAGASCAR	226.143	134.390	211.280
MALAWI	0.856	0.396	0.812
MALAYSIA	2.217	1.201	2.177
MALI	217.663	252.244	211.280
MARTINIQUE	4.436	3.775[f]	4.226
MAURITANIA	45.441	36.058	45.914
MAURITIUS	7.534	6.123	7.684
MEXICO	26.579	14.223	22.951
MONTSERRAT	2.959[a]	2.035[f]	2.700
MOROCCO	3.982	3.041	3.937
MOZAMBIQUE	35.212[a]	18.224	36.000
NEPAL	12.055	3.256	12.000
NETHERLANDS	1.980	2.633	1.988
NEW CALEDONIA	79.812	87.989[f]	76.829
NEW ZEALAND	1.030	1.053	1.027
NICARAGUA	10.781	3.897	10.050
NIGER	220.347	216.750	211.280
NIGERIA	0.627	0.667	0.547
NORWAY	5.227	6.280	4.939
OMAN	0.438	0.316	0.345
PAKISTAN	9.922	3.474	9.900

Table A.1 (contd.)

COUNTRY	Base Year (b = 1980) Exchange Rates		
	Atlas (e_b^*)	PPP	AAER(e_b)
PANAMA	1.012	0.690	1.000
PAPUA NEW GUINEA	0.703	0.371	0.670
PARAGUAY	136.380	94.959	126.000
PERU	334.964[c]	137.421	288.860
PHILIPPINES	7.864	3.550	7.511
PORTUGAL	51.951	33.292	50.062
PUERTO RICO	1.006	0.893[f]	1.000
QATAR	4.298	4.869[f]	3.657
RWANDA	88.425	55.416	92.840
SAO TOME & PRINCIPE	34.963	18.512[f]	34.771
SAUDI ARABIA	3.983	3.765	3.327
SENEGAL	218.800	151.615	211.280
SEYCHELLES	7.035	4.818[f]	6.392
SIERRA LEONE	1.077	0.766	1.050
SINGAPORE	2.177	1.730	2.141
SOLOMON ISLANDS	0.897[a]	0.567[f]	0.830
SOMALIA	9.369	4.452	6.295
SOUTH AFRICA	0.935	0.511	0.778
SPAIN	76.071	66.252	71.702
SRI LANKA	17.293	3.868	16.534
ST KITTS-NEVIS	2.746[a]	1.908[f]	2.700
ST LUCIA	2.885	1.908[f]	2.700
ST VINCENT	2.691	1.844[f]	2.700
SUDAN	0.474	0.328	0.500
SURINAME	1.829	1.334	1.785
SWAZILAND	0.903	0.681	0.778
SWEDEN	4.404	7.129	4.230
SWITZERLAND	1.607	2.664	1.676
SYRIAN ARAB REPUBLIC	3.995	1.939	3.925
TAIWAN	38.973[a]	27.741	36.015
TANZANIA	8.159	6.141	8.197
THAILAND	21.574	8.704	20.476
TOGO	211.565	148.310	211.280
TONGA	0.914[a]	0.613[f]	0.898
TRINIDAD & TOBAGO	2.749	2.027	2.400
TUNISIA	0.421	0.300	0.405
TURKEY	68.066	42.034	76.038
TUVALU	0.891	0.599[f]	0.878
UGANDA	74.548	28.461	7.424
UNITED ARAB EMIRATES	3.733	4.218	3.707
UNITED KINGDOM	0.523	0.503	0.430
UNITED STATES	1.000	1.000	1.000
URUGUAY	11.178	7.046	9.099
VANUATU	70.101[a]	46.635[f]	68.292
VENEZUELA	4.893	3.677	4.293
VIRGIN ISLANDS (U.S.)	1.000	1.145[f]	1.000
WESTERN SAMOA	0.945[a]	0.650[f]	0.920
YEMEN, ARAB REPUBLIC	4.675	1.948	4.563
YEMEN, P.D.R.	0.361	0.184[f]	0.345
ZAIRE	2.451	2.653	2.800
ZAMBIA	0.839	0.736	0.789
ZIMBABWE	0.701	0.544	0.643

Sources:

Unless otherwise stated, World Bank atlas exchange rates (Atlas) are from O'Connor, John C. Personal communication. Chief, Comparative Analysis & Data Division, Economic Analysis & Projections Department, The World Bank, Washington D.C., 15 May 1987; purchasing power parity (PPP) indices are from Summers, Robert, and Alan Heston. "A New Set of International Comparisons of Real Product and Price Levels Estimates for 130 Countries, 1950–1985." *Review of Income and Wealth* Series 34, No. 1 (March 1988): 1–25; and average annual exchange rates (AAER) are generally the IMF's rf or (inverted) rh rate taken from various issues of the IMF *International Financial Statistics Yearbook.*

Notes:

Atlas:

a. Due to the unavailability of a World Bank Atlas exchange rate, an Atlas rate was estimated using the approach described by equation 4.1, Chapter 4, Part I. IMF and World Bank annual average exchange rates and implicit GDP deflators were taken from Pardey, Roseboom, Kang and Folger (1988).

b. Due to the unavailability of a World Bank Atlas exchange rate, and the annual average exchange rates or implicit GDP deflators required to estimate an Atlas rate, the 1980 annual average exchange rate (AAER) was used.

c. Original Atlas rate scaled up by a factor of 1000 to be commensurable with research expenditure data expressed in 1980 current LCUs.

d. Original Atlas rate scaled up by a factor of 1,000,000 to be commensurable with research expenditure data expressed in 1980 current LCUs.

e. Original Atlas rate scaled up by a factor of 10,000,000 to be commensurable with research expenditure data expressed in 1980 current LCUs.

PPP:

f. Calculated by multiplying the base year annual average exchange rate (e_b) by an appropriately estimated GDP price level, where the base year is represented by b, and the GDP price level is given by $P_{Ib} = PPP_b/e_b$. Where necessary, P_{Ib} was estimated by first ranking the 118 countries for which both P_{Ib} and atlas-converted 1980 per capita GDP (indexed on U.S. = 100) are available, dividing them into 7 per capita income groups, and estimating the simple average P_{Ib} for each income group using the Summers and Heston (1988) price level data. These average price levels were then assigned to each country, depending on their respective per capita income groupings.

g. Original PPP rate scaled down by a factor of 100 to be commensurable with research expenditure data expressed in 1980 current LCUs.

Table A.2: Implicit GDP deflators for 1960 – 1986; base year 1980 = 100

COUNTRY	1960	1961	1962	1963	1964	1965	1966	1967	1968	1969
AFGHANISTAN	23.21	23.50	22.99	31.09	32.41	40.00	55.04	63.94	50.66	51.17
ALGERIA	22.00	19.80	20.20	20.80	22.20	23.20	24.40	25.60	27.30	28.40
ANGOLA	11.87	11.41	12.19	12.50	12.65	12.79	12.94	13.19	14.30	15.55
ANTIGUA	28.14	26.90	27.67	27.36	28.22	28.91	29.61	30.47	34.11	35.19
ARGENTINA	0.007	0.008	0.010	0.012	0.016	0.020	0.025	0.032	0.036	0.038
AUSTRALIA	26.90	27.20	27.40	28.20	29.20	29.90	31.20	32.20	33.10	34.10
AUSTRIA	37.30	39.20	40.60	42.00	42.60	44.90	46.60	48.10	49.70	51.40
BAHAMAS	..	..	..	..	..	..	38.20	40.30	42.10	45.90
BANGLADESH	15.20	14.80	15.80	14.40	15.50	16.50	18.60	17.30	19.60	19.80
BARBADOS	26.49	25.63	24.26	28.79	27.15	26.01	27.02	27.18	28.81	29.94
BELGIUM	36.80	37.20	37.60	38.80	40.60	42.60	44.50	45.90	47.10	49.10
BELIZE	..	..	..	..	..	..	..	..	..	..
BENIN	33.82	34.18	35.45	36.36	36.24	36.96	37.39	37.45	38.66	39.99
BERMUDA	..	..	..	..	..	..	..	..	..	..
BOLIVIA	9.70	10.40	11.10	11.20	12.10	12.90	13.30	14.10	13.90	14.40
BOTSWANA	22.57	22.79	22.97	23.67	24.49	24.76	25.91	25.60	26.96	28.37
BRAZIL	0.10	0.13	0.20	0.36	0.67	1.03	1.43	1.85	2.36	2.84
BRUNEI	..	..	..	..	..	..	..	..	..	..
BURKINA FASO	25.60	30.00	30.40	32.50	32.40	32.60	33.40	32.10	33.10	36.30
BURMA	49.30	50.30	43.80	44.10	44.20	43.80	46.10	51.00	53.50	55.60
BURUNDI	22.26	26.71	25.75	26.97	28.51	28.28	29.15	28.80	29.92	31.45
CAMEROON	24.70	25.95	26.74	27.40	28.65	29.63	29.70	31.93	33.84	35.74
CANADA	32.10	32.60	33.20	33.80	34.70	35.20	36.80	38.20	39.50	41.30
CAPE VERDE	20.83	18.43	20.34	20.23	20.52	20.89	19.99	19.74	20.34	21.43
CENTRAL AFRICAN REP.	28.60	30.30	31.50	32.90	35.40	37.60	38.70	41.10	40.90	40.70
CHAD	26.60	27.80	28.30	29.80	32.70	34.00	36.70	37.70	38.60	40.00
CHILE	0.0007	0.0007	0.0010	0.0014	0.0020	0.0027	0.0037	0.0044	0.0061	0.0085
CHINA	85.40	99.20	98.50	96.30	96.40	98.00	95.80	96.80	98.50	94.30
COLOMBIA	4.92	5.32	5.67	7.00	8.13	8.93	10.23	11.09	12.10	13.10
COMOROS	..	..	..	..	..	..	..	..	..	..
CONGO	23.50	24.50	23.10	23.60	24.20	24.60	27.10	28.00	27.70	31.10
COOK ISLANDS	19.90	19.90	20.80	21.40	22.10	22.50	22.60	23.60	24.50	26.80
COSTA RICA	21.60	22.30	22.40	22.80	23.20	23.00	23.30	23.80	24.30	25.40
COTE D'IVOIRE	25.33	26.24	26.16	26.49	27.24	27.81	27.98	29.47	30.13	32.37
CUBA	..	..	..	..	..	..	..	..	..	..
CYPRUS	31.70	34.70	35.90	36.00	36.00	36.50	37.50	37.70	38.80	40.70
DENMARK	21.80	22.90	24.40	25.80	27.10	29.00	30.60	32.40	34.70	37.10
DOMINICA	16.22	17.15	16.95	17.29	17.89	18.22	18.82	19.48	21.74	22.41
DOMINICAN REPUBLIC	35.20	34.20	37.00	39.50	42.20	40.30	40.10	40.50	41.30	43.00
ECUADOR	19.80	20.70	21.20	22.10	22.80	20.60	21.90	22.90	23.90	25.70
EGYPT	27.43	27.48	27.63	28.27	30.04	30.98	32.41	32.86	33.10	33.60
EL SALVADOR	37.30	36.20	36.60	37.40	37.70	38.20	37.70	37.60	37.60	37.80
ETHIOPIA	50.30	52.20	54.00	55.10	57.90	60.80	60.80	62.00	64.00	63.80
FIJI	24.15	24.30	24.68	24.38	25.21	27.17	27.85	26.87	27.77	29.51
FINLAND	19.20	20.20	21.00	22.10	23.70	24.90	26.10	28.00	31.40	32.70
FRANCE	26.40	27.30	28.60	30.40	31.70	32.40	33.30	34.40	35.60	38.20
FRENCH POLYNESIA	..	..	..	..	..	..	..	..	..	..
GABON	12.86	13.73	14.46	15.05	16.00	16.80	17.46	18.55	19.72	20.75
GAMBIA	32.19	33.91	34.19	34.62	34.98	35.63	36.49	35.05	34.84	37.28
GERMANY, FED.REP.	41.40	43.40	45.10	46.50	47.90	49.70	51.30	52.00	53.10	55.40
GHANA	2.84	2.94	2.99	3.20	3.51	4.11	4.45	4.27	4.81	5.34
GREECE	20.30	20.60	21.60	21.90	22.70	23.60	24.70	25.30	25.80	26.60

Table A.2 (contd.)

COUNTRY	1970	1971	1972	1973	1974	1975	1976	1977	1978	1979
AFGHANISTAN	62.34	68.47	60.80	59.34	68.47	72.99	74.16	82.77	88.83	97.59
ALGERIA	29.50	30.90	32.70	35.60	53.00	55.90	62.30	69.30	81.20	92.10
ANGOLA	16.65	17.32	18.65	22.34	30.25	37.85	48.82	61.27	68.96	79.24
ANTIGUA	37.52	40.31	43.10	48.37	57.52	64.11	69.53	75.46	76.20	89.40
ARGENTINA	0.040	0.055	0.089	0.15	0.19	0.57	2.96	7.68	19.77	49.75
AUSTRALIA	35.30	37.40	40.20	45.10	52.90	61.50	69.90	76.40	81.90	89.60
AUSTRIA	54.20	57.60	62.00	66.90	73.30	78.00	82.40	86.80	91.30	95.10
BAHAMAS	48.70	50.90	54.40	57.40	65.00	71.60	74.70	77.10	81.80	98.20
BANGLADESH	20.80	22.20	34.00	48.50	84.40	65.00	61.10	71.40	80.60	90.60
BARBADOS	32.45	34.09	34.99	40.98	55.07	62.98	62.00	64.52	71.27	79.50
BELGIUM	50.20	53.00	56.30	60.40	68.00	76.30	82.20	88.40	92.20	96.40
BELIZE	39.30	41.30	40.80	46.80	59.80	69.30	64.60	68.90	73.90	87.20
BENIN	42.04	43.74	45.92	48.88	55.90	60.50	69.09	74.83	79.13	86.99
BERMUDA	..	..	..	..	..	..	..	..	..	..
BOLIVIA	14.90	15.60	18.80	26.60	42.00	44.80	48.40	53.70	60.90	72.00
BOTSWANA	30.34	32.94	43.37	50.73	58.76	65.24	73.72	71.37	91.10	107.47
BRAZIL	3.38	4.02	4.78	5.90	7.88	10.66	15.63	22.62	32.61	51.40
BRUNEI	..	..	..	40.36	42.57	44.96	48.75	47.66	53.75	..
BURKINA FASO	38.20	40.40	42.80	46.20	51.60	56.10	60.40	75.10	86.30	94.50
BURMA	54.40	53.40	53.20	57.80	70.70	82.80	90.80	92.20	100.00	100.00
BURUNDI	32.03	32.48	33.89	35.68	40.17	48.46	52.40	59.80	66.98	84.21
CAMEROON	38.57	39.75	42.90	45.80	50.72	60.25	65.37	72.21	79.43	89.29
CANADA	43.20	44.20	46.50	50.70	58.50	65.20	70.90	76.40	81.50	90.00
CAPE VERDE	21.71	25.20	27.35	31.95	43.75	52.99	54.95	59.42	83.23	92.74
CENTRAL AFRICAN REP.	42.40	45.30	48.20	50.60	55.90	62.90	68.50	79.30	83.70	96.00
CHAD	45.20	48.10	49.20	51.00	56.20	61.40	65.00	68.90	77.70	89.40
CHILE	0.012	0.014	0.026	0.13	1.07	4.73	16.60	33.80	52.91	77.40
CHINA	91.20	91.90	91.80	92.00	92.20	90.80	90.40	91.40	92.70	96.40
COLOMBIA	14.38	15.93	18.00	21.63	27.11	33.30	41.79	53.95	63.17	78.39
COMOROS	..	..	..	..	..	..	..	..	..	..
CONGO	31.20	34.90	37.40	36.80	50.80	57.70	63.30	72.90	81.00	84.30
COOK ISLANDS	29.30	33.70	37.10	40.30	42.60	59.20	70.00	66.90	75.80	86.30
COSTA RICA	27.30	27.90	29.70	34.10	42.10	52.40	61.10	71.50	77.10	84.20
COTE D'IVOIRE	34.19	34.02	34.44	39.07	49.59	52.32	62.17	82.78	87.25	96.11
CUBA	..	..	..	..	..	..	..	..	..	..
CYPRUS	42.40	43.40	42.60	49.50	55.00	57.30	62.60	69.20	76.30	87.50
DENMARK	40.10	43.20	47.10	51.50	58.20	65.50	71.40	78.20	85.90	92.40
DOMINICA	24.87	25.00	26.13	29.65	39.03	44.02	51.26	66.49	73.07	87.43
DOMINICAN REPUBLIC	43.80	44.30	47.90	50.00	59.00	68.90	70.80	78.30	79.20	88.00
ECUADOR	28.00	30.10	30.80	32.70	45.70	50.30	56.80	66.80	72.10	83.70
EGYPT	34.34	35.18	36.06	40.21	44.75	49.33	55.50	61.08	67.74	82.88
EL SALVADOR	39.60	39.80	40.20	44.20	49.20	52.90	64.80	76.80	77.40	88.20
ETHIOPIA	67.80	68.50	65.80	67.60	74.00	73.50	78.10	88.30	95.00	99.20
FIJI	31.55	32.60	37.43	43.40	56.30	69.96	75.47	75.47	78.79	85.21
FINLAND	33.90	36.50	39.60	45.20	55.30	63.30	71.30	78.50	84.60	91.60
FRANCE	40.40	42.70	45.30	48.90	54.30	61.60	67.60	74.00	81.50	89.80
FRENCH POLYNESIA	..	..	..	..	..	..	71.23	77.71	82.34	90.24
GABON	21.91	22.72	23.59	27.47	45.43	47.41	54.35	59.68	61.43	73.05
GAMBIA	46.95	41.15	41.86	42.94	52.83	63.23	71.68	87.24	91.54	90.32
GERMANY, FED.REP.	59.60	64.40	67.80	72.10	77.20	81.80	84.80	88.00	91.70	95.40
GHANA	5.49	5.78	6.68	8.07	10.05	13.01	16.67	27.87	48.30	66.62
GREECE	27.70	28.50	30.00	35.80	43.30	48.60	56.10	63.40	71.60	85.00

Table A.2 (contd.)

COUNTRY	1980	1981	1982	1983	1984	1985	1986
AFGHANISTAN	100.00	104.38	..	..	..	..	..
ALGERIA	100.00	112.17	116.96	124.69	132.59	..	..
ANGOLA	100.00	108.23	115.33	118.75	..	..	..
ANTIGUA	100.00	109.10	115.30	127.00	137.70	..	..
ARGENTINA	100.00	207.60	606.80	2730.90	23558.40	180928.70	..
AUSTRALIA	100.00	109.40	121.30	131.90	141.30	149.80	160.80
AUSTRIA	100.00	106.30	112.90	117.30	123.00	127.00	131.60
BAHAMAS	100.00	111.10	117.80	122.60	127.40	133.20	..
BANGLADESH	100.00	112.70	118.40	137.80	159.90	174.80	184.50
BARBADOS	100.00	114.21	124.60	132.08	139.20	149.00	..
BELGIUM	100.00	105.10	112.40	119.50	126.10	132.60	..
BELIZE	100.00	106.80	102.00	105.80	109.80	..	..
BENIN	100.00	109.50	123.71	143.68	151.72	..	..
BERMUDA	..	..	..	..	..	..	..
BOLIVIA	100.00	126.00	331.30	1367.40	20056.50	2564342.00	..
BOTSWANA	100.00	102.28	104.94	111.64	115.84	119.52	..
BRAZIL	100.00	195.20	387.50	928.40	2893.10	9679.70	..
BRUNEI	100.00	109.00	103.80	91.90	92.60	..	..
BURKINA FASO	100.00	105.70	123.50	131.40	137.40	147.20	..
BURMA	100.00	106.00	109.30	110.30	111.60	110.70	113.10
BURUNDI	100.00	94.00	102.00	110.00	124.00	135.00	141.00
CAMEROON	100.00	107.20	119.90	136.00	149.00	149.90	..
CANADA	100.00	110.50	121.70	128.00	135.30	139.80	143.10
CAPE VERDE	100.00	109.95	122.88	131.48	146.32	163.81	..
CENTRAL AFRICAN REP.	100.00	110.40	119.60	132.50	140.80	152.30	..
CHAD	100.00	112.30	123.70	144.10	153.20	164.70	..
CHILE	100.00	112.20	127.10	160.92	183.90	244.30	291.30
CHINA	100.00	101.90	— 102.80	..	..	..	..
COLOMBIA	100.00	122.80	153.20	184.40	225.30	277.50	347.70
COMOROS	..	..	..	..	..	..	..
CONGO	100.00	121.20	135.20	147.20	167.40	180.10	..
COOK ISLANDS	100.00	119.40	129.70	142.70	150.70	172.30	..
COSTA RICA	100.00	141.10	259.90	327.30	391.00	457.10	543.70
COTE D'IVOIRE	100.00	100.90	106.20	112.80	129.50	135.10	..
CUBA	..	..	..	..	..	..	..
CYPRUS	100.00	111.47	122.88	129.59	141.06	146.36	..
DENMARK	100.00	110.10	121.70	131.70	139.10	146.10	..
DOMINICA	100.00	101.80	106.38	112.43	117.02	..	..
DOMINICAN REPUBLIC	100.00	105.30	113.80	117.60	146.30	202.30	221.40
ECUADOR	100.00	114.40	134.70	186.90	260.10	340.60	406.90
EGYPT	100.00	111.80	125.30	143.10	170.90	193.20	..
EL SALVADOR	100.00	105.70	116.10	130.50	146.50	176.60	242.60
ETHIOPIA	100.00	103.89	105.32	109.45	111.21	119.82	..
FIJI	100.00	100.96	109.22	114.30	117.96	127.60	..
FINLAND	100.00	111.40	121.50	132.10	143.90	152.60	159.90
FRANCE	100.00	111.40	124.40	136.60	146.70	155.20	162.40
FRENCH POLYNESIA	100.00	116.67	133.62	152.21	..	..	..
GABON	100.00	114.60	130.90	143.60	156.50	170.40	..
GAMBIA	100.00	105.16	109.65	137.63	164.63	172.10	..
GERMANY, FED.REP.	100.00	104.00	108.60	112.10	114.30	116.80	120.40
GHANA	100.00	172.30	221.30	467.90	643.30	809.70	..
GREECE	100.00	119.70	149.80	178.30	214.60	252.70	300.80

Table A.2 (contd.)

COUNTRY	1960	1961	1962	1963	1964	1965	1966	1967	1968	1969
GRENADA	..	..	..	..	..	..	..	..	..	..
GUADELOUPE	26.40	27.30	28.60	30.40	31.70	32.40	33.30	34.40	35.60	38.20
GUAM	..	..	..	..	..	..	..	..	..	..
GUATEMALA	39.20	38.80	39.80	40.10	39.30	38.70	38.30	38.50	39.20	39.90
GUINEA	48.90	49.90	49.60	49.40	51.10	52.20	52.90	55.10	56.50	61.40
GUINEA-BISSAU	34.68	34.11	35.38	35.24	35.24	35.87	38.55	36.36	36.92	37.59
GUYANA	30.28	31.10	30.86	31.58	32.56	31.45	31.66	32.80	34.67	36.03
HAITI	33.00	33.50	32.20	34.60	38.80	41.30	43.10	44.00	42.50	43.60
HONDURAS	33.30	34.70	35.90	36.70	38.70	39.60	40.60	41.90	42.80	44.00
HONG KONG	29.90	28.70	29.60	30.80	31.70	31.60	31.40	33.30	34.30	36.00
ICELAND	1.55	1.76	1.97	2.14	2.52	2.85	3.15	3.24	3.70	4.46
INDIA	23.60	24.30	25.30	27.60	30.10	32.90	37.30	40.30	40.40	42.00
INDONESIA	..	..	..	0.02	0.05	0.15	2.04	5.41	11.72	13.98
IRAN	15.30	15.60	15.50	15.70	16.40	16.00	15.90	15.80	14.80	16.90
IRAQ	14.90	15.20	15.90	15.50	16.10	16.00	16.80	19.40	17.70	18.30
IRELAND	15.40	15.90	16.60	17.00	18.60	19.30	20.20	21.00	21.80	23.90
ISRAEL	1.60	1.70	1.90	2.00	2.10	2.40	2.60	2.60	2.60	2.70
ITALY	14.10	14.40	15.30	16.20	17.70	18.80	18.80	19.40	19.70	20.50
JAMAICA	14.60	15.30	15.70	16.70	15.70	16.00	17.00	17.50	18.40	21.60
JAPAN	30.80	33.20	34.40	35.90	37.50	36.20	38.00	40.40	42.50	44.60
JORDAN	..	..	..	..	..	..	..	..	..	32.40
KENYA	34.39	37.42	37.42	36.70	37.67	36.94	37.55	38.52	39.13	39.00
KOREA, REPUBLIC OF	3.50	4.00	4.70	6.10	7.90	8.40	9.60	11.10	13.00	13.90
KUWAIT	..	..	8.04	8.04	8.05	8.77	9.32	10.08	9.83	9.62
LAOS, P.D.R.	..	..	..	..	..	..	..	..	..	..
LEBANON	23.76	23.76	24.00	24.41	24.41	24.41	25.08	26.03	25.84	27.03
LESOTHO	27.13	27.55	28.01	28.40	28.99	29.98	31.10	32.20	32.97	34.03
LIBERIA	38.63	36.43	37.48	38.23	42.89	43.11	42.32	42.41	43.16	44.30
LIBYA	31.40	29.10	15.10	12.60	11.20	11.10	11.70	12.30	11.90	11.80
MADAGASCAR	28.85	29.37	30.37	31.48	32.00	33.40	35.36	35.59	35.91	37.31
MALAWI	31.29	31.16	31.95	33.93	41.99	48.32	49.91	49.78	52.15	53.21
MALAYSIA	54.15	49.95	49.29	48.89	49.70	50.71	50.00	49.95	48.23	50.81
MALI	17.70	18.80	19.50	20.10	27.00	26.20	28.30	35.80	34.80	35.70
MARTINIQUE	26.40	27.30	28.60	30.40	31.70	32.40	33.30	34.40	35.60	38.20
MAURITANIA	33.02	34.08	35.96	37.72	37.96	37.37	38.78	39.84	40.31	40.78
MAURITIUS	13.70	16.70	17.00	19.20	17.80	18.00	18.40	18.80	20.10	20.50
MEXICO	13.60	13.80	14.30	14.30	14.60	15.60	16.40	16.70	17.10	17.80
MONTSERRAT	..	..	..	..	..	..	..	..	..	..
MOROCCO	56.60	46.30	46.50	44.30	44.30	46.60	47.70	48.30	48.20	48.10
MOZAMBIQUE	26.26	26.73	27.31	29.79	30.20	30.67	31.59	33.49	33.93	32.68
NEPAL	25.20	25.50	26.90	28.70	30.60	33.30	38.40	36.20	40.20	42.90
NETHERLANDS	28.70	29.40	30.50	31.90	34.60	36.70	38.90	40.60	42.30	45.10
NEW CALEDONIA	..	..	..	..	..	..	..	..	..	..
NEW ZEALAND	19.90	19.90	20.80	21.40	22.10	22.50	22.60	23.60	24.50	26.80
NICARAGUA	23.40	23.40	23.20	23.20	24.20	24.40	25.30	25.60	26.80	27.10
NIGER	33.50	34.60	34.50	33.60	34.70	35.70	29.50	29.60	28.40	31.50
NIGERIA	14.10	13.50	13.50	14.60	14.70	15.30	16.80	17.30	15.00	15.80
NORWAY	31.90	32.50	29.90	30.90	32.40	33.90	35.30	36.30	37.90	39.50
OMAN	7.44	7.59	7.78	7.93	8.18	8.33	8.48	8.18	9.12	9.22
PAKISTAN	21.10	22.20	22.20	22.30	23.70	25.10	26.40	29.50	30.90	31.70
PANAMA	41.80	42.10	42.30	43.20	44.40	44.70	45.60	46.40	46.70	47.30
PAPUA NEW GUINEA	29.97	29.97	30.05	30.46	31.03	31.77	34.07	37.03	39.00	40.89

Table A.2 (contd.)

COUNTRY	1970	1971	1972	1973	1974	1975	1976	1977	1978	1979
GRENADA	..	29.30	29.50	35.00	48.60	53.00	58.60	64.10	76.10	87.40
GUADELOUPE	40.40	42.70	45.30	48.90	54.30	61.60	67.60	74.00	81.50	89.80
GUAM	42.16	44.01	45.47	49.49	57.02	62.50	63.78	64.35	109.38	83.74
GUATEMALA	41.90	41.30	40.80	46.70	54.00	61.10	68.10	79.30	83.70	90.90
GUINEA	68.30	67.60	70.90	76.70	78.80	84.80	85.90	85.10	87.70	90.80
GUINEA-BISSAU	39.00	39.52	41.62	45.58	49.50	53.44	58.77	68.46	80.38	93.00
GUYANA	37.42	38.59	42.39	46.95	64.04	75.46	74.61	73.84	81.01	88.84
HAITI	45.50	47.00	47.60	61.50	52.50	62.60	74.50	82.60	81.40	83.70
HONDURAS	44.50	46.60	47.60	51.20	57.20	63.00	68.30	83.60	80.40	88.10
HONG KONG	39.20	42.00	45.70	52.20	58.20	60.50	66.00	68.10	73.30	86.60
ICELAND	5.20	5.86	6.87	8.99	12.63	17.54	23.42	31.61	45.70	65.41
INDIA	43.40	45.70	50.80	60.40	71.20	69.10	73.60	76.30	77.80	89.80
INDONESIA	15.75	16.27	18.48	24.58	36.20	40.72	46.60	52.67	58.44	77.44
IRAN	14.70	16.40	17.60	23.70	38.10	42.10	47.60	56.00	62.70	78.60
IRAQ	19.30	19.90	21.20	19.80	43.30	44.70	49.20	56.00	57.00	73.40
IRELAND	26.90	29.80	33.80	38.80	41.30	50.60	61.30	69.40	76.70	87.20
ISRAEL	2.90	3.30	3.90	4.80	6.50	8.80	11.50	16.20	25.30	44.70
ITALY	21.90	23.50	25.00	27.90	33.00	38.70	45.70	54.50	62.00	71.90
JAMAICA	22.80	24.20	24.80	29.30	38.50	46.50	51.60	57.90	72.80	84.80
JAPAN	48.00	50.70	53.60	60.50	73.10	78.70	84.30	89.20	93.50	96.30
JORDAN	34.30	36.10	38.70	43.10	51.50	57.70	64.30	73.70	78.90	90.00
KENYA	45.08	40.83	41.43	45.57	52.98	59.17	70.35	82.26	84.81	90.40
KOREA, REPUBLIC OF	15.80	18.00	21.00	23.80	30.80	38.80	46.60	64.20	66.10	80.10
KUWAIT	9.78	13.44	12.52	14.70	40.41	42.83	44.53	49.91	48.55	68.03
LAOS, P.D.R.	..	..	..	..	..	..	..	..	..	66.74
LEBANON	27.03	27.46	28.86	30.76	34.08	38.92	49.46	58.11	65.24	82.24
LESOTHO	35.19	37.33	39.90	43.21	45.50	56.12	62.17	68.75	76.95	90.08
LIBERIA	43.86	43.99	45.84	49.54	59.30	72.15	71.93	81.13	83.68	91.68
LIBYA	11.50	16.20	20.80	23.60	51.70	50.00	51.10	55.70	55.90	70.80
MADAGASCAR	39.90	41.34	42.58	47.65	58.54	61.25	67.36	73.14	78.09	86.95
MALAWI	57.17	63.86	61.03	62.77	71.54	79.17	85.79	91.15	91.36	96.06
MALAYSIA	50.61	50.35	50.56	59.56	67.16	65.08	73.38	78.39	82.44	93.78
MALI	36.20	37.80	38.70	43.00	45.80	50.40	54.80	64.80	76.20	88.00
MARTINIQUE	40.40	42.70	45.30	48.90	54.30	61.60	67.60	74.00	81.50	89.80
MAURITANIA	41.48	44.77	46.18	54.88	62.87	72.97	77.67	82.02	84.02	94.59
MAURITIUS	20.90	22.20	25.30	29.30	51.50	54.60	50.80	56.80	60.80	78.50
MEXICO	19.70	20.80	22.10	25.00	30.70	35.50	42.50	55.40	64.60	77.70
MONTSERRAT	..	..	..	..	..	49.50	52.90	61.00	66.30	72.80
MOROCCO	49.90	51.50	53.70	56.80	69.30	73.00	75.40	80.90	85.90	92.50
MOZAMBIQUE	33.97	38.52	42.12	44.63	51.77	59.85	62.33	69.50	77.41	88.59
NEPAL	45.90	47.30	53.30	51.40	62.20	79.70	80.10	77.30	84.50	92.90
NETHERLANDS	47.60	51.70	56.60	61.90	67.50	74.40	81.10	86.50	91.30	94.70
NEW CALEDONIA	46.36	49.82	53.73	57.19	64.39	69.01	73.53	77.80	83.57	88.81
NEW ZEALAND	29.30	33.70	37.10	40.30	42.60	59.20	70.00	66.90	75.80	86.30
NICARAGUA	27.70	28.10	28.60	33.60	41.30	42.60	46.50	50.20	52.70	72.90
NIGER	32.10	33.50	37.20	41.10	44.10	43.70	55.60	67.30	75.70	83.60
NIGERIA	17.80	19.00	19.20	28.20	43.00	51.30	58.60	64.30	76.40	89.20
NORWAY	44.60	47.60	49.90	54.50	60.10	66.10	71.10	77.00	82.00	87.40
OMAN	9.52	11.06	11.35	15.91	47.84	48.98	49.58	52.60	54.64	72.68
PAKISTAN	32.50	34.10	36.40	42.10	51.80	63.40	71.10	78.70	85.80	90.50
PANAMA	49.10	50.50	53.00	57.50	64.20	70.20	73.40	76.80	82.90	90.60
PAPUA NEW GUINEA	43.19	44.66	47.87	57.31	62.40	63.88	76.44	82.10	82.35	93.35

Table A.2 (contd.)

COUNTRY	1980	1981	1982	1983	1984	1985	1986
GRENADA	100.00	113.20	119.30	129.50	131.60	..	..
GUADELOUPE	100.00	111.40	124.40	136.60	146.70	155.20	162.40
GUAM	100.00	120.41	..	..	..	..	..
GUATEMALA	100.00	108.50	113.90	121.40	126.40	150.10	212.00
GUINEA	100.00	112.70	122.80	131.70	140.90	151.60	..
GUINEA-BISSAU	100.00	109.54	121.70	128.96	140.58	157.04	..
GUYANA	100.00	108.96	104.97	111.46	124.15	142.79	..
HAITI	100.00	106.00	110.20	120.00	133.30	147.20	163.40
HONDURAS	100.00	107.50	113.70	119.40	124.40	130.20	135.70
HONG KONG	100.00	113.90	123.30	127.20	137.10	144.80	..
ICELAND	100.00	150.90	234.60	417.50	526.00	690.00	871.00
INDIA	100.00	109.30	117.90	128.50	136.80	146.20	..
INDONESIA	100.00	118.50	124.60	141.00	157.80	169.90	..
IRAN	100.00	118.40	132.80	150.10	163.70	169.20	..
IRAQ	100.00	87.90	99.30	92.00	113.70	111.20	..
IRELAND	100.00	117.40	135.80	151.20	160.60	168.70	
ISRAEL	100.00	233.20	491.50	1179.00	5769.80	19719.50	..
ITALY	100.00	118.50	137.70	158.80	175.00	190.40	205.80
JAMAICA	100.00	108.20	118.80	137.50	187.80	236.40	..
JAPAN	100.00	103.20	105.10	105.90	107.20	108.90	110.70
JORDAN	100.00	107.70	115.70	121.50	126.20	130.00	130.00
KENYA	100.00	110.50	122.10	132.50	146.40	157.50	..
KOREA, REPUBLIC OF	100.00	115.70	123.90	127.50	133.10	138.00	141.00
KUWAIT	100.00	115.63	123.06	105.34	104.03	103.95	..
LAOS, P.D.R.	100.00	133.26	205.34	318.28	388.50	..	..
LEBANON	100.00	..	..	..	..	..	..
LESOTHO	100.00	110.11	123.17	136.19	152.17	162.83	..
LIBERIA	100.00	100.50	103.70	100.60	103.90	101.80	..
LIBYA	100.00	104.90	98.20	99.90	100.70	101.80	..
MADAGASCAR	100.00	123.40	159.60	191.20	209.60	231.10	..
MALAWI	100.00	119.32	119.41	132.69	161.01	163.35	206.40
MALAYSIA	100.00	101.10	103.60	108.40	115.00	113.30	103.80
MALI	100.00	110.10	114.20	123.10	138.10	155.00	..
MARTINIQUE	100.00	111.40	124.40	136.60	146.70	155.20	162.40
MAURITANIA	100.00	119.10	132.80	134.40	142.70	159.30	..
MAURITIUS	100.00	111.60	122.10	128.80	139.70	148.00	..
MEXICO	100.00	127.30	205.10	394.10	637.60	984.60	..
MONTSERRAT	100.00	108.80	155.50	125.40	..	..	..
MOROCCO	100.00	110.50	122.80	126.60	138.90	147.40	..
MOZAMBIQUE	100.00	112.20	128.60	146.60	171.20	203.00	..
NEPAL	100.00	107.90	118.00	132.50	139.10	147.70	170.40
NETHERLANDS	100.00	105.50	112.10	114.10	115.50	118.80	118.90
NEW CALEDONIA	100.00	114.03	125.04	140.59	..	..	..
NEW ZEALAND	100.00	119.40	129.70	142.70	150.70	172.30	..
NICARAGUA	100.00	111.80	130.40	149.20	195.40	623.50	..
NIGER	100.00	108.00	128.40	133.60	145.70	149.60	..
NIGERIA	100.00	102.70	109.90	112.70	118.60	124.10	..
NORWAY	100.00	114.00	125.60	133.20	141.80	149.20	147.00
OMAN	100.00	113.68	119.39	122.36	122.26	..	..
PAKISTAN	100.00	110.80	120.80	127.70	140.00	148.20	155.60
PANAMA	100.00	104.60	109.40	111.40	116.80	119.90	..
PAPUA NEW GUINEA	100.00	97.95	102.05	112.64	123.56	128.69	..

Table A.2 (contd.)

COUNTRY	1960	1961	1962	1963	1964	1965	1966	1967	1968	1969
PARAGUAY	21.30	23.20	25.00	26.10	26.60	27.40	28.40	28.30	28.70	29.70
PERU	2.70	2.90	3.00	3.10	3.60	4.00	4.50	5.00	5.90	6.30
PHILIPPINES	16.00	16.40	17.50	19.00	19.90	20.70	21.80	23.20	24.40	25.50
PORTUGAL	17.80	18.20	18.10	18.70	19.00	19.70	20.30	21.10	21.10	22.70
PUERTO RICO	36.10	36.40	37.30	37.80	38.40	39.40	40.80	41.90	44.00	46.40
QATAR	..	..	..	..	..	..	..	..	..	..
RWANDA	29.30	30.40	31.00	31.80	34.10	35.60	37.40	39.40	41.80	44.50
SAO TOME &										
PRINCIPE	20.18	21.19	20.81	21.26	21.00	21.29	21.45	22.12	22.38	23.26
SAUDI ARABIA	..	..	..	12.23	12.02	11.91	12.16	12.40	12.89	13.28
SENEGAL	37.13	38.30	39.28	39.37	40.72	41.08	41.35	41.79	41.61	44.39
SEYCHELLES	..	..	..	..	..	..	..	..	..	..
SIERRA LEONE	27.52	28.03	24.57	27.31	28.93	28.81	29.87	31.23	31.77	33.18
SINGAPORE	49.70	49.70	50.10	50.60	51.00	51.60	46.80	52.70	53.30	54.50
SOLOMON ISLANDS	32.04	32.54	32.83	33.33	34.84	35.70	35.49	36.93	39.15	41.02
SOMALIA	17.50	19.21	19.04	19.67	22.29	25.14	24.34	24.29	25.09	26.74
SOUTH AFRICA	20.80	21.10	21.30	22.00	22.60	23.30	24.30	25.30	26.20	27.90
SPAIN	12.20	12.50	13.20	14.30	16.40	18.00	19.40	20.90	21.90	22.90
SRI LANKA	28.00	27.50	28.20	27.90	27.70	28.10	27.60	28.20	32.00	32.30
ST KITTS-NEVIS	25.87	25.95	25.64	25.95	27.27	27.89	28.75	29.60	34.27	35.28
ST LUCIA	24.91	25.12	24.91	25.47	26.23	26.64	27.54	28.37	31.56	32.66
ST VINCENT	21.64	22.22	22.22	22.37	23.24	23.60	25.13	28.03	28.98	30.57
SUDAN	17.50	19.03	19.28	20.18	21.07	20.56	20.95	23.24	23.88	24.39
SURINAME	..	..	..	..	..	..	..	..	..	..
SWAZILAND	24.90	25.20	25.50	26.10	27.20	28.20	28.10	28.70	29.40	30.30
SWEDEN	26.40	27.30	28.50	28.80	30.10	31.80	34.00	35.70	36.50	37.40
SWITZERLAND	39.40	41.00	43.40	45.50	47.90	49.70	52.10	54.40	56.10	57.50
SYRIAN										
ARAB REPUBLIC	25.90	26.70	26.20	26.60	28.10	29.00	31.00	31.40	32.10	30.80
TAIWAN	..	27.97	28.57	29.55	30.78	30.59	31.42	32.85	35.08	37.32
TANZANIA	28.35	30.55	30.72	32.21	33.14	31.99	31.90	32.00	32.60	33.60
THAILAND	32.90	34.10	34.20	33.60	34.60	36.20	38.80	38.50	38.20	39.00
TOGO	30.60	31.20	31.80	32.30	33.60	33.60	34.50	32.70	32.60	35.00
TONGA	..	..	..	..	..	..	..	..	..	..
TRINIDAD &										
TOBAGO	17.10	16.10	16.40	17.30	16.80	16.60	15.90	16.30	16.40	18.40
TUNISIA	35.90	34.90	35.30	37.20	38.00	41.20	42.00	43.70	44.90	41.30
TURKEY	3.10	3.23	3.54	3.74	3.83	4.00	4.25	4.53	4.71	4.96
TUVALU	26.90	27.20	27.40	28.20	29.20	29.90	31.20	32.20	33.10	34.10
UGANDA	6.20	6.40	6.20	6.30	6.60	7.30	7.30	7.30	7.60	7.70
UNITED ARAB										
EMIRATES	..	..	..	..	..	..	..	..	..	..
UNITED KINGDOM	17.90	18.50	19.20	19.60	20.30	21.30	22.30	23.00	23.90	25.20
UNITED STATES	36.10	36.40	37.30	37.80	38.40	39.40	40.80	41.90	44.00	46.40
URUGUAY	0.025	0.030	0.034	0.040	0.057	0.096	0.158	0.29	0.62	0.79
VANUATU	..	..	..	..	..	..	..	..	..	..
VENEZUELA	16.00	15.70	15.20	15.30	15.70	15.80	16.30	16.30	17.20	17.00
VIRGIN ISLANDS										
(U.S.)	36.10	36.40	37.30	37.80	38.40	39.40	40.80	41.90	44.00	46.40
WESTERN SAMOA	..	26.00	26.60	26.80	28.30	28.80	29.80	29.60	30.20	31.20
YEMEN,										
ARAB REPUBLIC	..	..	..	..	..	..	..	..	..	..
YEMEN, P.D.R.	..	..	..	..	..	..	..	..	..	..
ZAIRE	0.53	0.59	0.73	1.41	1.45	1.98	2.08	3.31	4.95	5.76
ZAMBIA	41.69	41.53	40.83	41.45	41.77	46.54	47.25	48.83	58.46	68.99
ZIMBABWE	41.51	41.84	42.07	40.36	42.39	43.50	42.53	43.22	44.65	46.13

Table A.2 (contd.)

COUNTRY	1970	1971	1972	1973	1974	1975	1976	1977	1978	1979
PARAGUAY	30.50	32.40	35.20	42.50	52.60	56.10	58.90	64.30	71.00	85.60
PERU	6.80	7.10	7.50	8.60	10.00	12.00	16.20	22.40	36.20	64.60
PHILIPPINES	29.20	32.80	35.00	41.30	54.10	58.60	64.00	68.70	75.10	86.50
PORTUGAL	23.30	24.50	26.40	28.90	34.30	39.90	46.10	57.80	70.50	84.90
PUERTO RICO	49.00	51.80	54.20	57.80	63.00	69.20	73.60	78.50	84.30	91.70
QATAR	..	..	..	..	..	..	..	..	..	93.60
RWANDA	27.20	46.50	47.50	49.30	57.60	60.60	69.70	76.70	79.20	85.90
SAO TOME & PRINCIPE	25.91	24.90	23.95	31.48	43.57	45.52	50.53	85.20	86.40	91.51
SAUDI ARABIA	12.80	14.73	15.75	18.87	40.17	56.31	61.10	66.15	68.63	71.22
SENEGAL	45.29	46.73	48.61	52.38	61.17	68.25	70.76	76.59	83.59	89.69
SEYCHELLES	21.90	25.10	30.40	35.90	44.70	53.00	60.90	70.00	78.20	88.10
SIERRA LEONE	32.38	33.31	35.13	42.17	47.96	52.57	63.94	75.87	86.10	96.01
SINGAPORE	55.40	57.90	61.00	68.50	79.10	81.10	82.60	83.80	85.80	90.30
SOLOMON ISLANDS	43.61	45.47	48.71	50.22	59.63	60.27	63.22	71.84	78.59	85.92
SOMALIA	26.91	27.42	30.67	34.26	38.71	44.41	51.65	57.01	66.36	76.11
SOUTH AFRICA	29.00	30.40	33.70	40.10	46.00	50.40	55.70	62.10	69.10	79.60
SPAIN	24.50	26.40	28.70	32.10	37.40	43.70	51.00	62.60	75.20	87.80
SRI LANKA	32.60	33.50	33.70	39.30	48.80	52.20	56.80	65.90	72.20	83.30
ST KITTS-NEVIS	37.37	38.62	42.27	45.38	60.06	66.12	71.79	77.70	83.22	88.34
ST LUCIA	36.61	30.52	38.34	43.32	47.68	57.85	63.04	69.20	75.16	87.47
ST VINCENT	30.57	31.88	33.91	42.56	53.38	60.93	67.18	72.62	83.59	93.17
SUDAN	27.84	28.35	31.67	37.04	46.74	50.32	51.98	54.53	63.73	80.08
SURINAME	48.29	48.41	46.55	48.97	57.00	59.75	63.66	74.42	79.23	91.33
SWAZILAND	31.70	34.30	35.90	39.80	46.00	53.80	60.70	67.10	72.70	84.10
SWEDEN	39.80	42.60	45.60	48.80	53.40	61.20	68.50	75.70	82.90	89.50
SWITZERLAND	60.30	65.80	72.20	78.10	83.50	89.50	91.90	92.10	95.50	97.40
SYRIAN ARAB REPUBLIC	32.80	35.20	34.80	39.40	49.90	52.45	57.90	64.20	71.30	82.40
TAIWAN	38.61	39.79	42.11	48.37	63.97	65.46	69.09	73.51	73.45	86.18
TANZANIA	35.30	36.20	38.60	43.90	52.30	58.90	63.50	71.00	76.90	85.00
THAILAND	38.80	39.40	42.80	51.40	61.10	62.80	65.30	70.90	77.00	85.90
TOGO	36.90	39.20	40.10	44.10	60.10	59.00	66.90	81.70	83.50	89.60
TONGA	35.96	36.73	38.34	47.29	54.14	59.63	58.68	70.01	81.16	87.54
TRINIDAD & TOBAGO	18.90	20.20	22.40	27.10	42.40	52.50	56.60	64.10	66.10	81.70
TUNISIA	42.60	45.20	46.20	50.10	61.70	65.90	69.90	75.90	80.10	87.40
TURKEY	5.53	6.52	7.59	9.26	11.83	13.76	16.19	20.17	28.97	49.45
TUVALU	35.30	37.40	40.20	45.10	52.90	61.50	69.90	76.40	81.90	89.60
UGANDA	8.70	9.10	10.00	11.50	14.10	20.00	23.50	42.50	55.30	76.60
UNITED ARAB EMIRATES	..	..	38.60	26.30	62.70	74.90	84.10	89.20	87.30	92.10
UNITED KINGDOM	27.10	29.60	32.10	34.30	39.50	50.20	57.60	65.60	72.90	83.50
UNITED STATES	49.00	51.80	54.20	57.80	63.00	69.20	73.60	78.50	84.30	91.70
URUGUAY	0.88	1.05	1.84	3.78	6.50	11.04	16.43	25.59	37.75	66.25
VANUATU	..	..	..	..	..	..	76.70	81.10	86.30	89.90
VENEZUELA	18.10	20.20	22.50	27.70	43.00	48.80	52.80	59.70	66.10	80.20
VIRGIN ISLANDS (U.S.)	49.00	51.80	54.20	57.80	63.00	69.20	73.60	78.50	84.30	91.70
WESTERN SAMOA	32.10	33.60	36.20	40.60	50.70	55.10	57.90	66.20	67.70	75.20
YEMEN, ARAB REPUBLIC	25.91	26.18	28.93	31.74	37.84	45.05	52.99	66.30	77.58	90.10
YEMEN, P.D.R.	..	..	..	..	..	73.00	74.20	76.80	77.60	87.10
ZAIRE	5.69	6.00	6.62	7.80	9.22	10.35	16.30	22.37	32.73	66.16
ZAMBIA	54.90	53.02	52.82	58.85	60.12	51.57	58.45	65.14	73.38	89.47
ZIMBABWE	46.17	48.85	51.25	55.21	59.69	66.19	72.28	79.94	80.44	91.70

Table A.2 (contd.)

COUNTRY	1980	1981	1982	1983	1984	1985	1986
PARAGUAY	100.00	116.60	122.20	139.80	177.50	222.30	292.40
PERU	100.00	166.90	272.70	564.30	1201.20	3091.00	..
PHILIPPINES	100.00	111.00	120.30	134.30	201.30	236.90	241.20
PORTUGAL	100.00	116.40	142.80	172.80	216.00	260.80	..
PUERTO RICO	100.00	109.60	116.70	121.20	125.90	130.10	133.50
QATAR	100.00	108.50	114.70	117.90	119.20	..	..
RWANDA	100.00	108.70	112.60	118.10	136.80	144.80	..
SAO TOME & PRINCIPE	100.00	98.09	105.42	125.34	136.82	147.50	..
SAUDI ARABIA	100.00	125.03	123.97	109.92	98.59	93.69	88.88
SENEGAL	100.00	107.62	118.57	129.60	143.23	155.50	..
SEYCHELLES	100.00	110.60	109.70	116.30	120.90	122.00	..
SIERRA LEONE	100.00	119.07	199.65	234.93	346.85	515.92	..
SINGAPORE	100.00	106.40	111.40	116.00	116.80	114.50	110.20
SOLOMON ISLANDS	100.00	114.37	125.14	135.06	..	..	..
SOMALIA	100.00	142.90	171.90	248.20	447.30	627.20	..
SOUTH AFRICA	100.00	109.30	124.10	141.40	158.80	183.40	213.10
SPAIN	100.00	112.00	127.40	142.20	157.70	171.50	..
SRI LANKA	100.00	116.70	128.00	149.50	175.80	178.90	..
ST KITTS-NEVIS	100.00	111.27	119.74	120.75	123.39	..	..
ST LUCIA	100.00	111.35	114.39	117.02	118.34	..	..
ST VINCENT	100.00	116.56	125.78	129.99	135.00	..	..
SUDAN	100.00	125.30	144.80	191.80	230.00	271.30	..
SURINAME	100.00	105.04	115.23	116.44	116.44	120.50	..
SWAZILAND	100.00	112.75	120.84	134.71	152.01	179.38	..
SWEDEN	100.00	109.50	118.90	130.50	140.50	150.20	160.90
SWITZERLAND	100.00	106.90	114.70	118.50	122.00	125.10	130.20
SYRIAN ARAB REPUBLIC	100.00	117.00	119.00	124.50	132.80	139.30	..
TAIWAN	100.00	112.11	115.97	118.20	119.18	119.43	..
TANZANIA	100.00	118.20	138.90	164.70	195.10	250.40	..
THAILAND	100.00	108.00	111.60	115.10	116.70	119.10	121.40
TOGO	100.00	114.20	125.00	136.00	141.00	143.70	..
TONGA	100.00	110.44	115.44	117.89	..	..	..
TRINIDAD & TOBAGO	100.00	105.10	118.10	126.00	144.50	147.40	..
TUNISIA	100.00	111.38	129.09	139.95	149.68	161.89	..
TURKEY	100.00	142.00	181.90	235.10	351.30	506.60	..
TUVALU	100.00	109.40	121.30	131.90	141.30	149.80	160.80
UGANDA	100.00	115.20	137.60	168.20	209.90	288.30	..
UNITED ARAB EMIRATES	100.00	104.10	105.10	104.80	99.50	..	..
UNITED KINGDOM	100.00	111.50	120.10	126.20	131.40	139.10	144.50
UNITED STATES	100.00	109.60	116.70	121.20	125.90	130.10	133.50
URUGUAY	100.00	130.30	151.20	235.10	373.00	659.00	1126.00
VANUATU	100.00	127.50	135.30	137.60	145.20	146.30	..
VENEZUELA	100.00	114.40	117.60	128.20	152.20	163.40	..
VIRGIN ISLANDS (U.S.)	100.00	109.60	116.70	121.20	125.90	130.10	133.50
WESTERN SAMOA	100.00	120.50	142.60	166.00	185.70	202.60	..
YEMEN, ARAB REPUBLIC	100.00	95.00	99.00	112.00	118.00	..	..
YEMEN, P.D.R.	100.00	104.90	117.10	126.50	..	..	..
ZAIRE	100.00	134.10	179.70	336.60	657.30	782.10	..
ZAMBIA	100.00	107.16	113.74	197.36	224.22	318.11	..
ZIMBABWE	100.00	111.08	126.40	144.26	160.14	160.74	..

Sources:

Directorate-General of Budget, Accounting & Statistics. *Statistical Yearbook of the Republic of China 1986.* Taiwan: Executive Yuan, November 1986. [E]

IMF. *International Financial Statistical Yearbook, 1986,* Vol. 39. Washington D.C.: IMF, 1986. [B]

IMF. *International Financial Statistics Yearbook,* 1987, Vol. 40. Washington D.C.: IMF, 1987. [C]

UN. *National Accounts Statistics: Analysis of Main Aggregates, 1983/84.* New York: United Nations, 1987 [D]

UN. "Gross Domestic Product by Broad Economic Sector: Deflators." Office for Development Research and Policy Analysis of the United Nations Secretariat, July 1988. Mimeo. [F]

World Bank. "World Bank Deflators: 1960–1984." World Bank, Washington D.C., 1984. Mimeo. [A]

Notes:

Whenever possible the series represents implicit GDP deflators – denoted by GDP in the 'Deflator' column of the Source Code Key presented below – otherwise the following deflators were included: GNP is Gross National Product; NI is National Income; and CPI is Consumer Price Index.

To obtain a deflator series back to 1960 it was sometimes necessary to splice together data from different sources. Where possible the splicing was centered on the base year of 1980. The Source Key below indicates the sources used for each sub-period within a country. For instance, "A:60–80; B:80–84" indicates that source A was used for the 1960–80 period, source B for the 1980–84 period, and the 1980 base year deflator represents a simple average of the source A and B figure.

The Taiwan deflator was constructed on the basis of a series of GDP at current market prices and GDP at constant market prices obtained from source [E].

Source Key for Implicit GDP Deflators

COUNTRY	DEFLATOR	SOURCE
AFGHANISTAN	GDP	A:60-81
ALGERIA	GDP	B:60-80; A:80-84
ANGOLA	GDP	A:60-83
ANTIGUA	GDP	A:60-77; B:77-81; C:82-84
ARGENTINA	GDP	A:60-80; F:80-85
AUSTRALIA	GDP	B:60-80; C:81-86
AUSTRIA	GDP	B:60-81; C:82-86
BAHAMAS	CPI	B:66-85
BANGLADESH	GDP	F:60-85; C:86
BARBADOS	GDP	F:60-85
BELGIUM	GNP	B:60-81; C:82-85
BELIZE	GDP	A:70-80; B:80-84
BENIN	GDP	A:60-84
BERMUDA	–	–
BOLIVIA	GDP	B:60-80; F:80-85
BOTSWANA	GDP	F:80-85
BRAZIL	GDP	A:60-80; F:80-85
BRUNEI	GDP	A:74-81; D:82-84
BURKINA FASO	GDP	F:60-85
BURMA	GDP	F:60-85; C:86
BURUNDI	GDP	A:60-80; C:80-86
CAMEROON	GDP	A:60-80; F:80-85
CANADA	GDP	C:60-86
CAPE VERDE	GDP	F:60-85
CENTRAL AFRICAN REPUBLIC	GDP	F:60-85
CHAD	GDP	F:60-85
CHILE	GDP	A:60-80; B:80-85; C:86
CHINA	NI	C:60-82
COLOMBIA	GDP	A:60-80; B:80-83; C:84-86

COUNTRY	DEFLATOR	SOURCE
COMOROS	–	–
CONGO	GDP	F:60-85
COOK ISLANDS	GDP	See New Zealand
COSTA RICA	GDP	B:60-83; C:84-86
CUBA	–	–
CYPRUS	GDP	B:60-80; F:80-85
DENMARK	GDP	B:60-85
DOMINICA	GDP	A:60-84
DOMINICAN REPUBLIC	GDP	B:60-84; C:85-86
ECUADOR	GDP	B:60-83; C:84-86
EGYPT	GDP	A:60-80; F:80-85
EL SALVADOR	GDP	B:60-82; C:83-86
ETHIOPIA	GDP	B:60-80; F:80-85
FIJI	GDP	A:60-80; F:80-85
FINLAND	GDP	B:60-84; C:85-86
FRANCE	GDP	B:60-76; C:77-86
FRENCH POLYNESIA	GDP	A:76-83
GABON	GDP	A:60-80; F:80-85
GAMBIA	GDP	A:60-80; F:80-85
GERMANY, FED.REP. OF	GNP	B:60-83; C:84-86
GHANA	GDP	A:60-80; F:80-85
GREECE	GDP	B:60-80; C:81-86
GRENADA	GDP	A:71-84
GUADELOUPE	GDP	See France
GUAM	GDP	A:70-81
GUATEMALA	GDP	B:60-81; C:82-86
GUINEA	GDP	F:60-85
GUINEA-BISSAU	GDP	F:60-85
GUYANA	GDP	F:60-85
HAITI	GDP	B:60-65; C:66-86
HONDURAS	GDP	C:60-86
HONG KONG	GDP	A:60-80; F:80-85
ICELAND	GNP	A:60-80; B:80-83; C:84-86
INDIA	GDP	B:60-80; C:81-85
INDONESIA	GDP	A:63-80; C:80-85
IRAN	GDP	B:60-84; C:85
IRAQ	GDP	F:60-85
IRELAND	GDP	B:60-80; C:81-85
ISRAEL	GDP	A:60-80; F:80-85
ITALY	GDP	C:60-86
IVORY COAST	GDP	A:60-80; F:80-85
JAMAICA	GDP	B:60-83; C:84-85
JAPAN	GNP	B:60-83; C:84-86
JORDAN	GDP	B:69-84; C:85-86
KENYA	GDP	A:60-80; C:80-85
KOREA, REPUBLIC OF	GDP	B:60-84; C:85-86
KUWAIT	GDP	F:62-85
LAOS, P.D.R.	GDP	A:79-84
LEBANON	GDP	A:60-80
LESOTHO	GDP	A:60-80; F:80-85
LIBERIA	GDP	A:60-80; C:80-85
LIBYAN ARAB REPUBLIC	GDP	F:60-85
MADAGASCAR	GDP	A:60-80; F:80-85
MALAWI	GDP	F:60-85
MALAYSIA	GDP	A:60-80; C:80-86
MALI	GDP	F:60-85
MARTINIQUE	GDP	See France
MAURITANIA	GDP	A:60-80; F:80-85

COUNTRY	DEFLATOR	SOURCE
MAURITIUS	GDP	B:60-80; F:80-85
MEXICO	GDP	B:60-83; C:84-85
MONTSERRAT	GDP	D:75-83
MOROCCO	GDP	F:60-85
MOZAMBIQUE	GDP	A:60-80; F:80-85
NEPAL	GDP	B:60-84; C:85-86
NETHERLANDS	GNP	B:60-82; C:83-86
NEW CALEDONIA	GDP	A:70-83
NEW ZEALAND	GDP	B:60-75; C:76-85
NICARAGUA	GDP	B:60-80; F:80-85
NIGER	GDP	F:80-85
NIGERIA	GDP	B:60-67; C:68-80; F:80-85
NORWAY	GDP	B:60-82; C:83-86
OMAN	GDP	A:60-84
PAKISTAN	GDP	C:60-86
PANAMA	GDP	B:60-83; C:84-85
PAPUA NEW GUINEA	GDP	A:60-84; F:85
PARAGUAY	GDP	B:60-85; C:86
PERU	GDP	B:60-80; F:80-85
PHILIPPINES	GNP	B:60-80; C:80-86
PORTUGAL	GDP	C:60-80; F:80-85
PUERTO RICO	GNP	See USA
QATAR	CPI	B:79-84
RWANDA	GDP	B:60-80; F:80-84
SAO TOME & PRINCIPE	GDP	F:60-85
SAUDI ARABIA	GDP	A:63-68; C:68-86
SENEGAL	GDP	A:60-84; F:85
SEYCHELLES	CPI	B:70-85
SIERRA LEONE	GDP	F:60-85
SINGAPORE	GDP	C:60-86
SOLOMON ISLANDS	GDP	A:60-83
SOMALIA	GDP	A:60-80; F:80-85
SOUTH AFRICA	GDP	B:60-82; C:83-86
SPAIN	GDP	B:60-80; F:80-85
SRI LANKA	GDP	B:60-80; F:80-85
ST KITTS-NEVIS	GDP	A:60-84
ST LUCIA	GDP	A:60-84
ST VINCENT	GDP	A:60-84
SUDAN	GDP	A:60-80; F:80-85
SURINAME	GDP	F:70-85
SWAZILAND	GDP	A:60-80; F:80-85
SWEDEN	GDP	B:60-83; C:84-86
SWITZERLAND	GDP	B:60-84; C:85-86
SYRIAN ARAB REPUBLIC	GDP	A:60-75; B:75-85
TAIWAN	GDP	E:61-85
TANZANIA	GDP	A:60-65; B:65-80; F:80-85
THAILAND	GDP	B:60-82; C:83-86
TOGO	GDP	F:60-85
TONGA	GDP	A:70-83
TRINIDAD & TOBAGO	GDP	F:60-85
TUNISIA	GDP	B:60-80; F:80-85
TURKEY	GNP	A:60-80; F:80-85
TUVALU	GDP	See Australia
UGANDA	GDP	F:60-85
UNITED ARAB EMIRATES	GDP	B:72-80; C:80-84
UNITED KINGDOM	GDP	B:60-78; C:79-86
UNITED STATES	GNP	B:60-82; C:83-86
URUGUAY	GDP	A:60-80; B:80-83; C:84-86

COUNTRY	DEFLATOR	SOURCE
VANUATU	CPI	B:76-85
VENEZUELA	GDP	F:60-85
VIRGIN ISLANDS (U.S.)	GNP	See USA
WESTERN SAMOA	CPI	B:61-85
YEMEN, ARAB REPUBLIC	GDP	B:70-80; C:81-84
YEMEN, P.D.R.	GDP	A:75-83
ZAIRE	GDP	A:60-80; F:80-85
ZAMBIA	GDP	F:80-85; C:86
ZIMBABWE	GDP	A:60-80; F:80-85

Printed in the United States
By Bookmasters